INDUSTRIAL VENTILATION

20th Edition

A Manual of Recommended Practice

1988

COMMITTEE ON INDUSTRIAL VENTILATION
P. O. BOX 16153
LANSING, MICHIGAN 48901 USA

Sales
American Conference of Governmental Industrial Hygienists
6500 Glenway Avenue, Bldg. D-7
Cincinnati, Ohio 45211 USA

1st Edition — 1951
2nd Edition — 1952
3rd Edition — 1954
4th Edition — 1956
5th Edition — 1958
6th Edition — 1960
7th Edition — 1962
8th Edition — 1964
9th Edition — 1966
10th Edition — 1968
11th Edition — 1970
12th Edition — 1972
13th Edition — 1974
14th Edition — 1976
15th Edition — 1978
16th Edition — 1980
17th Edition — 1982
18th Edition — 1984
19th Edition — 1986
20th Edition — 1988

Second Printing

Published in the United States of America by
American Conference of Governmental Industrial Hygienists, Inc.
6500 Glenway Avenue, Building D-7
Cincinnati, Ohio 45211-4438
1-513-661-7881

Library of Congress Catalog Card Number: 62-12929

ISBN: 0-936712-79-1

Printed in the United States

CONTENTS

FOREWORD		vii
ACKNOWLEDGMENTS		viii
DEFINITIONS		ix
ABBREVIATIONS		xi
CHAPTER 1	GENERAL PRINCIPLES OF VENTILATION	1-1
	1.1 Introduction	
	1.2 Supply Systems	
	1.3 Exhaust Systems	
	1.4 Basic Definitions	
	1.5 Principles of Air Flow	
	1.6 Acceleration of Hood and Entry Losses	
	1.7 Duct Losses	
	1.8 Multiple-Hood Exhaust Systems	
	1.9 Air Flow Characteristics of Blowing and Exhausting	
	References	
CHAPTER 2	GENERAL INDUSTRIAL VENTILATION	2-1
	2.1 Introduction	
	2.2 Dilution Ventilation Principles	
	2.3 Dilution Ventilation for Health	
	2.4 Mixtures — Dilution Ventilation for Health	
	2.5 Dilution Ventilation for Fire and Explosion	
	2.6 Fire Dilution Ventilation for Mixtures	
	2.7 Ventilation for Heat Control	
	2.8 Heat Balance and Exchange	
	2.9 Adaptive Mechanism of the Body	
	2.10 Acclimatization	
	2.11 Acute Heat Disorders	
	2.12 Heat Stress Measurement	
	2.13 Heat Stress Indices	
	2.14 Ventilation Control	
	2.15 Ventilation Systems	
	2.16 Velocity Cooling	
	2.17 Radiant Heat Control	
	2.18 Protective Suits for Short Exposures	
	2.19 Respiratory Heat Exchangers	
	2.20 Refrigerated Suits	
	2.21 Enclosures	
	2.22 Insulation	
	References	
CHAPTER 3	LOCAL EXHAUST HOODS	3-1
	3.1 Introduction	
	3.2 Contaminant Characteristics	
	3.3 Hood Types	

3.4 Hood Design Factors
3.5 Hood Losses
3.6 Minimum Duct Velocity
3.7 Special Hood Requirements
3.8 Push-Pull Ventilation
3.9 Hot Processes
References

CHAPTER 4 AIR CLEANING DEVICES **4-1**
4.1 Introduction
4.2 Selection of Dust Collection Equipment
4.3 Dust Collector Types
4.4 Additional Aids in Duct Collector Selection
4.5 Control of Mist, Gas and Vapor Contaminants
4.6 Gaseous Contaminant Collectors
4.7 Unit Collectors
4.8 Dust Collection Equipment Cost
4.9 Selection of Air Filtration Equipment
4.10 Radioactive and High Toxicity Operations
4.11 Explosion Venting
References

CHAPTER 5 EXHAUST SYSTEM DESIGN PROCEDURE **5-1**
5.1 Introduction
5.2 Preliminary Steps
5.3 Design Procedure
5.4 Design Methods
5.5 Aids to Calculations
5.6 Distribution of Air Flow
5.7 Plenum Type Exhaust Systems
5.8 Fan Pressure Ratings
5.9 Corrections for Velocity Changes
5.10 Sample System Design
5.11 Corrections for Different Duct Materials
5.12 Friction Loss for Non-Circular Ducts
5.13 Corrections for Temperature, Moisture, and Altitude
5.14 Air Cleaning Equipment
5.15 Evasé Discharge
5.16 Exhaust Stack Outlets
5.17 Air Bleed-Ins
5.18 Optimum Economic Velocity
References

CHAPTER 6 FANS **6-1**
6.1 Introduction
6.2 Basic Definitions
6.3 Fan Selection
6.4 Fan Installation and Maintenance
References

CHAPTER 7 REPLACEMENT AND RECIRCULATED AIR **7-1**
7.1 Introduction
7.2 Replacement Air
7.3 Replacement Air Flow Rate
7.4 Environmental Control
7.5 Environmental Control Air Flow Rate
7.6 Air Changes

7.7 Air Supply Temperatures
7.8 Air Supply vs. Plant Heating Costs
7.9 Replacement Air Heating Equipment
7.10 Cost of Heating Replacement Air
7.11 Air Conservation
7.12 Evaluation of Employee Exposure Levels
References

CHAPTER 8 CONSTRUCTION GUIDELINES FOR LOCAL EXHAUST SYSTEMS 8-1
8.1 Introduction
8.2 General
8.3 Materials
8.4 Construction
8.5 System Details
8.6 Codes
8.7 Other Types of Duct Materials
8.8 Testing
References

CHAPTER 9 TESTING OF VENTILATION SYSTEMS 9-1
9.1 Introduction
9.2 Pressure Measurement
9.3 Volumetric Flow Measurement
9.4 Air Velocity Instruments
9.5 Calibration of Air Measuring Instruments
9.6 Evaluating Exhaust Systems
9.7 Difficulties Encountered in Field Measurement
References

CHAPTER 10 SPECIFIC OPERATIONS 10-1

BIBLIOGRAPHY 11-1

APPENDICES 12-1
A Threshold Limit Values for Chemical Substances in the Work Environment with Intended Changes for 1988-1989
B Physical Constants/Conversion Factors
C Metric Supplement

INDEX 13-1

FOREWORD

INDUSTRIAL VENTILATION: A Manual of Recommended Practice is the outgrowth of years of experience by Committee members and a compilation of research data and information on design, maintenance, and evaluation of industrial exhaust ventilation systems. The Manual attempts to present a logical method for designing and testing these systems. It has found wide acceptance as a guide for official agencies, as a standard for industrial ventilation designers, and as a textbook for industrial hygiene courses.

The Manual is not intended to be used as law, but rather as a guide. Because of new information on industrial ventilation becoming available through research projects, reports from engineers, and articles in various periodicals and journals, review and revision of each section of the manual is an ongoing Committee project. The Manual is available in a hard bound edition only and includes a metric supplement.

In this Twentieth Edition, the Committee has reviewed and rewritten or modified every text chapter — virtually the entire Manual. Over the past four years, the contents of each section of the Manual have been reviewed in depth and revised as needed to increase the depth or breadth of coverage. Our policy of presenting the material in a practical, concise, easy-to-understand manner continues. Where appropriate, limited theoretical discussions have been included to show the basis for the practical equations, especially in areas where the Committee has expanded on subject matter. The Manual is sufficiently complete so that industrial exhaust ventilation systems can be designed without reference to other texts. In our constant effort to present the latest techniques and data, the Committee desires, welcomes, and actively seeks comments and suggestions on the accuracy and adequacy of the information presented herein.

One comprehensive change to the Manual is that it has been reorganized, and it is important to understand the new order. Each component of the Manual is now called a "Chapter" instead of a "Section" as in previous editions. The relationship of the Chapters of this edition to the Sections of older editions is as follows:

Chapter 1 was Section 1
Chapter 2 was Sections 2 and 3
Chapter 3 was Section 4
Chapter 4 was Section 11
Chapter 5 was Section 6
Chapter 6 was Section 10
Chapter 7 was Section 7
Chapter 8 was Section 8
Chapter 9 was Section 9
Chapter 10 was Section 5

The primary reason for this reorganization was to collect the design theory of local exhaust ventilation systems in a contiguous block. Chapter 1 was expanded to show the principles of air flow in somewhat more detail. Sections 2 and 3 were combined in Chapter 2 and expanded. Chapters 3, 4, 5, and 6 now contain all the relevant design procedures for the local exhaust system and present hoods, air cleaners, duct systems, and fans in that order. Chapters 7, 8, and 9 maintain their original order, but each was reviewed and changed. Chapter 10 gives the hood data for specific operations.

All the graphics for Chapters 1 through 9 were done on a computer to facilitate future changes. The drawings of Chapter 10 were too extensive to computerize in the time available, so they appear in the same form as in previous editions. They will be revised and converted to the new style drawings for the next edition.

This publication is designed to present accurate and authoritative information with regard to the subject matter covered. It is distributed with the understanding that neither the Committee nor its members collectively or individually assume any responsibility for any inadvertent misinformation nor for omissions nor for the results in the use of this publication.

ACKNOWLEDGMENTS

Industrial Ventilation is a true Committee effort. It brings into focus in one source useful, practical ventilation data from all parts of the country. The committee membership of industrial ventilation and industrial hygiene engineers represents a diversity of experience and interest that ensures a well-rounded cooperative effort.

From the First Edition in 1951, this effort has been successful as witnessed by the acceptance of the "Ventilation Manual" throughout industry, by governmental agencies, and as a world-wide reference and text.

The present Committee is grateful for the faith and firm foundation provided by past Committees and members listed below.

Special acknowledgment is made to the Division of Occupational Health, Michigan Department of Health, for contributing their original field manual which was the basis for the First Edition, and to Mr. Knowlton J. Caplan who supervised the preparation of that manual.

The Committee is grateful also to those consultants who have contributed so greatly to the preparation of this and previous editions of Industrial Ventilation and to Mrs. Norma Donovan, Secretary to the Committee, for her untiring zeal in our efforts.

To many other individuals and agencies who have made specific contributions and have provided support, suggestions, and constructive criticism, our special thanks.

COMMITTEE ON INDUSTRIAL VENTILATION

Previous Members

A.B. Apol, 1984-present
H. Ayer, 1962-1966
R.E. Bales, 1954-1960
J. Baliff, 1950-1956; Chair, 1954-1956
J.T. Barnhart, Consultant, 1986-present
J.C. Barrett, 1956-1976; Chair, 1960-1968
J.L. Beltran, 1964-1966
D. Bonn, Consultant, 1958-1968
D.J. Burton, 1988-present
K.J. Caplan, 1974-1978; Consultant, 1980-1986
W.M. Cleary, 1976-present; Chair, 1978-1984
L. Dickie, 1984-present; Consultant, 1968-1984
B. Feiner, 1956-1968
S.E. Guffey, 1984-present
G.M. Hama, 1950-1984; Chair, 1956-1960
R.P. Hibbard, 1968-present
R.T. Hughes, 1976-present
H.S. Jordan, 1960-1962
J. Kane, Consultant, 1950-1952
J. Kayse, Consultant, 1956-1958
J.F. Keppler, 1950-1954, 1958-1960
G.W. Knutson, Consultant, 1986-present
J.J. Loeffler, 1980-present; Chair, 1984-present
J. Lumsden, 1962-1968
J.R. Lynch, 1966-1976
G. Michaelson, 1958-1960
K.M. Morse, 1950-1951; Chair, 1950-1951
R.T. Page, 1954-1956
O.P. Petrey, Consultant, 1978-present
G.S. Rajhans, 1978-present
K.E. Robinson, 1950-1954; Chair, 1952-1954
A. Salazar, 1952-1954
E.L. Schall, 1956-1958
M.M. Schuman, 1962-present; Chair, 1968-1978
J.C. Soet, 1950-1960
A.L. Twombly, Consultant, 1986-present
J. Willis, Consultant, 1952-1956
R. Wolle, 1966-1974
J.A. Wunderle, 1960-1964

DEFINITIONS

Aerosol: An assemblage of small particles, solid or liquid, suspended in air. The diameter of the particles may vary from 100 microns down to 0.01 micron or less, e.g., dust, fog, smoke.

Air Cleaner: A device designed for the purpose of removing atmospheric airborne impurities such as dusts, gases, vapors, fumes, and smoke. (Air cleaners include air washers, air filters, eletrostatic precipitators, and charcoal filters.)

Air Filter: An air cleaning device to remove light particulate loadings from normal atmospheric air before introduction into the building. Usual range: loadings up to 3 grains per thousand cubic feet (0.003 grains per cubic foot). Note: Atmospheric air in heavy industrial areas and in-plant air in many industries have higher loadings than this, and dust collectors are then indicated for proper air cleaning.

Air Horsepower: The theoretical horsepower required to drive a fan if there were no loses in the fan, that is, if its efficiency were 100%.

Air, Standard: Dry air at 70 F and 29.92 in (Hg) barometer. This is substantially equivalent to 0.075 lb/ft^3. Specific heat of dry air = 0.24 btu/lb/F.

Aspect Ratio: The ratio of the width to the length; AR = W/L.

Aspect Ratio of an Elbow: The width (W) along the axis of the bend divided by depth (D) in plane of bend; AR = W/D.

Blast Gate: Sliding damper.

Blow (throw): In air distribution, the distance an air stream travels from an outlet to a position at which air motion along the axis reduces to a velocity of 50 fpm. For unit heaters, the distance an air stream travels from a heater without a perceptible rise due to temperature difference and loss of velocity.

Brake Horsepower: The horsepower actually required to drive a fan. This includes the energy losses in the fan and can be determined only by actual test of the fan. (This does not include the drive losses between motor and fan.)

Capture Velocity: The air velocity at any point in front of the hood or at the hood opening necessary to overcome opposing air currents and to capture the contaminated air at that point by causing it to flow into the hood.

Coefficient of Entry: The actual rate of flow caused by a given hood static pressure compared to the theoretical flow which would result if the static pressure could be converted to velocity pressure with 100% efficiency. It is the ratio of actual to theoretical flow.

Comfort Zone (Average): The range of effective temperatures over which the majority (50% or more) of adults feel comfortable.

Convection: The motion resulting in a fluid from the differences in density and the action of gravity. In heat transmission, this meaning has been extended to include both forced and natural motion or circulation.

Density: The ratio of the mass of a specimen of a substance to the volume of the specimen. The mass of a unit volume of a substance. When weight can be used without confusion, as synonymous with mass, density is the weight of a unit volume of a substance.

Density Factor: The ratio of actual air density to density of standard air. The product of the density factor and the density of standard air (0.075 lb/ft^3) will give the actual air density in pounds per cubic foot; d × 0.075 = actual density of air, lbs/ft^3.

Dust: Small solid particles created by the breaking up of larger particles by processes such as crushing, grinding, drilling, explosions, etc. Dust particles already in existence in a mixture of materials may escape into the air through such operations as shoveling, conveying, screening, sweeping, etc.

Dust Collector: An air cleaning device to remove heavy particulate loadings from exhaust systems before discharge to outdoors. Usual range: loadings 0.003 grains per cubic foot and higher.

Entry Loss: Loss in pressure caused by air flowing into a duct or hood (inches H_2O.)

Fumes: Small, solid particles formed by the condensation of vapors of solid materials.

Gases: Formless fluids which tend to occupy an entire space uniformly at ordinary temperatures and pressures.

Gravity, Specific: The ratio of the mass of a unit volume of a substance to the mass of the same volume of a standard substance at a standard temperature. Water at 39.2 F is the standard substance usually referred to. For gases, dry air, at the same temperature and pressure as the gas, is often taken as the standard substance.

Hood: A shaped inlet designed to capture contaminated air and conduct it into the exhaust duct system.

Humidity, Absolute: The weight of water vapor per unit volume, pounds per cubic foot or grams per cubic centimeter.

Humidity, Relative: The ratio of the actual partial pressure of the water vapor in a space to the saturation pressure of pure water at the same temperature.

Inch of Water: A unit of pressure equal to the pressure exerted by a column of liquid water one inch high at a standard temperature.

Lower Explosive Limit: The lower limit of flammability or explosibility of a gas or vapor at ordinary ambient temperatures expressed in percent of the gas or vapor in air by volume. This limit is assumed constant for temperatures up to 250 F. Above these temperatures, it should be decreased by a factor of 0.7 since explosibility increases with higher temperatures.

Manometer: An instrument for measuring pressure; essentially a U-tube partially filled with a liquid, usually water, mercury or a light oil, so constructed that the amount of displacement of the liquid indicates the pressure being exerted on the instrument.

Micron: A unit of length, the thousandth part of 1 mm or the millionth of a meter (approximately 1/25,000 of an inch.)

Minimum Design Duct Velocity: Minimum air velocity required to move the particulates in the air stream, fpm.

Mists: Small droplets of materials that are ordinarily liquid at normal temperature and pressure.

Plenum: Pressure equalizing chamber.

Pressure, Static: The potential pressure exerted in all directions by a fluid at rest. For a fluid in motion, it is measured in a direction normal to the direction of flow. Usually expressed in inches water gauge when dealing with air. (The tendency to either burst or collapse the pipe.)

Pressure, Total: The algebraic sum of the velocity pressure and the static pressure (with due regard to sign).

Pressure, Vapor: The pressure exerted by a vapor. If a vapor is kept in confinement over its liquid so that the vapor can accumulate above the liquid, the temperature being held constant, the vapor pressure approaches a fixed limit called the maximum or saturated vapor pressure, dependent only on the temperature and the liquid. The term vapor pressure is sometimes used as synonymous with saturated vapor pressure.

Pressure, Velocity: The kinetic pressure in the direction of flow necessary to cause a fluid at rest to flow at a given velocity. Usually expressed in inches water gauge.

Radiation, Thermal (Heat) Radiation: The transmission of energy by means of electromagnetic waves of very long wave length. Radiant energy of any wave length may, when absorbed, become thermal energy and result in an increase in the temperature of the absorbing body.

Replacement Air: A ventilation term used to indicate the volume of controlled outdoor air supplied to a building to replace air being exhausted.

Slot Velocity: Linear flow rate of contaminated air through slot, fpm.

Smoke: An air suspension (aerosol) of particles, usually but not necessarily solid, often originating in a solid nucleus, formed from combustion or sublimation.

Temperature, Effective: An arbitrary index which combines into a single value the effect of temperature, humidity, and air movement on the sensation of warmth or cold felt by the human body. The numerical value is that of the temperature of still, saturated air which would induce an identical sensation.

Temperature, Wet-Bulb: Thermodynamic wet-bulb temperature is the temperature at which liquid or solid water, by evaporating into air, can bring the air to saturation adiabatically at the same temperature. Wet-bulb temperature (without qualification) is the temperature indicated by a wet-bulb psychrometer constructed and used according to specifications.

Threshold Limit Values (TLVs): The values for airborne toxic materials which are to be used as guides in the control of health hazards and represent time-weighted concentrations to which nearly all workers may be exposed 8 hours per day over extended periods of time without adverse effects (see Appendix).

Transport (Conveying) Velocity: See Minimum Design Duct Velocity.

Vapor: The gaseous form of substances which are normally in the solid or liquid state and which can be changed to these states either by increasing the pressure or decreasing the temperature.

ABBREVIATIONS

A . . . . Area
AHP . . . . Air horsepower
acfm . . . . actual cfm
AR . . . . Aspect ratio
B . . . . barometric pressure
bhp . . . . brake horsepower
bhp_a . . . . brake horsepower, actual
bhp_s . . . . brake horsepower standard air
btu . . . . British thermal unit
btuh . . . . btu/hr
C_e . . . . Coefficient of entry
cfm . . . . cubic feet per minute
CLR . . . . Centerline Radius
ft^3 . . . . cubic foot
d_f . . . . density factor
D . . . . diameter
ET . . . . effective temperature
F . . . . degree, Fahrenheit
fpm . . . . feet per minute
fps . . . . feet per second
g . . . . gravitational force, ft/sec/sec
gpm . . . . gallons per minute
gr . . . . grains
h_e . . . . hood entry loss
HEPA . . . . High Efficient Particulate Air filters
hp . . . . horsepower
hr . . . . hour
in . . . . inch
″wg . . . . inches water gauge
lb . . . . pound
LEL . . . . lower explosive limit
ME . . . . mechanical efficiency
mg . . . . milligram
MRT . . . . mean radiant temperature
mm . . . . millimeter
min . . . . minute
MW . . . . molecular weight
ppm . . . . parts per million
psi . . . . pounds per square inch
PWR . . . . power
Q . . . . quantity of air, cfm
R . . . . degree, Rankin
RH . . . . relative humidity
ρ . . . . density of air in lb/ft^3
rpm . . . . revolutions per minute
SFM . . . . surface feet per minute
ft^2 . . . . square foot
in^2 . . . . square inch
SP . . . . static pressure
SP_a . . . . actual SP, air other than standard
SP_h . . . . hood static pressure
SP_s . . . . SP, system handling standard air
scfm . . . . cfm at standard air conditions
sp gr . . . . specific gravity
STP . . . . standard temperature and pressure
TLV . . . . Threshold Limit Value
TP . . . . total pressure
V . . . . velocity, fpm
VP . . . . velocity pressure

Chapter 1
GENERAL PRINCIPLES OF VENTILATION

1.1 INTRODUCTION 1-2

1.2 SUPPLY SYSTEMS 1-2

1.3 EXHAUST SYSTEMS 1-2

1.4 BASIC DEFINITIONS 1-3

1.5 PRINCIPLES OF AIR FLOW 1-5

1.6 ACCELERATION OF AIR AND HOOD ENTRY LOSSES 1-5

1.7 DUCT LOSSES 1-7

1.7.1 FRICTION LOSSES 1-7

1.7.2 FITTING LOSSES 1-9

1.8 MULTIPLE-HOOD EXHAUST SYSTEMS 1-9

1.9 AIR FLOW CHARACTERISTICS OF BLOWING AND EXHAUSTING 1-9

REFERENCES 1-10

1.1 INTRODUCTION

The importance of clean uncontaminated air in the industrial work environment is well known. Modern industry with its complexity of operations and processes uses an increasing number of chemical compounds and substances, many of which are highly toxic. The use of such materials may result in particulates, gases, vapors and/or mists in the workroom air in concentrations which exceed safe levels. Heat stress can also result in unsafe or uncomfortable work environments. Effective, well designed ventilation offers a solution to these problems where worker protection is needed. Ventilation can also serve to control odor, moisture and other undesirable environmental conditions.

The health hazard potential of an airborne substance is characterized by the *threshold limit value (TLV)*. The TLV is defined as that airborne concentration of a substance which it is believed that nearly all workers may be exposed to day after day without developing adverse health effects. The *time-weighted average (TWA)*, defined as the time-weighted average concentration for a normal 8-hour workday and a 40-hour workweek which will produce no adverse health effects for nearly all workers, is usually used to determine a safe exposure level. TLV values are published by the American Conference of Governmental Industrial Hygienists, with revisions made yearly as more evidence accrues on the toxicity of the substance. Chapter 12 of this Manual lists the current TLV list as of the date of publication.

Ventilation systems used in industrial plants are of two generic types. The SUPPLY system is used to supply air, usually tempered, to a work space. The EXHAUST system is used to remove the contaminants generated by an operation in order to maintain a healthful work environment.

A complete ventilation program must consider both the supply and the exhaust system. If the overall quantity of air exhausted from a work space is greater than the quantity of outside air supplied to the space, the plant interior will experience a lower pressure than the local atmospheric pressure. This may be desirable when using a dilution ventilation system to control or isolate contaminants in a specific area of the overall plant. Often, this condition occurs simply because local exhaust systems are installed and consideration is not given to the corresponding replacement air systems. Air will then enter the plant in an uncontrolled manner through cracks, walls, windows and doorways. This typically results in 1) employee discomfort in winter months for those working near the plant perimeter, 2) exhaust system performance degradation, possibly leading to loss of contaminant control and a potential health hazard, and 3) higher heating and cooling costs. Chapter 7 of this Manual discusses these points in more detail.

1.2 SUPPLY SYSTEMS

Supply systems are used for two purposes: 1) to create a comfortable environment in the plant (the HVAC system); and 2) to replace air exhausted from the plant (the REPLACEMENT system). Many times, supply and exhaust systems are coupled, as in dilution control systems (see section 1.3 and Chapter 2.)

A well-designed supply system will consist of an air inlet section, filters, heating and/or cooling equipment, a fan, ducts, and register/grilles for distributing the air within the work space. The filters, heating and/or cooling equipment and fan are often combined into a complete unit called an airhouse or air supply unit. If part of the air supplied by a system is recirculated, a RETURN system is used to bring the air back to the airhouse.

1.3 EXHAUST SYSTEMS

Exhaust ventilation systems are classified in two generic groups: 1) the GENERAL exhaust system and 2) the LOCAL exhaust system.

The general exhaust system can be used for heat control and/or removal of contaminants generated in a space by flushing out a given space with large quantities of air. When used for heat control, the air may be tempered and recycled. When used for contaminant control (the dilution system), enough outside air must be mixed with the contaminant so that the average concentration is reduced to a safe level. The contaminated air is then typically discharged to the atmosphere. A supply system is usually used in conjunction with a general exhaust system to replace the air exhausted.

Dilution ventilation systems are normally used for contaminant control only when local exhaust is impractical, as the large quantities of tempered replacement air required to offset the air exhausted can lead to high operating costs. Chapter 2 describes the basic features of general ventilation systems and their application to contaminant and fire hazard control.

Local exhaust ventilation systems operate on the principle of capturing a contaminant at or near its source. It is the preferred method of control because it it more effective and the smaller exhaust flow rate results in lower heating costs compared to high flow rate general exhaust requirements. The present emphasis on air pollution control stresses the need for efficient air cleaning devices on industrial ventilation systems, and the smaller flow rates of the local exhaust system result in lower costs for air cleaning devices.

Local exhaust systems are comprised of up to four basic elements: the hood(s), the duct system (including the exhaust stack and/or recirculation duct), the air cleaning device and the fan. The purpose of the hood is to collect the contaminant generated in an air stream directed toward the hood. A duct system must then transport the contaminated air to the air cleaning device, if present, or to the fan. In the air cleaner, the contaminant is removed from the air stream. The fan must overcome all the losses due to friction, hood entry, and fittings in the system while producing the intended flow rate. Duct on the fan outlet usually discharges the air to the atmosphere in such a way that it will not be re-entrained by the replacement and/or HVAC systems. In some situations, the cleaned air is returned to the plant. Chapter 7 discusses whether this is possible and how it may be accomplished.

This Manual deals with the design aspects of exhaust ventilation systems, but the principles described also apply to supply systems.

1.4 BASIC DEFINITIONS

The following basic definitions are used to describe air flow and will be used extensively in the remainder of the Manual.

The density (ρ) of the air is defined as its mass per unit volume and is normally expressed in pounds mass per cubic foot (lbm/ft^3). At standard atmospheric pressure (14.7 psia), room temperature (70 F) and 0 water content, its value is normally taken to be 0.075 lbm/ft^3, as calculated from the perfect gas equation of state relating pressure, density and temperature:

$$p = \rho RT \qquad [1.1]$$

where:

p = the absolute pressure in pounds per square foot absolute (psfa)

ρ = the density, lbm/ft^3

R = the gas constant for air and equals 53.35 ft-lb/lbm-degrees Rankine

T = the absolute temperature of the air in degrees Rankine

Note that degrees Rankine = degrees Fahrenheit + 459.7.

From the above equation, density varies inversely with temperature when pressure is held constant. Therefore, for any *dry air* situation (see Chapter 5 for moist air calculations),

$$\rho T = (\rho T)_{STD}$$

or

$$\rho = \rho_{STD} \frac{T_{STD}}{T} = 0.075 \frac{530}{T} \qquad [1.2]$$

For example, the density of dry air at 250 F would be

$$\rho = 0.075 \frac{530}{460 + 250} = 0.056 \text{ lbm/ft}^3$$

The volumetric flow rate, many times referred to as "volume," is defined as the volume or quantity of air that passes a given location per unit of time. It is related to the average velocity and the flow cross-sectional area by the equation

$$Q = VA \qquad [1.3]$$

where:

Q = volumetric flow rate, cfm

V = average velocity, fpm

A = cross-sectional area, ft^2

Given any two of these three quantities, the third can readily be determined.

Air or any other fluid will always flow from a region of higher total pressure to a region of lower total pressure in the absence of work addition (a fan). There are three different but mathematically related pressures associated with a moving air stream.

Static pressure (SP) is defined as the pressure in the duct that tends to burst or collapse the duct and is expressed in inches of water gage ("wg). It is usually measured with a water manometer, hence the units. SP can be positive or negative with respect to the local atmospheric pressure, but must be measured perpendicular to the air flow. The holes in the side of a pitot tube (see Figure 9-3) or a small hole carefully drilled to avoid internal burrs that disturb the air flow (never punched) into the side of a duct will yield SP.

Velocity pressure (VP) is defined as that pressure required to accelerate air from zero velocity to some velocity (V) and is proportional to the kinetic energy of the air stream. The relationship between V and VP is given by

$$V = 1096 \sqrt{\frac{VP}{\rho}}$$

or

$$VP = \rho \left(\frac{V}{1096} \right)^2 \qquad [1.4]$$

where:

V = velocity, fpm

VP = velocity pressure, "wg

If standard air is assumed to exist in the duct with a density of 0.075 lbm/ft^3, this equation reduces to

$$V = 4005 \sqrt{VP}$$

or

$$VP = \left(\frac{V}{4005} \right)^2 \qquad [1.5]$$

VP will only be exerted in the direction of air flow and is *always* positive. Figure 1-1 shows graphically the difference

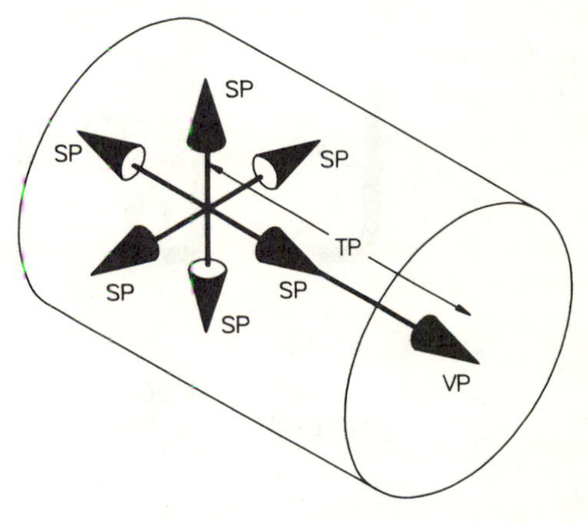

FIGURE 1-1. SP, VP AND TP AT A POINT.

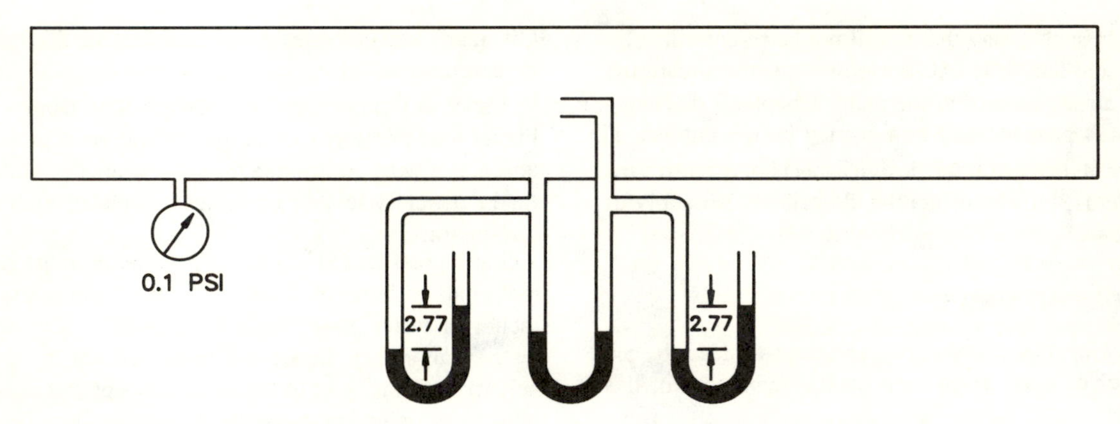

FIGURE 1-2. MEASUREMENT OF SP, VP AND TP IN A PRESSURIZED DUCT.

between SP and VP.

Total pressure (TP) is defined as the algebraic sum of the static and velocity pressures or

$$TP = SP + VP \qquad [1.6]$$

Total pressure can be positive or negative with respect to atmospheric pressure and is a measure of the energy content of the air stream, always dropping as the flow proceeds downstream through a duct. The only place it will rise is across the fan.

Total pressure can be measured with an impact tube pointing directly upstream and connected to a manometer. It will vary across a duct due to the change of velocity across a duct and therefore single readings of TP will not be representative of the energy content. Chapter 9 illustrates procedures for measurement of all pressures in a duct system.

The significance of these pressures can be illustrated as follows. Assume a duct segment with both ends sealed was pressurized to a static pressure of 0.1 psi above the atmospheric pressure as shown in Figure 1-2. If a small hole (typically 1⁄16" to 3⁄32") were drilled into the duct wall and connected to one side of a U-tube manometer, the reading would be approximately 2.77 "wg. Note the way the left-hand manometer is deflected. If the water in the side of the manometer exposed to the atmosphere is higher than the water level in the side connected to the duct, then the pressure read by the gauge is positive (greater than atmospheric). Because there is no velocity, the velocity pressure is 0 and SP = TP. A probe which faces the flow is called an impact tube and will measure TP. In this example, a manometer connected to an impact tube (the one on the right) will also read 2.77 "wg. Finally, if one side of a manometer were connected to the impact tube and the other side were connected to the static pressure opening (the center one), the manometer would read the difference between the two pressures. As VP = TP – SP, a manometer so connected would

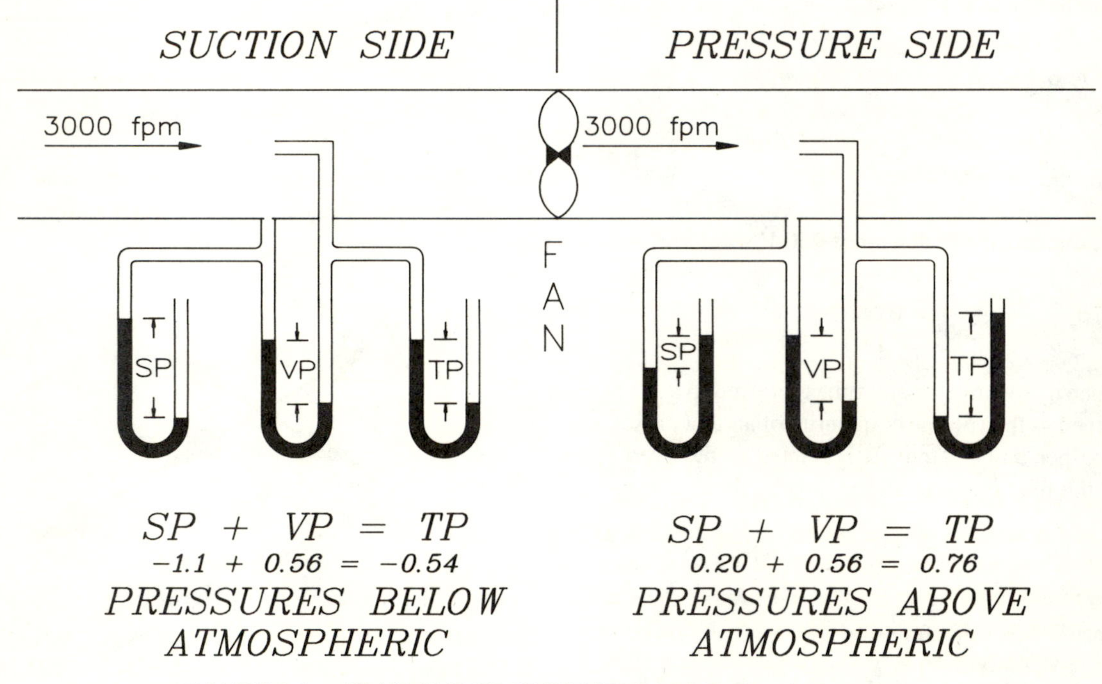

FIGURE 1-3. SP, VP AND TP AT POINTS IN A VENTILATION SYSTEM.

read VP directly. In this example, there is no flow and hence VP = 0 as indicated by the lack of manometer deflection.

If the duct ends were removed and a fan placed midway in the duct, the situation might change to the one shown on Figure 1-3. Upstream of the fan, SP and TP are negative (less than atmospheric). This is called the *suction side*. Downstream of the fan, both SP and TP are positive. This is called the *pressure side*. Regardless of which side of the fan is considered, VP is always positive. Note that the direction in which the manometers are deflected shows whether SP and TP are positive or negative with respect to the local atmospheric pressure.

1.5 PRINCIPLES OF AIR FLOW

Two basic principles of fluid mechanics govern the flow of air in industrial ventilation systems: conservation of mass and conservation of energy. These are essentially bookkeeping laws which state that all mass and all energy must be completely accounted for. A coverage of fluid mechanics is not in the purview of this manual; reference to any standard fluid mechanics textbook will show the derivation of these principles. However, it is important to know what simplifying assumptions are included in the principles discussed below. They include:

1. Heat transfer effects are neglected. If the temperature inside the duct is significantly different from the air temperature surrounding the duct, heat transfer will occur. This will lead to changes in the duct air temperature and hence in the volumetric flow rate.
2. Compressibility effects are neglected. If the overall pressure drop from the start of the system to the fan is greater than about 20 "wg, then the density will change by about 5% and the volumetric flow rate will also change (see Chapter 5).
3. The air is assumed to be dry. Water vapor in the airstream will lower the air density and correction for this effect, if present, should be made. Chapter 5 describes the necessary psychrometric analysis.
4. The weight and volume of the contaminant in the air stream is ignored. This is permissible for the contaminant concentrations in typical exhaust ventilation systems. For high concentrations of solids or significant amounts of gases other than air, corrections for this effect should be included.

Conservation of mass requires that the net change of mass flow rate must be zero. If the effects discussed above are negligible, then the density will be constant and the net change of volumetric flow rate (Q) must be zero. Therefore, the flow rate that enters a hood must be the same as the flow rate that passes through the duct leading from the hood. At a branch entry (converging wye) fitting, the sum of the two flow rates that enter the fitting must leave it. At a diverging wye, the flow rate entering the wye must equal the sum of the flow rates that leave it. Figure 1-4 illustrates these concepts.

Conservation of energy means that all energy changes must be accounted for as air flows from one point to another. In terms of the pressures previously defined, this principle can be expressed as:

$$TP_1 = TP_2 + h_L$$

$$SP_1 + VP_1 = SP_2 + VP_2 + h_L \quad [1.7]$$

where:

subscript 1 = some upstream point
subscript 2 = some downstream point
h_L = the energy losses encountered by the air as it flows from the upstream to the downstream point.

Note that, according to this principle, the *total pressure must fall in the direction of flow.*

The application of these principles will be demonstrated by an analysis of the simple system shown in Figure 1-5. The normally vertical exhaust stack is shown laying horizontally to facilitate graphing the variation of static, total and velocity pressures. The grinder wheel hood requires 300 cfm and the duct diameter is constant at 3.5 inches (0.0668 ft^2 area).

1.6 ACCELERATION OF AIR AND HOOD ENTRY LOSSES

Air flows from the room (point 1 of Figure 1-5) through the hood to the duct (point 2 of Figure 1-5) where the velocity can be calculated by the basic equation:

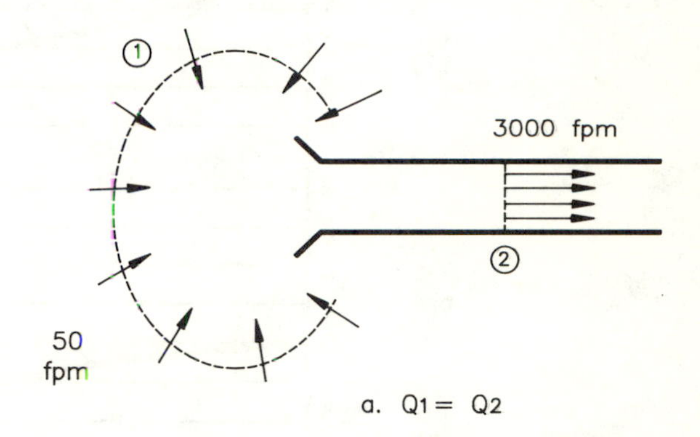

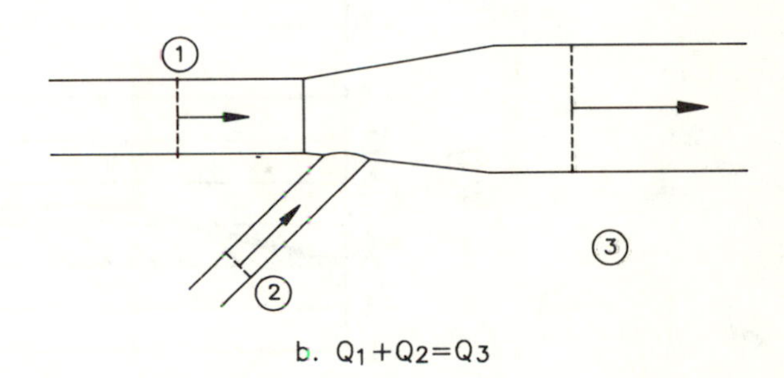

FIGURE 1-4. VOLUMETRIC FLOW RATES IN VARIOUS SITUATIONS. A. FLOW THROUGH A HOOD; B. FLOW THROUGH A BRANCH ENTRY.

$$V = \frac{Q}{A}$$

$$= \frac{300}{0.0668} = 4490 \text{ fpm}$$

This velocity corresponds to a velocity pressure of 1.26 "wg, assuming standard air.

If there are no losses associated with entry into a hood, then applying the energy conservation principle (Equation 1.7) to the hood yields

$$SP_1 + VP_1 = SP_2 + VP_2$$

This is the well known Bernoulli principle of fluid mechanics. Subscript l refers to the room conditions where the static pressure is atmospheric ($SP_1 = 0$) and the air velocity is assumed to be very close to zero ($VP_1 = 0$). Therefore, the energy principle yields

$$SP_2 = -VP_2 = -1.26 \text{ "wg}$$

Even if there were no losses, the *static pressure must decrease due to the acceleration of air to the duct velocity.*

In reality, there are losses as the air enters the hood. These entry losses (h_e) are normally expressed as a loss coefficient (F_h) multiplied by the duct velocity pressure, so $h_e = F_hVP$. The energy conservation principle then becomes

$$SP_2 = -(VP_2 + h_e) \quad \textbf{[1.8]}$$

The absolute value of SP_2 is known as the hood static suction (SP_h). Then

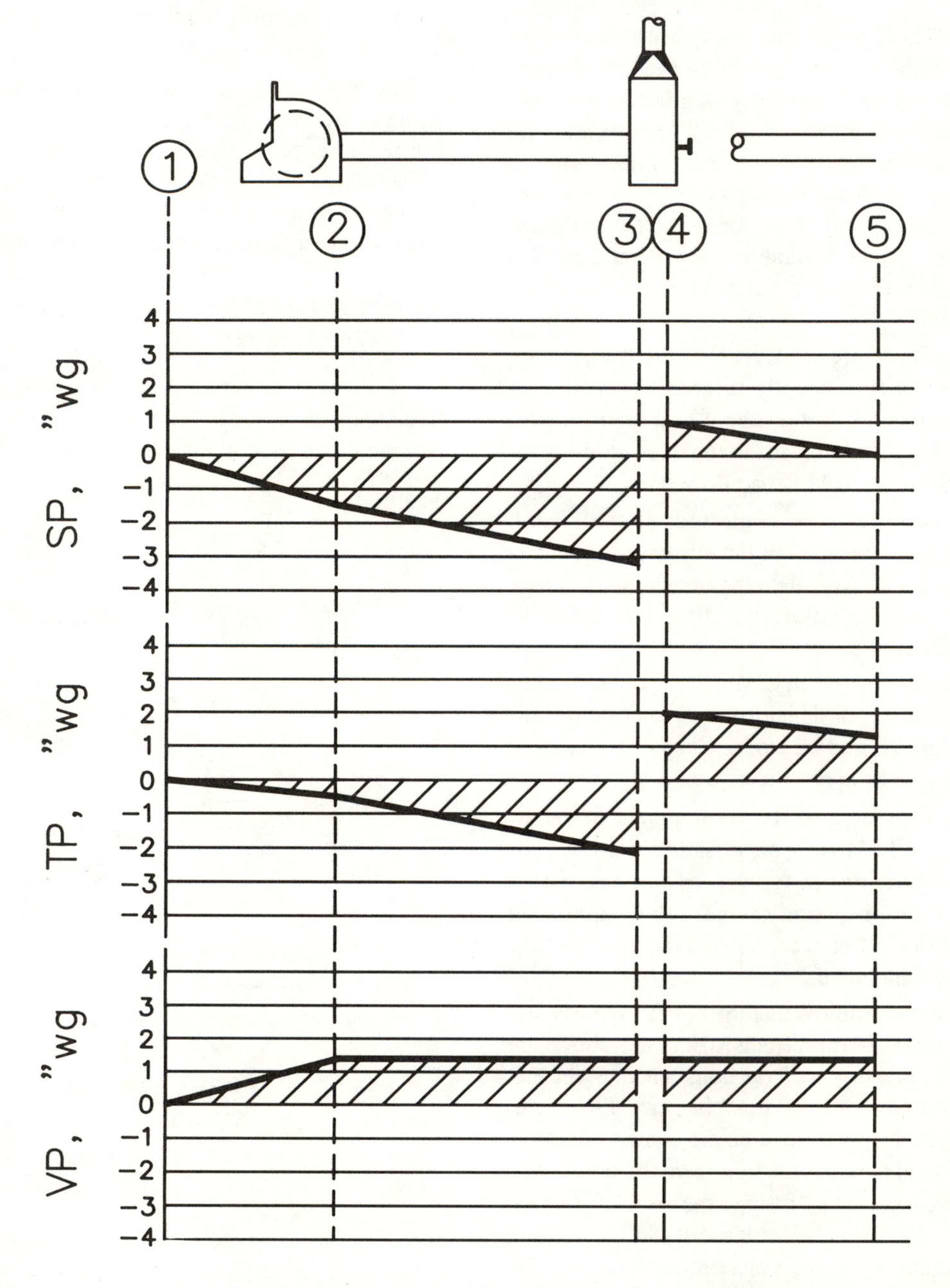

FIGURE 1-5. VARIATION OF SP, VP AND TP THROUGH A VENTILATION SYSTEM.

$$SP_h = -SP_2 = VP_2 + h_e \quad [1.9]$$

For the example in Figure 1-5, assuming an entry loss coefficient of 0.40,

$$SP_h = VP_2 + F_hVP_2$$
$$= 1.26 + (0.40)(1.26)$$
$$= 1.26 + 0.50$$
$$= 1.76 \text{ "wg}$$

In summary, the static pressure downstream of the hood is negative (less than atmospheric) due to two effects:

1. Acceleration of air to the duct velocity; and
2. Hood entry losses.

From the graph, note that $TP_2 = -h_e$, which confirms the premise that total pressure decreases in the flow direction.

An alternate method of describing hood entry losses is by the hood entry coefficient (C_e). This coefficient is defined as the square root of the ratio of duct velocity pressure to hood static suction, or

$$C_e = \sqrt{\frac{VP}{SP_h}} \quad [1.10]$$

If there were no losses, then $SP_h = VP$ and $C_e = 1.00$. However, as hoods always have some losses, C_e is always less than 1.00. In Figure 1-5,

$$C_e = \sqrt{\frac{VP}{SP_h}}$$
$$= \sqrt{\frac{1.26}{1.76}}$$
$$= 0.845$$

An important feature of C_e is that it is a constant for any given hood. It can, therefore, be used to determine the flow rate if the hood static suction is known. This is because

$$Q = VA = 1096\,A\sqrt{\frac{VP}{\rho}} = 1096\,A\,C_e\sqrt{\frac{SP_h}{\rho}} \quad [1.11]$$

For standard air, this equation becomes

$$Q = 4005\,A\,C_e\sqrt{SP_h} \quad [1.12]$$

For the example in Figure 1-5,

$$Q = 4005(0.0668)(0.845)\sqrt{1.76} = 300 \text{ cfm}$$

By use of C_e and a measurement of SP_h, the flow rate of a hood can be quickly determined and corrective action can be taken if the calculated flow rate does not agree with the design flow rate.

1.7 DUCT LOSSES

There are two components to the overall total pressure losses in a duct run: 1) friction losses and 2) fitting losses.

1.7.1 Friction Losses: Losses due to friction in ducts are a complicated function of duct velocity, duct diameter, air density, air viscosity, and duct surface roughness. The effects of velocity, diameter, density and viscosity are combined into the *Reynolds number (Re),* as given by

$$Re = \frac{\rho d v}{\mu} \quad [1.13]$$

where:

d = diameter, ft

v = velocity, ft/sec

μ = the air viscosity, lbm/s-ft

The effect of surface roughness is typically given by the *relative roughness,* which is the ratio of the absolute surface roughness height (k), defined as the average height of the roughness elements on a particular type of material, to the duct diameter. Some standard values of absolute surface roughness used in ventilation systems are given in Table 1-1.

TABLE 1-1. Absolute Surface Roughness

Duct Material	Surface Roughness (k), feet
Galvanized metal	0.0005
Black iron	0.00015
Aluminum	0.00015
Stainless steel	0.00015
Flexible duct (wires exposed)	0.01
Flexible duct (wires covered)	0.003

The above roughness heights are design values. It should be noted that significant variations from these values may occur, depending on the manufacturing process.

L. F. Moody[1.1] combined these effects into a single chart commonly called the *Moody diagram* (see Figure 1-6). With a knowledge of both the Reynolds number and the relative roughness, the *friction factor (f),* can be found.

Once determined, the friction factor is used in the *Darcy-Weisbach friction factor equation* to determine the overall duct friction losses:

$$h_L = f\,\frac{L}{d}\,VP$$

where:

h_L = friction losses in a duct, "wg

f = Moody diagram friction factor (dimensionless)

L = duct length, ft

d = duct diameter, ft

VP = duct velocity pressure, "wg

There are many equations available for computer solutions to the Moody diagram. One of these is that of Churchill,[1.2] which gives accurate (to within a few percent) results over the entire range of laminar, critical, and turbulent flow, all in a single equation. This equation is:

$$f = 8\left[\left(\frac{8}{Re}\right)^{12} + (A + B)^{-3/2}\right]^{1/12} \quad [1.15]$$

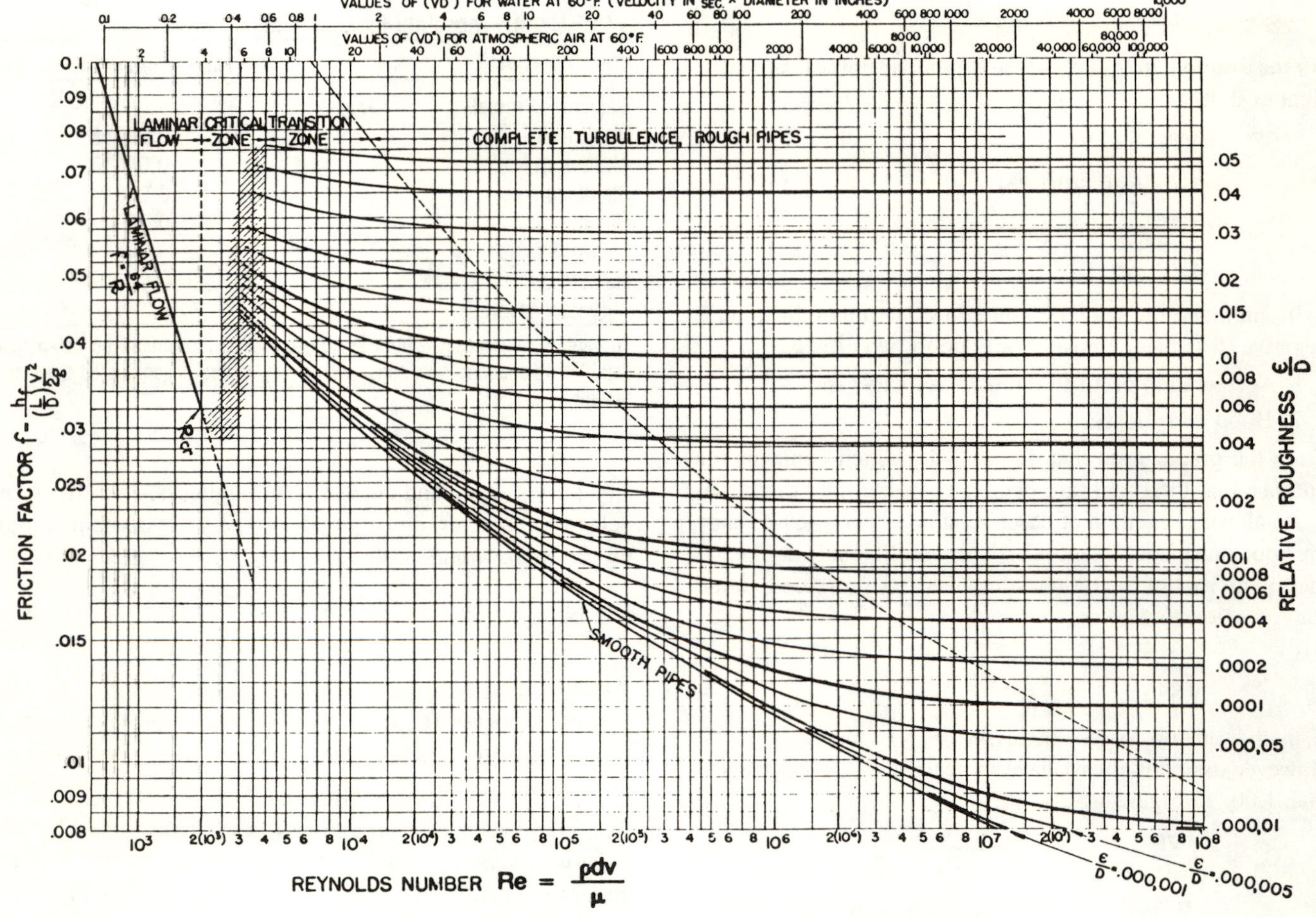

FIGURE 1-6. MOODY DIAGRAM (adapted from reference 1.1).

where:

$$A = \left\{-2.457 \ln\left[\left(\frac{7}{Re}\right)^{0.9} + \left(\frac{k}{3.7D}\right)\right]\right\}^{16}$$

$$B = \left(\frac{37{,}530}{Re}\right)^{16}$$

While useful, this equation is quite difficult to use without a computer. Several attempts have been made to simplify the determination of friction losses for specialized situations. For many years, charts based on the Wright[(1.3)] equation have been used in ventilation system design:

$$h_L = 2.74\frac{(V/1000)^{1.9}}{D^{1.22}} \qquad [1.16]$$

where:
 V = duct velocity, fpm
 D = duct diameter, inches

This equation gives the friction losses, expressed as "wg per 100 feet of pipe, for standard air of 0.075 lbm/ft^3 density flowing through average, clean, round galvanized pipe having approximately 40 slip joints per 100 feet (k = 0.0005 ft) and is the basis for the *equivalent length* friction chart shown on Figure 5-16.

The later work by Loeffler[(1.4)] presented equations for use in the *"velocity pressure"* calculation method. Using the standard values of surface roughness, equations were obtained that could be used with the Darcy-Weisbach equation in the form:

$$h_L = \left(12\,\frac{f}{D}\right)L\,VP = H_f\,L\,VP \qquad [1.17]$$

where the "12" is used to convert the diameter D in inches to feet.

Simplified equations were determined for the flow of standard air through various types of duct material with good accuracy (less than 5% error.) The equations thus resulting were:

$$H_f = 12\frac{f}{D} = \frac{aV^b}{Q^c} \qquad [1.18]$$

where the constant "a" and the exponents "b" and "c" vary as a function of the duct material as shown in Table 1-2. Note that no correlation with the extremely rough flexible duct with wires exposed was made. This equation, using the constants from Table 1-2 for galvanized sheet duct, were used to develop the friction chart of Figure 5-18. *Note that the value obtained from the chart or from equation 1.18 must be multiplied by both the length of duct and the velocity pressure.*

TABLE 1-2. Correlation Equation Constants

Duct Material	k, Ft	a	b	c
Aluminum, black iron, stainless steel	0.00015	0.0425	0.465	0.602
Galvanized sheet duct	0.0005	0.0307	0.533	0.612
Flexible duct, fabric wires covered	0.003	0.0311	0.604	0.639

1.7.2 Fitting Losses: The fittings (elbows, entries, etc.) in a duct run will also produce a loss in total pressure. These losses are given by one of two methods: 1) the velocity pressure method and 2) the equivalent length method.

In the velocity pressure method, the fitting losses are given by a loss coefficient (F) multiplied by the duct velocity pressure. Thus,

$$h_L = F\ VP \qquad [1.19]$$

In contractions, entries, or expansions, there are several different velocity pressures. The proper one to use with the loss coefficient will be identified where the coefficients are listed.

In the equivalent length method, the fitting is considered to be equivalent to a length of straight duct that will yield the same loss as the fitting. This loss is a function of duct size and velocity pressure. Figure 5-20 corresponds to the losses expected for duct velocities of about 4000 fpm.

In Figure 1-5, 15 feet of straight, constant diameter galvanized duct connects the hood to a fan inlet. Because the duct area is constant, the velocity, and therefore the velocity pressure, is also constant for any given flow rate. The energy principle is:

$$SP_2 + VP_2 = SP_3 + VP_3 + h_L$$

where subscript 3 refers to the fan inlet location. Because $VP_2 = VP_3$, the losses will appear as a reduction in static pressure (there will, of course, be a corresponding reduction in total pressure). The friction loss can be found from Equation 1.17 with the aid of Equation 1.18:

$$H_f = 0.0307\frac{V^{0.533}}{Q^{0.612}}$$

$$= 0.0307\frac{4490^{0.533}}{300^{0.612}}$$

$$= 0.0828$$

From Equation 1.17, $h_L = (0.0828)(15)(1.26) = 1.56$ "wg. Using this in the energy principle,

$$SP_3 = SP_2 - h_L = -1.76\text{ "wg} - 1.56\text{ "wg} = -3.32\text{ "wg}$$

Another 10 feet of straight duct is connected to the discharge side of the fan. The losses from the fan to the end of the system would be about 1.04 "wg. Because the static pressure at the end of the duct must be atmospheric ($SP_5 = 0$), the energy principle results in

$$SP_4 = SP_5 + h_L = 0\text{ "wg} + 1.04\text{ "wg} = 1.04\text{ "wg}$$

Therefore, the static pressure at the fan outlet must be higher than atmospheric by an amount equal to the losses in the discharge duct.

1.8 MULTIPLE-HOOD EXHAUST SYSTEMS

Most exhaust systems are more complicated than the preceding example. It is usually more economical to purchase a single fan and air cleaner to service a series of similar operations than to create a complete system for each operation. For example, the exhaust from 10 continuously used grinders can be combined into a single flow which leads to a common air cleaner and fan. This situation is handled similarly to a simple system, but with some provision to ensure that the air flow from each hood is as desired (see Chapter 5).

1.9 AIR FLOW CHARACTERISTICS OF BLOWING AND EXHAUSTING

Air blown from a small opening retains its directional effect for a considerable distance beyond the plane of the opening. However, if the flow of air *through the same opening* were reversed so that it operated as an exhaust opening handling the same volumetric flow rate, the flow would become almost non-directional and its range of influence would be greatly reduced. For this reason, local exhaust must

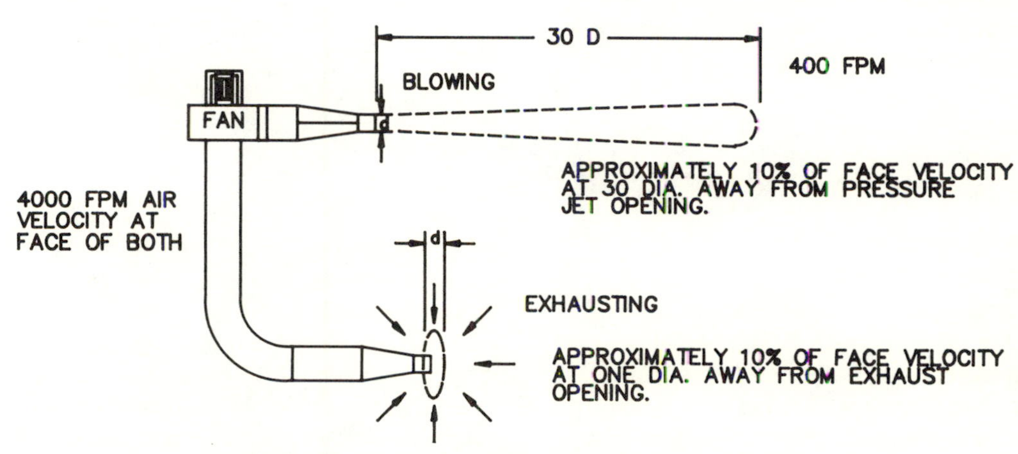

FIGURE 1-7. BLOWING VERSUS EXHAUSTING.

not be contemplated for any process which cannot be conducted in the *immediate* vicinity of the hood. Also, because of this effect, every effort should be made to enclose the operation as much as possible. Figure 1-7 illustrates the fundamental difference between blowing and exhausting.

This effect also shows how the supply or replacement air discharge grilles can influence an exhaust system. If care is not taken, the discharge pattern from a supply grille could seriously affect the flow pattern in front of an exhaust hood.

REFERENCES

1.1. L.F. Moody: "Friction Factors for Pipe Flow." *ASME Trans.* 66:672 (1944).

1.2. S.W. Churchill: "Friction Factor Equation Spans All Fluid Flow Regimes." *Chemical Engineering*, Vol. 84 (1977).

1.3. D.K. Wright, Jr.: "A new Friction Chart for Round Ducts." *ASHVE Trans.*, Vol. 51, Appendix I, p. 312 (1945).

1.4. J.J. Loeffler: "Simplified Equations for HVAC Duct Friction Factors." *ASHRAE J.*, p. 76 (January 1980).

Chapter 2
GENERAL INDUSTRIAL VENTILATION

2.1 INTRODUCTION 2-2

2.2 DILUTION VENTILATION PRINCIPLES 2-2

2.3 DILUTION VENTILATION FOR HEALTH . 2-2
- 2.3.1 General Dilution Ventilation Equation 2-2
- 2.3.2 Calculating Dilution Ventilation for Steady State Concentration 2-5
- 2.3.3 Contaminant Concentration Buildup 2-5
- 2.3.4 Rate of Purging 2-6

2.4 MIXTURES — DILUTION VENTILATION FOR HEALTH 2-6

2.5 DILUTION VENTILATION FOR FIRE AND EXPLOSION 2-7

2.6 FIRE DILUTION VENTILATION FOR MIXTURES 2-8

2.7 VENTILATION FOR HEALTH CONTROL 2-8

2.8 HEAT BALANCE AND EXCHANGE 2-8
- 2.8.1 Convection 2-9
- 2.8.2 Radiation 2-9
- 2.8.3 Evaporation 2-9

2.9 ADAPTIVE MECHANISM OF THE BODY 2-9

2.10 ACCLIMATIZATION 2-9

2.11 ACUTE HEAT DISORDERS 2-10
- 2.11.1 Heatstroke 2-10
- 2.11.2 Heat Exhaustion 2-10
- 2.11.3 Heat Cramps 2-10
- 2.11.4 Heat Rash 2-10

2.12 HEAT STRESS MEASUREMENT 2-10
- 2.12.1 Dry Build (Air Temperature) 2-10
- 2.12.2 Natural Wet Bulb Temperature 2-10
- 2.12.3 Psychrometric Wet Bulb Temperature 2-11
- 2.12.4 Air Velocity 2-11
- 2.12.5 Radiant Heat 2-11
- 2.12.6 Betabolic Heat Estimates 2-11

2.13 HEAT STRESS INDICES 2-11
- 2.13.1 WBGT Index 2-11
- 2.13.2 Wet Globe Temperature Index (WGT Index) 2-12

2.14 VENTILATION CONTROL 2-12

2.15 VENTILATION SYSTEMS 2-12

2.16 VELOCITY COOLING 2-14

2.17 RADIANT HEAT CONTROL 2-15

2.18 PROTECTIVE SUITS FOR SHORT EXPOSURES 2-15

2.19 RESPIRATORY HEAT EXCHANGERS 2-15

2.20 REFRIGERATED SUITS 2-15

2.21 ENCLOSURES 2-15

2.22 INSULATION 2-15

REFERENCES 2-16

2.1 INTRODUCTION

"General industrial ventilation" is a broad term which refers to the supply and exhaust of air with respect to an area, room or building. It can be divided further into specific functions as follows:

1. *Dilution Ventilation* — is the dilution of contaminated air with uncontaminated air for the purpose of controlling potential airborne health hazards, fire and explosive conditions, odors, and nuisance type contaminants. Dilution ventilation also can include the control of airborne contaminants (vapors, gases and particulates) generated within tight buildings.

   Dilution ventilation is not as satisfactory for health hazard control as is local exhaust ventilation. Circumstances may be found in which dilution ventilation provides an adequate amount of control more economically than a local exhaust system. One should be careful, however, not to base the economical considerations entirely upon the first cost of the system since dilution ventilation frequently exhausts large amounts of heat from a building which may greatly increase the energy cost of the operation.
2. *Heat Control Ventilation* — is the control of indoor atmospheric conditions associated with hot industrial environments such as are found in foundries, laundries, bakeries, etc., for the purpose of preventing acute discomfort or injury.

2.2 DILUTION VENTILATION PRINCIPLES

The principles of dilution ventilation system design are as follows:

1. Select from available data the amount of air required for satisfactory dilution of the contaminant. The values tabulated on Table 2-1 assume perfect distribution and dilution of the air and solvent vapors. These values must be multiplied by the selected K value (see Section 2.3.1).
2. Locate the exhaust openings near the sources of contamination, if possible, in order to obtain the benefit of "spot ventilation."
3. Locate the air supply and exhaust outlets such that the air passes through the zone of contamination. The operator should remain between the air supply and the source of the contaminant.
4. Replace exhausted air by use of a replacement air system. This replacement air should be heated during cold weather. Dilution ventilation systems usually handle large quantities of air by means of low pressure fans. Replacement air must be provided if the system is to operate satisfactorily.
5. Avoid re-entry of the exhausted air by discharging the exhaust high above the roof line or by assuring that no window, outside air intakes, or other such openings are located near the exhaust discharge.

2.3 DILUTION VENTILATION FOR HEALTH

The use of dilution ventilation for health has four limiting factors: 1) the quantity of contaminant generated must not be too great or the air flow rate necessary for dilution will be impractical; 2) workers must be far enough away from the contaminant source or evolution of contaminant must be in sufficiently low concentrations so that workers will not have an exposure in excess of the established TLV; 3) the toxicity of the contaminant must be low; and 4) the evolution of contaminants must be reasonably uniform.

Dilution ventilation is used most often to advantage to control the vapors from organic liquids with a TLV of 100 ppm or higher. In order to successfully apply the principles of dilution to such a problem, factual data are needed on the rate of vapor generation or on the rate of liquid evaporation. Usually such data can be obtained from the plant if any type of adequate records on material consumption are kept.

2.3.1 General Dilution Ventilation Equation: The ventilation rate needed to maintain a constant concentration at a uniform generation rate is derived by starting with a fundamental material balance and assuming no contaminant in the air supply,

$$\text{Rate of Accumulation} = \text{Rate of Generation} - \text{Rate of Removal}$$

or

$$VdC = Gdt - Q'\,Cdt \qquad [2.1]$$

where:

V = volume of room
G = rate of generation
Q' = effective volumetric flow rate
C = concentration of gas or vapor
t = time

At a steady state,

$$dC = 0$$

$$Gdt = Q'\,Cdt$$

$$\int_{t_1}^{t_2} Gdt = \int_{t_1}^{t_2} Q'\,Cdt$$

At a constant concentration, C, and uniform generation rate, G,

$$G(t_2 - t_1) = Q'\,C\,(t_2 - t_1)$$

$$Q' = \frac{G}{C} \qquad [2.2]$$

Due to incomplete mixing, a K value is introduced to the rate of ventilation; thus:

$$Q' = \frac{Q}{K} \qquad [2.3]$$

TABLE 2-1. Dilution Air Volumes for Vapors

The following values are tabulated using the TLV values shown in parentheses, parts per million. TLV values are subject to revision if further research or experience indicates the need. If the TLV value has changed, the dilution air requirements must be recalculated. The values on the table must be multiplied by the evaporation rate (pts/min) to yield the effective ventilation rate (Q′) (see Equation 2.5).

Liquid (TLV in ppm)*	Ft^3 of Air (STP) Required for Dilution to TLV** Per Pint Evaporation
Acetone (750)	7,350
n-Amyl acetate (100)	27,200
Benzene (10)	NOT RECOMMENDED
n-Butanol (butyl alcohol) (50)	88,000
n-Butyl acetate (150)	20,400
Butyl Cellosolve (2-butoxyethanol) (25)	NOT RECOMMENDED
Carbon disulfide (10)	NOT RECOMMENDED
Carbon tetrachloride (5)	NOT RECOMMENDED
Cellosolve (2-ethoxyethanol) (5)	NOT RECOMMENDED
Cellosolve acetate (2-ethoxyethyl acetate) (5)	NOT RECOMMENDED
Chloroform (10)	NOT RECOMMENDED
1-2 Dichloroethane (10) (ethylene dichloride)	NOT RECOMMENDED
1-2 Dichloroethylene (200)	26,900
Dioxane (25)	NOT RECOMMENDED
Ethyl acetate (400)	10,300
Ethyl alcohol (1000)	6,900
Ethyl ether (400)	9,630
Gasoline (300)	REQUIRES SPECIAL CONSIDERATION
Isoamyl alcohol (100)	37,200
Isopropyl alcohol (400)	13,200
Isopropyl ether (250)	11,400
Methyl acetate (200)	25,000
Methyl alcohol (200)	49,100
Methyl n-butyl ketone (5)	NOT RECOMMENDED
Methyl Cellosolve (2-methoxyethanol) (5)	NOT RECOMMENDED
Methyl Cellosolve acetate (2-methoxyethyl acetate) (5)	NOT RECOMMENDED
Methyl chloroform (350)	11,390
Methyl ethyl ketone (200)	22,500
Methyl isobutyl ketone (50)	64,600
Methyl propyl ketone (200)	19,900
Naptha (coal tar)	REQUIRES SPECIAL CONSIDERATION
Naptha VM & P (300)	REQUIRES SPECIAL CONSIDERATION
Nitrobenzene (1)	NOT RECOMMENDED
n-Propyl acetate (200)	17,500
Stoddard solvent (100)	30,000-35,000
1,1,2,2-Tetrachloroethane (1)	NOT RECOMMENDED
Tetrachloroethylene (50)	79,200
Toluene (100)	38,000
Trichloroethylene (50)	90,000
Xylene (100)	33,000

*See Threshold Limit Values 1988-89 in Appendix A.

**The tabulated dilution air quantities must be multiplied by the selected K value.

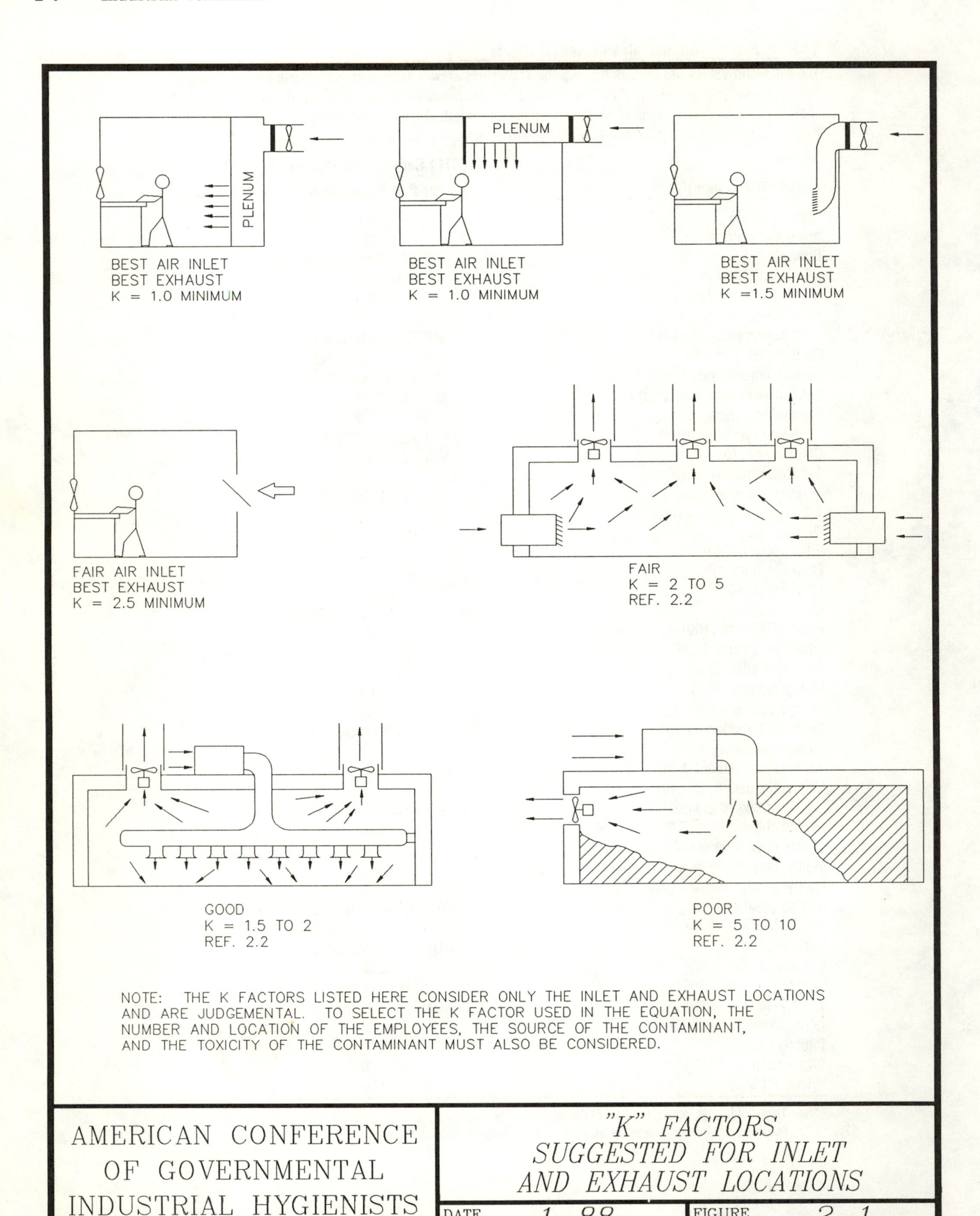
PLENUM
BEST AIR INLET
BEST EXHAUST
K = 1.0 MINIMUM
PLENUM
BEST AIR INLET
BEST EXHAUST
K = 1.0 MINIMUM
BEST AIR INLET
BEST EXHAUST
K =1.5 MINIMUM
FAIR AIR INLET
BEST EXHAUST
K = 2.5 MINIMUM
FAIR
K = 2 TO 5
REF. 2.2
GOOD
K = 1.5 TO 2
REF. 2.2
POOR
K = 5 TO 10
REF. 2.2
NOTE: THE K FACTORS LISTED HERE CONSIDER ONLY THE INLET AND EXHAUST LOCATIONS AND ARE JUDGEMENTAL. TO SELECT THE K FACTOR USED IN THE EQUATION, THE NUMBER AND LOCATION OF THE EMPLOYEES, THE SOURCE OF THE CONTAMINANT, AND THE TOXICITY OF THE CONTAMINANT MUST ALSO BE CONSIDERED.
AMERICAN CONFERENCE OF GOVERNMENTAL INDUSTRIAL HYGIENISTS
"K" FACTORS SUGGESTED FOR INLET AND EXHAUST LOCATIONS
DATE 1-88
FIGURE 2-1

where:

Q = actual ventilation rate, cfm

Q′ = effective ventilation rate, cfm

K = a factor to allow for incomplete mixing

Equation 2.2 then becomes:

$$Q = \left(\frac{G}{C}\right) K \qquad [2.4]$$

This K factor is based on several considerations:

1. The efficiency of mixing and distribution of replacement air introduced into the room or space being ventilated (see Figure 2-1).
2. The toxicity of the solvent. Although TLV and toxicity are not synonymous, the following guidelines have been suggested for choosing the appropriate K value:

   Slightly toxic material: TLV > 500 ppm

   Moderately toxic material: TLV 100 = 500 ppm

   Highly toxic material: TLV < 100 ppm
3. A judgement of any other circumstances which the industrial hygienist determined to be of importance — based on experience and the individual problem. Included in these criteria are such considerations as:
   a. Duration of the process, operational cycle and normal locations of workers relative to sources of contamination.
   b. Location and number of points of generation of the contaminant in the workroom or area.
   c. Seasonal changes in the amount of natural ventilation.
   d. Reduction in operational effectiveness of mechanical air moving devices.
   e. Other circumstances which may affect the concentration of hazardous material in the breathing zone of the workers.

The K value selected, depending on the above considerations, ranges from 1 to 10.

2.3.2 Calculating Dilution Ventilation for Steady State Concentration: The concentration of a gas or vapor at a steady state can be expressed by the material balance equation

$$Q' = \frac{G}{C}$$

Therefore, the rate of flow of uncontaminated air required to maintain the atmospheric concentration of a hazardous material at an acceptable level can be easily calculated if the generation rate can be determined. Usually, the acceptable concentration (C) expressed in parts per million (ppm) is considered to be the Threshold Limit Value (TLV). For liquid solvents, the rate of generation is

$$G = \frac{403 \times SG \times ER}{MW}$$

where:

G = generation rate, cfm

403 = the volume in ft^3 that 1 pt of liquid, when vaporized, will occupy at STP, ft^3/pt

SG = specific gravity of volatile liquid

ER = evaporation rate of liquid, pts/min

MW = molecular weight of liquid

Thus, Q′ = G/C can be expressed as

$$Q' = \frac{403 \times 10^6 \times SG \times ER}{MW \times C} \qquad [2.5]$$

EXAMPLE PROBLEM

Methyl chloroform is lost by evaporation from a tank at a rate of 1.5 pints per 60 minutes. What is the effective ventilation rate (Q′) and the actual ventilation rate (Q) required to maintain the vapor concentration at the TLV?

TLV = 350 ppm, SG = 1.32, MW = 133.4, Assume K = 5

Assuming perfect dilution, the effective ventilation rate (Q′) is

$$Q' = \frac{(403)(10^6)(1.32)(1.5/60)}{(133.4)(350)} = 285 \text{ cfm}$$

Due to incomplete mixing the actual ventilation rate (Q) is

$$Q = \frac{(403)(10^6)(1.32)(1.5/60)(5)}{(133.4)(350)} = 1425 \text{ cfm}$$

2.3.3 Contaminant Concentration Buildup (see Figure 2-2): The concentration of a contaminant can be calculated after any change of time. Rearranging the differential material balance results in

$$\frac{dC}{G - Q'C} = \frac{dt}{V}$$

which can be integrated to yield

$$\ln\left(\frac{G - Q'C_2}{G - Q'C_1}\right) = -\frac{Q'(t_2 - t_1)}{V} \qquad [2.6]$$

where subscript 1 refers to the initial condition and subscript 2 refers to the final condition. If it is desired to calculate the time required to reach a given concentration, then rearranging $t_2 - t_1$, or Δt, gives

$$\Delta t = -\frac{V}{Q'}\left[\ln\left(\frac{G - Q'C_2}{G - Q'C_1}\right)\right] \qquad [2.7]$$

If $C_1 = 0$, then the equation becomes

$$\Delta t = -\frac{V}{Q'}\left[\ln\left(\frac{G - Q'C_2}{G}\right)\right] \qquad [2.8]$$

Note: the concentration C_2 is ppm or parts/10^6 (e.g., 200 ppm is 200/10^6).

If it is desired to determine the concentration level (C_2) after a certain time interval, $t_2 - t_1$ or Δt, and if $C_1 = 0$, then the equation becomes

$$C_2 = \frac{G\left[1 - e^{\left(-\frac{Q'\Delta t}{V}\right)}\right]}{Q'} \quad \text{[2.9]}$$

Note: to convert C_2 to ppm, multiply the answer by 10^6.

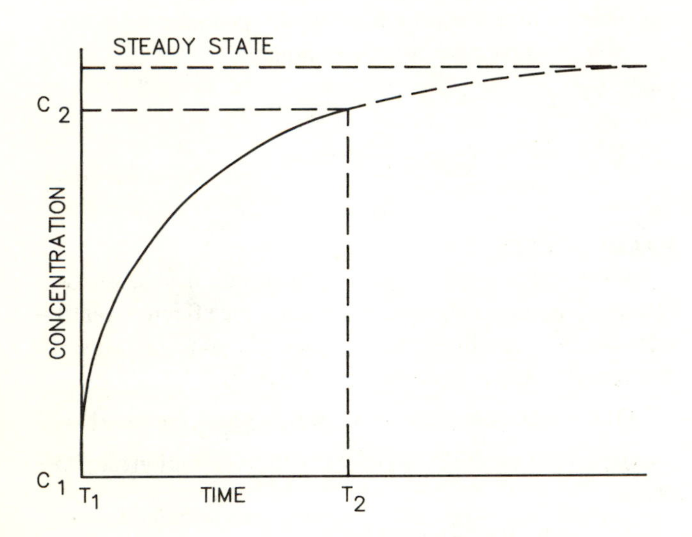

FIGURE 2-2.

EXAMPLE

Methyl chloroform vapor is being generated under the following conditions: G = 1.2 cfm; Q′ = 2,000 cfm; V = 100,000 cu ft; C_1 = 0; K = 3. How long before the concentration (C_2) reaches 200 ppm or $200/10^6$?

$$\Delta t = -\frac{V}{Q'}\left[\ln\left(\frac{G - Q' C_2}{G}\right)\right] = 20.3 \text{ min}$$

Using the same values as in the preceding example, what will be the concentration after 60 minutes?

$$C_2 = \frac{G\left[1 - e^{\left(-\frac{Q'\Delta t}{V}\right)}\right]}{Q'} \times 10^6 = 419 \text{ ppm}$$

2.3.4 Rate of Purging (see Figure 2-3): Where a quantity of air is contaminated but where further contamination or generation has ceased, the rate of decrease of concentration over a period of time is as follows:

$$V dC = -Q' C dt$$

$$\int_{C_1}^{C_2} \frac{dC}{C} = -\frac{Q'}{V}\int_{t_1}^{t_2} dt$$

$$\ln\left(\frac{C_2}{C_1}\right) = -\frac{Q'}{V}(t_2 - t_1)$$

or,

$$C_2 = C_1 e^{\left[-\frac{Q'(t_2 - t_1)}{V}\right]} \quad \text{[2.10]}$$

EXAMPLE

In the room of the example in Section 2.3.3, assume that

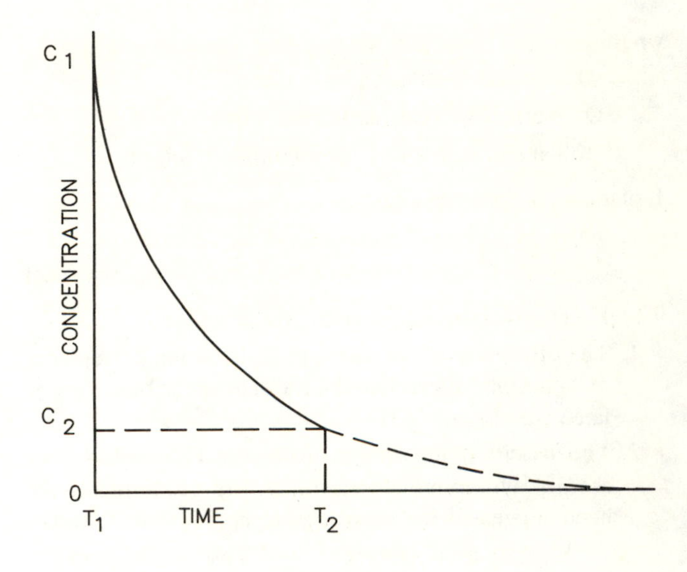

FIGURE 2-3.

ventilation continues at the same rate (Q′ = 2000 cfm), but that the contaminating process is interrupted. How much time is required to reduce the concentration from 100 (C_1) to 25 (C_2) ppm?

$$t_2 - t_1 = -\frac{V}{Q}\ln\left(\frac{C_2}{C_1}\right) = 69.3 \text{ min}$$

In the problem above, if the concentration (C_1) at t_1 is 100 ppm, what will concentration (C_1) be after 60 minutes (Δt)?

$$C_2 = C_1 e^{\left(-\frac{Q'\Delta t}{V}\right)} = 30.1 \text{ ppm}$$

2.4 MIXTURES — DILUTION VENTILATION FOR HEALTH

In many cases the parent liquid for which dilution ventilation rates are being designed will consist of a mixture of solvents. The common procedure used in such instances is as follows.

When two or more hazardous substances are present, their combined effect, rather than that of either individually, should be given primary consideration. *In the absence of information to the contrary, the effects of the different hazards should be considered as additive.* That is, if the sum of the following fractions,

$$\frac{C_1}{TLV_1} + \frac{C_2}{TLV_2} + \ldots + \frac{C_n}{TLV_n} \quad \text{[2.11]}$$

exceeds unity, then the threshold limit of the mixture should be considered as being exceeded. "C" indicates the observed atmospheric concentration and TLV the corresponding threshold limit. In the absence of information to the contrary, the dilution ventilation therefore should be calculated on the basis that the effect of the different hazards is additive. The air quantity required to dilute each component of the mixture to the required safe concentration is calculated and the *sum* of the air quantities is used as the required dilution ventilation for the mixture.

Exceptions to the above rule may be made when there is good reason to believe that the chief effects of the different harmful substances are not additive but independent, as when purely local effects on different organs of the body are produced by the various components of the mixture. In such cases, the threshold limit ordinarily is exceeded only when at least one member of the series itself has a value exceeding unity, e.g.,

$$\frac{C_1}{TLV_1} \text{ or } \frac{C_2}{TLV_2}, \text{ etc.}$$

Therefore, where two or more hazardous substances are present and it is known that the effects of the different substances are not additive but act independently on the different organs of the body, the required dilution ventilation for each component of the mixture should be calculated and the highest cfm thus obtained used as the dilution ventilation rate.

EXAMPLE PROBLEM

A cleaning and gluing operation is being performed; methyl ethyl ketone (MEK) and toluene are both being released. Both have narcotic properties and the effects are considered additive. Air samples disclose concentrations of 150 ppm MEK and 50 ppm toluene. Using the equation given, the sum of the fractions [(150/200) + (50/100) = 1.25] is greater than unity and the TLV of the mixture is exceeded. The volumetric flow rate at standard conditions required for dilution of the mixture to the TLV would be as follows:

Assume 2 pints of each is being released each 60 min. Select a K value of 4 for MEK and a K value of 5 for toluene; sp gr for MEK = 0.805, for toluene = 0.866; MW for MEK = 72.1, for toluene = 92.13.

$$Q \text{ for MEK} = \frac{(403)(0.805)(10^6)(4)(2/60)}{72.1 \times 200}$$

$$= 3000 \text{ cfm}$$

$$Q \text{ for toluene} = \frac{(403)(0.866)(10^6)(5)(2/60)}{92.13 \times 100}$$

$$= 6313 \text{ cfm}$$

$$Q \text{ for mixture} = 3000 + 6313 = 9313 \text{ cfm}$$

2.5 DILUTION VENTILATION FOR FIRE AND EXPLOSION

Another function of dilution ventilation is to reduce the concentration of vapors within an enclosure to below the lower explosive limit. It should be stressed that this concept is never applied in cases where workers are exposed to the vapor. In such instances, dilution rates for health hazard control are always applied. The reason for this will be apparent when comparing TLVs and lower explosive limits (LELs).

The TLV of xylene is 100 ppm. The LEL of xylene is 1% or 10,000 ppm. An atmosphere of xylene safe-guarded against fire and explosion usually will be kept below 25% of the LEL or 2500 ppm. Exposure to such an atmosphere may cause severe illness or death. However, in baking and drying ovens, in enclosed air drying spaces, within ventilation ductwork, etc., dilution ventilation for fire and explosion is used to keep the vapor concentration to below the LEL.

Equation 2.5 can be modified to yield air quantities to dilute below the LEL. By substituting LEL for TLV:

$$Q = \frac{(403)(\text{sp gr liquid})(100)(ER)(S_f)}{(\text{MW liquid})(LEL)(B)} \quad \text{(for Standard Air)} \qquad [2.12]$$

Note: 1. Since LEL is expressed in percent (parts per 100) rather than ppm (parts per million as for the TLV), the factor of 1,000,000 becomes 100.

2. S_f is a safety factor which depends on the percentage of the LEL necessary for safe conditions. In most ovens and drying enclosures, it has been found desirable to maintain vapor concentrations at not more than 25% of the LEL at all times in all parts of the oven. In properly ventilated continuous ovens, a S_f factor of 4 (25% of the LEL) is used. In batch ovens, with good air distribution, the existence of peak drying rates requires a S_f factor of 10 or 12 to maintain safe concentrations at all times. In non-recirculating or improperly ventilated batch or continuous ovens, larger S_f factors may be necessary.

3. B is a constant which takes into account the fact that the lower explosive limit of a solvent vapor or air mixture decreases at elevated temperatures. B = 1 for temperatures up to 250 F; B = 0.7 for temperatures above 250 F.

EXAMPLE PROBLEM

A batch of enamel dipped shelves is baked in a recirculating oven at 350 F for 60 minutes. Volatiles in the enamel applied to the shelves consist of two pints of xylene. What oven ventilation rate, in cfm, is required to dilute the xylene vapor concentration within the oven to a safe limit at all times?

For xylene, the LEL = 1.0%; sp gr = 0.88; MW = 106; S_f = 10; B = 0.7. From Equation 2.12:

$$Q = \frac{(403)(0.88)(2/60)(100)(10)}{(106)(1.0)(0.7)} = 159 \text{ cfm}$$

Since the above equation is at standard conditions, the air flow rate must be converted from 70 F to 350 F (operating conditions):

$$Q_A = (\text{cfm}_{STP}) \text{ (Ratio of Absolute Temperature)}$$

$$= (\text{cfm}_{STP}) \frac{(460 \text{ F} + 350 \text{ F})}{(460 \text{ F} + 70 \text{ F})}$$

$$Q_A = 159 \left(\frac{810}{530}\right)$$

$$= 243 \text{ cfm}$$

EXAMPLE PROBLEM

In many circumstances, solvent evaporation rate is non-uniform due to the process temperature or the manner of solvent use.

A 6-ft diameter muller is used for mixing resin sand on a 10-minute cycle. Each batch consists of 400 pounds of sand, 19 pounds of resin, and 8 pints of ethyl alcohol (the ethyl alcohol evaporates in the first two minutes). What ventilation rate is required?

For ethyl alcohol, LEL = 3.28%; sp gr = 0.789; MW = 46.07; S_f = 4; B = 1

$$Q = \frac{(403)(0.789)(8/2)(100)(4)}{(46.07)(3.28)(1)} = 3367 \text{ scfm}$$

Another source of data is the National Board of Fire Underwriters' Pamphlet #86, *Standard for Class A Ovens and Furnaces*.(2.3) This contains a more complete list of solvents and their properties. In addition, it lists and describes a number of safeguards and interlocks which must always be considered in connection with fire dilution ventilation. See also Reference 2.4.

2.6 FIRE DILUTION VENTILATION FOR MIXTURES

It is common practice to regard the entire mixture as consisting of the components requiring the highest amount of dilution per unit liquid volume and to calculate the required air quantity on that basis. (This component would be the one with the highest value for sp gr/(MW)(LEL).)

2.7 VENTILATION FOR HEAT CONTROL

Ventilation for heat control in a hot industrial environment is a specific application of general industrial ventilation. The primary function of the ventilation system is to prevent the acute discomfort or possible injury of those working in or generally occupying a designated hot industrial environment. Heat-induced occupational illnesses, injuries, or reduced productivity may occur in situations where the total heat load may exceed the defenses of the body and result in a heat stress situation. It follows, therefore, that a heat control ventilation system or other engineering control method must follow a physiological evaluation in terms of potential heat stress for the occupant in the hot industrial environment.

Due to the complexity of conducting a physiological evaluation, the criteria presented here are limited to general considerations. It is strongly recommended, however, that the NIOSH Publication No. 86-113, *Criteria for a Recommended Standard, Occupational Exposure to Hot Environments*,(2.5) be reviewed thoroughly in the process of developing the heat control ventilation system.

The development of a ventilation system for a hot industrial environment usually includes the control of the ventilation air flow rate, velocity, temperature, humidity, and air flow path through the space in question. This may require inclusion of certain phases of mechanical air-conditioning engineering design which is outside the scope of this manual. The necessary engineering design criteria that may be required are available in appropriate publications of the American Society of Heating, Refrigeration and Air-Conditioning Engineers (ASHRAE) handbook series.

2.8 HEAT BALANCE AND EXCHANGE

An essential requirement for continued normal body function is that the deep body core temperature be maintained within the acceptable range of about 37 C (98.6 F) ± 1 C (1.8 F). To achieve this, body temperature equilibrium requires a constant exchange of heat between the body and the environment. The rate and amount of the heat exchange are governed by the fundamental laws of thermodynamics of heat exchange between objects. The amount of heat that must be exchanged is a function of 1) the total heat produced by the body (metabolic heat), which may range from about 1 kilocalorie (kcal) per kilogram (kg) of body weight per hour (1.16 watts) at rest to 5 kcal/kg body weight/hour (7 watts) for moderately hard industrial work; and 2) the heat gained, if any, from the environment. The rate of heat exchange with the environment is a function of air temperature and humidity, skin temperature, air velocity, evaporation of sweat, radiant temperature , and type, amount, and characteristics of the clothing worn. Respiratory heat loss is generally of minor consequence except during hard work in a very dry environment.

The basic heat balance equation is:

$$\Delta S = (M - W) \pm C \pm R - E \qquad [2.13]$$

where:

ΔS = change in body heat content
$(M - W)$ = total metabolism − external work performed
C = convective heat exchange
R = radiative heat exchange
E = evaporative heat loss

To solve the equation, measurement of metabolic heat production, air temperature, air water vapor pressure, wind velocity, and mean radiant temperature are required.

The major modes of heat exchange between man and the environment are convection, radiation, and evaporation. Other than for brief periods of body contact with hot tools, equipment, floors, etc., which may cause burns, conduction plays a minor role in industrial heat stress.

The equations for calculating heat exchange by convection, radiation, and evaporation are available in Standard International (SI) units, metric units, and English units. In SI units heat exchange is in watts per square meter of body surface (W/m^2). The heat exchange equations are available in both metric and English units for both the seminude individual and the worker wearing conventional long-sleeved workshirt and trousers. The values are in kcal/h or British thermal units per hour (Btu/h) for the "standard worker" defined as one who weighs 70 kg (154 lbs) and has a body surface area of 1.8 m^2 (19.4 ft^2).

2.8.1 Convection: The rate of convective heat exchange between the skin of a person and the ambient air immediately surrounding the skin is a function of the difference in temperature between the ambient air (t_a), the mean weighted skin temperature (t_{sk}) and the rate of air movement over the skin (V_a). This relationship is stated algebraically for the "standard worker" wearing the customary one layer work clothing ensemble as:

$$C = 0.65\, V_a^{0.6}\,(t_a - t_{sk}) \quad [2.14]$$

where:

C = convective heat exchange, Btu/h
V_a = air velocity, fpm
t_a = air temperature, F
t_{sk} = mean weighted skin temperature, usually assumed to be 95 F

When $t_a > 95$ F there will be a gain in body heat from the ambient air by convection. When $t_a < 95$ F, heat will be lost from the body to the ambient air by convection.

2.8.2 Radiation: The rate of radiative heat exchange between the skin of a person and the radiant heat source is a function of the fourth power of the absolute temperature of the solid surroundings less the skin absolute temperature, $(T_w - T_{sk})^4$, but an acceptable approximation for the customary one layer clothed individual is:

$$R = 15.0\,(t_w - t_{sk}) \quad [2.15]$$

where:

R = radiant heat exchange, Btu/h
t_w = mean radiant temperature, F
t_{sk} = mean weighted skin temperature

2.8.3 Evaporation: The evaporation of water (sweat) from the skin surface results in a heat loss from the body. The maximum evaporative capacity (and heat loss) is a function of air motion (V_a) and the water vapor pressure difference between the ambient air (ρ_a) and the wetted skin at skin temperature (ρ_{sk}). The equation for this relationship is for the customary one layer clothed worker:

$$E = 2.4 V_a^{0.6}\,(\rho_{sk} - \rho_a) \quad [2.16]$$

where:

E = evaporative heat loss, Btu/h
V_{6a} = air velocity, fpm
ρ_a = water vapor pressure of ambient air, mmHg
ρ_{sk} = water vapor pressure on the skin, assumed to be 42 mmHg at a 95 F skin temperature

2.9 ADAPTIVE MECHANISM OF THE BODY

The human body, by a very complex mechanism, can attain perfect adaptation to environmental conditions through a narrow range. When this is true the storage factor is zero and optimum comfort is attained. The chief adaptive mechanisms of the body are peripheral blood circulation, sweating and change in metabolism.

In cold environments the skin surface has reduced blood circulation while in a warm environment the blood circulation to the skin is increased. This increase in blood at the skin surface is at the expense of the internal organs and brain and, if excessive, can result in heat exhaustion.

Sweating increases sharply in warm environments. However, in order to have a cooling effect, the humidity must be low and/or the air velocity must be high since the cooling effect is due to the evaporation of the liquid content of the sweat. Profuse sweating depletes the body salt content of unacclimatized persons and extra salt may be given when heavy work must be carried out under hot dry conditions.

The rate of metabolism is fairly constant in a temperate environment. There is an increase, however, in metabolism at low temperatures and at high temperatures. At elevated temperatures a sharp increase in metabolism denotes the beginning of the breakdown of the regulative process (see Figure 2-4).

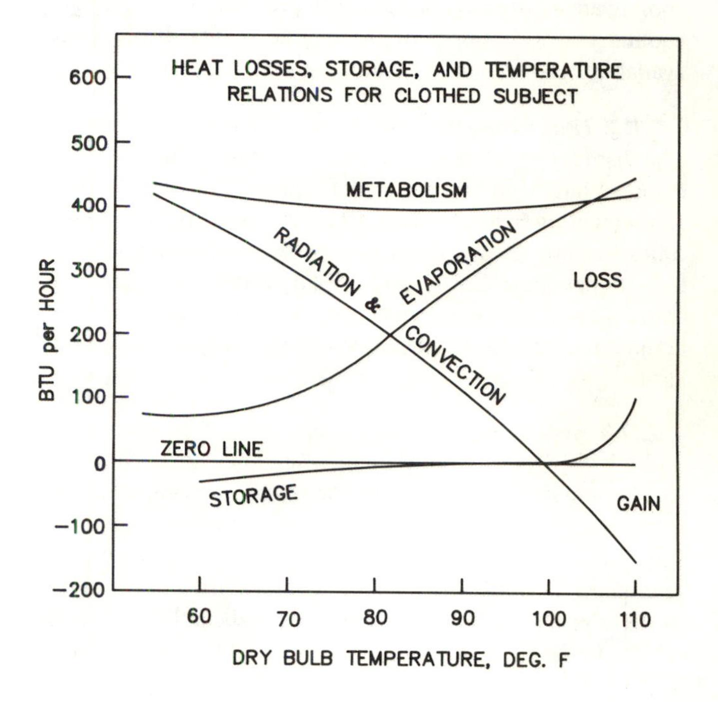

FIGURE 2–4

2.10 ACCLIMATIZATION

Acclimatization of personnel exposed to heat for extended periods of time is well demonstrated. Over a period of two weeks or so, the capacity of the individual to withstand heat is considerably increased. The acclimatization starts with a decrease in heat production as the individual adjusts by using energy more efficiently and relaxing when the work situation permits. During the first few days an increase in sweating develops. The heat regulating mechanism apparently becomes more sensitive and enables the individual to better react to rapidly changing environmental conditions. The blood volume is increased, as is the volume of extracellular fluid; at the same time there is a marked drop in pulse rate response. The concentration of salt in the sweat decreases to a point where it is virtually impossible for a chloride deficit to be produced even by hot work in a hot dry environment.

2.11 ACUTE HEAT DISORDERS

A variety of heat disorders can be distinguished clinically when individuals are exposed to excessive heat. A brief description of these disorders follows.

2.11.1 Heatstroke: Heatstroke includes 1) a major disruption of central nervous function (unconsciousness or convulsions), 2) a lack of sweating, and 3) a rectal temperature in excess of 41 C (105.8 F). Heatstroke is a *MEDICAL EMERGENCY*, and any procedure from the moment of onset which will cool the patient improves the prognosis. Placing the patient in a shady area, removing outer clothing, wetting the skin and increasing air movement to enhance evaporative cooling are all urgently needed until professional methods of cooling and assessment of the degree of the disorder are available.

2.11.2 Heat Exhaustion: Heat exhaustion is a mild form of heat disorder which readily yields to prompt treatment. This disorder has been encountered frequently in experimental assessment of heat tolerance. Characteristically, it is sometimes but not always accompanied by a small increase in body temperature (100.4–102.2 F). The symptoms of headache, nausea, vertigo, weakness, thirst, and giddiness are common to both heat exhaustion and the early stage of heatstroke.

2.11.3 Heat Cramps: Heat cramps are not uncommon in individuals who work hard in the heat. They are attributable to a continued loss of salt in the sweat, accompanied by copious intake of water without appropriate replacement of salt. Other electrolytes, such as magnesium, calcium and potassium, may also be involved. Cramps often occur in the muscles principally used during work and can be alleviated readily by rest, the ingestion of water, and the correction of any fluid electrolyte imbalance.

2.11.4 Heat Rash: Heat rash is prickly heat (miliaria rubra), which appears as red papules, usually in areas where the clothing is restrictive, and gives rise to a prickling sensation, particularly as sweating increases. It occurs in skin that is persistently wetted by unevaporated sweat, apparently because the keratinous layers of the skin absorb water, swell, and mechanically obstruct the sweat ducts. The papules may become infected unless they are treated.

2.12 HEAT STRESS MEASUREMENT

The assessment of heat stress may be conducted by measuring the climatic and physical factors of the environment and then evaluating their effects on the human body by using an appropriate heat stress index. Environmental factors which are of concern in industrial heat stress determinations follow.

2.12.1 Dry Bulb (Air) Temperature (t_a) – is the simplest to measure of the climatic factors. It is the temperature of the ambient air as measured with a thermometer. Temperature units proposed by the International Standards Organization (ISO) are degrees Celsius, C = (F 32) × 5/9 and degrees Kelvin, K = C + 273.

2.12.2 Natural Wet Bulb Temperature (t_{nwb}) – is the temperature measured by a thermometer which has its sensor covered by a wetted cotton wick and which is exposed only to the natural prevailing air movement.

TABLE 2-2. Estimating Energy Cost of Work by Task Analysis

A. Body position and movement		kcal/min*
Sitting		0.3
Standing		0.6
Walking		2.0 – 3.0
Walking uphill		Add 0.8 / meter rise
B. Type of work	**Average kcal/min**	**Range kcal/min**
Hand work — light	0.4	0.2 – 1.2
heavy	0.9	
Work one arm — light	1.0	0.7 – 2.5
heavy	1.8	
Work both arms — light	1.5	1.0 – 3.5
heavy	2.5	
Work whole body — light	3.5	2.5 – 9.0
moderate	5.0	
heavy	7.0	
very heavy	9.0	
C. Basal metabolism	1.0	
D. Sample calculation**		
Assembling work with heavy hand tools		
1. Standing	0.6	
2. Two-arm work	3.5	
3. Basal metabolism	1.0	
TOTAL	5.1 kcal/min	

*For standard worker of 70 kg body weight (154 lbs) and 1.8 m^2 body surface (19.4 ft^2).

**Example of measuring metabolic heat production of a worker when performing initial screening.

2.12.3 Psychrometric Wet Bulb Temperature (t_{wb}) – is obtained when the wetted wick covering the sensor is exposed to a high forced air movement. The t_{wb} is commonly measured with a psychrometer which consists of two mercury in glass thermometers mounted beside each other on the frame of the psychrometer. One thermometer is used to measure the t_{wb} by covering its bulb with a clean cotton wick wetted with water, and the second measures the dry bulb temperature (t_a). The air movement is obtained manually with a sling psychrometer or mechanically with a motor driven psychrometer.

2.12.4 Air Velocity – is the rate, in feet per minute (fpm) or meters per second (m/sec), at which the air moves and is important in heat exchange between the human body and the environment because of its role in convective and evaporative heat transfer.

2.12.5 Radiant Heat: is the thermal load of solar and infrared radiation on the human body. It is measured by the use of a black globe thermometer which consists of a 6-inch hollow copper globe painted matte black with a thermometer inserted to the center of the globe. The instrument measures directly the radiant heat temperature used to determine the wet bulb globe temperature (WBGT). (See Figure 2-5.)

2.12.6 Metabolic Heat Estimates – can be made using tables of energy expenditure or task analysis and can be applied to both short and long duration activities. A training program for using the tables of energy expenditure and task analysis is necessary to produce a fair degree of accuracy. Use of the task analysis procedure to estimate the metabolic heat of an activity is summarized in Table 2-2.

2.13 HEAT STRESS INDICES

Several attempts have been made to develop an empirical index of heat stress. These include several adaptations of Effective Temperature (ET), Belding Hatch Stress Index, Skin Wettedness, and others. The most frequently used and recommended are the Wet Bulb Globe Thermometer Index (WBGT Index) and the Wet Globe Temperature Index (WGT Index).

2.13.1 WBGT Index – is the National Institute for Occupational Safety and Health (NIOSH) recommended index for heat stress alert limits as developed in the criteria document.[2.5] Its advantages are that measurements are few and easy to make; the instrumentation is simple, relatively inexpensive and rugged; and the calculations are straight forward. The calculation of the WBGT for indoor air is:

$$WBGT = 0.7\,t_{nwb} + 0.3\,t_g \qquad [2.17]$$

where:

t_{nwb} = natural wet bulb temperature

t_g = radiant or globe temperature

For outdoor air considerations, the WBGT calculation is:

$$WBGT = 0.7\,t_{nwb} + 0.2\,t_g + 0.1\,t_a \qquad [2.18]$$

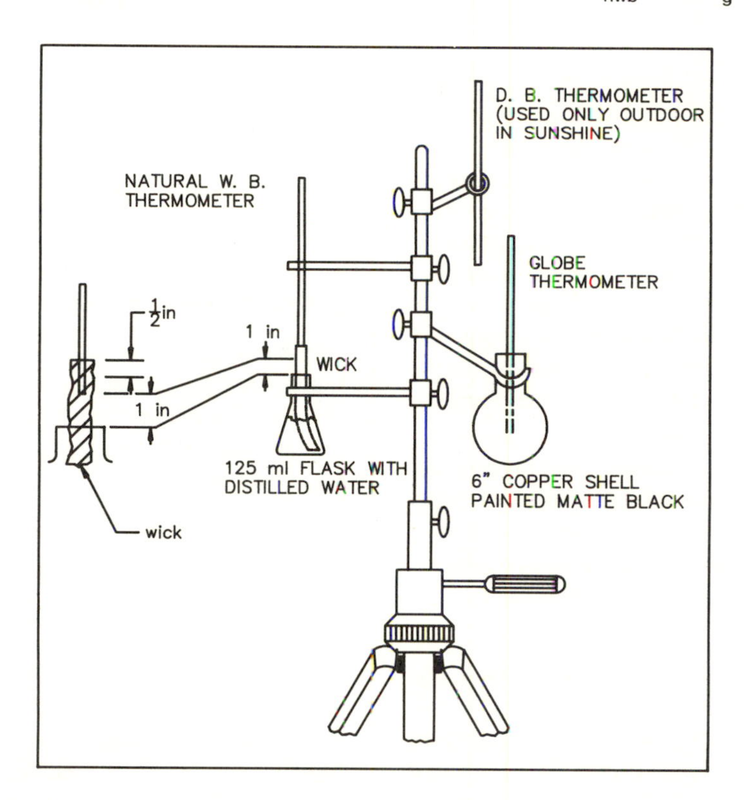

FIGURE 2–5

where:

t_a = ambient air temperature

The WBGT Index combines the effect of humidity and air movement, air temperature and radiation, and air temperature as a factor in outdoor situations in the presence of sunshine. If there is a radiant heat load, not sunshine, the globe temperature reflects the effects of air velocity and air temperature. The NIOSH suggested instrumentation used to determine the WBGT is shown in Figure 2-5. There are commercially available instruments that will provide instantaneous readings of the individual components in the WBGT or a combined and integrated digital readout.

Once the WBGT has been determined together with the metabolic heat being produced during a specific task, the evaluation of the potential heat stress situation can be addressed. The recommended limits recognize unacclimatized and acclimatized workers, the effect of clothing and specific ceiling limits is as follows:

1. Unacclimatized workers: Total heat exposure to workers should be controlled so that unprotected healthy workers who are not acclimatized to working in hot environments are not exposed to combinations of metabolic and environmental heat greater than the applicable RALs given in Figure 2-6.
2. Acclimatized workers: Total heat exposure to workers should be controlled so that unprotected healthy workers who are acclimatized to working in hot environments are not exposed to combinations of metabolic and environmental heat greater than the applicable RELs given in Figure 2-7.
3. Effect of clothing: The recommended limits given in Figures 2-6 and 2-7 are for healthy workers who are physically and medically fit for the level of activity required by their job and who are wearing the customary one layer work clothing ensemble consisting of not more than long-sleeved work shirts and trousers (or equivalent). The REL and RAL values given in Figures 2-6 and 2-7 may not provide adequate protection if workers wear clothing with lower air and vapor permeability or insulation values greater than those for the customary one layer work clothing ensemble discussed above.
4. Ceiling limits: No worker shall be exposed to combinations of metabolic and environmental heat exceeding the applicable Ceiling Limits of Figures 2-6 or 2-7 without being provided with and properly using appropriate and adequate heat protective clothing and equipment.

2.13.2 Wet Globe Temperature Index (WGT Index): The wet globe thermometer is the simplest, most easily read and the most portable of the environmental heat measuring devices. It consists of a hollow 3-inch copper sphere covered with a black cloth which is kept at 100% wettedness from a water reservoir. The sensing element of a thermometer is located inside the sphere at the center and the temperature is read on the dial gauge at the end of the stem. Presumably, the wet sphere exchanges heat with the environment in a manner similar to humans and the heat exchange by convection, radiation, and evaporation are integrated into a single instrument reading.

Over the past several years, the WGT has been used in various situations and compared with the WBGT. In general, the WGT and the WBGT have a high correlation but this is not constant for all combinations of environmental factors. A simple approximation of the relationship is as follows:

$$\text{WBGT} = \text{WGT} + 3.6\ \text{F} \qquad [2.19]$$

This approximation assumes moderate radiant heat and humidity and is probably adequate for general monitoring in industry. The WGT does not provide data for solving the equations for heat exchange but does provide a simple and rapid indicator of the level of heat stress.

2.14 VENTILATION CONTROL

The control method presented here is limited to a general engineering approach. Due to the complexity of evaluating a potential heat stress producing situation, it is essential that the accepted industrial hygiene method of recognition, evaluation, and control be utilized to its fullest extent. In addition to the usual time limited exposures, it may be necessary to specify additional protection which may include insulation, baffles, shields, partitions, personal protective equipment, administrative control, and other measures to prevent possible heat stress. Ventilation control measures may require a source of cooler replacement air, an evaporative or mechanically cooled source, a velocity cooling method, or any combination thereof. Specific guidelines, texts, and other publications or sources should be reviewed for the necessary data to develop the ventilation system.

2.15 VENTILATION SYSTEMS

Exhaust ventilation may be used to remove excessive heat and/or humidity if a replacement source of cooler air is available. If it is possible to enclose the heat source, such as in the case of ovens or certain furnaces, a gravity or forced air stack may be all that is necessary to prevent excessive heat from entering the workroom. If a partial enclosure or local hood is indicated, control velocities, as shown in Chapters 3 and 10, may be used to determine the volume of air to be exhausted.

In the case of many operations which do not lend themselves to local exhaust, general ventilation may be indicated. In order to arrive at the volumetric flow rate required, it is necessary to estimate the summation of all sources of both sensible and latent heat, as well as to determine in advance the temperature rise or humidity rise which will be acceptable. The volumetric flow rate required for sensible heat may be estimated from the following equation:

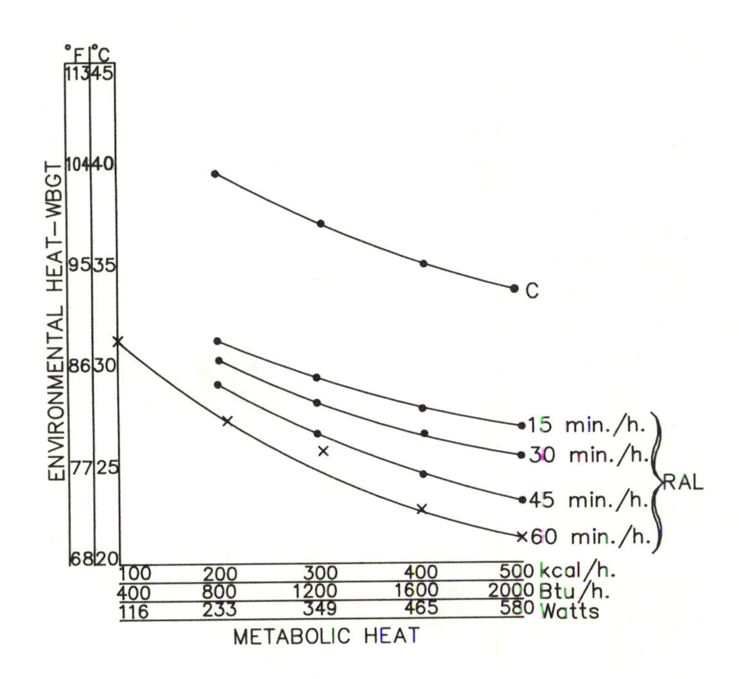

C = CEILING LIMIT
RAL = RECOMMENDED ALERT LIMIT
*FOR "STANDARD WORKER" OF 70 kg (154 lbs) BODY WEIGHT AND 1.8 m^2 (19.4 ft^2) BODY SURFACE.

FIGURE 2–6 RECOMMENDED HEAT–STRESS ALERT LIMITS HEAT–UNACCLIMATIZED WORKERS.

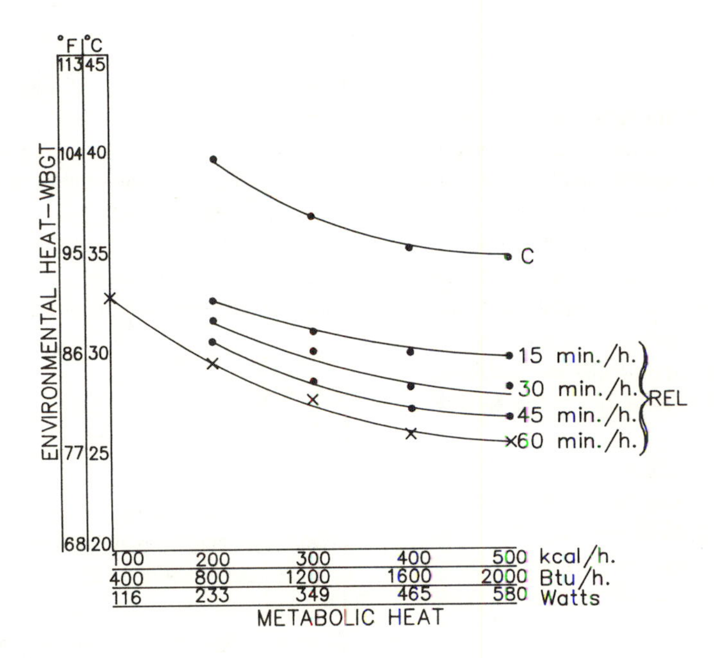

C = CEILING LIMIT
REL = RECOMMENDED EXPOSURE LIMIT
*FOR "STANDARD WORKER" OF 70 kg (154 lbs) BODY WEIGHT AND 1.8 m^2 (19.4 ft^2) BODY SURFACE.

FIGURE 2–7 RECOMMENDED HEAT–STRESS EXPOSURE LIMITS HEAT–ACCLIMATIZED WORKERS.

$$Q = \frac{\text{Total Btu/hr sensible heat}}{1.08 \times \text{temp rise, F}} \quad [2.20]$$

In order to use this equation, it is necessary to first estimate the heat load. This will include sun load, people, lights, and motors as well as other particular sources of heat. Of these, sun load, lights, and motors are all completely sensible. The people heat load is part sensible and part latent. In the case of hot processes which give off both sensible and latent heat, it will be necessary to estimate the amounts or percentages of each. In using the above equation for sensible heat, one must decide the amount of temperature rise which will be permitted. Thus, in a locality where 90 F outside dry bulb may be expected, if it is desired that the inside temperature not exceed 100 F, or a 10 degree rise, a certain air flow rate will be necessary. If an inside temperature of 95 F is required, the air flow rate will be doubled.

For latent heat load, the procedure is similar although more difficult. If the total amount of steam evaporated is known, the heat load may be estimated by multiplying the pounds of steam per hour by 1000. Then

$$Q = \frac{\text{Btu/hr latent heat}}{0.67 \times \text{Grains/lb difference}} \quad [2.21]$$

When the amount of water released is known, the following formula is used:

$$Q = \frac{\text{116.7 Pounds/hr water released as vapor}}{\text{Grains/lb difference} \times \rho} \quad [2.22]$$

where:

ρ = density of air, lbs/ft^3

The term "grains per pound difference" is taken from the psychrometric chart or tables and represents the difference in moisture content of the outside air and the conditions acceptable to the engineer designing the exhaust system. The air quantities calculated from the above two equations should not be added to arrive at the required quantity. Rather, the higher quantity should be used since both sensible and latent heat are absorbed simultaneously. Furthermore, in the majority of cases the sensible heat load far exceeds the latent heat load so the design can be calculated only on the basis of sensible heat.

The ventilation should be designed to flow through the hot environment in a manner that will control the excess heat by removing it from that environment. Figures 2-8 and 2-9 illustrate this principle.

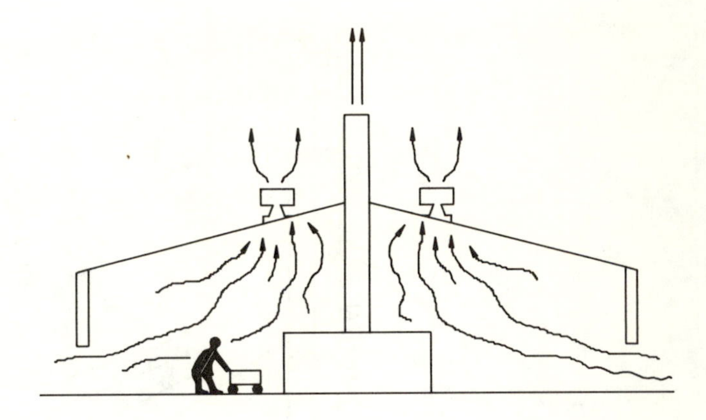

FIGURE 2—8. NATURAL VENTILATION.

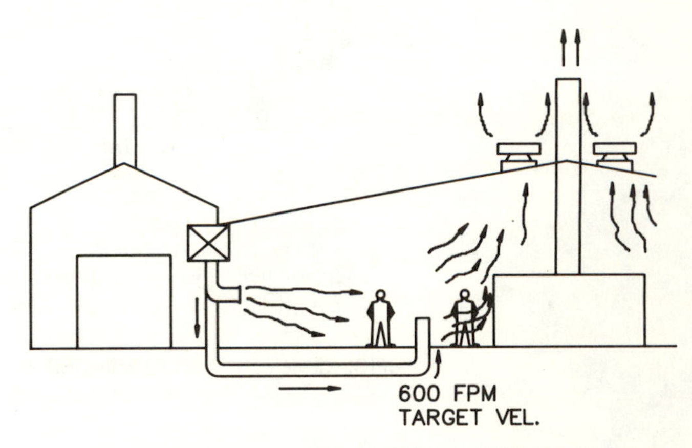

FIGURE 2—9. MECHANICAL VENTILATION.

2.16 VELOCITY COOLING

If the air dry bulb or wet bulb temperatures are lower than 95 – 100 F, the worker may be cooled by convection or evaporation. When the dry bulb temperature is higher than 95 – 100 F, increased air velocity may add heat to the worker by convection; if the wet bulb temperature is high also, evaporative heat loss may not increase proportionately and the net result will be an increase in the worker's heat burden. Many designers consider that supply air temperature should not exceed 80 F for practical heat relief.

Current practice indicates that air velocities in Table 2-3 can be used successfully for direct cooling of workers. For best results provide directional control of the air supply (Figure 2-10) to accommodate daily and seasonal variations in heat exposure and supply air temperature.

TABLE 2-3. Acceptable Comfort Air Motion at the Worker

	Air Velocity, fpm*
Continuous Exposure	
Air conditioned space	50 – 75
Fixed work station, general ventilation or spot cooling: Sitting	75 – 125
Standing	100 – 200
Intermittent Exposure, Spot Cooling or Relief Stations	
Light heat loads and activity	1000 – 2000
Moderate heat loads and activity	2000 – 3000
High heat loads and activity	3000 – 4000

*Note: Velocities greater than 1000 fpm may seriously disrupt the performance of nearby local exhaust systems. Care must be taken to direct air motion to prevent such interference.

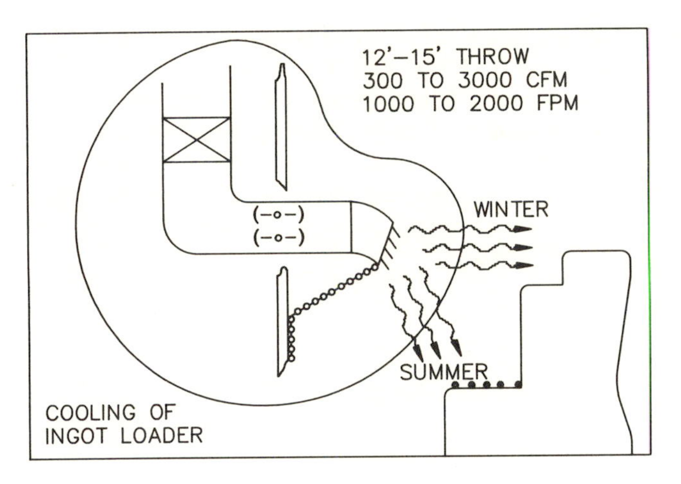

FIGURE 2–10 SPOT COOLING WITH VOLUME AND DIRECTIONAL CONTROL.

2.17 RADIANT HEAT CONTROL

Since radiant heat is a form of heat energy which needs no medium for its transfer, radiant heat cannot be controlled by any of the above means. Painting or coating the surface of hot bodies with materials having low radiation emission characteristics is one method of reducing radiation.

For materials such as molten masses of metal or glass which cannot be controlled directly, radiation shields are effective. These shields can consist of metal plates, screens, or other material interposed between the source of radiant heat and the workers. Shielding reduces the radiant heat load by reflecting the major portion of the incident radiant heat away from the operator and by re-emitting to the operator only a portion of that radiant heat which has been absorbed. Table 2-4 indicates the percentage of both reflection and emission of radiant heat associated with some common shielding materials. Additional ventilation will control the sensible heat load but will have only a minimal effect, if any, upon the radiant heat load. See Figure 2-11.

TABLE 2-4. Relative Efficiencies of Common Shielding Materials

Surface of Shielding	Reflection of Radiant Heat Incident Upon Surface	Emission of Radiant Heat from Surface
Aluminum, bright	95	5
Zinc, bright	90	10
Aluminum, oxidized	84	16
Zinc, oxidized	73	27
Aluminum paint, new, clean	65	35
Aluminum paint, dull, dirty	40	60
Iron, sheet, smooth	45	55
Iron, sheet, oxidized	35	65
Brick	20	80
Lacquer, black	10	90
Lacquer, white	10	90
Asbestos board	6	94
Lacquer, flat black	3	97

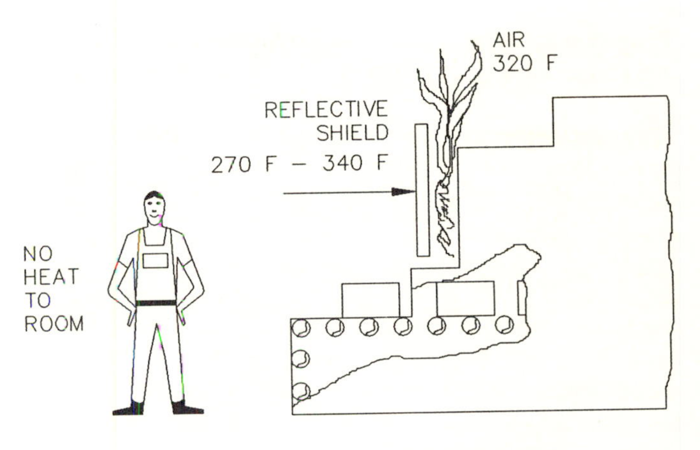

FIGURE 2–11 HEAT SHIELDING

2.18 PROTECTIVE SUITS FOR SHORT EXPOSURES

For brief exposures to very high temperatures, insulated aluminized suits and other protective clothing may be worn. These suits reduce the rate of heat gain by the body but provide no means of removing body heat; therefore, only short exposures may be tolerated.

2.19 RESPIRATORY HEAT EXCHANGERS

For brief exposure to air of good quality but high temperature, a heat exchanger on a half-mask respirator facepiece is available. This device will bring air into the respiratory passages at a tolerable temperature but will not remove contaminants nor furnish oxygen in poor atmospheres.

2.20 REFRIGERATED SUITS

Where individuals must move about, cold air may be blown into a suit or hood worn as a portable enclosure. The usual refrigeration methods may be used with insulated tubing to the suit. It may be difficult, however, to deliver air at a sufficiently low temperature. If compressed air is available, cold air may be delivered from a vortex tube worn on the suit. Suits of this type are commercially available.

2.21 ENCLOSURES

In certain hot industries, such as in steel mills, it is unnecessary and impractical to attempt to control the heat from the process. If the operation is such that remote control is possible, an air conditioned booth or cab can be utilized to keep the operators reasonably comfortable in an otherwise intolerable atmosphere.

2.22 INSULATION

If the source of heat is a surface giving rise to convection, insulation at the surface will reduce this form of heat transfer.

Insulation by itself, however, will not usually be sufficient if the temperature is very high or if the heat content is high.

REFERENCES

2.1 U.S. Department of Health, Education and Welfare, PHS, CDC, NIOSH: *The Industrial Environment–Its Evaluation and Control*, 1973.

2.2 Air Force AFOSH Standard 161.2.

2.3 National Board of Fire Underwriters: Pamphlet #86, *Standards for Class A Ovens and Furnaces*.

2.4 B. Feiner and L. Kingsley: "Ventilation of Industrial Ovens." *Air Conditioning, Heating and Ventilating*, December 1956, pp. 82 89.

2.5 U.S. Department of Health and Human Services, PHS, CDC, NIOSH: *Occupational Exposure to Hot Environments*, Revised Criteria, 1986.

Chapter 3
LOCAL EXHAUST HOODS

3.1 INTRODUCTION **3-2**

3.2 CONTAMINANT CHARACTERISTICS **3-2**
3.2.1 Inertial Effects **3-2**
3.2.2 Effect of Specific Gravity **3-2**

3 HOOD TYPES **3-2**
3.3.1 Enclosing Hoods **3-2**
3.3.2 Exterior Hoods **3-2**

3.4 HOOD DESIGN FACTORS **3-2**
3.4.1 Capture Velocity **3-6**
3.4.2 Hood Flow Rate Determination **3-6**
3.4.3 Effect of Flanges and Baffles **3-7**
3.4.4 Air Distribution **3-8**
3.4.5 Rectangular and Round Hoods **3-8**

3.5 HOOD LOSSES **3-8**
3.5.1 Simple Hoods **3-8**
3.5.2 Compound Hoods **3-15**

3.6 MINIMUM DUCT VELOCITY **3-17**

3.7 SPECIAL HOOD REQUIREMENTS **3-17**
3.7.1 Ventilation of Radioactive and High Toxicity Processes **3-17**
3.7.2 Laboratory Operations **3-17**

3.8 PUSH-PULL VENTILATION **3-18**
3.8.1 Push Jet **3-18**
3.8.2 Pull Hood **3-19**
3.8.3 Push-Pull System Design **3-19**

3.9 HOT PROCESSES **3-19**
3.9.1 Circular High Canopy Hoods **3-19**
3.9.2 Rectangular High Canopy Hoods **3-20**
3.9.3 Low Canopy Hoods **3-20**

REFERENCES **3-20**

3.1 INTRODUCTION

Local exhaust systems are designed to capture and remove process emissions prior to their escape into the workplace environment. The local exhaust hood is the point of entry into the exhaust system and is defined herein to include all suction openings regardless of their physical configuration. The primary function of the hood is to create an air flow field which will effectively capture the contaminant and transport it into the hood. Figure 3-1 provides nomenclature associated with local exhaust hoods.

3.2 CONTAMINANT CHARACTERISTICS

3.2.1 Inertial Effects: Gases, vapors, and fumes will not exhibit significant inertial effects. Also, fine dust particles, 20 microns or less in diameter (which includes respirable particles), will not exhibit significant inertial effects. These materials will move solely with respect to the air in which they are mixed. In such cases, the hood needs to generate an air flow pattern and capture velocity sufficient to control the motion of the contaminant-laden air plus extraneous air currents caused by room cross-drafts, vehicular traffic, etc.

3.2.2 Effective Specific Gravity: Frequently, the location of exhaust hoods is mistakenly based on a supposition that the contaminant is "heavier than air" or "lighter than air." In most health hazard applications, this criterion is of little value (see Figure 3-2). Hazardous fine dust particles, fumes, vapors, and gases are truly airborne, following air currents and are not subject to appreciable motion either upward or downward because of their own density. Normal air movement will assure an even mixture of these contaminants. Exception to these observations may occur with very hot or very cold operations or where a contaminant is generated at very high levels and control is achieved before the contaminant becomes diluted.

3.3 HOOD TYPES

Hoods may be of a wide range of physical configuration but can be grouped into two general categories: enclosing and exterior. The type of hood to be used will be dependent on the physical characteristics of the process equipment, the contaminant generation mechanism and the operator/equipment interface (see Figure 3-3).

3.3.1 Enclosing Hoods: Enclosing hoods are those which completely or partially enclose the process or contaminant generation point. A complete enclosure would be a laboratory glove box or similar type of enclosure where only minimal openings exist. A partial enclosure would be a laboratory hood or paint spray booth. An inward flow of air through the enclosure opening will contain the contaminant within the enclosure and prevents its escape into the work environment.

The enclosing hood is preferred wherever the process configuration and operation will permit. If complete enclosure is not feasible, partial enclosure should be used to the maximum extent possible (see Figure 3-3.)

3.3.2 Exterior Hoods: Exterior hoods are those which are located adjacent to an emission source without enclosing it. Examples of exterior hoods are slots along the edge of the tank or a rectangular opening on a welding table.

Where the contaminant is a gas, vapor or fine particulate and is not emitted with any significant velocity, the hood orientation is not critical. However, if the contaminant contains large particulates which are emitted with a significant velocity, the hood should be located in the path of the emission. An example would be a grinding operation (see VS-411).

If the process emits hot contaminated air, it will rise due to thermal buoyancy. Use of a side draft exterior hood (located horizontally from the hot process) may not provide satisfactory capture due to the inability of the hood induced air flow to overcome the thermally induced air flow. This will be especially true for very high temperature processes such as a melting furnace. In such cases a canopy hood located over the process may be indicated (see Section 3.9).

A variation of the exterior hood is the push-pull system (Section 3.8). In this case, a jet of air is pushed across a contaminant source into the flow field of a hood. Contaminant control is primarily achieved by the jet. The function of the exhaust hood is to receive the jet and remove it. The advantage of the push-pull system is that the push jet can travel in a controlled manner over much greater distances than air can be drawn by an exhaust hood alone. The push-pull system is used successfully for some plating and open surface vessel operations but has potential application for many other processes. However, the push portion of the system has potential for increasing operator exposure if not properly designed, installed or operated. Care must be taken to ensure proper design, application and operation.

3.4 HOOD DESIGN FACTORS

Capture and control of contaminants will be achieved by the inward air flow created by the exhaust hood. Air flow toward the hood opening must be sufficiently high to maintain control of the contaminant until it reaches the hood. External air motion may disturb the hood-induced air flow and require higher air flow rates to overcome the disturbing effects. Elimination of sources of external air motion is an important factor in achieving effective control without the need for excessive air flow and its associated cost. Important sources of air motion are

- Thermal air currents, especially from hot processes or heat-generating operations.
- Motion of machinery, as by a grinding wheel, belt conveyor, etc.
- Material motion, as in dumping or container filling.

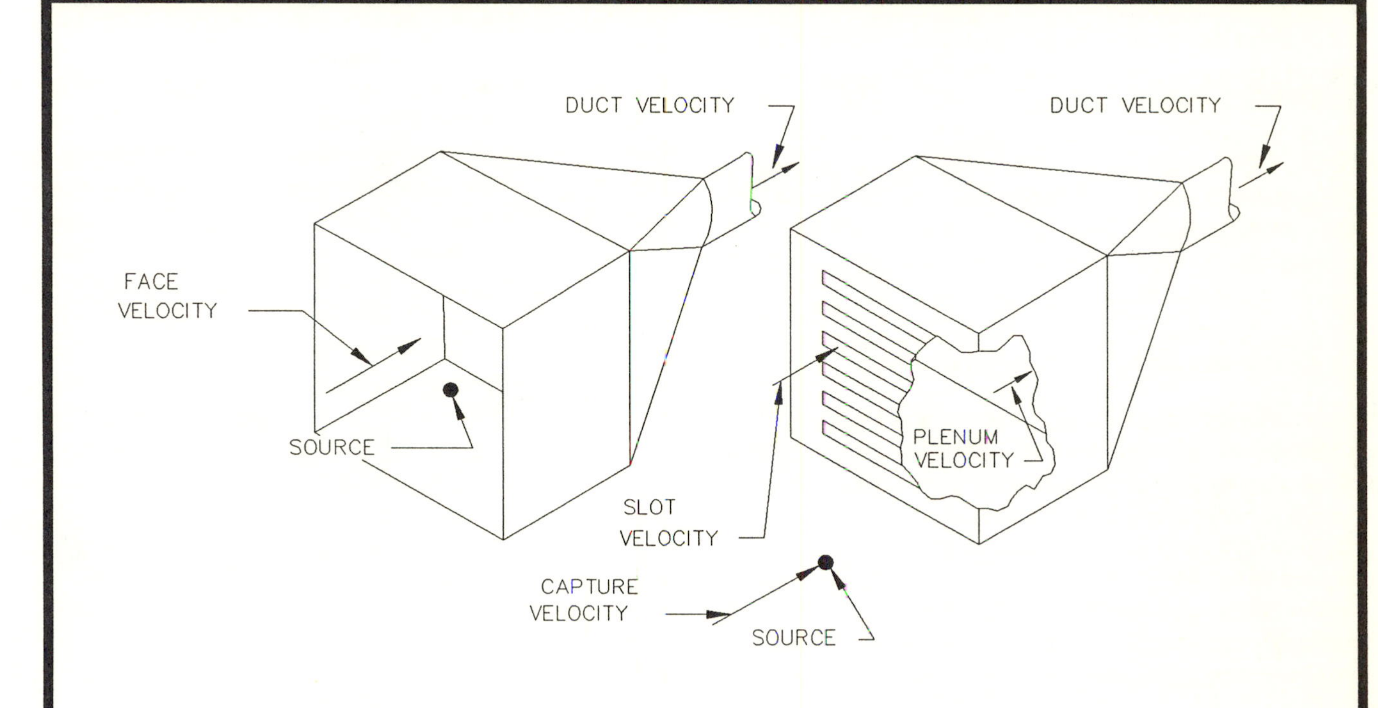

CAPTURE VELOCITY— AIR VELOCITY AT ANY POINT IN FRONT OF THE HOOD OR AT THE HOOD OPENING NECESSARY TO OVERCOME OPPOSING AIR CURRENTS AND TO CAPTURE THE CONTAMINATED AIR AT THAT POINT BY CAUSING IT TO FLOW INTO THE HOOD.

FACE VELOCITY— AIR VELOCITY AT THE HOOD OPENING.

SLOT VELOCITY— AIR VELOCITY THROUGH THE OPENINGS IN A SLOT—TYPE HOOD. IT IS USED PRIMARILY AS A MEANS OF OBTAINING UNIFORM AIR DISTRIBUTION ACROSS THE FACE OF THE HOOD.

PLENUM VELOCITY— AIR VELOCITY IN THE PLENUM. FOR GOOD AIR DISTRIBUTION WITH SLOT—TYPES OF HOODS, THE MAXIMUM PLENUM VELOCITY SHOULD BE 1/2 OF THE SLOT VELOCITY OR LESS.

DUCT VELOCITY— AIR VELOCITY THROUGH THE DUCT CROSS SECTION. WHEN SOLID MATERIAL IS PRESENT IN THE AIR STREAM, THE DUCT VELOCITY MUST BE EQUAL TO OR GREATER THAN THE MINIMUM AIR VELOCITY REQUIRED TO MOVE THE PARTICLES IN THE AIR STREAM.

AMERICAN CONFERENCE OF GOVERNMENTAL INDUSTRIAL HYGIENISTS	*HOOD NOMENCLATURE LOCAL EXHAUST*			
	DATE	*1—88*	FIGURE	*3—1*

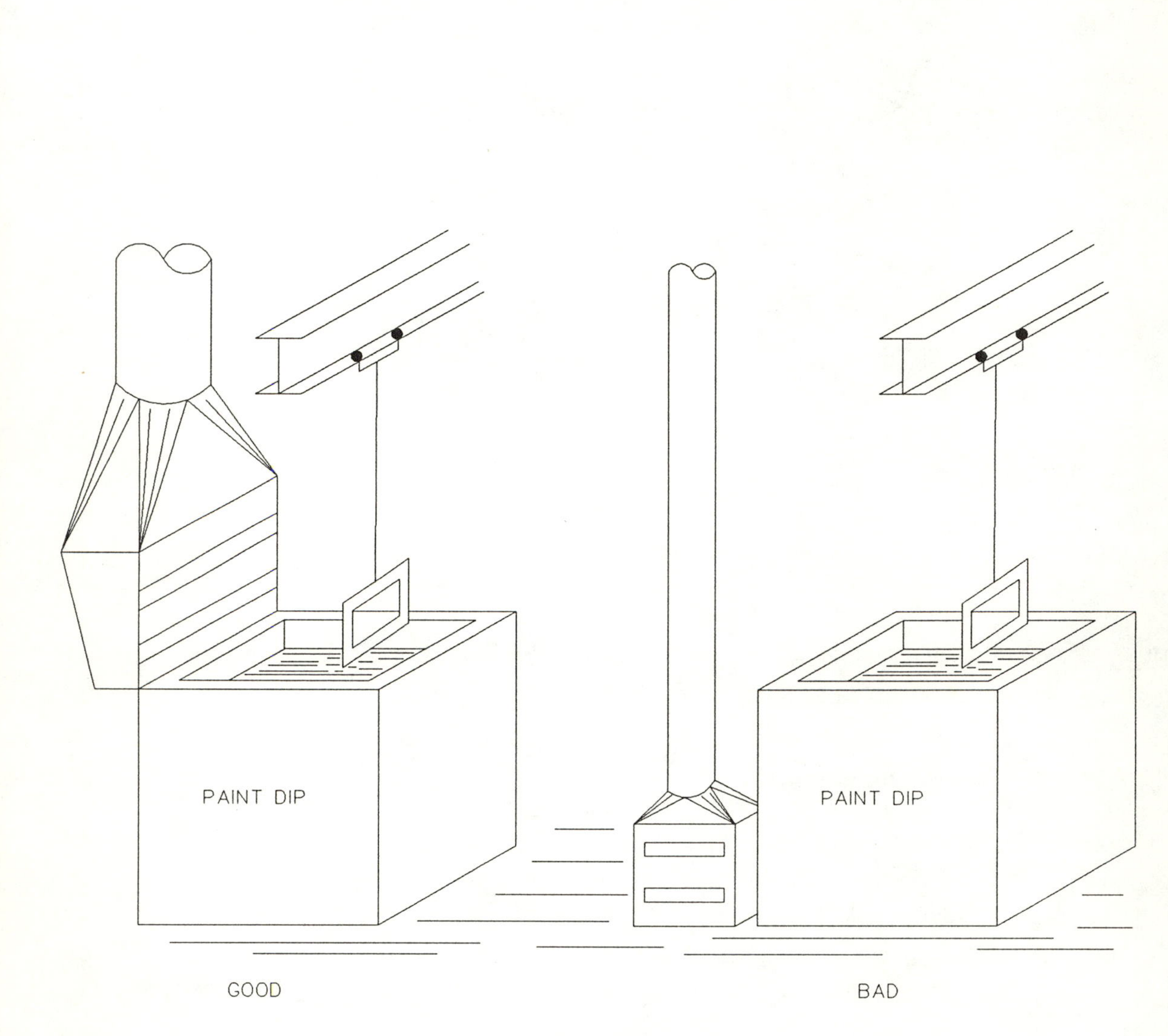

SOLVENT VAPORS IN HEALTH HAZARD CONCENTRATIONS ARE NOT APPRECIABLY HEAVIER THAN AIR.
EXHAUST FROM THE FLOOR USUALLY GIVES FIRE PROTECTION ONLY.

AMERICAN CONFERENCE OF GOVERNMENTAL INDUSTRIAL HYGIENISTS	*EFFECTS OF SPECIFIC GRAVITY*	
	DATE *1-88*	FIGURE *3-2*

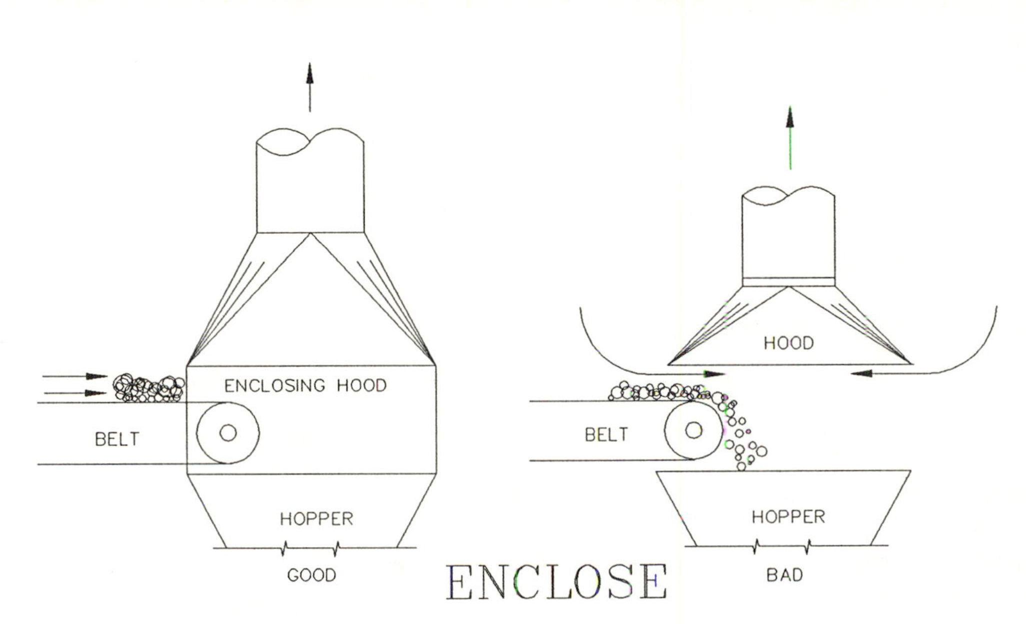

ENCLOSE

ENCLOSE THE OPERATION AS MUCH AS POSSIBLE. THE MORE COMPLETELY ENCLOSED THE SOURCE, THE LESS AIR REQUIRED FOR CONTROL.

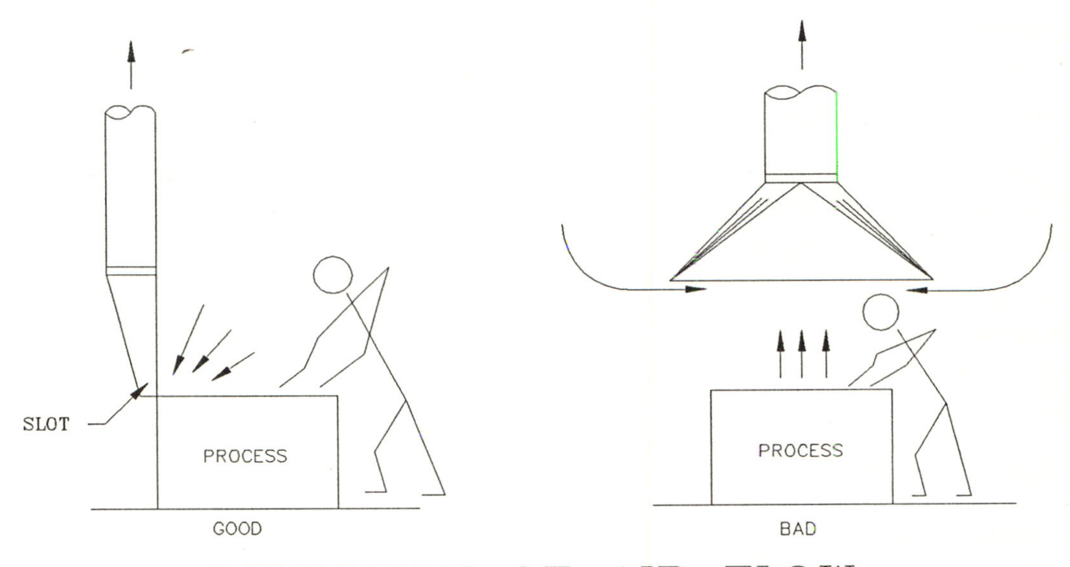

DIRECTION OF AIR FLOW

LOCATE THE HOOD SO THE CONTAMINANT IS REMOVED AWAY FROM THE BREATHING ZONE OF THE OPERATOR.

AMERICAN CONFERENCE OF GOVERNMENTAL INDUSTRIAL HYGIENISTS	*ENCLOSURE AND OPERATOR/ EQUIPMENT INTERFACE*
	DATE 1-88 FIGURE 3-3

- Movements of the operator.
- Room air currents (which are usually taken at 50 fpm minimum and may be much higher).
- Rapid air movement caused by spot cooling and heating equipment.

The shape of the hood, its size, location, and rate of air flow are important design considerations.

3.4.1 Capture Velocity: The minimum hood-induced air velocity necessary to capture and convey the contaminant into the hood is referred to as capture velocity. This velocity will be a result of the hood air flow rate and hood configuration.

Exceptionally high air flow hoods (example, large foundry side-draft shakeout hoods) may require less air flow than would be indicated by the capture velocity values recommended for small hoods. This phenomenon may be ascribed to:

- The presence of a large air mass moving into the hood.
- The fact that the contaminant is under the influence of the hood for a much longer time than is the case with small hoods.
- The fact that the large air flow rate affords considerable dilution as described above.

Table 3-1 offers capture velocity data. Additional information is found in Chapter 10.

3.4.2 Hood Flow Rate Determination: Within the bounds of flanges, baffles, adjacent walls, etc., air will move into an opening under suction from all directions. For an enclosure, the capture velocity at the enclosed opening(s) will be the exhaust flow rate divided by the opening area. The capture velocity at a given point in front of the exterior hood will be established by the hood air flow through the geometric surface which contains the point.

As an example, for a theoretical unbounded point suction source, the point in question would be on the surface of a sphere whose center is the suction point (Figure 3-4).

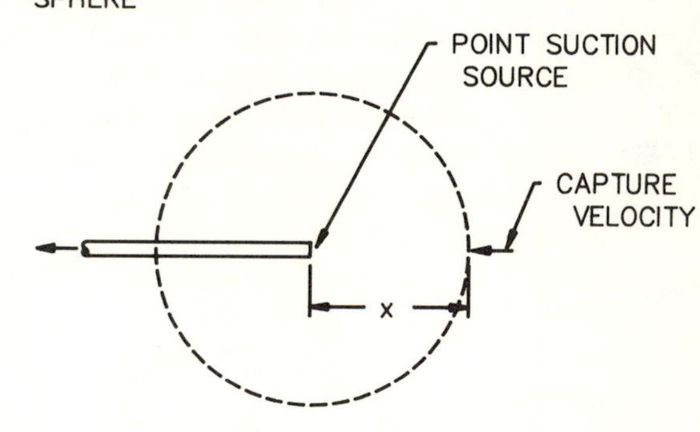

FIGURE 3–4 POINT SUCTION SOURCE

The surface area of a sphere is $4\pi X^2$. Using $V = Q/A$ (Equation 1.3), the velocity at point X on the sphere's surface can be given by

$$Q = V(4\pi X^2) = 12.57VX^2 \quad [3.1]$$

where:

Q = air flow into suction point, cfm

V = velocity at distance X, fpm

$A = 4\pi X^2$ = area of sphere, ft^2

X = radius of sphere, ft

Similarly, if an unbounded line source were considered, the surface would be that of a cylinder and the flow rate (neglecting end effects) would be

$$Q = V(2\pi XL) = 6.28VXL \quad [3.2]$$

where:

L = length of line source, ft

Equations 3.1 and 3.2 illustrate, on a theoretical basis, the

TABLE 3-1. Range of Capture Velocities[3.1,3.2]

Condition of Dispersion of Contaminant	Example	Capture Velocity, fpm
Released with practically no velocity into quiet air.	Evaporation from tanks; degreasing, etc.	50–100
Released at low velocity into moderately still air.	Spray booths; intermittent container filling; low speed conveyor transfers; welding; plating; pickling.	100–200
Active generation into zone of rapid air motion.	Spray painting in shallow booths; barrel filling; conveyor loading; crushers.	200–500
Released at high initial velocity into zone at very rapid air motion.	Grinding; abrasive blasting; tumbling	500–2000

In each category above, a range of capture velocity is shown. The proper choice of values depends on several factors:

Lower End of Range

1. Room air currents minimal or favorable to capture.
2. Contaminants of low toxicity or of nuisance value only.
3. Intermittent, low production.
4. Large hood-large air mass in motion.

Upper End of Range

1. Disturbing room air currents.
2. Contaminants of high toxicity.
3. High production, heavy use.
4. Small hood-local control only.

relationship between distance, flow and capture velocity and can be used for gross estimation purposes. In actual practice, however, suction sources are not points or lines, but rather have physical dimensions which cause the flow surface to deviate from the standard geometric shape. Velocity contours have been determined experimentally. Flow[3.3] for round hoods, and rectangular hoods which are essentially square, can be approximated by

$$Q = V(10X^2 + A) \quad [3.3]$$

where:

Q = air flow, cfm

V = centerline velocity at X distance from hood, fpm

X = distance outward along axis in ft. (NOTE: equation is accurate only for limited distance of X, where X is within 1.5 D)

A = area of hood opening, ft^2

D = diameter of round hoods or side of essentially square hoods, ft

Where distances of X are greater than 1.5 D, the flow rate increases less rapidly with distance than Equation 3.3 indicates.[3.4-3.5]

It can be seen from Equation 3.3 that velocity decreases inversely with the square of the distance from the hood (see

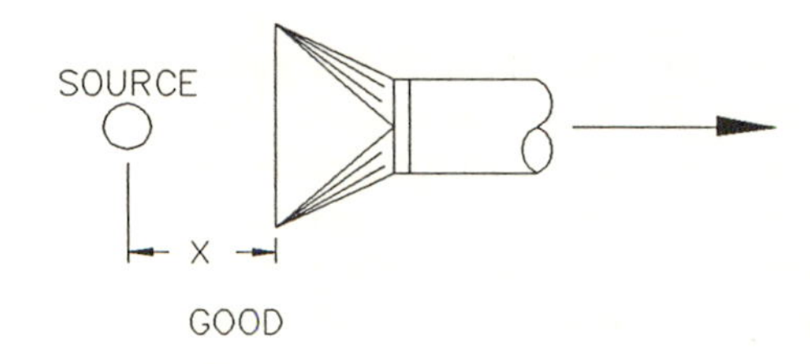

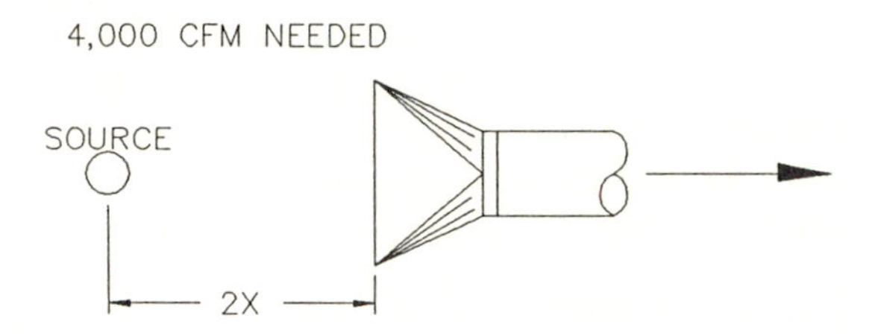

LOCATION

PLACE HOOD AS CLOSE TO THE SOURCE OF CONTAMINANT AS POSSIBLE. THE REQUIRED VOLUME VARIES WITH THE SQUARE OF THE DISTANCE FROM THE SOURCE.

FIGURE 3-5 FLOW RATE AS DISTANCE FROM HOOD

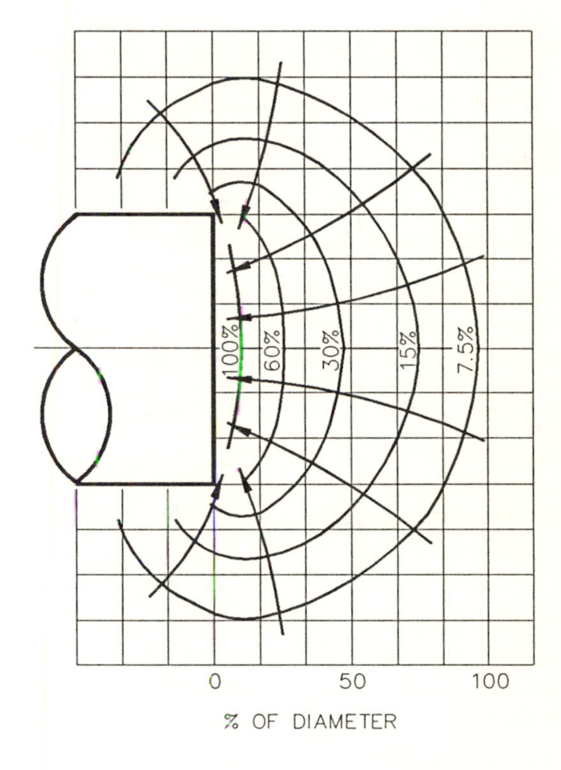

FIGURE 3-6 VELOCITY CONTOURS – PLAIN CIRCULAR OPENING – % OF OPENING VELOCITY

Figure 3-5.)

Figures 3-6 and 3-7 show flow contours and streamlines for plane and flanged circular hood openings. Flow contours are lines of equal velocity in front of a hood. Similarly, streamlines are lines perpendicular to velocity contours. (The tangent to a streamline at any point indicates the direction of air flow at that point.)

Flow capture velocity equations for various hood configurations are provided in Figures 3-8, 3-9 and 3-10.

3.4.3 Effects of Flanges and Baffles: A flange is a surface at and parallel to the hood face which provides a barrier to unwanted air flow from behind the hood. A baffle is a surface but which provides a barrier to unwanted air flow from the front or sides of the hood.

If the suction source were located on a plane, the flow area would be reduced (½ in both cases), thereby decreasing the flow rate required to achieve the same velocity. A flange around a hood opening has the same effect of decreasing the required flow rate to achieve a given capture velocity. In practice, flanging can decrease flow rate (or increase velocity) by approximately 25% (see Figures 3-6, 3-7, and 3-11). For most applications the flange width should be equal to the square root of the hood area ($\sqrt{A}$).

Baffles can provide a similar effect. The magnitude of the effort will depend on the baffle location and size.

Figure 3-11 illustrates several hood types and gives the velocity/flow formulas which apply.

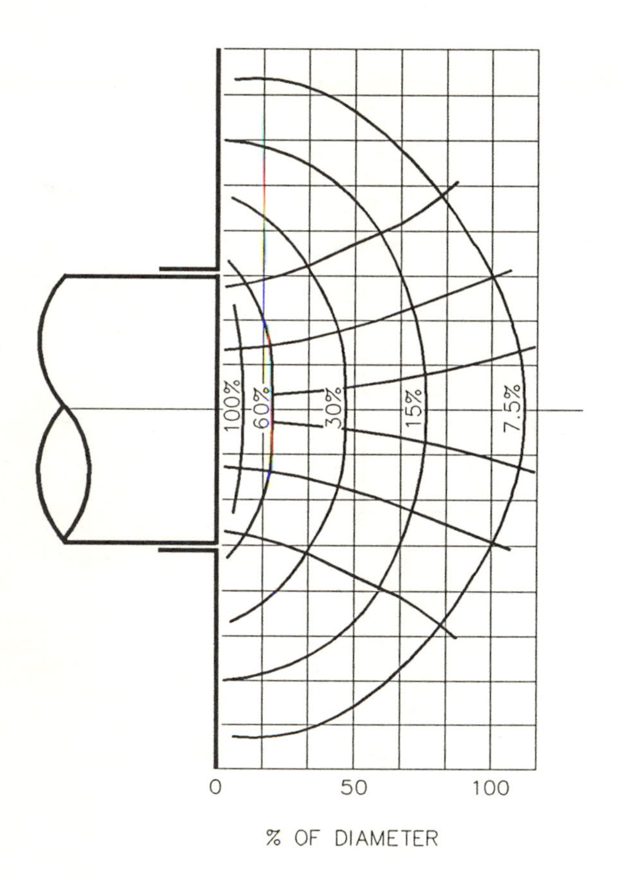

FIGURE 3-7 VELOCITY CONTOURS–FLANGED CIRCULAR OPENING–% OF OPENING VELOCITY

3.4.4 Air Distribution: Slot hoods are defined as hoods with an opening width to length ratio (W/L) of 0.2 or less. Slot hoods are most commonly used to provide uniform exhaust air flow and an adequate capture velocity over a finite length of contaminant generation, e.g. an open tank or over the face of a large hood such as a side-draft design. *The function of the slot is solely to provide uniform air distribution.* Slot velocity does not contribute toward capture velocity. A high slot velocity simply generates high pressure losses. Note that the capture velocity equation (Figure 3-11) shows that capture velocity is related to the exhaust volume and the slot length, not to the slot velocity.

Slot hoods usually consist of a narrow exhaust opening and a plenum chamber. Uniform exhaust air distribution across the slot is obtained by sizing slot width and plenum depth so that velocity through the slot is much higher than in the plenum. Splitter vanes may be used in the plenum; however, in most industrial exhaust systems, vanes are subject to corrosion and/or erosion and provide locations for material to accumulate. Adjustable slots can be provided but are subject to tampering and maladjustment. The most practical hood is the fixed slot and unobstructed plenum type. The design of the slot and plenum is such that the pressure loss through the slot is high compared with the pressure loss through the plenum. Thus, all portions of the slot are subjected to essentially equal suction and the slot velocity will be essentially uniform.

There is no straightforward method for calculating the pressure drop from one end to the other of a slot-plenum combination. A very useful approximation, applicable to most hoods, is to design for a maximum plenum velocity equal to one-half of the slot velocity. For most slot hoods a 2,000 fpm slot velocity and 1,000 fpm plenum velocity is a reasonable choice for uniformity of flow and moderate pressure drop. Centered exhaust take-off design results in the smallest practical plenum size since the air approaches the duct from both directions. Where large, deep plenums are possible, as with foundry shake-out hoods, the slot velocity may be as low as 1,000 fpm with a 500 fpm plenum velocity.

3.4.5 Rectangular and Round Hoods: Air distribution for rectangular and round hoods is achieved by air flow within the hood rather than by pressure drop as for the slot hood. The plenum (length of hood from face to tapered hood to duct connection) should be as long as possible. The hood take-off should incorporate a 60° to 90° total included tapered angle. Multiple take-offs may be required for long hoods. End take-off configurations require large plenum sizes because all of the air must pass in one direction.

Figures 3-12 and 3-13 provide a number of distribution techniques.

3.5 HOOD LOSSES

Plain duct openings, flanged duct openings, canopies, and similar hoods have only one significant energy loss. As air enters the duct, a vena contracta is formed and a small energy loss occurs first in the conversion of static pressure to velocity pressure (see Figure 3-14). As the air passes through the vena contracta, the flow area enlarges to fill the duct and velocity pressure converts to static pressure. At this point the uncontrolled slow down of the air from the vena contracta to the downstream duct velocity results in the major portion of the entry loss. The more pronounced the vena contracta, the greater will be the energy loss and hood static pressure. The hood entry loss (h_e) can be expressed, therefore, in terms of a hood loss factor (F_h) which, when multiplied by the duct velocity pressure (VP), will give the entry loss (h_e) in inches of water. Figure 3-15 gives hood entry loss factors for several typical hood types.

3.5.1 Simple Hood: In a simple hood (Figure 3-16), the hood static pressure is equal to the velocity pressure in the duct plus the hood entry loss (see Chapter 1, Section 1.6, "Acceleration of Air and Hood Entry Losses"). The velocity pressure represents the pressure necessary to accelerate the air from rest to the duct velocity; the hood entry loss represents the energy necessary to overcome the loss as the air enters the duct. This may be expressed as:

$$SP_h = h_{ed} + VP_d \quad [3.4]$$

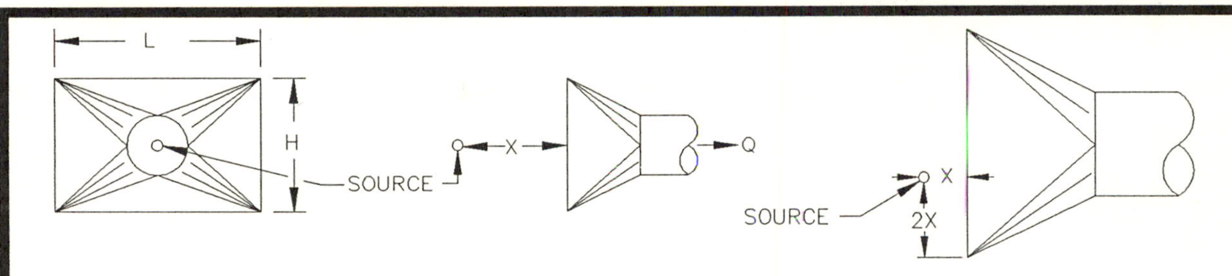

FREELY SUSPENDED HOOD

$Q = V(10X^2 + A)$

LARGE HOOD

LARGE HOOD, X SMALL--MEASURE X PERPENDICULAR TO HOOD FACE, NOT LESS THAN 2X FROM HOOD EDGE.

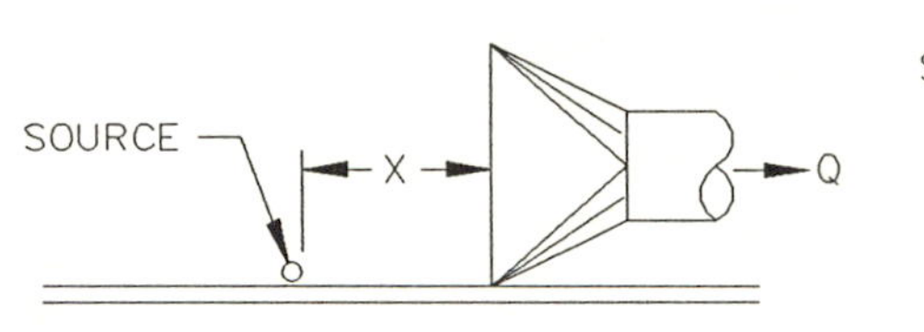

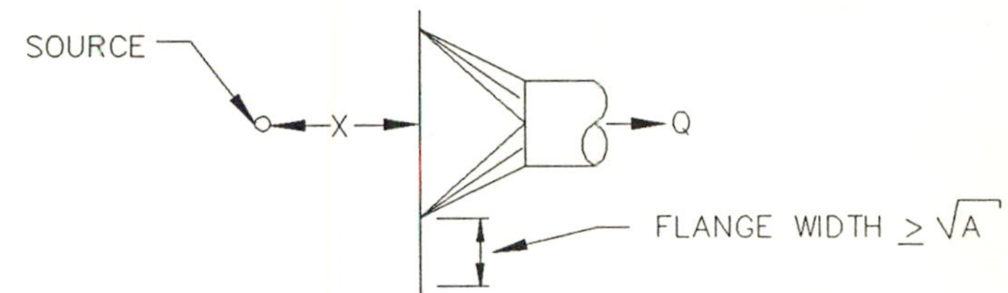

HOOD ON BENCH OR FLOOR

$Q = V(5X^2 + A)$

HOOD WITH WIDE FLANGE

$Q = V\ 0.75(10X^2 + A)$

SUSPENDED HOODS
(SMALL SIDE-DRAFT HOODS)

Q = REQUIRED EXHAUST AIR FLOW, CFM.
X = DISTANCE FROM HOOD FACE TO FARTHEST POINT OF CONTAMINANT RELEASE, FEET.
A = HOOD FACE AREA, FT^2
V = CAPTURE VELOCITY, FPM, AT DISTANCE X.

NOTE: AIR FLOW RATE MUST INCREASE AS THE SQUARE OF DISTANCE OF THE SOURCE FROM THE HOOD. BAFFLING BY FLANGING OR BY PLACING ON BENCH, FLOOR, ECT. HAS A BENEFICIAL EFFECT.

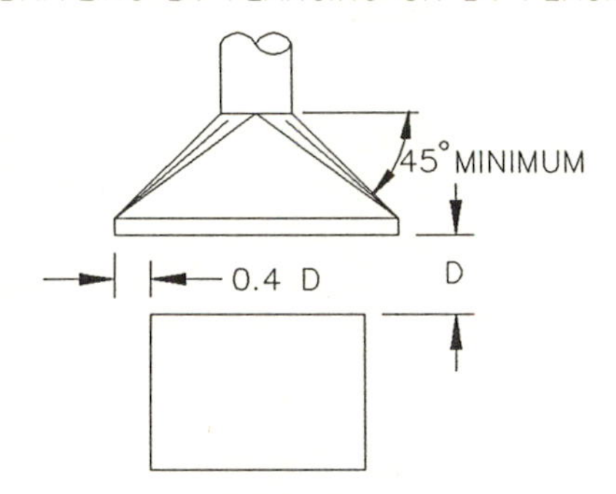

CANOPY HOOD

Q = 1.4 PDV(P=PERIMETER OF TANK, FEET).
NOT RECOMMENDED IF WORKERS MUST BEND OVER SOURCE. V RANGES FROM 50 TO 500 FPM DEPENDING ON CROSSDRAFTS. SIDE CURTAINS ON TWO OR THREE SIDES TO CREATE A SEMI-BOOTH OR BOOTH ARE DESIRABLE.

AMERICAN CONFERENCE OF GOVERNMENTAL INDUSTRIAL HYGIENISTS	*FLOW/CAPTURE VELOCITY*	
	DATE *1-88*	FIGURE *3-8*

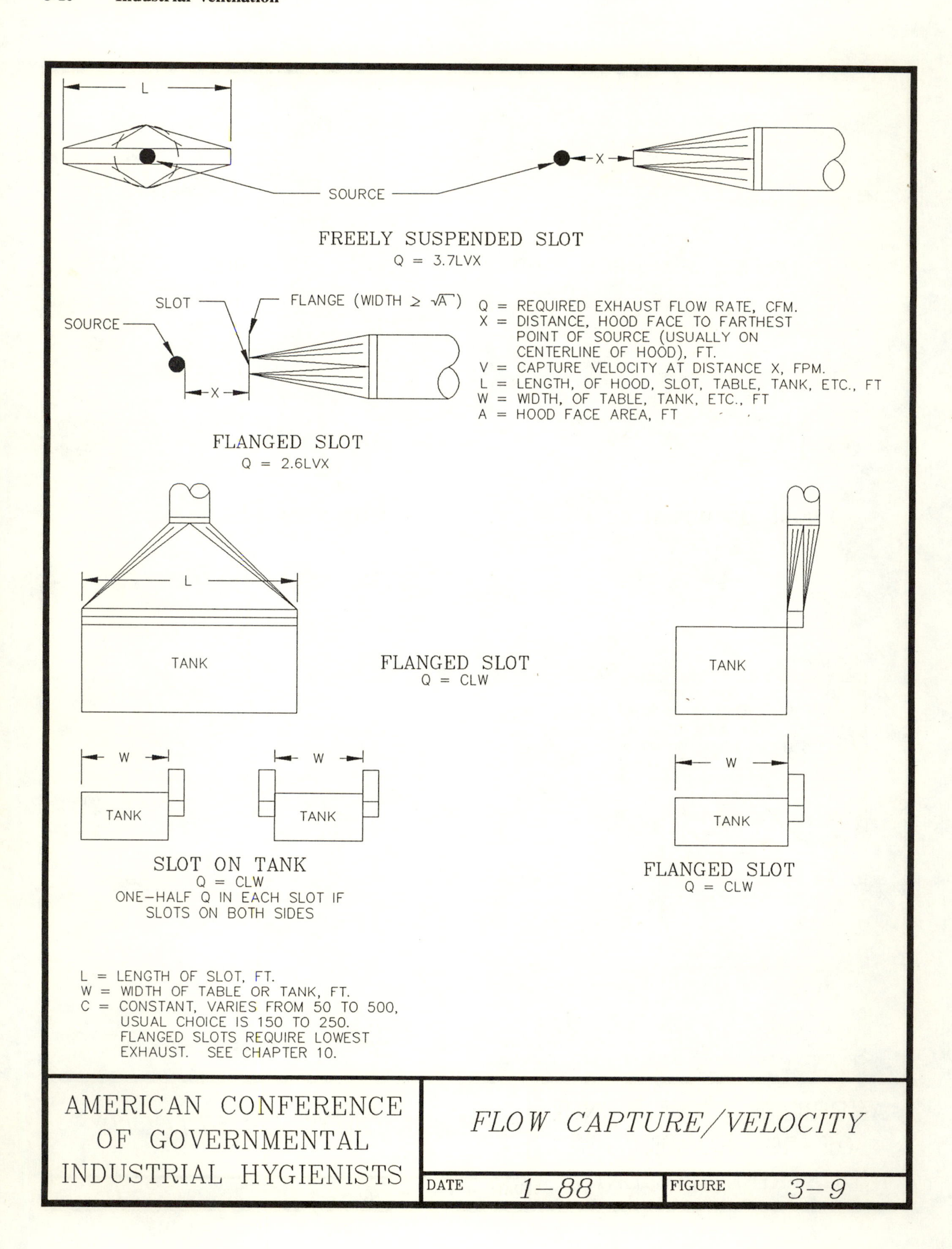

L
SOURCE
X
FREELY SUSPENDED SLOT
Q = 3.7LVX
SLOT
FLANGE (WIDTH ≥ √A)
SOURCE
X
FLANGED SLOT
Q = 2.6LVX
Q = REQUIRED EXHAUST FLOW RATE, CFM.
X = DISTANCE, HOOD FACE TO FARTHEST POINT OF SOURCE (USUALLY ON CENTERLINE OF HOOD), FT.
V = CAPTURE VELOCITY AT DISTANCE X, FPM.
L = LENGTH, OF HOOD, SLOT, TABLE, TANK, ETC., FT
W = WIDTH, OF TABLE, TANK, ETC., FT
A = HOOD FACE AREA, FT
L
TANK
FLANGED SLOT
Q = CLW
TANK
W
TANK
W
TANK
SLOT ON TANK
Q = CLW
ONE-HALF Q IN EACH SLOT IF SLOTS ON BOTH SIDES
W
TANK
FLANGED SLOT
Q = CLW
L = LENGTH OF SLOT, FT.
W = WIDTH OF TABLE OR TANK, FT.
C = CONSTANT, VARIES FROM 50 TO 500, USUAL CHOICE IS 150 TO 250. FLANGED SLOTS REQUIRE LOWEST EXHAUST. SEE CHAPTER 10.
AMERICAN CONFERENCE OF GOVERNMENTAL INDUSTRIAL HYGIENISTS
FLOW CAPTURE/VELOCITY
DATE 1-88
FIGURE 3-9

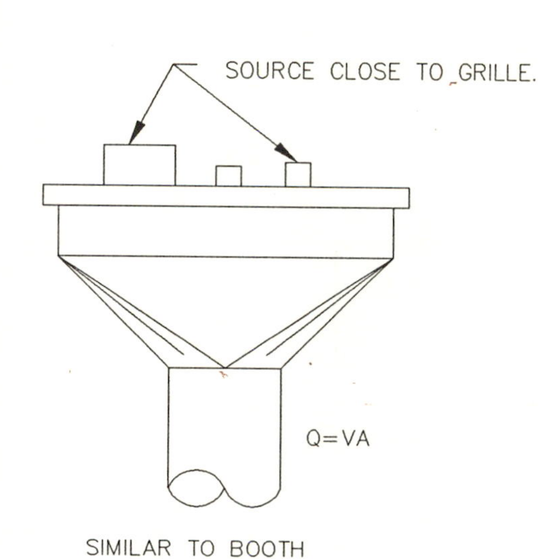

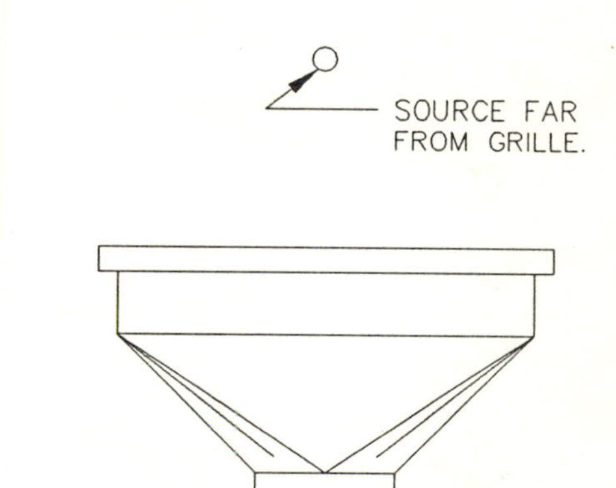

$Q=(10X^2 + A)V$

SIMILAR TO SUSPENDED HOOD

DOWNDRAFT HOODS

NOT RECOMMENDED FOR HOT OR HEAT-PRODUCING OPERATIONS IF DOWNDRAFT AREA IS LARGE, SEE "CAPTURE VELOCITY" IN THIS SECTION.

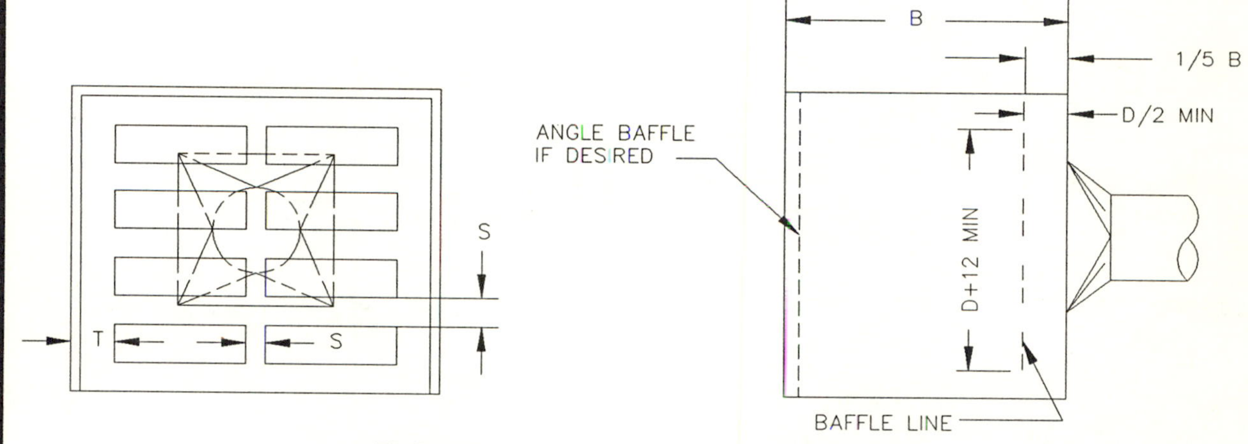

BOOTH-TYPE HOODS

Q=AV (A=FACE AREA, FT^2; V=FACE VELOCITY, FPM.).
BAFFLES ARE OPTIONAL FOR AIR DISTRIBUTION; NOT REQUIRED IF A WATER WALL BOOTH OR IF OTHER MEANS FOR DISTRIBUTION IS PROVIDED.
S VARIES FROM 4 INCHES TO 8 INCHES, DEPENDING ON SIZE OF BOOTH.
T VARIES FROM 6 INCHES TO 12 INCHES, DEPENDING ON SIZE OF BOOTH.
INCREASE THE NUMBER OF PANELS WITH SIZE OF BOOTH.

AMERICAN CONFERENCE OF GOVERNMENTAL INDUSTRIAL HYGIENISTS	*FLOW/CAPTURE VELOCITY*	
	DATE *1-88*	FIGURE *3-10*

HOOD TYPE	DESCRIPTION	ASPECT RATIO, W/L	AIR FLOW
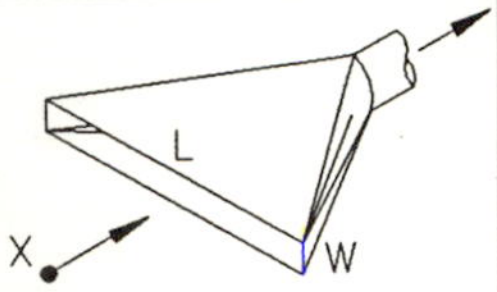	SLOT	0.2 OR LESS	$Q = 3.7\ LVX$
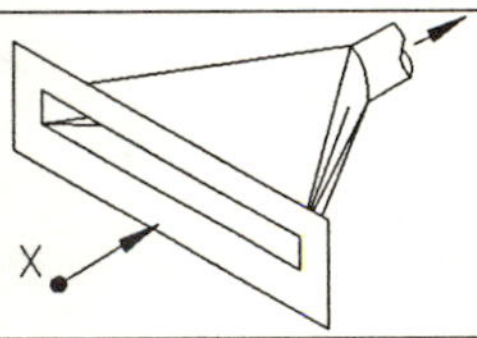	FLANGED SLOT	0.2 OR LESS	$Q = 2.6\ LVX$
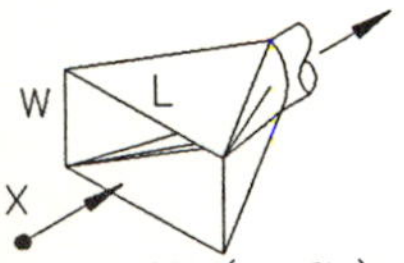 A = WL (sq.ft.)	PLAIN OPENING	0.2 OR GREATER AND ROUND	$Q = V(10X^2 + A)$
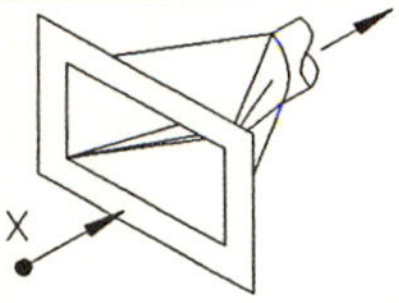	FLANGED OPENING	0.2 OR GREATER AND ROUND	$Q = 0.75V(10X^2 + A)$
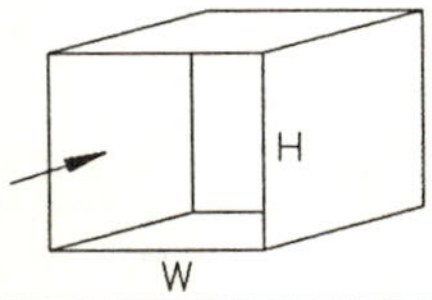	BOOTH	TO SUIT WORK	$Q = VA = VWH$
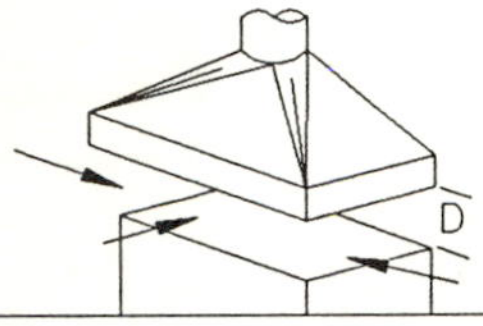	CANOPY	TO SUIT WORK	$Q = 1.4\ PVD$ SEE VS-903 P = PERIMETER D = HEIGHT ABOVE WORK
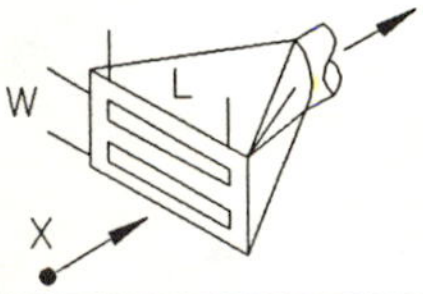	PLAIN MULTIPLE SLOT OPENING 2 OR MORE SLOTS	0.2 OR GREATER	$Q = V(10X^2 + A)$
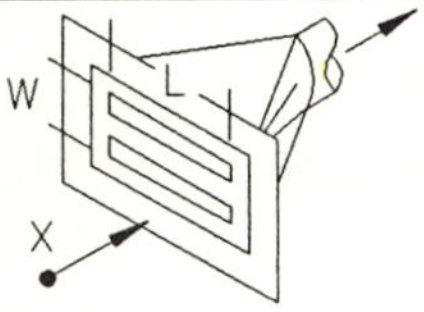	FLANGED MULTIPLE SLOT OPENING 2 OR MORE SLOTS	0.2 OR GREATER	$Q = 0.75V(10X^2 + A)$

AMERICAN CONFERENCE OF GOVERNMENTAL INDUSTRIAL HYGIENISTS

HOOD TYPES

DATE 1-88 FIGURE 3-11

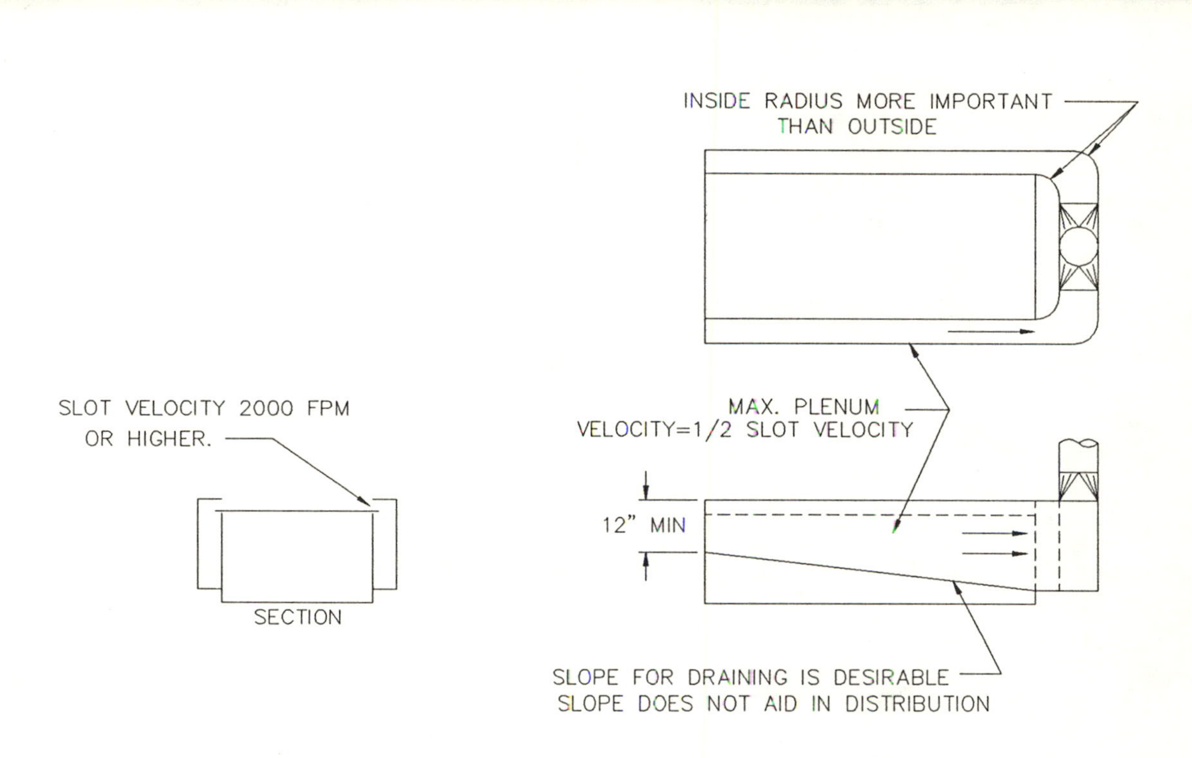

DISTRIBUTION BY SLOT RESISTANCE

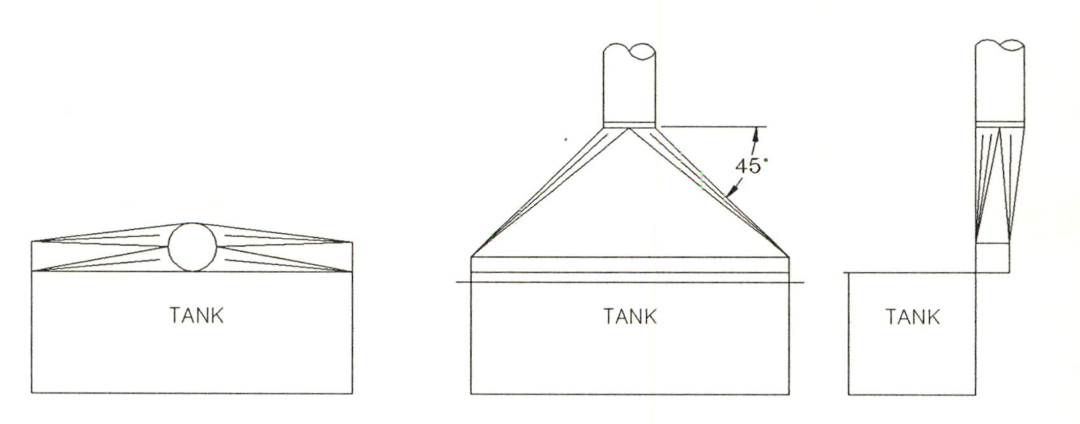

DISTRIBUTION BY FISH TAIL

WITH LOW PLENUM VELOCITIES AND HIGH SLOT VELOCITIES, GOOD DISTRIBUTION IS OBTAINED.
SLOTS OVER 10 FEET TO 12 FEET IN LENGTH USUALLY NEED MUTIPLE TAKE-OFFS.

AMERICAN CONFERENCE OF GOVERNMENTAL INDUSTRIAL HYGIENISTS	*DISTRIBUTION TECHNIQUES*	
	DATE *1-88*	FIGURE *3-12*

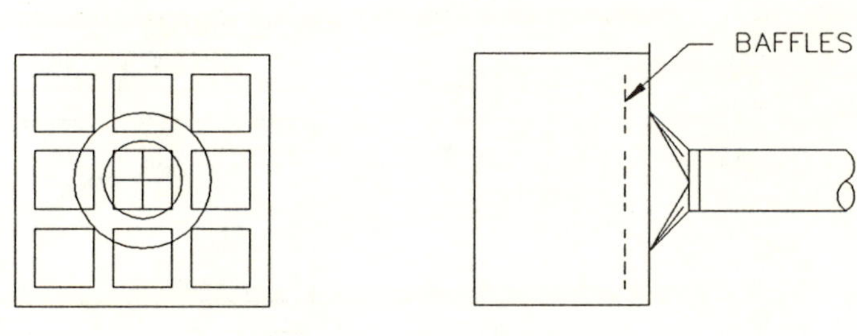

DISTRIBUTION BY BAFFLES
SEE FIG. 3-10

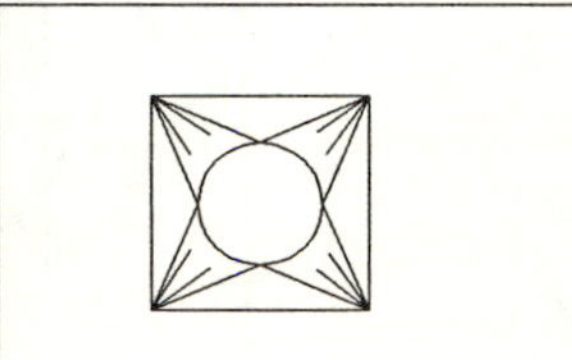

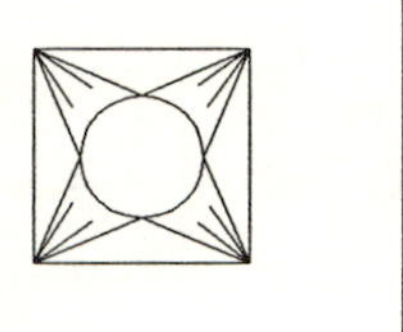

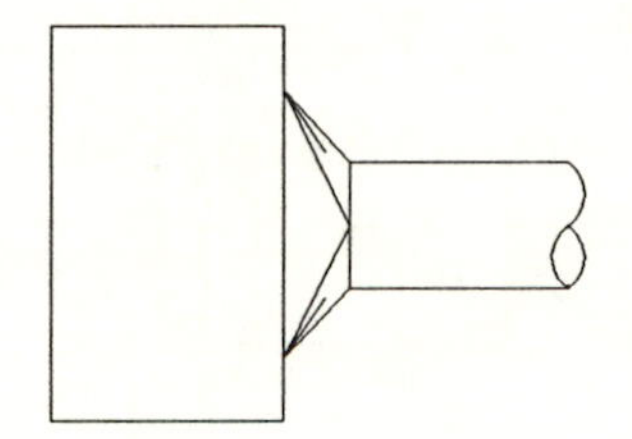

LONG BOOTHS – DISTRIBUTION BY MULTIPLE TAKE-OFFS AND TAPERS

BOOTH CANOPY

(SAME PRINCIPLES APPLY TO CANOPY TYPE)

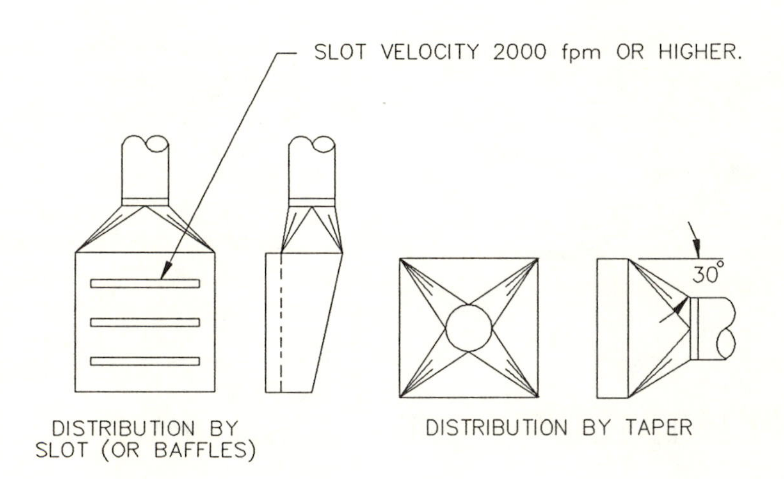

SIDE-DRAFTS AND SUSPENDED HOODS

AMERICAN CONFERENCE OF GOVERNMENTAL INDUSTRIAL HYGIENISTS	*DISTRIBUTION TECHNIQUES*			
	DATE	*1-88*	FIGURE	*3-13*

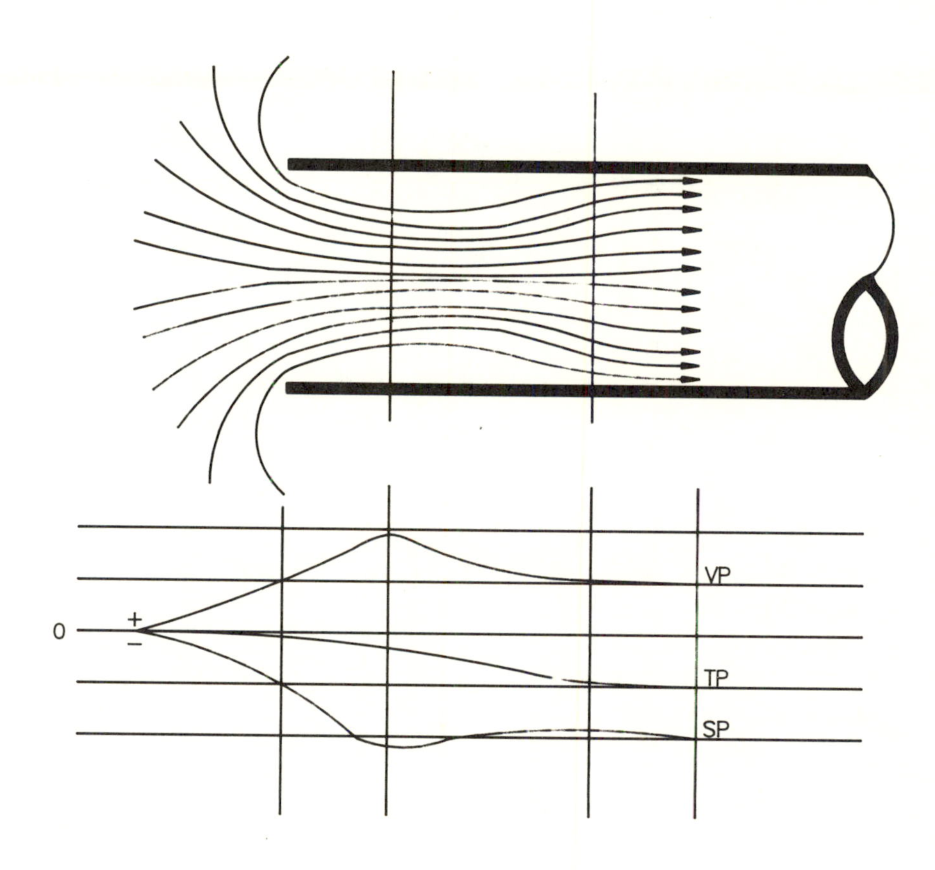

FIGURE 3–14 AIRFLOW AT THE VENA CONTRACTA

where:

SP_h = hood static pressure, "wg

h_{ed} = entry loss of transition (see Figure 5-15) = $F_h \times VP_d$

VP_d = Duct velocity pressure, "wg

Equation 3.4 is used when the face velocity is less than 1000 fpm. When the face velocity is higher than 1000 fpm, the hood should be treated as a compound hood.

EXAMPLE PROBLEM

$$\text{Given: Face Velocity } (V_f) = \frac{Q}{A_{face}} = 250 \text{ fpm}$$

$$\text{Duct Velocity } (V_d) = \frac{Q}{A_{face}} = 2{,}000 \text{ fpm}$$

$$VP_d = \left(\frac{V_d}{4005}\right)^2 = 0.56 \text{ "wg}$$

F_h = 0.25 as shown in Figure 5-15

$$SP_h = h_{ed} + VP_d$$
$$= (0.25 \times 0.56) + 0.56$$
$$= 0.70 \text{ "wg}$$

3.5.2 Compound Hoods: Compound hoods are hoods which have two or more points of significant energy loss and must be considered separately and added together to arrive at the total loss for the hood. Common examples of hoods having double entry losses are slot type hoods and multiple opening, lateral draft hoods commonly used on plating, paint dipping and degreasing tanks, and foundry side-draft shake-out ventilation.

Figure 3-17 illustrates a double entry loss hood. This is a single slot hood with a plenum and a transition from the plenum to the duct. The purpose of the plenum is to give uniform velocity across the slot opening. Air enters the slot, in this case a sharp-edged orifice, and loses energy due to the vena contracta at this point. The air then continues through the plenum where the greater portion of the slot velocity is retained because the air stream projects itself across the plenum in a manner similar to the "blowing" supply stream shown in Figure 1-7. (The retention of velocity in the plenum is characteristic of most local exhaust hoods because of the short plenum length. In the case of very large hoods or exhausted closed rooms, however, the velocity loss must be taken into account. Finally, the air converges into the duct through the transition where the second significant energy loss occurs.

In Figure 3-17, air enters the slot and a slot entry loss occurs as the air is accelerated to the slot velocity. The air passes through the plenum and enters the hood transition where another entry loss occurs as the air is accelerated further to the higher duct velocity.

The hood static pressure for a double entry hood can be

HOOD TYPE	DESCRIPTION	HOOD ENTRY LOSS FACTOR (F_h)
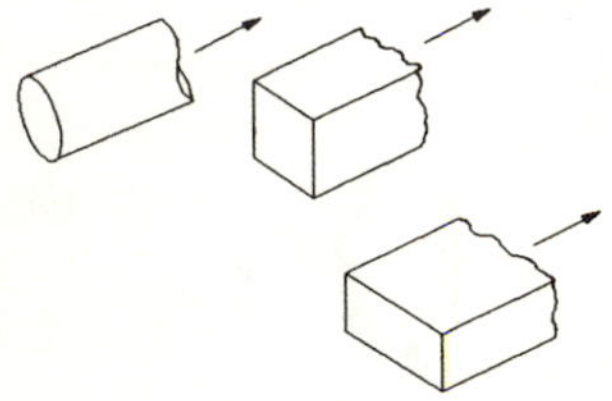	PLAIN OPENING	0.93
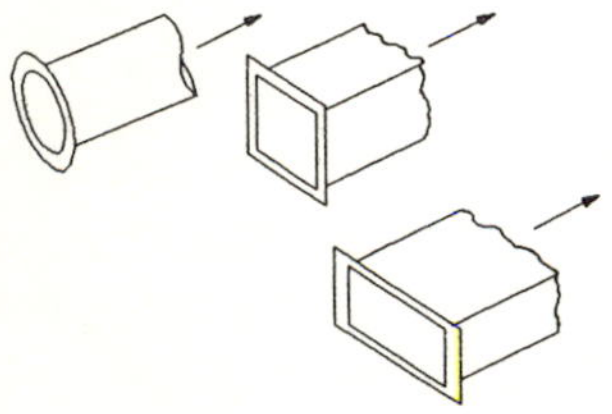	FLANGED OPENING	0.49
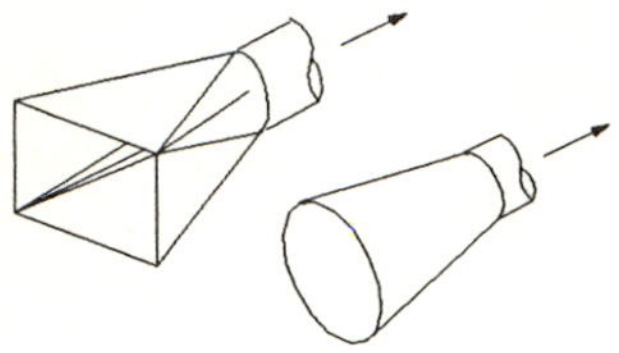	TAPER OR CONE HOOD	SEE FIGURE 5–15
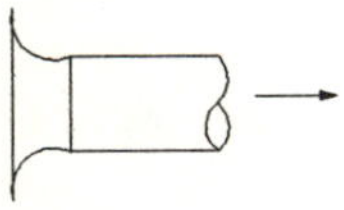	BELL MOUTH INLET	0.04
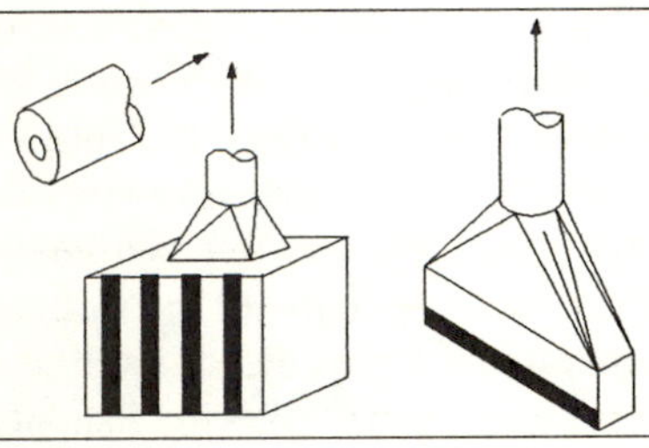	ORIFICE	SEE FIGURE 5–15
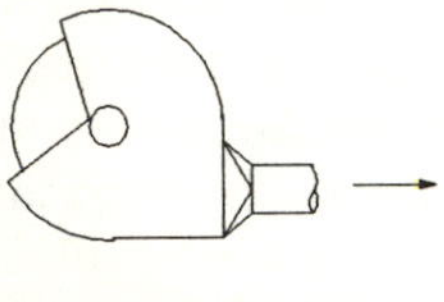	TYPICAL GRINDING HOOD	(STRAIGHT TAKEOFF) 0.65 (TAPERED TAKEOFF) 0.40

AMERICAN CONFERENCE OF GOVERNMENTAL INDUSTRIAL HYGIENISTS

HOOD LOSS FACTORS

DATE *7–89* FIGURE *3–15*

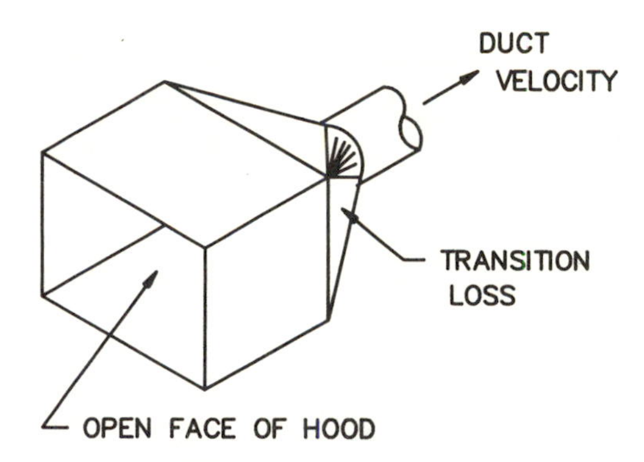

FIGURE 3–16 SIMPLE HOOD

expressed as:

$$SP_h = h_{es} + h_{ed} + VP_d \quad [3.5]$$

when duct velocity is greater than slot velocity.

EXAMPLE PROBLEM

Given: Slot Velocity = 2,000 fpm

VP_s = 0.25 "wg

$h_{es} = VP_s$

Duct Velocity = 3,500 fpm

VP_d = 0.76 "wp

$h_{ed} = 0.25\ VP_d$

$SP_h = h_{es} + h_{ed} + VP_d$

$= 1.78\ (0.25) + 0.25\ (0.76) + 0.76$

= 1.40 "wg

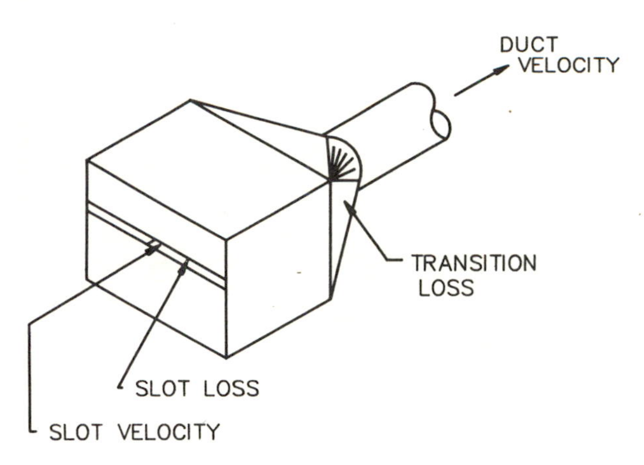

FIGURE 3–17 COMPOUND HOOD

3.6 MINIMUM DUCT VELOCITY

The velocity pressure, VP_d, utilized to determine hood losses in the previous examples is determined from the air velocity in the duct immediately downstream of the hood to duct connection. This velocity is determined by the type of material being transported in the duct.

For systems handling particulate, a minimum design velocity is required to prevent settling and plugging of the duct. On the other hand, excessively high velocities are wasteful of power and may cause rapid abrasion of ducts.[3.6-3.13] Minimum recommended design velocities are higher than theoretical and experimental values to protect against practical contingencies such as:

1. Plugging or closing one or more branch will reduce the total flow rate in the system and correspondingly will reduce the velocities in at least some sections of the duct system.
2. Damage to ducts, by denting for example, will increase the resistance and decrease the flow rate and velocity in the damaged portion of the system.
3. Leakage of ducts will increase flow rate and velocity downstream of the leak but will decrease airflow upstream and in other parts of the system.
4. Corrosion or erosion of the fan wheel or slipping of a fan drive belt will reduce flow rates and velocities.
5. Velocities must be adequate to pick up or re-entrain dust which may have settled due to improper operation of the exhaust system.

The designer is cautioned that for some conditions such as sticky materials, condensing conditions in the presence of dust, strong electrostatic effects, etc., velocity alone may not be sufficient to prevent plugging and other special measures may be necessary.

Some typical duct velocities are provided in Table 3-2. The use of minimum duct velocity is treated in detail in Chapter 5.

3.7 SPECIAL HOOD REQUIREMENTS

3.7.1 Ventilation of Radioactive and High Toxicity Processes: Ventilation of radioactive and high toxicity processes requires a knowledge of the hazards, the use of extraordinarily effective control methods, and adequate maintenance which includes monitoring. Only the basic principles can be covered here. For radioactive processes, reference should be made to the standards and regulations of the nuclear regulatory agencies.

Local exhaust hoods should be of the enclosing type with the maximum enclosure possible. Where complete or nearly complete enclosure is not possible, control velocities from 50 to 100% higher than the minimum standards in this manual should be used. If the enclosure is not complete and an operator must be located at an opening, such as in front of a laboratory hood, the maximum control velocity should not exceed 125 fpm. Air velocities higher than this value will create eddies in front of the operator which may pull contaminant from the hood into the operator's breathing zone. Replacement air should be introduced at low velocity and in a direction which does not cause disruptive cross drafts at the hood opening.

3.7.2 For Laboratory Operations: Glove boxes should be used for high activity alpha or beta emitters and highly toxic

TABLE 3-2. Range of Minimum Duct Design Velocities*

Nature of Contaminant	Examples	Design Velocity
Vapors, gases, smoke	All vapors, gases and smoke	Any desired velocity (economic optimum velocity usually 1000–2000 fpm)
Fumes	Welding	2000–2500
Very fine light dust	Cotton lint, wood flour, litho powder	2500–3000
Dry dusts & powders	Fine rubber dust, Bakelite molding powder dust, jute lint, cotton dust, shavings (light), soap dust, leather shavings	3000–4000
Average industrial dust	Grinding dust, buffing lint (dry), wool jute dust (shaker waste), coffee beans, shoe dust, granite dust, silica flour, general material handling, brick cutting, clay dust, foundry (general), limestone dust, packaging and weighing asbestos dust in textile industries	3500–4000
Heavy dusts	Sawdust (heavy and wet), metal turnings, foundry tumbling barrels and shake-out, sand blast dust, wood blocks, hog waste, brass turnings, cast iron boring dust, lead dust	4000–4500
Heavy or moist	Lead dusts with small chips, moist cement dust, asbestos chunks from transite pipe cutting machines, buffing lint (sticky), quick-line dust	4500 and up

and biological materials. An exhaust flow rate of from 35 to 50 cfm is usually sufficient for a glove box. The air locks used with the glove box should be exhausted if they open directly to the room.

For low activity radioactive laboratory work, a laboratory fume hood may be acceptable. For such hoods, an average face velocity of 125 fpm is recommended. See VS-202 to 204, 204.1, 205, and 205.2.

For new buildings it is frequently necessary to estimate the air conditioning early — before the detailed design and equipment specifications are available. For early estimating, the guidelines provided in VS-204.1 for hood air flow and replacement air flow can be used.

3.8 PUSH-PULL VENTILATION

Push-pull ventilation consists of a push nozzle and an exhaust hood to receive and remove the push jet. Push-pull is used most commonly on open surface vessels such as plating tanks(3.14) but may be effectively used elsewhere (see VS-504. The advantage of push-pull is that a push jet will maintain velocity over large distances, 20-30 ft or more, whereas the velocity in front of an exhaust hood decays very rapidly as the distance from the hood increases. Properly used, the push jet intercepts contaminated air and carries it relatively long distances into the exhaust hood thus providing control where it may be otherwise difficult or impossible.

3.8.1 Push Jet: Ambient air is entrained in the push jet and results in a jet flow at the exhaust hood several times greater than the push nozzle flow rate. The jet velocity will decay with distance from the nozzle. The entrainment ratio for a long thin slot (or pipe) type nozzle may be approximated by:(3.15)

$$\frac{Q_x}{Q_o} = 1.2\sqrt{\left(\frac{ax}{b_o}\right) + 0.41} \qquad [3.6]$$

The velocity ratio may be approximated by:(3.15)

$$\frac{V_x}{V_o} = \frac{1.2}{\sqrt{\left(\frac{ax}{b_o}\right) + 0.41}} \qquad [3.7]$$

where:

Q_o = the push nozzle supply flow

Q_x = the jet flow rate at a distance x from the nozzle

V_o = the push nozzle exit air velocity

V_x = the peak push jet velocity at a distance x

a = a factor characteristic of the nozzle (0.13 for slots and pipes)

x = distance from the nozzle

b_o = the slot width*

[*If the nozzle is freely suspended (free plane jet), b_o is equal to one-half the total slot width. If the nozzle is positioned on or very near a plane surface (wall jet), b_o is equal to the full slot width. For pipes with holes, b_o is the width of a slot with equivalent area.]

Typical jet velocity profiles are shown in Figure 3-18.

Obstructions in the jet path should be minimized near the jet. Objects with small cross-sections, such as parts hangers, will not cause serious problems; however, large flat surface objects should be avoided. At further distances from the nozzle where the jet has expanded, larger objects may be acceptable if they are located within the jet.

The nozzle may be constructed as a long thin slot, a pipe with holes or individual nozzles. The total nozzle exit area should not exceed 25% of the nozzle plenum cross-sectional area to assure even flow distribution. Slot width can range from 0.125 to 0.25 inch for short push length such as plating tanks (4 to 8 ft). Hole size should be 0.25 inch on 3 to 8 diameter spacing. The nozzle momentum factor, which is proportional to nozzle exit flow per foot of nozzle length times nozzle exit velocity ($Q_o \times V_o$) must be sufficient to result in an effective jet, but not so strong so that the exhaust hood is overpowered. A Q_oV_o range should be approximately 50,000 to 75,000 per foot of nozzle length for short distances of 4 to 8 feet.

3.8.2 Pull Hood: The pull hood will accept and remove the push jet flow. The same design considerations regarding flow distribution, hood entry losses, etc., used for a normal pull only hood should be used. The hood pull flow should be approximately 1.5 to 2.0 times the push flow which reaches the hood. If design criteria specifying pull flow rate is not available, Equation 3.6 can be used.

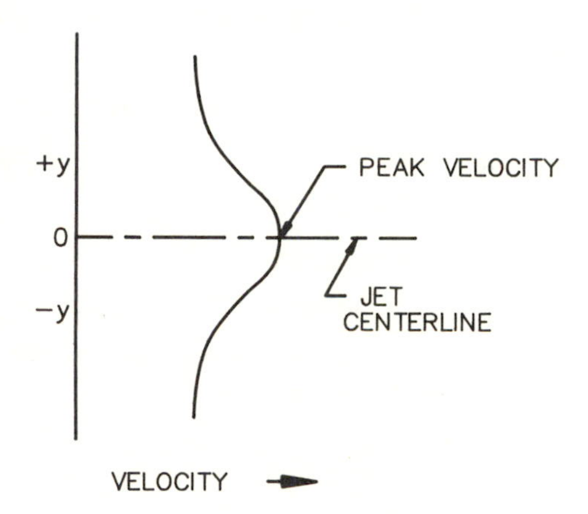

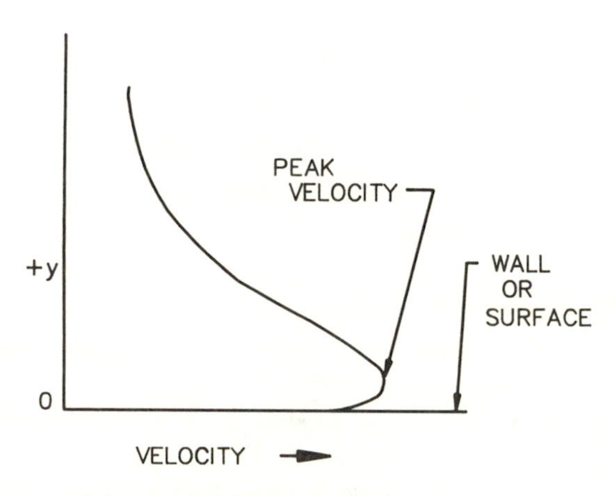

FIGURE 3–18 JET VELOCITY PROFILE

The hood opening height should be the same as the width of the expanded jet, if possible. However, smaller opening heights are acceptable if the hood flow rate meets the 1.5 to 2.0 times jet flow criteria.

Each push-pull application will necessitate special attention. Wherever possible, a pilot system should be evaluated prior to final installation.

3.8.3 Push-Pull System Design: Specific design criteria have been developed experimentally for plating, cleaning, or other open surface vessels and are provided in VS-504, VS-504.1, and VS-504.2. Where such specific design criteria are not available, the criteria provided in sections 3.8.1 and 3.8.2 can be used. When designing with Equation 3.7, a push jet velocity (Vx) of 150 to 200 fpm at the exhaust hood face should be specified.

3.9 HOT PROCESSES

Design of hooding for hot processes requires different considerations than design for cold processes.[3.16] When significant quantities of heat are transferred to the air above and around the process by conduction and convection, a thermal draft is created which causes an upward air current with air velocities as high as 400 fpm. The design of the hood and exhaust rate must take this thermal draft into consideration.

3.9.1 Circular High Canopy Hoods: As the heated air rises, it mixes turbulently with the surrounding air. This results in an increasing air column diameter and volumetric flow rate. The diameter of the column (see Figure 3-19) can be approximated by:

$$D_c = 0.5X_c^{0.88} \qquad [3.8]$$

where:

D_c = column diameter at hood face

X_c = y + z = the distance from the hypothetical point source to the hood face, ft

y = distance from the process surface to the hood face, ft

z = distance from the process surface to the hypothetical point source, ft.

"z" can be calculated from:

$$z = (2\ D_s)^{1.138} \qquad [3.9]$$

where:

D_s = diameter of hot source, ft.

The velocity of the rising hot air column can be calculated from:

$$V_f = 8(A_s)^{0.33}\ \frac{(\Delta t)^{0.42}}{X_c^{0.25}} \qquad [3.10]$$

where:

V_f = velocity of hot air column at the hood face, fpm

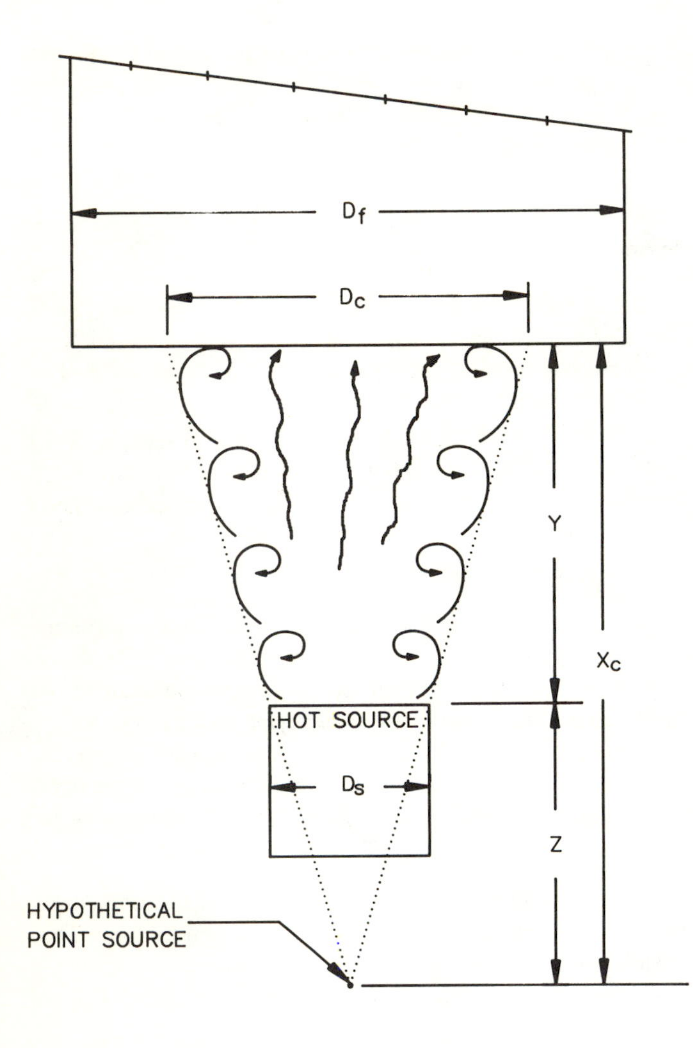

FIGURE 3–19 DIMENSIONS USED TO DESIGN HIGH–CANOPY HOODS FOR HOT SOURCES (HEMEON, 1955).

A_s = area of the hot source, ft^2

Δt = the temperature difference between the hot source and the ambient air, F

$X_c = y + z$ = the distance from the hypothetical point source to the hood face, ft.

The diameter of the hood face must be larger than the diameter of the rising hot air column to assure complete capture. The hood diameter is calculated from

$$D_f = D_c + 0.8\,y \qquad [3.11]$$

where:

D_f = diameter of the hood face, ft.

Total hood airflow rate is

$$Q_t = V_f A_c + V_r(A_f - A_c) \qquad [3.12]$$

where:

Q_t = total volume entering hood, cfm

V_f = velocity of hot air column at the hood face, fpm

A_c = area of the hot air column at the hood face, ft^2

V_r = the required air velocity through the remaining hood area, fpm

A_f = total area of hood face, ft^2.

3.9.2 *Rectangular High Canopy Hoods:* Hot air column from sources which are not circular may be better controlled by a rectangular canopy hood. Hood air flow calculations are performed in the same manner as for circular hoods except the dimensions of the hot air column at the hood (and the hood dimensions) are determined by considering both the length and width of the source. Equations 3.8, 3.9, and 3.11 are used individually to determine length and width of the hot air column and the hood. The remaining values are calculated in the same manner as for the circular hood.

3.9.3 *Low Canopy Hoods:* If the distance between the hood and the hot source does not exceed approximately the diameter of the source or 3 ft, whichever is smaller, the hood may be considered a low canopy hood. Under such conditions, the diameter or cross-section of the hot air column will be approximately the same as the source. The diameter or side dimensions of the hood therefore need only be 1 ft larger than the source.

The total flow rate for a circular low canopy hood is

$$Q_t = 4.7\,(D_f)^{2.33}\,(\Delta t)^{0.42} \qquad [3.13]$$

where:

Q_t = total hood air flow, cfm

D_f = diameter of hood, ft

Δt = difference between temperature of the hot source, and the ambient, F.

The total flow rate for a rectangular low hood is

$$\frac{Q_t}{L} = 6.2\,b^{1.33}\,\Delta t^{0.42} \qquad [3.14]$$

where:

Q_t = total hood air flow, cfm

L = length of the rectangular hood, ft

b = width of the rectangular hood, ft

Δt = difference between temperature of the hot source and the ambient, F.

REFERENCES

3.1. A.D. Brandt: *Industrial Health Engineering*, John Wiley and Sons, New York (1947).

3.2. J.M. Kane: "Design of Exhaust Systems." *Health and Ventilating* 42:68 (November 1946).

3.3. J.M. Dalla Valle: *Exhaust Hoods*. Industrial Press, New York (1946).

3.4. L. Silverman: "Velocity Characteristics of Narrow Exhaust

Slots." *J. Ind. Hg. Toxicol.* 24:267 (November 1942).

3.5. L. Silverman: "Center-line Characteristics of Round Openings Under Suction." *J. Ind. Hyg. Toxicol.* 24:259 (November 1942).

3.6. American Society of Mechanical Engineers: Power Test Code 19.2.4, "Liquid Column Gages" (1942).

3.7. W.C.L. Hemeon: *Plant and Process Ventilation*. Industrial Press, New York (1963).

3.8. J.L. Alden: *Design of Industrial Exhaust System*. Industrial Press, New York (1939).

3.9. G.S. Rajhans and R.W. Thompkins: "Critical Velocities of Mineral Dusts." *Canadian Mining J.* (October 1967).

3.10. O.T. Djamgowz and S.A.A. Ghoneim: "Determining the Pick-Up Air Velocity of Mineral Dusts." Canadian Mining J. (July 1974).

3.11. J.L. Baliff Greenburg and A.C. Stern: "Transport Velocities for Industrial Dusts — An Experimental Study." Ind. Hyg. Q. (December 1948).

3.12. J.M. Dalla Valle: "Determining Minimum Air Velocities for Exhaust Systems. *Heating, Piping and Air Conditioning* (1932).

3.13. T.F. Hatch: *Economy in the Design of Exhaust Systems*

3.14. R.T. Hughes: "Design Criteria for Plating Tank Push-Pull Ventilation." *Ventilation '86*. Elsiever Press, Amsterdam (1986).

3.15. V.V. Baturin: *Fundamentals Industrial Ventilation*. Pergamon Press, New York (1972).

3.16. U.S. Public Health Service: *Air Pollution Engineering Manual*. Publication No. 999-AP-40 (1973).

Chapter 4
AIR CLEANING DEVICES

4.1 INTRODUCTION 4-2
4.2 SELECTION OF DUST COLLECTION EQUIPMENT 4-2
4.2.1 Contaminant Concentration 4-2
4.2.2 Efficiency Required 4-2
4.2.3 Gas Stream Characteristics 4-3
4.2.4 Contaminant Characteristics 4-3
4.2.5 Energy Considerations 4-3
4.2.6 Dust Disposal 4-3
4.3 DUST COLLECTOR TYPES 4-3
4.3.1 Electrostatic Precipitators 4-3
4.3.2 Fabric Collectors 4-9
4.3.3 Wet Collectors 4-17
4.3.4 Dry Centrifugal Collectors 4-18
4.4 ADDITIONAL AIDS IN DUST COLLECTOR SELECTION 4-23
4.5 CONTROL OF MIST, GAS AND VAPOR CONTAMINANTS 4-23
4.6 GASEOUS CONTAMINANT COLLECTORS 4-26
4.6.1 Absorbers 4-26
4.6.2 Adsorbers 4-26
4.6.3 Thermal Oxidizers 4-26
4.6.4 Direct Combustors 4-26
4.6.5 Catalytic Oxidizers 4-26
4.7 UNIT COLLECTORS 4-28
4.8 DUST COLLECTING EQUIPMENT COST 4-28
4.8.1 Price versus Capacity 4-28
4.8.2 Accessories Included 4-28
4.8.3 Installation Cost 4-28
4.8.4 Special Construction 4-28
4.9 SELECTION OF AIR FILTRATION EQUPMENT 4-28
4.9.1 Training 4-28
4.9.2 Impingement 4-32
4.9.3 Interception 4-32
4.9.4 Diffusion 4-32
4.9.5 Electrostatic 4-32
4.10 RADIOACTIVE AND HIGH TOXICITY OPERATIONS 4-33
4.11 EXPLOSION VENTING 4-32
REFERENCES 4-32

4.1 INTRODUCTION

Air cleaning devices remove contaminants from an air or gas stream. They are available in a wide range of designs to meet variations in air cleaning requirements. Degree of removal required, quantity and characteristics of the contaminant to be removed and conditions of the air or gas stream will all have a bearing on the device selected for any given application. In addition, fire safety and explosion control must be considered in all selections. (See NFPA publications.)

For particulate contaminants, air cleaning devices are divided into two basic groups: AIR FILTERS and DUST COLLECTORS. Air filters are designed to remove low dust concentrations of the magnitude found in atmospheric air. They are typically used in ventilation, air-conditioning, and heating systems where dust concentrations seldom exceed 1.0 grains per thousand cubic feet of air and are usually well below 0.1 grains per thousand cubic feet of air. (One pound contains 7,000 grains. A typical atmospheric dust concentration in an urban area is 87 micrograms per cubic meter or 0.038 grains per thousand cubic feet of air.)

Dust collectors are usually designed for the much heavier loads from industrial processes where the air or gas to be cleaned originates in local exhaust systems or process stack gas effluents. Contaminant concentrations will vary from less than 0.1 to 20 grains or more *for each cubic foot* of air or gas. Therefore, dust collectors are, and must be, capable of handling concentrations 100 to 20,000 times greater than those for which air filters are designed.

Small, inexpensive versions of all categories of air cleaning devices are available. The principles of selection, application and operation are the same as for larger equipment. However, due to the structure of the market which focuses on small, quickly available and inexpensive equipment, much of the available equipment is of light duty design and construction. One of the major economies of unit collectors implies recirculation, for which such equipment may or may not be suitable. For adequate prevention of health hazards, fires and explosions, application engineering is just as essential for unit collectors as it is for major systems.

4.2 SELECTION OF DUST COLLECTION EQUIPMENT

Dust collection equipment is available in numerous designs utilizing many different principles and featuring wide variations in effectiveness, first cost, operating and maintenance cost, space, arrangement and materials of construction. Consultation with the equipment manufacturer is the recommended procedure in selecting a collector for any problem where extensive previous plant experience on the specific dust problem is not available. Factors influencing equipment selection include:

4.2.1 Contaminant Concentrations: Contaminants in exhaust systems cover an extreme range in concentration and particle size. Concentrations can range from less than 0.1 to much more than 20.0 grains of dust per cubic foot of air. In low pressure conveying systems, the dust ranges from 0.5 to 100 or more microns in size. Deviation from mean size (the range over and under the mean) will also vary with the material.

4.2.2 Efficiency Required: The degree of collection required will depend on the specific problem under consideration and whether the cleaned air will be recirculated to the plant (see Chapter 7) or discharged outside. Evaluation will consider the need for high efficiency-high cost equipment requiring minimum energy such as high voltage electrostatic precipitators; high efficiency-moderate cost equipment such as fabric or wet collectors; or the lower cost primary units such as the dry centrifugal group. If either of the first two groups is selected, the question of combination with primary collectors should be considered.

When the cleaned air is to be discharged outside, the required degree of collection can depend on plant location; comparison of quantities of material released to atmosphere with different types of collection; nature of contaminant (its salvage value and its potential as a health hazard, public nuisance or ability to damage property); and the regulations of governmental agencies. In remote locations, damage to farms or contribution to air pollution problems of distant cities can influence the need for and importance of effective collection equipment. Many industries, originally located away from residential areas, failed to anticipate the construction of residential building which frequently develops around a plant. Such lack of foresight has required installation of air cleaning equipment at greater expense than initially would have been necessary. Today, the remotely located plant must comply, in most cases, with the same regulations as the plant located in an urban area. With present emphasis on public nuisance, public health, and preservation and improvement of community air quality, management can continue to expect criticism for excessive emissions of air contaminants whether located in a heavy industry section of a city or in an area closer to residential zones.

The mass rate of emission will also influence equipment selection. For a given concentration, the larger the exhaust volumetric flow rate, the greater the need for better equipment. Large central steam generating stations might select high efficiency electrostatic precipitators or fabric collectors for their pulverized coal boiler stacks while a smaller industrial pulverized fuel boiler might be able to use slightly less efficient collectors.

A safe recommendation in equipment selection is to select the collector that will allow the least possible amount of contaminant to escape and is reasonable in first cost and maintenance while meeting all prevailing air pollution regulations. For some applications even the question of reasonable cost and maintenance must be sacrificed to meet established standards for air pollution control or to prevent damage to health or property.

It must be remembered that visibility of an effluent will be a function of the light reflecting surface area of the escaping material. Surface area per pound increases inversely as the square of particle size, which means that the removal of 80% or more of the dust on a weight basis may remove only the coarse particles without altering the stack appearance.

4.2.3 Gas Stream Characteristics: The characteristics of the carrier gas stream can have a marked bearing on equipment selection. Temperature of the gas stream may limit the material choices in fabric collectors. Condensation of water vapor will cause packing and plugging of air or dust passages in dry collectors. Corrosive chemicals can attack fabric or metal in dry collectors and when mixed with water in wet collectors can cause extreme damage.

4.2.4 Contaminant Characteristics: The contaminant characteristics will also affect equipment selection. Chemicals emitted may attack collector elements or corrode wet type collectors. Sticky materials, such as metallic buffing dust impregnated with buffing compounds, can adhere to collector elements, plugging collector passages. Linty materials will adhere to certain types of collector surfaces or elements. Abrasive materials in moderate to heavy concentrations will cause rapid wear on dry metal surfaces. Particle size, shape, and density will rule out certain designs. For example, the parachute shape of particles like the "bees wings" from grain will float through centrifugal collectors because their velocity of fall is less than the velocity of much smaller particles having the same specific gravity but a spherical shape. The combustible nature of many finely divided materials will require specific collector designs to assure safe operation.

4.2.5 Energy Considerations: The cost and availability of energy makes essential the careful consideration of the total energy requirement for each collector type which can achieve the desired performance. An electrostatic precipitator, for example, might be a better selection at a significant initial cost penalty because of the energy savings through its inherently lower pressure drop.

4.2.6 Dust Disposal: Methods of removal and disposal of collected materials will vary with the material, plant process, quantity involved, and collector design. Dry collectors can be unloaded continuously or in batches through dump gates, trickle valves, and rotary locks to conveyors or containers. Dry materials can create a secondary dust problem if careful thought is not given to dust-free material disposal or to collector dust bin locations suited to convenient material removal. See Figures 4-1, 4-2, and 4-3 for some typical discharge arrangements and valves.

Wet collectors can be arranged for batch removal or continual ejection of dewatered material. Secondary dust problems are eliminated although disposal of wet sludge can be a material handling problem. Solids carry-over in waste water can create a sewer or stream pollution problem if waste water is not properly clarified.

Material characteristics can influence disposal problems. Packing and bridging of dry materials in dust hoppers, floating or slurry forming characteristics in wet collectors are examples of problems that can be encountered.

4.3 DUST COLLECTOR TYPES

The four major types of dust collectors for particulate contaminants are Electrostatic Precipitators, Fabric Collectors, Wet Collectors, and Dry Centrifugal Collectors.

4.3.1 Electrostatic Precipitators: In electrostatic precipitation, a high potential electric field is established between discharge and collecting electrodes of opposite electrical charge. The discharge electrode is of small cross-sectional area, such as a wire or a piece of flat stock, and the collection electrode is large in surface area such as a plate.

The gas to be cleaned passes through an electrical field that develops between the electrodes. At a critical voltage, the gas molecules are separated into positive and negative ions. This is called "ionization" and takes place at, or near, the surface of the discharge electrode. Ions having the same polarity as the discharge electrode attach themselves to neutral particles in the gas stream as they flow through the precipitator. These charged particles are then attracted to a collecting plate of opposite polarity. Upon contact with the collecting surface, dust particles lose their charge and then can be easily removed by washing, vibration or gravity.

The electrostatic process consists of:

1. Ionizing the gas
2. Charging the dust particles
3. Transporting the particles to the collecting surface
4. Neutralizing, or removing the charge from the dust particles
5. Removing the dust from the collecting surface.

The two basic types of electrostatic precipitators are "Cottrell," or single-stage, and "Penny," or two-stage (see Figures 4-4 and 4-5).

The "Cottrell," single-stage, precipitator (Figure 4-4) combines ionization and collection in a single stage. Because it operates at ionization voltages from 40,000 to 70,000 volts DC, it may also be called a high voltage precipitator and is used extensively for heavy duty applications such as utility boilers, larger industrial boilers, and cement kilns. Some precipitator designs use sophisticated voltage control systems and rigid electrodes instead of wires to minimize maintenance problems.

The "Penny," or two-stage, precipitator (Figure 4-5) uses DC voltages from 11,000 to 14,000 for ionization and is

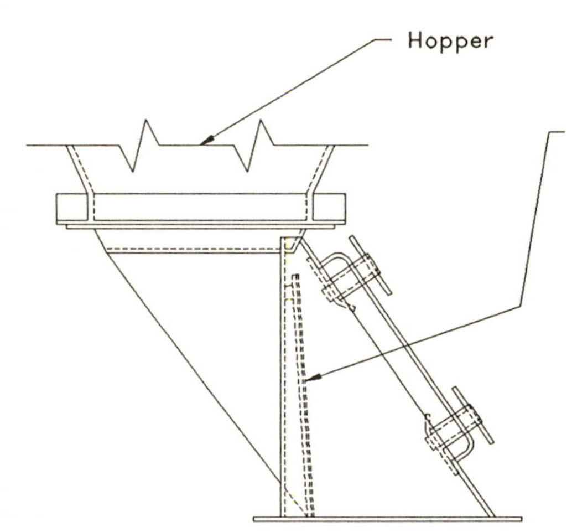

For continuous removal of collected dust where hopper is under negative pressure. Curtain is kept closed by pressure differential until collected material builds up sufficient height to overcome pressure.

TRICKLE VALVE

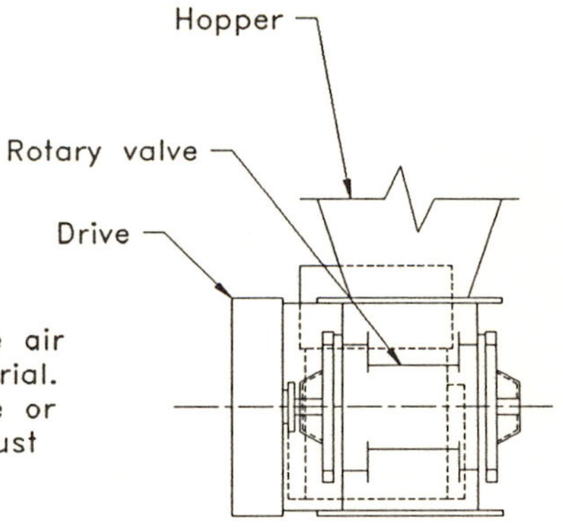

Motor driven multiple blade rotary valve provide air lock while continuously dumping collected material. Can be used with hoppers under either positive or negative pressure. Flanged for connection to dust disposal chute.

ROTARY LOCK

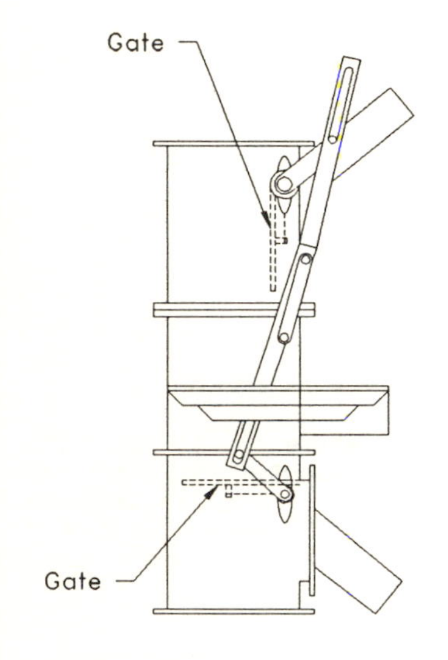

Motor driven, double gate valve for continuous removal of collected dust. Gates are sequenced so only one is open at a time in order to provide air seal. Flanged for connection to dust disposal chute.

DOUBLE DUMP VALVE

AMERICAN CONFERENCE OF GOVERNMENTAL INDUSTRIAL HYGIENISTS	*DRY TYPE DUST COLLECTORS DISCHARGE VALVES*
	DATE *1-88* FIGURE *4-3*

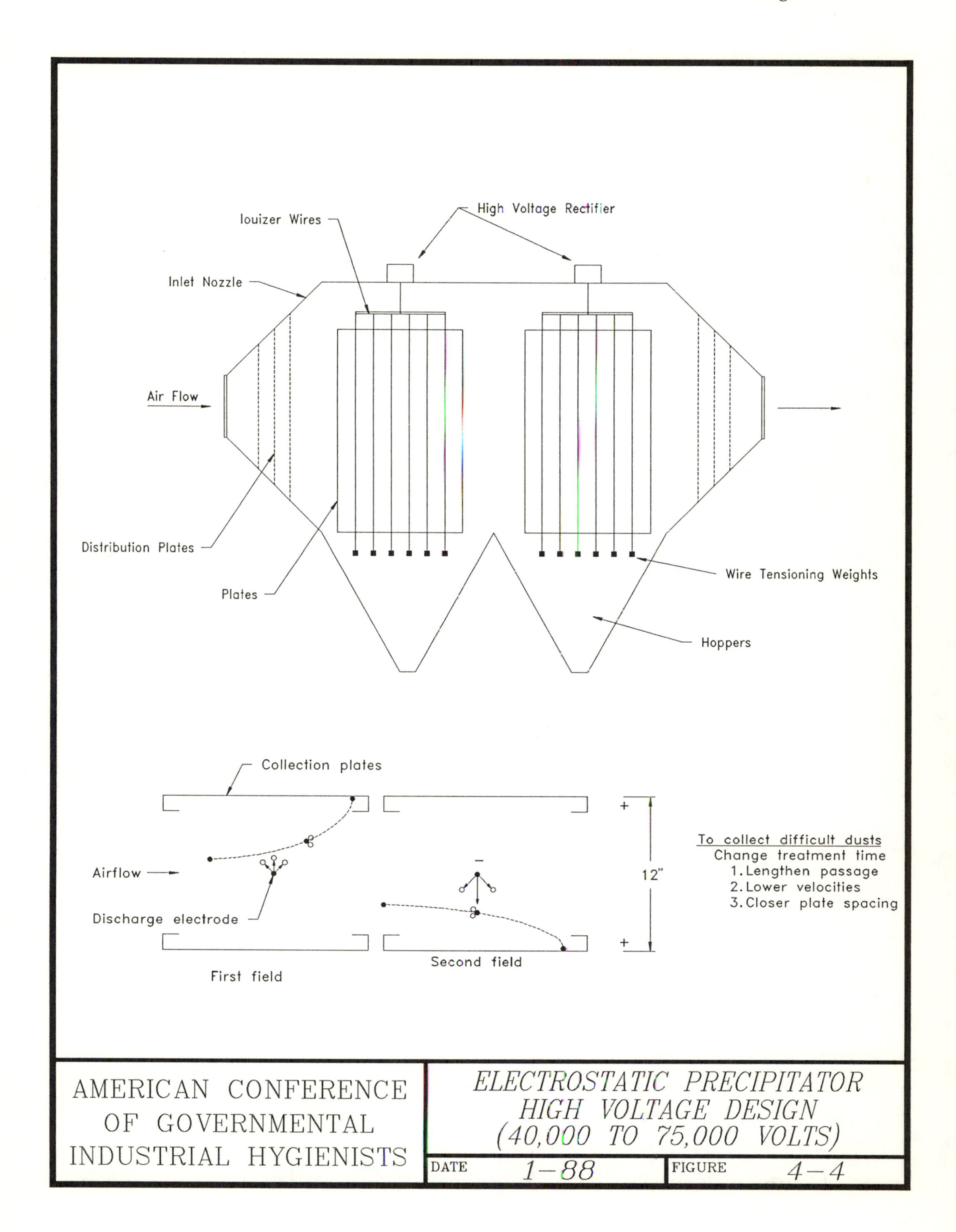
Iouizer Wires
High Voltage Rectifier
Inlet Nozzle
Air Flow
Distribution Plates
Plates
Wire Tensioning Weights
Hoppers
Collection plates
Airflow
Discharge electrode
First field
Second field
12"
To collect difficult dusts
Change treatment time
1. Lengthen passage
2. Lower velocities
3. Closer plate spacing
AMERICAN CONFERENCE OF GOVERNMENTAL INDUSTRIAL HYGIENISTS
ELECTROSTATIC PRECIPITATOR
HIGH VOLTAGE DESIGN
(40,000 TO 75,000 VOLTS)
DATE 1-88
FIGURE 4-4

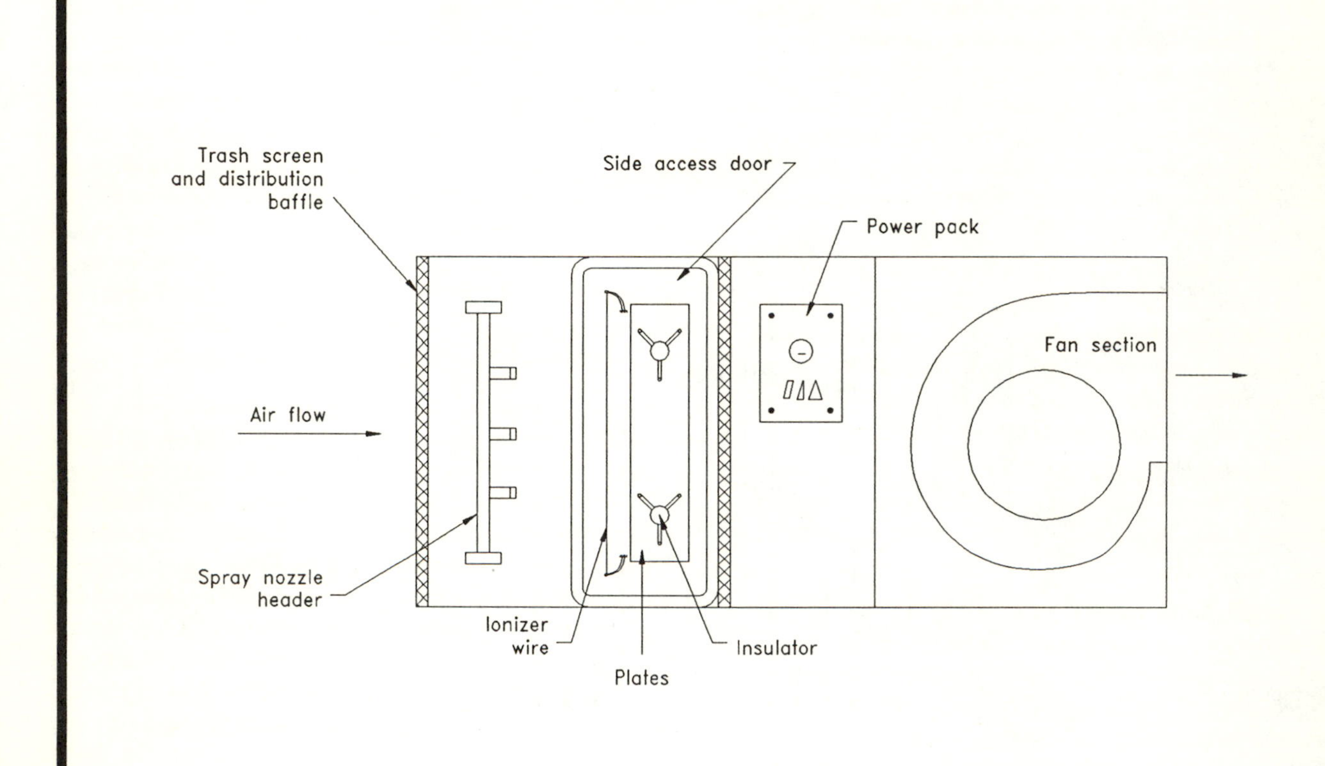

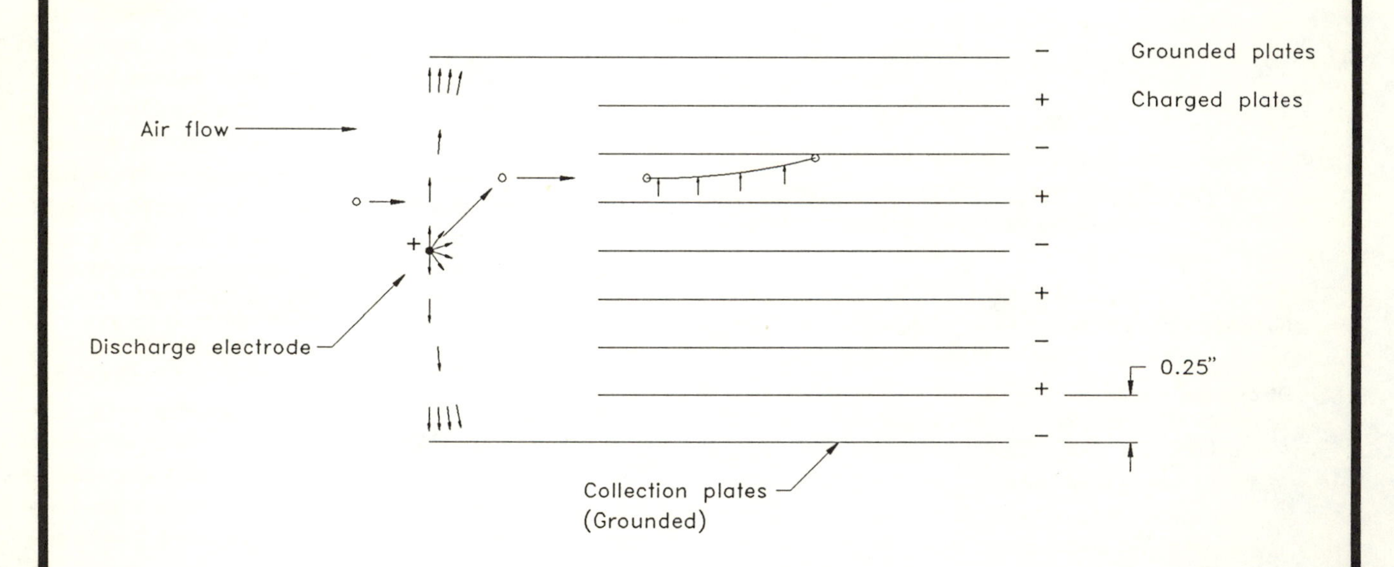

AMERICAN CONFERENCE OF GOVERNMENTAL INDUSTRIAL HYGIENISTS

ELECTROSTATIC PRECIPITATOR LOW VOLTAGE DESIGN (11,000 TO 15,000 VOLTS)

DATE 1-88 FIGURE 4-5

frequently referred to as a low voltage precipitator. Its use is limited to low concentrations, normally not exceeding 0.025 grains per cubic foot. It is the most practical collection technique for the many hydrocarbon applications where an initially clear exhaust stack turns into a visible emission as vapor condenses. Some applications include plasticizer ovens, forge presses, die-casting machines, and various welding operations. Care must be taken to keep the precipitator inlet temperature low enough to insure that condensation has already occurred.

For proper results, the inlet gas stream should be evaluated and treated where necessary to provide proper conditions for ionization. For high-voltage units a cooling tower is sometimes necessary. Low voltage units may use wet scrubbers, evaporative coolers, heat exchangers or other devices to condition the gas stream for best precipitator performance.

The pressure drop of an electrostatic precipitator is extremely low, usually less than 1 "wg; therefore, the energy requirement is significantly less than for other techniques.

4.3.2 Fabric Collectors: Fabric collectors remove particulate by straining, impingement, interception, diffusion, and electrostatic charge. The "fabric" may be constructed of any fibrous material, either natural or man-made, and may be spun into a yarn and woven or felted by needling, impacting or bonding. Woven fabrics are identified by thread count and weight of fabric per unit area. Non-woven (felts) are identified by thickness and weight per unit area. Regardless of construction, the fabric represents a porous mass through which the gas is passed unidirectionally such that dust particles are retained on the dirty side and the cleaned gas passed on through.

The ability of the fabric to pass air is stated as "permeability" and is defined as the cubic feet of air that is passed through one square foot of fabric each minute at a pressure drop of 0.5 "wg. Typical permeability values for commonly used fabrics range from 25 to 40 cfm/ft^2.

A non-woven (felted) fabric is more efficient than a woven fabric of identical weight because the void areas or pores in the non-woven fabric are smaller. A specific type of fabric can be made more efficient by using smaller fiber diameters, a greater weight of fiber per unit area and by packing the fibers more tightly. For non-woven construction, the use of finer needles for felting also improves efficiency. While any fabric is made more efficient by these methods, the cleanability and permeability are reduced. A highly efficient fabric that cannot be cleaned represents an excessive resistance to air flow and is not an economical engineering solution. Final fabric selection is generally a compromise between efficiency and permeability.

Choosing a fabric with better cleanability or greater permeability but lower inherent efficiency is not as detrimental as it may seem. The efficiency of the fabric as a filter is meaningful only when new fabric is first put into service. Once the fabric has been in service any length of time, collected particulate which is in contact with the fabric acts as a filter aid, improving collection efficiency. Depending on the amount of particulate and the time interval between fabric reconditioning, it may well be that virtually all filtration is accomplished by the previously collected particulate — or dust cake — as opposed to the fabric itself. Even immediately after cleaning, a residual and/or redeposited dust cake provides additional filtration surface and higher collection efficiency than obtainable with new fabric. While the collection efficiency of new, clean fabric is easily determined by laboratory test and the information is often published, it is not representative of operating conditions and therefore is of little importance in selecting the proper collector.

Fabric collectors are not 100% efficient, but well-designed, adequately sized, and properly operated fabric collectors can be expected to operate at efficiencies in excess of 99%, and often as high as 99.9+ % on a weight basis. The inefficiency, or penetration, that does occur is greatest during or immediately after reconditioning. Fabric collector inefficiency is frequently a result of by-pass due to damaged fabric, faulty seals or sheet metal leaks rather than penetration of the fabric. Where extremely high collection efficiency is essential, the fabric collector should be leak tested for mechanical leaks by feeding a fluorescent dye powder to the collector, then checking for leaks with an ultraviolet light.

The combination of fabric and collected dust becomes increasingly efficient as the dust cake accumulates on the fabric surface. At the same time, the resistance to air flow increases. Unless the air moving device is adjusted to compensate for the increased resistance, the gas flow rate will be reduced. Figure 4-6 shows how efficiency, resistance to flow and flow rate change with time as dust accumulates on the fabric. Fabric collectors are suitable for service on relatively heavy dust concentrations. The amount of dust collected on a single square yard of fabric may exceed five pounds per hour. In virtually all applications, the amount of dust cake accumulated in just a few hours will represent sufficient resistance to flow to cause an unacceptable reduction in air flow.

In a well-designed fabric collector system, the fabric or filter mat is cleaned or reconditioned before the reduction in air flow is critical. The cleaning is accomplished by mechanical agitation or air motion, which frees the excess accumulation of dust from the fabric surface and leaves a residual or base cake. The residual dust cake does not have the same characteristics of efficiency or resistance to air flow as new fabric.

Commercially available fabric collectors employ fabric configured as bags or tubes, envelopes (flat bags), or pleated cartridges. Most of the available fabrics, whether woven or non-woven, are employed in either bag or envelope configuration. The pleated cartridge arrangement uses a paper-like fiber in either a cylindrical or flat panel configuration. It features extremely high efficiency on light concentrations of

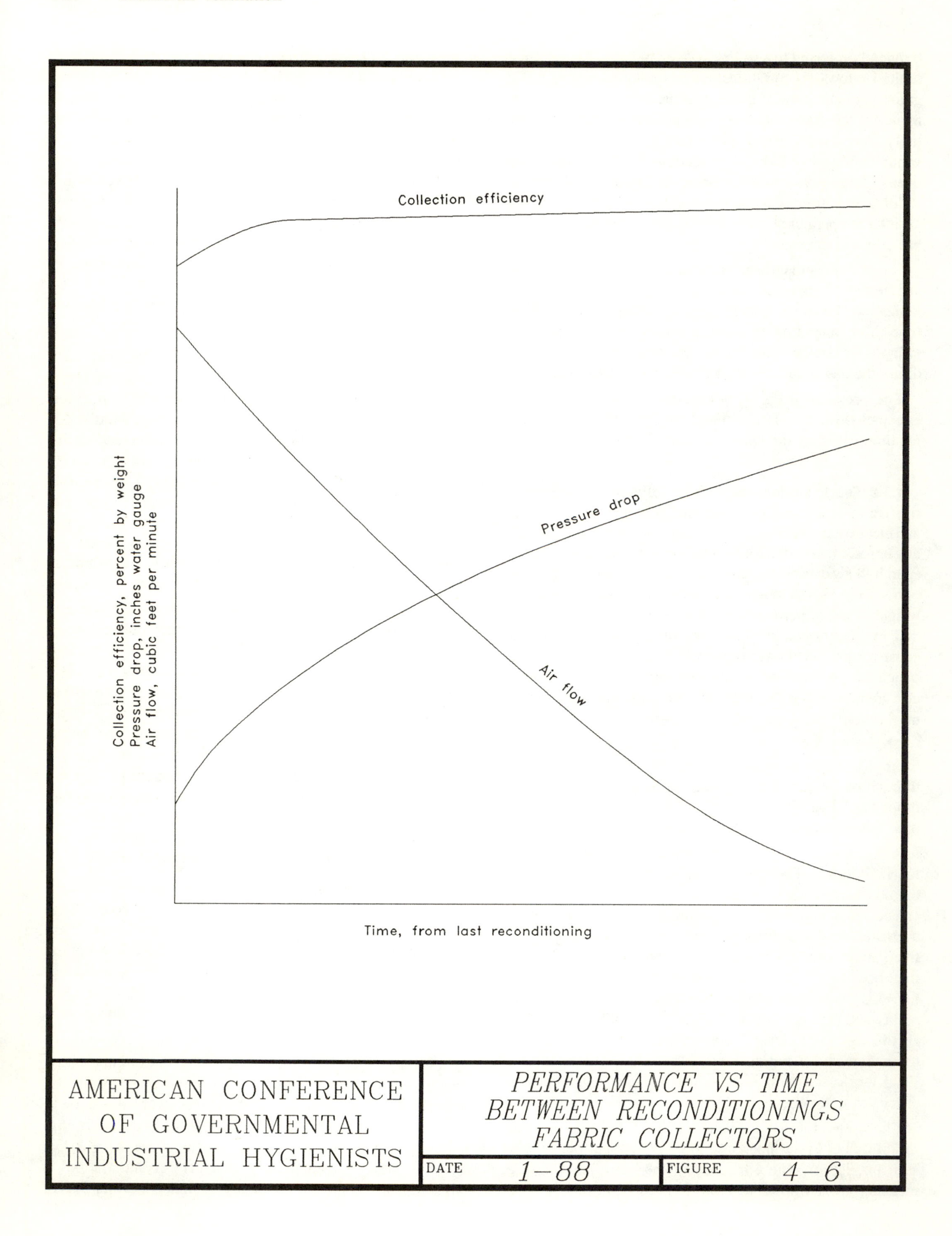
Collection efficiency
Pressure drop
Air flow
Collection efficiency, percent by weight
Pressure drop, inches water gauge
Air flow, cubic feet per minute
Time, from last reconditioning
AMERICAN CONFERENCE OF GOVERNMENTAL INDUSTRIAL HYGIENISTS
PERFORMANCE VS TIME BETWEEN RECONDITIONINGS FABRIC COLLECTORS
DATE 1-88
FIGURE 4-6

TABLE 4-1. Characteristics of Filter Fabrics*

Generic Name	Example Trade Name Fabrics**	Max. Temp, F		Resistance to Physical Action					Resistance to Chemicals				
		Con-tinuous	Inter-mittent	Dry Heat	Moist Heat	Abra-sion	Shaking	Flex-ing	Mineral Acid	Organic Acid	Alkalies	Oxi-dizing	Sol-vents
Cotton	Cotton	180	—	G	G	F	G	G	P	G	F	F	E
Polyester	Dacron;[1] Fortrel;[2] Vycron;[3] Kodel;[4] Enka Polyester[5]	275	—	G	F	G	E	E	G	G	F	G	E
Acrylic	Orlon;[1] Acrilan;[6] Creslan;[7] Dralong T;[8] Zefran[9]	275	285	G	G	G	G	E	G	G	F	G	E
Modacrylic	Dynel;[10] Verel[4]	160	—	F	F	F	P–F	G	G	G	G	G	G
Nylon (Polyamide)	Nylon 6,6;[1,2,6] Nylon 6;[11,5,12] Nomex[1]	225 400	— 450	G E	G E	E E	E E	E E	P P–F	F E	G G	F G	E E
Polypropylene	Herculon;[13] Reevon;[14] Vectra[15]	200	250	G	F	E	E	G	E	E	E	G	G
Teflon (Fluorocarbon)	Teflon TFE[1] Teflon FEP[1]	500 450	550 —	E E	E E	P–F P–F	G G	G G	E E	E E	E E	E E	E E
Vinyon	Vinyon;[16] Clevylt[17]	350	—	F	F	F	G	G	E	E	G	G	P
Glass	Glass	500	600	E	E	P	P	F	E	E	F	E	E
Wool	Wool	215	250	F	F	G	F	G	F	F	P	P	F

*E = excellent; G = good; F = fair; P = poor.
**Registered Trademarks
(1) Du Pont
(2) Celanese
(3) Beaunit
(4) Eastman
(5) American Enka
(6) Chemstrand
(7) American Cyanamid
(8) Farbenfabriken Boyer AG
(9) Dow Chemical
(10) Union Carbide
(11) Allied Chemical
(12) Firestone
(13) Hercules
(14) Alamo Polymer
(15) National Plastic
(16) FMC
(17) Societe Rhovyl

dry, rounded dust such as from abrasive cleaning or powder coating operations. Although conventional fabric, such as polyester or polypropylene, is sometimes used in pleated cartridge collectors, this is usually done because the paper-like fabric cannot handle temperature, moisture, or other gas stream conditions. For such operation, there is a reduction in efficiency and fabric reconditioning capability.

The variable design features of the many fabric collectors available are:

1. Type of fabric (woven or non-woven)
2. Fabric configuration (bags or tubes, envelopes, cartridges)
3. Intermittent or continuous service
4. Type of reconditioning (shaker, pulse-jet, reverse-air)
5. Housing configuration (single compartment, multiple compartment)

At least two of these features will be interdependent. For example, non-woven fabrics are more difficult to recondition and therefore require high pressure cleaning.

A given fabric is selected for its mechanical, chemical, or thermal characteristics. Table 4-1 lists those characteristics for some common filter fabrics.

Fabric collectors are sized to provide a sufficient area of filter media to allow operation without excessive pressure drop. The amount of filter area required depends on many factors, including:

1. Release characteristics of dust
2. Porosity of dust cake
3. Concentration of dust in carrier gas stream
4. Type of fabric and surface finish, if any
5. Type of reconditioning
6. Reconditioning interval
7. Air flow pattern within the collector
8. Temperature and humidity of gas stream

Because of the many variables and their range of variation, fabric collector sizing is a judgment based on experience. The sizing is usually made by the equipment manufacturer, but at times may be specified by the user or a third party where first-hand experience exists from duplicate or very similar applications. Where no experience exists, a pilot installation is the only reliable way to determine proper size.

The sizing or rating of a fabric collector is expressed in terms of air flow rate versus fabric media area. The resultant ratio is called "air to cloth ratio" with units of cfm per square foot of fabric. This ratio represents the average velocity of the gas stream through the filter media. The expression filtration velocity is used synonymously with air to cloth ratio for rating fabric collectors. For example, an air to cloth ratio of 7:1 (7 cfm/ft^2) is equivalent to a filtration velocity of 7 fpm.

The term filter drag is sometimes used to define fabric performance, especially in large installations such as utility boilers. This refers to the pressure drop per cfm per square foot of filter media and is analogous to the resistance of an element in an electrical circuit, i.e., the ratio of filter pressure to filter velocity.

Table 4-2 compares the various characteristics of fabric collectors. The different types will be described in detail later. Inspection of Table 4-2 now may make the subsequent discussion more meaningful. The first major classification of fabric collectors is intermittent or continuous duty. Intermittent duty fabric collectors cannot be reconditioned while in operation. By design, they require that the gas flow be interrupted while the fabric is agitated to free accumulated dust cake. Continuous duty collectors do not require shut down for reconditioning.

Intermittent Duty Fabric Collectors may use a tube, cartridge, or envelope configuration of woven fabric and will generally employ shaking or vibration for reconditioning. Figure 4-7 shows both tube and envelope shaker collector

TABLE 4-2. Summary of Fabric Type Collectors and Their Characteristics

	INTERRUPTABLE OPERATION Light to Moderate Loading	INTERRUPTABLE OPERATION Heavy Loading		CONTINUOUS OPERATION Any Loading	
Fabric Reconditioning Requirement	Intermittent	Continuous			
Type of Reconditioning	Shaker	Shaker	Reverse Air (Low Pressure)	Reverse Pulse - (High Pressure) Pulse Jet or Fan Pulse	
Collector Configuration	Single Compartment	Multiple Compartments with inlet or outlet dampers for each		Single Compartment	
Fabric Configuration	Tube or Envelope	Tube or Envelope	Tube	Tube or Envelope	Pleated-Cartridge
Type Fabric	Woven	Woven		Non-Woven (Felt)	Non-Woven (Paper Mat)
Air Flow	Highly Variable	Slightly Variable		Virtually Constant	Virtually Constant
Normal Rating (filtration velocity, fpm)	1 to 6 fpm	1 to 3 fpm	1 to 2 pm	5 to 12 fpm	<1 to 2.5 fpm

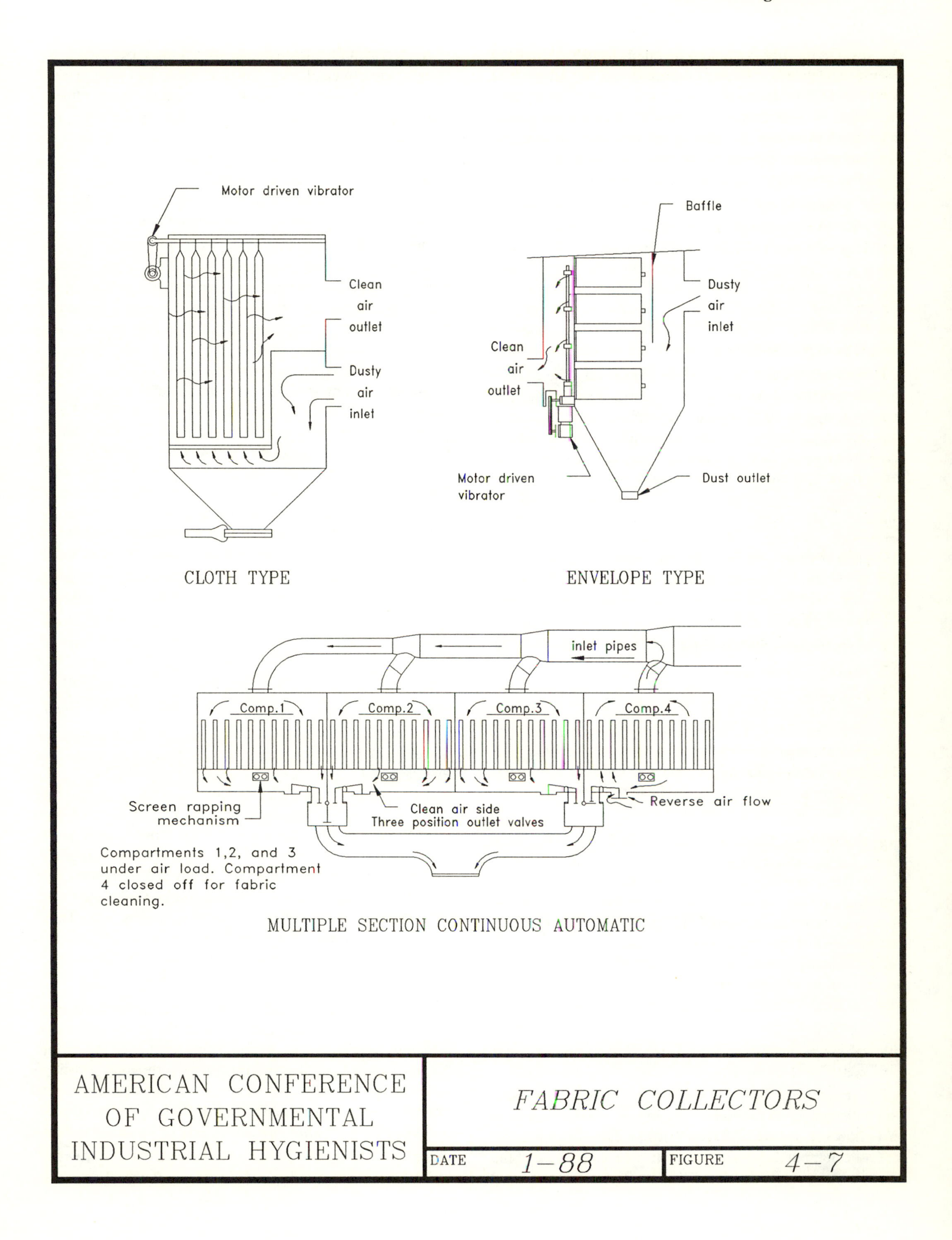
Motor driven vibrator
Clean
air
outlet
Dusty
air
inlet
CLOTH TYPE
Baffle
Dusty
air
inlet
Clean
air
outlet
Motor driven
vibrator
Dust outlet
ENVELOPE TYPE
inlet pipes
Comp.1
Comp.2
Comp.3
Comp.4
Screen rapping
mechanism
Clean air side
Three position outlet valves
Reverse air flow
Compartments 1,2, and 3
under air load. Compartment
4 closed off for fabric
cleaning.
MULTIPLE SECTION CONTINUOUS AUTOMATIC
AMERICAN CONFERENCE
OF GOVERNMENTAL
INDUSTRIAL HYGIENISTS
FABRIC COLLECTORS
DATE 1-88
FIGURE 4-7

designs. For the tube type, dirty air enters the open bottom of the tube and dust is collected on the inside of the fabric. The bottoms of the tubes are attached to a tube sheet and the tops are connected to a shaker mechanism. Since the gas flow is from inside to outside, the tubes tends to inflate during operation and no other support of the fabric is required.

Gas flow for envelope type collectors is from outside to inside; therefore, the envelopes must be supported during operation to prevent collapsing. This is normally done by inserting wire mesh or fabricated wire cages into the envelopes. The opening of the envelope from which the cleaned air exits is attached to a tube sheet and, depending on design, the other end may be attached to a support member or cantilevered without support. The shaker mechanism may be located in either the dirty air or cleaned air compartments.

Periodically (usually at 3- to 6-hour intervals) the air flow must be stopped to recondition the fabric. Figure 4-8 illustrates the system air flow characteristics of an intermittent-duty fabric collector. As dust accumulates on the fabric, resistance to flow increases and air flow decreases until the fan is turned off and the fabric reconditioned. Variations in air flow due to changing pressure losses is sometimes a disadvantage and, when coupled with the requirement to periodically stop the air flow, may preclude the use of intermittent collectors. Reconditioning seldom requires more than two minutes but must be done without air flow through the fabric. If reconditioning is attempted with air flowing it will be less effective and the flexing of the woven fabric will allow a substantial amount of dust to escape to the clean air side.

The filtration velocity for intermittent duty fabric collectors seldom exceeds 6 fpm and normal selections are in the 2 fpm to 4 fpm range. Lighter dust concentrations and the ability to recondition more often allow the use of higher filtration velocities. Ratings are usually selected so that the pressure drop across the fabric will be in the 2 to 5 "wg range between start and end of operating cycle.

With Multiple-Section, Continuous-Duty, Automatic Fabric Collectors the disadvantage of stopping the air flow to permit fabric reconditioning and the variations in air flow with dust cake buildup can be overcome. The use of sections or compartments, as indicated in Figure 4-7, allows continuous operation of the exhaust system because automatic dampers periodically remove one section from service for fabric reconditioning while the remaining compartments handle the total gas flow. The larger the number of compartments, the more constant the pressure loss and air flow. Either tubes or envelopes may be used and fabric reconditioning is usually accomplished by shaking or vibrating.

Figure 4-8 shows air flow versus time for a multiple-section collector. Each individual section or compartment has an air flow versus time characteristic like that of the intermittent collector, but the total variation is reduced because of the multiple compartments. Note the more constant air flow characteristic of the five compartment unit as opposed to the three-compartment design. Since an individual section is out of service only a few minutes for reconditioning and remaining sections handle the total gas flow during that time, it is possible to clean the fabric more frequently than with the intermittent type. This permits the multiple-section unit to handle higher dust concentrations. Compartments are reconditioned in fixed sequence with the ability to adjust the time interval between cleaning of individual compartments.

One variation of this design is the low-pressure, reverse-air collector which does not use shaking for fabric reconditioning. Instead, a compartment is isolated for cleaning and the tubes collapsed by means of a low pressure secondary blower, which draws air from the compartment in a direction opposite to the primary air flow. This is a "gentle" method of fabric reconditioning and was developed primarily for the fragile glass cloth used for high temperature operation. The reversal of air flow and tube deflation is accomplished very gently to avoid damage to the glass fibers. The control sequence usually allows the deflation and re-inflation of tubes several times for complete removal of excess dust. Tubes are 6 to 11 inches in diameter and can be as long as 30 feet. For long tubes, stainless rings may be sewn on the inside to help break up the dust cake during deflation. A combination of shaking and reverse air flow has also been utilized.

When shaking is used for fabric reconditioning, the filtration velocity usually is in the 1 fpm to 4 fpm range. Reverse air collapse type reconditioning generally necessitates lower filtration velocities since reconditioning is not as complete. They are seldom rated higher than 3 fpm. The air to cloth ratio or filtration velocity is based on *net* cloth area available when a compartment is out of service for reconditioning.

Reverse-Jet, Continuous-Duty, Fabric Collectors may use envelopes or tubes of non-woven (felted) fabric or pleated cartridges of non-woven mat (paper-like) in cylindrical or flat panel configuration. They differ from the low pressure reverse air type in that they employ a brief burst of high pressure air to recondition the fabric. Woven fabric is not used because it allows excessive dust penetration during reconditioning. The most common designs use compressed air at 80 to 100 psig, while others use an integral pressure blower at a lower pressure but higher secondary flow rate. Those using compressed air are generally called pulse-jet collectors and those using pressure blowers are called fan-pulse collectors.

All designs, even the tube type, collect dust on the outside and have air flow from outside to inside the fabric. All recondition the media by introducing the pulse of cleaning air into the opening where cleaned air exits from the tube, envelope or cartridge. In many cases, a venturi shaped fitting is used at this opening to provide additional cleaning by inducing additional air flow. The venturi also directs or focuses the cleaning pulse for maximum efficiency.

Figure 4-9 shows a typical pulse-jet collector. Under nor-

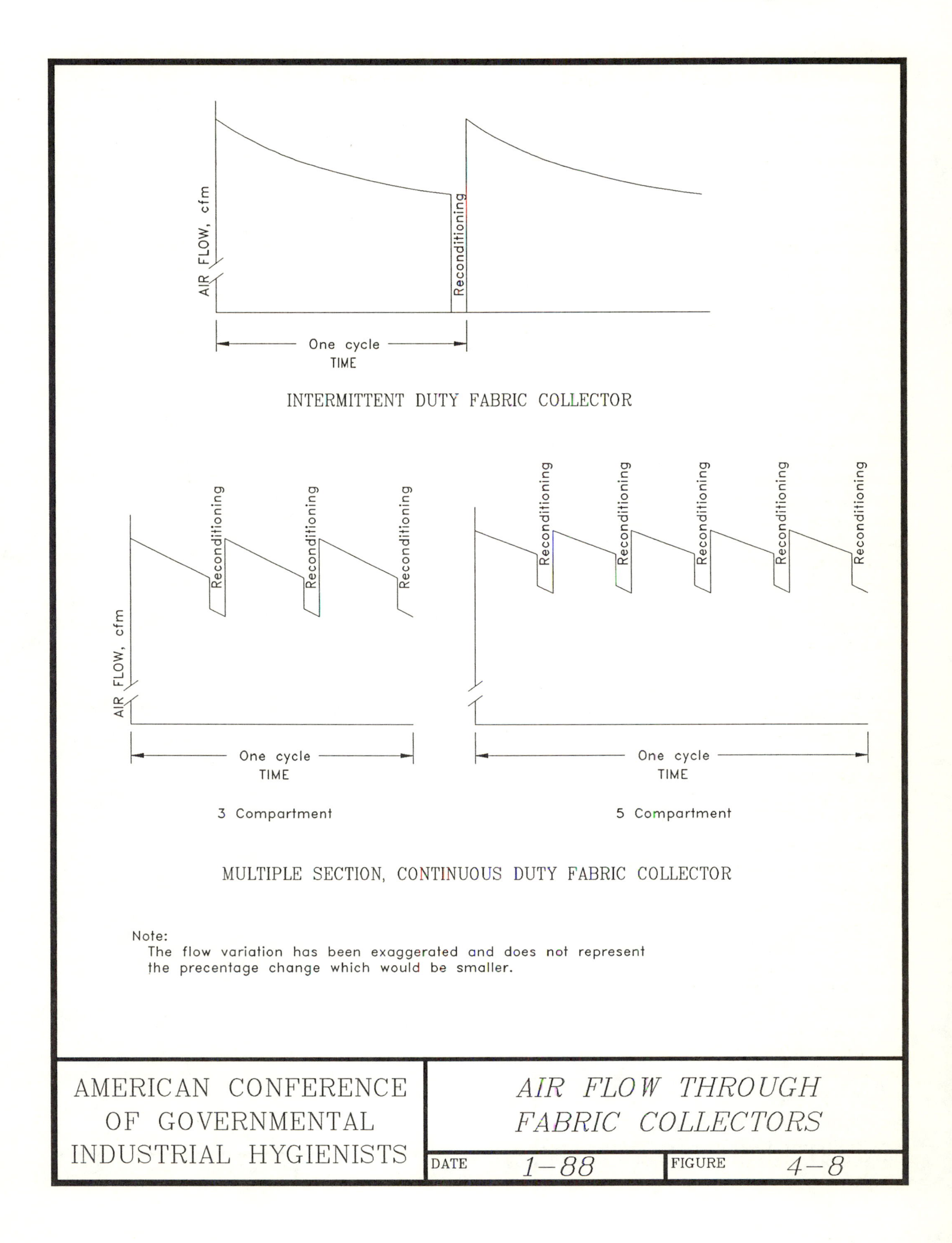
AIR FLOW, cfm
Reconditioning
One cycle
TIME
INTERMITTENT DUTY FABRIC COLLECTOR
AIR FLOW, cfm
Reconditioning
Reconditioning
Reconditioning
One cycle
TIME
3 Compartment
Reconditioning
Reconditioning
Reconditioning
Reconditioning
Reconditioning
One cycle
TIME
5 Compartment
MULTIPLE SECTION, CONTINUOUS DUTY FABRIC COLLECTOR
Note:
The flow variation has been exaggerated and does not represent
the precentage change which would be smaller.
AMERICAN CONFERENCE
OF GOVERNMENTAL
INDUSTRIAL HYGIENISTS
AIR FLOW THROUGH
FABRIC COLLECTORS
DATE 1-88
FIGURE 4-8

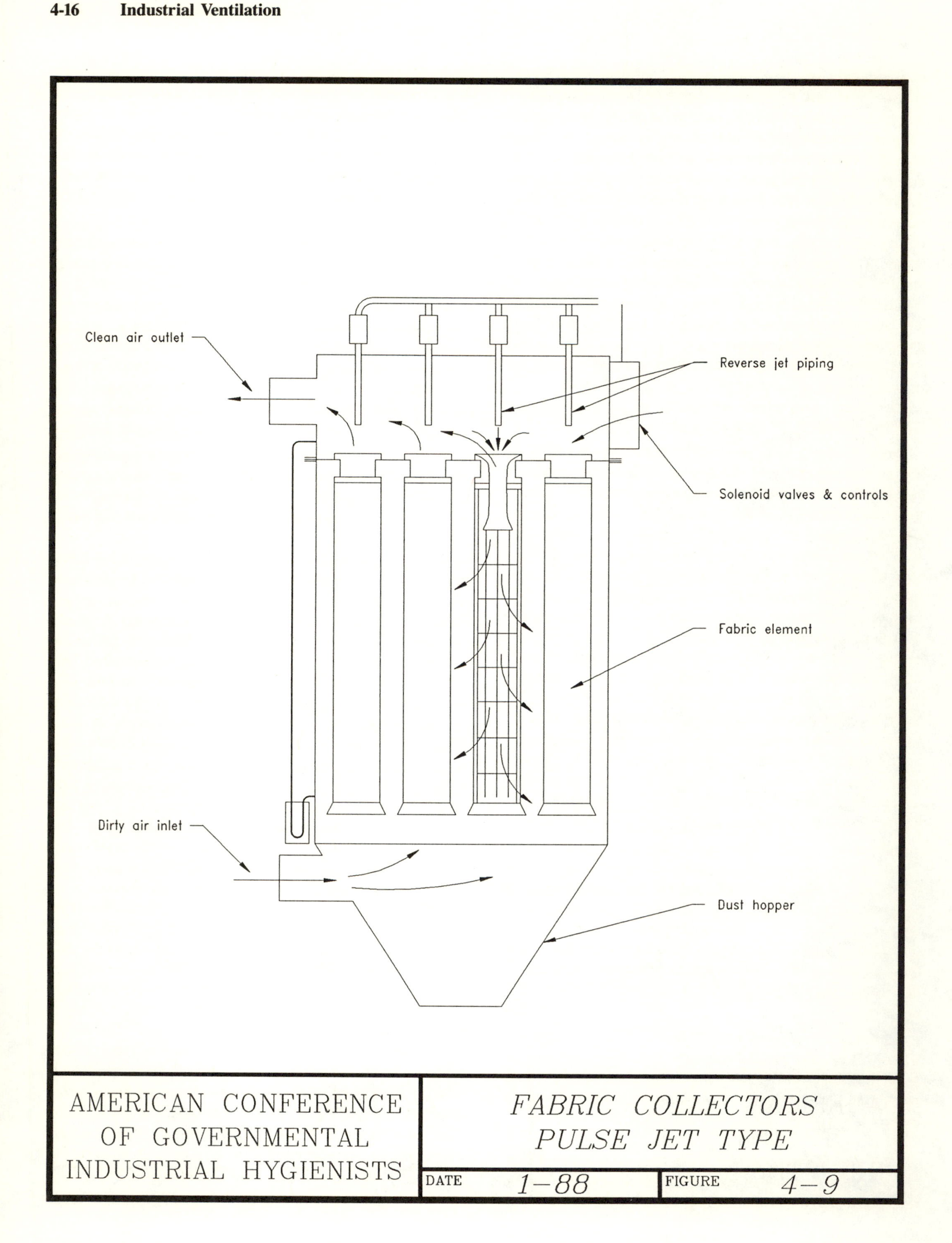
Clean air outlet
Reverse jet piping
Solenoid valves & controls
Fabric element
Dirty air inlet
Dust hopper
AMERICAN CONFERENCE OF GOVERNMENTAL INDUSTRIAL HYGIENISTS
FABRIC COLLECTORS PULSE JET TYPE
DATE 1-88
FIGURE 4-9

mal operation (air flow from outside to inside) the fabric shape would tend to collapse, therefore, a support cage is required. The injection of a short pulse of high pressure air induces a secondary flow from the clean air compartment in a direction opposite to the normal air flow. Reconditioning is accomplished by the pulse of high pressure air which stops forward air flow, then rapidly pressurizes and inflates the tube causing it to snap away from the supporting cage, breaking up the dust cake and freeing accumulated dust from the fabric. The secondary or induced air acts as a damper, preventing flow in the normal direction during reconditioning. The entire process, from injection of the high pressure pulse and initiation of secondary flow until the secondary flow ends, takes place in approximately one second. Solenoid valves which control the pulses of compressed air may be open for a tenth of a second or less. An adequate flow rate of clean and dry compressed air of sufficient pressure must be supplied to ensure effective reconditioning.

Reverse-jet collectors normally clean no more than 10% of the fabric at any one time. Because such a small percentage is cleaned at any one time and because the induced secondary flow blocks normal flow during that time, reconditioning can take place while the collector is in service and without the need for compartmentation and dampers. The cleaning intervals are adjustable and are considerably more frequent than the intervals for shaker or reverse-air collectors. An individual tube may be pulsed and reconditioned as often as once a minute to every six minutes.

Due to this very short reconditioning cycle, higher filtration velocities are possible with reverse-jet collectors. However, with all reverse-jet collectors, the fabric is in the dirty air compartment and during reconditioning accumulated dust which is freed from one fabric surface may become reintrained and redeposited on an adjacent surface, or even on the original surface. This phenomenon of redeposition tends to limit filtration velocity to something less than might be anticipated with cleaning intervals of just a few minutes.

Laboratory tests[4.1] have shown that for a given collector design redeposition increases with filtration velocity. Other test work[4.2] indicates clearly that redeposition varies with collector design and especially with flow patterns in the dirty air compartment. EPA sponsored research[4.3] has shown that superior performance results from downward flow of the dirty air stream. This downward air flow reduces redeposition since it aids gravity in moving dust particles toward the hopper.

Filtration velocities of 5 to 12 fpm are normal for reverse-jet collectors. The pleated cartridge type of reverse-jet collector is limited to filtration velocities in the 1 to 2.5 fpm range because a pleat configuration produces very high approach velocities and greater redeposition.

4.3.3 Wet Collectors: Wet collectors, or scrubbers, are commercially available in many different designs, with pressure drops ranging from 1.5 "wg to as much as 100 "wg. There is a corresponding variation in collector performance. It is generally accepted that, for well-designed equipment, efficiency depends on the energy utilized in air to water contact and is independent of operating principle. Efficiency is a function of total energy input per cfm whether the energy is supplied to the air or to the water. This means that well-designed collectors by different manufacturers will provide similar efficiency if equivalent power is utilized.

Wet collectors have the ability to handle high temperature and moisture-laden gases. The collection of dust in a wetted form eliminates a secondary dust problem in disposal of collected material. Some dusts represent explosion or fire hazards when dry and wet collection eliminates, or at least reduces, the hazard. However, the use of water may introduce corrosive conditions within the collector and freeze protection may be necessary if collectors are located outside in cold climates. Space requirements are nominal. Pressure losses and collection efficiency vary widely for different designs.

Wet collectors, especially the high-energy types, are being used more frequently as the solution to air pollution problems. It should be realized that disposal of collected material in water without clarification or treatment may create water pollution problems.

Wet collectors have one characteristic not found in other collectors — the inherent ability to humidify. Humidification, the process of adding water vapor to the air stream through evaporation, may be either advantageous or disadvantageous depending on the situation. Where the initial air stream is at an elevated temperature and not saturated, the process of evaporation reduces the temperature and the volumetric flow rate of the gas stream leaving the collector. Assuming the fan is to be selected for operation on the clean air side of the collector, it may be smaller and will definitely require less power than if there had been no cooling through the collector. This is one of the obvious advantages of humidification; however, there are other applications where the addition of moisture to the gas stream is undesirable. For example, the exhaust of humid air to an air conditioned space normally places an unacceptable load on the air conditioning system. High humidity can also result in corrosion of finished goods. Therefore, humidification effects should be considered before designs are finalized. While all wet collectors humidify, the amount of humidification varies for different designs. Most manufacturers publish the humidifying efficiency for their equipment and will assist in evaluating the results.

Chamber or Spray Tower: Chamber or spray tower collectors consist of a round or rectangular chamber into which water is introduced by spray nozzles. There are many variations of design, but the principal mechanism is impaction of dust particles on the liquid droplets created by the nozzles. These droplets are separated from the air stream by centrifugal force or impingement on water eliminators.

The pressure drop is relatively low (on the order of 0.5 to

1.5 "wg), but water pressures range from 10 to 400 psig. The high pressure devices are the exception rather than the rule. In general, this type of collector utilizes low pressure supply water and operates in the lower efficiency range for wet collectors. Where water is supplied under high pressure, as with fog towers, collection efficiency can reach the upper range of wet collector performance.

For conventional equipment, water requirements are reasonable, with a maximum of about 5 gpm per thousand scfm of gas. Fogging types using high water pressure may require as much as 10 gpm per thousand scfm of gas.

Packed Towers: Packed towers (see Figure 4-10) are essentially contact beds through which gases and liquid pass concurrently, counter-currently, or in cross-flow. They are used primarily for applications involving gas, vapor, and mist removal. These collectors can capture solid particulate matter, but they are not used for that purpose because dust plugs the packing and requires unreasonable maintenance.

Water rates of 5 to 10 gpm per thousand scfm are typical for packed towers. Water is distributed over V-notched ceramic or plastic weirs. High temperature deterioration is avoided by using brick linings, allowing gas temperatures as high as 1600 F to be handled direct from furnace flues.

The air flow pressure loss for a 4-foot bed of packing, such as ceramic saddles, will range from 1.5 to 3.5 "wg. The face velocity (velocity at which the gas enters the bed) will typically be 200 to 300 fpm.

Wet Centrifugal Collectors: Wet centrifugal collectors (see Figure 4-11) comprise a large portion of the commercially available wet collector designs. This type utilizes centrifugal force to accelerate the dust particle and impinge it upon a wetted collector surface. Water rates are usually 2 to 5 gpm per thousand scfm of gas cleaned. Water distribution can be from nozzles, gravity flow or induced water pickup. Pressure drop is in the 2 to 6 "wg range.

As a group, these collectors are more efficient than the chamber type. Some are available with a variable number of impingement sections. A reduction in the number of sections results in lower efficiency, lower cost, less pressure drop, and smaller space. Other designs contain multiple collecting tubes. For a given air flow rate, a decrease in the tube size provides higher efficiency because the centrifugal force is greater.

Wet Dynamic Precipitator: The wet dynamic precipitator (see Figure 4-12) is a combination fan and dust collector. Dust particles in the dirty air stream impinge upon rotating fan blades which are wetted with spray nozzles. The dust particles impinge into water droplets and are trapped along with the water by a metal cone while the cleaned air makes a turn of 180 degrees and escapes from the front of the specially shaped impeller blades. Dirty water from the water cone goes to the water and sludge outlet and the cleaned air goes to an outlet section containing a water elimination device.

Orifice Type: In this group of wet collector designs (see Figure 4-12) the air flow through the collector is brought in contact with a sheet of water in a restricted passage. Water flow may be induced by the velocity of the air stream or maintained by pumps and weirs. Pressure losses vary from 1 "wg or less for a water wash paint booth to a range of 3 to 6 "wg for most of the industrial designs. Pressure drops as high as 20 "wg are used with some designs intended to collect very small particles.

Venturi: The venturi collector (see Figure 4-11) uses a venturi-shaped constriction to establish throat velocities considerably higher than those used by the orifice type. Gas velocities through venturi throats may range from 12,000 to 24,000 fpm. Water is supplied by piping or jets at or ahead of the throat at rates from 5 to 15 gpm per thousand scfm of gas.

The collection mechanism of the venturi is impaction. As is true for all well-designed wet collectors, collection efficiency increases with higher pressure drops. Specific pressure drops are obtained by designing for selected velocities in the throat. Some venturi collectors are made with adjustable throats allowing operation over a range of pressure drops for a given flow rate or over a range of flow rates with a constant pressure drop. Systems are available with pressure drops as low as 5 "wg for moderate collection efficiency and as high as 100 "wg for collection of extremely fine particles.

The venturi itself is a gas conditioner causing intimate contact between the particulates in the gas and the multiple jet streams of scrubbing water. The resulting mixture of gases, fume-dust agglomerates and dirty water must be channeled through a separation section for the elimination of entrained droplets as shown in Figure 4-11.

4.3.4 Dry Centrifugal Collectors: Dry centrifugal collectors separate entrained particulate from an air stream by the use or combination of centrifugal, inertial, and gravitational force. Collection efficiency is influenced by:

1. Particle size, weight and shape. Performance is improved as size and weight become larger and as the shape becomes more spherical.
2. Collector size and design. The collection of fine dust with a mechanical device requires equipment designed to best utilize mechanical forces and fit specific application needs.
3. Velocity. Pressure drop through a cyclone collector increases approximately as the square of the inlet velocity. There is, however, an optimum velocity that is a function of collector design, dust characteristics, gas temperature and density.
4. Dust Concentration. Generally, the performance of a mechanical collector increases as the concentration of dust becomes greater.

Gravity Separators: Gravity separators consist of a chamber or housing in which the velocity of the gas stream is made to drop rapidly so that dust particles settle out by gravity. Extreme space requirements and the usual presence of eddy

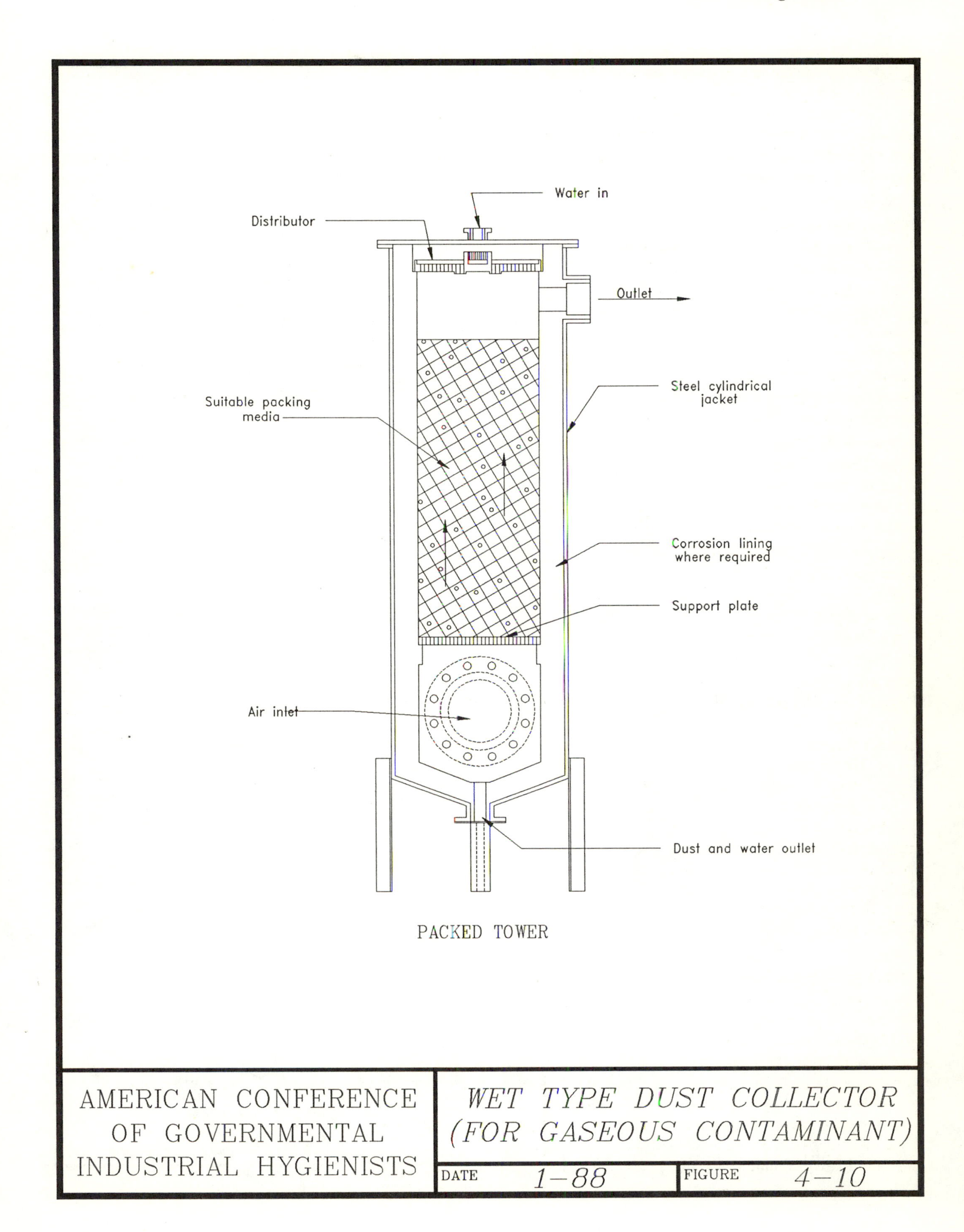
Water in
Distributor
Outlet
Steel cylindrical jacket
Suitable packing media
Corrosion lining where required
Support plate
Air inlet
Dust and water outlet
PACKED TOWER
AMERICAN CONFERENCE OF GOVERNMENTAL INDUSTRIAL HYGIENISTS
WET TYPE DUST COLLECTOR (FOR GASEOUS CONTAMINANT)
DATE 1-88
FIGURE 4-10

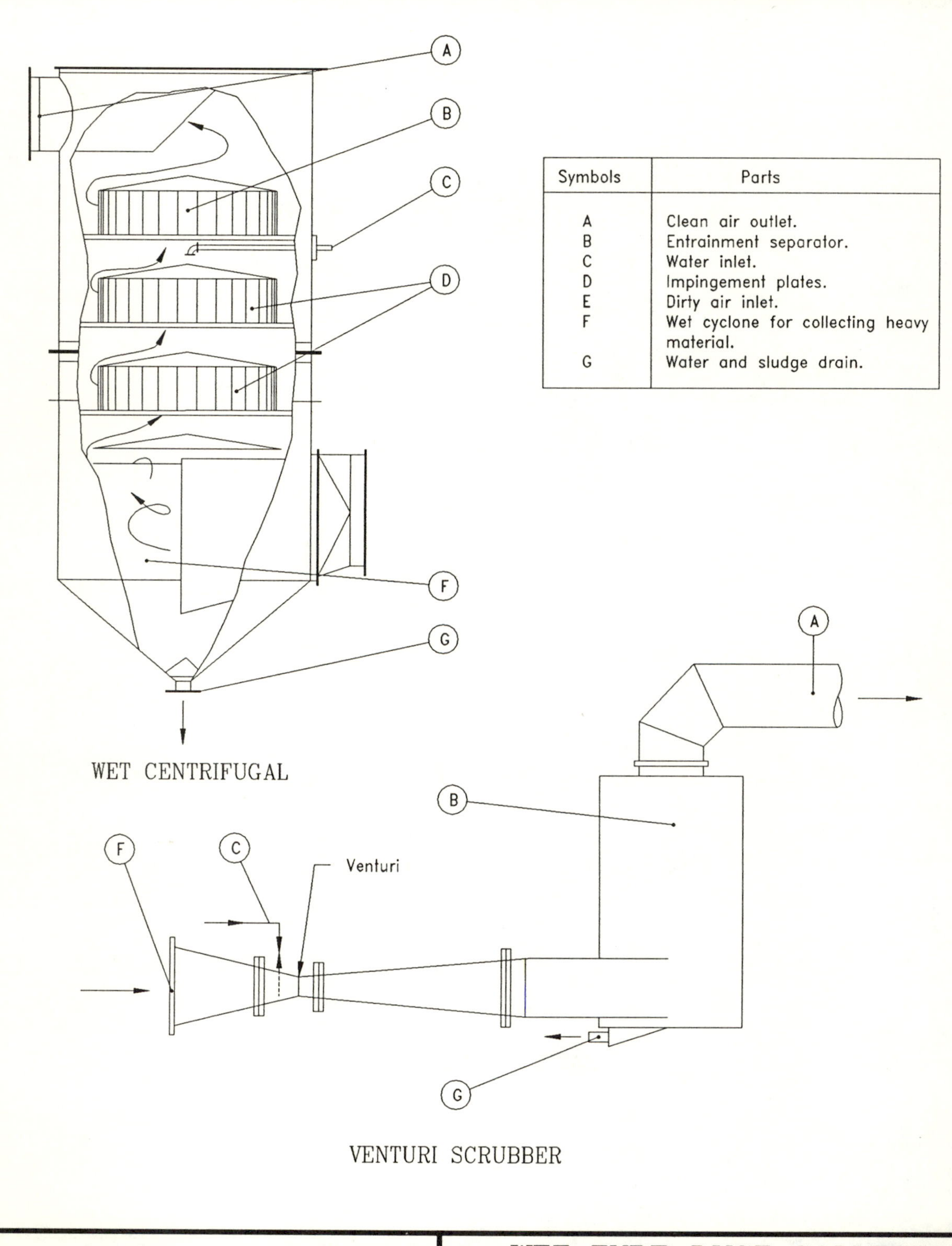

Symbols	Parts
A	Clean air outlet.
B	Entrainment separator.
C	Water inlet.
D	Impingement plates.
E	Dirty air inlet.
F	Wet cyclone for collecting heavy material.
G	Water and sludge drain.

AMERICAN CONFERENCE OF GOVERNMENTAL INDUSTRIAL HYGIENISTS	*WET TYPE DUST COLLECTORS (FOR PATICULATE CONTAMINANTS)*	
	DATE *1–88*	FIGURE *4–11*

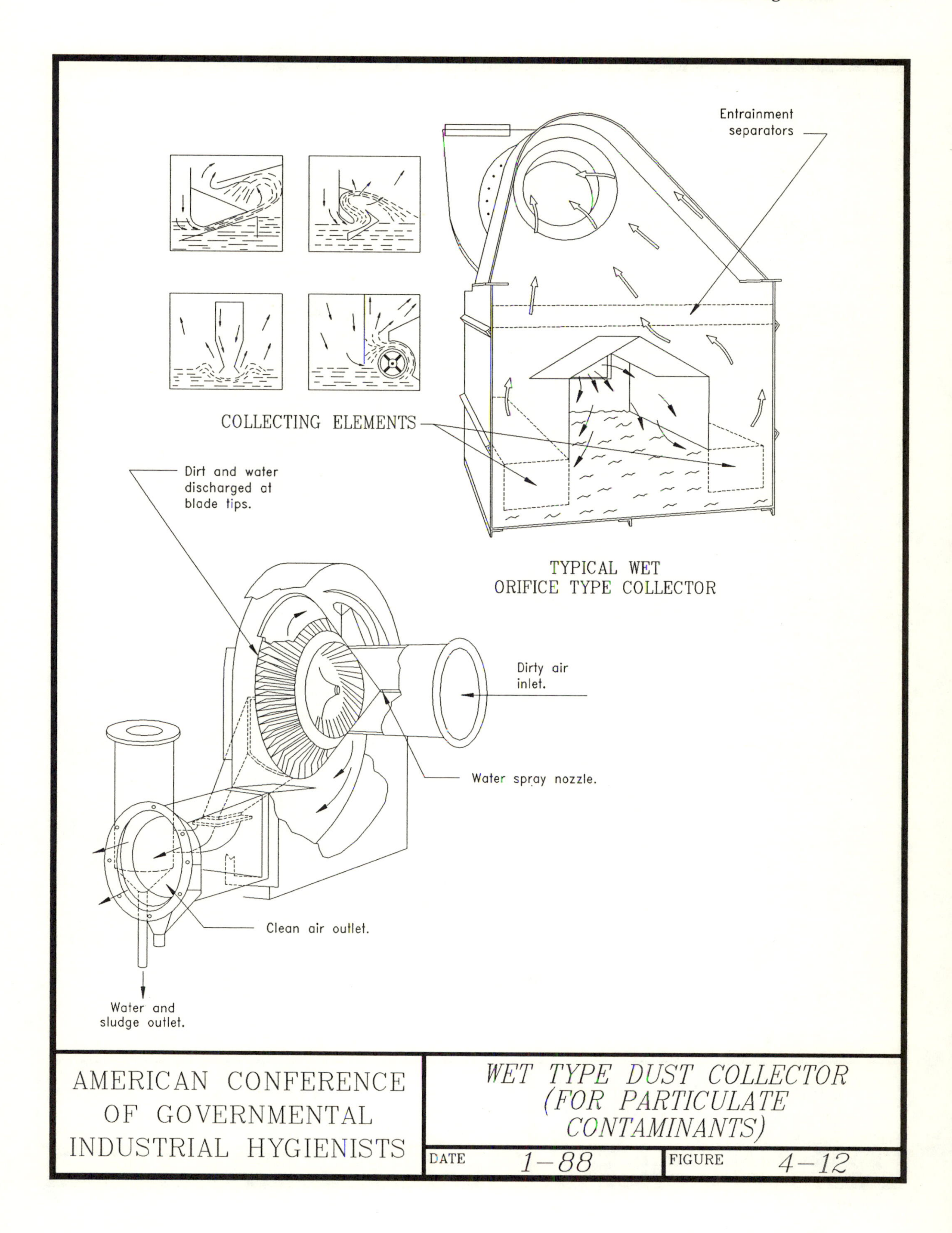
Entrainment separators
COLLECTING ELEMENTS
TYPICAL WET ORIFICE TYPE COLLECTOR
Dirt and water discharged at blade tips.
Dirty air inlet.
Water spray nozzle.
Clean air outlet.
Water and sludge outlet.
AMERICAN CONFERENCE OF GOVERNMENTAL INDUSTRIAL HYGIENISTS
WET TYPE DUST COLLECTOR (FOR PARTICULATE CONTAMINANTS)
DATE 1-88
FIGURE 4-12

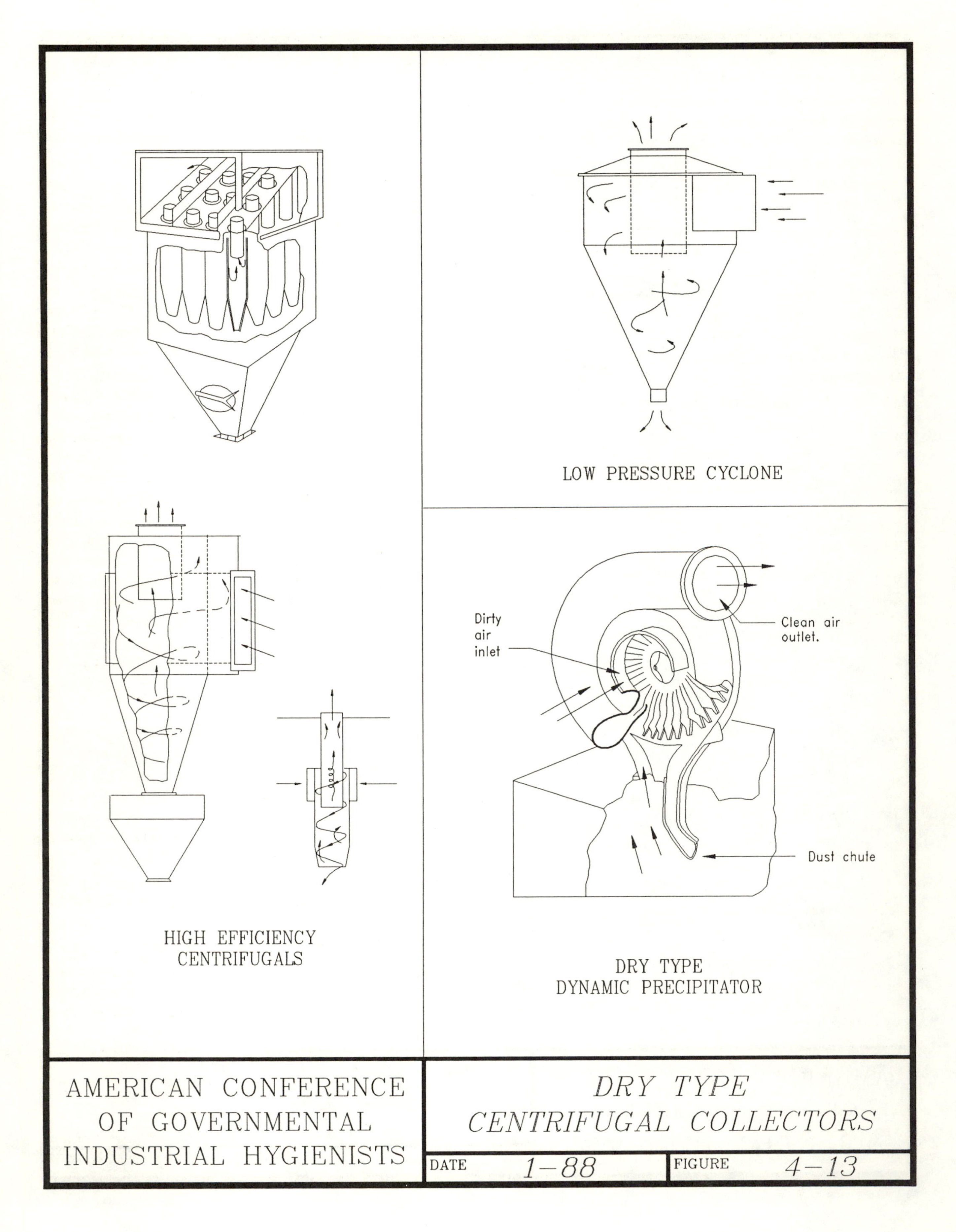

DRY TYPE CENTRIFUGAL COLLECTORS

currents nullify this method for removal of anything but extremely coarse particles.

Inertial Separators: Inertial separators depend on the inability of dust to make a sharp turn because its inertia is much higher than that of the carrier gas stream. Blades or louvers in a variety of shapes are used to require abrupt turns of 120 degrees or more. Well-designed inertial separators can separate particles in the 10 to 20 micron range with about 90% efficiency.

Dynamic Precipitator: The dynamic precipitator (see Figure 4-13) is a combined air mover and dust collector which has a large number of shaped blades attached to a concave disc. Impeller rotation creates the force necessary to draw in dust laden air and cause it to turn in an arc of nearly 180 degrees before passing through the scroll to the clean air outlet. Dust particles are thrown by centrifugal force to the blade tips where they are trapped and conveyed by a secondary circuit to the hopper below.

Cyclone Collector: The cyclone collector (see Figure 4-13) is commonly used for the removal of coarse dust from an air stream, as a precleaner to more efficient dust collectors and/or as a product separator in air conveying systems. Principal advantages are low cost, low maintenance, and relatively low pressure drops (in the 0.75 to 1.5 "wg range). It is not suitable for the collection of fine particles.

High Efficiency Centrifugals: High efficiency centrifugals (see Figure 4-13) exert higher centrifugal forces on the dust particles in a gas stream. Because centrifugal force is a function of peripheral velocity and angular acceleration, improved dust separation efficiency has been obtained by:

1. Increasing the inlet velocity
2. Making the cyclone body and cone longer
3. Using a number of small diameter cyclones in parallel
4. Placing units in series.

While high efficiency centrifugals are not as efficient on small particles as electrostatic, fabric, and wet collectors, their effective collection range is appreciably extended beyond that of other mechanical devices. Pressure losses of collectors in this group range from 3 to 8 "wg.

4.4 ADDITIONAL AIDS IN DUST COLLECTOR SELECTION

The collection efficiencies of the five basic groups of air cleaning devices have been plotted against mass mean particle size (Figure 4-14). The graphs were found through laboratory and field testing and were not compiled mathematically. The number of lines for each group indicates the range that can be expected for the different collectors operating under the same principle. Variables, such as type of dust, velocity of air, water rate, etc., will also influence the range for a particular application.

Deviation lines shown in the upper right hand corner of the chart allow the estimation of mass mean material size in the effluent of a collector when the inlet mean size is known. Space does not permit a detailed explanation of how the slopes of these lines were determined, but the following example illustrates how they are used. The deviation lines should not be used for electrostatic precipitators but can be used for the other groups shown at the bottom of the figure.

Example: A suitable collector will be selected for a lime kiln to illustrate the use of the chart. Referring to Figure 4-14, the concentration and mean particle size of the material leaving the kiln can vary between 3 and 10 grains per cubic foot, with 5 to 10 microns the range for mass mean particle size. Assume an inlet concentration of 7.5 grains per cubic foot and a mean inlet size of 9 microns. Projection of this point vertically downwardly to the collection efficiency portion of the chart will indicate that a low resistance cyclone will be less than 50% efficient; a high efficiency centrifugal will be 60 to 80% efficient and a wet collector, fabric arrester and electrostatic precipitator will be 97+ % efficient. A precleaner is usually feasible for dust concentrations over 5 grains per cubic foot unless it is undesirable to have the collected dust separated by size. For this example a high efficiency centrifugal will be selected as the precleaner. The average efficiency is 70% for this group, therefore the effluent from this collector will have a concentration of 7.5 $\times$ (1.00 $-$ 0.70) = 2.25 grains per cubic foot. Draw a line through the initial point with a slope parallel to the deviation lines marked "industrial dust." Where deviation is not known, the average of this group of lines normally will be sufficiently accurate to predict the mean particle size in the collector effluent. A vertical line from the point of intersection between the 2.25 grains per cubic foot horizontal and the deviation line to the base of the chart will indicate a mean effluent particle size of 6.0 microns.

A second high efficiency centrifugal in series would be less than 50% efficient on this effluent. A wet collector, fabric arrester, or electrostatic would have an efficiency of 94% or better. Assume that a good wet collector will be 98% efficient. The effluent would then be 2.25 $\times$ (1.00 $-$ 0.98) = 0.045 grains per cubic foot. Using the previous deviation line and its horizontal intersection of 0.045 grains per cubic foot yields a vertical line intersecting the mean particle size chart at 1.6 microns, which is the mean particle size of the wet collector effluent.

In Table 4-3, an effort has been made to report types of dust collectors used for a wide range of industrial processes. While many of the listings are purely arbitrary, they may serve as a guide in selecting the type of dust collector most frequently used.

4.5 CONTROL OF MIST, GAS AND VAPOR CONTAMINANTS

Previous discussion has centered on the collection of dust and fume or particulate existing in the solid state. Only the Packed Tower was singled out as being used primarily to

TABLE 4-3. Dust Collector Selection Guide

	Collector Types Used in Industry							
Operation	Concentration Note 1	Particle Sizes Note 2	Cyclone	Wet Collector	Fabric Collector	Low-Volt Electrostatic	Hi-Volt Electrostatic	See Remark No.
CERAMICS								
a. Raw product handling	light	fine	O	O	O	N	N	1
b. Fettling	light	fine to medium	S	S	O	N	N	2
c. Refractory sizing	heavy	coarse	N	S	O	N	N	3
d. Glaze & vitr. enamel spray	moderate	medium	N	O	O	N	N	
CHEMICALS								50
a. Material handling	light to moderate	fine to medium	O	O	O	N	N	4
b. Crushing, grinding	moderate to heavy	fine to coarse	O	O	O	N	N	5
c. Pneumatic conveying	very heavy	fine to coarse	O	S	O	N	N	6
d. Roasters, kilns, coolers	heavy	mid-coarse	O	O	O	O	N	7
COAL MINING AND POWER PLANT								50
a. Material handling	moderate	medium	O	O	O	N	N	8
b. Bunker ventilation	moderate	fine	S	S	O	N	N	9
c. Dedusting, air cleaning	heavy	med-coarse	S	O	O	N	N	10
d. Drying	moderate	fine	N	O	S	N	N	11
FLY ASH								
a. Coal burning—chain grate	light	fine	S	S	O	N	O	12
b. Coal burning—stoker fired	moderate	fine to coarse	S	S	O	N	O	
c. Coal burning—pulverized fuel	heavy	fine	S	S	O	N	O	13
d. Wood burning	varies	coarse	S	O	O	N	S	14
FOUNDRY								
a. Shakeout	light to moderate	fine	N	O	O	N	N	15
b. Sand handling	moderate	fine to medium	N	O	O	N	N	16
c. Tumbling mills	heavy	med-coarse	N	O	O	N	N	17
d. Abrasive cleaning	moderate to heavy	fine to medium	N	O	O	N	N	18
GRAIN ELEVATOR, FLOUR AND FEED MILLS								50
a. Grain handling	light	medium	O	S	O	N	N	19
b. Grain dryers	light	coarse	S	S	O	N	N	20
c. Flour dust	moderate	medium	O	S	O	N	N	21
d. Feed mill	moderate	medium	O	S	O	N	N	22
METAL MELTING								50
a. Steel blast furnace	heavy	varied	N	O	S	N	S	23
b. Steel open hearth	moderate	fine to coarse	N	O	S	N	S	24
c. Steel electric furnace	light	fine	N	S	O	N	S	25
d. Ferrous cupola	moderate	varied	N	O	O	N		26
e. Non-ferrous reverberatory	varied	fine	N	S	O	S	N	27
f. Non-ferrous crucible	light	fine	N	S	O	N	N	28
METAL MINING AND ROCK PRODUCTS								
a. Material handling	moderate	fine to medium	N	O	O	N	N	29
b. Dryers, kilns	moderate	med-coarse	N	O	O	N	O	30
c. Cement rock dryer	moderate	fine to medium	N	S	S	N	S	31
d. Cement kiln	heavy	fine to medium	N	N	O	N	S	32
e. Cement grinding	moderate	fine	N	N	O	N	N	33
f. Cement clinker cooler	moderate	coarse	N	N	O	N	N	34
METAL WORKING								50
a. Production grinding, scratch brushing, abrasive cut off	light	coarse	O	S	O	N	N	35
b. Portable and swing frame	light	medium	O	S	O	N	N	
c. Buffing	light	varied	S	S	O	N	N	36
d. Tool room	light	fine	S	O	S	N	N	37
e. Cast iron machining	moderate	varied	O	O	O	S	N	38
PHARMACEUTICAL AND FOOD PRODUCTS								50
a. Mixers, grinders, weighing, blending, bagging, packaging	light	medium	O	O	O	N	N	39
b. Coating pans,	varied	fine to medium	N	O	O	N	N	40

TABLE 4-3. Dust Collector Selection Guide (Con't)

	Collector Types Used in Industry								
Operation	Concentration Note 1	Particle Sizes Note 2	Cyclone	Wet Collector	Fabric Collector	Low-Volt Electrostatic	Hi-Volt Electrostatic	See Remark No.
PLASTICS								50
a. Raw material processing	(See comments	under Chemicals)	0	S	0	N	N	41
b. Plastic finishing	light to moderate	varied	S	S	0	N	N	42
RUBBER PRODUCTS								50
a. Mixers	moderate	fine	S	0	0	N	N	43
b. Batchout rolls	light	fine	S	0	S	S	N	44
c. Talc dusting and dedusting	moderate	medium	S	S	0	N	N	45
d. Grinding	moderate	coarse	0	0	0	N	N	46
WOODWORKING								50
a. Woodworking machines	moderate	varied	0	S	0	N	N	47
b. Sanding	moderate	fine	0	S	0	N	N	48
c. Waste conveying, hogs	heavy	varied	0	S	0	N	N	49

Note 1: Light: less than 2 gr/ft^3; Moderate: 2 to 5 gr/ft^3; Heavy: 5 gr/ft^3 and up.
Note 2: Fine: 50% less than 5 microns; Medium: 50% 5 to 15 microns; Coarse: 50% 15 microns and larger.
Note 3: 0 = often; S = seldom; N = never.

Remarks Referred to in Table 4-3

1. Dust released from bin filling, conveying, weighing, mixing, pressing, forming. Refractory products, dry pan, and screening operations more severe.
2. Operations found in vitreous enameling, wall and floor tile, pottery.
3. Grinding wheel or abrasive cut-off operation. Dust abrasive.
4. Operations include conveying, elevating, mixing, screening, weighing, packaging. Category covers so many different materials that recommendation will vary widely.
5. Cyclone and high efficiency centrifugals often act as primary collectors followed by fabric or wet type.
6. Cyclones used as product collector followed by fabric arrester for high over-all collection efficiency.
7. Dust concentration determines need for dry centrifugal; plant location, product value determines need for final collectors. High temperatures are usual and corrosive gases not unusual.
8. Conveying, screening, crushing, unloading.
9. Remote from other dust producing points. Separate collector usually.
10. Heavy loading suggests final high efficiency collector for all except very remote locations.
11. Difficult problem but collectors will be used more frequently with air pollution emphasis.
12. Public nuisance from boiler blow-down indicates collectors are needed.
13. Large installations in residential areas require electrostatic in addition to dry centrifugal.
14. Public nuisance from settled wood char indicates collectors are needed.
15. Hot gases and steam usually involved.
16. Steam from hot sand, adhesive clay bond involved.
17. Concentration very heavy at start of cycle.
18. Heaviest load from airless blasting due to higher cleaning speed. Abrasive shattering greater with sand than with grit or shot. Amounts removed greater with sand castings, less with forging scale removal, least when welding scale is removed.
19. Operations such as car unloading, conveying, weighing, storing.
20. Collection equipment expensive but public nuisance complaints becoming more frequent.
21. Operations include conveyors, cleaning rolls, sifters, purifiers, bins and packaging.
22. Operations include conveyors, bins, hammer mills, mixers, feeders, and baggers.
23. Primary dry trap and wet scrubbing usual. Electostatic is added where maximum cleaning required.
24. Collectors seldom installed in past. Air pollution emphasis indicates collector usage to increase.
25. Air pollution standards will probably require increased usage of fabric arresters.
26. Collectors not usually provided but air polution emphasis creating greater interest.
27. Zinc oxide loading heavy during zinc additions. Stack temperatures high.
28. Zinc oxide plume can be troublesome in certain plant locations.
29. Crushing, screening, conveying involved. Wet ores often introduce water vapor in exhaust air.
30. Dry centrifugals used as primary collectors, followed by final cleaner.
31. Collection equipment installed primarily to prevent public nuisance.
32. Collectors usually permit salvage of material and also reduce nuisance from settled dust in plant area.

Remarks Referred to in Table 4-3 (Con't)

33. Salvage value of collected material high. Same equipment used on raw grinding before calcining.
34. Coarse abrasive particles readily removed in primary collector types.
35. Roof discoloration, deposition on autos can occur with cyclones and less frequently with high efficiency dry centrifugal. Heavy duty air filters sometimes used as final cleaners.
36. Linty particles and sticky buffing compounds can cause pluggage and fire hazard in dry collectors.
37. Unit collectors extensively used, especially for isolated machine tools.
38. Dust ranges from chips to fine floats including graphitic carbon.
39. Materials vary widely. Collector selection depends on salvage value, toxicity, sanitation yardsticks.
40. Controlled temperature and humidity of supply air to coating pans makes recirculation desirable.
41. Plastic manufacture allied to chemical industry and vary with operations involved.
42. Operations and collector selection similar to woodworking. See Item 13.
43. Concentration is heavy during feed operation. Carbon black and other fine additions make collection and dust-free disposal difficult.
44. Often no collector used where dispersion from or location of exhaust stack favorable.
45. Salvage of collected material often dictates type of high efficiency collector.
46. Fire hazard from some operations must be considered.
47. Bulking material. Collected material storage and bridging from splinters and chips can be a problem.
48. Dry centrifugals not too effective on heavy concentration of fine particles from production sanding.
49. Primary collector invariably indicated with concentration and particle size range involved, wet or fabric collectors when used are employed as final collectors.
50. See NFPA publications for fire hazards, e.g., zirconium, magnesium, aluminum, woodworking, plastics, etc.

---

collect mist, gas, or vapor. The character of a mist aerosol is very similar, aerodynamically, to that of a dust or fume aerosol, and the mist can be removed from an air stream by applying the principles that are used to remove solid particulate.

Many of the devices previously described, or equipment quite similar, are used to collect mist. Standard wet collectors are used to collect many types of mists. Specially designed electrostatic precipitators are frequently employed to collect sulfuric acid or oil mist. Even fabric and centrifugal collectors, although not the types previously mentioned, are widely used to collect oil mist generated by high speed machining.

4.6 GASEOUS CONTAMINANT COLLECTORS

Equipment designed specifically to control gas or vapor contaminants can be classified as:

1. Absorbers
2. Adsorbers
3. Thermal oxidizers
4. Direct combustors
5. Catalytic oxidizers.

4.6.1 Absorbers: Absorbers remove soluble or chemically reactive gases from an air stream by contact with a suitable liquid. While all designs utilize intimate contact between the gaseous contaminant and the absorbent, different brands vary widely in configuration and performance. Removal may be by absorption if the gas solubility and vapor pressure promote absorption or chemical reaction. Water is the most frequently used absorbent, but additives are frequently required and occasionally other chemical solutions must be used. Packed Towers (Figure 4-10) are typical absorbers.

4.6.2 Adsorbers: Adsorbers remove contaminants by collection on a solid. No chemical reaction is involved as adsorption is a physical process where molecules of a gas adhere to surfaces of the solid adsorbent. Activated carbon or molecular sieves are popular adsorbents.

4.6.3 Thermal Oxidizers: Thermal Oxidizers, or afterburners, may be used where the contaminant is combustible. The contaminated air stream is introduced to an open flame or heating device followed by a residence chamber where combustibles are oxidized producing carbon dioxide and water vapor. Most combustible contaminants can be oxidized at temperatures between 1000 and 1500 F. The residence chamber must provide sufficient dwell time and turbulence to allow complete oxidation.

4.6.4 Direct Combustors: Direct combustors differ from thermal oxidizers by introducing the contaminated gases and auxiliary air directly into the burner as fuel. Auxiliary fuel, usually natural gas or oil, is generally required for ignition and may or may not be required to sustain burning.

4.6.5 Catalytic Oxidizers: Catalytic oxidizers may be used where the contaminant is combustible. The contaminated gas stream is preheated and then passed through a catalyst bed which promotes oxidation of the combustibles to carbon dioxide and water vapor. Metals of the platinum family are commonly used catalysts which will promote oxidation at

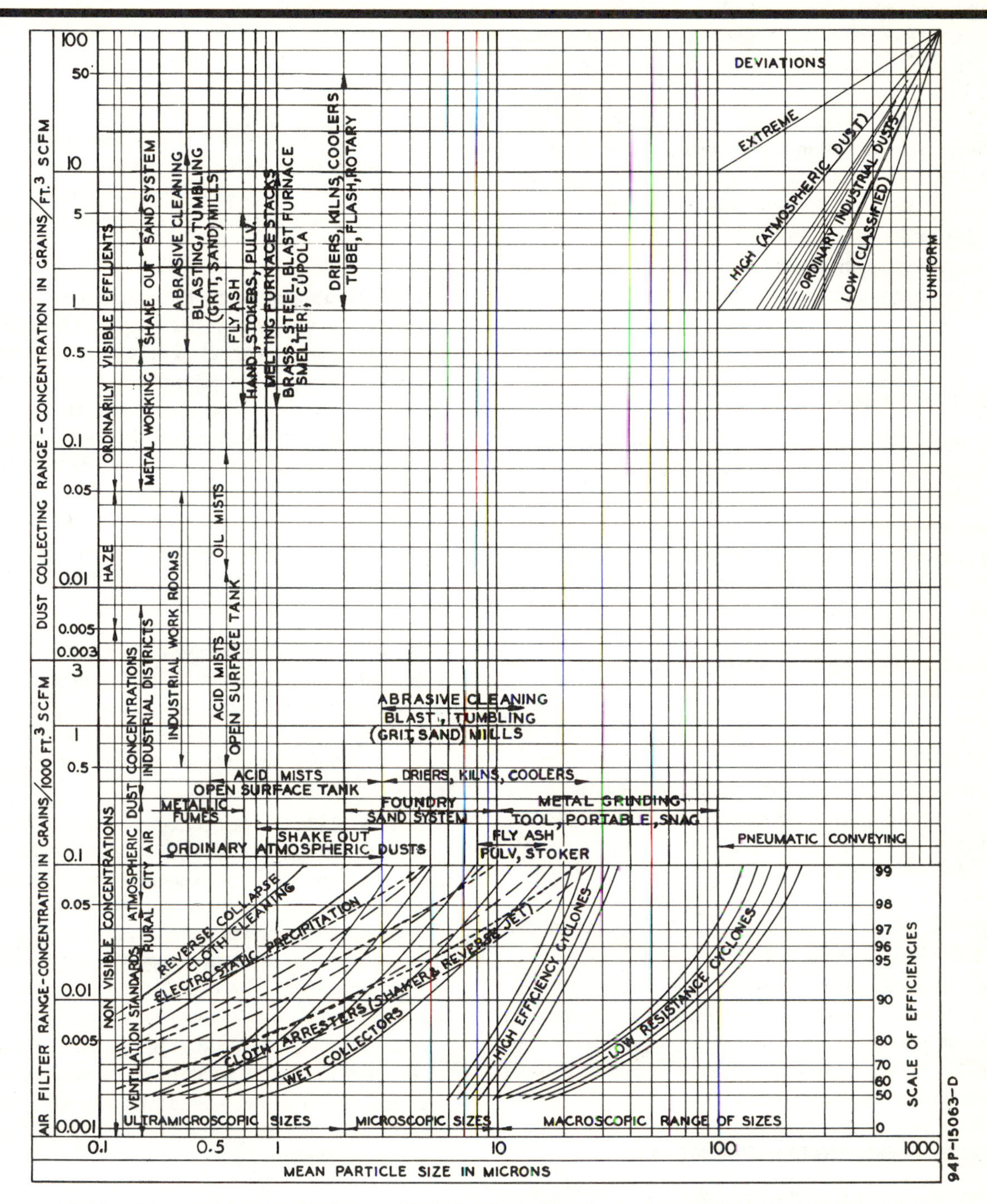

RANGE OF PARTICLE SIZES, CONCENTRATION, & COLLECTOR PERFORMANCE

COMPILED BY S. SYLVAN APRIL 1952 : COPYRIGHT 1952 AMERICAN AIR FILTER CO. INC.

ACKNOWLEDGEMENTS OF PARTIAL SOURCES OF DATA REPORTED:

1 FRANK W.G. - AMERICAN AIR FILTER - SIZE AND CHARACTERISTICS OF AIR BORNE SOLIDS - 1931
2 FIRST AND DRINKER - ARCHIVES OF INDUSTRIAL HYGIENE AND OCCUPATIONAL MEDICINE - APRIL 1952
3 TAFT INSTITUTE AND AAF LABORATORY TEST DATA - 1961 - '63
4 REVERSE COLLAPSE CLOTH CLEANING ADDED 1964

AMERICAN CONFERENCE OF GOVERNMENTAL INDUSTRIAL HYGIENISTS	*RANGE OF PARTICLE SIZE*	
	DATE *1-88*	FIGURE *4-14*

temperatures between 700 and 900 F.

To use either thermal or catalytic oxidation, the combustible contaminant concentration must be below the lower explosive limit. Equipment specifically designed for control of gaseous or vapor contaminants should be applied with caution when the air stream also contains solid particles. Solid particulates can plug absorbers, adsorbers, and catalysts and, if noncombustible, will not be converted in thermal oxidizers and direct combustors.

Air streams containing both solid particles and gaseous contaminants may require appropriate control devices in series.

4.7 UNIT COLLECTORS

Unit collector is a term usually applied to small fabric collectors having capacities in the 200-2000 cfm range. They have integral air movers, feature small space requirements and simplicity of installation. In most applications cleaned air is recirculated, although discharge ducts may be used if the added resistance is within the capability of the air mover. One of the primary advantages of unit collectors is a reduction in the amount of duct required, as opposed to central systems, and the addition of discharge ducts to unit collectors negates that advantage.

When cleaned air is to be recirculated, a number of precautions are required (see Chapter 7).

Unit collectors are used extensively, especially in the metal working industry, to fill the need for dust collection from isolated, portable, intermittently used or frequently relocated dust producing operations. Typically, a single collector serves a single dust source with the energy saving advantage that the collector need operate only when that particular dust producing machine is in operation.

Figure 4-15 shows a typical unit collector. Usually they are the intermittent duty, shaker-type in envelope configuration. Woven fabric is nearly always used. Automatic fabric cleaning is preferred since manual methods are unreliable without careful scheduling and supervision.

4.8 DUST COLLECTING EQUIPMENT COST

The variation in equipment cost, especially on an installed basis, is difficult to estimate and comparisons are difficult. Such comparisons can be misleading if these factors are not carefully evaluated.

4.8.1 Price versus Capacity: All dust collector prices per cfm of gas will vary with the gas flow rate. The smaller the flow rate, the higher the cost per cfm. The break point, where price per cfm cleaned tends to level off, will vary with the design. This can be noted on the typical curves shown on Figure 4-16.

4.8.2 Accessories Included: Careful analysis of components of equipment included is very important. Some collector designs include exhaust fan, motor, drive and starter. In other designs, these items and their supporting structure must be obtained by the purchaser from other sources. Likewise, while dust storage hoppers are integral parts of some dust collector designs, they are not provided in other types. Duct connections between elements may be included or omitted. Recirculating water pumps and/or settling tanks may be required and may not be included in the equipment price.

4.8.3 Installation Cost: The cost of installation can equal or exceed the cost of the collector. Actual cost will depend on the method of shipment (completely assembled, subassembled or completely knocked down), the location (which may require expensive rigging), and the need for expensive supporting steel and access platforms. The cost can also be measurably influenced by the need for water and drain connections, special or extensive electrical work, and expensive material handling equipment for collection material disposal. Items in the latter group will often also be variable, decreasing in cost per cfm as the flow rate of gas to be cleaned increases.

4.8.4 Special Construction: Prices shown in any tabulation must necessarily assume standard or basic construction. The increase in cost for corrosion resisting material, special high temperature fabrics, insulation, and/or weather protection for outdoor installations can introduce a multiplier of one to four times the standard cost.

A general idea of dust collector cost is provided in Figure 4-16. The additional notes and explanations included in these data should be carefully examined before they are used for estimating the cost of specific installations. For more accurate data, the equipment manufacturer or installer should be asked to provide estimates or a past history record for similar control problems utilized. Table 4-4 lists other characteristics that must be evaluated along with equipment cost.

Price estimates included in Figure 4-16 are for equipment of standard construction in normal arrangement. Estimates for exhausters and dust storage hoppers have been included, as indicated in Notes 1 and 2, where they are normally furnished by others.

4.9 SELECTION OF AIR FILTRATION EQUIPMENT

Air filtration equipment is available in a wide variety of designs and capability. Performance ranges from a simple throwaway filter for the home furnace to the "clean room" in the electronics industry, where the air must be a thousand times as clean as in a hospital surgical suite. Selection is based on efficiency, dust holding capacity, and pressure drop. There are five basic methods of air filtration.

4.9.1 Straining: Straining occurs when a particle is larger than the opening between fibers and cannot pass through. It is

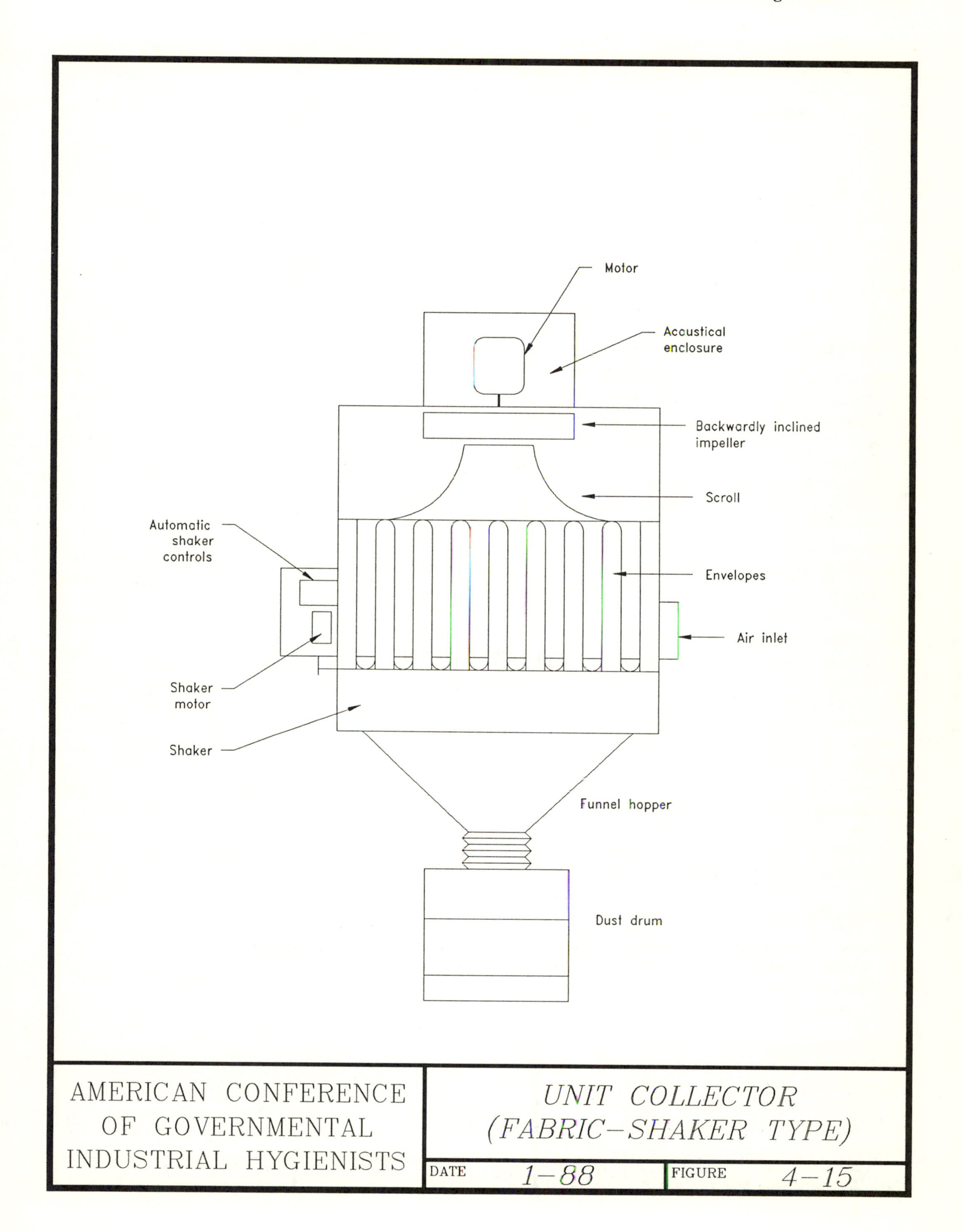
Motor
Acoustical enclosure
Backwardly inclined impeller
Scroll
Automatic shaker controls
Envelopes
Air inlet
Shaker motor
Shaker
Funnel hopper
Dust drum
AMERICAN CONFERENCE OF GOVERNMENTAL INDUSTRIAL HYGIENISTS
UNIT COLLECTOR (FABRIC–SHAKER TYPE)
DATE 1–88
FIGURE 4–15

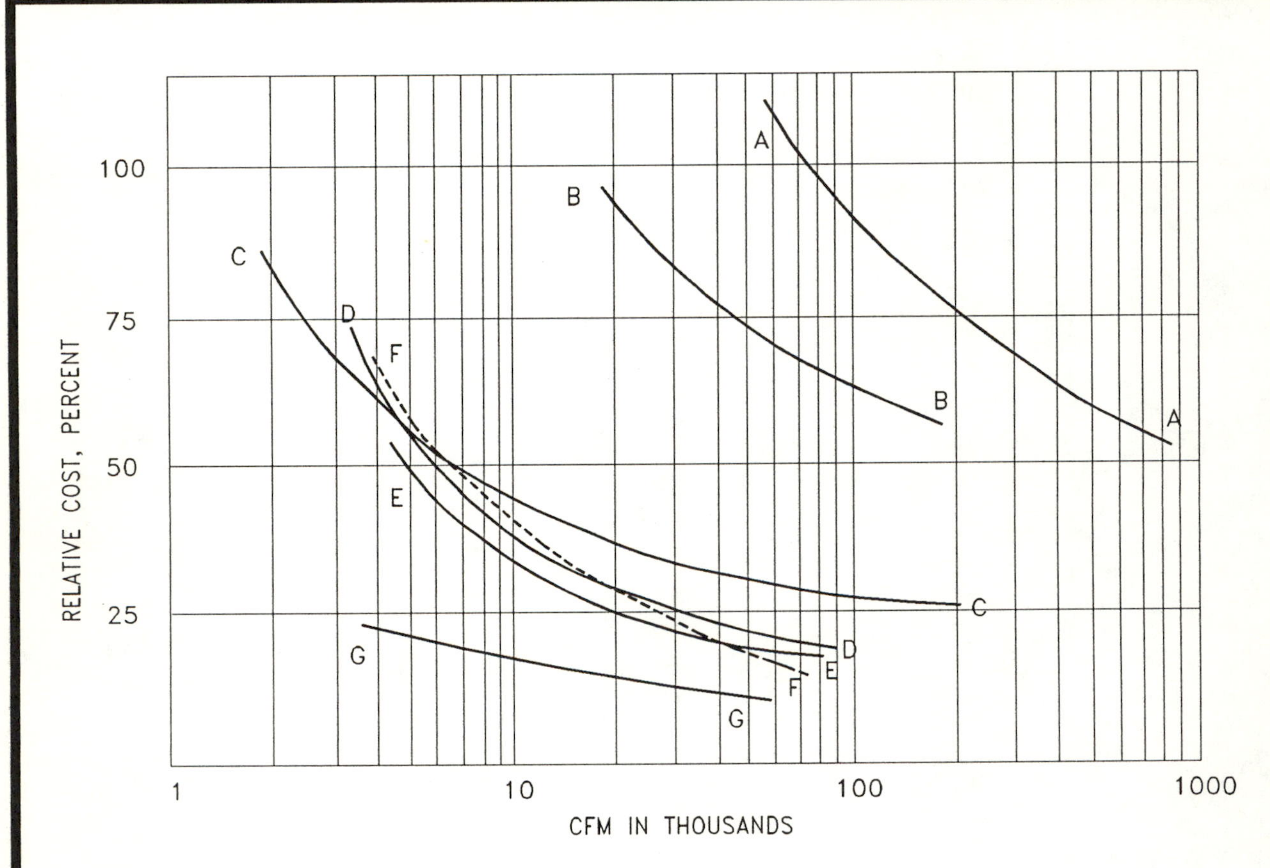

A. High voltage precipitator (minimum cost range)
B. Continuous duty high temperature fabric collector (2.0:1)
C. Continuous duty reverse pulse (8:1)
D. Wet collector
E. Intermittent duty fabric collector (2.0:1)
F. Low voltage precipitator
G. Cyclone

Note 1: Cost based on collector section only. Does not include ducts, dust disposal devices, pumps, exhausters or other accessories not an integral part of the collector.

Note 2: Price of high voltage precipitator will vary substantially with applications and efficiency requirements. Costs shown are for fly ash aplications where velocities of 200 to 300 fpm are normal.

AMERICAN CONFERENCE OF GOVERNMENTAL INDUSTRIAL HYGIENISTS	*COST ESTIMATES OF DUST COLLECTING EQUIPMENT*			
	DATE	*1–88*	FIGURE	*4–16*

TABLE 4-4. Comparison of Some Important Dust Collector Characteristics

Type	Higher efficiency Range on Particles Greater than Mean Size in Microns	Pressure Loss Inches	H_2O Gal per 1,000 cfm	Space	Sensitivity to cfm Change: Pressure	Sensitivity to cfm Change: Efficiency	Humid Air Influence	Max. Temp. F Standard Construction Note 4
Electrostatic	0.25	½	—	Large	Negligible	Yes	Improves Efficiency	500
Fabric								
Intermittent—Shaker	0.25	3-6	—	Large	As cfm	Negligible	May Make	See Table 4-1
Continuous—Shaker	0.25	3-6 Note 1	—	Large	As cfm	Negligible	Reconditioning	
Continuous—Reverse Air	0.25	3-6	—	Large	As cfm	Negligible	Difficult	
Continuous—Reverse Pulse	0.25	3-6	—	Moderate	As cfm	Negligible		
Glass, Reverse flow	0.25	3-8	—	Large	As cfm	Negligible		550
Wet:								
Packed Tower	1-5	1.5-3.5	5-10	Large	As cfm	Yes		
Wet Centrifugal	1-5	2.5-6	3-5	Moderate	As $(cfm)^2$	Yes		
Wet Dynamic	1-2	Note 2	½ to 1	Small	Note 2	No	None	Unlimited
Orifice Types	1-5	2½-6	10-40	Small	As cfm or less	Varies with design		
Higher Efficiency:								
Fog Tower	0.5-5	2-4	5-10	Moderate	As $(cfm)^2$	Slightly	None	Note 3
Venturi	0.5-2	10-100	5-15	Moderate	As $(cfm)^2$	Yes		Unlimited
Dry Centrifugal:								
Low Pressure Cyclone	20-40	0.75-1.5	—	Large	As $(cfm)^2$	Yes	May Cause	750
High Eff. Centrifugal	10-30	3-6	—	Moderate	As $(cfm)^2$	Yes	condensation	750
Dry Dynamic	10-20	Note 2	—	Small	Note 2	No	and plugging	

Note 1: Pressure loss is that for fabric and dust cake. Pressure losses associated with outlet connections to be added by system designer.
Note 2: A function of the mechanical efficiency of these combined exhausters and dust collectors.
Note 3: Precooling of high temperature gases will be necessary to prevent rapid evaporation of fine droplets.
Note 4: See NFPA requirements for fire hazards, e.g., zirconium, magnesium, aluminum, woodworking, etc.

a very ineffective method of filtration because the vast majority of particles are far smaller than the spaces between fibers. Straining will remove lint, hair, and other large particles.

4.9.2 Impingement: When air flows through a filter, it changes direction as it passes around each fiber. Larger dust particles, however, cannot follow the abrupt changes in direction because of their inertia. As a result, they do not follow the air stream and collide with a fiber. Filters using this method are often coated with an adhesive to help fibers retain the dust particles that impinge on them.

4.9.3 Interception: Interception is a special case of impingement where a particle is small enough to move with the air stream but, because its size is very small in relation to the fiber, makes contact with a fiber while following the tortuous air flow path of the filter. The contact is not dependent on inertia and the particle is retained on the fiber because of the inherent adhesive forces that exist between the particle and fiber. These forces, called van der Waals (J. D. van der Waals, 1837-1923) forces, enable a fiber to trap a particle without the use of inertia.

4.9.4 Diffusion: Diffusion takes place on particles so small that their direction and velocity are influenced by molecular collisions. These particles do not follow the air stream, but behave more like gases than particulate. They move across the direction of air flow in a random fashion. When a particle does strike a fiber, it is retained by the van der Waals forces existing between the particle and the fiber. Diffusion is the primary mechanism used by most extremely efficient filters.

4.9.5 Electrostatic: A charged dust particle will be attracted to a surface of opposite electrical polarity. Most dust particles are not electrically neutral, therefore, electrostatic attraction between dust particle and filter fiber aids the collection efficiency of all barrier type air filters. Electrostatic filters establish an ionization field to charge dust particles so that they can be collected on a surface that is grounded or of opposite polarity. This concept was previously discussed in Section 4.3.1.

TABLE 4-5. Media Velocity vs. Fiber Size

Filter Type	Filter Size (microns)	Media Velocity (fpm)	Filtration Mechanism
Panel Filters	25-50	250-625	Impingement
Automatic Roll Filters	25-50	500	Impingement
Extended Surface Filters	0.75-2.5	20-25	Interception
HEPA Filters	0.5-6.3	5	Diffusion

Table 4-5 shows performance versus filter fiber size for several filters. Note that efficiency increases as fiber diameter decreases because more small fibers are used per unit volume. Note also that low velocities are used for high efficiency filtration by diffusion.

The wide range in performance of air filters makes it necessary to use more than one method of efficiency testing. The industry-accepted methods are ASHRAE Arrestance, ASHRAE Efficiency, and DOP. For ASHRAE Arrestance, a measured quantity of 72% standardized air cleaner test dust, 23% carbon black, and 5% cotton lint is fed to the filter. The efficiency by weight on this specific test dust is the ASHRAE Arrestance. ASHRAE Efficiency is a measure of the ability of a filter to prevent staining or discoloration. It is determined by light reflectance readings taken before and after the filter in a specified test apparatus. Atmospheric dust is used for the test. Both ASHRAE tests are described in ASHRAE Publication 52-76.[4.4]

In a DOP Test, 0.3 micron particles of dioctylphthalate (DOP) are drawn through a HEPA (High Efficiency Particulate Air) filter. Efficiency is determined by comparing the downstream and upstream particle counts. To be designated as a HEPA filter, the filter must be at least 99.97% efficient, i.e., only three particles of 0.3 micron size can pass for every ten thousand particles fed to the filter. Unlike both ASHRAE tests, the DOP test is not destructive, so it is possible to repair leaks and retest a filter that has failed.

The three tests are not directly comparable; however, Figure 4-17 shows the general relationship. Table 4-6 compares several important characteristics of commonly used air fil-

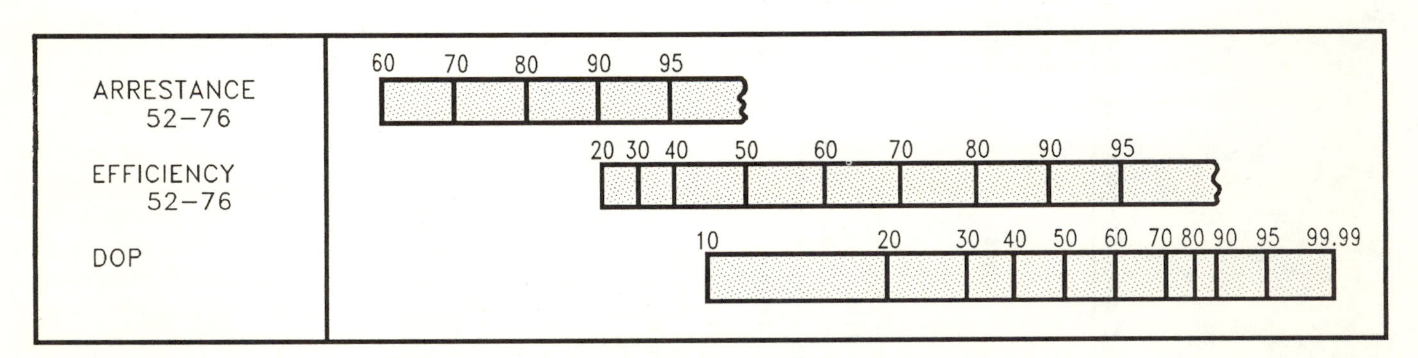

FIGURE 4-17. COMPARISON BETWEEN VARIOUS METHODS OF MEASURING AIR CLEANING CAPABILITY.

TABLE 4-6. Comparison of Some Important Air Filter Characteristics*

Type	Pressure Drop in wg (Notes 1 & 2) Initial	Final	ASHRAE Performance (Note 4) Arrestance	Efficiency	Face Velocity fpm	Maintenance (Note 5) Labor	Material
Low/Medium Efficiency							
1. Glass Throwaway (2″ deep)	0.1	0.5	77%	NA Note 6	300	High	High
2. High Velocity (permanent units) (2″ deep)	0.1	0.5	73%	NA Note 6	500	High	Low
3. Automatic (viscous)	0.4	0.4	80%	NA Note 6	500	Low	Low
Medium/High Efficiency							
1. Extended Surface (dry)	0.15–0.60	0.5–1.25	90-99%	25-95%	300–625	Medium	Medium
2. Electrostatic:							
a. Dry Agglomerator/ Roll Media	0.35	0.35	NA Note 7	90%	500	Medium	Low
b. Dry Agglomerator/ Extended Surface Media	0.55	1.25	NA Note 7	95% +	530	Medium	Medium
c. Automatic Wash Type	0.25	0.25	NA Note 7	85–95%	400–600	Low	Low
Ultra High Efficiency							
1. HEPA	0.5–1.0	1.0–3.0	Note 3	Note 3	250–500	High	High

Note 1: Pressure drop values shown constitute a range or average, whichever is applicable.
Note 2: Final pressure drop indicates point at which filter or filter media is removed and the media is either cleaned or replaced. All others are cleaned in place, automatically, manually or media renewed automatically. Therefore, pressure drop remains approximately constant.
Note 3: 95-99.97% by particle count, DOP test.
Note 4: ASHRAE Standard 52-76 defines (a) Arrestance as a measure of the ability to remove injected synthetic dust, calculated as a percentage on a weight basis and (b) Efficiency as a measure of the ability to remove atmospheric dust determined on a light-transmission (dust spot) basis.
Note 5: Compared to other types within efficiency category.
Note 6: Too low to be meaningful.
Note 7: Too high to be meaningful.

*Air filters should be used only for supply air systems or other applications where dust loading does not exceed 1 grain per 1000 cubic feet of air.

ters. Considerable life extension of an expensive final filter can be obtained by the use of one or more cheaper, less efficient, prefilters. For example, the life of a HEPA filter can be increased 25% with a throwaway prefilter. If the throwaway filter is followed by a 90% efficient extended surface filter, the life of the HEPA filter can be extended nearly 900%. This concept of "progressive filtration" allows the final filters in clean rooms to remain in place for ten years or more.

4.10 RADIOACTIVE AND HIGH TOXICITY OPERATIONS

There are three major requirements for air cleaning equipment to be utilized for radioactive or high toxicity applications:

1. High efficiency
2. Low maintenance
3. Safe disposal

High efficiency is essential because of extremely low tolerances for the quantity and concentration of stack effluent and the high cost of the materials handled. Not only must the efficiency be very good, it must also be verifiable because of the legal requirement to account for all radioactive material.

The need for low maintenance is of special importance when exhausting any hazardous material. For many radioactive processes, the changing of bags in a conventional fabric collector may expend the daily radiation tolerances of 20 or more persons, so infrequent, simple, and rapid maintenance requirements are vital. Another important factor is the desirability of low residual build up of material in the collector since dose rates increase with the amount of material and reduce the allowable working time.

Disposal of radioactive or toxic materials by air, water or land is a serious and very difficult problem. For example, scalping filters loaded with radioactive dust are usually incinerated to reduce the quantity of material that must be disposed of in special burial grounds. The incinerator will require an air cleaning device, such as a wet collector of very special design, to avoid unacceptable pollution of air and water.

With these factors involved, it is necessary to select an air cleaning device that will meet efficiency requirements without causing too much difficulty in handling and disposal.

Filter units especially designed for high efficiency and low maintenance are available. These units feature quick changeout through a plastic barrier which is intended to encapsulate spent filters, thereby eliminating the exposure of personnel to radioactive or toxic material. A filtration efficiency of 99.97% by particle count on 0.3 micron particles is standard for this type of unit.

For further information on this subject, see reference 4.5.

4.11 EXPLOSION VENTING

Many dusts are combustible and, in certain concentration ranges, explosive. Dust collection equipment should be designed to reduce the risk of property damage and personal injury where explosive mixtures of dust and air are probable. Alternates are to design the dust collector housing to contain the considerable pressure rise resulting from an explosion or to equip it with explosion vents and design for the lower, but still significant, pressure rise that results from a vented explosion. Explosion vents are a normal optional accessory for dust collection equipment, although frequently employed without complete understanding of their purpose and limitations.

NFPA 68-1978, *Guide for Explosion Venting*,[4.6] is the most commonly referenced work on the subject and should be studied in detail by anyone responsible for design, selection, or purchase of dust collectors applied to potentially explosive dusts. As pointed out there, the purpose of explosion vents is to *limit* the maximum pressure developed during an explosion to a value less than the safety limit of the vessel (dust collector housing). Vents can be designed to open at any desired release pressure but the maximum pressure developed during venting can be very much higher than the pressure at which the vent opens.

NFPA 68-1978 contains monographs which allow the prediction of the maximum pressure occurring during venting. For example, for explosion vents having an area equal to 3 $ft^2/100\ ft^3$ of vessel volume and a release pressure of 1.5 psig, the predicted maximum pressure occurring as a result of a vented explosion could be 5 to 10 psig. Even though equipped with vents, most standard dust collectors require reinforcing to safely withstand this reduced maximum pressure.

REFERENCES

4.1. D. Leith, M.W. First and H. Feldman: "Performance of a Pulse-Jet at High Velocity Filtration II, Filter Cake Redeposition," *J. Air Pollut. Control Assoc.* 28:696 (July 1978).

4.2. E. Beake: "Optimizing Filtration Parameters." *J. Air Pollut. Control Assoc.* 24:1150 (1974).

4.3. D. Leith, D.D. Gibson and M.W. First: "Performance of Top and Bottom Inlet Pulse-Jet Fabric Filters." *J. Air Pollut. Control Assoc.* 24:1150 (1974).

4.4. American Society of Heating, Refrigerating, and Air-Conditioning Engineers: *Method of Testing Cleaning Devices Used in General Ventilation for Removing Particulate Matter.* ASHRAE Pub. No. 52-76. ASHRAE, 1791 Tullie Circle, NE, Atlanta, GA 30329 (May 1976).

4.5. National Council on Radiation Protection and Measurement: NCRP Report No. 39, *Basic Radiation Protection Criteria.* NCRP Report No. 39. NCRP Publications, 7910 Woodmont Ave., Suite 1016, Bethesda, MD 20814 (January 15, 1971).

4.6. National Fire Protection Association: *Guide for Explosion Venting.* NFPA 68-1978. NFPA, Batterymarch Park, Quincy, MA 02269 (1978).

Chapter 5
EXHAUST SYSTEM DESIGN PROCEDURE

5.1 INTRODUCTION 5-2

5.2 PRELIMINARY STEPS 5-2

5.3 DESIGN PROCEDURE 5-2

5.4 DESIGN METHODS 5-2
- 5.4.1 Velocity Pressure Method 5-2
- 5.4.2 Equivalent Foot Method 5-2

5.5 AIDS TO CALCULATIONS 5-3

5.6 DISTRIBUTION OF AIR FLOW 5-3
- 5.6.1 Balance by Design Method 5-6
- 5.6.2 Blast Gate Method 5-6
- 5.6.3 Choice of Methods 5-6
- 5.6.4 Balance by Design Procedure 5-6
- 5.6.5 Blast Gate Procedure 5-7

5.7 PLENUM TYPE EXHAUST SYSTEMS 5-7
- 5.7.1 Choice of Systems 5-7
- 5.7.2 Design 5-7

5.8 FAN PRESSURE RATINGS 5-7
- 5.8.1 Fan Total Pressure 5-7
- 5.8.2 Fan Static Pressure 5-7

5.9 CORRECTIONS FOR VELOCITY CHANGES 5-10
- 5.9.1 Branch Entries to Main Ducts 5-10
- 5.9.2 Duct Contractions and Enlargements 5-10

5.10 SAMPLE SYSTEM DESIGN 5-11

5.11 CORRECTIONS FOR DIFFERENT DUCT MATERIALS 5-11

5.12 FRICTION LOSS FOR NON-CIRCULAR DUCTS 5-11

5.13 CORRECTIONS FOR TEMPERATURE, MOISTURE AND ALTITUDE 5-23
- 5.13.1 Variable Temperature and/or Different Altitude 5-23
- 5.13.2 Elevated Moisture 5-23
- 5.13.3 Psychrometric Principles 5-23
- 5.13.4 Density Determination 5-24
- 5.13.5 Hood Volumetric Flow Rate Changes with Density 5-24

5.14 AIR CLEANING EQUIPMENT 5-28

5.15 EVASE DISCHARGE 5-28

5.16 EXHAUST STACK OUTLETS 5-28

5.17 AIR BLEED-INS 5-29

5.18 OPTIMUM ECONOMIC VELOCITY 5-29

REFERENCES 5-29

5.1 INTRODUCTION

The calculations discussed below are essential to determining duct sizes and exhaust system pressure loss. The results, coupled with the exhaust volumetric flow rate, will determine the size and type of fan as well as its speed and power requirements.

5.2 PRELIMINARY STEPS

The designer should have the following data available:

1. A layout of the operations, workroom, building (if necessary), etc.
2. A line sketch of the duct system layout including plan and elevation dimensions, fan location, collector location, etc. Number, letter or otherwise identify each branch and section of main on the line sketch for convenience.
3. A rough design or sketch of the desired hood for each operation with direction and elevation of outlet for duct connection.
4. Information as to the details of the operation(s), toxicity of materials, physical and chemical characteristics, operational characteristics, etc.

5.3 DESIGN PROCEDURE

All exhaust systems, whether simple or complex, have in common the use of hoods, duct segments, and special fittings leading to an exhaust fan. In fact, a complex system is merely an arrangement of several simple exhaust systems connected to a common duct. In designing an exhaust system, start at the hood farthest away from the fan.

1. Select or design each exhaust hood to suit the operation to be controlled and determine its design flow rate (see Chapter 3).
2. Determine the minimum duct velocity based on the required transport velocity as determined from the data presented in Chapter 3.
3. Determine the branch duct size by dividing the design flow rate by the minimum duct velocity. For systems handling particulate matter, a commercially available duct size (see Table 5-5) with a *smaller* area should be selected to insure that the actual duct velocity is greater than the minimum required.
4. Using the line sketch, determine the design length for each duct segment and the number and type of any special fittings and elbows needed. A duct segment is defined as that constant diameter round duct that separates points of interest such as hoods, entry points, fan inlet, etc. Design length is the centerline distance along the duct.

5.4 DESIGN METHODS

1. Calculate the pressure losses for the exhaust system. The pressure losses due to friction and fittings can be calculated either by the velocity pressure method or the equivalent foot method. However, the velocity pressure method is preferred for several reasons:
   a. It is generally more rapid and puts all losses, including hood entry, on the same basis; and
   b. It offers the advantage of quick recalculation of branch duct sizes for the balanced duct design method (see Section 5.6.1).
2. Check for correct balance at entries and adjust volumetric flow rate, duct size, or hood design to obtain correct flow.
3. Select collector and fan based upon final volumetric flow rate and calculated system resistance.

5.4.1 Velocity Pressure Method: This method is based on the fact that all frictional and dynamic (fitting) losses in ducts and hoods are functions of the velocity pressure and can be calculated by a factor multiplied by the velocity pressure. Factors for hoods, straight ducts, elbows, branch entries, and many other fittings are shown in Figures 5-15 through 5-19. Thus in starting the design it is necessary to establish only one value for elbows and fittings. For convenience, loss factors for elbows and entries are presented on the calculation sheet.

A friction chart for this method is presented as Figures 5-18a and b. This chart gives the loss factor per foot of galvanized sheet metal duct. The equation for this chart is listed on these figures and also on the calculation sheet (see Figures 5-2, 5-9, and 5-10). The following steps will establish the overall pressure loss of a duct segment that starts at a hood:

1. Determine the *actual* velocity by dividing the flow rate by the area of the commercial duct size chosen. Then determine the corresponding velocity pressure from Table 5-4 or the equations in Chapter 1.
2. Determine the hood suction from the equations in Chapter 3.
3. Multiply the design duct length by the loss factor obtained from Figures 5-18a and b.
4. Determine the number and type of fittings in the duct segment. For each fitting type (see Figures 5-15, 5-16, 5-17, and 5-19), determine the loss factor and multiply by the number of fittings.
5. Add the results of Steps 3 and 4 above and multiply by the duct VP. This is the actual loss in inches of water for the duct segment.
6. Add the result of Step 5 to the hood suction. If there are any additional losses (expressed in inches of water), such as for an air cleaning device, add them in also. This establishes the cumulative energy required, expressed as static pressure, to move the design flow rate through the duct segment.

5.4.2 Equivalent Foot Method: This method is very similar to the velocity pressure method. The method differs in the manner by which the friction and fitting losses are calculated. The length of straight duct is determined as before. Fittings are "replaced" by a length of duct that will have an equivalent loss. These equivalent lengths are a function of the duct diameter and are listed on Figure 5-20. The equivalent length of all the fittings in the duct segment then are added to the straight duct length.

A friction chart for this method is presented as Figures 5-21a and b. This chart gives the loss in inches of water per 100 feet of galvanized sheet metal duct. The equation for this chart is given on the figure. To determine the loss, multiply the total length by the friction loss from the chart and divide by 100.

EXAMPLE

To illustrate the steps involved in calculating these pressure losses, Problem 1, Figure 5-1, is included. Note that the calculation of these losses is a straightforward procedure using the calculation sheet provided. Along with the drawing of the system there is also a pressure profile showing the magnitude and relationships of total, static, and velocity pressures on both the "suction" and the "pressure" sides of the fan. It should be noted that Velocity Pressure is always positive with respect to atmospheric pressure. Also, while Total and Static Pressure may be either negative or positive with respect to atmosphere, Total Pressure is always greater than Static Pressure (TP = SP + VP) on an absolute pressure basis.

5.5 AIDS TO CALCULATIONS

The use of a calculation sheet can be very beneficial when performing the calculations by hand. Figure 5-2 is a calculation sheet for the velocity pressure method which shows the results for the previous example.

Figure 5-3 shows the Equivalent Foot Method calculation sheet. It also shows a solution to this example.

As an alternative to performing these calculations manually, programmable calculators and computers can be used to provide automated design of entire systems. The Committee does not recommend any specific hardware or software. Many firms have developed their own software, and there are commercially available software packages on the market.

5.6 DISTRIBUTION OF AIR FLOW

As discussed earlier, a complex exhaust system is actually a group of simple exhaust systems connected to a common main duct. Therefore, in designing a system of multiple hoods and branches, the same rules apply as before. In a multiple branch system, however, it is necessary to provide a means of distributing air flow between the branches either by balanced design or by the use of blast gates.

The reason for this is that air will always take the path of least resistance. A natural balance at each junction will occur; that is, the exhaust volumetric flow rate will distribute itself automatically according to the pressure losses of the available flow paths. To provide distribution that will result in the design air flow at each hood, the designer must make sure that *all flow paths (ducts) entering a junction will have equal calculated static pressure requirements.*

To accomplish this, the designer has a choice of two methods. The object of both methods is the same: to obtain the desired volumetric flow rate at each hood in the system while maintaining the desired velocity in each branch and main.

The two methods, labeled Balance by Design Method and Blast Gate Method, are outlined below. Their relative advantages and disadvantages can be found in Table 5-1.

TABLE 5-1. Relative Advantages and Disadvantages of the Two Methods

Balance by Design Method	Blast Gate Method
1. Volumetric flow rates cannot be changed easily by workers or at the whim of the operator.	1. Volumetric flow rates may be changed relatively easily. Such changes are desirable where pickup of unnecessary quantities of material may affect the process.
2. There is little degree of flexibility for future equipment changes or additions. The duct is "tailor made" for the job.	2. Depending on the fan and motor selected, there is somewhat greater flexibility for future changes or additions.
3. The choice of exhaust volumetric flow rates for a new operation may be incorrect. In such cases, some duct revision may be necessary.	3. Correcting improperly estimated exhaust volumetric flow rates is relatively easy within certain ranges.
4. No unusual erosion or accumulation problems will occur.	4. Partially closed blast gates may cause erosion thereby changing resistance or causing particulate accumulation.
5. Duct will not plug if velocities are chosen wisely.	5. Duct may plug if the blast gate insertion depth has been adjusted.
6. Total volumetric flow rate may be greater than design due to higher air requirements.	6. Balance may be achieved with design volumetric flow rate; however, the net energy required is usually greater than for the Balance by Design Method.
7. The system layout must be in complete detail with all obstructions cleared and length of runs accurately determined. Installation must follow layout exactly.	7. Moderate variations in duct layout are possible.

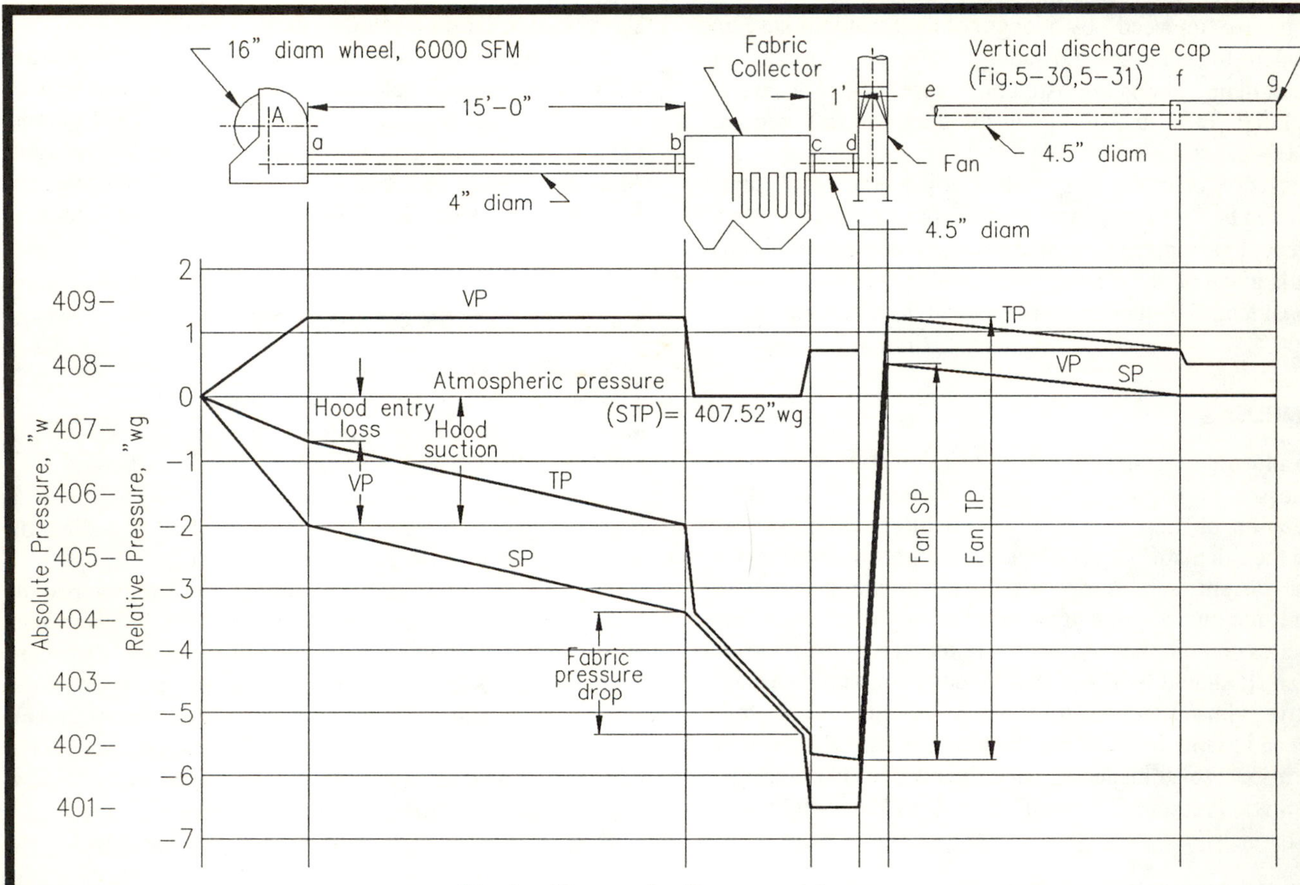

Details of Operation

NO.		HOOD NO.	VS-PRINT	REQUIRED AIR FLOW, cfm
1	16" Diameter Grinding wheel, 2" Wide	A	411	390

Dimensions

No. of Branch or Main	Straight Run, Ft	CFM Required	Elbows	Entries
ab	15	390	--	--
bc	Collector	390	--	--
cd	1	390	--	--
ef	10	390	--	--
fg	Stack Head	390	--	--

AMERICAN CONFERENCE OF GOVERNMENTAL INDUSTRIAL HYGIENISTS	*PROBLEM 1*
	DATE *1-88* FIGURE *5-1*

Velocity Pressure Method Calculation Sheet

Plant Name: PROBLEM 1 Elevation: ________ Date: 4-88

Location: ________ Temperature: ________ Drawing No.: ________

Department: ________ Factor: ________ Designer: ________

No.			Item	Reference	Units						
1			Duct Segment Identification			A-b	b-c	c-d	e-f		
2			Volumetric Flowrate		cfm	390	390	390	390		
3			Minimum Transport Velocity		fpm	4500					
4			Duct Diameter		inches	4		4.5	4.5		
5			Duct area		sq. ft.	0.0873		0.1104	0.1104		
6			Actual Duct Velocity		fpm	4467		3531	3531		
7			Duct Velocity Pressure		"wg	1.24		0.78	0.78		
8	H		Slot Area		sq. ft.						
9	O	S	Slot Velocity		fpm						
10	O	L	Slot Velocity Pressure		"wg						
11	D	O	Slot Loss Factor	Fig. 5-15 or Chap. 10							
12		T	Acceleration Factor		0 or 1						
13	S	S	Plenum loss per VP	Items 11 + 12							
14	U		Plenum SP	Items 10 x 13	"wg						
15	C		Duct Entry Loss Factor	Fig. 5-15 or Chap. 10		0.65		0.49			
16	T		Acceleration Factor		1 or 0	1		1			
17	I		Duct Entry Loss per VP	Items 15 + 16		1.65		1.49			
18	O		Duct Entry Loss	Items 7 x 17		2.05		1.16			
19	N		Other Loss		"wg		2.00				
20			Hood Static Pressure	Items 14 + 18 + 19	"wg	2.05		1.16			
21			Straight Duct Length		feet	15		1	10		
22			Friction Factor (H_f)	Fig. 5-18 or equation		0.0703		0.062	0.062		
23			Friction Loss per VP	Items 21 x 22		1.05		0.062	0.62		
24			No. of 90° Elbows								
25			Elbow Loss per VP	Item 24 x Loss Factor							
26			No. Entries								
27			Entry Loss per VP	Item 26 x Loss Factor							
28			Special Fittings Loss Factors								
29			Duct Loss per VP	Items 23 + 25 + 27 + 28		1.05		0.062	0.62		
30			Duct Loss	Items 7 x 29	"wg	1.30		0.05	0.48		
31			Duct SP Loss	Items 20 + 30	"wg	3.35	2.00	1.21			
32			Cumulative Static Pressure		"wg	-3.35	-5.35	-6.56	0.48		
33			Governing Static Pressure		"wg						
34			Corrected Volumetric Flowrate		cfm						
35			Resultant Velocity Pressure		"wg						

PERTINENT EQUATIONS:

$$Q_{corr} = Q_{design}\sqrt{\frac{SP_{gov.}}{SP_{duct}}}$$

$$H_f = 0.0307\frac{V^{0.533}}{Q^{0.612}} = \frac{0.4937}{(Q^{0.079})(D^{1.066})}$$

$$VP_r = \left[\frac{Q_1 + Q_2}{4005(A_1 + A_2)}\right]^2$$

$$FSP = SP_{outlet} - SP_{inlet} - VP_{inlet}$$

90° ROUND ELBOW LOSS FACTORS

C.L. R/D	Factor
Mitered	1.25
1.5	0.39
2.0	0.27
2.5	0.22

60° elbow = 2/3 loss
45° elbow = 1/2 loss

BRANCH ENTRY LOSS FACTORS

Angle	Factor
15°	0.09
30°	0.18
45°	0.28
60°	0.44
90°	1.00

CALCULATIONS:

FSP = +0.48 - (-6.56) - 0.78 = 6.26 "wg

FIGURE 5-2.

Equivalent Foot Method
Calculation Sheet

Committee on Industrial Ventilation
P. O. Box 16153
Lansing, Michigan 48901

Plant name PROBLEM 1	Refer to	Elevation	Factor	Remarks
Location		Temperature		
Department				

1	2	3	4	5	6	7	8		9	10	11	12	13	14	15	16	17	18	19
									From Fig. 5-20	Col. 7 plus Col. 9	From Fig. 5-21	Col. 10 x Col. 11 / 100	From Col. 6 & Table 5-4	From Fig. 5-15	1.00 plus Col. 14	Col. 13 times Col. 15	Col. 12 plus Col. 16	At junction	
			Air volume CFM			Length of duct in feet					Friction loss, inches of water					Pressure, inches of water			
No. of br. or main	Dia. duct in in.	Area duct sq. ft.	in branch	in main	Vel. in FPM	straight runs	Number of elbows	Number of entries	equiv. length	total length	per 100	of run	one VP	entry loss (VP)	hood suct (VP)	hood suct.	static press.	gov. SP	corrected CFM
A	4	0.087	390		4470								1.25	0.65	1.65	2.06			
a-b	4				4470	15				15	8.5	1.28	1.25				3.34		
Collector							SP drop across fabric					2.00					5.34		
c	4 1/2	0.111			3510								0.77	0.49	1.49	1.15	6.49		
c-d	4 1/2				3510	1				1	4.5	0.05					6.54		
Fan																			
e-f	4 1/2				3510	10				10	4.5	0.45							

Fan SP = $SP_o - SP_i - VP_i$

= 0.45 − (−6.54) − 0.77

Fan SP = 6.22 "wg

Calculated fan characteristics for standard conditions		Corrected for temperature and elevation		
Capacity ( ) CFM	Fan, type and size	RPM	CFM	Motor
Fan TP in WG		BHP	TP	V Belt
Fan SP	RPM BHP		SP	

FIGURE 5-3.

5.6.1 *Balance By Design Method:* This procedure (see Section 5.10) provides for achievement of desired air flow (a "balanced" system) without the use of blast gates. It is often called the "Static Pressure Balance Method." In this type of design, the calculation usually begins at the hood farthest from the fan and proceeds, branch to main, and section of main to section of main, to the fan. *At each junction, the static pressure necessary to achieve desired flow in one stream must equal the static pressure in the joining air stream.* The static pressures are balanced by suitable choice of duct sizes, elbow radii, etc., as detailed below.

5.6.2 *Blast Gate Method:* The design procedure depends on the use of blast gates which must be adjusted after installation in order to achieve the desired flow at each hood. At each junction, the flow rates of two joining ducts are added.

5.6.3 *Choice of Methods:* The Balance by Design Method will normally be selected where highly toxic materials are controlled to safeguard against tampering with blast gates and consequently subjecting personnel to potentially excessive exposures. This method is mandatory where explosives and radioactive dusts are exhausted as the possibility of accumulations in the system caused by a blast gate obstruction is eliminated.

5.6.4 *Balance by Design Procedure:* The pressure loss of each branch is calculated, based on design data, and totaled for the length running from exhaust hood to junction with the next branch. At each junction the SP for each parallel path of air flow must be the same. Where the ratio of the lower SP to the higher SP is less than 0.8, redesign of the branch with lower pressure loss should be considered. This may include a change of duct size, selection of different fittings, and/or modifications to the hood design. Where static pressures of parallel paths are unequal, balance can be obtained by increasing the air flow through the run with the lower resistance. This change in flow rate is calculated by noting that pressure losses vary with the velocity pressure and therefore as the square of the volumetric flow rate, so:

$$Q_{Corrected} = Q_{Design}\sqrt{\frac{\text{SP chosen for the junction}}{\text{SP calculated for this branch}}} \quad [5.1]$$

Where the static pressure ratio is between 0.95 and 1.0, the flow rate correction will be less than 2.5% and is usually ignored. This small error is treated as if the paths were in complete balance at the SP corresponding to the duct run with the higher loss (the governing static pressure).

5.6.5 Blast Gate Procedure: Data and calculations involved are the same as for the balanced design method except that the duct sizes, fittings, and volumetric flow rates are not adjusted; the blast gates are set after installation to provide the design volumetric flow rates. It should be noted that a change in any of the the blast gate settings will change the flow rates in all of the other branches. Readjusting the blast gates during the system balancing process sometimes can result in increases to the actual fan static pressure and increased fan power requirements.

5.7 PLENUM TYPE EXHAUST SYSTEMS

Plenum type systems differ from the designs illustrated earlier (see Figure 5-4). Minimum transport velocities are maintained only in the branch ducts to prevent settling of particulate matter; the main duct is oversized and velocities are allowed to decrease far below normal values. The function of the main duct is to provide a low pressure loss path for air flow from the various branches to the air cleaner or the fan. This helps to maintain balanced exhaust in all of the branches and often provides a minimum operating power.

Advantages of the plenum type exhaust system include:

1. Branch ducts can be added, removed or relocated at any convenient point along the main duct.
2. Branch ducts can be closed off and the volumetric flow rate in the entire system reduced so long as minimum transport velocities are maintained in the remaining branches.
3. The main duct can act as a primary separator (settling chamber) for large particulate matter and refuse material which might be undesirable in the air cleaner or fan.

Limitations of this design include:

1. Sticky, linty materials, such as buffing dust, are difficult to handle and tend to clog the main duct. It may be expected that greatest difficulty will be encountered with the drag chain type of cleaning, but the other types will be susceptible to buildup as well.
2. Materials which are subject to direct or spontaneous combustion must be handled with care. Wood dust has been handled successfully in systems of this type; buffing dust and lint are subject to this limitation and are not recommended. Explosive dusts, such as magnesium or titanium as well as grain dusts, should not be handled in systems of this type.

5.7.1 Choice of Systems: Various types of plenum exhaust systems are used in industry (see Figure 5-5). They include both self-cleaning and manual-cleaning designs. Self-cleaning types include pear-shaped designs which incorporate a drag chain conveyor in the bottom of the duct to convey the dust to a chute, tote box, or enclosure for disposal. Another self-cleaning design uses a rectangular main with a belt conveyor. In these types, the conveyors may be run continuously or on periodic cycles to empty the main duct before considerable buildup and clogging occur. A third type[5.3] of self-cleaning design utilizes a standard conveying main duct system to remove the collected material from a hopper type of main duct above. Such a system is usually run continuously to avoid clogging of the pneumatic air circuit. Manual-cleaning designs may be built into the floor or may be large enclosures behind the equipment to be ventilated. Experience indicates that these should be generously oversized, particularly the underfloor designs, to permit added future exhaust capacity as well as convenient housekeeping intervals.

5.7.2 Design: Control volumetric flow rates, hoods, and duct sizes for all branches are calculated in the same manner as with the balanced by design and blast gate methods. The branch segment with the greatest pressure loss will govern the static pressure required in the main duct. Other branches will be designed to operate at this static pressure or locking dampers can be used to adjust their pressure loss to the same static pressure as the governing branch. Where the main duct is relatively short or where the air cleaners or fans can be spaced along the duct, static pressure losses due to air flow in the main duct can be ignored. For extremely long ducts, it is necessary to calculate the static pressure loss along the main in a manner similar to that used in the balanced and blast gate methods. However, only approximate results will be obtained. Duct connections to air cleaners, fans and discharge to outside are handled in the normal manner.

5.8 FAN PRESSURE RATINGS

Exhaust system calculations are based on static pressure; that is, all hood static pressures and balancing or governing pressures at the duct junctions are given as static pressures which can be measured directly as described in Chapter 9. Most fan rating tables are based on Fan Static Pressure. An additional calculation is required to determine Fan Static Pressure before selecting the fan.

5.8.1 Fan Total Pressure (FTP) is the increase in total pressure through or across the fan and can be expressed by the equation:

$$FTP = TP_{outlet} - TP_{inlet} \quad [5.2]$$

Some fan manufacturers base catalog ratings on Fan Total Pressure. To select a fan on this basis the Fan Total Pressure is calculated as follows:

$$FTP = TP_{outlet} - TP_{inlet}$$

$$FTP = (SP_{outlet} + VP_{outlet}) - (SP_{inlet} + VP_{inlet}) \quad [5.3]$$

5.8.2 Fan Static Pressure: The Air Movement and Control Association Test Code [5.1] defines the Fan Static Pres-

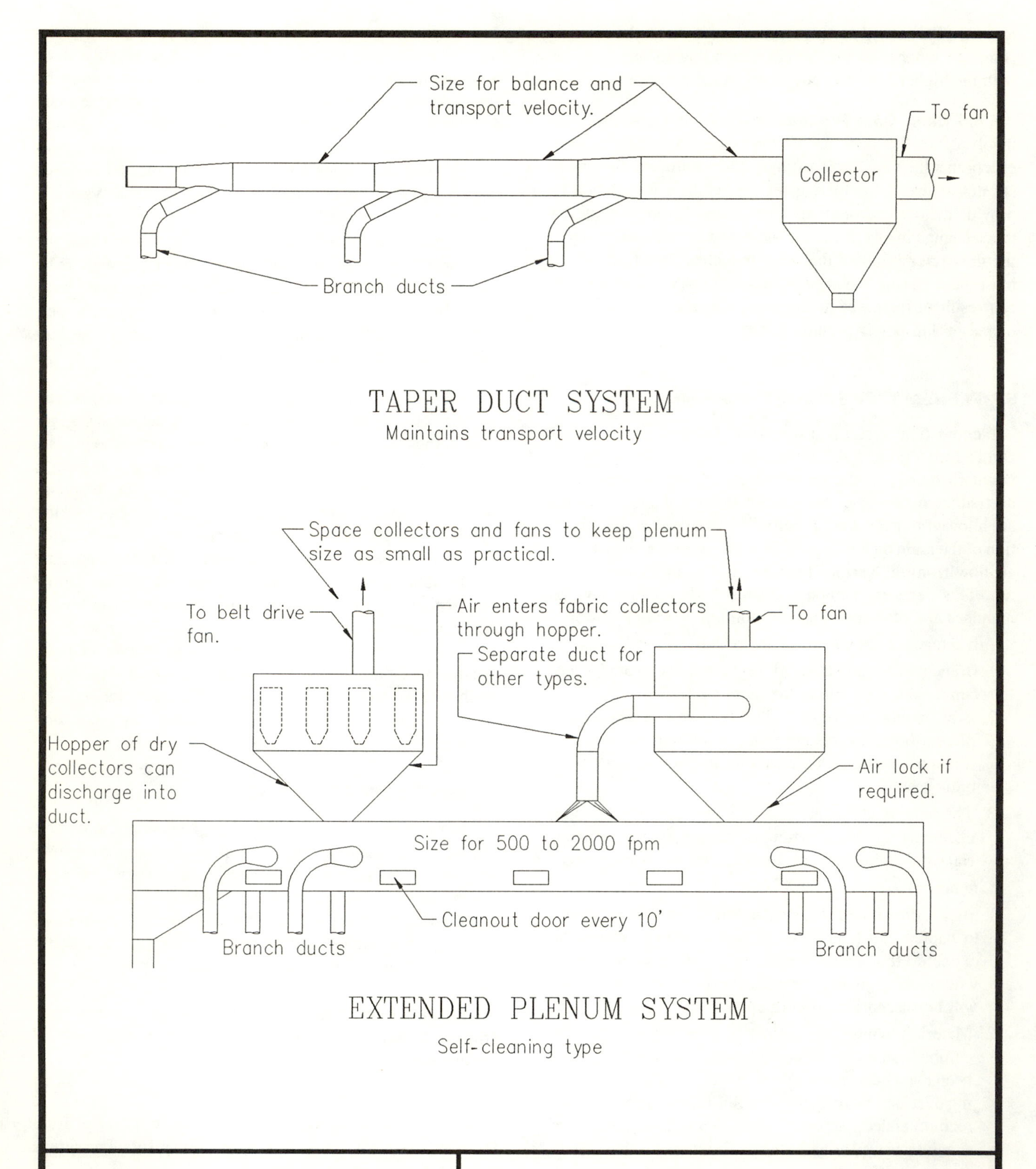

AMERICAN CONFERENCE OF GOVERNMENTAL INDUSTRIAL HYGIENISTS	*PLENUM VS CONVENTIONAL SYSTEM*		
	DATE 1-88	FIGURE 5-4	

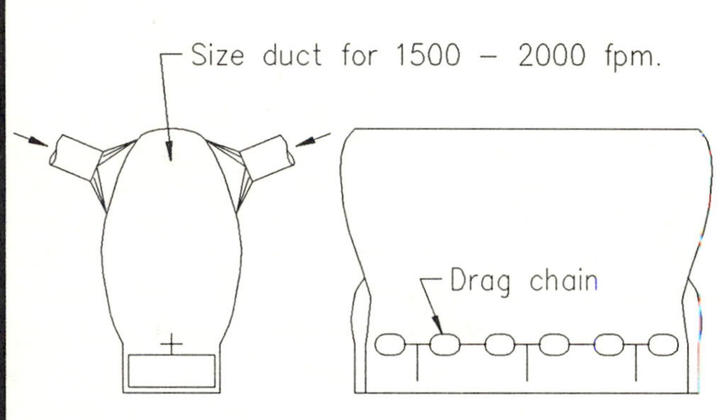

1. Self-cleaning main – drag chain

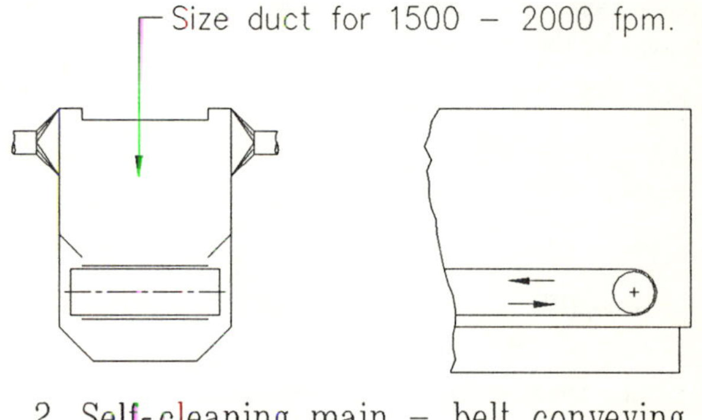

2. Self-cleaning main – belt conveying

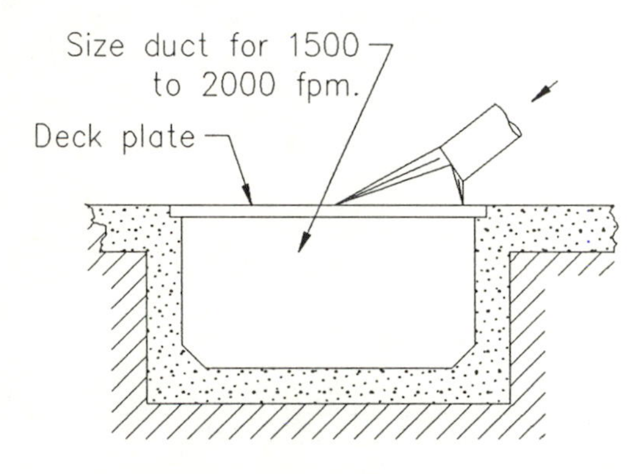

3. Under floor – manual cleaning

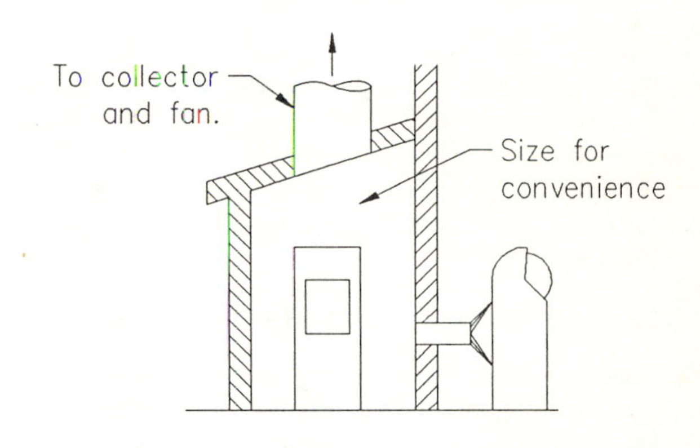

4. Large plenum – manual cleaning

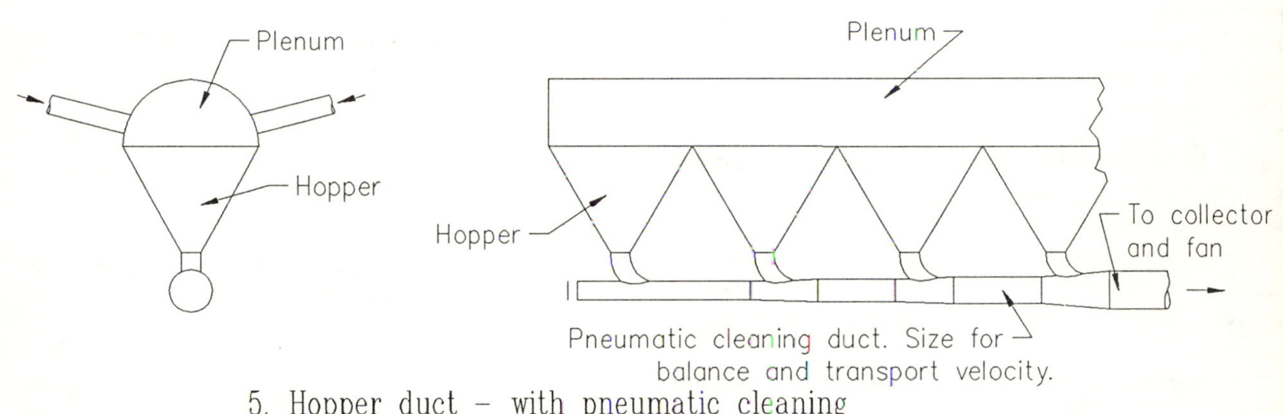

5. Hopper duct – with pneumatic cleaning

Reference 5.3

AMERICAN CONFERENCE OF GOVERNMENTAL INDUSTRIAL HYGIENISTS	*TYPES OF PLENUMS*		
	DATE 1–88	FIGURE 5–5	

sure (FSP) as follows: the static pressure of the fan is the total pressure diminished by the fan velocity pressure. The fan velocity pressure is defined as the pressure corresponding to the air velocity at the fan outlet. Fan Static Pressure can be expressed by the equation:

$$FSP = FTP - VP_{outlet} \quad [5.4]$$

or

$$FSP = TP_{outlet} - TP_{inlet} - VP_{outlet}$$

Since TP = SP + VP, the equation can be rewritten:

$$FSP = (SP_{outlet} + VP_{outlet}) - (SP_{inlet} + VP_{inlet}) - VP_{outlet}$$

Combining terms leads to a final equation:

$$FSP = SP_{outlet} - SP_{inlet} - VP_{inlet} \quad [5.5]$$

In selecting a fan from catalog ratings, the rating tables should be examined to determine whether they are based on Fan Static Pressure or Fan Total Pressure. The proper pressure rating can then be calculated keeping in mind the proper algebraic signs; i.e., VP is always positive (+), SP_{inlet} is usually negative (−) and SP_{outlet} is usually positive (+).

5.9 CORRECTIONS FOR VELOCITY CHANGES

Variations in duct velocity occur at many locations in exhaust systems because of necessary limitations of available standard duct sizes (area) or due to duct selections based on balanced system design. As noted earlier, small accelerations and decelerations are usually compensated automatically in the system where good design practices and proper fittings are used. There are times, however, when special circumstances require the designer to have a knowledge of the energy losses and regains which occur since these may work to his advantage or disadvantage in the final performance of the system.

5.9.1 Branch Entries to Main Ducts: Sometimes the final main duct velocity exceeds the higher of the two velocities in the branches entering the main. If the difference is great, additional static pressure is required to produce the increased velocity. A difference of 0.10 "wg or greater between the main VP and the resultant VP of the two branches should be corrected.

The correction is made by first computing the resultant velocity pressure (VP_r) corresponding to a pseudo velocity of the two volumetric flow rates entering the junction. This is accomplished by applying the basic velocity pressure equation, $VP = (V/4005)^2$ using the summation of the two volumetric flow rates and the summation of the two duct areas:

$$VP_r = \left[\frac{Q_1 + Q_2}{4005\,(A_1 + A_2)}\right]^2 \quad [5.6]$$

where:

VP_r = resultant velocity pressure of the combined branches

Q_1 = volumetric flow rate in branch #1

Q_2 = volumetric flow rate in branch #2

A_1 = area of branch duct #1

A_2 = area of branch duct #2

It is assumed that branches #1 and #2 are balanced at the junction so that $SP_1 = SP_2$. If VP_3 is less than VP_r, a deceleration has occurred and no correction is made. If VP_3 is greater than VP_r an acceleration has occurred, and the difference between VP_3 and VP_r is the necessary loss in SP required to produce the increase in kinetic energy between VP_3 and VP_r. The correction is made as follows:

$$SP_3 = SP_1 - (VP_3 - VP_r) \quad [5.7]$$

where:

SP_3 = SP in main #3

SP_1 = SP at branch #1 = SP at branch #2

VP_3 = velocity pressure in main #3

EXAMPLE

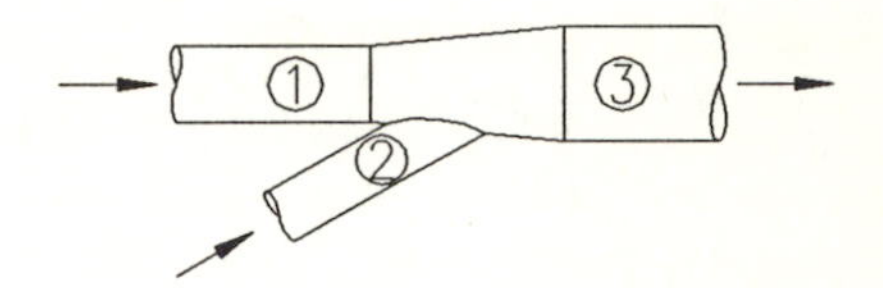

Duct No.	Dia.	Area	Q	V	VP	SP
(1)	10	0.545	1935	3550	0.79	−2.11
(2)	4	0.087	340	3890	0.94	−2.11
Main (3)	10	0.545	2275	4170	1.08	−

$$VP_r = \left[\frac{Q_1 + Q_2}{4005\,(A_1 + A_2)}\right]^2 = \left[\frac{1935 + 340}{4005\,(0.545 + 0.087)}\right]^2$$

$$= \left[\frac{2275}{4005 \times 0.632}\right]^2 = 0.81 \text{ "wg}$$

$$SP_3 = SP_1 - (VP_3 - VP_r) = -2.11 - (1.08 - 0.81)$$

$$= -2.11 - 0.27$$

$SP_3 = -2.38$ "wg (corrected SP to provide for increased velocity)

FIGURE 5-6. BRANCH ENTRY VELOCITY CORRECTION.

5.9.2 Duct Contractions and Enlargements: Duct contractions are used where it is necessary to reduce the size of the duct to fit into tight places, to fit equipment, or to provide a high discharge velocity at the end of the stack. Duct enlargements are used to fit a particular piece of equipment or to reduce the energy consumed in the system by reducing velocity and friction. Enlargements are not desirable in transport systems since the duct velocity may become less than the minimum transport velocity and material may settle in the ducts.

Regain of pressure in a duct system is possible because static pressure and velocity pressure are mutually convertible. This conversion is accompanied by some energy loss; the amount of this loss is a function of the geometry of the transition piece (the more abrupt the change in velocity, the greater the loss) and depends on whether air is accelerated or decelerated. Loss is expressed as a percentage of the difference between the velocity pressures at the entrance and the exit of the transition piece. One minus the loss is the efficiency of the energy conversion or regain.

A perfect contraction or enlargement (no loss) would cause no change in the total pressure in the duct. There would be an increase (decrease) in static pressure corresponding exactly to the decrease (increase) in velocity pressure of the air. In practice, the enlargement or contraction will not be perfect, and there will be a change in total pressure (see Figure 5-7). In each example, total pressure and static pressure are plotted in order to show their relationship at various points in each system.

5.10 SAMPLE SYSTEM DESIGN

A discussion of the calculations for either taper duct method can best be done by a typical example using the exhaust system shown in Figure 5-8. Typical calculation sheets illustrate an orderly and concise arrangement of data and calculation (see Figures 5-9 through 5-12).

The problem considered is a foundry sand handling and shake-out system. A minimum conveying velocity of 3,500 fpm is used throughout the problem. The operations, hood designations on the diagram, VS-print references, and required volumetric flow rates are presented in Table 5-2.

5.11 CORRECTIONS FOR DIFFERENT DUCT MATERIALS

The friction loss charts, Figures 5-18 (a & b) and 5-21 (a & b), provide average values for clean, round galvanized metal ducts having approximately 40 joints per 100 feet based on the flow of standard air of 0.075 pounds per cubic foot density. The values obtained can be used with no significant error for the majority of designs; occasionally, however, more precise calculations of duct friction are required.

Figure 5-22 presents correction factors depending on duct smoothness, size and velocity. These are applied directly to the values obtained from Figures 5-18 (a & b) and 5-21 (a & b).

EXAMPLE

4″ diameter duct; 4469 fpm (VP = 1.25 ″wg); 390 cfm. Friction factor (Figure 5-18a) = 0.07. Then friction loss for "standard" duct would be 0.07 × 1.25 = 0.0875 ″wg per foot of duct. Special construction: 10′ duct lengths, welded and polished longitudinal seam. From Figure 5-22, consider as medium smooth, 5″ duct is closest. Factor = 0.85. Actual friction = 0.85 × 0.0875 = 0.074 ″wg per foot.

Table 5-2. Details of Operation

No.	Hood No.	VS-Print	Minimum Exhaust, cfm
1. Vibrating shake out 4′ × 6′ grate	1	110, 112	9600
2. Shake out hopper	2	112	960
3. Vibrating pan feeder 24″	3	112, 306	700
4. Incline sand belt 24″ × 28′	5	306	700
5. Magnetic pulley			
6. Tramp iron box			
7. Bucket elevator 24″ × 30″ casing,	7a(lower) 7b(upper)	305	500
8. Vibrating screen 34′	8	307	1200
9. Sand bin 600 ft^3 capacity, 18″ × 20′ opening	9	304	500
10. Waste sand box 44″ × 54″, clearance 6″	10	903 (V = 150 fpm)	1225
11. Sand weigh hopper	11	108	900
12. Sand muller 6′ dia.	12		
13. Wet dust collector			

DIMENSIONS

No. of Branch or Main	CFM Required, Minimum	Straight Run, ft	Elbows	Entries
1-A	9,600	13	1-90°	1-30°
2-B	960	3	1-60°	1-30°
3-B	700	4	1-90° + 1-60°	1-30°
B-A	1,600	18	2-90°	1-30°
A-C	11,260	34		
5-D	700	7	1-30° + 1-60°	1-30°
7a-D	250	5		
D-C	1,200	14	1-90° + 1-60°	1-30°
C-E	12,360	6.5		
8-F	1,200	11	2-90°	
9-F	500	4	1-90° + 6-60°	1-30°
F-G	1,700	5		
7b-G	250	15	1-60°	1-30°
G-E	2,200	6	1-60°	1-30°
E-H	14,660	3.5		
10-J	1,225	6	1-45°	
12-J	900	2.5	1-30°	1-30°
J-H	2,125	8	1-90° + 1-60°	1-30°
H-K	16,785	9	2-45°	
13	16,785			
14-L	16,785	20		

In addition to Figure 5-22, various manufacturers have developed data for special duct materials, nonmetallic flexible ducts, and metallic flexible ducts. These values should be obtained from the manufacturer.

5.12 FRICTION LOSS FOR NON-CIRCULAR DUCTS

Round ducts are preferred for industrial exhaust systems

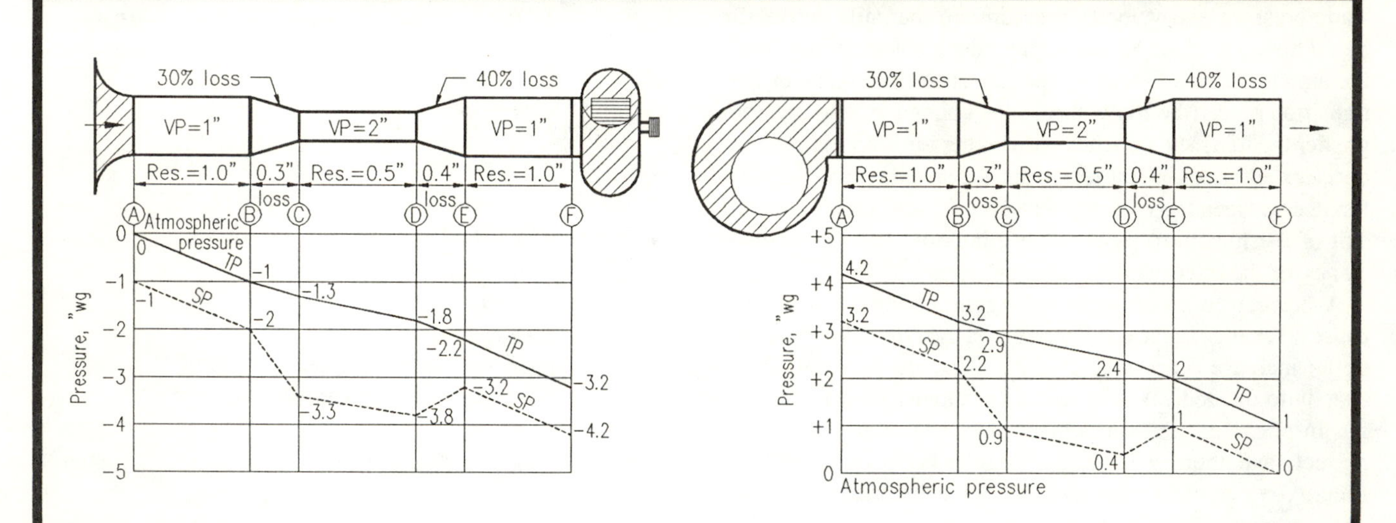

EXAMPLE — DUCT LOCATED ON SUCTION SIDE OF FAN. Velocity changes as indicated. Since all the ductwork is on the suction side of the fan, SP at the fan inlet (point F) is equal to VP at the fan inlet plus the total duct resistance up to that point. This equals −4.2 SP since static pressure on the suction side of the fan is always negative. The duct system is the same as was used in Example 2 and therefore has the same overall resistance of 3.2. If it is again assumed that the inlet and discharge of the fan are equal areas, the total pressure across the fan will be the same as in Example 1 and, in each case, the fan will deliver the same air horsepower when handling equal volumes of air.

Static pressure conversion between B and C follows contraction formula (Figure 5-19). There must be sufficient SP at B to furnish the additonal VP required at C. In addition, the energy transfer between these two points is accompanied by a loss of 0.3. Since SP at B = −2, SP at C = −2.0 + (−1.0) + (−0.3) = −3.3

Static pressure regain between D and E follows the regain formulae (Figure 5-19). If there were no loss in the transition piece, the difference of 1 in velocity pressure would be regained as static pressure at E, and SP at that point would be −2.8. However, the transition is only 60% efficient (0.4 loss) so the SP at E = −2.8 + (−0.4) = −3.2.

EXAMPLE — DUCT LOCATED ON DISCHARGE SIDE OF FAN. Velocity changes as indicated. The ductwork is located all on the discharge of side of the fan. Total pressure at the fan discharge (point A) is equal to the velocity pressure at the discharge end of the duct (point F) plus the accumulated resistances. These add up to 1.0 + 1.0 + 0.4 + 0.5 + 0.3 + 1.0 = 4.2.

Static pressure regain between D and E follows the regain formulae (Figure 5-19). If there were no energy loss in the transition piece, static pressure at D would be 0 because the difference in VP of 1 would show up as static pressure regain. However, the transition is only 60% efficient which means a loss of 0.4, so SP at point D = 0 + 0.4 = 0.4.

Conversion of static pressure into veclocity pressure between B anc C follows contraction formulae (Figure 5-19). There must be sufficient static pressure at B to furnish the additional velocity pressure required at C. In addition, transformation of energy between these two points is accompanied by a loss of 0.3. Since SP at C = 0.9, SP at B = 0.9 + 0.3 + 1.0 = 2.2. Since there is no ductwork on the suction side of the fan, total pressure against which the fan is operating is 4.2".

AMERICAN CONFERENCE OF GOVERNMENTAL INDUSTRIAL HYGIENISTS

EXPANSIONS AND CONTRACTIONS

DATE *1-88* FIGURE *5-7*

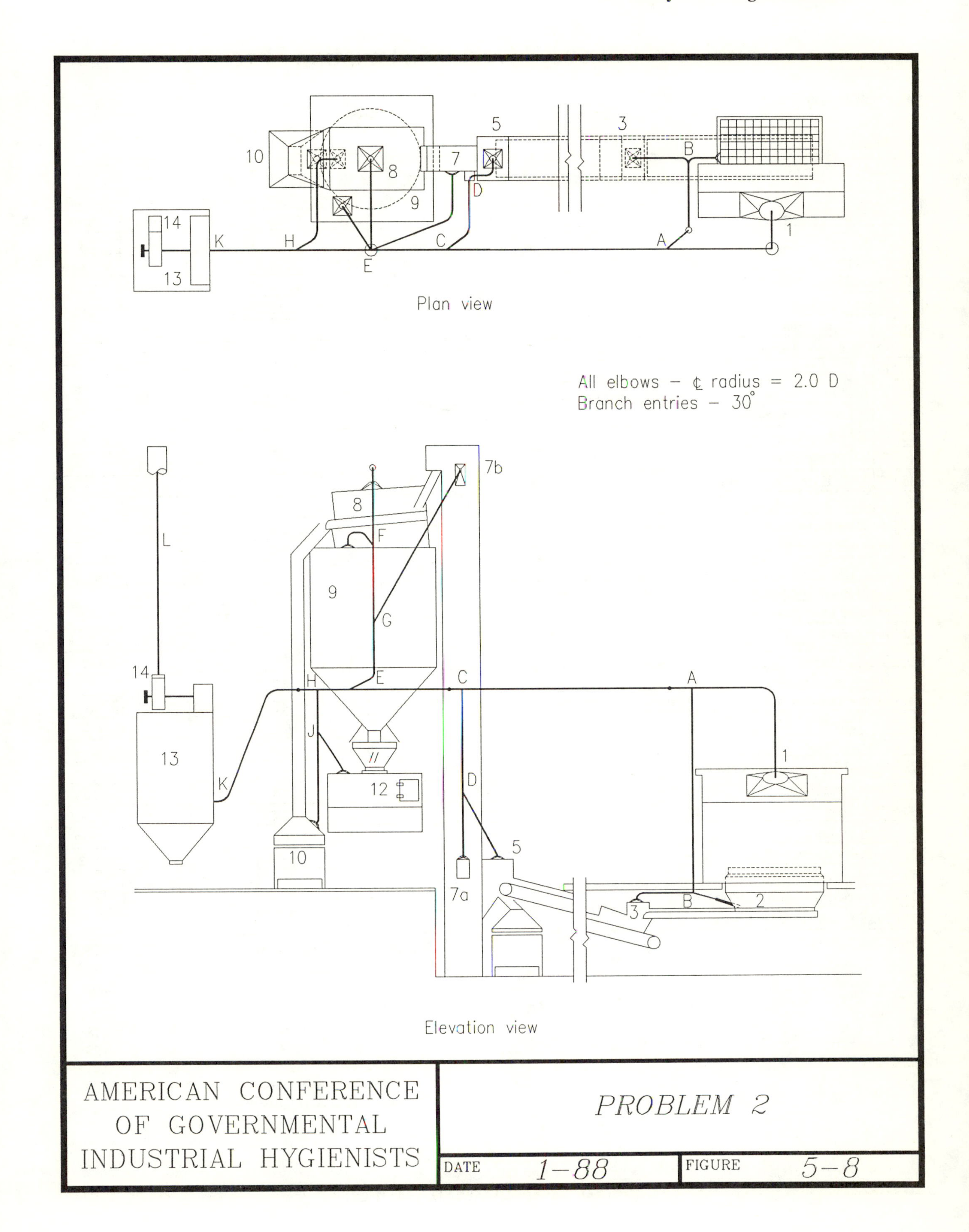

10
8
9
7
5
3
B
D
14
13
K
H
C
E
A
1
Plan view
All elbows – ℄ radius = 2.0 D
Branch entries – 30°
7b
8
F
L
9
G
14
H
E
C
A
J
13
K
12
D
1
5
10
7a
3
B
2
Elevation view
AMERICAN CONFERENCE
OF GOVERNMENTAL
INDUSTRIAL HYGIENISTS
PROBLEM 2
DATE 1–88
FIGURE 5–8

Velocity Pressure Method Calculation Sheet

Plant Name: PROBLEM 2
Location:
Department:
Elevation:
Temperature:
Factor:
Date: 4-88
Drawing No.: Fig. 5-8
Designer:

1			Duct Segment Identification			1-A	2-B	3-B	B-A	1-A	A-C
2			Volumetric Flowrate		cfm	9600	960	700	1743	9600	11429
3			Minimum Transport Velocity		fpm	3500	3500	3500	3500	3500	3500
4			Duct Diameter		inches	22	7	6	9	18	24
5			Duct area		sq. ft.	2.64	0.2673	0.1963	0.4418	1.767	3.1416
6			Actual Duct Velocity		fpm	3637	3592	3565	3945	5432	3638
7			Duct Velocity Pressure		"wg	0.82	0.80	0.79	0.97	1.84	0.82
8	H		Slot Area		sq. ft.	10				10	
9	O	S	Slot Velocity		fpm	960				960	
10	O	L	Slot Velocity Pressure		"wg	0.06				0.06	
11	D	O	Slot Loss Factor	Fig. 5-15 or Chap. 10		1.78				1.78	
12		T	Acceleration Factor		0 or 1	0				0	
13	S	S	Plenum loss per VP	Items 11 + 12		1.78				1.78	
14	U		Plenum SP	Items 10 x 13	"wg	0.11				0.11	
15	C		Duct Entry Loss Factor	Fig. 5-15 or Chap. 10		0.25	0.25	0.25		0.25	
16	T		Acceleration Factor		1 or 0	1	1	1		1	
17	I		Duct Entry Loss per VP	Items 15 + 16		1.25	1.25	1.25		1.25	
18	O		Duct Entry Loss	Items 7 x 17		1.03	1.00	0.99		2.30	
19	N		Other Loss		"wg				0.09 ③		
20			Hood Static Pressure	Items 14 + 18 + 19	"wg	1.03	1.00	0.99		2.30	
21			Straight Duct Length		feet	13	3	4	18	13	34
22			Friction Factor (H_f)	Fig. 5-18 or equation		0.0089	0.036	0.0436	0.0263	0.011	0.008
23			Friction Loss per VP	Items 21 x 22		0.115	0.11	0.17	0.47	0.14	0.27
24			No. of 90° Elbows			1	0.67	1.67	2	1	
25			Elbow Loss per VP	Item 24 x Loss Factor		0.27	0.18	0.45	0.54	0.27	
26			No. Entries				1	1	1		
27			Entry Loss per VP	Item 26 x Loss Factor			0.18	0.18	0.18		
28			Special Fittings Loss Factors								
29			Duct Loss per VP	Items 23 + 25 + 27 + 28		0.385	0.47	0.80	1.19	0.41	0.27
30			Duct Loss	Items 7 x 29	"wg	0.32	0.38	0.64	1.15	0.75	0.22
31			Duct SP Loss	Items 20 + 30	"wg	1.46	1.38	1.63	1.24	3.16	0.22
32			Cumulative Static Pressure		"wg	-1.46	-1.38	-1.63	-2.87	-3.16	-3.38
33			Governing Static Pressure		"wg	-2.87	-1.63		-3.16		
34			Corrected Volumetric Flowrate		cfm	④	1043 ①		1829		
35			Resultant Velocity Pressure		"wg		0.88 ②		1.66		

PERTINENT EQUATIONS:

$$Q_{corr} = Q_{design}\sqrt{\frac{SP_{gov.}}{SP_{duct}}}$$

$$H_f = 0.0307\frac{V^{0.533}}{Q^{0.612}} = \frac{0.4937}{(Q^{0.079})(D^{1.066})}$$

$$VP_r = \left[\frac{Q_1 + Q_2}{4005(A_1 + A_2)}\right]^2$$

$$FSP = SP_{outlet} - SP_{inlet} - VP_{inlet}$$

90° ROUND ELBOW LOSS FACTORS

C.L. R/D	Factor
Mitered	1.25
1.5	0.39
2.0	0.27
2.5	0.22

60° elbow = 2/3 loss
45° elbow = 1/2 loss

BRANCH ENTRY LOSS FACTORS

Angle	Factor
15°	0.09
30°	0.18
45°	0.28
60°	0.44
90°	1.00

FIGURE 5-9. BALANCED DESIGN METHOD

Remarks: ____

1	5-D	7a-D	5-D	D-C	C-E	8-F	9-F	F-G	7b-G	F-G	G-E	E-H	1
2	700	250	700	959	12419	1200	500	1700	250	1700	2086	14574	2
3	3500	3500	3500	3500	3500	3500	3500	3500	3500	3500	3500	3500	3
4	6	3.5	5.5	7	24	8	5	9	3.5	8	10	28	4
5	0.1963	0.0668	0.1650	0.2673	3.1416	0.3491	0.1364	0.4418	0.0668	0.3491	0.5454	4.276	5
6	3565	3742	4243	3588	3953	3438	3667	3848	3742	4870	3825	3408	6
7	0.79	0.87	1.12	0.80	0.97	0.74	0.84	0.92	0.87	1.48	0.91	0.72	7
8													8
9													9
10													10
11													11
12													12
13													13
14													14
15	0.25	1.0	0.25			0.5	0.25		1.0				15
16	1	1	1			1	1		1				16
17	1.25	2.0	1.25			1.5	1.25		2.0				17
18	0.99	1.74	1.40			1.11	1.05		1.74				18
19					0.15 ⑥			0.16		0.72			19
20	0.99	1.74	1.40			1.11	1.05		1.74				20
21	7	5	7	14	6.5	11	4	5	15	5	6	3.5	21
22	0.0436	0.084	0.0478	0.0361	0.0079	0.0307	0.0543	0.0264	0.084	0.03	0.023	0.0066	22
23	0.31	0.42	0.33	0.50	0.05	0.34	0.22	0.13	1.26	0.15	0.14	0.02	23
24	1		1	1.67		2	1.67		0.67		0.67		24
25	0.27		0.27	0.45		0.54	0.45		0.18		0.18		25
26	1		1	1			1		1		1		26
27	0.18		0.18	0.18			0.18		0.18		0.18		27
28													28
29	0.76	0.42	0.78	1.13	0.05	0.88	0.85	0.13	1.62	0.15	0.50	0.02	29
30	0.60	0.37	0.87	0.90	0.05	0.65	0.71	0.12	1.41	0.22	0.46	0.02	30
31	1.59	2.11	2.27	0.90	0.19	1.76	1.76	0.28	3.15	0.94	0.46	0.02	31
32	-1.59	-2.11	-2.27	-3.17	-3.58	-1.76	-1.76	-2.04	-3.15	-2.70	-3.61	-3.63	32
33	-2.11	-2.27		-3.38	-3.61			-3.15		-3.15			33
34	⑤	259		990	12488			⑦		1836			34
35		1.07		0.83	0.97	0.76	0.76			1.57			35

NOTES:

1. $\frac{SP_{3-B}}{SP_{2-B}} = 1.18$ (< 20%) so increase flowrate in 2-B; $Q_{2-B} = 960\sqrt{\frac{1.63}{1.38}} = 1043$

2. $VP_r = \left[\frac{1043 + 700}{4005(0.2673 + 0.1963)}\right]^2 = 0.88$ "wg

3. $VP_{B-A} - VP_r = 0.97 - 0.88 = 0.09$ "wg

4. $\frac{SP_{B-A}}{SP_{1-A}} = \frac{2.87}{1.46} = 1.97$ therefore, must resize 1-A to increase its SP loss

5. $\frac{SP_{7a-D}}{SP_{5-D}} = \frac{2.11}{1.59} = 1.33$ therefore resize 5-D

FIGURE 5-9. BALANCED DESIGN METHOD (continued)

Velocity Pressure Method Calculation Sheet

Plant Name: PROBLEM 2 (Contd.) Elevation: ______ Date: 4-88
Location: ______ Temperature: ______ Drawing No.: Fig. 5-8
Department: ______ Factor: ______ Designer: ______

1			Duct Segment Identification			10-J	12-J	10-J	J-H	J-H	H-K
2			Volumetric Flowrate		cfm	1225	900	1225	2137	2137	16747
3			Minimum Transport Velocity		fpm	3500	3500	3500	3500	3500	3500
4			Duct Diameter		inches	8	6	7	10	9	30
5			Duct area		sq. ft.	0.3491	0.1963	0.2673	0.5454	0.4418	4.909
6			Actual Duct Velocity		fpm	3509	4585	4583	3918	4837	3411
7			Duct Velocity Pressure		"wg	0.77	1.31	1.31	0.96	1.46	0.77
8	H		Slot Area		sq. ft.						
9	O	S	Slot Velocity		fpm						
10	O	L	Slot Velocity Pressure		"wg						
11	D	O	Slot Loss Factor	Fig. 5-15 or Chap. 10							
12		T	Acceleration Factor		0 or 1						
13	S	S	Plenum loss per VP	Items 11 + 12							
14	U		Plenum SP	Items 10 x 13	"wg						
15	C		Duct Entry Loss Factor	Fig. 5-15 or Chap. 10		0.25	0.25	0.25			
16	T		Acceleration Factor		1 or 0	1	1	1			
17	I		Duct Entry Loss per VP	Items 15 + 16		1.25	1.25	1.25			
18	O		Duct Entry Loss	Items 7 x 17		0.96	1.64	1.64			
19	N		Other Loss		"wg					0.14	
20			Hood Static Pressure	Items 14 + 18 + 19	"wg	0.96	1.64	1.64			
21			Straight Duct Length		feet	6	2.5	6	8	8	9
22			Friction Factor (H_f)	Fig. 5-18 or equation		0.0307	0.0427	0.0354	0.0231	0.0259	0.0061
23			Friction Loss per VP	Items 21 x 22		0.18	0.11	0.21	0.19	0.21	0.05
24			No. of 90° Elbows			0.5	0.33	0.5	1.67	1.67	1
25			Elbow Loss per VP	Item 24 x Loss Factor		0.14	0.09	0.14	0.45	0.45	0.27
26			No. Entries				1		1	1	
27			Entry Loss per VP	Item 26 x Loss Factor			0.18		0.18	0.18	
28			Special Fittings Loss Factors								
29			Duct Loss per VP	Items 23 + 25 + 27 + 28		0.32	0.38	0.35	0.82	0.84	0.32
30			Duct Loss	Items 7 x 29	"wg	0.25	0.50	0.46	0.79	1.23	0.23
31			Duct SP Loss	Items 20 + 30	"wg	1.21	2.14	2.10	0.79	1.37	0.23
32			Cumulative Static Pressure		"wg	-1.21	-2.14	-2.10	-2.93	-3.51	-3.86
33			Governing Static Pressure		"wg	-2.14		-2.14	-3.63	-3.63	
34			Corrected Volumetric Flowrate		cfm	(8)		1237	(9)	2173	
35			Resultant Velocity Pressure		"wg			1.32		0.80	

PERTINENT EQUATIONS:

$$Q_{corr} = Q_{design}\sqrt{\frac{SP_{gov.}}{SP_{duct}}}$$

$$H_f = 0.0307\frac{V^{0.533}}{Q^{0.612}} = \frac{0.4937}{(Q^{0.079})(D^{1.066})}$$

$$VP_r = \left[\frac{Q_1 + Q_2}{4005(A_1 + A_2)}\right]^2$$

$$FSP = SP_{outlet} - SP_{inlet} - VP_{inlet}$$

90° ROUND ELBOW LOSS FACTORS

C.L. R/D	Factor
Mitered	1.25
1.5	0.39
2.0	0.27
2.5	0.22

60° elbow = 2/3 loss
45° elbow = 1/2 loss

BRANCH ENTRY LOSS FACTORS

Angle	Factor
15°	0.09
30°	0.18
45°	0.28
60°	0.44
90°	1.00

FIGURE 5-9. BALANCED DESIGN METHOD (continued)

Remarks: ______________________________

1	13	14 in	14-L										1
2	16747	16747	16747										2
3													3
4		35.5	32										4
5	C	6.874	5.5850										5
6	O	2436	3000										6
7	L	0.37	0.56										7
8	L												8
9	E												9
10	C												10
11	T												11
12	O												12
13	R												13
14													14
15													15
16													16
17													17
18													18
19	4.5												19
20													20
21			20										21
22			0.0057										22
23			0.11										23
24													24
25													25
26													26
27													27
28													28
29			0.11										29
30			0.06										30
31	4.50		0.06										31
32	-8.36		0.06										32
33													33
34													34
35													35

NOTES:

6. Correction for acceleration = 0.97 - 0.82 = 0.15 "wg

7. $\frac{SP_{7b-G}}{SP_{F-G}} = \frac{3.15}{2.04} = 1.54$ therefore, resize F-G

8. $\frac{SP_{12-J}}{SP_{10-J}} = \frac{2.14}{1.21} = 1.77$ therefore, resize 10-J

9. $\frac{SP_{E-H}}{SP_{J-H}} = \frac{3.63}{2.93} = 1.24$ therefore, resize J-H

10. Fan SP = $SP_o - SP_i - VP_i$ = +0.06 - (-8.36) - 0.37 = 8.05 "wg

Fan TP = Fan SP + VP_o = 8.05 + 0.56 = 8.61 "wg

FIGURE 5-9. BALANCED DESIGN METHOD (continued)

Velocity Pressure Method Calculation Sheet

Plant Name: PROBLEM 2 Elevation: ______ Date: 4-88

Location: ______ Temperature: ______ Drawing No.: Fig. 5-8

Department: ______ Factor: ______ Designer: ______

No.			Item	Reference	Units						
1			Duct Segment Identification			1-A	2-B	3-B	B-A	A-C	5-D
2			Volumetric Flowrate		cfm	9600	960	700	1660	11260	700
3			Minimum Transport Velocity		fpm	3500	3500	3500	3500	3500	3500
4			Duct Diameter		inches	22	7	6	9	24	6
5			Duct area		sq. ft.	2.64	0.2673	0.1963	0.4418	3.1416	0.1963
6			Actual Duct Velocity		fpm	3637	3592	3565	3757	3584	3565
7			Duct Velocity Pressure		"wg	0.82	0.80	0.79	0.88	0.80	0.79
8	H		Slot Area		sq. ft.	10					
9	O	S	Slot Velocity		fpm	960					
10	O	L	Slot Velocity Pressure		"wg	0.06					
11	D	O	Slot Loss Factor	Fig. 5-15 or Chap. 10		1.78					
12		T	Acceleration Factor		0 or 1	0					
13	S	S	Plenum loss per VP	Items 11 + 12		1.78					
14	U		Plenum SP	Items 10 x 13	"wg	0.11					
15	C		Duct Entry Loss Factor	Fig. 5-15 or Chap. 10		0.25	0.25	0.25			0.25
16	T		Acceleration Factor		1 or 0	1	1	1			1
17	I		Duct Entry Loss per VP	Items 15 + 16		1.25	1.25	1.25			1.25
18	O		Duct Entry Loss	Items 7 x 17		1.03	1.00	0.99			0.99
19	N		Other Loss		"wg						
20			Hood Static Pressure	Items 14 + 18 + 19	"wg	1.03	1.00	0.99			0.99
21			Straight Duct Length		feet	13	3	4	18	34	7
22			Friction Factor (H_f)	Fig. 5-18 or equation		0.0089	0.036	0.0436	0.0264	0.0080	0.0436
23			Friction Loss per VP	Items 21 x 22		0.115	0.11	0.17	0.48	0.27	0.31
24			No. of 90° Elbows			1	0.67	1.67	2		1
25			Elbow Loss per VP	Item 24 x Loss Factor		0.27	0.18	0.45	0.54		0.27
26			No. Entries				1	1	1		1
27			Entry Loss per VP	Item 26 x Loss Factor			0.18	0.18	0.18		0.18
28			Special Fittings Loss Factors								
29			Duct Loss per VP	Items 23 + 25 + 27 + 28		0.385	0.47	0.80	1.20	0.27	0.76
30			Duct Loss	Items 7 x 29	"wg	0.32	0.38	0.64	1.06	0.22	0.60
31			Duct SP Loss	Items 20 + 30	"wg	1.46	1.38	1.63	1.06	0.22	1.59
32			Cumulative Static Pressure		"wg	-1.46	-1.38	-1.63	-2.69	-2.91	-1.59
33			Governing Static Pressure		"wg	-2.69	-1.63				-2.11
34			Corrected Volumetric Flowrate		cfm	GATE	GATE				GATE
35			Resultant Velocity Pressure		"wg						

PERTINENT EQUATIONS:

$$Q_{corr} = Q_{design}\sqrt{\frac{SP_{gov.}}{SP_{duct}}}$$

$$H_f = 0.0307\frac{V^{0.533}}{Q^{0.612}} = \frac{0.4937}{(Q^{0.079})(D^{1.066})}$$

$$VP_r = \left[\frac{Q_1 + Q_2}{4005(A_1 + A_2)}\right]^2$$

$$FSP = SP_{outlet} - SP_{inlet} - VP_{inlet}$$

90° ROUND ELBOW LOSS FACTORS

C.L. R/D	Factor
Mitered	1.25
1.5	0.39
2.0	0.27
2.5	0.22

60° elbow = 2/3 loss
45° elbow = 1/2 loss

BRANCH ENTRY LOSS FACTORS

Angle	Factor
15°	0.09
30°	0.18
45°	0.28
60°	0.44
90°	1.00

FIGURE 5-10. BLAST GATE METHOD

Remarks: ______________________________

1	7a-D	D-C	C-E	8-F	9-F	F-G	7b-G	G-E	E-H	10-J	12-J	J-H	1
2	250	950	12210	1200	500	1700	250	1950	14160	1225	900	2125	2
3	3500	3500	3500	3500	3500	3500	3500	3500	3500	3500	3500	3500	3
4	3.5	7	24	8	5	9	3.5	10	26	8	6	10	4
5	0.0668	0.2673	3.1416	0.3491	0.1364	0.4418	0.0668	0.5454	3.687	0.3491	0.1963	0.5454	5
6	3742	3554	3887	3438	3667	3848	3742	3575	3841	3509	4585	3896	6
7	0.87	0.79	0.94	0.74	0.84	0.92	0.87	0.80	0.92	0.77	1.31	0.95	7
8													8
9													9
10													10
11													11
12													12
13													13
14													14
15	1.0			0.5	0.25		1.0			0.25	0.25		15
16	1			1	1		1			1	1		16
17	2.0			1.5	1.25		2.0			1.25	1.25		17
18	1.74			1.11	1.05		1.74			0.96	1.64		18
19			0.14			0.16							19
20	1.74			1.11	1.05		1.74			0.96	1.64		20
21	5	14	6.5	11	4	5	15	6	3.5	6	2.5	8	21
22	0.084	0.0361	0.0079	0.0307	0.0543	0.0264	0.084	0.023	0.0072	0.0307	0.0427	0.0232	22
23	0.42	0.51	0.05	0.34	0.22	0.13	1.26	0.14	0.025	0.18	0.11	0.19	23
24		1.67		2	1.67		0.67	0.67		0.5	0.33	1.67	24
25		0.45		0.54	0.45		0.18	0.18		0.14	0.09	1.45	25
26		1			1		1	1			1	1	26
27		0.18			0.18		0.18	0.18			0.18	0.18	27
28													28
29	0.42	1.14	0.05	0.88	0.85	0.13	1.62	0.50	0.025	0.32	0.38	0.82	29
30	0.37	0.90	0.05	0.65	0.71	0.12	1.41	0.40	0.02	0.25	0.50	0.78	30
31	2.11	0.90	0.19	1.76	1.76	0.28	3.15	0.40	0.02	1.21	2.14	0.78	31
32	-2.11	-3.09	-3.28	-1.76	-1.76	-2.04	-3.15	-3.55	-3.57	-1.21	-2.14	-2.92	32
33			-3.55			-3.15				-2.14		-3.57	33
34			GATE			GATE				GATE		GATE	34
35				0.76	0.76								35

NOTES:

FIGURE 5-10. BLAST GATE METHOD (continued).

Velocity Pressure Method Calculation Sheet

Plant Name: PROBLEM 2 (Contd.) Elevation: ______ Date: 4-88
Location: ______ Temperature: ______ Drawing No.: Fig. 5-8
Department: ______ Factor: ______ Designer: ______

			Item	Reference	Units						
1			Duct Segment Identification			H-K	13	14 in	14-L		
2			Volumetric Flowrate		cfm	16285	16285	16285	16285		
3			Minimum Transport Velocity		fpm	3500					
4			Duct Diameter		inches	28		35.5	32		
5			Duct area		sq. ft.	4.276	C	6.874	5.585		
6			Actual Duct Velocity		fpm	3808	O	2369	2916		
7			Duct Velocity Pressure		"wg	0.90	L	0.35	0.53		
8	H		Slot Area		sq. ft.		L				
9	O	S	Slot Velocity		fpm		E				
10	O	L	Slot Velocity Pressure		"wg		C				
11	D	O	Slot Loss Factor	Fig. 5-15 or Chap. 10			T				
12		T	Acceleration Factor		0 or 1		O				
13	S	S	Plenum loss per VP	Items 11 + 12			R				
14	U		Plenum SP	Items 10 x 13	"wg						
15	C		Duct Entry Loss Factor	Fig. 5-15 or Chap. 10							
16	T		Acceleration Factor		1 or 0						
17	I		Duct Entry Loss per VP	Items 15 + 16							
18	O		Duct Entry Loss	Items 7 x 17							
19	N		Other Loss		"wg		4.5				
20			Hood Static Pressure	Items 14 + 18 + 19	"wg						
21			Straight Duct Length		feet	9			20		
22			Friction Factor (H_f)	Fig. 5-18 or equation		0.0066			0.0057		
23			Friction Loss per VP	Items 21 x 22		0.06			0.11		
24			No. of 90° Elbows			1					
25			Elbow Loss per VP	Item 24 x Loss Factor		0.27					
26			No. Entries								
27			Entry Loss per VP	Item 26 x Loss Factor							
28			Special Fittings Loss Factors								
29			Duct Loss per VP	Items 23 + 25 + 27 + 28		0.33			0.11		
30			Duct Loss	Items 7 x 29	"wg	0.30			0.06		
31			Duct SP Loss	Items 20 + 30	"wg	0.30	4.50		0.06		
32			Cumulative Static Pressure		"wg	-3.87	-8.37		0.06		
33			Governing Static Pressure		"wg						
34			Corrected Volumetric Flowrate		cfm						
35			Resultant Velocity Pressure		"wg						

PERTINENT EQUATIONS:

$$Q_{corr} = Q_{design}\sqrt{\frac{SP_{gov.}}{SP_{duct}}} \qquad H_f = 0.0307\frac{V^{0.533}}{Q^{0.612}} = \frac{0.4937}{(Q^{0.079})(D^{1.066})}$$

$$VP_r = \left[\frac{Q_1 + Q_2}{4005(A_1 + A_2)}\right]^2 \qquad FSP = SP_{outlet} - SP_{inlet} - VP_{inlet}$$

90° ROUND ELBOW LOSS FACTORS

C.L. R/D	Factor
Mitered	1.25
1.5	0.39
2.0	0.27
2.5	0.22

60° elbow = 2/3 loss
45° elbow = 1/2 loss

BRANCH ENTRY LOSS FACTORS

Angle	Factor
15°	0.09
30°	0.18
45°	0.28
60°	0.44
90°	1.00

CALCULATIONS:
FSP = +0.06 - (-8.37) - 0.35 = 8.08 "wg

FIGURE 5-10. BLAST GATE METHOD (continued).

Equivalent Foot Method
Calculation Sheet

Committee on Industrial Ventilation
P O Box 16153
Lansing, Michigan 48901

Plant name PROBLEM 2	Refer to Figure 5-8	Factor	Remarks
Location	Elevation		
Department	Temperature		

1	2	3	4	5	6	7	8	8	9	10	11	12	13	14	15	16	17	18	19
									From Fig. 5-20	Col. 7 plus Col. 9	From Fig. 5-21	Col.10xCol.11 / 100	From Col. 6 & Table 5-4	From Fig. 5-15	1.00 plus Col. 14	Col. 13 times Col. 15	Col. 12 plus Col. 16	At junction	
			Air volume CFM			Length of duct in feet					Friction loss, inches of water					Pressure, inches of water			
No. of br. or main	Dia. duct in in.	Area duct sq. ft.	in branch	in main	Vel. in FPM	straight runs	Number of elbows	Number of entries	equiv. length	total length	per 100	of run	one VP	entry loss(VP)	hood suct(VP)	hood suct.	static press.	gov. SP	corrected CFM
1-A	Slots	10.0	9600		960								0.06	1.78	--	0.11			
	22	2.64			3635	13	90°		36	49	0.70	0.34	0.82	0.25	1.25	1.02	1.47		
2-B	7	0.267	960		3590	3	60°	30°	12	15	2.8	0.42	0.80	0.25	1.25	1.00	1.42	1.65	1030
3-B	6	0.196	700		3570	4	90°+60°	30°	16	20	3.3	0.66	0.79	0.25	1.25	0.99	1.65		
													$Q = 960 \sqrt{\frac{1.65}{1.42}} = 1030$ cfm						
	Branches 2 and 3 are within 15% of balance at "B". Air volume in 2 will increase																		
B-A	9	0.442	1730		3920	18	2-90°	30°	32	50	2.5	1.25+0.09	0.96				2.99	3.22	1790
	Velocity pressure in B-A is higher than Branch 2 or 3. See text.																		
	$VP_{2-3} = \left[\frac{(Q_2 + Q_3)}{4005(A_2 + A_3)}\right]^2 = \left[\frac{(1030 + 700)}{4005(0.267 + 0.196)}\right]^2 = 0.87$ Increase = 0.96 - 0.87 = 0.09"																		
	At jct A, branch B-A requires 2.99" SP; branch 1-A requires 1.47". Refigure 1-A.																		
1-A	Slots	10.0	9600	960									0.06	1.78	--	0.11			
	18	1.77			5420	13	90°		28	41	2.0	0.82	1.83	0.25	1.25	2.29	3.22		
	Branches 1-A and B-A are within 8% of balance at "A". Air volume in B-A will increase:																		
	$Q = 1730 \sqrt{\frac{3.22}{2.99}} = 1790$ cfm																		
A-C	24	3.14		11390	3625	34				34	0.62	0.21					3.43		
7a-D	5	0.136	500		3670	5				5	4.5	0.23	0.84	1.0	2.0	1.68	1.91		
5-D	6	0.196	700		3570	7	90°	30°	12	19	3.3	0.63	0.79	0.25	1.25	0.99	1.62	1.91	760
	Branches 5 and 7a are within 18% of balance at "D". Air volume will increase: $Q = 700 \sqrt{\frac{1.91}{1.62}} = 760$ cmf																		
D-C	8	0.349	1260		3610	14	90°+60°	30°	24	38	2.5	0.95					2.86	3.43	1380
	Branches D-C and A-C are within 20% of balance at jct. "C". Air volume in D-C will increase: $Q = 1260 \sqrt{\frac{3.43}{2.86}} = 1380$ cfm																		
C-E	26	3.69		12770	3460	6.5				6.5	0.60	0.04					3.47		
8-F	8	0.349	1200		3440	11	2-90°		20	31	2.2	0.68	0.74	0.50	1.50	1.11	1.79	1.86	OK*
9-F	5	0.136	500		3670	4	90°+60°	30°	14	18	4.5	0.81	0.84	0.25	1.25	1.05	1.86		
	*Actually Q will increase slightly and duct velocity in branch 8 will be $\frac{1225}{0.349} = 3500$ fpm																		
F-G	9	0.442	1725		3900	5				5	2.4	0.12+0.16	0.95				2.14		
	Velocity pressure in F-G is higher than branch 8 or 9. See text.																		
	$VP_{8-9} = \frac{(Q_8 + Q_9)^2}{[4005(A_8 + A_9)]^2} = \frac{(1725)^2}{(4005 \times 0.485)^2} = 0.79$". Increase = 0.95 - 0.79 = 0.16".																		
7b-G	5	0.136	500		3670	15	60°	30°	8	23	4.5	1.04	0.84	1.0	2.0	1.68	2.72	2.82	OK
	At jct G, branch F-G requires 2.15" SP, branch 7b-G requires 2.72" SP. Refigures F-G.																		
F-G	8	0.349	1725		4940	5				5	4.5	0.23+0.73	1.52				2.82		
	Velocity pressure in F-G is higher than branch 8 or 9. Increase = 1.52" - VP_{8-9} = 1.52 - 0.79 = 0.73.																		
G-E	10	0.545	2225		4090	6	60°	30°	19	25	2.3	0.58	1.04				3.40	3.47	OK
E-H	28	4.28		14995	3500	3.5				3.5	0.50	0.02					3.49		
10-J	8	0.349	1225		3500	6	45°		5	11	2.3	0.25	0.76	0.25	1.25	0.95	1.20		
12-J	6	0.196	900		4590	2.5	30°	30°	7	9.5	5.5	0.52	1.31	0.25	1.25	1.64	2.16		
	At jct "J" branch 12-J requires 2.16" SP, branch 10-J requires 1.20". Refigure 10-J.																		
10-J	7	0.267	1225		4590	6	45°		4.5	10.5	4.4	0.46	1.26	0.25	1.25	1.58	2.04	2.16	1260
	Branches 10 and 12 are within % of balance. Air volume in branch 10 will increase: $Q = 1225 \sqrt{\frac{2.16}{2.04}} = 1260$ cfm																		
J-H	10	0.545	2160		3960	8	90°+60°	30°	32	40	2.1	0.84	0.98				3.00	3.49	2320
At jct "H" branch J-H and main E-H are within 17% of balance. Air volume in J-H will increase: $Q = 2160 \sqrt{\frac{3.49}{3.00}} = 2330$ cfm																			
H-K	30	4.91		17325	3530	9	2-45°		51	60	0.45	0.27					3.76		
13	Collector resistance, including fan inlet.											4.5					8.26		
14	Backward curved fan. Inlet = 35½ I.D. Outlet = 5.15 ft^2.																		
Inlet	35½	6.87		17325	2520								0.40						
L	32	5.59		17325	3100	20				20	0.33	0.07					8.33		
	Fan SP = $SP_o - SP_i - VP_i$ = 0.07" - (-8.26") - 0.40" = 8.33" - 0.40" = 7.93"																		

Calculated fan characteristics for standard conditions		Corrected for temperature and elevation	
Capacity( ) 17325 CFM	Fan, type and size	RPM CFM	Motor
Fan TP in WG		BHP TP	V Belt
Fan SP 8"	RPM BHP	SP	

FIGURE 5-11. BALANCED SYSTEM DESIGN.

Equivalent Foot Method
Calculation Sheet

Committee on Industrial Ventilation
P O Box 16153
Lansing, Michigan 48901

Plant name PROBLEM 2 · Location · Department FOUNDRY · Refer to Figure 5-8 · Elevation · Temperature · Factor · Remarks

1	2	3	4	5	6	7	8		9	10	11	12	13	14	15	16	17	18	19
									From Fig. 5-20	Col. 7 plus Col. 9	From Fig. 5-21	Col. 10 x Col. 11 / 100	From Col. 6 & Table 5-4	From Fig. 5-15	1.00 plus Col. 14	Col. 13 times Col. 15	Col. 12 plus Col. 16	At junction	
			Air volume CFM			Length of duct in feet					Friction loss, inches of water					Pressure, inches of water			
No. of br. or main	Dia. duct in in.	Area duct sq. ft.	in branch	in main	Vel. in FPM	straight runs	Number of elbows	Number of entries	equiv. length	total length	per 100	of run	one VP	entry loss (VP)	hood suct (VP)	hood suct.	static press.	gov. SP	corrected CFM
1-A	Slots	10.0	9600		960								0.06	1.78	--	0.11			
	22	2.64			3635	13	90°		36	49	0.70	0.34	0.82	0.25	1.25	1.02	1.47	2.75	GATE
2-B	7	0.267	960		3590	3	60°	30°	12	15	2.8	0.42	0.80	0.25	1.25	1.00	1.42	1.65	GATE
3-B	6	0.196	700		3570	4	90°+60°	30°	16	20	3.3	0.66	0.79	0.25	1.25	0.99	1.65		GATE
B-A	9	0.442	1660		3750	18	2-90°	30°	32	50	2.2	1.10	0.88				2.75		
A-C	24	3.14		11260	3580	34				34	0.61	0.21					2.96		
7a-D	5	0.136	500		3670	5				5	4.5	0.23	0.84	1.0	2.0	1.68	1.91		GATE
5-D	6	0.196	700		3570	7	90°	30°	12	19	3.3	0.63	0.79	0.25	1.25	0.99	1.62	1.91	GATE
D-C	8	0.349	1200		3440														
	Increase air volume in 5-D to give minimum transport velocity in D-C.																		
D-C	8	0.349	1225		3500	14	90°+60°	30°	24	38	2.3	0.87					2.78	2.96	
C-E	24	3.14		12485	3980	6.5				6.5	0.75	0.05+0.20	0.99				3.21	3.30	
	Velocity pressure in C-E is higher than Branch A-C or D-C. See text.																		
	$VP_{A-C} = \frac{(Q_A + Q_C)^2}{[4005(A_A + A_C)]^2} = \frac{(11260 + 1200)^2}{(4005 \times 3.489)^2} = 0.79$ Increase = 0.99 - 0.79 = 0.20".																		
8-F	8	0.349	1200		3440														
	Increase Q to 1225 to maintain minimum transport velocity.																		
8-F	8	0.349	1225		3500	11	2-90°		20	31	2.3	0.71	0.76	0.50	1.50	1.14	1.85	1.86	GATE
9-F	5	0.136	500		3670	4	90°+60°	30°	14	18	4.5	0.81	0.84	0.25	1.25	1.05	1.86		GATE
F-G	9	0.442	1725		3900	5				5	2.4	0.12+0.16	0.95				2.14	2.72	
	Velocity pressure in F-G is higher than branch 8 and 9. See text.																		
	$VP_{8-9} = \frac{(Q_8 + Q_9)^2}{[(4005(A_8 + A_9)]^2} = \frac{(1725)^2}{(4005 \times 0.485)^2} = 0.79''$. Increase = 0.95 - 0.79 = 0.16".																		
7b-G	5	0.136	500		3670	15	60°	30°	8	23	4.5	1.04	0.84	1.0	2.0	1.68	2.72		GATE
G-E	10	0.545	2225		4090	6	60°	30°	19	25	2.3	0.58	1.04				3.30		
E-H	26	3.69		14710	3990	3.5				3.5	0.69	0.02	0.99				3.32		
10-J	8	0.349	1225		3500	6	45°		5	11	2.3	0.25	0.76	0.25	1.25	0.95	1.20	2.16	GATE
12-J	6	0.196	900		4590	2.5	30°	30°	7	9.5	5.5	0.52	1.31	0.25	1.25	1.64	2.16		GATE
J-H	10	0.545	2125		3900	8	90°+60°	30°	32	40	2.0	0.80					2.96	3.32	
H-K	30	4.91		16835	3430														
	Increase air volume at selected hoods to maintain minimum transport velocity																		
H-K	30	4.91		17200	3500	9	2-45°		51	60	0.45	0.27					3.59		
13	Collector resistance, including fan inlet											4.5					8.09		
14	Backward curved fan, inlet = 35½" I.D., outlet = 5.15 ft².																		
Inlet	35½	6.87		17200	2500								0.39						
L	32	5.59		17200	3080	20				20	0.33	0.07					8.16		
	All blast gates to be locked in place after system is balanced.																		
	Fan SP = $SP_o - SP_i - VP_i$																		
	= 0.07" - (-8.09") - 0.39"																		
	=8.16" - 0.39" = 7.77"																		

Calculated fan characteristics for standard conditions		Corrected for temperature and elevation	
Capacity ( ) 17,200 CFM	Fan, type and size	RPM / CFM	Motor
Fan TP in WG		BHP / TP	V Belt
Fan SP 7 3/4"	RPM / BHP	SP	

FIGURE 5-12. BLAST GATE DESIGN.

because of a more uniform air velocity to resist settling of material and an ability to withstand higher static pressure. At times, however, the designer must use other duct shapes.

Rectangular duct friction can be calculated by using the friction charts, Figures 18 (a & b), and 21 (a & b), in conjunction with Table 5-6 to obtain rectangular equivalents for circular ducts on the basis of equal friction loss. It should be noted that, on this basis, the area of the rectangular duct will be larger than the equivalent round duct; consequently, the actual air velocity in the duct will be reduced. Therefore, it is necessary to use care to maintain minimum transport velocities.

Occasionally the designer will find it necessary to estimate the air handling ability of odd-shaped ducts. The following procedure[5.2] will be helpful in determining the frictional pressure losses for such ducts.

1. Find duct cross sectional area, ft^2 .... A
2. Find wetted perimeter, ft. .... P
3. Calculate hydraulic radius, ft. .... R ($R = A/P$)
4. Convert R to inches .... r ($r = 12R$)
5. Calculate equivalent diameter, in ...... D ($D = 4r$)
6. Use friction chart based on the equivalent diameter and volumetric flow rate (or velocity).

5.13 CORRECTIONS FOR TEMPERATURE, MOISTURE, AND ALTITUDE

Fan tables and exhaust volumetric flow rate requirements assume standard atmospheric conditions. These assumptions fix the air density at 0.075 lbm/ft^3. Where appreciable variation occurs, the change in air density must be considered. Factors for different temperatures and elevations are listed in Table 5-7. As noted, correction for temperatures between 40 F and 100 F and/or elevations between −1,000 feet and +1,000 feet are seldom required with the permissible variations in usual exhaust system design.

The density variation equations of Chapter 1 (Section 1.4) demonstrate that if temperature increases or absolute pressure decreases, the density will decrease but the *mass* flow rate at the hood(s) must remain the same; therefore, the volumetric flow rate must change if density changes. It is helpful to remember that a fan connected to a given system will exhaust the same VOLUMETRIC flow rate regardless of air density. The mass of air moved, however, will be a function of the density.

5.13.1 Variable Temperature and/or Different Altitude: Consider an exhaust system at sea level where 5,000 cfm of air at 70 F is drawn into a hood. The air is then heated to 600 F and the density of the air leaving the heater becomes 0.0375 lbm/cu ft. The volumetric flow rate downstream of the heater would be 10,000 actual cubic feet per minute (acfm) at the new density of 0.0375 lbm/cu ft. This is true because the 50% decrease in density *must* correspond to a twofold increase in the volumetric flow rate since the mass flow rate has remained constant.

If this temperature effect is ignored and a fan selected for 5,000 cfm is placed in the system, the hood flow rate will be well below that required to maintain contaminant control. The exact operating point of such a system would have to be recalculated based upon the operating point of the incorrectly sized fan.

5.13.2 Elevated Moisture: When air temperature is under 100 F, no correction for humidity is necessary. When air temperature exceeds 100 F and moisture content is greater than 0.02 lbs H_2O per pound of dry air, correction is required to determine fan operating RPM and power. Correction factors may be read from the psychrometric charts such as those illustrated in Figures 5-23 through 5-26.

5.13.3 Psychrometric Principles: The properties of moist air are presented on the psychrometric chart at a single pressure. These parameters define the physical properties of an air/water vapor mixture. *The actual gas volumetric flow rate and the density of the gas stream at the inlet of the fan must be known in order to select the fan.* The psychrometric chart provides the information required to calculate changes in the volumetric flow rate and density of the gas as it passes through the various exhaust system components. These properties are:

- *Dry Bulb Temperature* is the temperature observed with an ordinary thermometer. Expressed in degrees Fahrenheit, it may be read directly on the chart and is indicated on the bottom horizontal scale.
- *Wet Bulb Temperature* is the temperature at which liquid or solid water, by evaporating into air, can bring the air to saturation adiabatically at the same temperature. Expressed in degrees Fahrenheit, it is read directly at the intersection of the constant enthalpy line with the 100% saturation curve.
- *Dew Point Temperature* is that temperature at which the air in an air/vapor mixture becomes saturated with water vapor and any further reduction of dry bulb temperature causes the water vapor to condense or deposit as drops of water. Expressed in degrees Fahrenheit, it is read directly at the intersection of the saturation curve with a horizontal line representing constant moisture content.
- *Percent Saturation* curves reflect the mass of moisture actually in the air as a percentage of the total amount possible at the various dry bulb and moisture content combinations. Expressed in percent, it may be read directly from the curved lines on the chart.
- *Density Factor* is a dimensionless quantity which expresses the ratio of the actual density of the mixture to the density of standard air (0.075 lbm/ft^3). The lines representing density factor typically do not appear on low temperature psychrometric charts when relative humidity or percent saturation curves are presented. A method of calculating the density of the gas defined by a point on the chart (when

density factor curves are not presented) is discussed in section 5.13.4.

- *Moisture Content*, or weight of water vapor, is the amount of water which has been evaporated into the air. In ordinary air it is very low pressure steam and has been evaporated into the air at a temperature corresponding to the boiling point of water at that low pressure. Moisture content is expressed in grains of water vapor per pound of dry air (7,000 grains = one pound) or pounds of water vapor per pound of dry air and is read directly from a vertical axis.
- *Enthalpy (Total Heat)* as shown on the psychrometric chart is the sum of the heat required to raise the temperature of a pound of air from 0 F to the dry bulb temperature, plus the heat required to raise the temperature of the water contained in that pound of air from 32 F to the dew point temperature, plus the latent heat of vaporization, plus the heat required to superheat the vapor in a pound of air from the dew point temperature to the dry bulb temperature. Expressed in BTUs per pound of dry air, it is shown by following the diagonal wet bulb temperature lines.
- *Humid Volume* is the volume occupied by the air/vapor mixture per pound of dry air and is expressed in cubic feet of mixture per pound of dry air. It is most important to understand the dimensions of this parameter and realize that *the reciprocal of humid volume is not density*. Humid volume is the parameter used most frequently in determining volumetric flow rate changes within a system as a result of mixing gases of different properties or when evaporative cooling occurs within the system.

5.13.4 Density Determination: When the quality of an air/vapor mixture is determined by a point on a psychrometric chart having a family of density factor curves, all that must be done to determine the actual density of the gas at the pressure reference for which the chart is drawn is to multiply the density factor taken from the chart by the density of standard air (0.075 lbm/ft^3). Should relative humidity curves be presented on the chart in lieu of density factor curves, information available through dimensional analysis must be used to determine the actual density of the mixture. This can be done quite easily as follows: The summation of one pound of dry air plus the mass of the moisture contained within that pound of dry air divided by the humid volume will result in the actual density of the mixture.

$$\rho = \frac{1 + W}{HV} \quad [5.8]$$

where:

ρ = density of the mix (lbm/ft^3)
W = moisture content (lbm H_2O/lbm dry air)
HV = humid volume (ft^3 mix/lbm dry air)

5.13.5 Hood Volumetric Flow Rate Changes With Density: If the density of the air entering a hood is different from standard density due to changes in elevation, ambient pressure, temperature, or moisture, the volumetric flow rate through the hood should be changed to keep the mass flow rate the same as for standard air. This can be accomplished by multiplying the hood flow rate required for standard air by the ratio of the density of standard air to the actual ambient density.

The following example illustrates the effect of elevated moisture and temperature and a method of calculation:

EXAMPLE

Given: The exit volumetric flow rate from a 60″ × 24′ dryer is 16,000 scfm plus removed moisture. Exhaust air temperature is 500 F. The dryer delivers 60 tons/hr of dried material with capacity to remove 5% moisture. Required suction at the dryer hood is −2.0 ″wg; minimum conveying velocity must be 4000 fpm.

It has been determined that the air pollution control system should include a cyclone for dry product recovery and a high energy wet collector. These devices have the following operating characteristics:

- *Cyclone:* Pressure loss is 4.5 ″wg at rated volumetric flow rate of 35,000 scfm. The pressure loss across any cyclone varies directly with any change in density and as the square of any change in volumetric flow rate from the rated conditions.

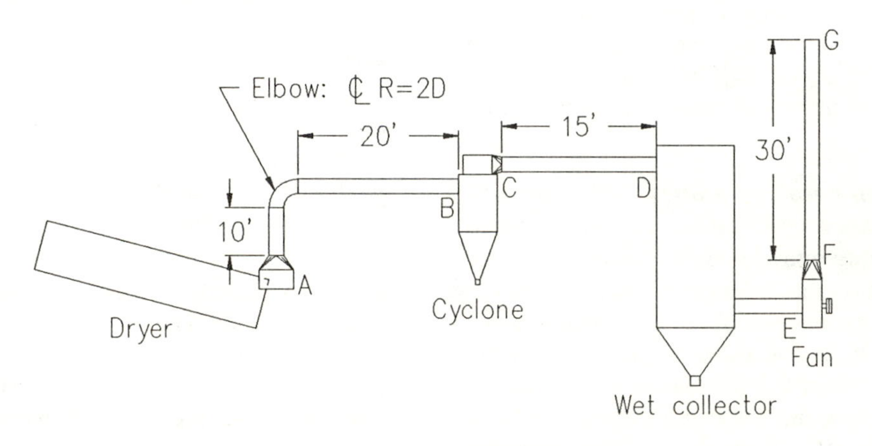

FIGURE 5-13. SYSTEM LAYOUT.

TABLE 5-3. Fan Rating Table

Fan size Nº 34										*Inlet diameter = 34"*									*Max. safe rpm = 1700*			
CFM VOLUME	20" SP		22" SP		24" SP		26" SP		28" SP		30" SP		32" SP		34" SP		36" SP		38" SP		40" SP	
	RPM	BHP	RPM	BHP	RPM	BHP	RPM	BHP	RPM	BHP	RPM	BHP	RPM	BHP	RPM	BHP	RPM	BHP	RPM	BHP	RPM	BHP
14688	1171	73.3	1225	81.4	1277	89.8	1326	98.3	1374	107	1421	116	1466	125	1510	134	1552	143	1594	153	1634	162
16524	1181	81.8	1234	90.2	1286	98.8	1335	107	1382	116	1428	126	1472	135	1516	145	1557	155	1600	165	1639	175
18360	1191	90.2	1244	99.5	1294	108	1344	118	1391	127	1437	137	1481	146	1524	157	1565	167	1606	178	1645	188
20196	1204	99.9	1256	109	1306	119	1354	129	1400	139	1446	149	1490	160	1532	170	1574	181	1615	191	1654	202
22032	1217	110	1268	120	1318	130	1366	141	1412	151	1456	162	1499	173	1542	184	1584	196	1624	207	1663	218
23868	1230	120	1282	131	1331	142	1378	154	1424	165	1468	176	1511	187	1553	199	1594	211	1633	223	1672	235
25704	1245	131	1296	143	1345	155	1391	167	1437	179	1481	191	1524	203	1565	215	1606	227	1645	239	1683	252
27540	1261	143	1311	156	1359	168	1406	181	1450	193	1494	206	1537	219	1578	232	1618	245	1658	258	1695	271
29376	1277	156	1327	169	1374	182	1421	196	1465	209	1508	222	1550	236	1591	249	1631	263	1670	277		
31212	1295	170	1344	184	1391	197	1436	211	1480	225	1523	239	1564	253	1605	268	1644	282	1683	297		
33048	1313	184	1361	198	1407	213	1453	228	1496	242	1538	257	1580	272	1620	287	1659	302	1697	317		
34884	1331	198	1379	214	1425	229	1469	245	1513	260	1555	276	1595	291	1635	307	1674	323				

- *High Energy Wet Scrubber:* The manufacturer has determined that a pressure loss of 20 "wg is required in order to meet existing air pollution regulations and has sized the collector accordingly. The humidifying efficiency of the wet collector is 90%.
- *Fan:* A size #34 "XYZ" fan with the performance shown in Table 5-3 has been recommended:

REQUIRED

Size the duct and select fan RPM and motor size.

SOLUTION

Step 1

Find the actual gas flow rate that must be exhausted from the dryer. This volumetric flow rate must include both the air used for drying and the water, as vapor, which has been removed from the product. Since it is actual flow rate, it must be corrected from standard air conditions to reflect the actual moisture, temperature and pressures which exist in the duct.

Step 1A

Find the amount (weight) of water vapor exhausted.

Dryer Discharge = 60 tons/hr of dried material (given)

Since the dryer has capacity to remove 5% moisture the Dryer Discharge = (0.95) (dryer feed).

60 tons/hr dried material = (0.95) (dryer feed)

$$\text{dryer feed} = \frac{60 \text{ tons/hr}}{0.95} = 63.2 \text{ tons/hr}$$

Moisture removed = (feed rate) − (discharge rate)
= 63.2 tons/hr − 60 tons/hr
= 6400 lbs/hr or 106.6 lbs/min

Step 1B

Find the amount (weight) of dry air exhausted.

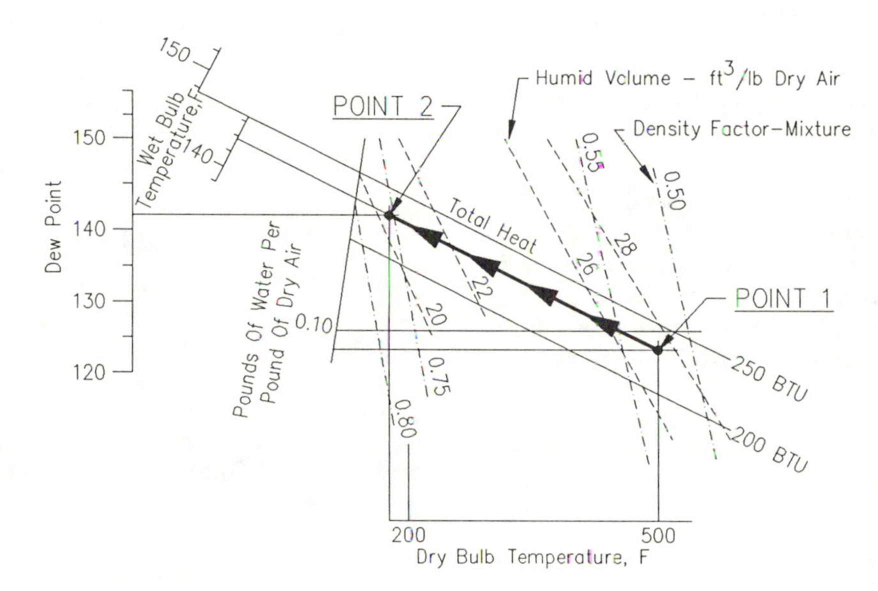

FIGURE 5-14. PSYCHROMETRIC CHART FOR HUMID AIR. (See Figures 5-23 through 5-26.)

Dry air exhausted = 16,000 scfm at 70 F and 29.92" Hg (0.075 lbs/ft^3density)

Exhaust rate, lbs/min = (16,000 scfm)(0.075 lbs/ft^3)

= 1200 lbs/min dry air

Step 1C

Knowing the water to dry air ratio and the temperature of the mixture, it is possible to determine other quantities of the air to water mixture. This can be accomplished by the use of psychrometric charts (see Figures 5-23 to 5-26) which are most useful tools when working with humid air.

$$W = \frac{106.6 \text{ lbs } H_2O}{1200 \text{ lbs dry air/min}}$$

$$= 0.089 \text{ lbs } H_2O\text{/lb dry air}$$

Dry bulb temperature = 500 F (given)

The intersection of the 500 F dry bulb temperature line and the 0.089 lbs H_2O/lb dry air line can be located on the psychrometric chart (see Figure 5-14). This point, Point #1, completely defines the quality of the air and water mixture. Other data relative to this specific mixture can be read as follows:

Dew Point Temperature	122 F
Wet Bulb Temperature	145 F
Humid volume, ft^3 of mix/lb of dry air	27.5 ft^3/lb dry air
Enthalpy, BTU/lb of dry air	235 BTU/lb dry air
Density factor, df	0.53

Step 1D

Find actual gas volumetric flow rate, (acfm).

Exhaust volumetric flow rate, acfm = (humid volume)(weight of dry air/min). Humid volume, HV, was found in Step 1C as 27.5 ft^3/lb. Weight of dry air/min was found in Step 1B as 1200 lb/min. Exhaust volumetric flow rate = (27.5 ft^3/lb)(1200 lb/min) = 33,000 acfm.

Step 2

Size the duct. Minimum conveying velocity of 4000 fpm was given. Suction at the dryer exit of −2.0"wg corresponds to hood suction.

The duct area equals the actual flow rate divided by the minimum duct velocity, or A = 33,000/4,000 = 8.25 ft^2. A 38" diameter duct with a cross-sectional area of 7.876 ft^2 should be chosen as this is the largest size available with an area smaller than calculated. Then the actual duct velocity would be 33,000 actual ft^3/min ÷ 7.876 ft^2 = 4,190 fpm.

Step 2A

The velocity pressure in the duct cannot be found using the equation VP = $(V/4005)^2$, as this equation is for standard air only. The actual velocity pressure in the duct is given by

$$VP_{actual} = (df)(VP_{std})$$

where:

df = density factor

As the density factor was determined in Step 1C, the actual velocity pressure in the duct will be

$$VP = (df)(VP_{std}) = (0.53)(1.09 \text{ "wg}) = 0.58 \text{ "wg}$$

Step 3

Calculate the pressure loss from A to B and determine static pressure at Point B.

The data from Figures 5-16 through 5-19 can be used directly. The static pressure loss through the duct can be found by multiplying the length of duct by the friction factor, adding the elbow loss factor, and multiplying the result by the duct velocity pressure:

$$\text{SP loss} = [(0.0045)(30) + 0.27][0.58]$$

$$= (0.405)(0.58) = 0.23 \text{ "wg}$$

Then the static pressure at the inlet to the cyclone should be −2.23 "wg (hood suction plus friction and fitting losses).

Step 4

The pressure loss of the cyclone is provided by the manufacturer. In this example, the cyclone pressure loss is 4.5 "wg at a rated flow of 35,000 scfm. The pressure loss through a cyclone, as with duct, varies as the square of the change in volumetric flow rate and directly with change in density.

Therefore, the actual loss through the cyclone would be

$$(4.5)\left[\frac{33{,}000}{35{,}000}\right]^2(0.53) = -2.12 \text{ "wg}$$

and the static pressure at the cyclone outlet would be − 4.35 "wg.

Step 5

The calculation from Point C to D is the same as from A to B in Step 3. Thus, the static pressure at the wet collector inlet would be

$$-4.35 - (0.0045)(15)(0.58) = -4.39 \text{ "wg}$$

NOTE: Information for Steps 6 and 7 which involve calculation of changes in volumetric flow rate, density, etc., across the wet collector should be provided by the equipment manufacturer.

Step 6

An important characteristic of wet collectors is their ability to humidify a gas stream. The humidification process is generally assumed to be adiabatic (without gain or loss of heat to the surroundings). Therefore, water vapor is added to the mixture but the enthalpy, expressed in BTU/lb dry air, remains unchanged. During the process of humidification, the point on the psychrometric chart that defines the quality of the mixture moves to the left, along a line of constant enthalpy, toward saturation.

All wet collectors do not have the same ability to humidify. If a collector is capable of taking an air stream to complete adiabatic saturation, it is said to have a humidifying efficiency of 100%. The humidifying efficiency of a given device may be expressed by either of the following equations:

$$\eta_n = \frac{t_i - t_o}{t_i - t_s} \times 100 \qquad [5.9]$$

where:

η_n = humidifying efficiency, %
t_i = dry bulb temperature at collector inlet, F
t_o = dry bulb temperature at collector outlet, F
t_s = adiabatic saturation temperature, F

or

$$\eta_n = \frac{W_i - W_o}{W_i - W_s} \times 100 \qquad [5.10]$$

where:

W_i = moisture content in lb H_2O/lb dry air at inlet
W_o = moisture content in lb H_2O/lb dry air at outlet
W_s = moisture content in lb H_2O/lb dry air at adiabatic saturation conditions

Step 6A

Find the quality of the air to water mixture at Point 2, the collector outlet.

Humidifying Efficiency = 90% (given). Dry Bulb Temperature at Collector Inlet = 500 F (given). Adiabatic saturation temperature = 145 F from inspection of Psychrometric Chart.

$$90 = \frac{(500 - t_o)}{(500 - 145)} \times 100$$

$$t_o = 180 \text{ F}$$

Then the air leaving the collector will have a dry bulb temperature of 180 F and an enthalpy of 235 BTU/lb of dry air as the humidifying process does not change the total heat or enthalpy.

The point of intersection of 180 F dry bulb and 235 BTU/lb dry air on the psychrometric chart defines the quality of the air leaving the collector and allows other data to be read from the chart as follows:

Dew Point Temperature	143 F
Wet Bulb Temperature	145 F
Humid Volume, ft^3/lb dry air	20.5 ft^3/lb dry air
Enthalpy, BTU/lb dry air	235 BTU/lb dry air
Density factor, df	0.76

Step 7

What is the exhaust volumetric flow rate in acfm and the density factor at the collector outlet?

Step 7A

Exhaust volumetric flow rate = (humid volume)(weight of dry air/min). Humid Volume from Step 6 is 20.5 ft^3/lb dry air. Weight of dry air/min from Step 1B is 1200 lbs/min. Volumetric flow rate = (20.5 ft^3/lb)(1200 lbs/min) = 24,600 acfm.

As the wet collector loss was stated to be 20 "wg, the static pressure at the wet collector outlet would be −24.39 "wg.

Step 7B

On low pressure exhaust systems, where the negative pressure at the fan inlet is less than 20 "wg, the effect of the negative pressure is usually ignored. However, as the pressures decrease, or the magnitude of negative pressures increases, it is understood that gases expand to occupy a larger volume. Unless this larger volume is anticipated and the fan sized to handle the larger volumetric flow rate, it will have the effect of reducing the amount of air that is pulled into the hood at the beginning of the system. From the characteristic equation for the ideal gas laws, PQ = wRT (where w = the mass flow rate in lbm/min), the pressure volumetric flow rate relationship is

$$P_1 Q_1 = P_2 Q_2 \text{ or } \frac{P_1}{P_2} = \frac{Q_2}{Q_1}$$

Up to this point the air has been considered to be at standard atmospheric pressure which is 14.7 PSI A, 29.92 "Hg or 407 "wg. The pressure within the duct at Point E is −24.4 "wg and minus or negative only in relation to the pressure outside the duct which is 407 "wg. Therefore, the absolute pressure within the duct is 407 "wg − 24.4 "wg = 382.6 "wg.

$$\frac{407}{382.6} = \frac{Q_2}{24{,}600 \text{ cfm}}$$

$$Q_2 = 26{,}170 \text{ acfm}$$

Step 7C

Pressure also affects the density of the air. From PQ = wRT the relationship

$$\frac{(w_1/Q_1)RT_1}{(w_2/Q_2)RT_2} = \frac{P_1}{P_2}$$

can be derived. Density factor is directly proportional to the density and the equation can be rewritten

$$\frac{P_1}{P_2} = \frac{df_1}{df_2}$$

Substitute

$$\frac{407}{382.6} = \frac{0.76}{df_2}$$

(df_1 was determined to be 0.76 in Step 6.)

$$df_2 = 0.71$$

Step 7D

The duct from the wet collector to the fan can now be sized. The flow rate leaving the wet collector was 26,170 acfm. As the fan selected has a 34-inch diameter inlet (area = 6.305 ft^2) it is logical to make the duct from the wet collector to the fan a 34-inch diameter. Thus, the velocity through the duct would be 26,170/6.305 = 4,151 fpm. The VP would be $(0.71)(4151 \div 4005)^2 = (0.71)(1.07) = 0.76$ "wg.

Step 7E

The duct pressure loss, based on 26,170 cfm and a 34-inch

diameter duct, would be (0.0052)(5)(0.76) = 0.02 "wg. Therefore, the SP at the fan inlet would be −24.41 "wg.

Step 8

Calculate the pressure loss from fan discharge F to stack discharge G. Since the air is now on the discharge side of the fan, the pressure is very near atmospheric. No pressure correction is needed. The flow rate and density factor are 24,600 acfm and 0.76, respectively.

Assuming that the fan discharge area is nearly the same as at the fan inlet, the same 34" diameter duct would result in a velocity of 3902 fpm. The velocity pressure would be $(0.76)(3902 \div 4005)^2 = 0.72$ "wg.

From Figure 5-18, the friction factor is 0.0052 and the frictional pressure loss for the 30′ high stack would be (0.0052)(30)(0.72) = 0.11 "wg. As the static pressure at the exit of the stack must be atmospheric, the static pressure at the fan exit will be positive.

Step 9

Determine actual fan static pressure.

$$\begin{aligned} \text{Actual FSP} &= SP_{out} - SP_{in} - VP_{in} \\ &= +0.11 - (-24.41) - 0.76 \\ &= 23.76 \text{ "wg} \end{aligned}$$

Step 10

Determine equivalent fan static pressure in order to enter fan rating table. Equivalent fan static pressure is determined by dividing the actual fan static pressure by the density factor at the fan inlet. This is necessary since fan rating tables are based on standard air.

$$\text{Equivalent FSP} = 23.76 \div 0.71 = 33.46 \text{ "wg.}$$

Step 11

Select fan from rating table using the equivalent fan SP and the fan inlet flow rate. Interpolating the fan rating table for 26,200 cfm at 33.5 "wg yields a fan speed of 1559 RPM at 217 BHP.

Step 12

Determine the actual required fan power. Since actual density is less than standard air density the actual required power is determined by multiplying by the density factor, or (217 BHP)(0.71) = 154 BHP. If a damper is installed in the duct to prevent overloading of the motor, at cold start the motor need only be a 200 HP (see Chapter 6).

5.14 AIR CLEANING EQUIPMENT

Dusts, fume, and toxic or corrosive gases should not be discharged to the atmosphere. Each exhaust system handling such materials should be provided with an adequate air cleaner as outlined in Chapter 4. As a rule the exhaust fan should be located on the clean air side of such equipment. An exception is in the use of cyclone cleaners where the hopper discharge is not tightly sealed and better performance is obtained by putting the fan ahead of the collector.

5.15 EVASE DISCHARGE

An evasé discharge is a gradual enlargement at the outlet of the exhaust system (see Figure 5-19). The purpose of the evasé is to reduce the air discharge velocity efficiently; thus, the available velocity pressure can be regained and credited to the exhaust system instead of being wasted. Practical considerations usually limit the construction of an evasé to approximately a 10° angle (5° side angle) and a discharge velocity of about 2,000 fpm (0.25 "wg velocity pressure) for normal exhaust systems. Further streamlining or lengthening the evasé yields diminishing returns.

It should be noted, however, that for optimum vertical dispersion of contaminated air many designers feel that the discharge velocity from the stack should not be less than 3,000-3,500 fpm. When these considerations prevail, the use of an evasé is questionable.

The following example indicates the application of the evasé fitting. It is not necessary to locate the evasé directly after the outlet of the fan. It should be noted that, depending upon the evasé location, the static pressure at the fan discharge may be below atmospheric, i.e., negative (−), as shown in this example.

EXAMPLE

	Duct No.	Dia.	Q	V	VP	SP
1	Fan Inlet	20	8300	3800	0.90	−7.27
2	Fan Discharge = 16.5 × 19.5			3765	0.88	
3	Round Duct Connection		20	3800	0.90	
4	Evasé Outlet		28	1940	0.23	0

To calculate the effect of the evasé, see Figure 5-19, where the Diameter Ratio, $D_4/D_3 = 28/20 = 1.4$ and Taper length L/D = 40/20 = 2.

R = 0.52 × 70% (since the evasé is within 5 diameters of the fan outlet)

SP_4 = 0 (since the end of the duct is at atmospheric pressure)

$$\begin{aligned} VP_3 = 0.9 \text{ as given } SP_3 &= SP_4 - R(VP_3) \\ &= 0 - (0.52)(0.70)(0.90") \\ &= -0.33 \text{ "wg} \end{aligned}$$

$$\begin{aligned} FSP &= SP_{outlet} - SP_{inlet} - VP_{inlet} \\ &= -0.33 - (-7.27) - 0.9 = 6.04 \text{ "wg} \end{aligned}$$

5.16 EXHAUST STACK OUTLETS

The type and location of exhaust stacks are important to permit good dispersion of contaminated air from exhaust systems even when an efficient air cleaner is used. Poor discharge conditions result in low level contamination which can re-enter the building due to wind effect (building turbulence), negative pressure, or the action of mechanical air supply systems. Figures 5-30 and 5-31 illustrate the principles of good stack design; Figure 5-32 shows the influence of building turbulence and stack height. Whenever there is doubt about the proper height and location of outlets, simple observations and tests with smoke candles will be helpful in determining the air flow pattern across the building roof.

5.17 AIR BLEED-INS

Bleed-ins are used at the ends of branch ducts to provide additional air volumetric flow rates to transport heavy material loads as in woodworking at saws and jointers or at the ends of a main duct to maintain minimum transport velocity when the system has been oversized deliberately to provide for future expansion. Some designers use bleed-ins also to introduce additional air to an exhaust system to reduce air temperature and to assist in balancing the system.

EXAMPLE

End cap bleed-in. Consider as an orifice, Figure 5-15; h_e = 1.78 VP.

1. Calculate SP for branch duct to junction (X).
2. Determine volumetric flow rate in main duct according to design or future capacity or determine Q bleed in directly from temperature or moisture considerations.
3. Q bleed-in = (Q main duct) − (Q branch)
4. SP bleed-in = SP branch as calculated
5. VP, bleed-in = $\dfrac{\text{SP bleed-in}}{h_e + 1\text{VP}} = \dfrac{\text{SP}}{(1.78 + 1.0)} = \dfrac{\text{SP}}{2.78}$
6. Velocity, bleed-in from VP and Fig. 5-16.
7. Area bleed-in = $\dfrac{\text{Q bleed-in}}{\text{V bleed-in}}$

5.18 OPTIMUM ECONOMIC VELOCITY

In systems which are intended to carry dust, a minimum conveying velocity is necessary to ensure that the dust will not settle in the duct. Also, when a system is installed in a quiet area, it may be necessary to keep velocities below some maximum to avoid excessive duct noise. When axial flow fans are used, duct velocities of 1,000 to 1,500 fpm are preferred. In a gas or vapor exhaust system installed in a typical factory environment where none of these restrictions apply, the velocity may be selected to yield the lowest annual operating cost.

To determine the optimum economic velocity, the system must first be designed at any assumed velocity and the total initial costs of duct material, fabrication, and installation estimated.[5.4]

This optimum economic velocity may range from under 2000 fpm to over 4000 fpm. Lengthy expected service periods and system operating times tend to lower the optimum while high interest rates and duct costs tend to raise the optimum. In general, a velocity of 2500 to 3000 fpm will not result in equivalent total annual costs much in excess of the true optimum.

REFERENCES

5.1 Air Movement and Control Association, Inc.: *AMCA Standard 210-74*. 30 West University Drive, Arlington Heights, IL 60004.

5.2 J.A. Constance: "Estimating Air Friction in Triangular Ducts," *Air Conditioning, Heating and Ventilating*, 60, 6, June 1963, pp. 85 86.

5.3 The Kirk and Blum Mfg. Co.: *Woodworking Plants*, pp. W-9, Cincinnati, OH.

5.4 J.R. Lynch, "Computer Design of Industrial Exhaust Systems," *Heating, Piping and Air Conditioning*, September 1968.

5.5 J.J. Loeffler: "Simplified Equations for HVAC Duct Friction Factors," *ASHRAE Journal*, January 1980, pp. 76-79.

5.6 D.K. Wright, Jr.: "A New Friction Chart for Round Ducts," *ASHVE Transactions*, Vol. 51, 1945, p. 303.

5.7 J.H. Clarke: "Air Flow Around Buildings," *Heating, Piping and Air Conditioning*, 39, 5, May 1967, pp. 145-154.

5.8 R.D. Madison and W.R. Elliot: "Friction Charts for Gases Including Correction for Temperature, Viscosity and Pipe Roughness," *Heating, Piping and Air Conditioning*, 18, 107, October 1946.

5.9 American Society of Heating, Refrigerating and Air Conditioning Engineers. *Heating, Ventilating, Air Conditioning Guide*, 37th ed., 1959. ASHRAE, Atlanta, GA.

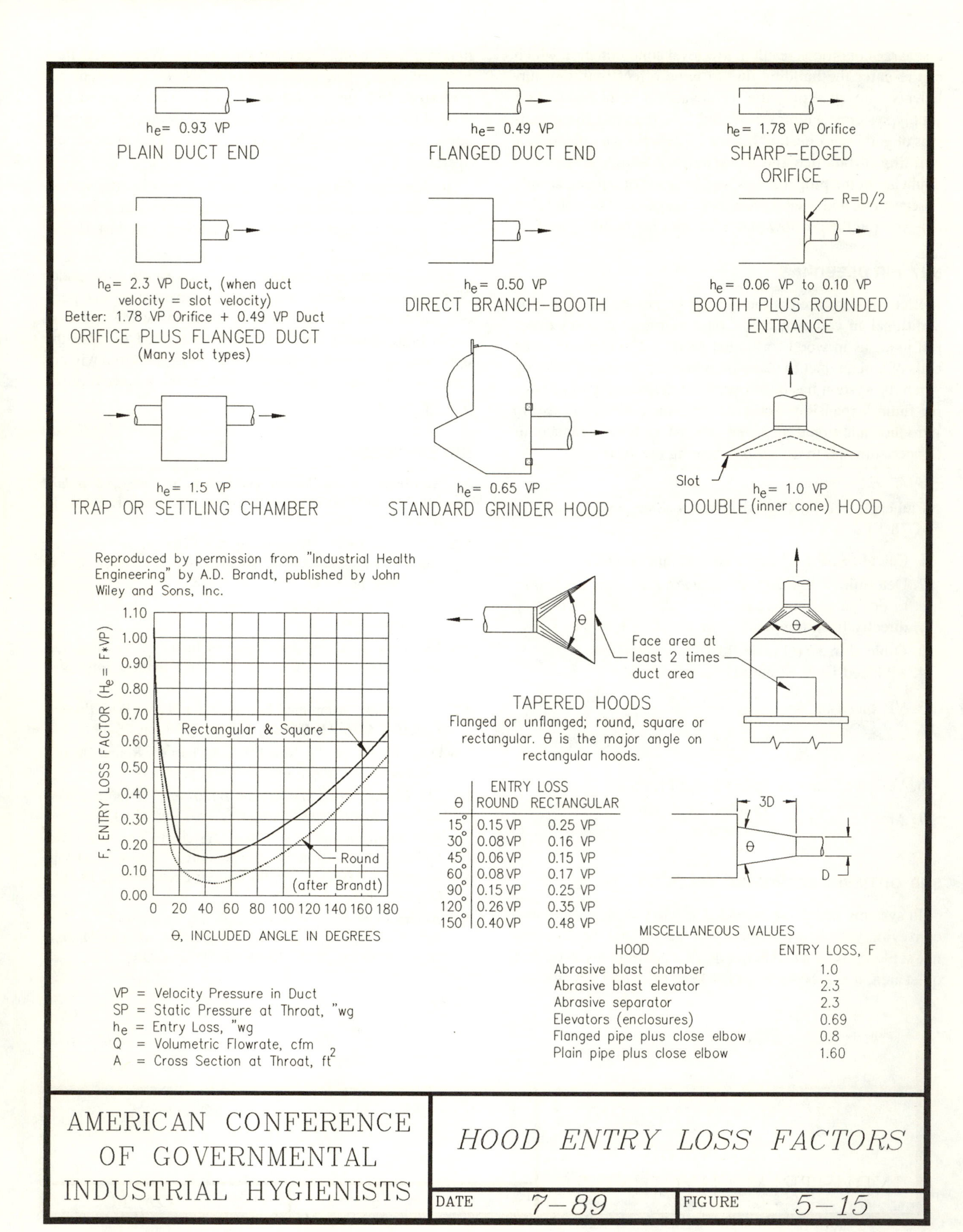

θ	ENTRY LOSS ROUND	ENTRY LOSS RECTANGULAR
15°	0.15 VP	0.25 VP
30°	0.08 VP	0.16 VP
45°	0.06 VP	0.15 VP
60°	0.08 VP	0.17 VP
90°	0.15 VP	0.25 VP
120°	0.26 VP	0.35 VP
150°	0.40 VP	0.48 VP

MISCELLANEOUS VALUES

HOOD	ENTRY LOSS, F
Abrasive blast chamber	1.0
Abrasive blast elevator	2.3
Abrasive separator	2.3
Elevators (enclosures)	0.69
Flanged pipe plus close elbow	0.8
Plain pipe plus close elbow	1.60

VP = Velocity Pressure in Duct
SP = Static Pressure at Throat, "wg
h_e = Entry Loss, "wg
Q = Volumetric Flowrate, cfm
A = Cross Section at Throat, ft^2

AMERICAN CONFERENCE OF GOVERNMENTAL INDUSTRIAL HYGIENISTS

HOOD ENTRY LOSS FACTORS

DATE 7-89 FIGURE 5-15

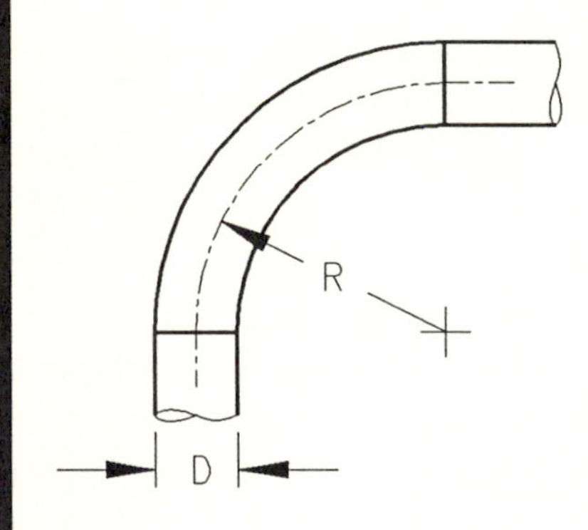

R/D	Loss Fraction of VP
2.75	0.26
2.50	0.22
2.25	0.26
2.00	0.27
1.75	0.32
1.50	0.39
1.25	0.55

ROUND ELBOWS

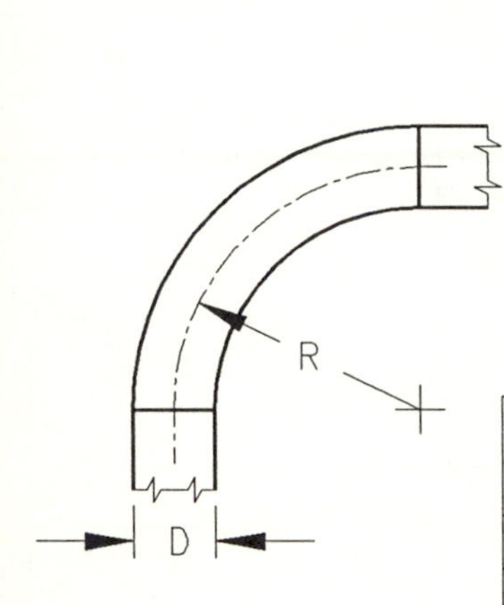

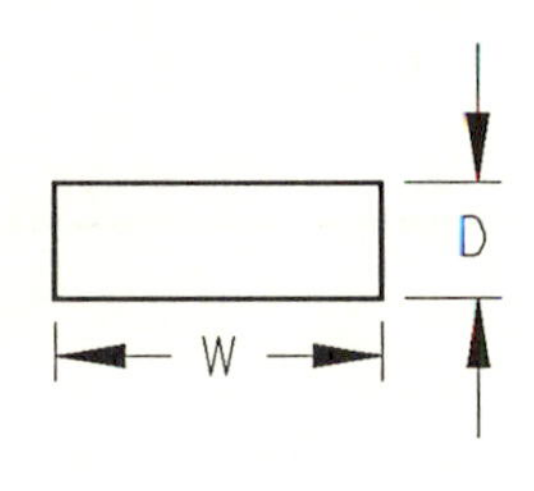

R/D	Aspect Ratio, W/D					
	0.25	0.5	1.0	2.0	3.0	4.0
0.0(Mitre)	1.50	1.32	1.15	1.04	0.92	0.86
0.5	1.36	1.21	1.05	0.95	0.84	0.79
1.0	0.45	0.28	0.21	0.21	0.20	0.19
1.5	0.28	0.18	0.13	0.13	0.12	0.12
2.0	0.24	0.15	0.11	0.11	0.10	0.10
3.0	0.24	0.15	0.11	0.11	0.10	0.10

SQUARE & RECTANGULAR ELBOWS

ELBOW LOSSES

AMERICAN CONFERENCE OF GOVERNMENTAL INDUSTRIAL HYGIENISTS

DUCT DESIGN DATA

DATE *1-88* FIGURE *5-16*

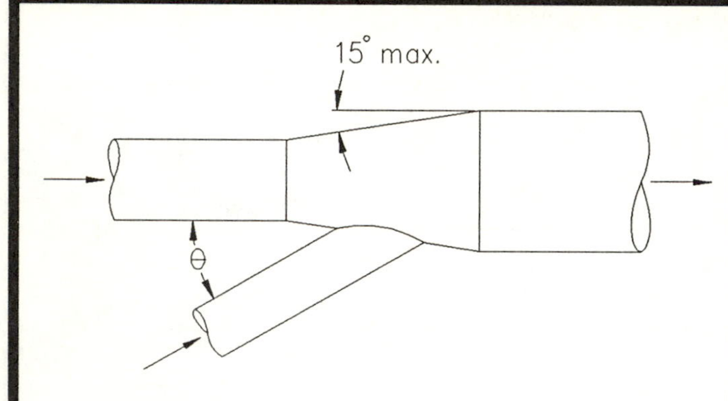

Angle θ Degrees	Loss Fraction of VP in Branch
10	0.06
15	0.09
20	0.12
25	0.15
30	0.18
35	0.21
40	0.25
45	0.28
50	0.32
60	0.44
90	1.00

Note: Branch entry loss assumed to occur in branch and is so calculated.

Do not include an enlargement regain calculation for branch entry enlargements.

BRANCH ENTRY LOSSES

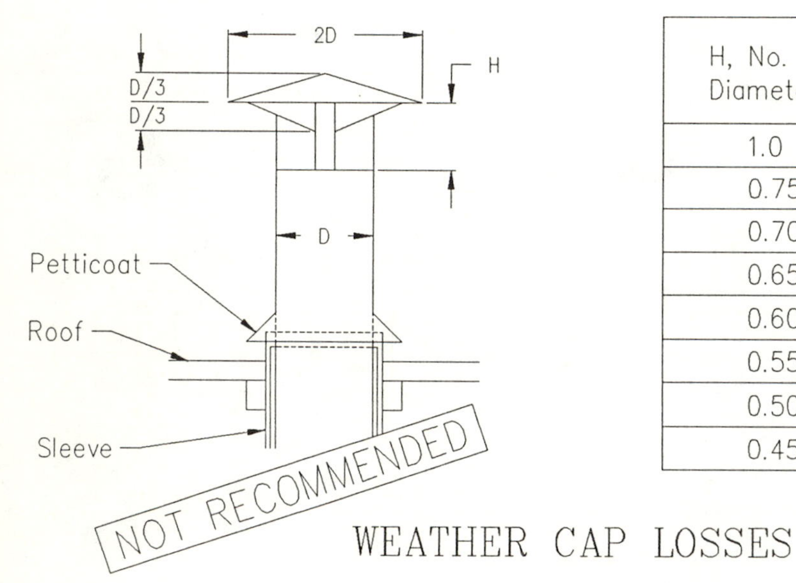

H, No. of Diameters	Loss Fraction of VP
1.0 D	0.10
0.75 D	0.18
0.70 D	0.22
0.65 D	0.30
0.60 D	0.41
0.55 D	0.56
0.50 D	0.73
0.45 D	1.0

WEATHER CAP LOSSES

See Fig. 5-30

AMERICAN CONFERENCE OF GOVERNMENTAL INDUSTRIAL HYGIENISTS

DUCT DESIGN DATA

DATE *1-88* FIGURE *5-17*

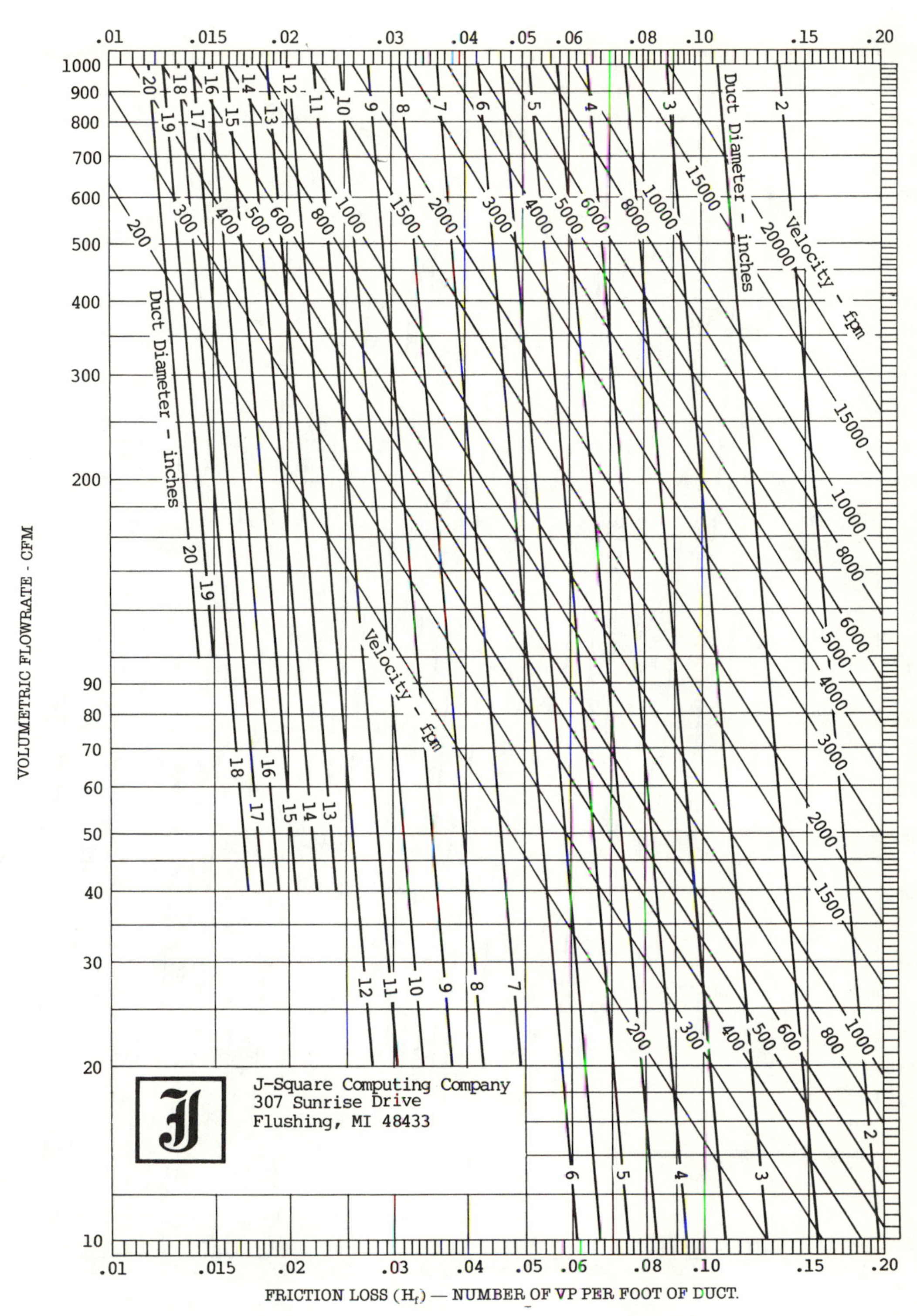

Based on Standard Air of 0.075 lb/ft³ density flowing through clean, round, galvanized metal ducts (equivalent sand grain roughness = 0.0005 ft).[5.5]

$$H_f = 0.0307 \frac{V^{0.533}}{Q^{0.612}} = \frac{0.4937}{Q^{0.079}D^{1.066}}$$

FIGURE 5-18a.

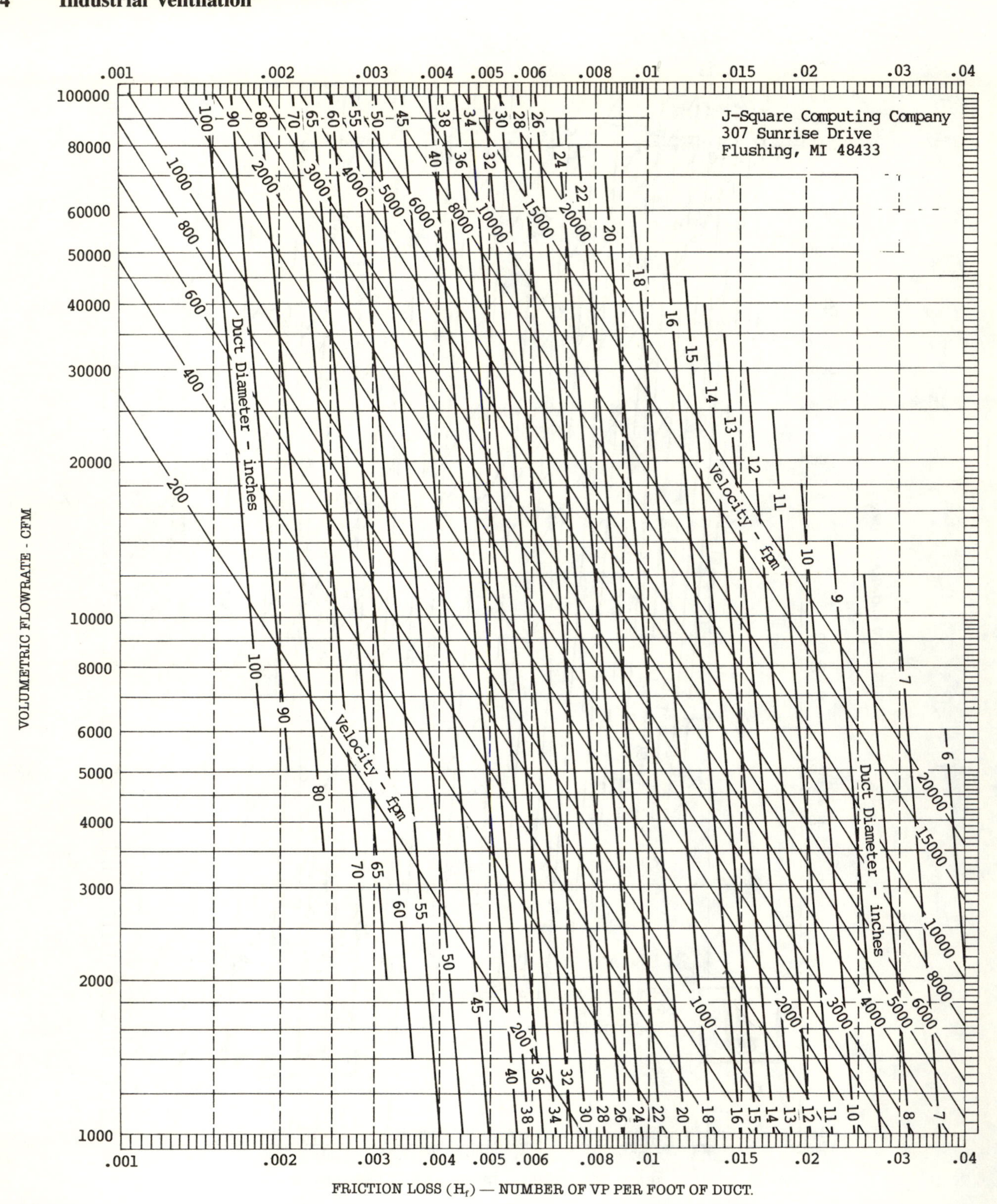

Based on Standard Air of 0.075 lb/ft^3 density flowing through clean, round, galvanized metal ducts (equivalent sand grain roughness = 0.0005 ft).(5.5)

$$H_f = 0.0307 \frac{V^{0.533}}{Q^{0.612}} = \frac{0.4937}{Q^{0.079}D^{1.066}}$$

FIGURE 5-18b.

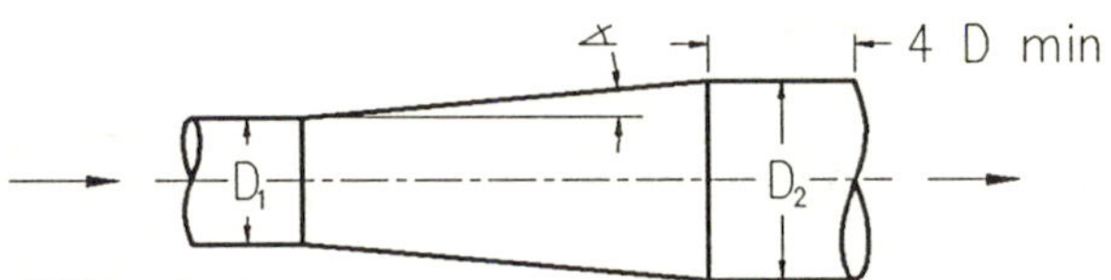

Within duct

Regain (R), fraction of VP difference					
Taper angle degrees	Diameter ratios D_2/D_1				
	1.25:1	1.5:1	1.75:1	2:1	2.5:1
3 1/2	0.92	0.88	0.84	0.81	0.75
5	0.88	0.84	0.80	0.76	0.68
10	0.85	0.76	0.70	0.63	0.53
15	0.83	0.70	0.62	0.55	0.43
20	0.81	0.67	0.57	0.48	0.43
25	0.80	0.65	0.53	0.44	0.28
30	0.79	0.63	0.51	0.41	0.25
Abrupt 90	0.77	0.62	0.50	0.40	0.25

Where: $SP_2 = SP_1 + R(VP_1 - VP_2)$

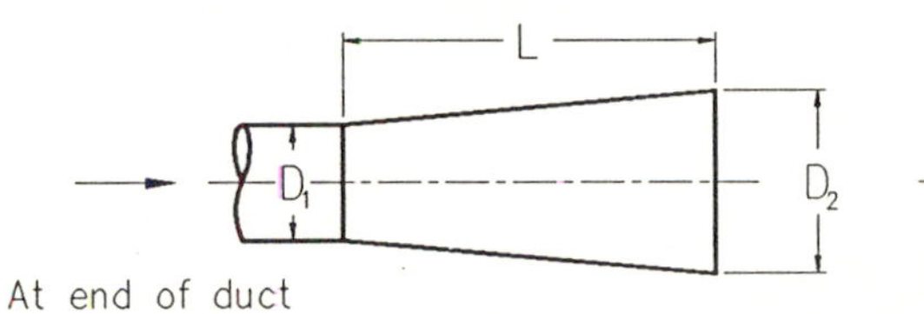

At end of duct

Regain (R), fraction of inlet VP						
Taper length to inlet diam L/D	Diameter ratios D_2/D_1					
	1.2:1	1.3:1	1.4:1	1.5:1	1.6:1	1.7:1
1.0:1	0.37	0.39	0.38	0.35	0.31	0.27
1.5:1	0.39	0.46	0.47	0.46	0.44	0.41
2.0:1	0.42	0.49	0.52	0.52	0.51	0.49
3.0:1	0.44	0.52	0.57	0.59	0.60	0.59
4.0:1	0.45	0.55	0.60	0.63	0.63	0.64
5.0:1	0.47	0.56	0.62	0.65	0.66	0.68
7.5:1	0.48	0.58	0.64	0.68	0.70	0.72

Where: $SP_1 = SP_2 - R(VP_1)$*

*When $SP_2 = 0$ (atmosphere) SP_1 will be (−)

The regain (R) will only be 70% of value shown above when expansion follows a disturbance or elbow (including a fan) by less than 5 duct diameters.

STATIC PRESSURE LOSSES FOR CONTRACTIONS

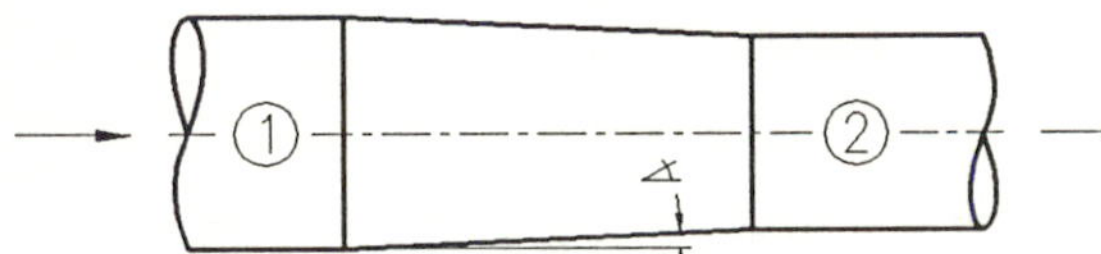

Tapered contraction

$SP_2 = SP_1 - (VP_2 - VP_1) - L(VP_2 - VP_1)$

Taper angle degrees	L(loss)
5	0.05
10	0.06
15	0.08
20	0.10
25	0.11
30	0.13
45	0.20
60	0.30
over 60	Abrupt contraction

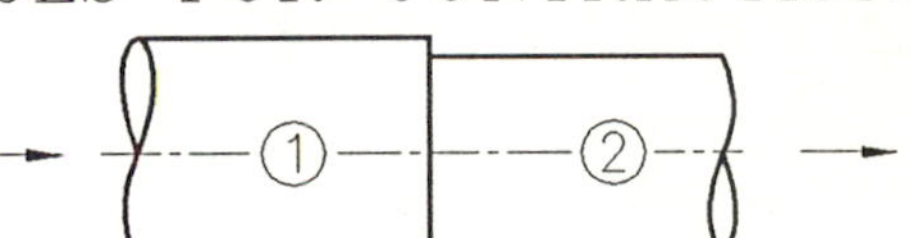

Abrupt contraction

$SP_2 = SP_1 - (VP_2 - VP_1) - K(VP_2)$

Ratio A_2/A_1	K
0.1	0.48
0.2	0.46
0.3	0.42
0.4	0.37
0.4	0.32
0.6	0.26
0.7	0.20

A= duct area, ft^2

Note:

In calculating SP for expansion or contraction use algebraic signs: VP is (+), and usually SP is (+) in discharge duct from fan, and SP is (−) in inlet duct to fan.

AMERICAN CONFERENCE OF GOVERNMENTAL INDUSTRIAL HYGIENISTS	*DUCT DESIGN DATA*		
	DATE *1-88*	FIGURE *5-19*	

EQUIVALENT RESISTANCE IN FEET OF STRAIGHT PIPE

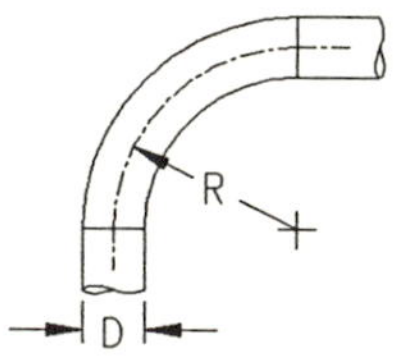

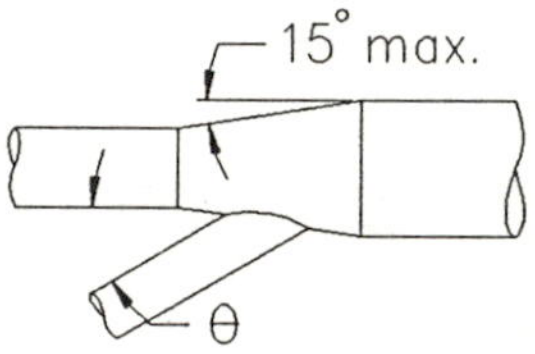

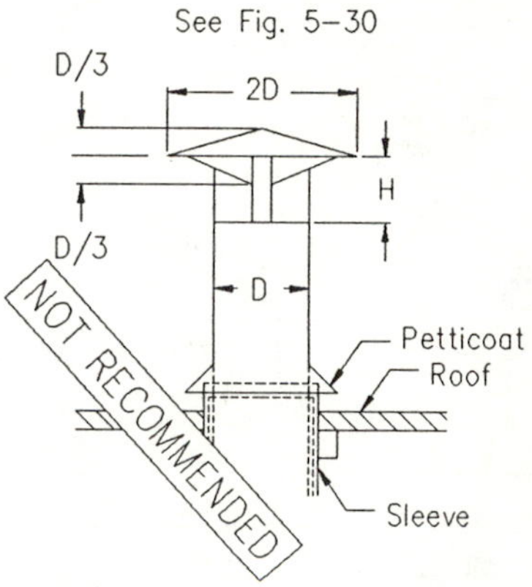

See Fig. 5-30

Pipe Dia.	90° Elbow * Centerline Radius			Angle of Entry		H, No of Diameters		
D	1.5 D	2.0 D	2.5 D	30°	45°	1.0 D	.75 D	.5 D
3"	5	3	3	2	3	2	2	9
4"	6	4	4	3	5	2	3	12
5"	9	6	5	4	6	2	4	16
6"	12	7	6	5	7	3	5	20
7"	13	9	7	6	9	3	6	23
8"	15	10	8	7	11	4	7	26
10"	20	14	11	9	14	5	9	36
12"	25	17	14	11	17	6	11	44
14"	30	21	17	13	21	7	13	53
16"	36	24	20	16	25	9	15	62
18"	41	28	23	18	28	10	18	71
20"	46	32	26	20	32	11	20	80
24"	57	40	32			13	24	92
30"	74	51	41			17	31	126
36"	93	64	52			22	39	159
40"	105	72	59					
48"	130	89	73					

* For 60° elbows – 0.67 x loss for 90°
45° elbows – 0.5 x loss for 90°
30° elbows – 0.33 x loss for 90°

AMERICAN CONFERENCE OF GOVERNMENTAL INDUSTRIAL HYGIENISTS

DUCT DESIGN DATA

DATE 1–88 | FIGURE 5–20

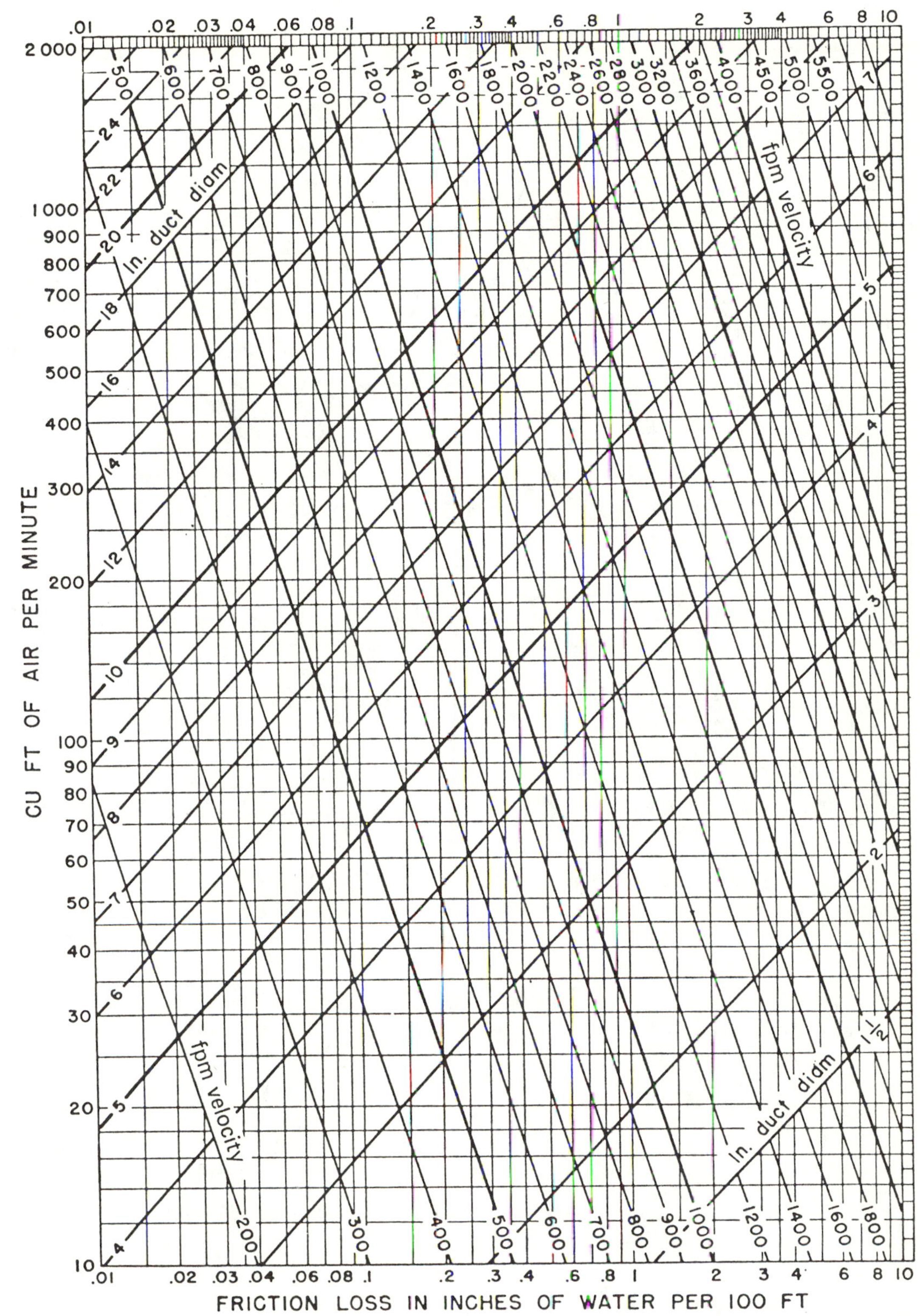

Based on Standard Air of 0.075 lb/ft³ density flowing through clean, round, galvanized metal ducts having approximately 40 joints per 100 ft.

$$\text{Friction Loss/100}' = \frac{2.74\left[\dfrac{V_{fpm}}{1000}\right]^{1.9}}{D_{inches}^{\;1.22}}$$

FIGURE 5.21a. FRICTION OF AIR IN STRAIGHT DUCTS FOR VOLUMETRIC FLOW RATES OF 10 TO 2000 cfm.(5.6) **Caution:** Do not extrapolate below chart. For propietary duct, obtain data from manufacturer.

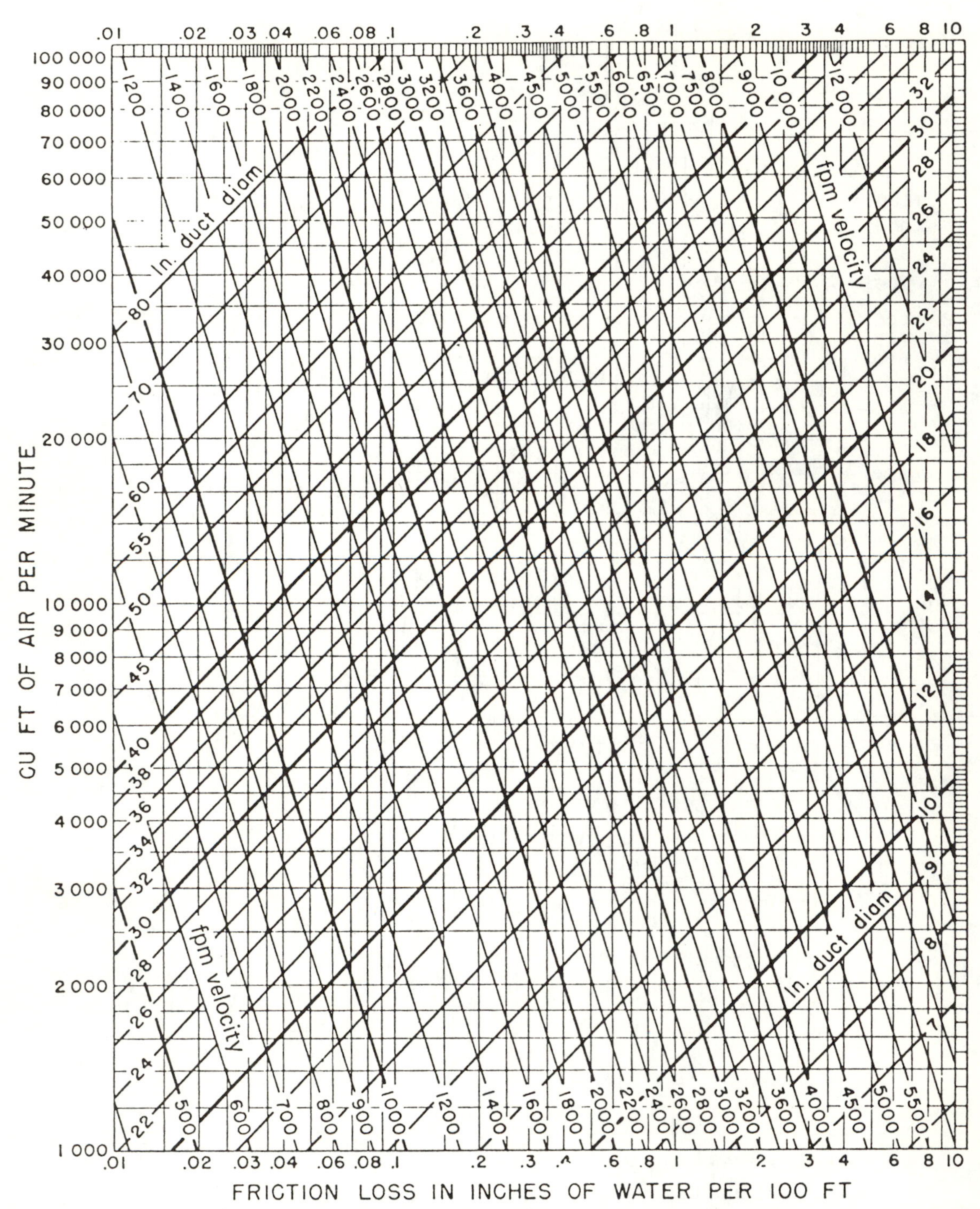

Based on Standard Air of 0.075 lb/ft^3 density flowing through average, clean, round, galvanized metal ducts having approximately 40 joints per 100 ft.

$$\text{Friction Loss/100}' = \frac{2.74\left[\frac{V_{fpm}}{1000}\right]^{1.9}}{D_{inches}^{\ 1.22}}$$

FIGURE 5.21b. FRICTION OF AIR IN STRAIGHT DUCTS FOR VOLUMETRIC FLOW RATES OF 10 TO 2000 cfm.[5.6] **Caution:** Do not extrapolate below chart. For propietary duct, obtain data from manufacturer. (Reprinted from Reference 5.9 by permission of the American Society of Heating, Refrigerating and Air-Conditioning Engineers.)

TABLE 5-4A. Velocity Pressure to Velocity Conversion — Standard Air

FROM: V = 4005 $\sqrt{VP}$

V = Velocity, fpm
VP = Velocity Pressure, "wg

VP	V	VP	V	VP	V	VP	V	VP	V	VP	V
0.01	401	0.51	2860	1.01	4025	1.51	4921	2.01	5678	2.60	6458
0.02	566	0.52	2888	1.02	4045	1.52	4938	2.02	5692	2.70	6581
0.03	694	0.53	2916	1.03	4065	1.53	4954	2.03	5706	2.80	6702
0.04	801	0.54	2943	1.04	4084	1.54	4970	2.04	5720	2.90	6820
0.05	896	0.55	2970	1.05	4104	1.55	4986	2.05	5734	3.00	6937
0.06	981	0.56	2997	1.06	4123	1.56	5002	2.06	5748	3.10	7052
0.07	1060	0.57	3024	1.07	4143	1.57	5018	2.07	5762	3.20	7164
0.08	1133	0.58	3050	1.08	4162	1.58	5034	2.08	5776	3.30	7275
0.09	1201	0.59	3076	1.09	4181	1.59	5050	2.09	5790	3.40	7385
0.10	1266	0.60	3102	1.10	4200	1.60	5066	2.10	5804	3.50	7493
0.11	1328	0.61	3128	1.11	4220	1.61	5082	2.11	5818	3.60	7599
0.12	1387	0.62	3154	1.12	4238	1.62	5098	2.12	5831	3.70	7704
0.13	1444	0.63	3179	1.13	4257	1.63	5113	2.13	5845	3.80	7807
0.14	1499	0.64	3204	1.14	4276	1.64	5129	2.14	5859	3.90	7909
0.15	1551	0.65	3229	1.15	4295	1.65	5145	2.15	5872	4.00	8010
0.16	1602	0.66	3254	1.16	4314	1.66	5160	2.16	5886	4.10	8110
0.17	1651	0.67	3278	1.17	4332	1.67	5176	2.17	5900	4.20	8208
0.18	1699	0.68	3303	1.18	4351	1.68	5191	2.18	5913	4.30	8305
0.19	1746	0.69	3327	1.19	4369	1.69	5206	2.19	5927	4.40	8401
0.20	1791	0.70	3351	1.20	4387	1.70	5222	2.20	5940	4.50	8496
0.21	1835	0.71	3375	1.21	4405	1.71	5237	2.21	5954	4.60	8590
0.22	1879	0.72	3398	1.22	4424	1.72	5253	2.22	5967	4.70	8683
0.23	1921	0.73	3422	1.23	4442	1.73	5268	2.23	5981	4.80	8775
0.24	1962	0.74	3445	1.24	4460	1.74	5283	2.24	5994	4.90	8865
0.25	2003	0.75	3468	1.25	4478	1.75	5298	2.25	6007	5.00	8955
0.26	2042	0.76	3491	1.26	4496	1.76	5313	2.26	6021	5.50	9393
0.27	2081	0.77	3514	1.27	4513	1.77	5328	2.27	6034	6.00	9810
0.28	2119	0.78	3537	1.28	4531	1.78	5343	2.28	6047	6.50	10211
0.29	2157	0.79	3560	1.29	4549	1.79	5358	2.29	6061	7.00	10596
0.30	2194	0.80	3582	1.30	4566	1.80	5373	2.30	6074	7.50	10968
0.31	2230	0.81	3604	1.31	4584	1.81	5388	2.31	6087	8.00	11328
0.32	2266	0.82	3627	1.32	4601	1.82	5403	2.32	6100	8.50	11676
0.33	2301	0.83	3649	1.33	4619	1.83	5418	2.33	6113	9.00	12015
0.34	2335	0.84	3671	1.34	4636	1.84	5433	2.34	6126	9.50	12344
0.35	2369	0.85	3692	1.35	4653	1.85	5447	2.35	6140	10.00	12665
0.36	2403	0.86	3714	1.36	4671	1.86	5462	2.36	6153	10.50	12978
0.37	2436	0.87	3736	1.37	4688	1.87	5477	2.37	6166	11.00	13283
0.38	2469	0.88	3757	1.38	4705	1.88	5491	2.38	6179	11.50	13582
0.39	2501	0.89	3778	1.39	4722	1.89	5506	2.39	6192	12.00	13874
0.40	2533	0.90	3799	1.40	4739	1.90	5521	2.40	6205	12.50	14160
0.41	2564	0.91	3821	1.41	4756	1.91	5535	2.41	6217	13.00	14440
0.42	2596	0.92	3841	1.42	4773	1.92	5549	2.42	6230	13.50	14715
0.43	2626	0.93	3862	1.43	4789	1.93	5564	2.43	6243	14.00	14985
0.44	2657	0.94	3883	1.44	4806	1.94	5578	2.44	6256	14.50	15251
0.45	2687	0.95	3904	1.45	4823	1.95	5593	2.45	6269	15.00	15511
0.46	2716	0.96	3924	1.46	4839	1.96	5607	2.46	6282	15.50	15768
0.47	2746	0.97	3944	1.47	4856	1.97	5621	2.47	6294	16.00	16020
0.48	2775	0.98	3965	1.48	4872	1.98	5636	2.48	6307	16.50	16268
0.49	2803	0.99	3985	1.49	4889	1.99	5650	2.49	6320	17.00	16513
0.50	2832	1.00	4005	1.50	4905	2.00	5664	2.50	6332	17.50	16754

TABLE 5-4B. Velocity to Velocity Pressure Conversion — Standard Air

FROM: $V = 4005\sqrt{VP}$

V = Velocity, fpm
VP = Velocity Pressure, "wg

V	VP	V	VP	V	VP	V	VP	V	VP	V	VP
400	0.01	2600	0.42	3850	0.92	4880	1.48	5690	2.02	6190	2.39
500	0.02	2625	0.43	3875	0.94	4900	1.50	5700	2.03	6200	2.40
600	0.02	2650	0.44	3900	0.95	4920	1.51	5710	2.03	6210	2.40
700	0.03	2675	0.45	3925	0.96	4940	1.52	5720	2.04	6220	2.41
800	0.04	2700	0.45	3950	0.97	4960	1.53	5730	2.05	6230	2.42
900	0.05	2725	0.46	3975	0.99	4980	1.55	5740	2.05	6240	2.43
1000	0.06	2750	0.47	4000	1.00	5000	1.56	5750	2.06	6250	2.44
1100	0.08	2775	0.48	4020	1.01	5020	1.57	5760	2.07	6260	2.44
1200	0.09	2800	0.49	4040	1.02	5040	1.58	5770	2.08	6270	2.45
1300	0.11	2825	0.50	4060	1.03	5060	1.60	5780	2.08	6280	2.46
1400	0.12	2850	0.51	4080	1.04	5080	1.61	5790	2.09	6290	2.47
1450	0.13	2875	0.52	4100	1.05	5100	1.62	5800	2.10	6300	2.47
1500	0.14	2900	0.52	4120	1.06	5120	1.63	5810	2.10	6310	2.48
1550	0.15	2925	0.53	4140	1.07	5140	1.65	5820	2.11	6320	2.49
1600	0.16	2950	0.54	4160	1.08	5160	1.66	5830	2.12	6330	2.50
1650	0.17	2975	0.55	4180	1.09	5180	1.67	5840	2.13	6340	2.51
1700	0.18	3000	0.56	4200	1.10	5200	1.69	5850	2.13	6350	2.51
1750	0.19	3025	0.57	4220	1.11	5220	1.70	5860	2.14	6360	2.52
1800	0.20	3050	0.58	4240	1.12	5240	1.71	5870	2.15	6370	2.53
1825	0.21	3075	0.59	4260	1.13	5260	1.72	5880	2.16	6380	2.54
1850	0.21	3100	0.60	4280	1.14	5280	1.74	5890	2.16	6390	2.55
1875	0.22	3125	0.61	4300	1.15	5300	1.75	5900	2.17	6400	2.55
1900	0.23	3150	0.62	4320	1.16	5320	1.76	5910	2.18	6410	2.56
1925	0.23	3175	0.63	4340	1.17	5340	1.78	5920	2.18	6420	2.57
1950	0.24	3200	0.64	4360	1.19	5360	1.79	5930	2.19	6430	2.58
1975	0.24	3225	0.65	4380	1.20	5380	1.80	5940	2.20	6440	2.59
2000	0.25	3250	0.66	4400	1.21	5400	1.82	5950	2.21	6450	2.59
2025	0.26	3275	0.67	4420	1.22	5420	1.83	5960	2.21	6460	2.60
2050	0.26	3300	0.68	4440	1.23	5440	1.84	5970	2.22	6470	2.61
2075	0.27	3325	0.69	4460	1.24	5460	1.86	5980	2.23	6480	2.62
2100	0.27	3350	0.70	4480	1.25	5480	1.87	5990	2.24	6490	2.63
2125	0.28	3375	0.71	4500	1.26	5500	1.89	6000	2.24	6500	2.63
2150	0.29	3400	0.72	4520	1.27	5510	1.89	6010	2.25	6550	2.67
2175	0.29	3425	0.73	4540	1.29	5520	1.90	6020	2.26	6600	2.72
2200	0.30	3450	0.74	4560	1.30	5530	1.91	6030	2.27	6650	2.76
2225	0.31	3475	0.75	4580	1.31	5540	1.91	6040	2.27	6700	2.80
2250	0.32	3500	0.76	4600	1.32	5550	1.92	6050	2.28	6750	2.84
2275	0.32	3525	0.77	4620	1.33	5560	1.93	6060	2.29	6800	2.88
2300	0.33	3550	0.79	4640	1.34	5570	1.93	6070	2.30	6900	2.97
2325	0.34	3575	0.80	4660	1.35	5580	1.94	6080	2.30	7000	3.05
2350	0.34	3600	0.81	4680	1.37	5590	1.95	6090	2.31	7100	3.14
2375	0.35	3625	0.82	4700	1.38	5600	1.96	6100	2.32	7200	3.23
2400	0.36	3650	0.83	4720	1.39	5610	1.96	6110	2.33	7300	3.32
2425	0.37	3675	0.84	4740	1.40	5620	1.97	6120	2.34	7400	3.41
2450	0.37	3700	0.85	4760	1.41	5630	1.98	6130	2.34	7500	3.51
2475	0.38	3725	0.87	4780	1.42	5640	1.98	6140	2.35	7600	3.60
2500	0.39	3750	0.88	4800	1.44	5650	1.99	6150	2.36	7700	3.70
2525	0.40	3775	0.89	4820	1.45	5660	2.00	6160	2.37	7800	3.79
2550	0.41	3800	0.90	4840	1.46	5670	2.00	6170	2.37	7900	3.89
2575	0.41	3825	0.91	4860	1.47	5680	2.01	6180	2.38	8000	3.99

TABLE 5-5. Area and Circumference of Circles

Diam. in Inches	AREA Square Inches	AREA Square Feet	CIRCUMFERENCE Inches	CIRCUMFERENCE Feet	Diam. in Inches	AREA Square Inches	AREA Square Feet	CIRCUMFERENCE Inches	CIRCUMFERENCE Feet
1	0.79	0.0055	3.14	0.2618	30	706.9	4.909	94.2	7.854
1.5	1.77	0.0123	4.71	0.3927	31	754.8	5.241	97.4	8.116
2	3.14	0.0218	6.28	0.5236	32	804.2	5.585	100.5	8.378
2.5	4.91	0.0341	7.85	0.6545	33	855.3	5.940	103.7	8.639
3	7.07	0.0491	9.42	0.7854	34	907.9	6.305	106.8	8.901
3.5	9.62	0.0668	11.00	0.9163	35	962.1	6.681	110.0	9.163
4	12.57	0.0873	12.57	1.0472	36	1017.9	7.069	113.1	9.425
4.5	15.90	0.1104	14.14	1.1781	37	1075.2	7.467	116.2	9.687
5	19.63	0.1364	15.71	1.3090	38	1134.1	7.876	119.4	9.948
5.5	23.76	0.1650	17.28	1.4399	39	1194.6	8.296	122.5	10.210
6	28.27	0.1963	18.85	1.5708	40	1256.6	8.727	125.7	10.472
6.5	33.18	0.2304	20.42	1.7017	41	1320.3	9.168	128.8	10.734
7	38.48	0.2673	21.99	1.8326	42	1385.4	9.621	131.9	10.996
7.5	44.18	0.3068	23.56	1.9635	43	1452.2	10.085	135.1	11.257
8	50.27	0.3491	25.13	2.0944	44	1520.5	10.559	138.2	11.519
8.5	56.75	0.3941	26.70	2.2253	45	1590.4	11.045	141.4	11.781
9	63.62	0.4418	28.27	2.3562	46	1661.9	11.541	144.5	12.043
9.5	70.88	0.4922	29.85	2.4871	47	1734.9	12.048	147.7	12.305
10	78.54	0.5454	31.42	2.6180	48	1809.6	12.566	150.8	12.566
10.5	86.59	0.6013	32.99	2.7489	49	1885.7	13.095	153.9	12.828
11	95.03	0.6600	34.56	2.8798	50	1963.5	13.635	157.1	13.090
11.5	103.87	0.7213	36.13	3.0107	52	2123.7	14.748	163.4	13.614
12	113.10	0.7854	37.70	3.1416	54	2290.2	15.904	169.6	14.137
13	132.73	0.9218	40.84	3.4034	56	2463.0	17.104	175.9	14.661
14	153.94	1.0690	43.98	3.6652	58	2642.1	18.348	182.2	15.184
15	176.71	1.2272	47.12	3.9270	60	2827.4	19.635	188.5	15.708
16	201.06	1.3963	50.27	4.1888	62	3019.1	20.966	194.8	16.232
17	226.98	1.5763	53.41	4.4506	64	3217.0	22.340	201.1	16.755
18	254.47	1.7671	56.55	4.7124	66	3421.2	23.758	207.3	17.279
19	283.53	1.9689	59.69	4.9742	68	3631.7	25.220	213.6	17.802
20	314.16	2.1817	62.83	5.2360	70	3848.5	26.725	219.9	18.326
21	346.36	2.4053	65.97	5.4978	72	4071.5	28.274	226.2	18.850
22	380.13	2.6398	69.12	5.7596	74	4300.8	29.867	232.5	19.373
23	415.48	2.8852	72.26	6.0214	76	4536.5	31.503	238.8	19.897
24	452.39	3.1416	75.40	6.2832	78	4778.4	33.183	245.0	20.420
25	490.87	3.4088	78.54	6.5450	80	5026.5	34.907	251.3	20.944
26	530.93	3.6870	81.68	6.8068	82	5281.0	36.674	257.6	21.468
27	572.56	3.9761	84.82	7.0686	84	5541.8	38.485	263.9	21.991
28	615.75	4.2761	87.96	7.3304	86	5808.8	40.339	270.2	22.515
29	660.52	4.5869	91.11	7.5922	88	6082.1	42.237	276.5	23.038

The usual sheet metal fabricator will have patterns for ducts in 0.5-inch steps through 5.5-inch diameter; 1 inch steps 6 inches through 20 inches and 2-inch steps 22 inches and larger diameters.

TABLE 5-6. Circular Equivalents of Rectangular Duct Sizes

A \ B	4.0	4.5	5.0	5.5	6.0	6.5	7.0	7.5	8.0	8.5	9.0	9.5	10.0	10.5	11.0	11.5	12.0	12.5	13.0	13.5	14.0	14.5	15.0	15.5	16.0
3.0	3.8	4.0	4.2	4.4	4.6	4.7	4.9	5.1	5.2	5.3	5.5	5.6	5.7	5.9	6.0	6.1	6.2	6.3	6.4	6.5	6.6	6.7	6.8	6.9	7.0
3.5	4.1	4.3	4.6	4.8	5.0	5.2	5.3	5.5	5.7	5.8	6.0	6.1	6.3	6.4	6.5	6.7	6.8	6.9	7.0	7.1	7.2	7.3	7.5	7.6	7.7
4.0	4.4	4.6	4.9	5.1	5.3	5.5	5.7	5.9	6.1	6.3	6.4	6.6	6.7	6.9	7.0	7.2	7.3	7.4	7.6	7.7	7.8	7.9	8.0	8.2	8.3
4.5	4.6	4.9	5.2	5.4	5.7	5.9	6.1	6.3	6.5	6.7	6.9	7.0	7.2	7.4	7.5	7.7	7.8	7.9	8.1	8.2	8.4	8.5	8.6	8.7	8.8
5.0	4.9	5.2	5.5	5.7	6.0	6.2	6.4	6.7	6.9	7.1	7.3	7.4	7.6	7.8	8.0	8.1	8.3	8.4	8.6	8.7	8.9	9.0	9.1	9.3	9.4
5.5	5.1	5.4	5.7	6.0	6.3	6.5	6.8	7.0	7.2	7.4	7.6	7.8	8.0	8.2	8.4	8.6	8.7	8.9	9.0	9.2	9.3	9.5	9.6	9.8	9.9

A \ B	6.0	7.0	8.0	9.0	10.0	11.0	12.0	13.0	14.0	15.0	16.0	17.0	18.0	19.0	20.0	22.0	24.0	26.0	28.0	30.0	32.0	34.0	36.0	38.0	40.0
6.0	6.6																								
7.0	7.1	7.7																							
8.0	7.6	8.2	8.7																						
9.0	8.0	8.7	9.3	9.8																					
10.0	8.4	9.1	9.8	10.4	10.9																				
11.0	8.8	9.5	10.2	10.9	11.5	12.0																			
12.0	9.1	9.9	10.7	11.3	12.0	12.6	13.1																		
13.0	9.5	10.3	11.1	11.8	12.4	13.1	13.7	14.2																	
14.0	9.8	10.7	11.5	12.2	12.9	13.5	14.2	14.7	15.3																
15.0	10.1	11.0	11.8	12.6	13.3	14.0	14.6	15.3	15.8	16.4															
16.0	10.4	11.3	12.2	13.0	13.7	14.4	15.1	15.7	16.4	16.9	17.5														
17.0	10.7	11.6	12.5	13.4	14.1	14.9	15.6	16.2	16.8	17.4	18.0	18.6													
18.0	11.0	11.9	12.9	13.7	14.5	15.3	16.0	16.7	17.3	17.9	18.5	19.1	19.7												
19.0	11.2	12.2	13.2	14.1	14.9	15.7	16.4	17.1	17.8	18.4	19.0	19.6	20.2	20.8											
20.0	11.5	12.5	13.5	14.4	15.2	16.0	16.8	17.5	18.2	18.9	19.5	20.1	20.7	21.3	21.9										
22.0	12.0	13.0	14.1	15.0	15.9	16.8	17.6	18.3	19.1	19.8	20.4	21.1	21.7	22.3	22.9	24.0									
24.0	12.4	13.5	14.6	15.6	16.5	17.4	18.3	19.1	19.9	20.6	21.3	22.0	22.7	23.3	23.9	25.1	26.2								
26.0	12.8	14.0	15.1	16.2	17.1	18.1	19.0	19.8	20.6	21.4	22.1	22.9	23.5	24.2	24.9	26.1	27.3	28.4							
28.0	13.2	14.5	15.6	16.7	17.7	18.7	19.6	20.5	21.3	22.1	22.9	23.7	24.4	25.1	25.8	27.1	28.3	29.5	30.6						
30.0	13.6	14.9	16.1	17.2	18.3	19.3	20.2	21.1	22.0	22.9	23.7	24.4	25.2	25.9	26.6	28.0	29.3	30.5	31.7	32.8					
32.0	14.0	15.3	16.5	17.7	18.8	19.8	20.8	21.8	22.7	23.5	24.4	25.2	26.0	26.7	27.5	28.9	30.2	31.5	32.7	33.9	35.0				
34.0	14.4	15.7	17.0	18.2	19.3	20.4	21.4	22.4	23.3	24.2	25.1	25.9	26.7	27.5	28.3	29.7	31.1	32.4	33.7	34.9	36.1	37.2			
36.0	14.7	16.1	17.4	18.6	19.8	20.9	21.9	22.9	23.9	24.8	25.7	26.6	27.4	28.2	29.0	30.5	32.0	33.3	34.6	35.9	37.1	38.2	39.4		
38.0	15.0	16.5	17.8	19.0	20.2	21.4	22.4	23.5	24.5	25.4	26.4	27.2	28.1	28.9	29.8	31.3	32.8	34.2	35.6	36.8	38.1	39.3	40.4	41.5	
40.0	15.3	16.8	18.2	19.5	20.7	21.8	22.9	24.0	25.0	26.0	27.0	27.9	28.8	29.6	30.5	32.1	33.6	35.1	36.4	37.8	39.0	40.3	41.5	42.6	43.7
42.0	15.6	17.1	18.5	19.9	21.1	22.3	23.4	24.5	25.6	26.6	27.6	28.5	29.4	30.3	31.2	32.8	34.4	35.9	37.3	38.7	40.0	41.3	42.5	43.7	44.8
44.0	15.9	17.5	18.9	20.3	21.5	22.7	23.9	25.0	26.1	27.1	28.1	29.1	30.0	30.9	31.8	33.5	35.1	36.7	38.1	39.5	40.9	42.2	43.5	44.7	45.8
46.0	16.2	17.8	19.3	20.6	21.9	23.2	24.4	25.5	26.6	27.7	28.7	29.7	30.6	31.6	32.5	34.2	35.9	37.4	38.9	40.4	41.8	43.1	44.4	45.7	46.9
48.0	16.5	18.1	19.6	21.0	22.3	23.6	24.8	26.0	27.1	28.2	29.2	30.2	31.2	32.2	33.1	34.9	36.6	38.2	39.7	41.2	42.6	44.0	45.3	46.6	47.9
50.0	16.8	18.4	19.9	21.4	22.7	24.0	25.2	26.4	27.6	28.7	29.8	30.8	31.8	32.8	33.7	35.5	37.2	38.9	40.5	42.0	43.5	44.9	46.2	47.5	48.8
54.0	17.3	19.0	20.6	22.0	23.5	24.8	26.1	27.3	28.5	29.7	30.8	31.8	32.9	33.9	34.9	36 8	38.6	40.3	41.9	43.5	45.1	46.5	48.0	49.3	50.7
58.0	17.8	19.5	21.2	22.7	24.2	25.5	26.9	28.2	29.4	30.6	31.7	32.8	33.9	35.0	36.0	38.0	39.8	41.6	43.3	45.0	46.6	48.1	49.6	51.0	52.4
62.0	18.3	20.1	21.7	23.3	24.8	26.3	27.6	28.9	30.2	31.5	32.6	33.8	34.9	36.0	37.1	39.1	41.0	42.9	44.7	46.4	48.0	49.6	51.2	52.7	54.1
66.0	18.8	20.6	22.3	23.9	25.5	26.9	28.4	29.7	31.0	32.3	33.5	34.7	35.9	37.0	38.1	40.2	42.2	44.1	46.0	47.7	49.4	51.1	52.7	54.2	55.7
70.0	19.2	21.1	22.8	24.5	26.1	27.6	29.1	30.4	31.8	33.1	34.4	35.6	36.8	37.9	39.1	41.2	43.3	45.3	47.2	49.0	50.8	52.5	54.1	55.7	57.3
74.0	19.6	21.5	23.3	25.1	26.7	28.2	29.7	31.2	32.5	33.9	35.2	36.4	37.7	38.8	40.0	42.2	44.4	46.4	48.4	50.3	52.1	53.8	55.5	57.2	58.8
78.0	20.0	22.0	23.8	25.6	27.3	28.8	30.4	31.8	33.3	34.6	36.0	37.2	38.5	39.7	40.9	43.2	45.4	47.5	49.5	51.4	53.3	55.1	56.9	58.6	60.2
82.0	20.4	22.4	24.3	26.1	27.8	29.4	31.0	32.5	33.9	35.4	36.7	38.0	39.3	40.6	41.8	44.1	46.4	48.5	50.6	52.6	54.5	56.4	58.2	59.9	61.6
86.0	20.8	22.9	24.8	26.6	28.3	30.0	31.6	33.1	34.6	36.1	37.4	38.8	40.1	41.4	42.6	45.0	47.3	49.6	51.7	53.7	55.7	57.6	59.4	61.2	63.0
90.0	21.2	23.3	25.2	27.1	28.9	30.6	32.2	33.8	35.3	36.7	38.2	39.5	40.9	42.2	43.5	45.9	48.3	50.5	52.7	54.8	56.8	58.8	60.7	62.5	64.3

TABLE 5-6. Circular Equivalents of Rectangular Duct Sizes (con't)

A \ B	42.0	44.0	46.0	48.0	50.0	54.0	58.0	62.0	66.0	70.0	74.0	78.0	82.0	86.0	90.0
6.0															
7.0															
8.0															
9.0															
10.0															
11.0															
12.0															
13.0															
14.0															
15.0															
16.0															
17.0															
18.0															
19.0															
20.0															
22.0															
24.0															
26.0															
28.0															
30.0															
32.0															
34.0															
36.0															
38.0															
40.0															
42.0	45.9														
44.0	47.0	48.1													
46.0	48.0	49.2	50.3												
48.0	49.1	50.2	51.4	52.5											
50.0	50.0	51.2	52.4	53.6	54.7										
54.0	52.0	53.2	54.4	55.6	56.8	59.0									
58.0	53.8	55.1	56.4	57.6	58.8	61.2	63.4								
62.0	55.5	56.9	58.2	59.5	60.8	63.2	65.5	67.8							
66.0	57.2	58.6	60.0	61.3	62.6	65.2	67.6	69.9	72.1						
70.0	58.8	60.3	61.7	63.1	64.4	67.1	69.6	72.0	74.3	76.5					
74.0	60.3	61.9	63.3	64.8	66.2	68.9	71.5	74.0	76.4	78.7	80.9				
78.0	61.8	63.4	64.9	66.4	67.9	70.6	73.3	75.9	78.4	80.7	83.0	85.3			
82.0	63.3	64.9	66.5	68.0	69.5	72.3	75.1	77.8	80.3	82.8	85.1	87.4	89.6		
86.0	64.7	66.3	67.9	69.5	71.0	74.0	76.8	79.6	82.2	84.7	87.1	89.5	91.8	94.0	
90.0	66.0	67.7	69.4	71.0	72.6	75.6	78.5	81.3	84.0	86.6	89.1	91.5	93.9	96.2	98.4

$$D_{equiv} = 1.3 \frac{(A \times B)^{0.625}}{(A + B)^{0.25}}$$

where:

D_{equiv} = equivalent round duct size for rectangular duct, in.

A = one side of rectangular duct, in.

B = adjacent side of rectangular duct, in.

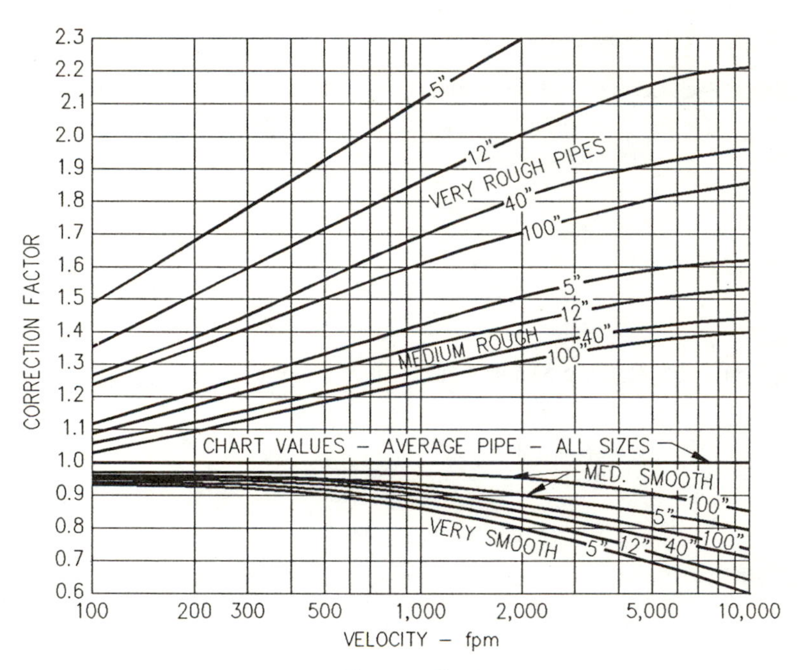

Figure 5–22
CORRECTIONS FOR NON–STANDARD MATERIALS.
From Chapter 41, Heating, Ventilating, Air Conditioning Guide, 1947. Used by Permission.
MADISON AND ELLIOTT (Ref. 5.8)

TABLE 5-7. Air Density Correction Factor, df

	ALTITUDE RELATIVE TO SEA LEVEL, ft															
	–5000	–4000	–3000	–2000	–1000	0	1000	2000	3000	4000	5000	6000	7000	8000	9000	10000
	Barometric Pressure															
"Hg	35.74	34.51	33.31	32.15	31.02	29.92	28.86	27.82	26.82	25.84	24.89	23.98	23.09	22.22	21.39	20.57
"w	486.74	469.97	453.67	437.84	422.45	407.50	392.98	378.89	365.21	351.93	339.04	326.54	314.42	302.66	291.26	280.21
Temp., F	Density Factor, df															
–40	1.51	1.46	1.40	1.36	1.31	1.26	1.22	1.17	1.13	1.09	1.05	1.01	0.97	0.94	0.90	0.87
0	1.38	1.33	1.28	1.24	1.19	1.15	1.11	1.07	1.03	1.00	0.96	0.92	0.89	0.86	0.82	0.79
40	1.27	1.22	1.18	1.14	1.10	1.06	1.02	0.99	0.95	0.92	0.88	0.85	0.82	0.79	0.76	0.73
70	1.19	1.15	1.11	1.07	1.04	1.00	0.96	0.93	0.90	0.86	0.83	0.80	0.77	0.74	0.71	0.69
100	1.13	1.09	1.05	1.02	0.98	0.95	0.91	0.88	0.85	0.82	0.79	0.76	0.73	0.70	0.68	0.65
150	1.04	1.00	0.97	0.93	0.90	0.87	0.84	0.81	0.78	0.75	0.72	0.70	0.67	0.65	0.62	0.60
200	0.96	0.93	0.89	0.86	0.83	0.80	0.77	0.75	0.72	0.69	0.67	0.64	0.62	0.60	0.57	0.55
250	0.89	0.86	0.83	0.80	0.77	0.75	0.72	0.69	0.67	0.64	0.62	0.60	0.58	0.55	0.53	0.51
300	0.83	0.80	0.78	0.75	0.72	0.70	0.67	0.65	0.62	0.60	0.58	0.56	0.54	0.52	0.50	0.48
350	0.78	0.75	0.73	0.70	0.68	0.65	0.63	0.61	0.59	0.57	0.54	0.52	0.50	0.49	0.47	0.45
400	0.74	0.71	0.69	0.66	0.64	0.62	0.59	0.57	0.55	0.53	0.51	0.49	0.48	0.46	0.44	0.42
450	0.70	0.67	0.65	0.63	0.60	0.58	0.56	0.54	0.52	0.50	0.48	0.47	0.45	0.43	0.42	0.40
500	0.66	0.64	0.61	0.59	0.57	0.55	0.53	0.51	0.49	0.48	0.46	0.44	0.43	0.41	0.39	0.38
550	0.63	0.61	0.58	0.56	0.54	0.52	0.51	0.49	0.47	0.45	0.44	0.42	0.40	0.39	0.38	0.36
600	0.60	0.58	0.56	0.54	0.52	0.50	0.48	0.46	0.45	0.43	0.42	0.40	0.39	0.37	0.36	0.34
700	0.55	0.53	0.51	0.49	0.47	0.46	0.44	0.42	0.41	0.39	0.38	0.37	0.35	0.34	0.33	0.31
800	0.50	0.49	0.47	0.45	0.44	0.42	0.41	0.39	0.38	0.36	0.35	0.34	0.32	0.31	0.30	0.29
900	0.47	0.45	0.43	0.42	0.40	0.39	0.38	0.36	0.35	0.34	0.32	0.31	0.30	0.29	0.28	0.27
1000	0.43	0.42	0.40	0.39	0.38	0.36	0.35	0.34	0.33	0.31	0.30	0.29	0.28	0.27	0.26	0.25

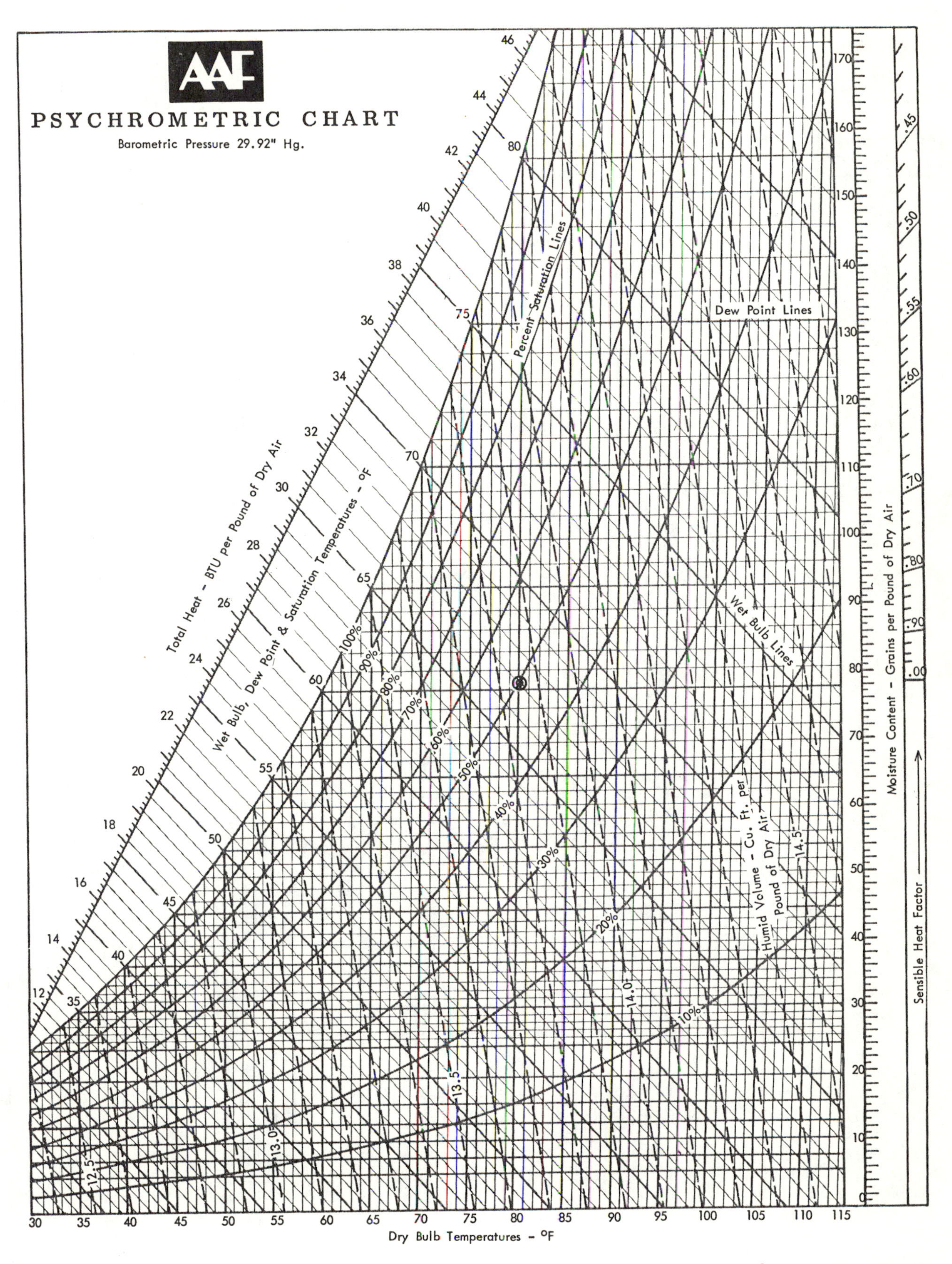
AAF
PSYCHROMETRIC CHART
Barometric Pressure 29.92" Hg.
Total Heat - BTU per Pound of Dry Air
Wet Bulb, Dew Point & Saturation Temperatures - °F
Percent Saturation Lines
Dew Point Lines
Wet Bulb Lines
Humid Volume - Cu. Ft. per Pound of Dry Air
Moisture Content - Grains per Pound of Dry Air
Sensible Heat Factor
Dry Bulb Temperatures - °F
© American Air Filter Co. Inc., 1959
Louisville, Ky. Form 1932
Total Heat Values - ASHAE Guide
End Points - Zimmerman & Lavine

FIGURE 5-23.

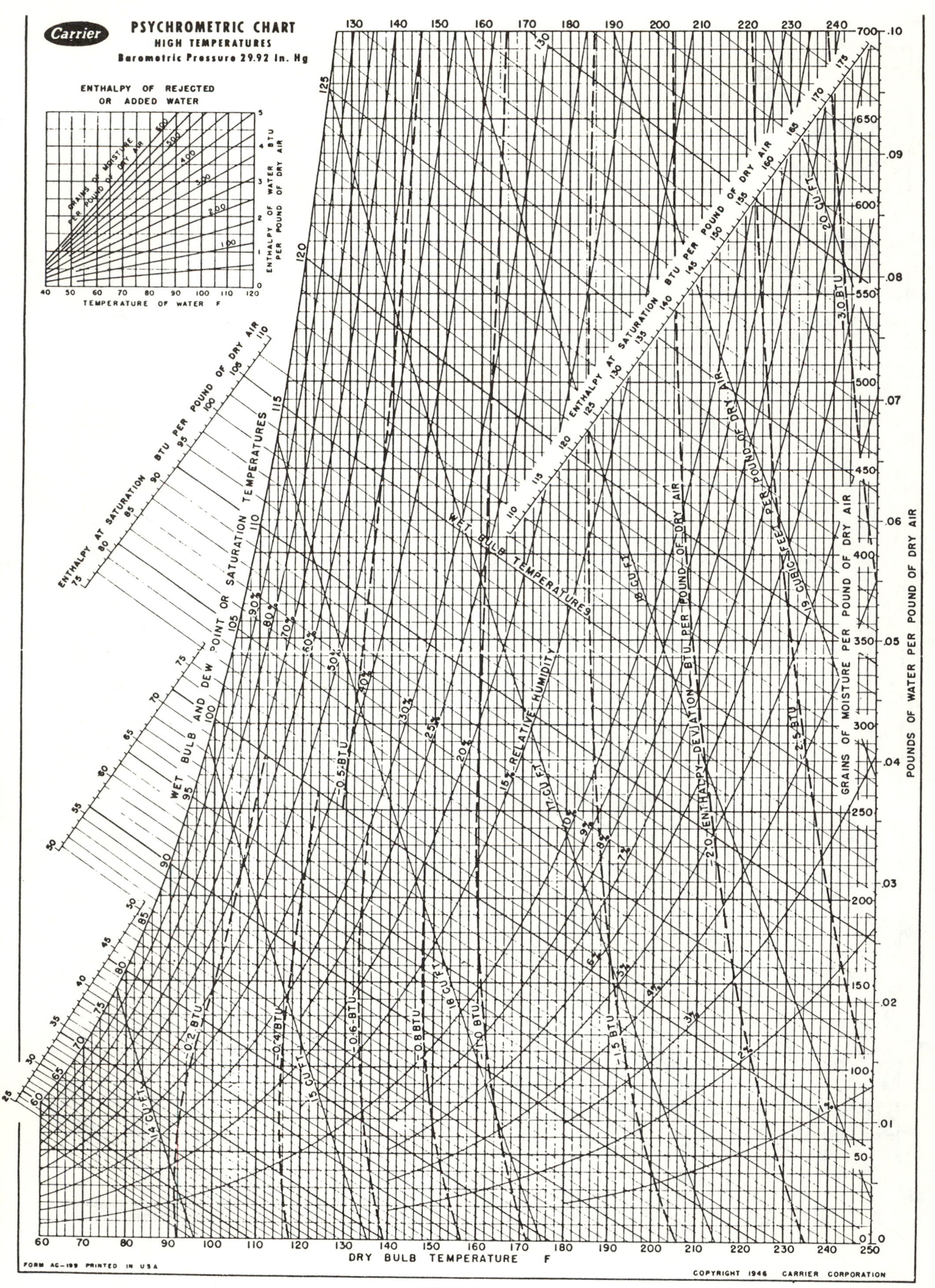
Carrier
PSYCHROMETRIC CHART
HIGH TEMPERATURES
Barometric Pressure 29.92 In. Hg
ENTHALPY OF REJECTED OR ADDED WATER
GRAINS OF MOISTURE PER POUND OF DRY AIR
ENTHALPY OF WATER BTU PER POUND OF DRY AIR
TEMPERATURE OF WATER F
ENTHALPY AT SATURATION BTU PER POUND OF DRY AIR
WET BULB AND DEW POINT OR SATURATION TEMPERATURES
WET BULB TEMPERATURES
RELATIVE HUMIDITY
ENTHALPY DEVIATION BTU PER POUND OF DRY AIR
CUBIC FEET PER POUND OF DRY AIR
GRAINS OF MOISTURE PER POUND OF DRY AIR
POUNDS OF WATER PER POUND OF DRY AIR
DRY BULB TEMPERATURE F
FORM AC-199 PRINTED IN U.S.A.
COPYRIGHT 1946 CARRIER CORPORATION

FIGURE 5-24.

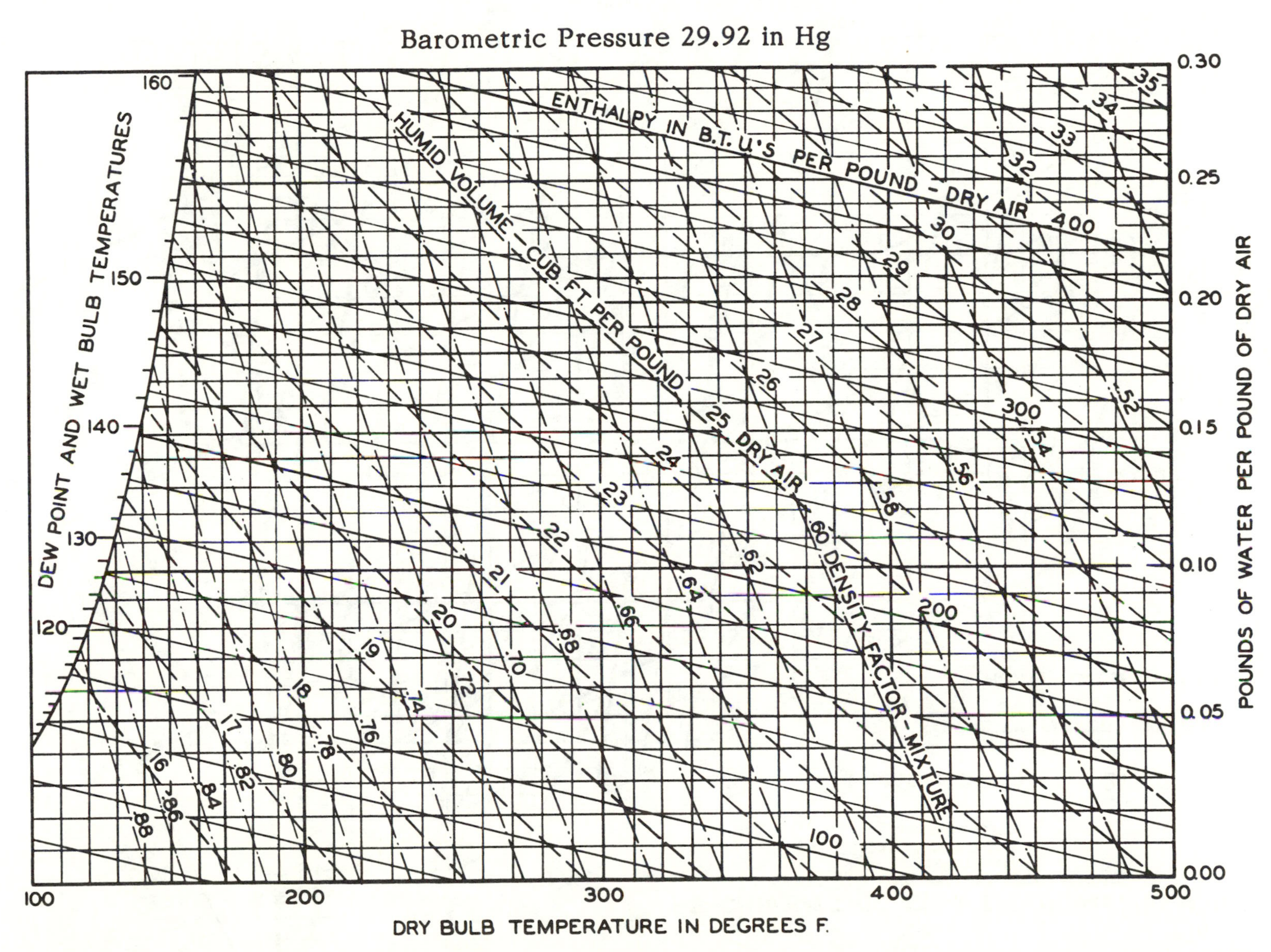

FIGURE 5-25. PSYCHROMETRIC CHART FOR HUMID AIR BASED ON ONE POUND DRY WEIGHT (© 1951 American Air Filter Co., Inc., Louisville, KY).

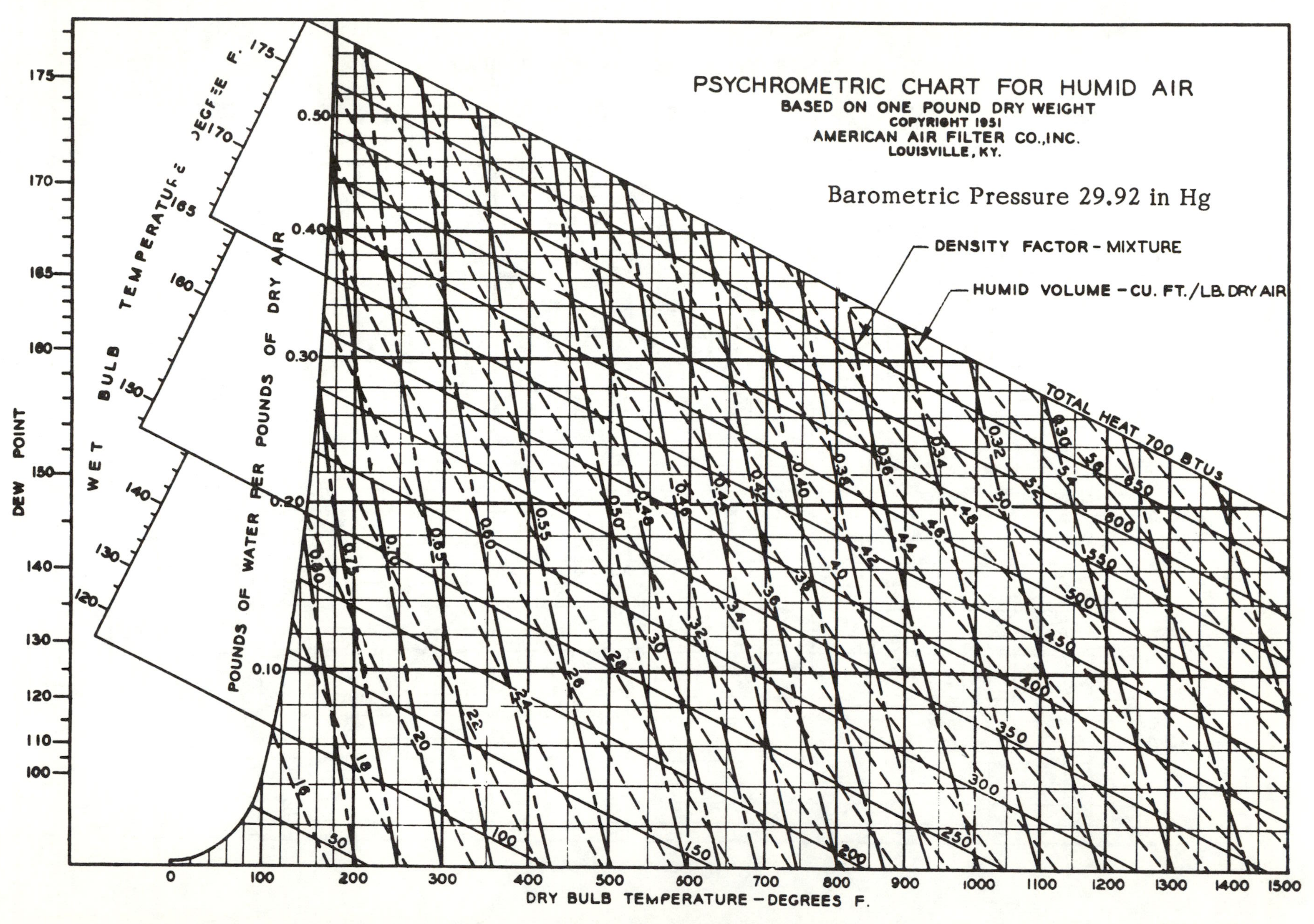
PSYCHROMETRIC CHART FOR HUMID AIR
BASED ON ONE POUND DRY WEIGHT
COPYRIGHT 1951
AMERICAN AIR FILTER CO., INC.
LOUISVILLE, KY.
Barometric Pressure 29.92 in Hg
DENSITY FACTOR - MIXTURE
HUMID VOLUME - CU. FT./LB. DRY AIR
TOTAL HEAT 700 BTUS
DRY BULB TEMPERATURE - DEGREES F.
WET BULB TEMPERATURE
POUNDS OF WATER PER POUNDS OF DRY AIR
DEW POINT

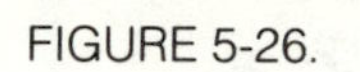
FIGURE 5-26.

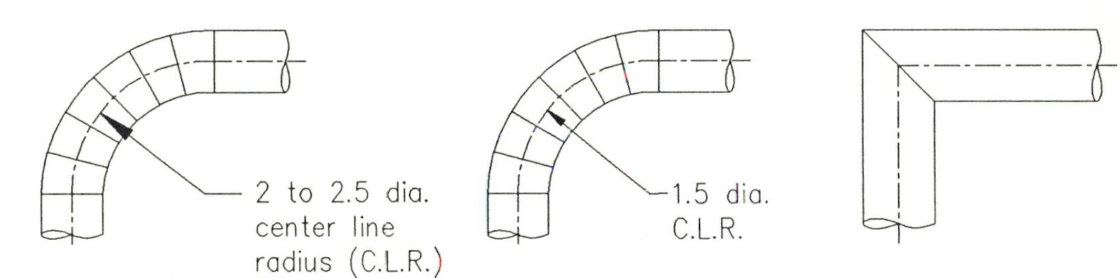

PREFERRED ACCEPTABLE AVOID

ELBOW RADIUS

Elbows should be 2 to 2.5 diameter centerline radius except where space does not permit. See Fig. 5-16 for loss factor.

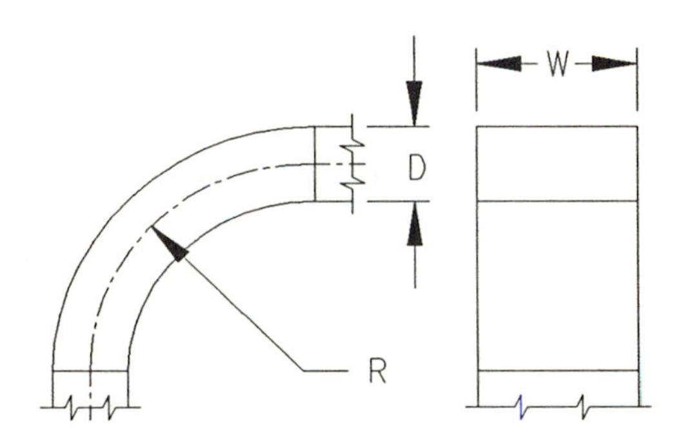

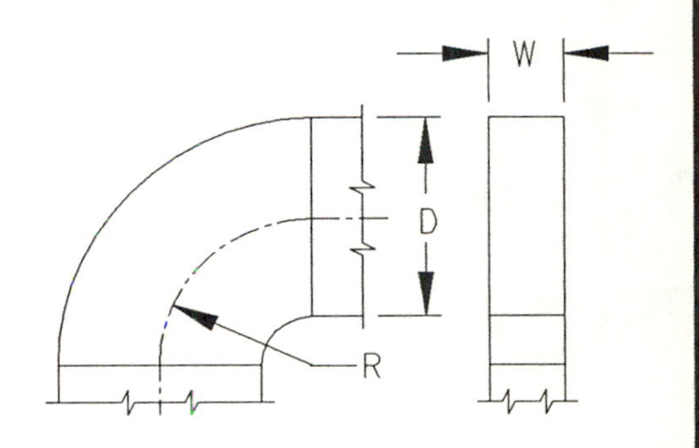

PREFERRED AVOID

ASPECT RATIO $(\frac{W}{D})$

Elbows should have $(\frac{W}{D})$ and $(\frac{R}{D})$ equal to or greater than (1). See Fig. 5-16 for loss factor.

Note: Avoid mitre elbows. If necessary, use only with clean air and provide turning vanes. Consult mfg. for turning vane loss factor.

AMERICAN CONFERENCE OF GOVERNMENTAL INDUSTRIAL HYGIENISTS	*PRINCIPLES OF DUCT DESIGN ELBOWS*	
	DATE *1-88*	FIGURE *5-27*

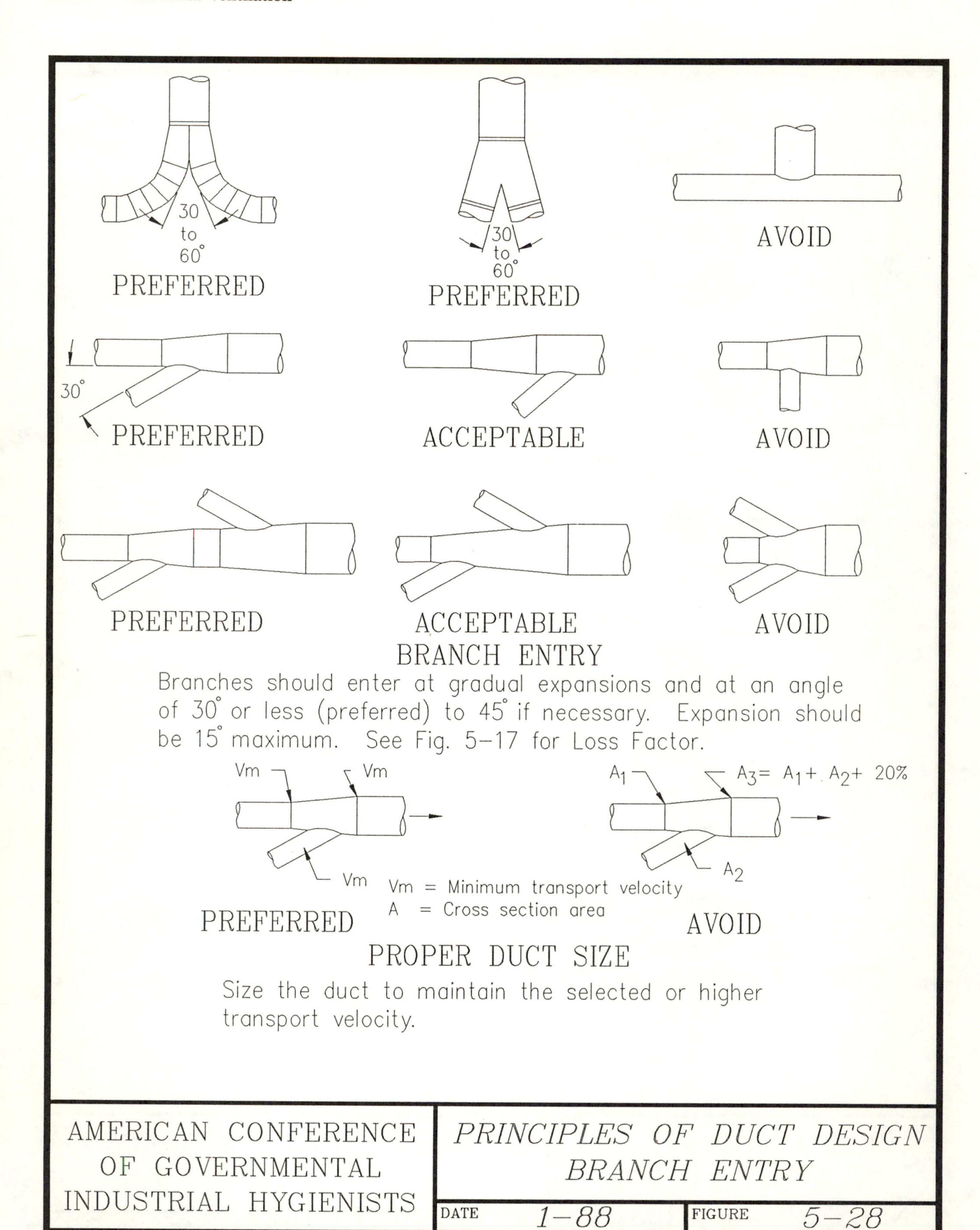
30
to
60°
PREFERRED
30
to
60°
PREFERRED
AVOID
30°
PREFERRED
ACCEPTABLE
AVOID
PREFERRED
ACCEPTABLE
AVOID
BRANCH ENTRY
Branches should enter at gradual expansions and at an angle of 30° or less (preferred) to 45° if necessary. Expansion should be 15° maximum. See Fig. 5-17 for Loss Factor.
Vm
Vm
Vm
A1
A3= A1+ A2+ 20%
A2
Vm = Minimum transport velocity
A = Cross section area
PREFERRED
AVOID
PROPER DUCT SIZE
Size the duct to maintain the selected or higher transport velocity.
AMERICAN CONFERENCE OF GOVERNMENTAL INDUSTRIAL HYGIENISTS
PRINCIPLES OF DUCT DESIGN
BRANCH ENTRY
DATE 1-88
FIGURE 5-28

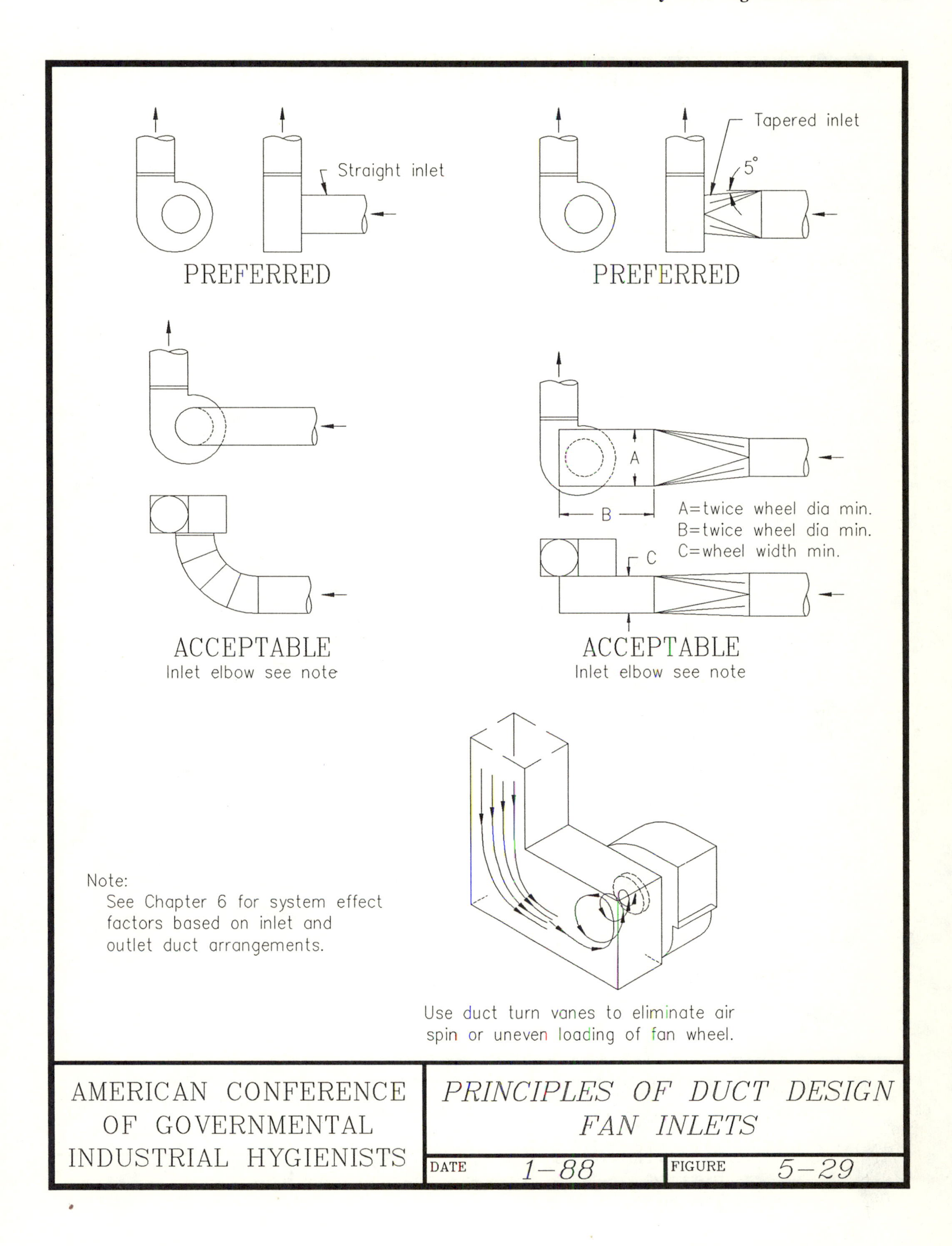
Straight inlet
PREFERRED
Tapered inlet
5°
PREFERRED
ACCEPTABLE
Inlet elbow see note
A
B
C
A=twice wheel dia min.
B=twice wheel dia min.
C=wheel width min.
ACCEPTABLE
Inlet elbow see note
Note:
See Chapter 6 for system effect factors based on inlet and outlet duct arrangements.
Use duct turn vanes to eliminate air spin or uneven loading of fan wheel.
AMERICAN CONFERENCE OF GOVERNMENTAL INDUSTRIAL HYGIENISTS
PRINCIPLES OF DUCT DESIGN
FAN INLETS
DATE 1-88
FIGURE 5-29

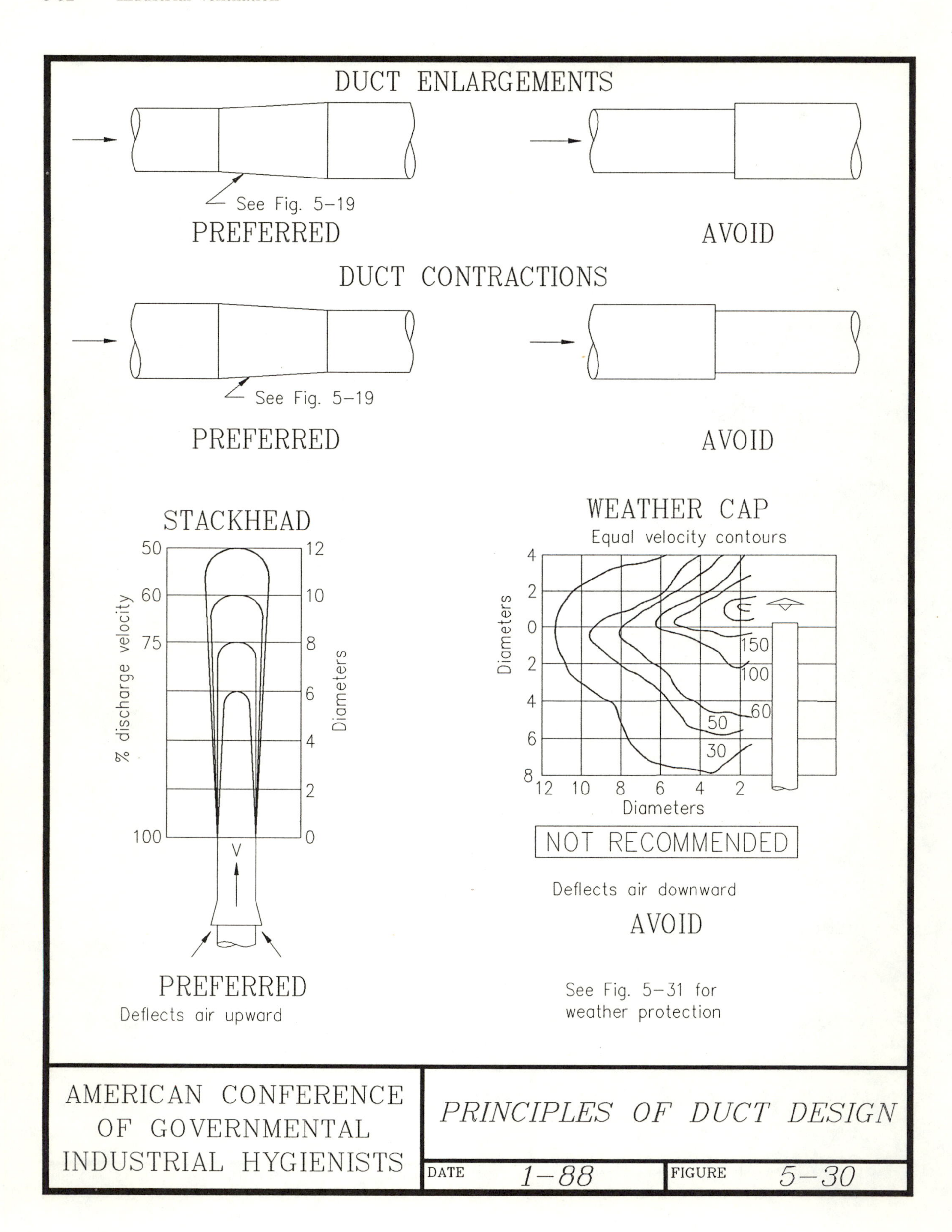
DUCT ENLARGEMENTS
See Fig. 5–19
PREFERRED
AVOID
DUCT CONTRACTIONS
See Fig. 5–19
PREFERRED
AVOID
STACKHEAD
% discharge velocity
50
60
75
100
Diameters
12
10
8
6
4
2
0
V
PREFERRED
Deflects air upward
WEATHER CAP
Equal velocity contours
Diameters
4
2
0
2
4
6
8
12 10 8 6 4 2
150
100
60
50
30
NOT RECOMMENDED
Deflects air downward
AVOID
See Fig. 5–31 for
weather protection
AMERICAN CONFERENCE OF GOVERNMENTAL INDUSTRIAL HYGIENISTS
PRINCIPLES OF DUCT DESIGN
DATE 1–88
FIGURE 5–30

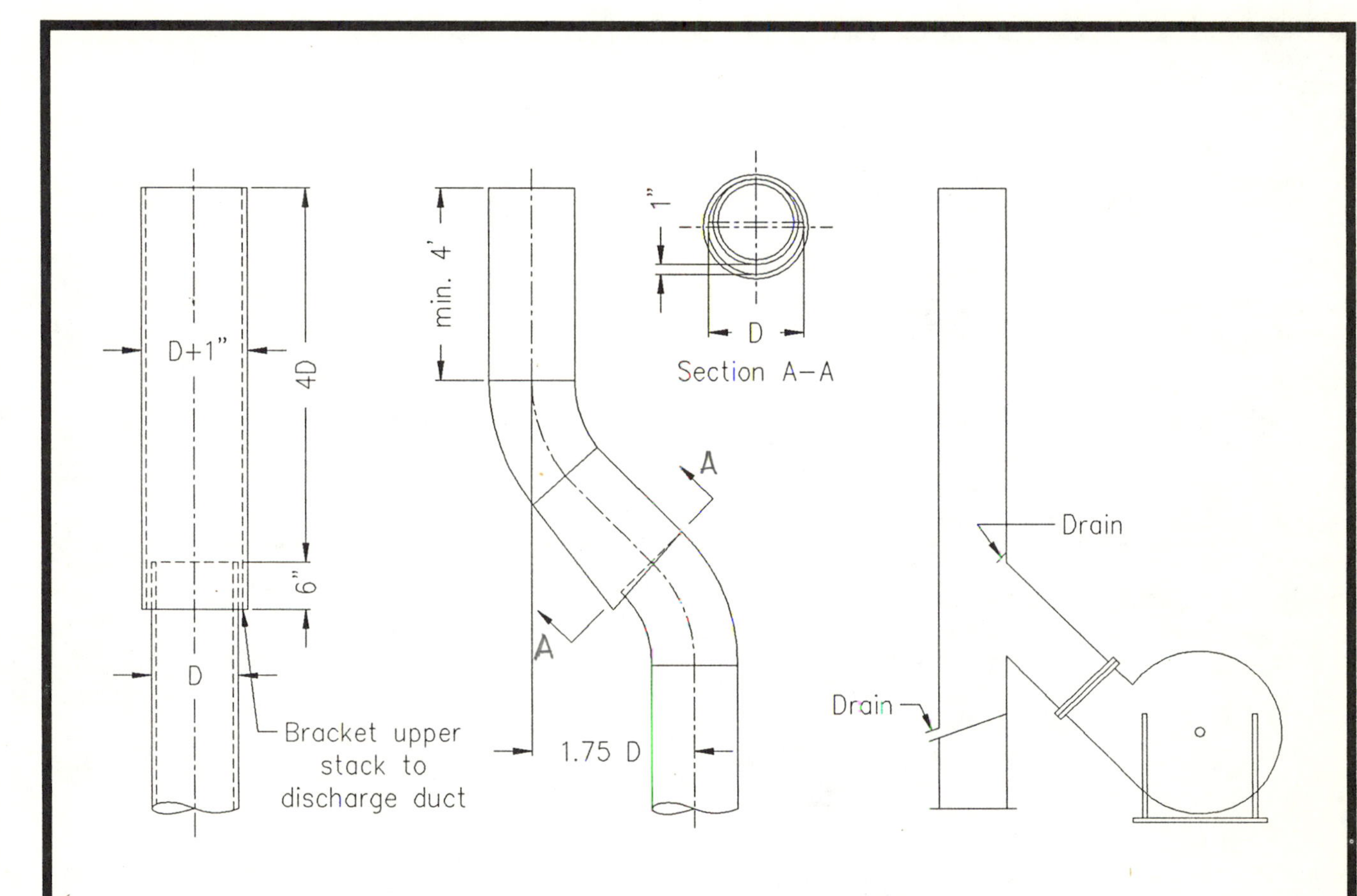

1. Rain protection is superior to a deflection cap 0.75 D above top of stack.

2. While the height of upper stack is related to the amount of rain protection afforded, excessive height may cause stack gas to flow out the drain.

AMERICAN CONFERENCE OF GOVERNMENTAL INDUSTRIAL HYGIENISTS

STACKHEAD DESIGNS

DATE 1–88

FIGURE 5–31

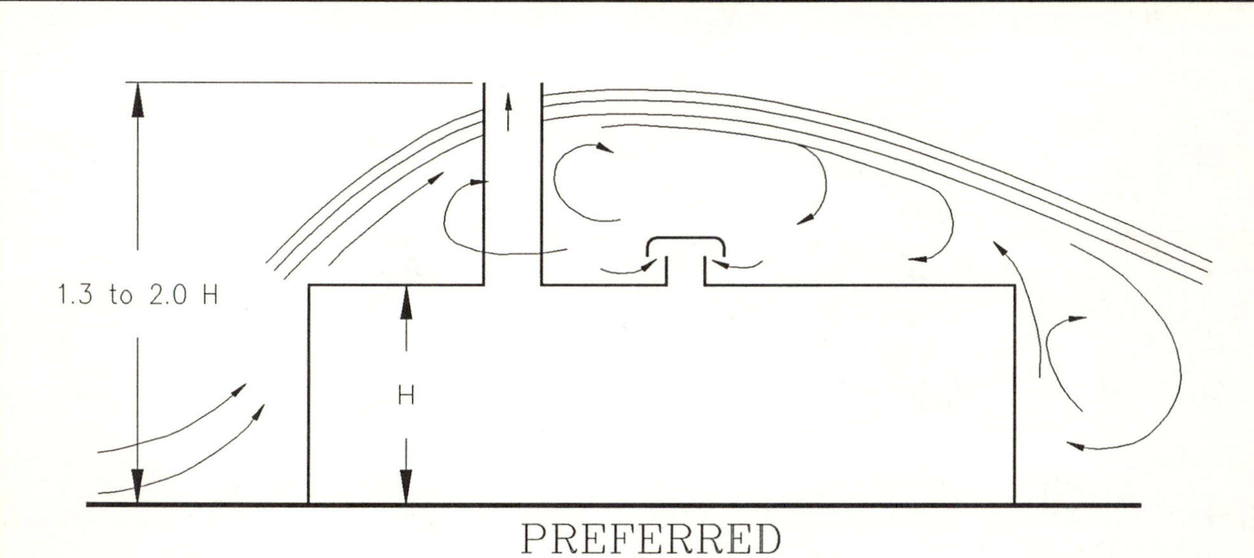

PREFERRED

High discharge stack relative to building height, air inlet on roof.

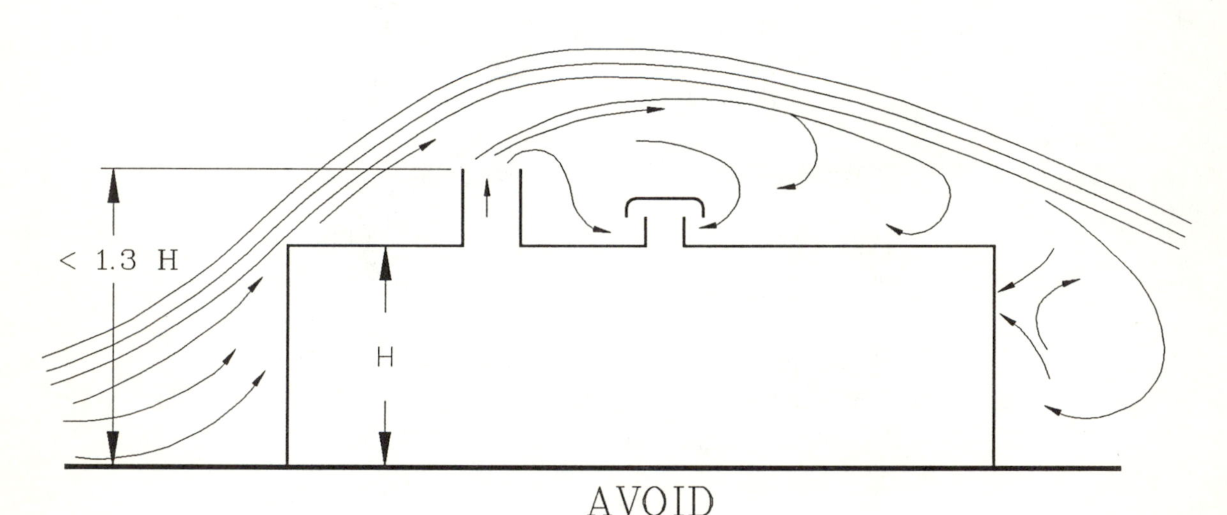

AVOID

Low discharge stack relative to building height and air inlets.

This applies only to the simple case of a low building without surrounding obstructions on reasonably level terain.

Note: Low pressure on the lee of a building may cause return of contaminants into the building through openings.

(Ref. 5.7)

AMERICAN CONFERENCE OF GOVERNMENTAL INDUSTRIAL HYGIENISTS	*STACK HEIGHT*	
	DATE *1-88*	FIGURE *5-32*

Chapter 6
FANS

6.1 INTRODUCTION 6-2

6.2 BASIC DEFINITIONS 6-2
- 6.2.1 Ejectors 6-2
- 6.2.2 Axial Fans 6-2
- 6.2.3 Centrifugal Fans 6-2
- 6.2.4 Special Type Fans 6-2

6.3 FAN SELECTION 6-6
- 6.3.1 Considerations for Fan Selection 6-6
- 6.3.2 Rating Tables 6-10
- 6.3.3 Point of Operation 6-13
- 6.3.4 Matching Fan Performance and System Requirements 6-13
- 6.3.5 Fan Laws 6-14
- 6.3.6 The Effect of Changing Rotation Rate or Gas Density 6-14
- 6.3.7 Limitations on the Use of Fan Laws 6-16
- 6.3.8 Fan Selection at Air Density Other Than Standard 6-16
- 6.3.9 Explosive or Flammable Materials 6-20

6.4 FAN INSTALLATION AND MAINTENANCE 6-20
- 6.4.1 System Effect 6-20
- 6.4.2 Inspection and Maintenance 6-21

REFERENCES 6-21

6.1 INTRODUCTION

To move air in a ventilation or exhaust system, energy is required to overcome the system losses. This energy can be in the form of natural convection or buoyancy. Most systems, however, require some powered air moving device such as a fan or an ejector.

This chapter will describe the various air moving devices that are used in industrial applications, provide guidelines for the selection of the air moving device for a given situation, and discuss the proper installation in the system to achieve desired performance.

Selection of an air moving device can be a complex task and the specifier is encouraged to take advantage of all available information from applicable trade associations as well as from individual manufacturers.

6.2 BASIC DEFINITIONS

Air moving devices can be divided into two basic classifications: ejectors and fans. Ejectors have low operating efficiencies and are used only for special material handling applications. Fans are the primary air moving devices used in industrial applications.

Fans can be divided into three basic groups: axial, centrifugal, and special types. As a general rule, axial fans are used for higher flow rates at lower resistances and centrifugal fans are used for lower flow rates at higher resistances.

6.2.1 Ejectors (see Figure 6-1): Ejectors sometimes are used when it is not desirable to have contaminated air pass directly through the air moving device. Ejectors are utilized for air streams containing corrosive, flammable, explosive, hot or sticky materials that might damage a fan, present a dangerous operating situation, or quickly degrade fan performance. Ejectors also may be used in pneumatic conveying systems.

6.2.2 Axial Fans: There are three basic types of axial fans: propeller, tubeaxial, and vaneaxial (see Figures 6-2 and 6-3).

Propeller Fans are used for moving air against low static pressures and are used commonly for general ventilation. Two types of blades are available: disc blade types when there is no duct present; narrow or propeller blade types for moving air against low resistances (less than 1 "wg). Performance is very sensitive to added resistance and a small increase will cause a marked reduction in flow rate.

Tubeaxial Fans (Duct Fans) contain narrow or propeller type blades in a short, cylindrical housing normally without any type of straightening vanes. Tubeaxial fans will move air against moderate pressures (less than 2 "wg).

Vaneaxial Fans have a propeller type configuration with a hub and airfoil blades mounted in cylindrical housings which normally incorporate straightening vanes on the discharge side of the impeller. Compared to other axial flow fans, vaneaxial fans are more efficient and will develop higher pressures (up to 8 "wg). They should be limited to clean air applications.

6.2.3 Centrifugal Fans (see Figures 6-4 and 6-7): These fans have three basic impeller designs: forward curved, radial, and backward inclined/backward curved.

Forward curved (commonly called "squirrel cages") impellers have blades which curve toward the direction of rotation. These fans have low space requirements, low tip speeds and are quiet in operation. They are used against low to moderate static pressures such as those encountered in heating and air conditioning work and replacement air systems. This type of fan is not recommended for dust or particulate that would adhere to the short curved blades and cause imbalance.

Radial impellers have blades which are straight or are in a radial direction from the hub. The housings are designed with their inlets and outlets sized to produce material conveying velocities. There is a variety of impeller types available ranging from "high efficiency minimum material" to "heavy impact resistance" designs. The radial blade shape will resist material buildup. This fan design is used for most exhaust system applications when particulate will pass through the fan. These fans usually have medium tip speeds and are used for a wide variety of exhaust systems which handle either clean or dirty air.

Backward Inclined/Backward Curved impeller blades are inclined opposite to the direction of fan rotation. This type usually has higher tip speeds and provides high fan efficiency and relatively low noise levels with "non-overloading" horsepower characteristics. In a non-overloading fan, the maximum horsepower occurs near the optimum operating point so any variation from that point due to a change in system resistance will result in a reduction in operating horsepower. The blade shape is condusive to material buildup so fans in this group should be limited as follows:

- *Single Thickness Blade:* Solid blades allow the unit to handle light dust loading or moisture. It should not be used with particulate that would build up on the underside of the blade surfaces.
- *Airfoil Blade*: Airfoil blades offer higher efficiencies and lower noise characteristics. Hollow blades erode more quickly with material and can fill with liquid in high humidity applications and should be limited to clean air service.

6.2.4 Special Type Fans (see Figure 6-4):

In-line Centrifugal fans have backward inclined blades with special housings which permit a straight line duct installation. Pressure versus flow rate performance curves are similar to a scroll type centrifugal fan of the same blade type. Space requirements are similar to vane axial fans.

Power Exhausters, Power Roof Ventilators are packaged units that can be either axial flow or centrifugal type. The centrifugal type does not use a scroll housing but discharges

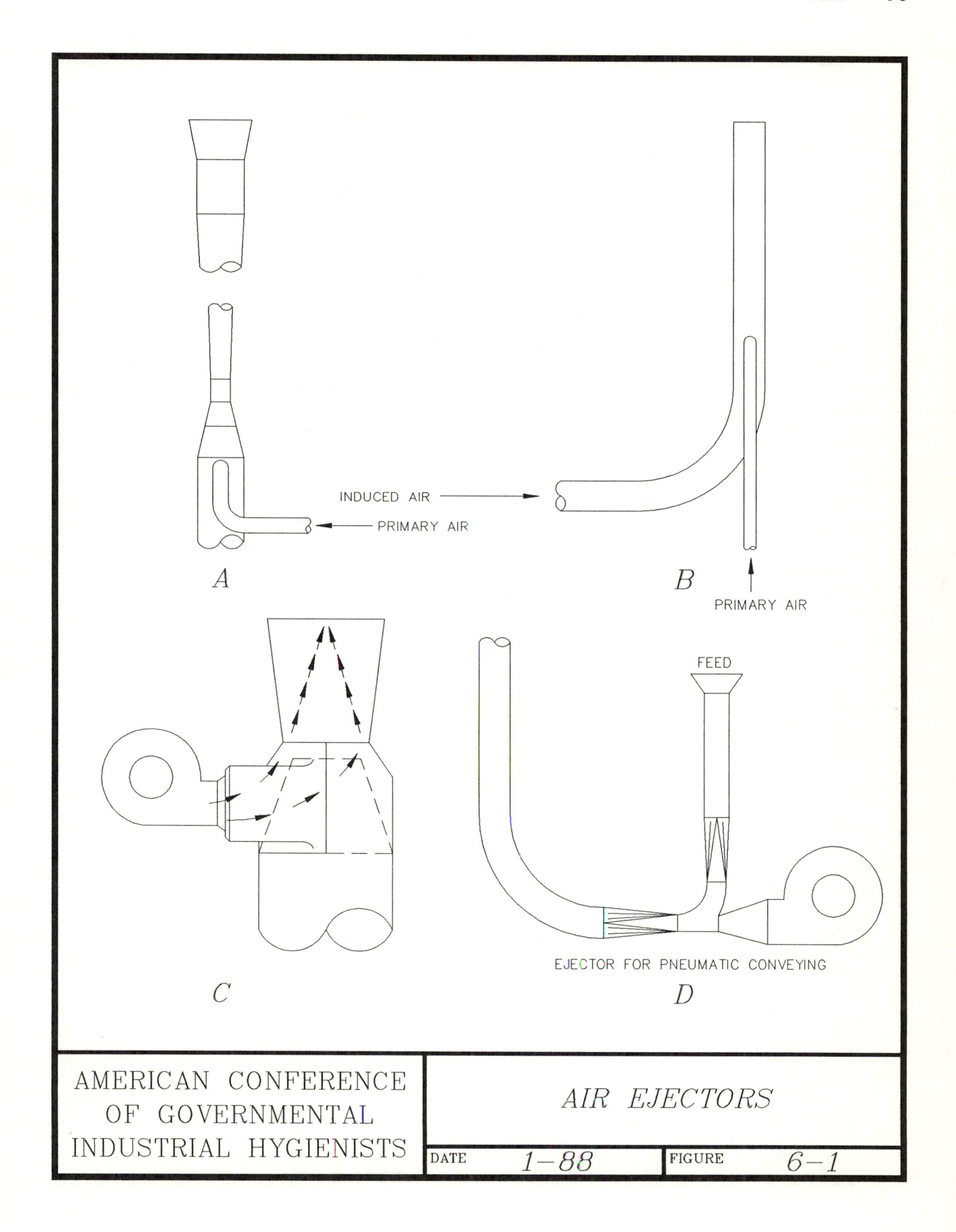
INDUCED AIR
PRIMARY AIR
A
B
PRIMARY AIR
FEED
C
EJECTOR FOR PNEUMATIC CONVEYING
D
AMERICAN CONFERENCE
OF GOVERNMENTAL
INDUSTRIAL HYGIENISTS
AIR EJECTORS
DATE 1-88
FIGURE 6-1

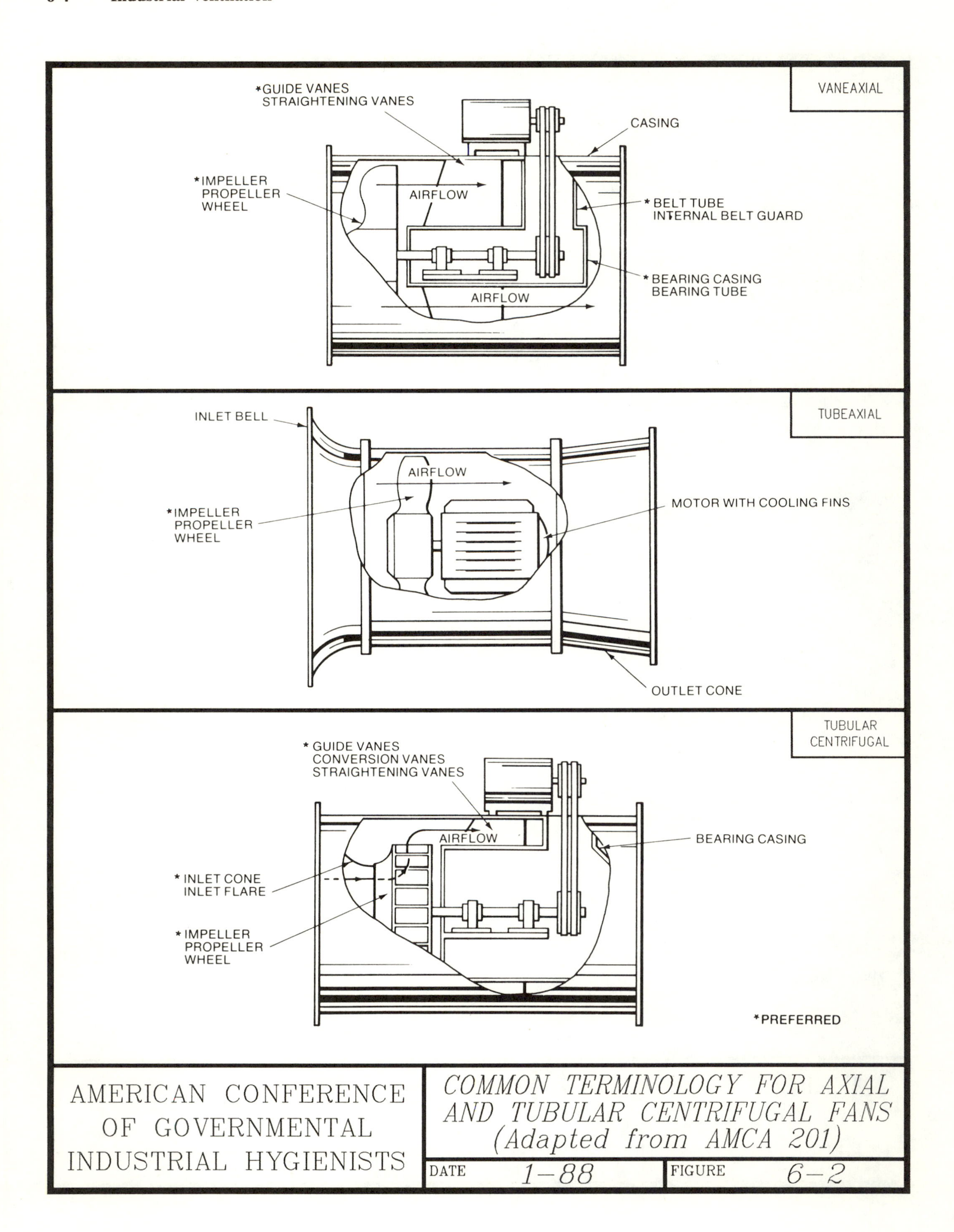

COMMON TERMINOLOGY FOR AXIAL AND TUBULAR CENTRIFUGAL FANS (Adapted from AMCA 201)

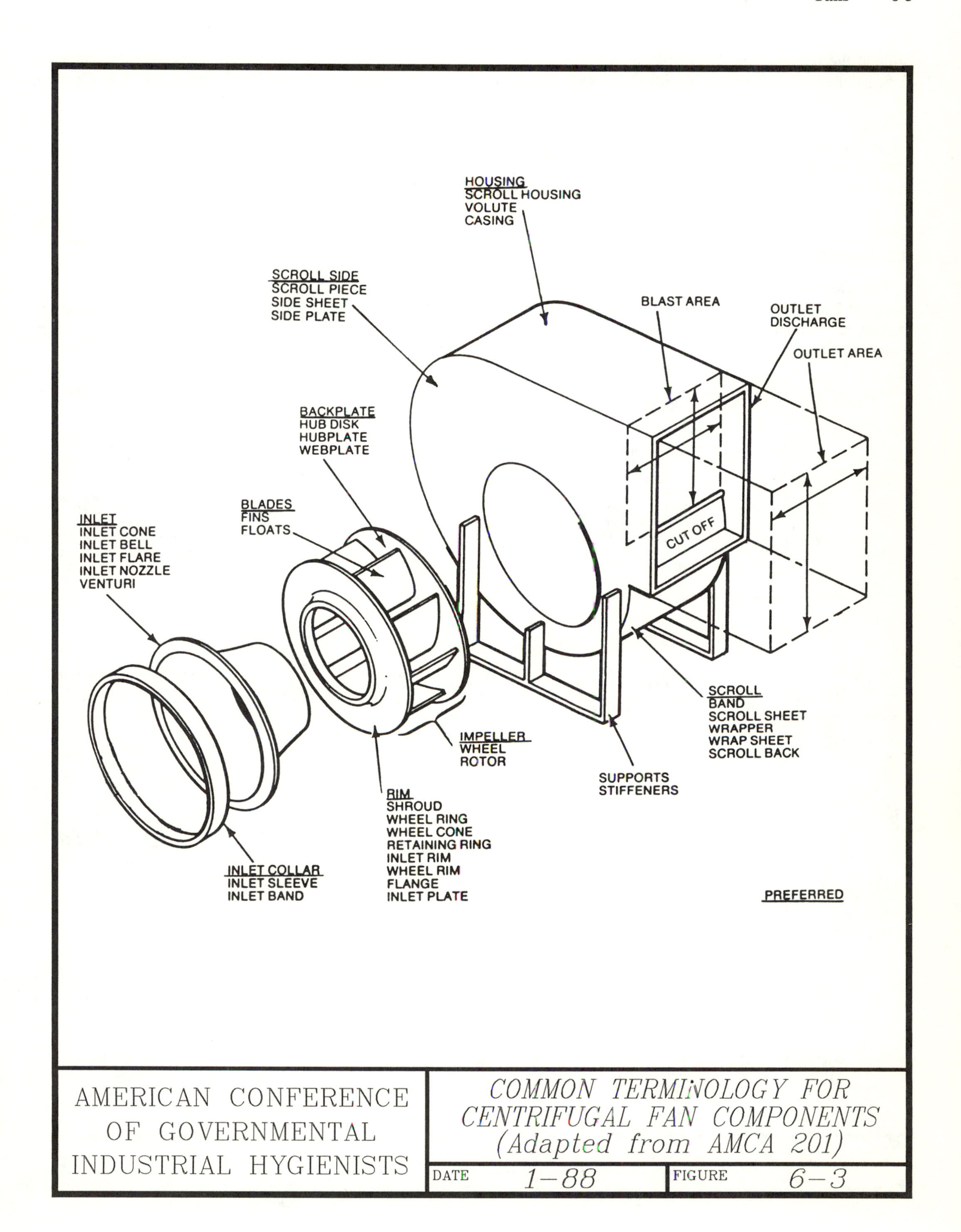
HOUSING
SCROLL HOUSING
VOLUTE
CASING
SCROLL SIDE
SCROLL PIECE
SIDE SHEET
SIDE PLATE
BLAST AREA
OUTLET
DISCHARGE
OUTLET AREA
BACKPLATE
HUB DISK
HUBPLATE
WEBPLATE
BLADES
FINS
FLOATS
INLET
INLET CONE
INLET BELL
INLET FLARE
INLET NOZZLE
VENTURI
CUT OFF
SCROLL
BAND
SCROLL SHEET
WRAPPER
WRAP SHEET
SCROLL BACK
IMPELLER
WHEEL
ROTOR
SUPPORTS
STIFFENERS
RIM
SHROUD
WHEEL RING
WHEEL CONE
RETAINING RING
INLET RIM
WHEEL RIM
FLANGE
INLET PLATE
INLET COLLAR
INLET SLEEVE
INLET BAND
PREFERRED
AMERICAN CONFERENCE OF GOVERNMENTAL INDUSTRIAL HYGIENISTS
COMMON TERMINOLOGY FOR CENTRIFUGAL FAN COMPONENTS (Adapted from AMCA 201)
DATE 1-88
FIGURE 6-3

around the periphery of the ventilator to the atmosphere. These units can be obtained with either downward deflecting or upblast discharges.

Fan and Dust Collector Combination: There are several designs in which fans and dust collectors are packaged in a unit. If use of such equipment is contemplated, the manufacturer should be consulted for proper application and performance characteristics.

6.3 FAN SELECTION

Fan selection involves not only finding a fan to match the required flow and pressure considerations but all aspects of an installation including the air stream characteristics, operating temperature, drive arrangement and mounting. Section 6.2 discussed the various fan types and why they might be selected. This section offers guidelines to fan selection; however, the exact performance and operating limitations of a particular fan should be obtained from the original equipment manufacturer.

6.3.1 Considerations for Fan Selection:

CAPACITY

Flow Rate (Q): Based on system requirements and expressed as actual cubic feet per minute (acfm) at the fan inlet.

Pressure Requirements: Based on system pressure requirements which normally are expressed as Fan Static Pressure (FSP) or Fan Total Pressure (FTP) in inches of water gauge at

TYPE		IMPELLER DESIGN	HOUSING DESIGN
CENTRIFUGAL FANS	AIRFOIL	Highest efficiency of all centrifugal fan designs. 9 to 16 blades of airfoil contour curved away from the direction of rotation. Air leaves the impeller at a velocity less than its tip speed and relatively deep blades provide for efficient expansion within the blade passages. For given duty, this will be the highest speed of the centrifugal fan designs.	Scroll-type, usually designed to permit efficient conversion of velocity pressure to static pressure, thus permitting a high static efficiency; essential that clearance and alignment between wheel and inlet bell be very close in order to reach the maxiumum efficiency capability. Concentric housings can also be used as in power roof ventilators, since there is efficient pressure conversion in the wheel.
	BACKWARD-INCLINED BACKWARD-CURVED	Efficiency is only slightly less than that of airfoil fans. Backward-inclined or backward-curved blades are single thickness. 9 to 16 blades curved or inclined away from the direction of rotation. Efficient for the same reasons given for the airfoil fan above.	Utilizes the same housing configuration as the airfoil design.
	RADIAL	Simplest of all centrifugal fans and least efficient. Has high mechanical strength and the wheel is easily repaired. For a given point of rating, this fan requires medium speed. This classification includes radial blades (R) and modified radial blades (M), usually 6 to 10 in number.	Scroll-type, usually the narrowest design of all centrifugal fan designs described here because of required high velocity discharge. Dimensional requirements of this housing are more critical than for airfoil and backward-inclined blades.
	FORWARD-CURVED	Efficiency is less than airfoil and backward-curved bladed fans. Usually fabricated of lightweight and low cost construction. Has 24 to 64 shallow blades with both the heel and tip curved forward. Air leaves wheel at velocity greater than wheel. Tip speed and primary energy transferred to the air is by use of high velocity in the wheel. For given duty, wheel is the smallest of all centrifugal types and operates at lowest speed.	Scroll is similar to other centrifugal-fan designs. The fit between the wheel and inlet is not as critical as on airfoil and backward-inclinded bladed fans. Uses large cut-off sheet in housing.

FIGURE 6-4. TYPES OF FANS: IMPELLER AND HOUSING DESIGNS (see facing page).

standard conditions (0.075 lbm/ft^3). If the required pressure is known only at non-standard conditions, a density correction (see Section 6.3.8) must be made.

AIRSTREAM

Material handled through the fan: When the exhaust air contains a small amount of smoke or dust, a backward inclined centrifugal or axial fan can be selected. With light dust, fume or moisture, a backward inclined or radial bladed fan would be the preferred selection. If the particulate loading is high, or when material is handled, the normal selection would be a radial fan.

Explosive or Flammable Material: Use spark resistant construction (explosion proof motor if the motor is in the airstream). Conform to the standards of the National Board of Fire Underwriters, the National Fire Protection Association and governmental regulations (see Section 6.3.9).

Corrosive Applications: May require a protective coating or special materials of construction (stainless, fiberglass, etc.)

Elevated Airstream Temperatures: Maximum operating temperature affects strength of materials and therefore must be known for selection of correct materials of construction, arrangement and bearing types.

PHYSICAL LIMITATIONS

Fan size should be determined by performance requirements. Inlet size and location, fan weight and ease of maintenance also must be considered. The most efficient fan size may not fit the physical space available.

PERFORMANCE CURVES	PERFORMANCE CHARACTERISTICS*	APPLICATIONS
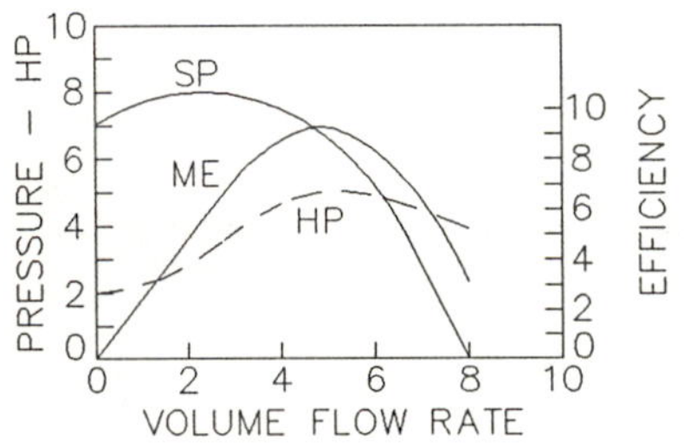	Highest efficiencies occur 50 to 65% of wide open volume. This is also the area of good pressure characteristics; the horsepower curve reaches a maximum near the peak efficiency area and becomes lower toward free delivery, a self-limiting power characteristic as shown.	General heating, ventilating and air-conditioning systems. Used in large sizes for clean air industrial applications where power savings are significant.
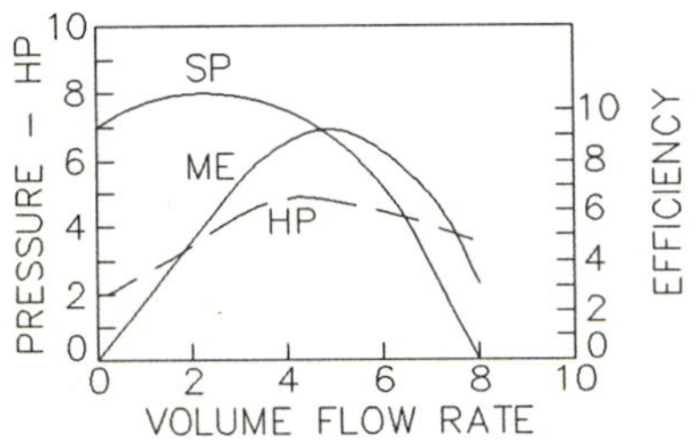	Operating characteristics of this fan are similar to the airfoil fan mentioned above. Peak efficiency for this fan is slightly lower than the airfoil fan. Normally unstable left of peak pressure.	Same heating, ventilating, and air-conditioning applications as the airfoil fan. Also used in some industrial applications where the airfoil blade is not acceptable because of corrosive and/or erosion environment.
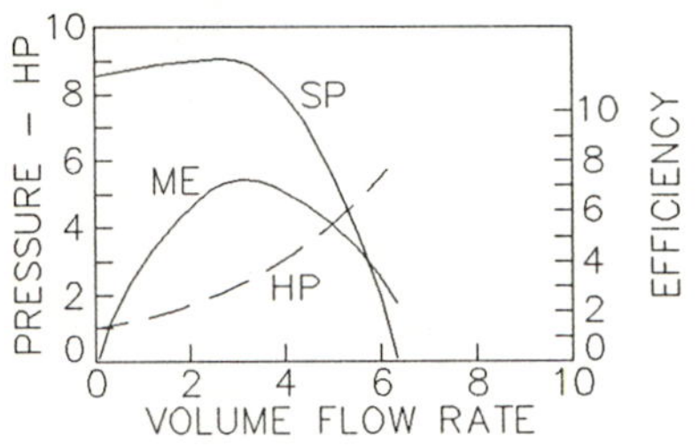	Higher pressure characteristics than the above mentioned fans. Power rises continually to free delivery.	Used primarily for material-handing applications in industrial plants. Wheel can be of rugged construction and is simple to repair in the field. Wheel is sometimes coated with special material. This design also used for high-pressure industrial requirements. Not commonly found in HVAC applications.
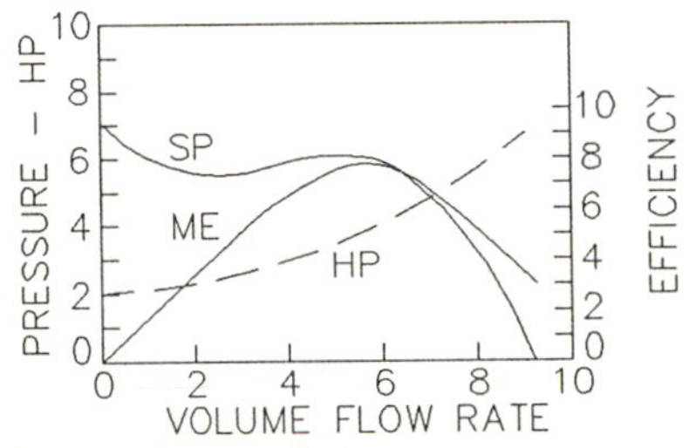	Pressure curve is less steep than that of backward-curved bladed fans. There is a dip in the pressure curve left of the peak pressure point and highest efficiency occurs to the right of peak pressure, 40 to 50% of wide open volume. Fan should be rated to the right of peak pressure. Power curve rises continually toward free delivery and this must be taken into account when motor is selected.	Used primarily in low-pressure heating, ventilating, and air-conditioning applications such as domestic furnaces, central station units, and packaged air-conditioning equipment from room air-conditioning units to roof top units.

TYPES OF FANS: PERFORMANCE CHARACTERISTICS AND APPLICATIONS. (*These performance curves reflect the general characteristics of various fans as commonly employed. They are not intended to provide complete selection criteria for application purposes, since other parameters, such as diameter and speed, are not defined.)

TYPE			IMPELLER DESIGN	HOUSING DESIGN
AXIAL FANS	PROPELLER	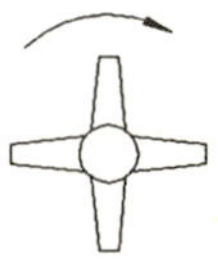	Efficiency is low. Impellers are usually of inexpensive construction and limited to low pressure applications. Impeller is of 2 or more blades, usually of single thickness attached to relatively small hub. Energy transfer is primarily in form of velocity pressure.	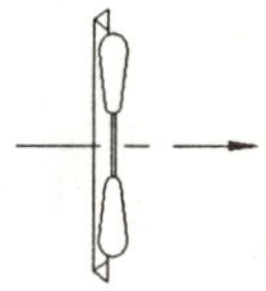 Simple circular ring, orifice plate, or venturi design. Design can substantially influence performance and optimum design is reasonably close to the blade tips and forms a smooth inlet flow contour to the wheel.
AXIAL FANS	TUBEAXIAL		Somewhat more efficient than propeller fan design and is capable of developing a more useful static pressure range. Number of blades usually from 4 to 8 and hub is usually less than 50% of fan tip diameter. Blades can be of airfoil or single thickness cross section.	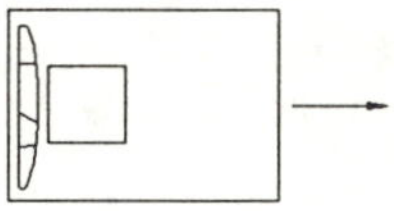 Cylindrical tube formed so that the running clearance between the wheel tip and tube is close. This results in significant improvement over propeller fans.
AXIAL FANS	VANEAXIAL	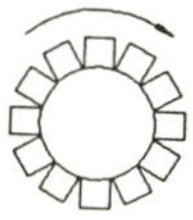	Good design of blades permits medium- to high-pressure capability at good efficiency. The most efficient fans of this type have airfoil blades. Blades are fixed or adjustable pitch types and hub is usually greater than 50% of fan tip diameter.	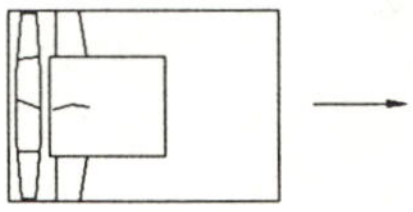 Cylindrical tube closely fitted to the outer diameter of blade tips and fitted with a set of guide vanes. Upstream or downstream from the impeller, guide vanes convert the rotary energy imparted to the air and increase pressure and efficiency of fan.
SPECIAL DESIGNS	TUBULAR	CENTRIFUGAL	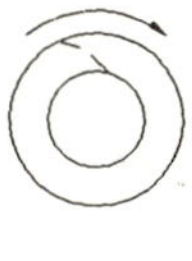 This fan usually has a wheel similar to the airfoil backward-inclined or backward-curved blade as described above. (However, this fan wheel type is of lower efficiency when used in fan of type.) Mixed flow impellers are sometimes used.	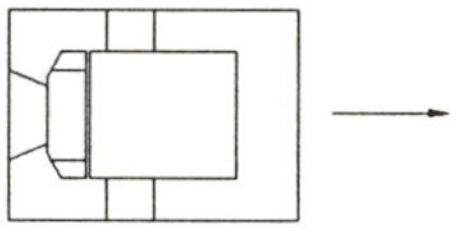 Cylindrical shell similar to a vaneaxial fan housing, except the outer diameter of the wheel does not run close to the housing. Air is discharged radially from the wheel and must change direction by 90 degrees to flow through the guide vane section.
SPECIAL DESIGNS	POWER ROOF VENTILATORS	CENTRIFUGAL	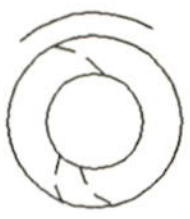 Many models use airfoil or backward-inclined impeller designs. These have been modified from those mentioned above to produce a low-pressure, high-volume flow rate characteristic. In addition, many special centrifugal impeller designs are used, including mixed-flow design.	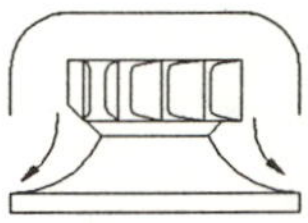 Does not utilize a housing in a normal sense since the air is simply discharged from the impeller in a 360 degree pattern and usually does not include a configuration to recover the velocity pressure component.
SPECIAL DESIGNS	POWER ROOF VENTILATORS	AXIAL	 A great variety of propeller designs are employed with the objective of high-volume flow rate at low pressure.	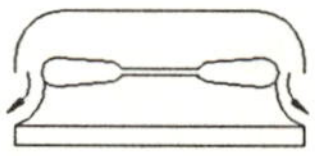 Essentially a propeller fan mounted in a supporting structure with a cover for weather protection and safety considerations. The air is discharged through the annular space around the bottom of the weather hood.

FIGURE 6-4 (continued). TYPES OF FANS: IMPELLER AND HOUSING DESIGN.

PERFORMANCE CURVES	PERFORMANCE CHARACTERISTICS*	APPLICATIONS
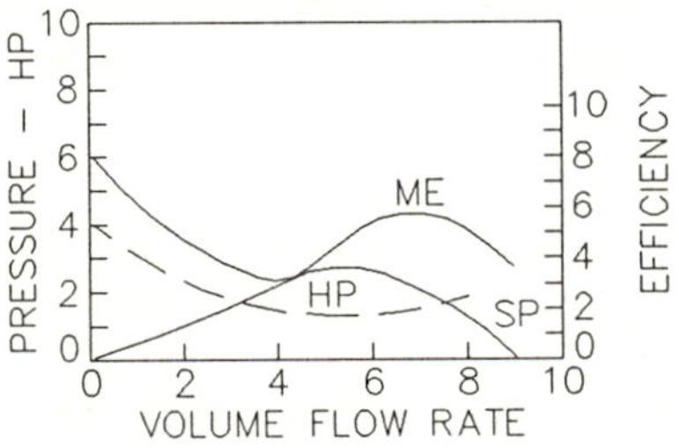	High flow rate but very low-pressure capabilities and maximumum efficiency is reached near free delivery. The discharge pattern of the air is circular in shape and the air stream swirls because of the action of the blades and the lack of straightening facilities.	For low-pressure, high-volume air moving applications such as air circulation within a space or ventilation through a wall without attached duct work. Used for replacement air applications.
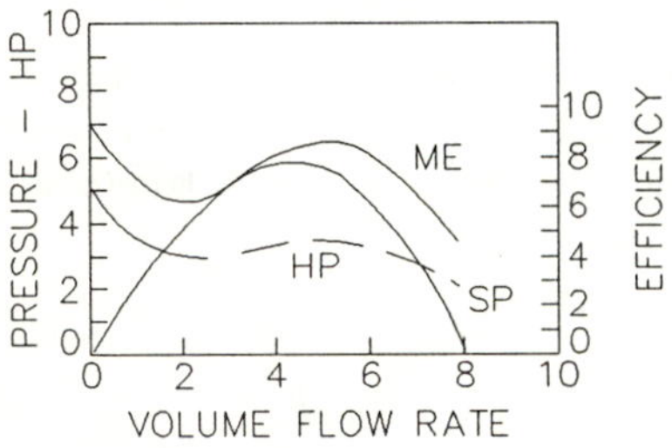	High flow-rate characteristics with medium-pressure capabilities. Performance curve includes a dip to the left of peak pressure which should be avoided. The discharge air pattern is circular and is rotating or whirling because of the propeller rotation and lack of guide vanes.	Low- and medium-pressure ducted heating, ventilating, and air-conditioning applications where air distribution on the downstream side is not critical. Also used in some industrial applications such as drying ovens, paint spray booths, and fume exhaust systems.
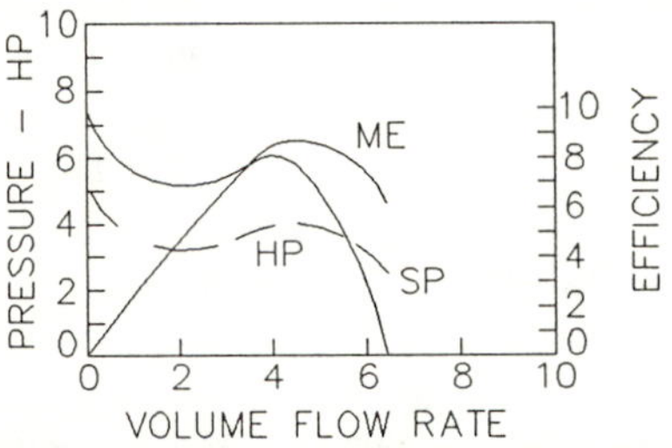	High-pressure characteristics with medium volume flow rate capabilities. Performance curve includes a dip caused by aerodynamic stall to the left of peak pressure, which should be avoided. Guide vanes correct the circular motion imparted to the air by the wheel and improve pressure characteristics and efficiency of the fan.	General heating, ventilating, and air-conditioning systems in low-, medium-, and high-pressure applications is of advantage where straight-through flow and compact instalation are required; air distribution on downstream side is good. Also used in industrial application similar to the tubeaxial fan. Relatively more compact than comparable centrifugal-type fans for same duty.
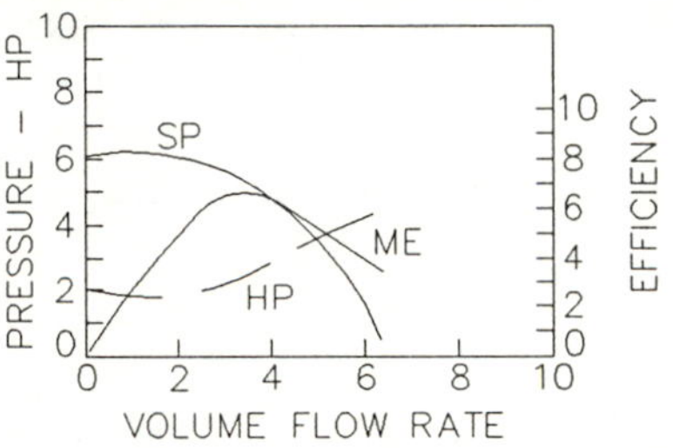	Performance is similar to backward-curved fan, except lower capacity and pressure because of the 90 degree change in direction of the air flow in the housing. The efficiency will be lower than the backward-curved fan. Some designs may have a dip in the curve similar to the axial-flow fan.	Used primarily for low-pressure return air systems in heating, ventilating, and air-conditioning applications. Has straight-through flow configuration.
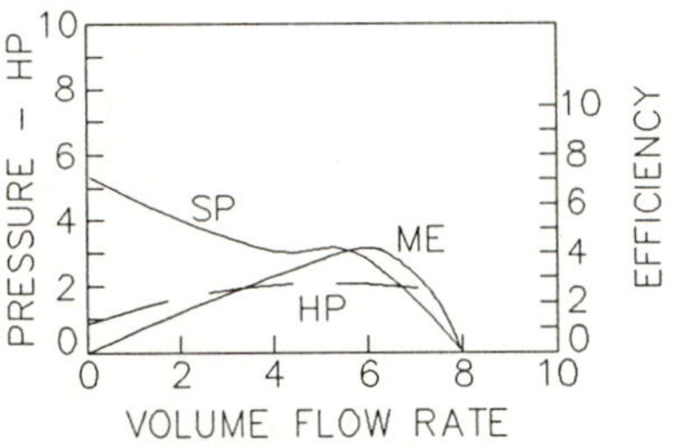	Usually intended to operate without attached ductwork and therefore to operate against a very low-pressure head. It is usually intended to have a rather high-volume flow rate characteristic. Only static pressure and static efficiency are shown for this type of product.	For low-pressure exhaust systems such as general factory, kitchen, warehouse, and commercial installations where the low-pressure rise limitation can be tolerated. Unit is low in first cost and low in operating cost and provides positive exhaust ventilation in the space which is a decided advantage over gravity-type exhaust units. The centrifugal unit is somewhat quieter than the axial unit desribed below.
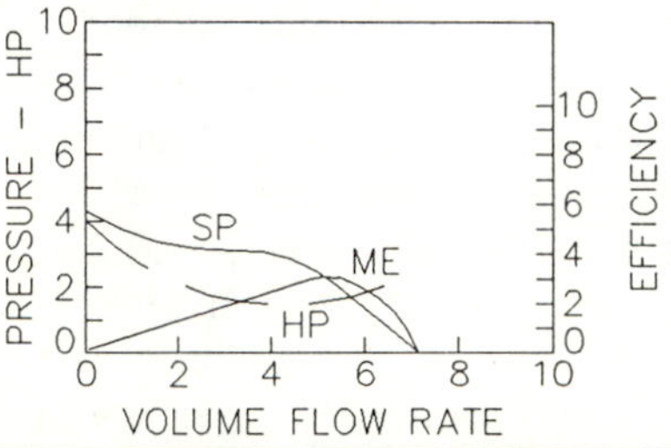	Usually intended to operate without attached ductwork and therefore to operate against very low-pressure head. It is usually intended to have a high-volume flow rate characteristic. Only static pressure and static efficiency are shown for this type of product.	For low-pressure exhaust systems such as general factory, kitchen, warehouse, and some commercial installations where the low-pressure rise limitations can be tolerated. Unit is low in first cost and low in operating cost and provides positive exhaust ventilation in the space which is a decided advantage over gravity-type exhaust units.

TYPES OF FANS: PERFORMANCE CHARACTERISTICS AND APPLICATIONS.

DRIVE ARRANGEMENTS

All fans must have some type of power source — usually an electric motor. On packaged fans, the motor is furnished and mounted by the manufacturer. On larger units, the motor is mounted separately and coupled directly to the fan or indirectly by a belt drive. A number of standard drive arrangements are shown in Figure 6-5.

Direct Drive offers a more compact assembly and assures constant fan speed. Fan speeds are limited to available motor speeds (except in the case of variable frequency controllers). Capacity is set during construction by variations in impeller geometry and motor speed.

Belt Drive offers flexibility in that fan speed can be changed by altering the drive ratio. This may be important in some applications to provide for changes in system capacity or pressure requirements due to changes in process, hood design, equipment location or air cleaning equipment. V-belt drives must be maintained and have some power losses which can be estimated from the chart in Figure 6-6.

NOISE

Fan noise is generated by turbulence within the fan housing and will vary by fan type, flow rate, pressure and fan efficiency. Because each design is different, noise ratings must be obtained from the fan manufacturer. Most fans produce a "white" noise which is a mixture of all frequencies. In addition to white noise, radial blade fans also produce a pure tone at a frequency equal to the blade passage frequency (BPF):

$$\text{BPF} = \text{RPM} \times \text{N} \times \text{CF} \qquad [6.1]$$

where:

BPF = blade passage frequency, Hz

RPM = rotational rate, rpm

N = number of blades

CF = conversion factor, $1/60$

This tone can be very noticeable in some installations and should be considered in the system design.

SAFETY AND ACCESSORIES

Safety Guards are required. Consider all danger points such as inlet, outlet, shaft, drive and cleanout doors. Construction should comply with applicable governmental safety requirements and attachment must be secure.

Accessories can help in the installation and in future maintenance requirements. Examples might include drains, cleanout doors, split housings and shaft seals.

Flow Control: Fan performance can be controlled by installing a damper directly on the fan inlet or outlet. This may be required on systems that vary throughout the day or for reduction in flow rate in anticipation of some future requirement. Dampers will build up with material and may not be acceptable on material handling fans. Two types of dampers are available:

- *Outlet Dampers* mount on the fan outlet to add resistance to the system when partially closed. These are available with both parallel and opposed blades. Selection depends on the degree of control required (opposed blade dampers will control the flow more evenly throughout the entire range from wide open to closed).
- *Inlet Dampers* mount on the fan inlet to pre-spin air into the impeller. This reduces fan output and lowers operating horsepower. Because of the power savings, inlet dampers should be considered when the fan will operate for long periods at reduced capacities.

6.3.2 Rating Tables: Fan size and operating RPM and BHP usually are obtained from a rating table based on required air flow and pressure. Tables are based on Fan Total Pressure or Fan Static Pressure:

$$\text{Fan TP} = (\text{SP}_{\text{outlet}} + \text{VP}_{\text{outlet}}) - (\text{SP}_{\text{inlet}} + \text{VP}_{\text{inlet}}) \qquad [6.2]$$

$$\text{Fan SP} = \text{SP}_{\text{outlet}} - \text{SP}_{\text{inlet}} - \text{VP}_{\text{inlet}} \qquad [6.3]$$

Fan Rating Tables are based on requirements for air at standard conditions (0.075 lbm/ft^3). If other than standard conditions exist, the actual pressure must be converted to standard conditions. See Section 6.3.8, "Selection at Air Densities Other Than Standard."

The most common form of table is a "multi-rating table" (see Table 6-1) which shows a range of capacities for a particular fan size. For a given pressure, the highest mechanical efficiency usually will be in the middle third of the "CFM" column. Some manufacturers show the rating of maximum efficiency for each pressure by underscoring or similar indicator. In the absence of such a guide, the design engineer must calculate the efficiency from the efficiency equation

$$\eta = \frac{\text{Q} \times \text{FTP}}{\text{CF} \times \text{PWR}} = \frac{\text{Q} \times (\text{FSP} + \text{VP}_{\text{outlet}})}{\text{CF} \times \text{PWR}} \qquad [6.4]$$

where:

η = Mechanical efficiency

Q = Volumetric flow rate, cfm

FTP = Fan Total Pressure, "wg

FSP = Fan Static Pressure, "wg

PWR = Power requirement, hp

CF = Conversion Factor, 6356

Even with a multi-rating table it is usually necessary to interpolate in order to select fan RPM and BHP for the exact conditions desired. In many cases a double interpolation will be necessary. Straight line interpolations throughout the multi-rating table will introduce negligible errors.

Certain types of fans may be offered in various Air Moving and Conditioning Association[6.1] construction classes identified as I through IV. A fan designated as meeting the requirements of a particular class must be physically capable of operating at any point within the performance limits for that class. Performance limits for each class are established

SW - Single Width **DW** - Double Width
SI - Single Inlet **DI** - Double Inlet

Arrangements (ARR.) 1, 3, 7, and 8 are also available with bearings mounted on pedestals or base set independent of the fan housing.

ARR. 1 SWSI For belt drive or direct connection. Impeller overhung. Two bearings on base.

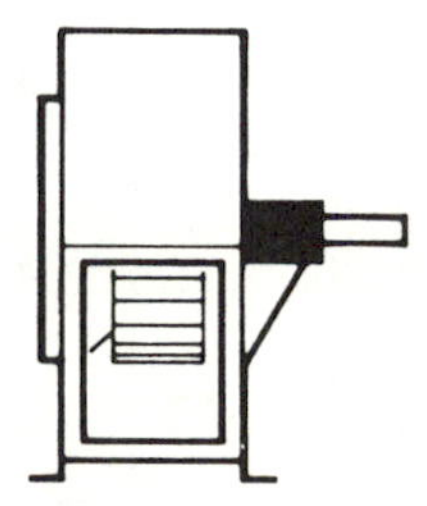

ARR. 2 SWSI For belt drive or direct connection. Impeller overhung. Bearings in bracket supported by fan housing.

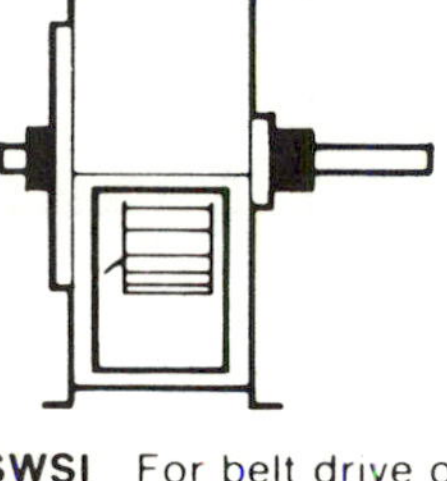
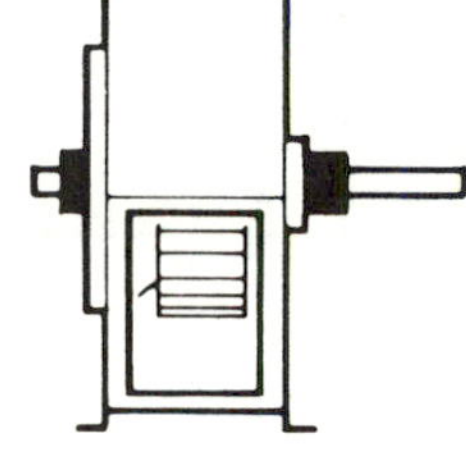

ARR. 3 SWSI For belt drive or direct connection. One bearing on each side and supported by fan housing.

ARR. 3 DWDI For belt drive or direct connection. One bearing on each side and supported by fan housing.

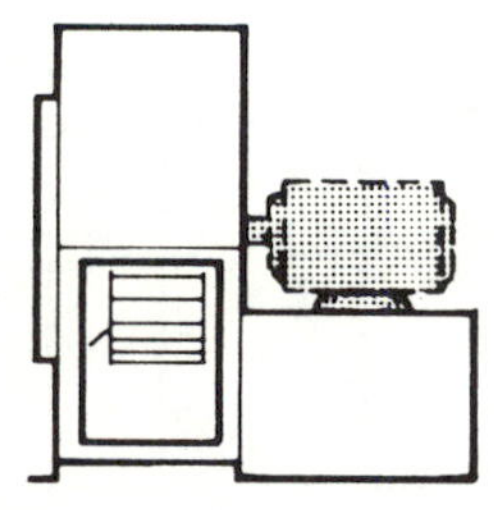

ARR. 4 SWSI For direct drive. Impeller overhung on prime mover shaft. No bearings on fan. Prime mover base mounted or integrally directly connected.

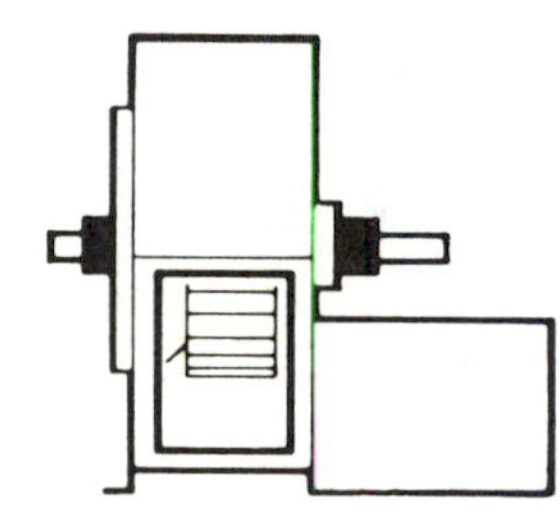

ARR. 7 SWSI For belt drive or direct connection. Arrangement 3 plus base for prime mover.

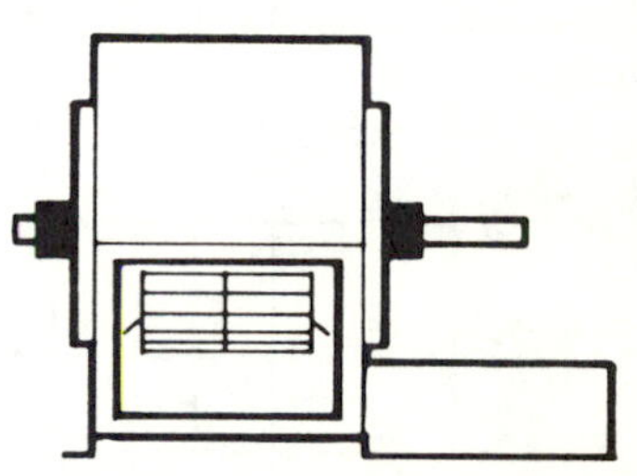

ARR. 7 DWDI For belt drive or direct connection. Arrangement 3 plus base for prime mover.

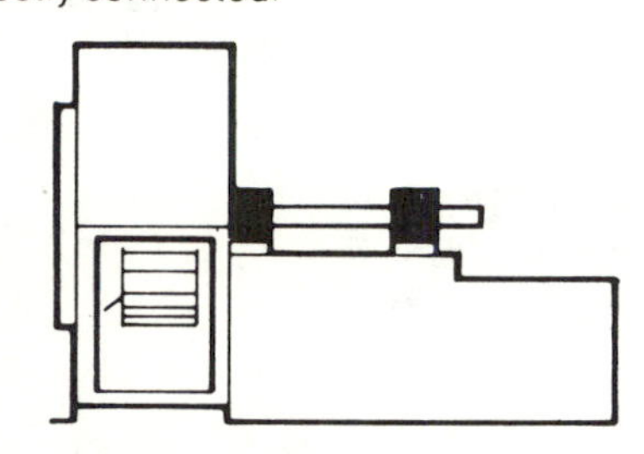

ARR. 8 SWSI For belt drive or direct connection. Arrangement 1 plus extended base for prime mover.

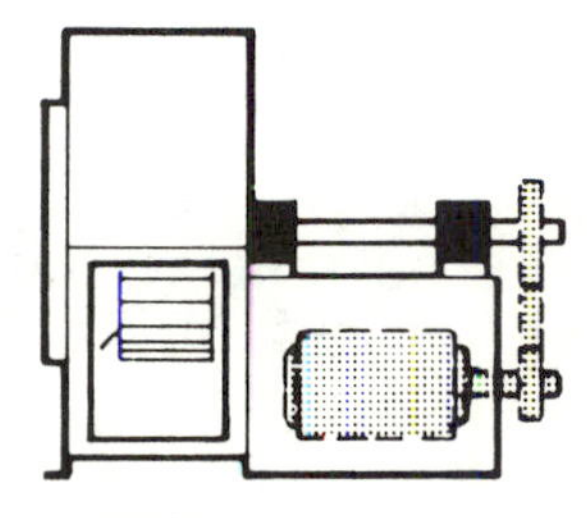

ARR. 9 SWSI For belt drive. Impeller overhung, two bearings, with prime mover outside base.

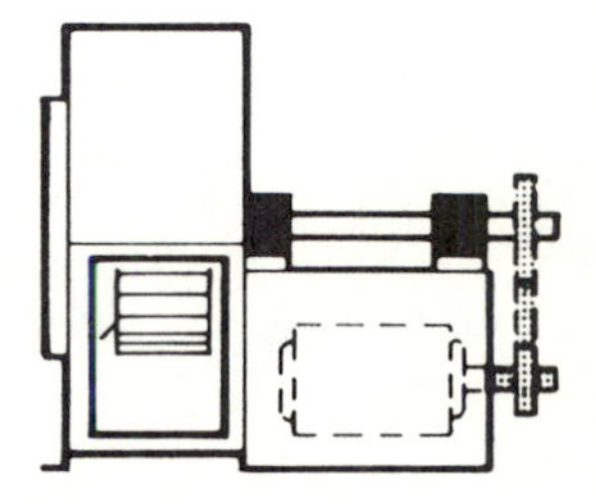

ARR. 10 SWSI For belt drive. Impeller overhung, two bearings, with prime mover inside base.

AMERICAN CONFERENCE OF GOVERNMENTAL INDUSTRIAL HYGIENISTS	*DRIVE ARRANGEMENTS FOR CENTRIFUGAL FANS (Adapted from AMCA 99-83)*	
	DATE *1-88*	FIGURE *6-5*

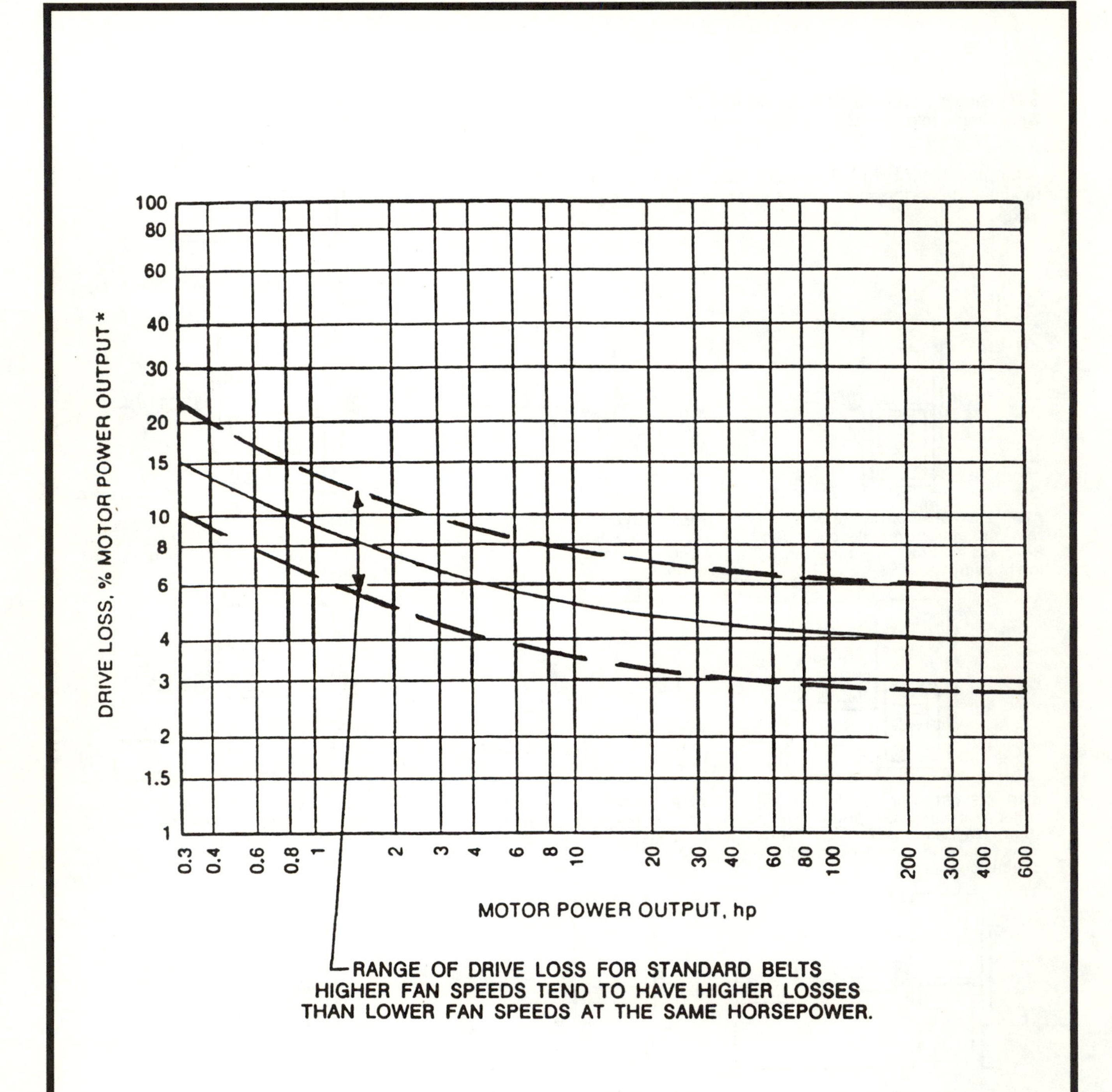

*Drive losses are based on the conventional V-belt.

AMERICAN CONFERENCE OF GOVERNMENTAL INDUSTRIAL HYGIENISTS	*ESTIMATED BELT DRIVE LOSS (Adapted from AMCA 203)*	
	DATE *1–88*	FIGURE *6–6*

TABLE 6-1. Example of Multi-Rating Table

Inlet diameter: 13" O.D. | Wheel diameter: 22⅜"
Outlet area: .930 sq. ft. inside | Wheel circumference: 5.92 ft.

CFM	OV	2"SP		4"SP		6"SP		8"SP		10"SP		12"SP		14"SP		16"SP		18"SP		20"SP		22"SP	
		RPM	BHP	RPM	BHP	RPM	BHP	RPM	BHP	RPM	BHP	RPM	BHP	RPM	BHP	RPM	BHP	RPM	BHP	RPM	BHP	RPM	BHP
930	1000	843	0.57	1176	1.21	1434	1.93	1653	2.75	1846	3.64	2021	4.59	2184	5.62	2333	6.68	2475	7.81	2610	9.01	2738	10.2
1116	1200	853	0.67	1183	1.35	1439	2.12	1656	2.98	1848	3.90	2022	4.89	2182	5.95	2333	7.07	2473	8.23	2606	9.45	2733	10.7
1302	1400	866	0.77	1191	1.51	1445	2.33	1660	3.22	1852	4.20	2025	5.23	2183	6.31	2333	7.47	2474	8.68	2606	9.95	2731	11.2
1488	1600	882	0.89	1201	1.69	1453	2.56	1668	3.50	1857	4.51	2030	5.59	2188	6.72	2337	7.92	2474	9.13	2606	10.4	2734	11.8
1674	1800	899	1.01	1213	1.88	1463	2.81	1676	3.81	1863	4.86	2035	5.98	2194	7.16	2340	8.38	2479	9.67	2610	11.0	2735	12.4
1860	2000	917	1.14	1227	2.09	1474	3.09	1685	4.13	1872	5.24	2040	6.39	2199	7.62	2344	8.89	2484	10.2	2613	11.6	2735	13.0
2046	2200	937	1.29	1242	2.32	1484	3.37	1694	4.48	1879	5.63	2048	6.84	2206	8.13	2351	9.43	2487	10.8	2618	12.2	2741	13.6
2232	2400	961	1.45	1257	2.56	1497	3.68	1704	4.85	1889	6.07	2056	7.33	2212	8.64	2357	10.0	2493	11.4	2622	12.8	2745	14.3
2418	2600	984	1.62	1275	2.81	1513	4.02	1717	5.25	1900	6.53	2065	7.84	2222	9.22	2364	10.6	2501	12.1	2631	13.6	2750	15.1
2790	3000	1038	2.02	1313	3.36	1543	4.73	1744	6.11	1924	7.52	2088	8.96	2241	10.4	2383	12.0	2517	13.5	2644	15.1	2766	16.7
3162	3400	1099	2.50	1358	3.99	1580	5.52	1775	7.05	1952	8.60	2115	10.2	2265	11.8	2405	13.4	2538	15.1	2665	16.8	2783	18.5
3534	3800	1164	3.07	1407	4.69	1620	6.37	1812	8.09	1984	9.79	2144	11.5	2290	13.3	2428	15.0	2562	16.8	2684	18.6	2803	20.5
3906	4200	1232	3.75	1462	5.48	1665	7.31	1851	9.19	2018	11.0	2174	12.9	2320	14.8	2458	16.8	2587	18.7	2708	20.6	2825	22.5
4278	4600	1306	4.56	1520	6.39	1717	8.38	1894	10.4	2058	12.4	2209	14.5	2355	16.5	2489	18.6	2614	20.6	2736	22.7	2852	24.8
4650	5000	1380	5.49	1582	7.41	1770	9.53	1941	11.7	2100	13.9	2247	16.1	2390	18.3	2521	20.5	2645	22.7	2766	25.0	2883	27.3
5022	5400	1457	6.56	1647	8.57	1827	10.8	1990	13.1	2146	15.5	2291	17.8	2428	20.2	2558	22.6	2681	25.0	2798	27.3		
5394	5800	1535	7.79	1719	9.93	1885	12.2	2045	14.7	2194	17.2	2334	19.7	2469	22.2	2594	24.7	2717	27.3	2830	29.8		

Performance shown is for fans with outlet ducts and with inlet ducts. BHP shown does not include belt drive losses.

in terms of outlet velocity and static pressure. Multi-rating tables usually will be shaded to indicate the selection zones for various classes or will state the maximum operating RPM. This can be useful in selecting equipment but class definition is only based on performance and will not indicate quality of construction.

Capacity tables which attempt to show the ratings for a whole series of homologous fans on one sheet cannot be used accurately unless the desired rating happens to be listed on the chart. Interpolation is practically impossible since usually only one point of the fan curve for a given speed is defined in such a table.

6.3.3 Point of Operation: Fans usually are selected for operation at some fixed condition or single "Point of Operation." Both the fan and the system have variable performance characteristics which can be represented graphically as curves depicting an array of operating points. The actual "point of operation" will be the one single point at the intersection of the fan curve and the system curve.

Fan Performance Curves: Certain fan performance variables usually are related to volumetric flow rate in graphic form to represent a fan performance curve. Figure 6-7 is a typical representation where pressure (P) and power requirement (PWR) are plotted against flow rate (Q). Other variables also may be included and more detailed curves representing various fan designs are provided in Figure 6-4. Pressure can be either fan static pressure (FSP) or fan total pressure (FTP). This depends on the manufacturer's method of rating.

It should be noted that a fan performance curve is always specific to a fan of given size operating at a single rotation rate (RPM). Even with size and rotation rate fixed, it should be obvious that pressure and power requirements vary over a range of flow rates.

System Requirement Curves: The duct system pressure also varies with volumetric flow rate. Figure 6-8 illustrates the variation of pressure (P) with flow rate (Q) for three different situations. The turbulent flow condition is representative of duct losses and is most common. In this case the pressure loss varies as the square of the flow rate.

The laminar flow condition is representative of the flow through low velocity filter media. Some wet collector designs operate at or close to a constant loss situation. The overall system curve results from the combined effects of the individual components.

6.3.4 Matching Fan Performance and System Requirement: A desired point of operation results from the process of

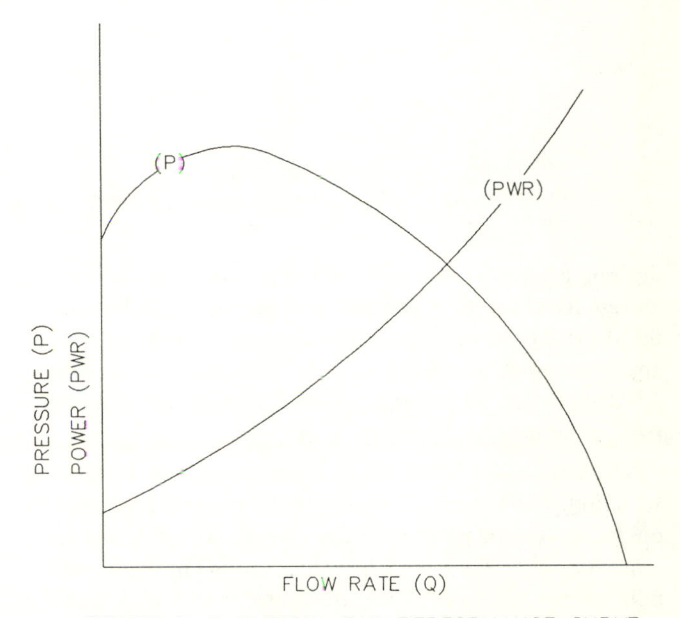

FIGURE 6-7. TYPICAL FAN PERFORMANCE CURVE.

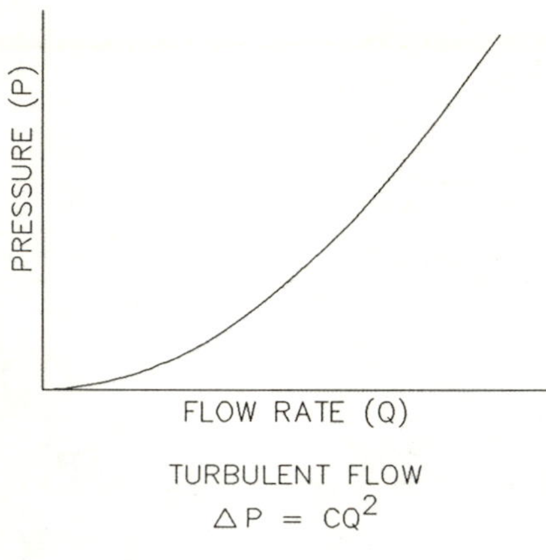

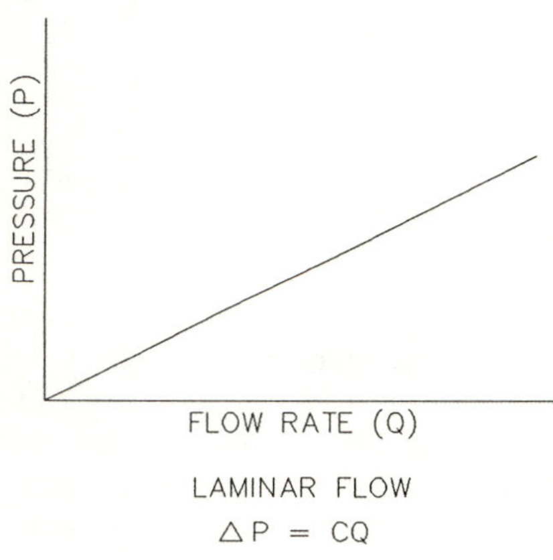

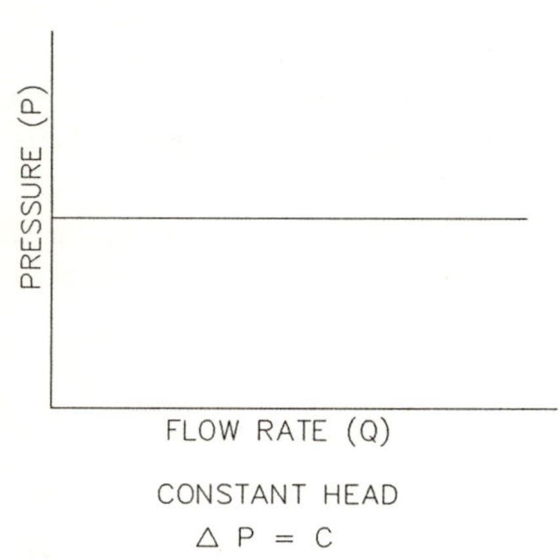

FIGURE 6-8. SYSTEM REQUIREMENT CURVES.

designing a duct system and selecting a fan. Considering the system requirement or fan performance curves individually, this desired point of operation has no special status relative to any other point of operation on the individual curve. Figure 6-9 depicts the four general conditions which can result from the system design fan selection process.

There are a number of reasons why the system design, fan selection, fabrication, and installation process can result in operation at some point other than design. When this occurs, it may become necessary to alter the system physically which will change the system requirement curve and/or cause a change in the fan performance curve. Because the fan performance curve is not only peculiar to a given fan but specific to a given rotation rate (RPM), a change of rotation rate can be relatively simple if a belt drive arrangement has been used. The "Fan Laws" are useful when changes of fan performance are required.

6.3.5 Fan Laws: These principles relate the performance variables for any homologous series of fans. A homologous series represents a range of sizes where all dimensional variables between sizes are proportional. The performance variables involved are fan size (SIZE), rotation rate (RPM), gas density (ρ), flow rate (Q), pressure (P), power requirement (PWR), and efficiency (η). Pressure (P) may be represented by total pressure (TP), static pressure (SP), velocity pressure (VP), fan static pressure (FSP), or fan total pressure (FTP).

At the same relative point of operation on any two performance curves in this homologous series, the efficiencies will be equal. The fan laws are mathematical expressions of these facts and establish the inter-relationship of the other variables. They predict the effect of changing size, speed, or gas density on capacity, pressure, and power requirement as follows:

$$Q_2 = Q_1\left(\frac{SIZE_2}{SIZE_1}\right)^3\left(\frac{RPM_2}{RPM_1}\right) \qquad [6.5]$$

$$P_2 = P_1\left(\frac{SIZE_2}{SIZE_1}\right)^2\left(\frac{RPM_2}{RPM_1}\right)^2\left(\frac{\rho_2}{\rho_1}\right) \qquad [6.6]$$

$$PWR_2 = PWR_1\left(\frac{SIZE_2}{SIZE_1}\right)^5\left(\frac{RPM_2}{RPM_1}\right)^3\left(\frac{\rho_2}{\rho_1}\right) \qquad [6.7]$$

As these expressions involve ratios of the variables, any convenient units may be employed so long as they are consistent. Size may be represented by any linear dimension since all must be proportional in a homologous series.

6.3.6 The Effect of Changing Rotation Rate or Gas Density: In practice, these principles are normally applied to determine the effect of changing only one variable. Most often the fan laws are applied to a given fan size and may be expressed in the simplified versions which follow:

- For changes of rotation rate:

  Flow varies directly with rotation rate; pressure varies as the square of the rotation rate; and power varies as the cube of the rotation rate.

$$Q_2 = Q_1\left(\frac{RPM_2}{RPM_1}\right) \qquad [6.8]$$

$$P_2 = P_1\left(\frac{RPM_2}{RPM_1}\right)^2 \qquad [6.9]$$

$$PWR_2 = PWR_1\left(\frac{RPM_2}{RPM_1}\right)^3 \qquad [6.10]$$

- For changes of gas density:

  Flow is not affected by a change in density; pressure and power vary directly with density:

$$Q_2 = Q_1 \qquad [6.11]$$

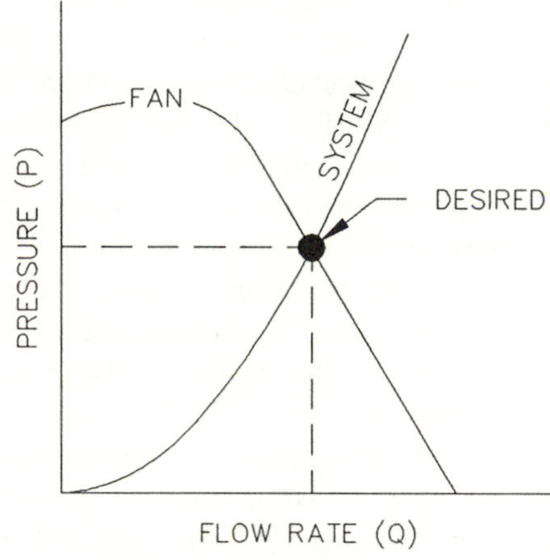

A. FAN AND SYSTEM MATCHED

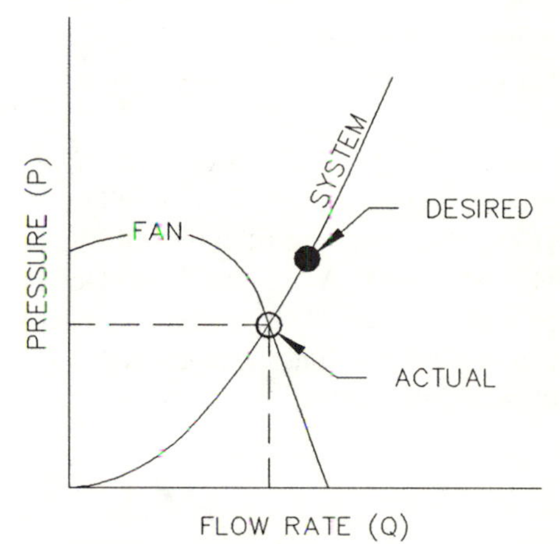

B. WRONG FAN.

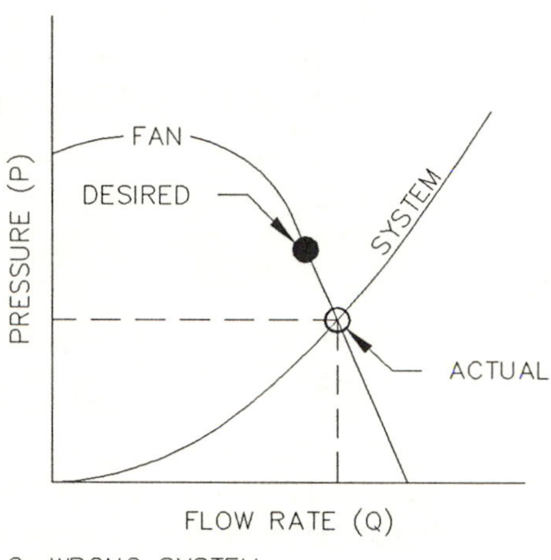

C. WRONG SYSTEM.

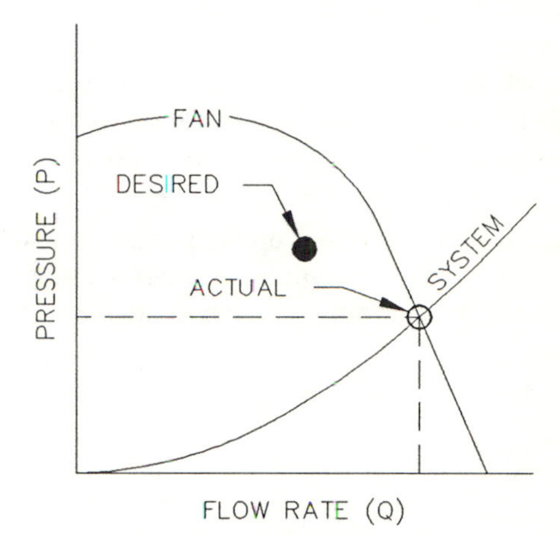

D. BOTH FAN AND SYSTEM WRONG

AMERICAN CONFERENCE OF GOVERNMENTAL INDUSTRIAL HYGIENISTS	*ACTUAL VERSUS DESIRED POINT OF OPERATION*
DATE *1–88*	FIGURE *6–9*

$$P_2 = P_1\left(\frac{\rho_2}{\rho_1}\right) \quad [6.12]$$

$$PWR_2 = PWR_1\left(\frac{\rho_2}{\rho_1}\right) \quad [6.13]$$

6.3.7 Limitations on the Use of Fan Laws: These expressions are equations which rely on the fact that the performance curves are homologous and that the ratios are for the *same relative points of rating* on each curve. Care must be exercised to apply the laws between the same relative points of rating.

Figure 6-10 contains a typical representation of two homologous fan performance curves, PQ_1 and PQ_2. These could be the performances resulting from two different rotation rates, RPM_1 and RPM_2. Assuming a point of rating indicated as A_1 on PQ_1 there is only one location on PQ_2 with the same relative point of rating and that is at A_2. The A_1 and A_2 points of rating are related by the expression

$$P_{A_2} = P_{A_1}\left(\frac{Q_{A_2}}{Q_{A_1}}\right)^2 \quad [6.14]$$

This equation can be used to identify every other point that would have the same relative point of rating as A_1 and A_2. The line passing through A_2,A_1 and the origin locates all conditions with the same relative points of rating. These lines are more often called "system lines" or "system curves." As discussed in Section 6.3.3, there are a number of exceptions to the condition where system pressure varies as the square of flow rate. These lines representing the same relative points of rating are "system lines" or "system curves" for turbulent flow conditions only.

Where turbulent flow conditions apply, it must be understood that the system curves or lines of relative points of rating represent a system having fixed physical characteristics. For example the, B_2,B_1 line defines another system which has lower resistance to flow than the A_2,A_1 system.

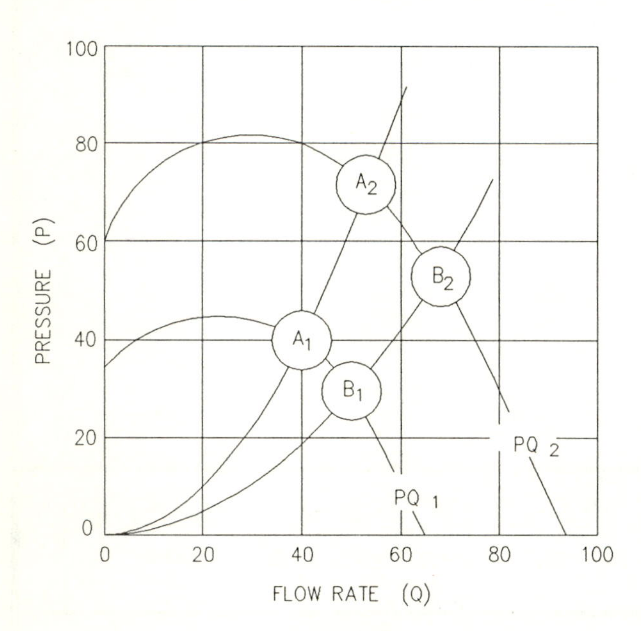

FIGURE 6–10 HOMOLGOUS PERFORMANCE CURVES

Special care must be exercised when applying the fan laws in the following cases:

1. Where any component of the system does not follow the "pressure varies as the square of the flow rate" rule.
2. Where the system has been physically altered or for any other reason operates on a different system line.

6.3.8 Fan Selection at Air Density Other Than Standard: As discussed in Section 6.3.6, fan performance is affected by changes in gas density. Variations in density due to normal fluctuations of ambient pressure, temperature and humidity are small and need not be considered. Where temperature, humidity, elevation, pressure, gas composition or a combination of two or more cause density to vary by more than 5 percent from the standard 0.075 lbm/ft^3, corrections should be employed.

Rating tables and performance curves as published by fan manufacturers are based on standard air. Performance variables are always related to conditions at the fan inlet. Fan characteristics are such that volumetric flow rate (Q) is unaffected but pressure (P) and power (PWR) vary directly with changes in gas density. Therefore, the selection process requires that rating tables are entered with *actual* volumetric flow rate but with a corrected or equivalent pressure.

The equivalent pressure is that pressure corresponding to standard density and is determined from Equation 6.12 as follows:

$$P_e = P_a\left(\frac{0.075}{\rho_a}\right)$$

where:

P_e = Equivalent Pressure

P_a = Actual Pressure

ρ_a = Actual density, lbm/ft^3

The pressures (P_e and P_a) can be either Fan Static Pressure or Fan Total Pressure in order to conform with the manufacturer's rating method.

The fan selected in this manner is to be operated at the rotation rate indicated in the rating table and actual volumetric flow rate is that indicated by the table. However, the pressure developed is not that indicated in the table but is the actual value. Likewise, the power requirement is not that of the table as it also varies directly with density. The actual power requirement can be determined from Equation 6.13 as follows:

$$PWR_a = PWR_t\left(\frac{\rho_a}{0.075}\right)$$

where:

PWR_a = Actual Power Requirement

PWR_t = Power Requirement in Rating Table

ρ_a = Actual Density, lbm/ft³

Fan selection at non-standard density requires knowledge of the actual volumetric flow rate at the fan inlet, the actual pressure requirement (either FSP or FTP, depending on the rating table used) and the density of the gas at the fan inlet. The determination of these variables requires that the system design procedure consider the effect of density as discussed in Chapter 5.

EXAMPLE

Consider the system illustrated in Figure 6-11 where the heater causes a change in volumetric flow rate and density. For simplicity, assume the heater has no resistance to flow and that the sum of friction losses will equal FSP. Using the Multi-Rating Table, Table 6-1, select the rotation rate and determine power requirements for fan locations ahead of and behind the heater.

Location 1: *Fan ahead of the heater* (side "A" to "B" in Figure 6-11).

Step 1. Determine actual FSP

$$\begin{aligned} FSP &= 1\ ''wg + 3\ ''wg \\ &= 4\ ''wg \text{ at } 0.075\ lbm/ft^3 \end{aligned}$$

Step 2a. Density at fan inlet is standard. Therefore, enter rating table with actual volumetric flow rate at fan inlet, 1000 acfm, and FSP of 4 ″wg.

b. Interpolation from Table 6-1 results in:

RPM = 1182 rpm

PWR = 1.32 bhp

Step 3. The fan should be operated at 1182 rpm and actual power requirement will be 1.32 bhp.

Location 2: *Fan behind the heater* (side "B" to "C" in Figure 6-11).

Step 1. Determine actual FSP

$$\begin{aligned} FSP &= 1\ ''wg + 3\ ''wg \\ &= 4\ ''wg \text{ at } 0.0375\ lbm/ft^3 \end{aligned}$$

Step 2a. Density at fan inlet is not standard and a pressure correction must be made (using Equation 6.12) to determine equivalent FSP.

$$\begin{aligned} FSP_e &= FSP_a\left(\frac{0.075}{\rho_a}\right) \\ &= 4\ ''wg\left(\frac{0.075}{0.0375}\right) \\ &= 8\ ''wg \end{aligned}$$

Now, enter rating table with actual volumetric flow rate at fan inlet, 2000 acfm, and equivalent FSP, 8 ″wg.

b. Interpolation from Table 6-1 results in:

RPM = 1692 rpm

PWR = 4.39 bhp

Step 3a. The fan should be operated at 1692 rpm, but actual power requirements will be affected by the density and can be determined by using Equation 6.13.

$$\begin{aligned} PWR_a &= PWR_t\left(\frac{\rho_a}{0.075}\right) \\ &= 4.39\left(\frac{0.0375}{0.075}\right) \\ &= 2.2\ bhp \end{aligned}$$

b. It also should be noted that a measurement of FSP will result in the value of 4 ″wg (actual) and not the equivalent value of 8 ″wg.

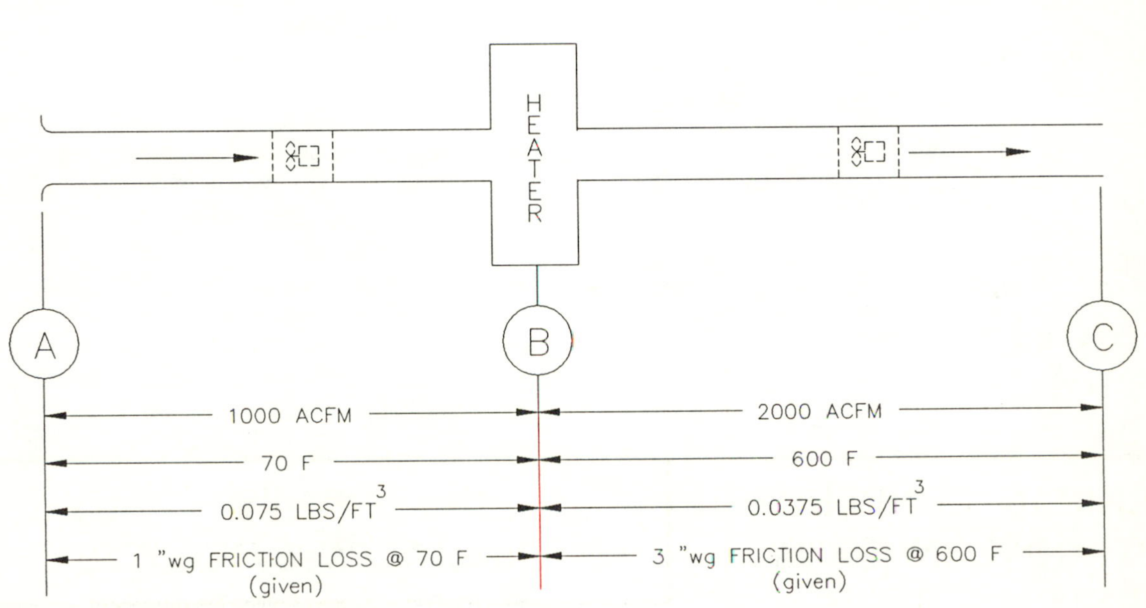

FIGURE 6–11 IN DUCT HEATER.

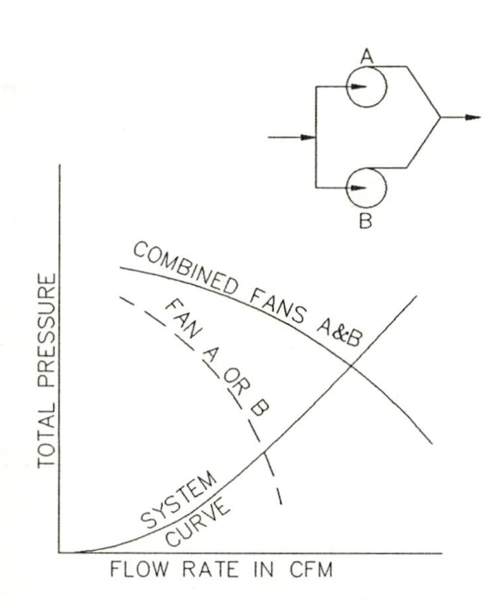

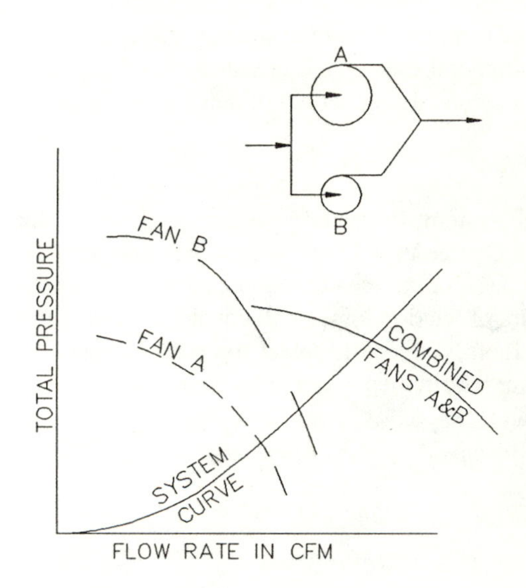

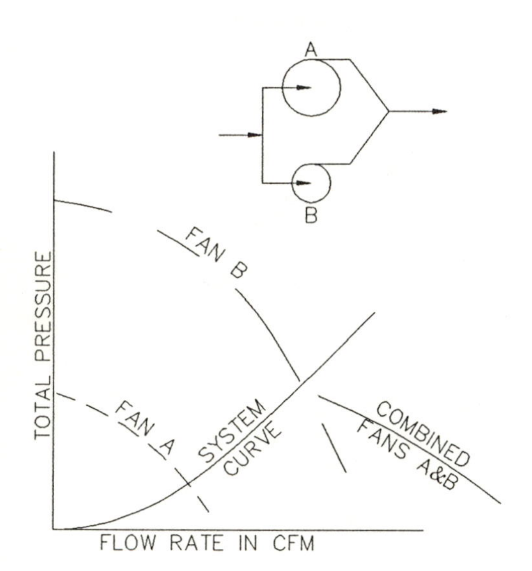

NOTES:
1. TO ESTABLISH COMBINED FAN CURVE, THE COMBINED AIR FLOW RATE, Q, IS THE SUM OF INDIVIDUAL FAN AIR VOLUMES AT POINTS OF EQUAL PRESSURE
2. TO ESTABLISH SYSTEM CURVE, INCLUDE LOSSES IN INDIVIDUAL FAN CONNECTIONS.
3. SYSTEM CURVE MUST INTERSECT COMBINED FAN CURVE OR HIGHER PRESSURE FAN MAY HANDLE MORE AIR ALONE.

WHEN SYSTEM CURVE DOES NOT CROSS COMBINED FAN CURVE, OR CROSSES PROJECTED COMBINED CURVE BEFORE FAN B, FAN B WILL HANDLE MORE AIR THAN FANS A AND B IN PARALLEL.

AMERICAN CONFERENCE OF GOVERNMENTAL INDUSTRIAL HYGIENISTS	*FANS PARALLEL OPERATION*			
	DATE	*1–88*	FIGURE	*6–12*

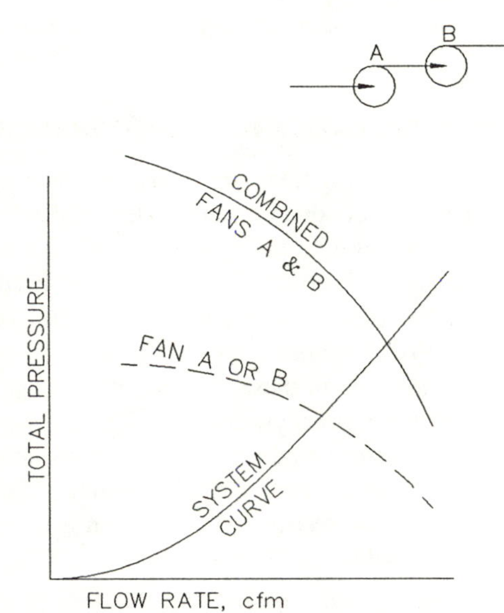

TWO IDENTICAL FANS
RECOMMENDED FOR BEST EFFICIENCY

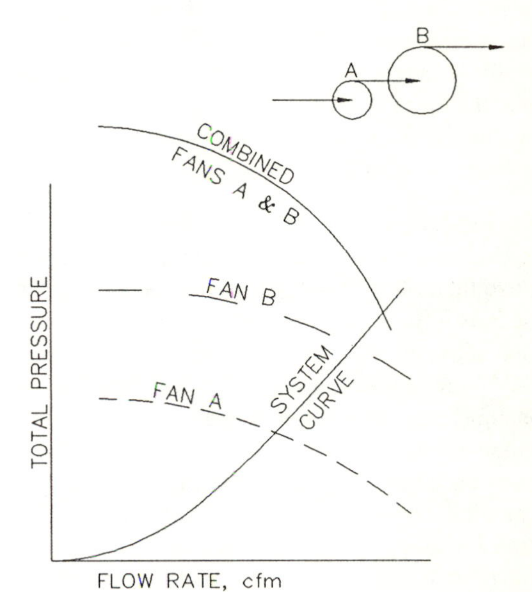

TWO DIFFERENT FANS
SATISFACTORY

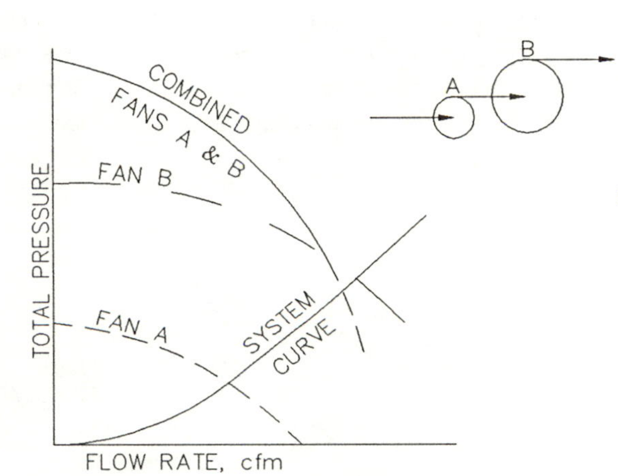

TWO DIFFERENT FANS
UNSATISFACTORY

WHEN SYSTEM CURVE DOES NOT INTERSECT COMBINED FAN CURVE, OR CROSSES PROJECTED COMBINED CURVE BEFORE FAN B CURVE, FAN B WILL MOVE MORE AIR THAN FAN A AND B IN SERIES.

NOTES:

1. TO ESTABLISH COMBINED FAN CURVE, THE COMBINED TOTAL PRESSURE IS THE SUM OF INDIVIDUAL FAN PRESSURES AT EQUAL AIR FLOW RATES, LESS THE PRESSURE LOSS IN THE FAN CONNECTIONS.

2. AIR FLOW RATE THROUGH EACH FAN WILL BE THE SAME, SINCE AIR IS CONSIDERED INCOMPRESSIBLE.

3. SYSTEM CURVE MUST INTERSECT COMBINED FAN CURVE OR LARGE FLOW RATE FAN MAY HANDLE MORE AIR ALONE.

AMERICAN CONFERENCE OF GOVERNMENTAL INDUSTRIAL HYGIENISTS	*FANS SERIES OPERATION*		
	DATE 1–88		FIGURE 6–13

It will be noted that regardless of location the fan will handle the same mass flow rate. Also, the actual resistance to flow is not affected by fan location. It may appear then that there is an error responsible for the differing power requirements of 1.32 bhp versus 2.2 bhp. In fact, the fan must work harder at the lower density to move the same mass flow rate. This additional work results in a higher temperature rise in the air from fan inlet to outlet.

6.3.9 Explosive or Flammable Materials: When conveying explosive or flammable materials, it is important to recognize the potential for ignition of the gas stream. This may be from airborne material striking the impeller or by the physical movement of the impeller into the fan casing. AMCA(6.1) and other associations offer guidelines for both the manufacturer and the user on ways to minimize this danger. These involve more permanent attachment of the wheel to the shaft and bearings and the use of buffer plates or spark resistant alloy construction. Because no single type of construction fits all applications, it is imperative that both the manufacturer and the user are aware of the dangers involved and agree on the type of construction and degree of protection that is being proposed.

NOTE: For many years aluminum alloy impellers have been specified to minimize sparking if the impeller were to contact other steel parts. This is still accepted but recent tests by the U. S. Bureau of Mines(6.2) and others have demonstrated that impact of aluminum with rusty steel creates a "Thermite" reaction and thus possible ignition hazards. Special care must be taken when aluminum alloys are used in the presence of steel.

6.4 FAN INSTALLATION AND MAINTENANCE

Fan rating tests for flow rate, static pressure and power requirements are conducted under ideal conditions which include uniform straight air flow at the fan inlet and outlet. However, if in practice duct connections to the fan cause non-uniform air flow, fan performance and operating efficiency will be affected. Location and installation of the fan must consider the location of these duct components to minimize losses. If adverse connections must be used, appropriate compensation must be made in the system calculations. Once the system is installed and operating, routine inspection and maintenance will be required if the system is to continue to operate at original design levels.

6.4.1 System Effect: System effect is defined as the estimated loss in fan performance from this non-uniform air flow. Figure 6-14 illustrates deficient fan system performance. The system pressure losses have been determined accurately and a suitable fan selected for operation at Point 1. However, no allowance has been made for the effect of the system connections on fan performance. The point of intersection between the resulting fan performance curve and the

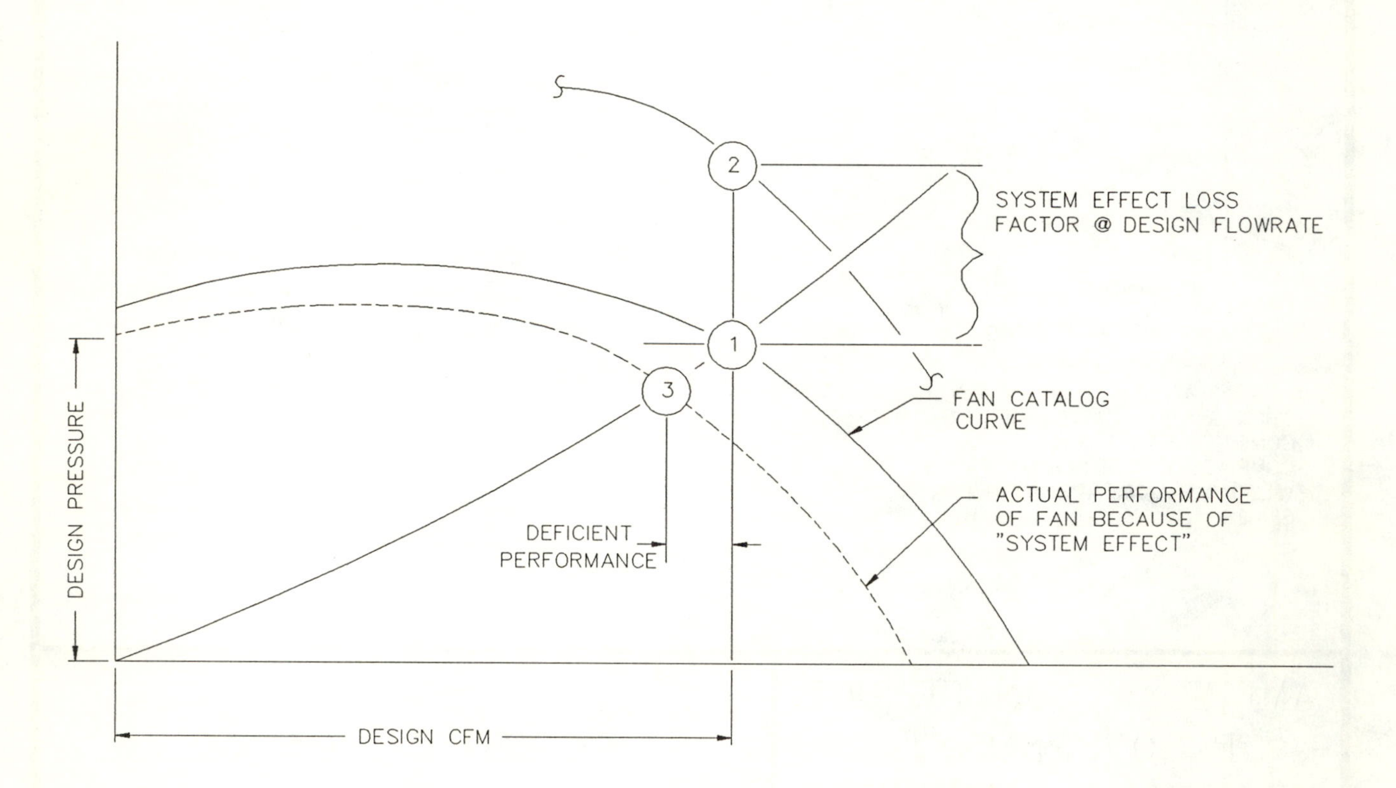

FIGURE 6–14. SYSTEM EFFECT FACTOR.

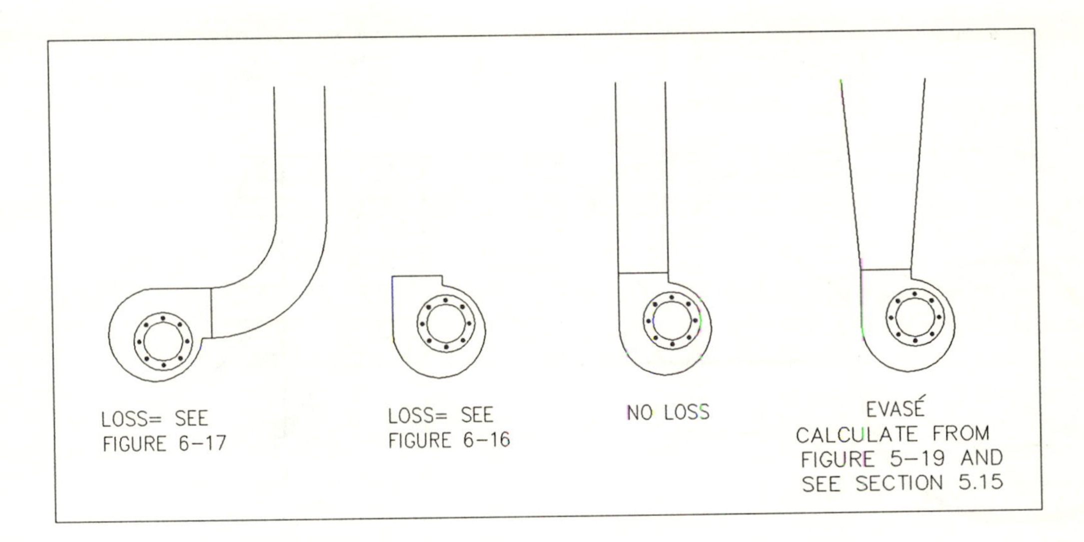

FIGURE 6-15. FAN DISCHARGE CONDITIONS.

actual system curve is Point 3. The resulting flow rate will, therefore, be deficient by the difference from 1 to 3. To compensate for this system effect, it will be necessary to add a "system effect factor" to the calculated system pressure. This will be equal to the pressure difference between Points 1 and 2 and will have to be added to the calculated system pressure losses. The fan then will be selected for this higher pressure (Point 2) but will operate at Point 1 due to loss in performance from system effects.

Figure 6-15 illustrates typical discharge conditions and the losses which may be anticipated. The magnitude of the change in system performance caused by elbows and other obstructions placed too close to a fan inlet or outlet can be estimated for the conditions shown on Figures 6-16 through 6-21 as follows:

Addition to System Static Pressure = System Effect Factor × VP

A vortex or spin of the air stream entering the fan inlet may be created by non-uniform flow conditions as illustrated in Figure 6-20. These conditions may be caused by a poor inlet box, multiple elbows or ducts near the inlet or by other spin producing conditions. Since the variations resulting in inlet spin are many, no System Effect Factors are tabulated. Where a vortex or inlet spin cannot be avoided or is discovered at an existing fan inlet, the use of turning vanes, splitter sheets or egg-crate straighteners will reduce the effect.

6.4.2 *Inspection and Maintenance:* Wear or accumulation on an impeller will cause weakening of the impeller structure and/or serious vibration. If these vibrations are severe, damage or failure also can occur at the bearings or fan structure.

Fan rotation often is reversed inadvertently during repair or alterations to wiring circuits or starters. As centrifugal fans do move a fraction of their rated capacity when running backward, incorrect rotation often goes unnoticed in spite of less effective performance of the exhaust system.

Scheduled inspection of fans is recommended. Items checked should include:

1. Bearings for proper operating temperature (greasing on an established schedule).
2. Excessive vibration of bearings or housing.
3. Belt drives for proper tension and minimum wear.
4. Correct coupling alignment.
5. Fan impeller for proper alignment and rotation.
6. Impeller free from excess wear or material accumulation.

REFERENCES

6.1. Air Movement and Control Association, Inc.: 30 W. University Dr., Arlington Heights, IL 60004.

6.2. N. Gibson, F.C. Lloyd and G.R. Perry: *Fire Hazards in Chemical Plants from Friction Sparks Involving the Thermite Reaction*. Symposium Series No. 25. Inst. Chem. Engrs., London (1968).

6.3. Air Movement and Control Association, Inc.: *Bulletin 201*. 30 W. University Dr., Arlington Heights, IL 60004.

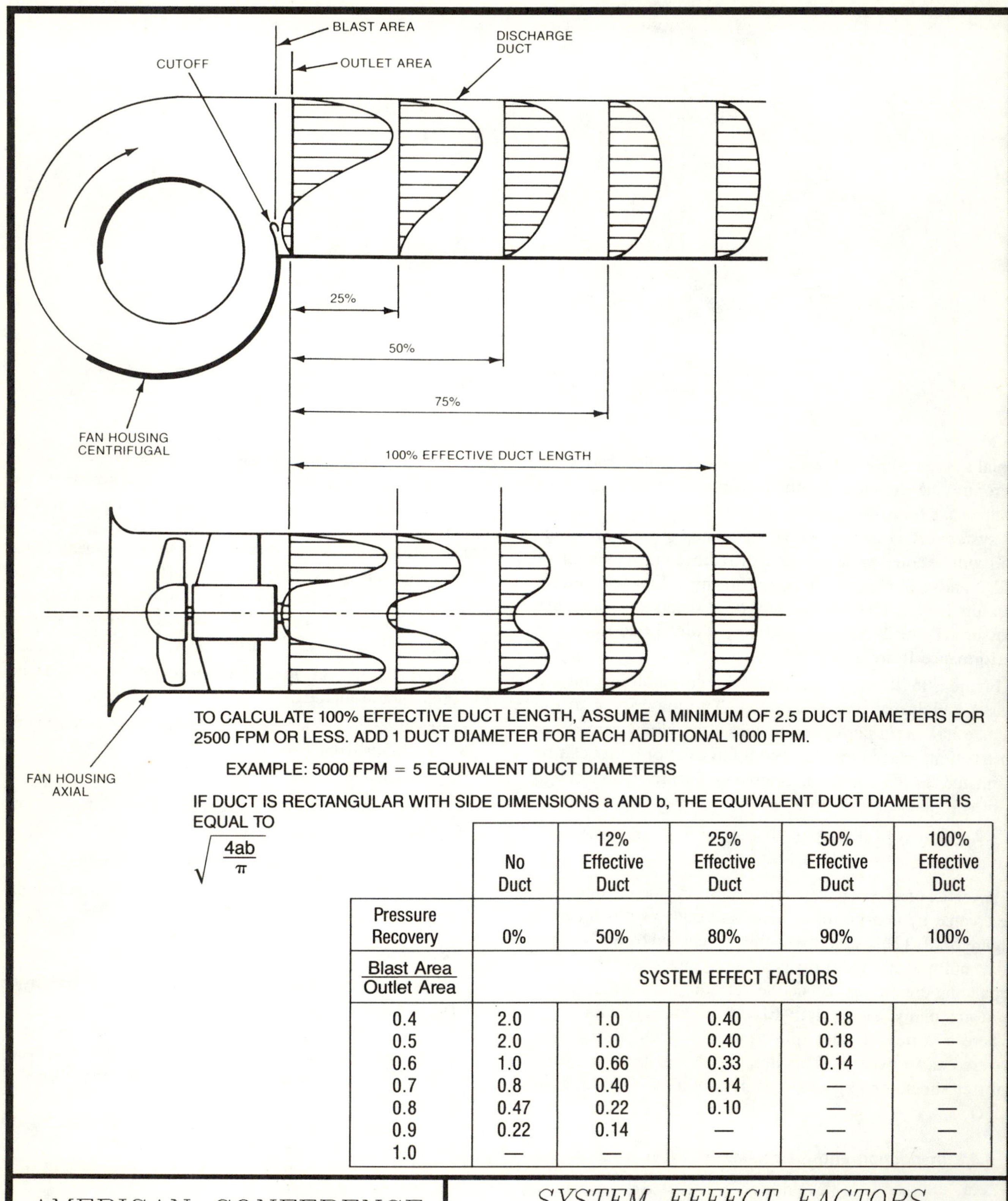

TO CALCULATE 100% EFFECTIVE DUCT LENGTH, ASSUME A MINIMUM OF 2.5 DUCT DIAMETERS FOR 2500 FPM OR LESS. ADD 1 DUCT DIAMETER FOR EACH ADDITIONAL 1000 FPM.

EXAMPLE: 5000 FPM = 5 EQUIVALENT DUCT DIAMETERS

IF DUCT IS RECTANGULAR WITH SIDE DIMENSIONS a AND b, THE EQUIVALENT DUCT DIAMETER IS EQUAL TO

$$\sqrt{\frac{4ab}{\pi}}$$

	No Duct	12% Effective Duct	25% Effective Duct	50% Effective Duct	100% Effective Duct
Pressure Recovery	0%	50%	80%	90%	100%
Blast Area / Outlet Area	SYSTEM EFFECT FACTORS				
0.4	2.0	1.0	0.40	0.18	—
0.5	2.0	1.0	0.40	0.18	—
0.6	1.0	0.66	0.33	0.14	—
0.7	0.8	0.40	0.14	—	—
0.8	0.47	0.22	0.10	—	—
0.9	0.22	0.14	—	—	—
1.0	—	—	—	—	—

AMERICAN CONFERENCE OF GOVERNMENTAL INDUSTRIAL HYGIENISTS

SYSTEM EFFECT FACTORS FOR OUTLET DUCTS (Adapted from AMCA 201)

DATE 1-88 FIGURE 6-16

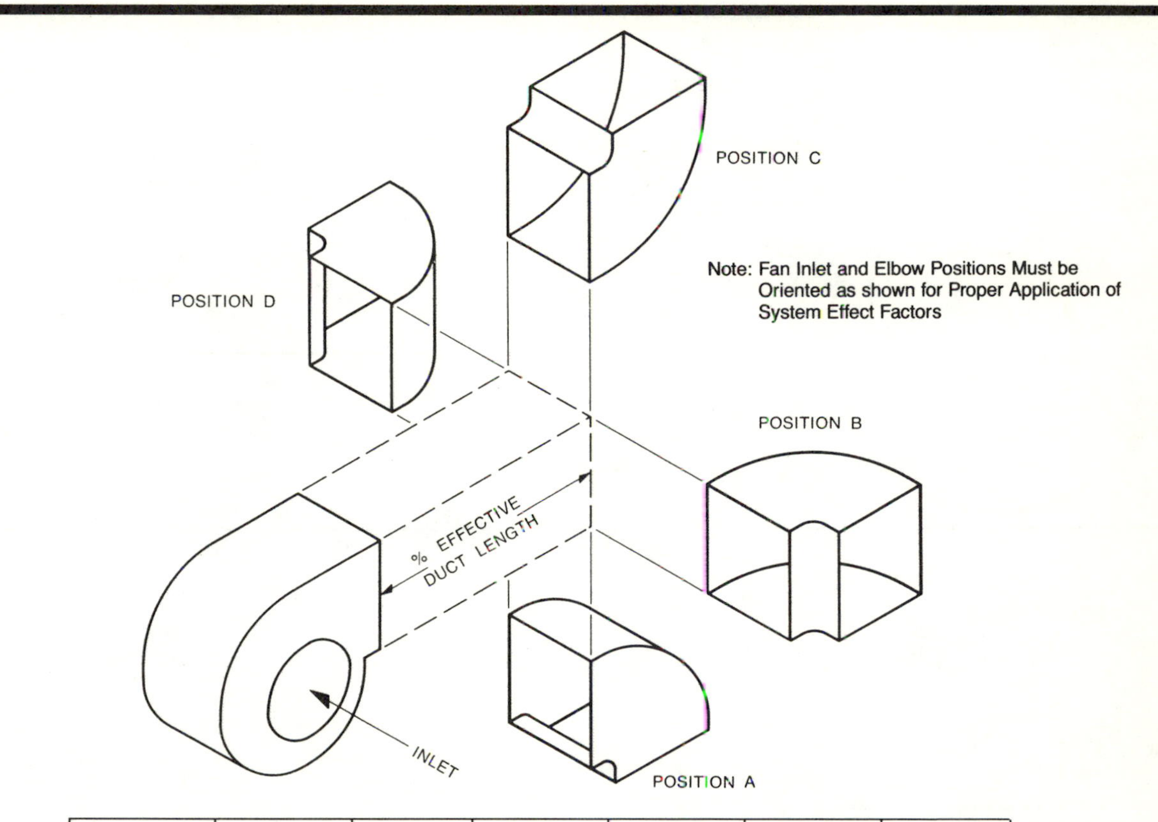

Blast Area / Outlet Area	Outlet Elbow Position	No Outlet Duct	12% Effective Duct	25% Effective Duct	50% Effective Duct	100% Effective Duct
0.4	A	3.2	2.5	1.8	0.8	NO SYSTEM EFFECT FACTOR
	B	4.6	3.9	2.5	1.2	
	C	5.5	4.6	3.2	1.6	
	D	5.5	4.6	3.2	1.6	
0.5	A	2.0	1.6	1.2	0.53	
	B	2.9	2.3	1.8	0.80	
	C	3.9	2.9	2.3	1.0	
	D	3.9	2.9	2.3	1.0	
0.6	A	1.6	1.4	1.0	0.40	
	B	2.0	1.6	1.2	0.53	
	C	2.9	2.3	1.8	0.80	
	D	2.5	2.0	1.4	0.66	
0.7	A	0.66	0.53	0.40	0.18	
	B	1.0	0.80	0.53	0.26	
	C	1.4	1.2	0.80	0.33	
	D	1.2	1.0	0.66	0.33	
0.8	A	0.8	0.66	0.47	0.22	
	B	1.2	1.0	0.66	0.33	
	C	1.6	1.4	1.0	0.40	
	D	1.4	1.2	0.8	0.33	
0.9	A	0.66	0.53	0.40	0.18	
	B	1.0	0.80	0.53	0.26	
	C	1.2	1.0	0.66	0.33	
	D	1.0	0.80	0.53	0.26	
1.0	A	1.0	0.80	0.53	0.26	
	B	0.66	0.53	0.40	0.18	
	C	1.0	0.80	0.53	0.26	
	D	1.0	0.80	0.53	0.26	

AMERICAN CONFERENCE OF GOVERNMENTAL INDUSTRIAL HYGIENISTS

SYSTEM EFFECT FACTORS FOR OUTLET ELBOWS (Adapted from AMCA 201)

DATE 1-88 | FIGURE 6-17

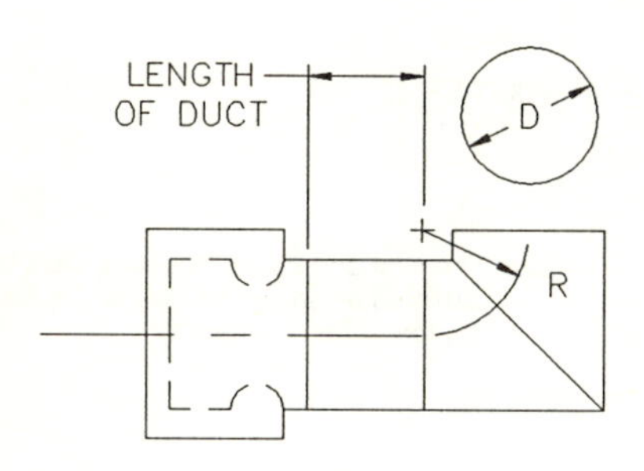

SYSTEM EFFECT FACTORS			
R/D	NO DUCT	2D DUCT	5D DUCT
–	3.2	2.0	1.0

A. TWO–PIECE MITERED 90° ROUND SECTION ELBOW –– NOT VANED.

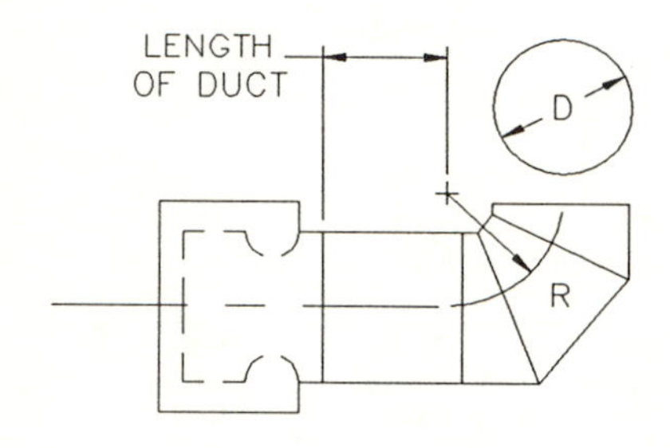

SYSTEM EFFECT FACTORS			
R/D	NO DUCT	2D DUCT	5D DUCT
0.5	2.5	1.6	0.8
0.75	1.6	1.0	0.47
1.0	1.2	0.66	0.33
2.0	1.0	0.53	0.33
3.0	0.8	0.47	0.26

B. THREE–PIECE MITERED 90° ROUND SECTION ELBOW –– NOT VANED.

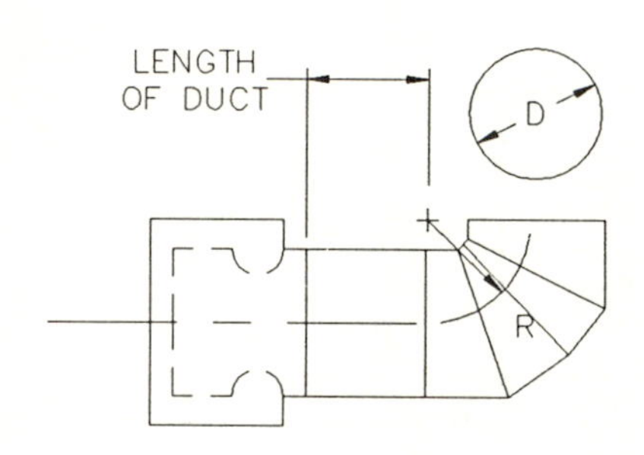

SYSTEM EFFECT FACTORS			
R/D	NO DUCT	2D DUCT	5D DUCT
0.5	1.8	1.0	0.53
0.75	1.4	0.8	0.40
1.0	1.2	0.66	0.33
2.0	1.0	0.53	0.33
3.0	0.66	0.40	0.22

C. FOUR OR MORE PIECE MITERED 90° ROUND SECTION ELBOW –– NOT VANED.

AMERICAN CONFERENCE OF GOVERNMENTAL INDUSTRIAL HYGIENISTS	*SYSTEM EFFECT FACTORS FOR VARIOUS MITERED ELBOWS WITHOUT TURNING VANES (Adapted from AMCA 201)*
	DATE *1–88* FIGURE *6–18*

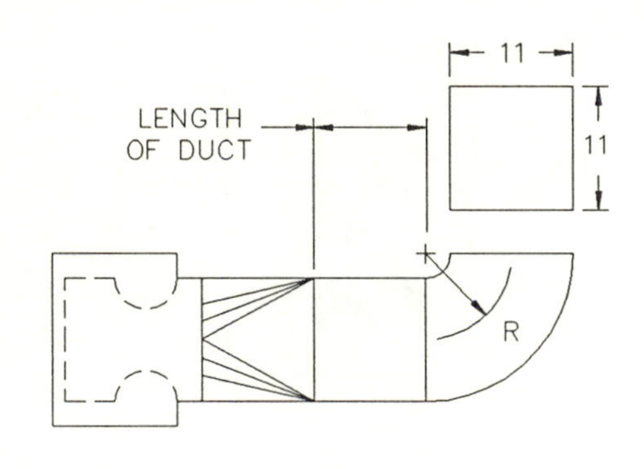

SYSTEM EFFECT FACTORS

R/D	NO DUCT	2D DUCT	5D DUCT
0.5	2.5	1.6	0.8
0.75	2.0	1.2	0.66
1.0	1.2	0.66	0.33
2.0	0.8	0.47	0.26

A. SQUARE ELBOW WITH INLET TRANSITION -- NO TURNING VANES.

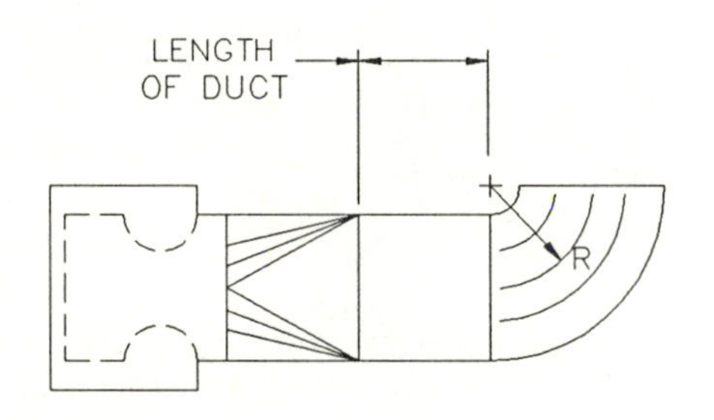

SYSTEM EFFECT FACTORS

R/D	NO DUCT	2D DUCT	5D DUCT
0.5	0.8	0.47	0.26
1.0	0.53	0.33	0.18
2.0	0.26	0.22	0.14

B. SQUARE ELBOW WITH INLET TRANSITION -- 3 LONG TURNING VANES.

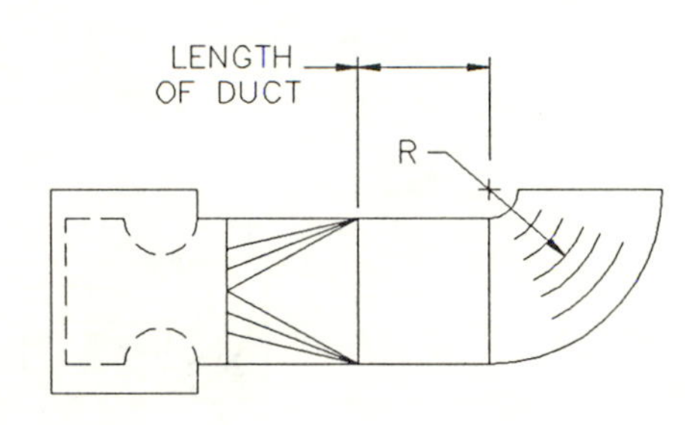

SYSTEM EFFECT FACTORS

R/D	NO DUCT	2D DUCT	5D DUCT
0.5	0.8	0.47	0.26
1.0	0.53	0.33	0.18
2.0	0.26	0.22	0.14

C. SQUARE ELBOW WITH INLET TRANSITION -- SHORT TURNING VANES.

$D = \frac{2H}{\sqrt{\pi}}$

THE INSIDE AREA OF THE SQUARE DUCT (H X H) IS EQUAL TO THE INSIDE AREA CIRCUMSCRIBED BY THE FAN INLET COLLAR. THE MAXIMUM PERMISSIBLE ANGLE OF ANY CONVERGING ELEMENT OF THE TRANSITION IS 15°, AND FOR A DIVERGING ELEMENT 7.5°.

AMERICAN CONFERENCE OF GOVERNMENTAL INDUSTRIAL HYGIENISTS	*SYSTEM EFFECT FACTORS FOR VARIOUS DUCT ELBOWS (Adapted from AMCA 201)*
	DATE 1-88 / FIGURE 6-19

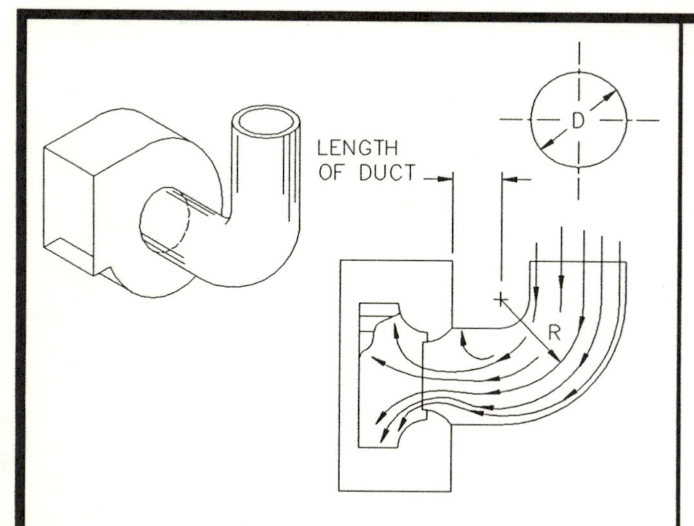

SYSTEM EFFECT FACTORS

R/D	NO DUCT	2D DUCT	5D DUCT
0.75	1.4	0.8	0.40
1.0	1.2	0.66	0.33
2.0	1.0	0.53	0.33
3.0	0.66	0.40	0.22

A. NON–UNIFORM FLOW INTO A FAN INLET BY A 90° ROUND SECTION ELBOW – NO TURNING VANES.

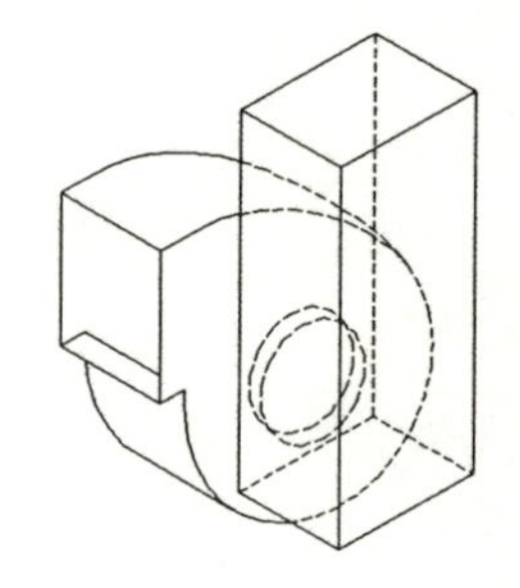

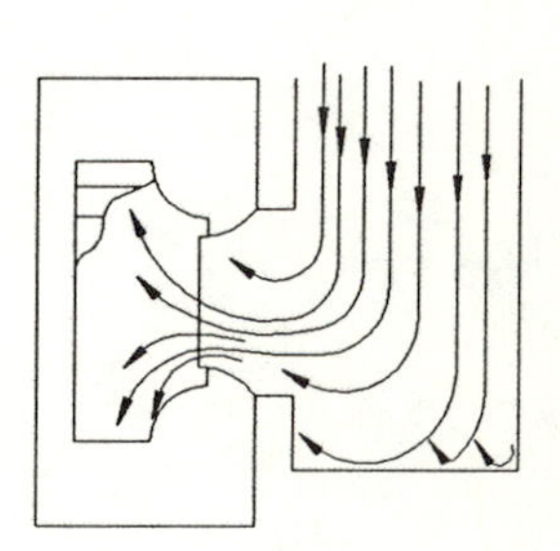

THE REDUCTION IN FLOW RATE AND PRESSURE FOR THIS TYPE OF INLET CONDITION IS IMPOSSIBLE TO TABULATE. THE MANY POSSIBLE VARIATIONS IN WIDTH AND DEPTH OF THE DUCT INFLUENCE THE REDUCTION IN PERFORMANCE TO VARYING DEGREES AND THEREFORE THIS INLET SHOULD BE AVOIDED. FLOW RATE LOSSES AS HIGH AS 45% HAVE BEEN OBSERVED. EXISTING INSTALLATIONS CAN BE IMPROVED WITH GUIDE VANES OR THE CONVERSION TO SQUARE OR MITERED ELBOWS WITH GUIDE VANES.

B. NON–UNIFORM FLOW INDUCED INTO FAN INLET BY A RECTANGULAR INLET DUCT.

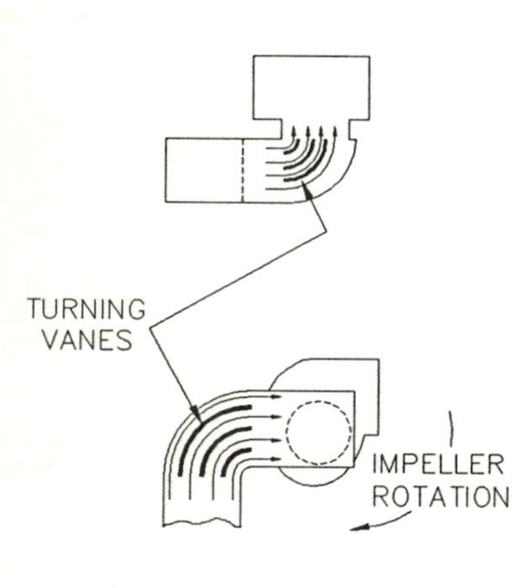

CORRECTED PRE–ROTATING SWIRL

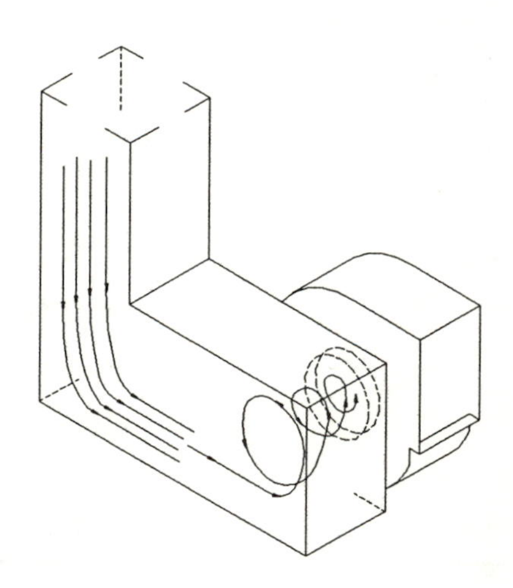

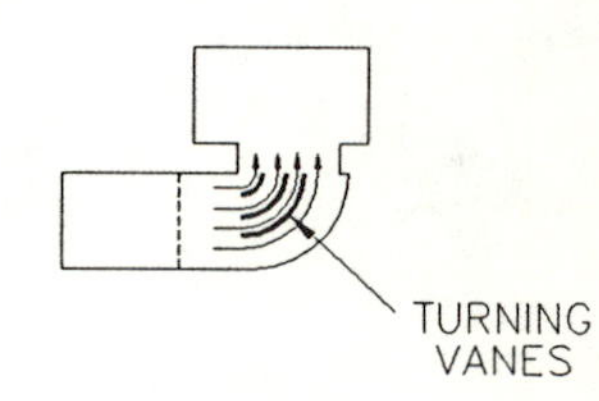

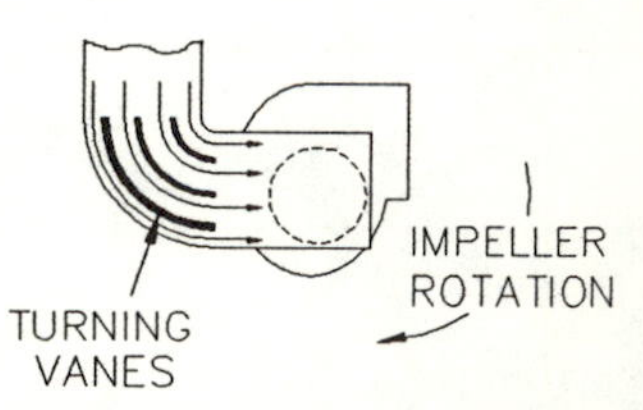

CORRECTED COUNTER–ROTATING SWIRL

C. NON–UNIFORM FLOW INTO A FAN INLET BY AN INDUCED VORTEX, SPIN OR SWIRL.

AMERICAN CONFERENCE OF GOVERNMENTAL INDUSTRIAL HYGIENISTS

NON–UNIFORM INLET CORRECTIONS (Adapted from AMCA 201)

DATE 1–88 FIGURE 6–20

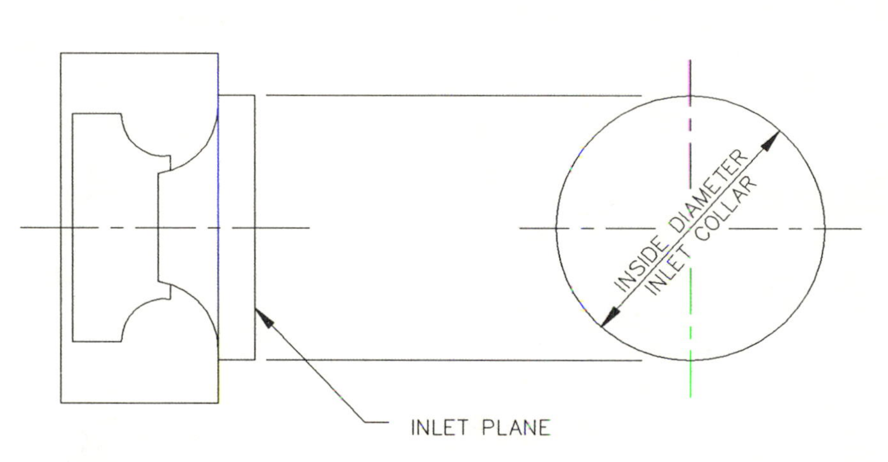

A. FREE INLET AREA PLANE -- FAN WITH INLET COLLAR.

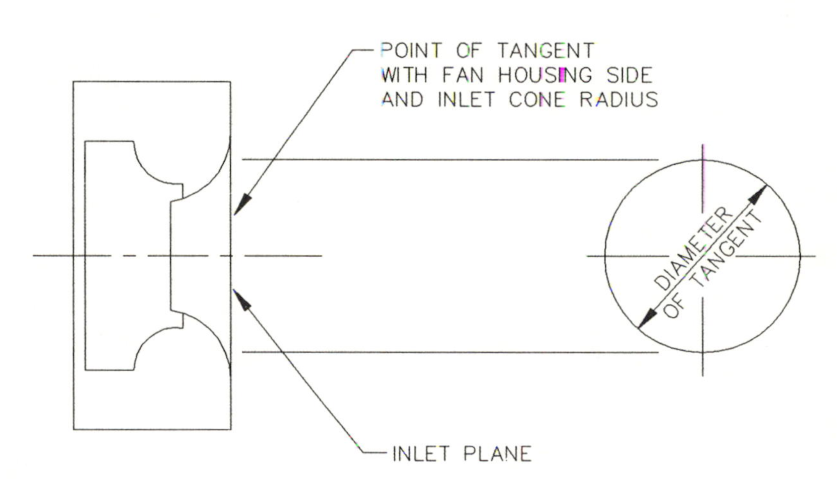

B. FREE INLET AREA PLANE -- FAN WITHOUT INLET COLLAR.

PERCENTAGE OF UNOBSTRUCTED INLET AREA	SYSTEM EFFECT FACTORS
100	NO LOSS
95	0.26
90	0.40
85	0.53
75	0.8
50	1.6
25	2.0

AMERICAN CONFERENCE OF GOVERNMENTAL INDUSTRIAL HYGIENISTS

SYSTEM EFFECT FACTORS FOR INLET OBSTRUCTIONS (Adapted from AMCA 201)

DATE 7-89 FIGURE 6-21

Chapter 7
REPLACEMENT AND RECIRCULATED AIR

7.1 INTRODUCTION 7-1

7.2 REPLACEMENT AIR 7-2

7.3 REPLACEMENT AIR FLOW RATE 7-4

7.4 ENVIRONMENTAL CONTROL 7-4

7.5 ENVIRONMENTAL CONTROL AIR FLOW RATE 7-5

7.6 AIR CHANGES 7-5

7.7 AIR SUPPLY TEMPERATURES 7-5

7.8 AIR SUPPLY VS. PLANT HEATING COSTS 7-6

7.9 REPLACEMENT AIR HEATING EQUIPMENT 7-8

7.10 COST OF HEATING REPLACEMENT AIR 7-12

7.11 AIR CONSERVATION 7-14

- 7.11.1 Reduced Flow Rate 7-14
- 7.11.2 Untempered Air Supply 7-15
- 7.11.3 Energy Recovery 7-15
- 7.11.4 Selection of Monitors 7-16

7.12 EVALUATION OF EMPLOYEE EXPOSURE LEVELS 7-12

REFERENCES 7-19

7.1 INTRODUCTION

Chapters 1 through 6 describe the purpose, function, and design of industrial exhaust systems. As mentioned in Chapter 1, Section 1.2, supply systems are used for two basic purposes: to create a comfortable environment and to replace air exhausted from the building. It is important to note that while properly designed exhaust systems will remove toxic contaminants they should not be relied upon to draw outside air into the building. If the amount of replacement air supplied to the building is lower than the amount of air exhausted, the pressure in the building will be lower than atmospheric. This condition is called "negative pressure" and results in air entering the building in an uncontrolled manner through window sashes, doorways, and walls. In turn, this may lead to many undesirable results such as high velocity drafts, backdrafting, difficulty in opening doors, etc.

In order to minimize these effects, mechanical air supply systems are needed to introduce sufficient outside air to avoid a negative pressure situation. A properly designed and installed air supply system can provide both replacement air and effective environmental control. Provided that important health and safety measures are taken, recirculation of the exhaust air may be an effective method that can substantially reduce heating and/or cooling costs.

7.2 REPLACEMENT AIR

Air will enter a building in an amount to equal the flow rate of exhaust air whether or not provision is made for this replacement. However, the actual exhaust flow rate will be less than the design value if the plant is under negative pressure. If the building perimeter is tightly sealed, thus blocking effective infiltration of outside air, a severe decrease of the exhaust flow rate will result. If, on the other hand, the building is relatively old with large sash areas, air infiltration may be quite pronounced and the exhaust system performance will decrease only slightly and other problems may occur.

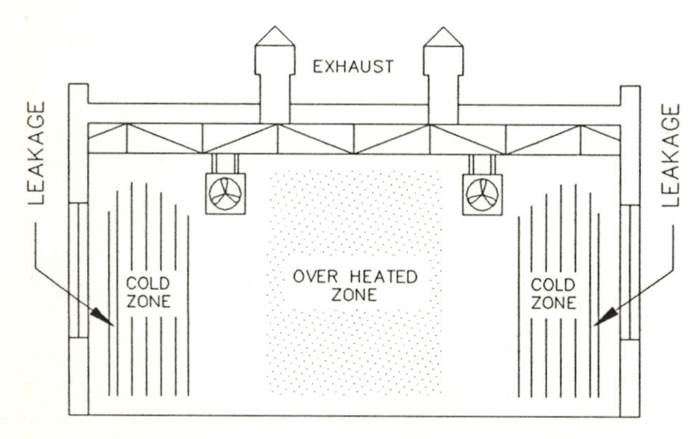

FIGURE 7–1 UNDER NEGATIVE PRESSURE CONDITIONS, WORKERS IN THE COLD ZONES TURNED UP THERMOSTATS IN AN ATTEMPT TO GET HEAT. BECAUSE THIS DID NOTHING TO STOP LEAKAGE OF COLD AIR, THEY REMAINED COLD WHILE CENTER OF PLANT WAS OVERHEATED.

TABLE 7-1. Negative Pressures and Corresponding Velocities Through Crack Openings (Calculated with air at room temperature, standard atmospheric pressure, C_e = 0.6.)

Negative Pressure, "wg	Velocity, fpm
0.004	150
0.008	215
0.010	240
0.014	285
0.016	300
0.018	320
0.020	340
0.025	380
0.030	415
0.040	480
0.050	540
0.060	590
0.080	680
0.100	760
0.150	930
0.200	1080
0.250	1200
0.300	1310
0.400	1520
0.500	1700
0.600	1860

When the building is relatively open the resultant in-plant environmental condition is often undesirable since the influx of cold outside air in the northern climates chills the perimeter of the building. Exposed workers are subjected to drafts, space temperatures are not uniform, and the building heating system is usually overtaxed (see Figure 7-1). Although the air may eventually be tempered to acceptable conditions by mixing as it moves to the building interior, this is an ineffective way of transferring heat to the air and usually results in fuel waste.

Experience has shown that replacement air is necessary for the following reasons:

1. *To insure that exhaust hoods operate properly:* A lack of replacement air and the attendant negative pressure condition results in an increase in the static pressure the exhaust fans must overcome. This can cause a reduction in exhaust flow rate from all fans and is particularly serious with low pressure fans such as wall fans and roof exhausters (see Figure 7-2).
2. *To eliminate high velocity cross-drafts through windows and doors.* Depending on the negative pressure created, cross-drafts may be substantial (see Table 7-1). Cross-drafts not only interfere with the proper operation of exhaust hoods, but also may disperse contaminated air from one section of the building to another and can interfere with the proper operation of process equip-

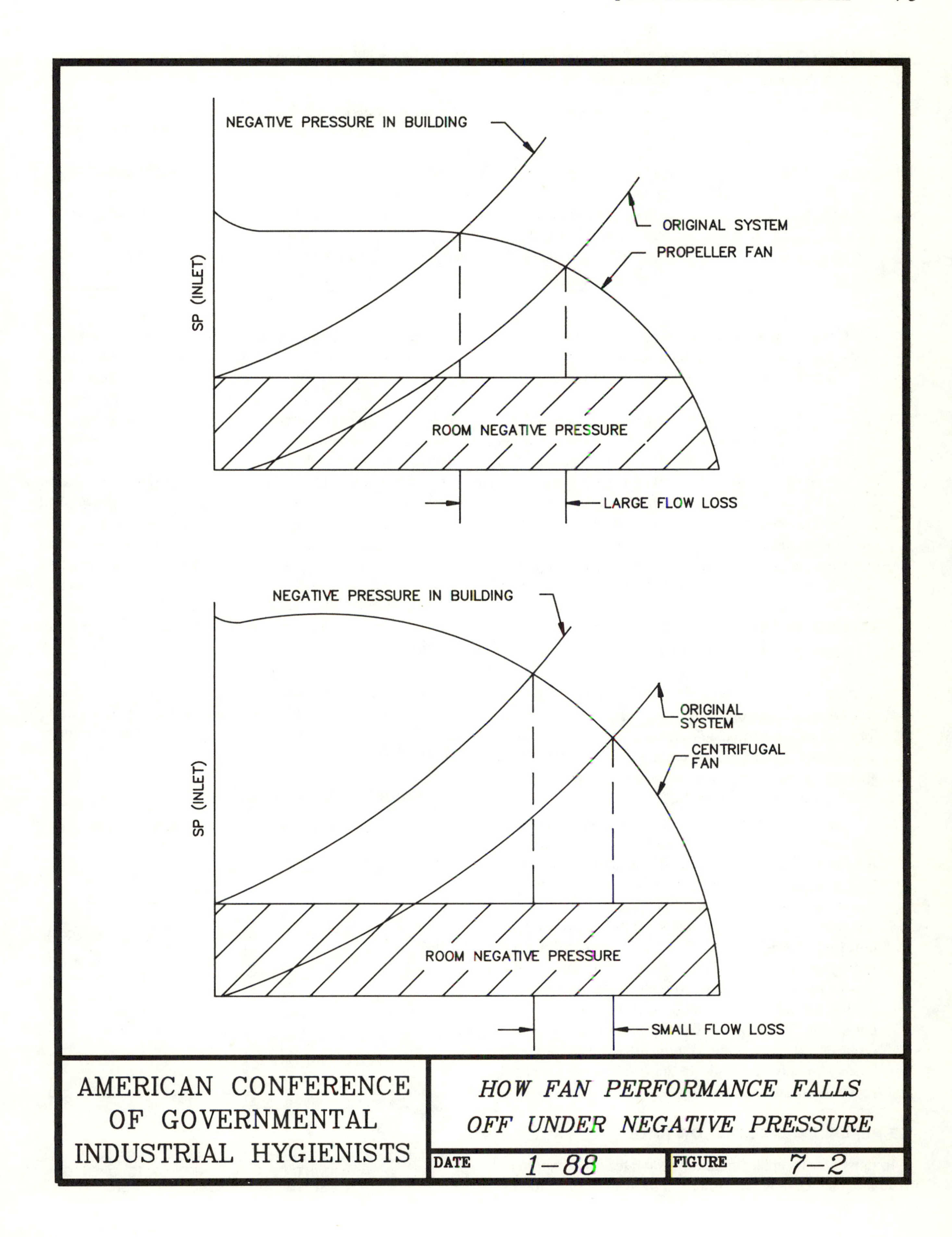
NEGATIVE PRESSURE IN BUILDING
ORIGINAL SYSTEM
PROPELLER FAN
SP (INLET)
ROOM NEGATIVE PRESSURE
LARGE FLOW LOSS
NEGATIVE PRESSURE IN BUILDING
ORIGINAL SYSTEM
CENTRIFUGAL FAN
SP (INLET)
ROOM NEGATIVE PRESSURE
SMALL FLOW LOSS
AMERICAN CONFERENCE OF GOVERNMENTAL INDUSTRIAL HYGIENISTS
HOW FAN PERFORMANCE FALLS OFF UNDER NEGATIVE PRESSURE
DATE 1-88
FIGURE 7-2

TABLE 7-2. Negative Pressures Which May Cause Unsatisfactory Conditions Within Buildings

Negative Pressure, "wg	Adverse Conditions
0.01 to 0.02	Worker Draft Complaints — High velocity drafts through doors and windows.
0.01 to 0.05	Natural Draft Stacks Ineffective — Ventilaton through roof exhaust ventilators, flow through stacks with natural draft greatly reduced.
0.02 to 0.05	Carbon Monoxide Hazard — Back drafting will take place in hot water heaters, unit heaters, furnaces and other combustion equipment not provided with induced draft.
0.03 to 0.10	General Mechanical Ventilation Reduced — Air flows reduced in propeller fans and low pressure supply and exhaust systems.
0.05 to 0.10	Doors Difficult to Open — Serious injury may result from non-checked, slamming doors.
0.10 to 0.25	Local Exhaust Ventilation Impaired — Centrifugal fan exhaust flow reduced.

ment such as open top solvent degreasers. In the case of dusty operations, settled material may be dislodged from surfaces and result in recontamination of the workroom.

3. *To insure operation of natural draft stacks such as combustion flues*. Moderate negative pressures can result in backdrafting of flues which may cause a dangerous health hazard from the release of combustion products, principally carbon monoxide, into the workroom. Back drafting may occur in natural draft stacks at negative pressures as low as 0.02 "wg (see Table 7-2). Secondary problems include difficulty in maintaining pilot lights in burners, poor operation of temperature controls, corrosion damage in stacks and heat exchangers due to condensation of water vapor in the flue gases.
4. *To eliminate cold drafts on workers*. Drafts not only cause discomfort and reduce working efficiency but also may result in lower overall ambient temperatures.
5. *To eliminate differential pressure on doors*. High differential pressures make doors difficult to open or shut and, in some instances, can cause personnel safety hazards when the doors move in an uncontrolled fashion (see Figure 7-3 and Table 7-2).
6. *To conserve fuel*. Without adequate replacement air, uncomfortable cold conditions near the building perimeter frequently lead to the installation of more heating equipment in those areas in an attempt to correct the problem. These heaters take excessive time to warm the air and the over-heated air moving toward the building interior makes those areas uncomfortably warm (see Figure 7-1). This in turn leads to the installation of more exhaust fans to remove the excess heat, further aggravating the problem. Heat is wasted without curing the problem. The fuel consumption with a replacement air heating system usually is lower than when attempts are made to achieve comfort without replacement air (see Section 7.10).

7.3 REPLACEMENT AIR FLOW RATE

In most cases, replacement air flow rate should approximate the total air flow rate of air removed from the building by exhaust ventilation systems, process systems and combustion processes. Determination of the actual flow rate of air removed usually requires an inventory of air exhaust locations and any necessary testing. When conducting the exhaust inventory it is necessary not only to determine the quantity of air removed, but also the *need* for a particular piece of equipment. At the same time, reasonable projections should be made of the total plant exhaust requirements for the next one to two years, particularly if process changes or plant expansions are contemplated. In such cases it can be practical to purchase a replacement air unit slightly larger than immediately necessary with the knowledge that the increased capacity will be required within a short time. The additional cost of a larger unit is relatively small and in most cases the fan drive can be regulated to supply only the desired quantity of air.

Having established the minimum air supply quantity necessary for replacement air purposes, many plants have found that it is wise to provide additional supply air flow rate to overcome natural ventilation leakage and further minimize drafts at the perimeter of the building.

7.4 ENVIRONMENTAL CONTROL

In addition to toxic contaminants which are most effec-

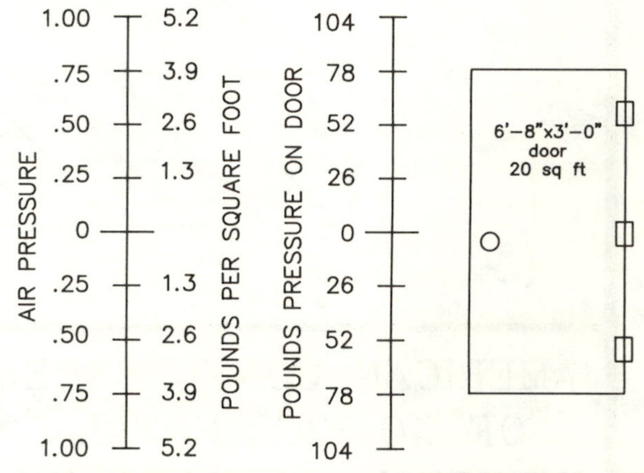

FIGURE 7–3 RELATIONSHIP BETWEEN AIR PRESSURE AND AMOUNT OF FORCE NEEDED TO OPEN OR CLOSE AN AVERAGE-SIZED DOOR.

TABLE 7-3. Air Exchanges Vs. Room Size

Room Size	Room ft^3	Air Changes/Minute	Air Changes/Hour
40 × 40 × 12 high	19,200	11,650/19,200 = 0.61	36
40 × 40 × 20 high	32,000	11,650/32,000 = 0.364	22

tively controlled by exhaust ventilation systems, industrial processes may create an undesirable heat load in the work space. Modern automated machining, conveying and transferring equipment require considerable horsepower. Precision manufacturing and assembling demand increasingly higher light levels in the plant with correspondingly greater heat release. The resulting in-plant heat burden raises indoor temperatures, often beyond the limits of efficient working conditions and, in some cases, beyond the tolerance limits for the product.

Many industrial processes release minor amounts of "nuisance" contaminants which, at low concentrations, have no known health effects but which are unpleasant or disagreeable to the workers or harmful to the product. The desire to provide a clean working environment for both the people and the product often dictates controlled air flow between rooms or entire departments.

Environmental control of these factors can be accomplished through the careful use of air supply systems. (It must be noted that radiant heat cannot be controlled by ventilation and methods such as shielding described in Chapter 2, are required.) Sensible and latent heat released by people and the process can be controlled to desired limits by proper use of ventilation.

7.5 ENVIRONMENTAL CONTROL AIR FLOW RATE

The flow rate of air needed depends on the factors which are to be considered and the degree of control necessary for a satisfactory environment. *Sensible heat* can be removed through simple air dilution (see Chapter 2 under ventilation).

"Nuisance" or undesirable contaminants can also be reduced by dilution with outside air. The control of odors from people at various conditions of rest and work can be accomplished with the outdoor air flow rate described in Chapter 2. However, these data apply mainly to offices, schools and similar types of environment and do not correspond well with the usual industrial or commercial establishment. Experience has shown that when the air supply is properly distributed into the working level (i.e., in the lower 8 to 10 feet of the space), an outdoor air supply of from 1 to 2 cfm/ft^2 of floor space will give good results.

7.6 AIR CHANGES

"Number of air changes per minute or per hour" is the ratio of the ventilation rate (per minute or per hour) to the room volume. "Air changes per hour" or "air changes per minute" is a poor basis for ventilation criteria where environmental control of hazards, heat, and/or odors is required. The required ventilation depends on the problem, not on the size of the room in which it occurs. For example, let us assume a situation where 11,650 cfm would be required to control solvent vapors by dilution. The operation may be conducted in either of two rooms, but in either case, 11,650 cfm is the required ventilation. The "air changes," however, would be quite different for the two rooms, as can be seen in Table 7-3. As can be seen, for the same "air change" rate, a high ceiling space will require more ventilation than a low ceiling space of the same floor area. Thus, there is little relationship between "air changes" and the required contaminant control.

The "air change" basis for ventilation does have some applicability for relatively standard situations such as office buildings and school rooms where a standard ventilation rate is reasonable. It is easily understood and reduces the engineering effort required to establish a design criteria for ventilation. It is this ease of application, in fact, which often leads to lack of investigation of the real engineering parameters involved and correspondingly poor results.

7.7 AIR SUPPLY TEMPERATURES

In the majority of cases, outside air will be supplied in the winter months at or slightly below desired space temperatures and during the summer at whatever temperature is available out-of-doors or if air conditioning is available, at desired space temperatures. Where high internal heat loads are to be controlled, however, the temperature of the air supply can be appreciably below that of the space by reducing the amount of heat supplied to the air during the winter months and by deliberately cooling the air in the summer. When a large air flow rate is delivered at approximately space temperatures or somewhat below, the distribution of the air becomes vitally important in order to maintain satisfactory environmental conditions for the persons in the space.

Maximum utilization of the supply air is achieved when the air is distributed in the "living zone" of the space, below the eight to ten foot level (see Figure 7-4). When delivered in this manner — where the majority of the people and processes are located — maximum ventilation results with minimum air handling. During the warm months of the year large air flow in the working space at relatively high velocities is welcomed by the workers. During the winter months, however, care must be taken to insure that air velocities over the person, except when extremely high heat loads are involved, are kept within acceptable values (see Chapter 2, Table 2-5). To accomplish this the air can be distributed uniformly in the space or where required for worker comfort. Heavy-duty, adjustable, directional grilles and louvers have proven to be very successful in allowing individual workers

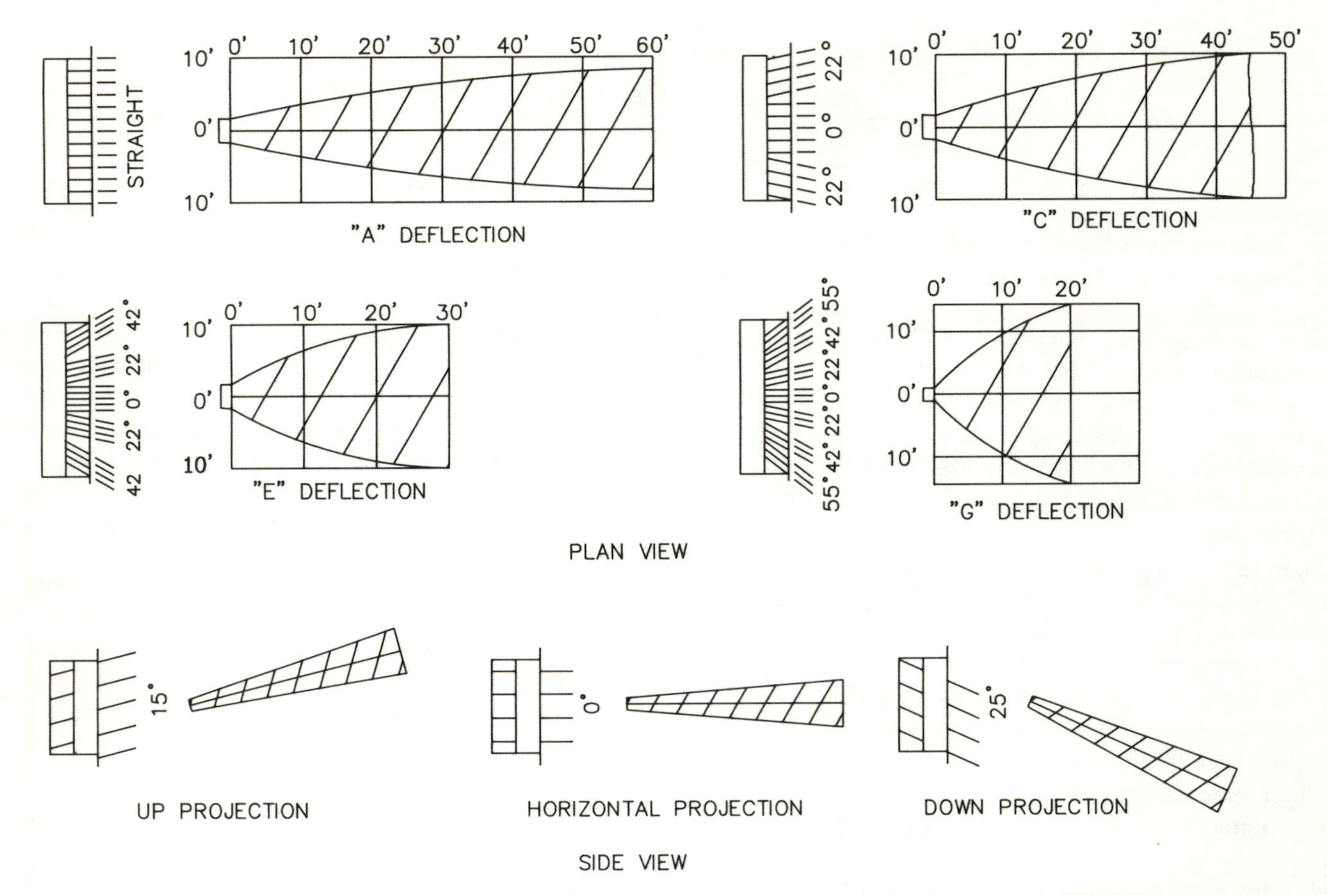

FIGURE 7-4. THROW PATTERNS AND DISTANCE FROM DIFFERENT REGISTER ADJUSTMENTS (Ref. 7.2).

to direct the air as needed.[7.1] Light gauge, stamped grilles intended for commercial use are not satisfactory. Suitable control must be provided to accommodate seasonal and even daily requirements with a minimum of supervision or maintenance attention.

Chapter 2 describes the relative comfort that can be derived through adequate air flow control. Published tables of data by register and diffuser manufacturers indicate the amount of throw (projection) and spread that can be achieved with different designs at different flow rates (see Figure 7-4). Terminal velocities at the throw distance can also be determined.

Multiple point distribution is usually best since it provides uniformity of air delivery and minimizes the re-entrainment of contaminated air that occurs when large volumes are "dumped" at relatively high velocities. Depending on the size and shape of the space and the amount of air to be delivered, various distributional layouts are employed. Single point distribution can be used; however, it is usually necessary to redirect the large volume of air with a baffle or series of baffles in order to reduce the velocity close to the outlet and minimize re-entrainment. In determining the number and types of registers or outlet points, it also is necessary to consider the effect of terminal air supply velocity on the performance of local exhaust hoods.

When large amounts of sensible heat are to be removed from the space during the winter months, it is most practical to plan for rapid mixing of the cooler air supply with the warmer air in the space. During the summer months the best distribution usually involves minimum mixing so that the air supply will reach the worker at higher velocities and with a minimum of heat pickup. These results can be obtained by providing horizontal distribution of winter air over the worker's head, mixing before it reaches the work area and directing the air toward the worker through register adjustment for the summer months (see Figure 7-5).

Delivered air temperatures during the winter usually range from 65 to 68 F for work areas without much process heat or vigorous work requirement downward to 60 F or even 55 F where hard work or significant heat sources are involved. For summer operation the temperature rise in indoor air can be estimated as described in Chapter 2. Evaporative cooling should be considered for summer operation. Although not as effective as mechanical refrigeration under all conditions, evaporative cooling significantly lowers the temperature of the outdoor air even in humid climates, improves the ability of the ventilation air to reduce heat stress, and costs much less to install and operate.

7.8 AIR SUPPLY VS. PLANT HEATING COSTS

Even if the supply air were drawn into the building simply

by the action of the exhaust fans, during the winter months there will be an added burden on the plant heating system and fuel costs will rise. Experience has shown, however, that when the same flow rate of outdoor air is introduced through properly designed replacement air heaters, the overall fuel cost does not exceed previous levels and often is decreased. A partial explanation of this savings is more efficient heat transfer. The most important factor, however, is that a well-designed air supply system is not dependent on the plant space heating system; rather, the two systems operate in an independent fashion. The air supply system and the plant heating system can be understood best by considering the building as a whole. In order for an equilibrium to be established, the heat outflow from the building must balance the

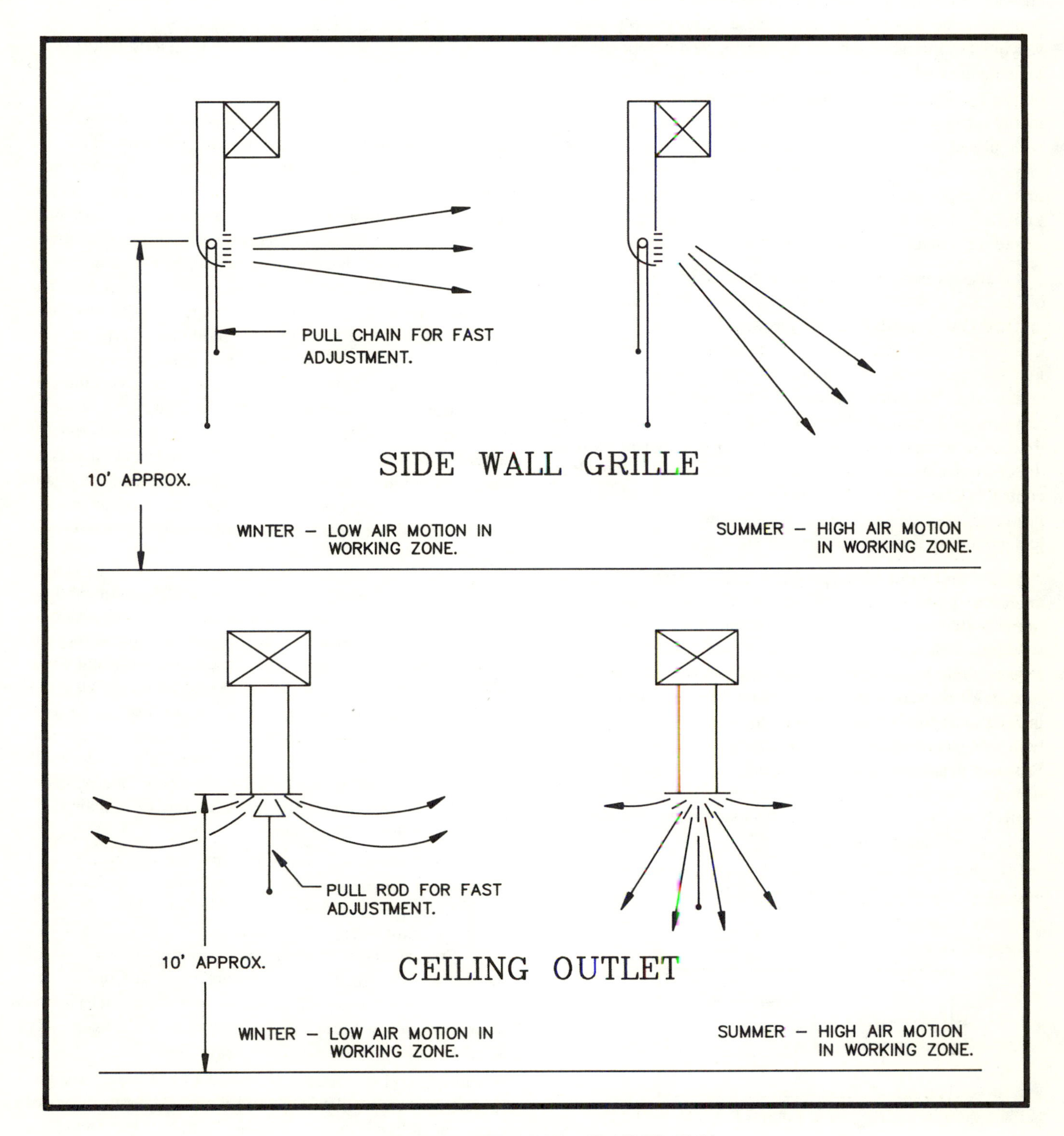

FIGURE 7–5 – SEASONAL AIR VENTILATION

heat inflow.

7.9 REPLACEMENT AIR HEATING EQUIPMENT

Replacement air heaters are usually designed to supply 100% outdoor air. The basic requirements for an air heater are that it be capable of continuous operation, constant delivered air flow rate and constant preselected discharge temperature. The heater must meet these requirements under varying conditions of service and accommodate outside air temperatures which vary as much as 40 F daily. Standard design heating and ventilating units are usually selected for mixed air applications, i.e., partial outdoor air and partial recirculated air; it is rare that their construction and operating capabilities will meet the requirements of industry. Such units are applicable in commercial buildings and institutional facilities where the requirements are less severe and where mixed air service is more common.

Air heaters are usually categorized according to the source of heat: steam and hot water units, indirect-fired gas and oil units, and direct-fired natural gas and Liquified Petroleum Gas (LPG) units. Each basic type is capable of meeting the first two requirements — constant operation and constant delivered air flow rate. Variations occur within each type in relation to the third requirement, that of constant preselected discharge temperature. One exception to this rule is the direct-fired air heater where the inherent design provides control over a wide range of temperatures. Each type of air heater has specific advantages and limitations which must be understood by the designer in making a selection.

Steam coil units were probably the earliest air heaters applied to general industry as well as commercial and institutional buildings (see Figure 7-6). When properly designed, selected, and installed, they are reliable and safe. They require a reliable source of clean steam at dependable pressure. For this reason they are applied most widely in large installations; smaller industrial plants often do not provide a boiler or steam capacity for operating a steam air heater. Principal disadvantages of steam units are potential damage from freezing or water hammer in the coils, the complexity of controls when close temperature limits must be maintained, high cost and excessive piping.

Freezing and water hammer are the result of poor selection and installation and can be minimized through careful

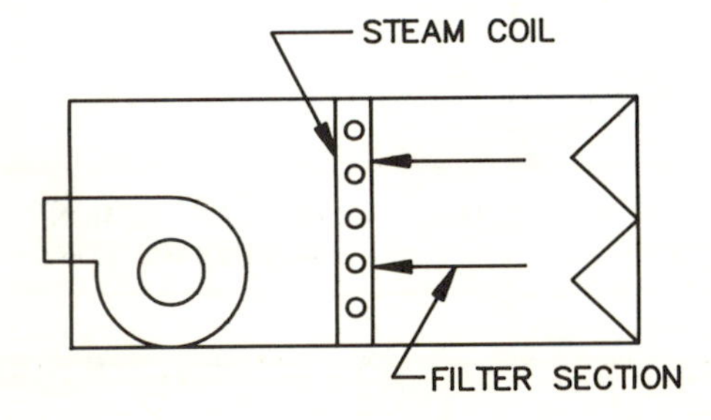

FIGURE 7—6 SINGLE STEAM COIL UNIT

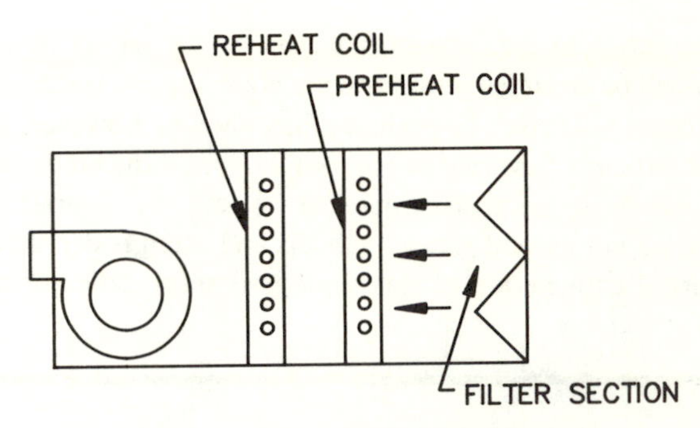

FIGURE 7—8 MULTIPLE COIL STEAM UNIT

application. The coil must be sized to provide desired heat output at the available steam pressure and flow. The coil preferably should be of the steam distributing type with vertical tubes. The traps and return piping must be sized for the maximum condensate flow at minimum steam pressure plus a safety factor. Atmospheric vents must be provided to minimize the danger of a vacuum in the coil which would hold up the condensate. Finally, the *condensate must never be lifted by steam pressure*. The majority of freeze-up and water hammer problems relate to the steam modulating type of unit which relies on throttling of the steam supply to achieve temperature control. When throttling occurs, a vacuum can be created in the coil and unless adequate venting is provided, condensate will not drain and can freeze rapidly under the influence of cold outside air. Most freeze-ups occur when outdoor air is in the range of 20 to 30 F and the steam control valve is partially closed, rather than when the outside air is a minimum temperature and full steam supply is on (see Figure 7-7).

"Safety" controls are often used to detect imminent danger from freeze-up. A thermostat in the condensate line or an extended bulb thermostat on the downstream side of the coil can be connected into the control circuit to shut the unit down when the temperature falls below a safe point. As an alternate, the thermostat can call for full steam flow to the coil with shutdown if a safe temperature is not maintained. An obvious disadvantage is that the plant air supply is reduced; if the building should be subjected to an appreciable negative pressure, unit freeze-up still may occur due to cold air leakage through the fresh air dampers.

The throttling range of a single coil unit can be extended by using two valves: one valve is usually sized for about two-thirds the capacity and the other valve one-third. Through suitable control arrangements both valves will provide 100% steam flow when fully opened and various combinations will provide a wide range of temperature control. Controls are complex in this type of unit and care must be taken to insure that pressure drop through the two valve circuits is essentially equal so as to provide expected steam flow.

Multiple coil steam units (Figure 7-8) and bypass designs (Figure 7-9) are available to extend the temperature control range and help minimize freeze-up. With multiple coil units, the first coil (preheat) is usually sized to raise the air tempera-

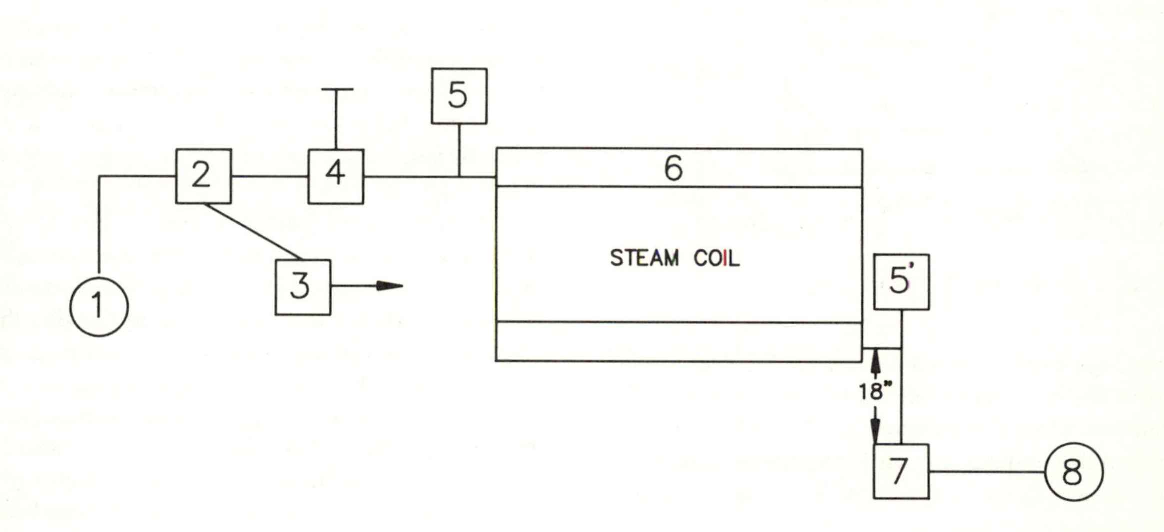

1. STEAM SUPPLY
   PROVIDE STEAM FROM A CLEAN SOURCE
   MAINTAIN CONSTANT PRESSURE WITH REDUCING VALVES IF REQUIRED
   PROVIDE TRAPPED DRIPS FOR SUPPLY LINES
   SIZE SUPPLY PIPING FOR FULL LOAD AT AVAILABLE PRESSURE
2. STRAINER
   1/32" DIAMETER MINIMUM PERFORATIONS
3. DRIP TRAP
   INVERTED BUCKET TRAP PREFERRED
4. CONTROL VALVE
   SIZE FOR MAXIMUM STEAM FLOW
   MAXIMUM PRESSURE DROP EQUAL TO 50% INLET STEAM PRESSURE
5. VACUUM BREAKER
   1/2" CHECK VALVE TO ATMOSPHERE

5'. ALTERNATE VACUUM BREAKER

6. STEAM COIL
   A. SIZE FOR DESIGN CAPACITY AT INLET STEAM PRESSURE (SUPPLY–VALVE DROP)
   B. VERTICAL COILS PREFFERED
   C. HORIZONTAL COILS MUST BE PITCHED 1/4" PER FOOT TOWARD DRAIN.
      6' MAXIMUM LENGTH RECOMMEMDED
7. CONDENSATE TRAP
   A. INVERTED BUCKET PREFERRED
   B. SIZE TRAP FOR THREE TIMES MAXIMUM CONDENSATE LOAD AT PRESSURE DROP EQUAL TO 50% INLET PRESSURE
   C. INDIVIDUAL TRAP FOR EACH COIL
8. CONDENSATE RETURN
   ATMOSPHERIC DRAIN ONLY

AMERICAN CONFERENCE OF GOVERNMENTAL INDUSTRIAL HYGIENISTS	*STEAM COIL PIPING*
	DATE *1–74* FIGURE *7–7*

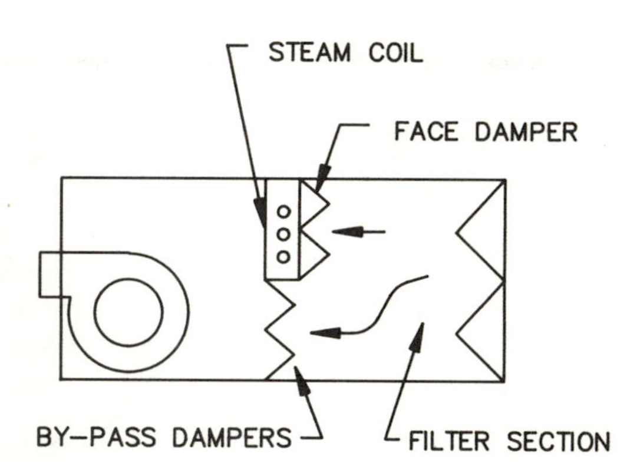

FIGURE 7-9 BY-PASS STEAM SYSTEM

ture from design outdoor temperature to at least 40 F. The coil is controlled with an on-off valve which will be fully open whenever outdoor temperature is below 40 F. The second (reheat) coil is designed to raise the air temperature from 40 F to the desired discharge condition. Temperature control will be satisfactory for most outdoor conditions, but overheating can occur when the outside air temperature approaches 40 F (39 F + the rise through the preheat coil can give temperatures of 79 to 89 F *entering* the reheat). Refined temperature control can be accomplished by using a second preheat coil to split the preheat load.

Bypass units incorporate dampers to direct the air flow. When maximum temperature rise is required, all air is directed through the coil. As the outdoor temperature rises, more and more air is diverted through the bypass section until finally all air is bypassed. Controls are relatively simple. The principal disadvantage is that the bypass is not always sized for full air flow at the same pressure drop as through the coil, thus (depending on the damper position) the unit may deliver differing air flow rates. Damper air flow characteristics are also a factor. An additional concern is that in some units the air coming through the bypass and entering the fan compartment may have a nonuniform flow and/or temperature characteristic which will affect the fan(s) ability to deliver air.

Another type of bypass design, called integral face and bypass (Figure 7-10), features alternating sections of coil and bypass. This design promotes more uniform mixing of the air stream, minimizes any nonuniform flow effect and, through carefully engineered damper design, permits minimum temperature pickup even at full steam flow and full bypass.

Hot water is an acceptable heating medium for air heaters. As with steam, there must be a dependable source of water at predetermined temperatures for accurate sizing of the coil. Hot water units are less susceptible to freezing than steam because of the forced convection which insures that the cooler water can be positively removed from the coil. Practical difficulties and pumping requirements thus far have limited the application of hot water to relatively small systems: for a 100 F air temperature rise and an allowable 100 F water temperature drop, 1 GPM of water will provide heat for only 450 cfm of air. This range can be extended with high temperature hot water systems.

Hybrid systems using an intermediate heat exchange fluid, such as ethylene glycol, have also been installed by industries with critical air supply problems and a desire to eliminate all freeze up dangers. A primary steam system provides the necessary heat to a converter which supplies a secondary closed loop of the selected heat exchange fluid. The added equipment cost is at least partially offset by the less complex control system.

Indirect-fired gas and oil units (Figure 7-11) are widely applied in small industrial and commercial applications. Economics appear to favor their use up to approximately 10,000 cfm; above this size the capital cost of direct-fired air heaters is lower. Indirect-fired heaters incorporate a heat

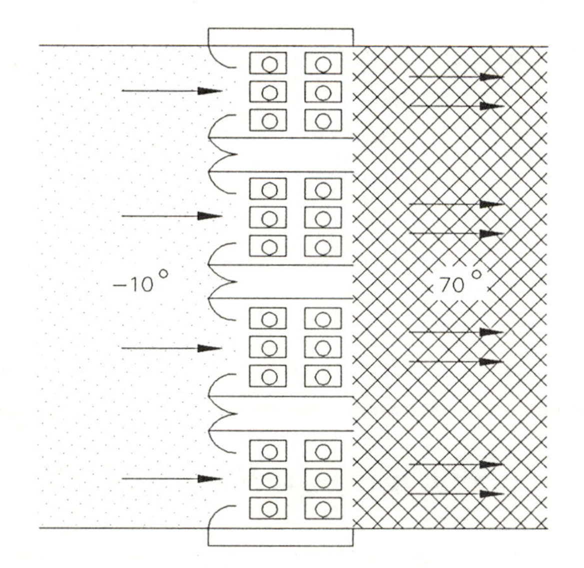

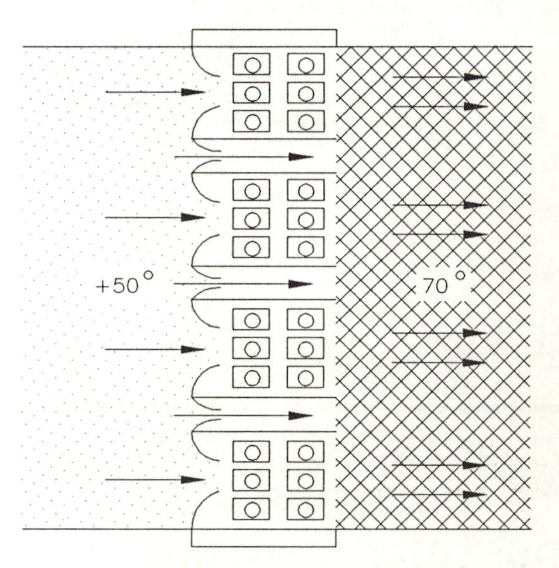

FIGURE 7-10 INTEGRAL FACE AND BY-PASS COIL (REF.7-4).

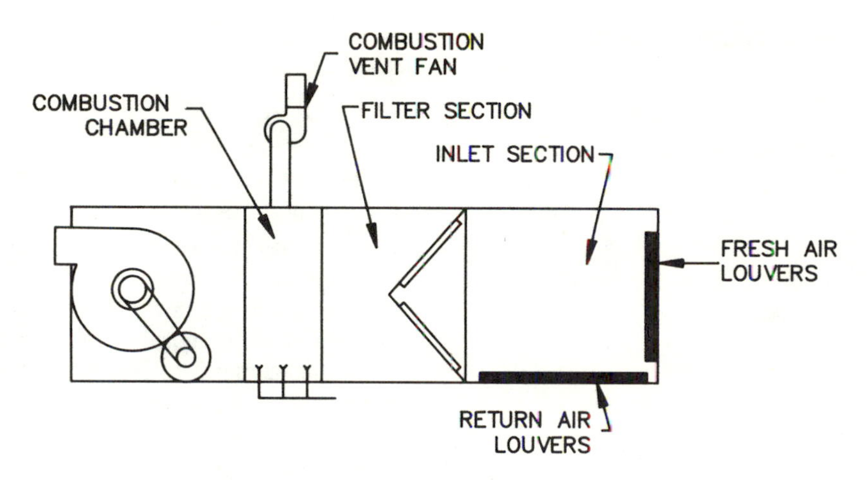

FIGURE 7–11. INDIRECT FIRED UNIT

exchanger, commonly stainless steel, which effectively separates the incoming air stream from the products of combustion of the fuel being burned. Positive venting of combustion products is usually accomplished with induced draft fans. These precautions are taken to minimize interior corrosion damage from condensation in the heat exchanger due to the chilling effect of the incoming cold air stream. The indirect-fired air heaters permit the use of oil as a heat source and room air recirculation is permitted with this type of unit since the air stream is separated from the products of combustion. A third major advantage is that this type of unit is economical in the smaller sizes and is widely applied as a "package" unit in small installations such as commercial kitchens and laundries.

Temperature control, "turn-down ratio," is limited to about 3:1 or 5:1 due to burner design limitations and the necessity to maintain minimum temperatures in the heat exchanger and flues. Temperature control can be extended through the use of a bypass system similar to that described for single coil steam air heaters. Bypass units of this design offer the same advantages and disadvantages as the steam bypass units.

Another type of indirect-fired unit incorporates a rotating heat exchanger. Temperature control can be as high as 20:1.

Direct-fired heaters wherein the fuel, natural or LPG gas, is burned directly in the air stream and the products of combustion are released in the air supply have been commercially available for some years (Figure 7-12). These units are economical to operate since all of the net heating value of the fuel is available to raise the temperature of the air resulting in a net heating efficiency approaching 90%. Commercially available burner designs provide turndown ratios from approximately 25:1 to as high as 45:1 and permit excellent temperature control. In sizes above 10,000 cfm the units are relatively inexpensive on a cost per cfm basis; below this capacity the costs of the additional combustion and safety controls weigh heavily against this design. A further disadvantage is that governmental codes often prohibit the recirculation of room air across the burner. Controls on these units are designed to provide a positive proof of air flow before the burner can ignite, a timed preignition purge to insure that any leakage gases will be removed from the housing, and constantly supervised flame operation which includes both flame controls and high temperature limits.

Concerns are often expressed with respect to potentially toxic concentrations of carbon monoxide, oxides of nitrogen, aldehydes, and other contaminants produced by combustion and released into the supply air stream. Practical field evaluations and detailed studies show that with a properly operated, adequately maintained unit, carbon monoxide concentrations will not be expected to exceed 5 ppm and that oxides of nitrogen and aldehydes are well within acceptable limits.[7.3]

A variation of this unit, known as a bypass design, has

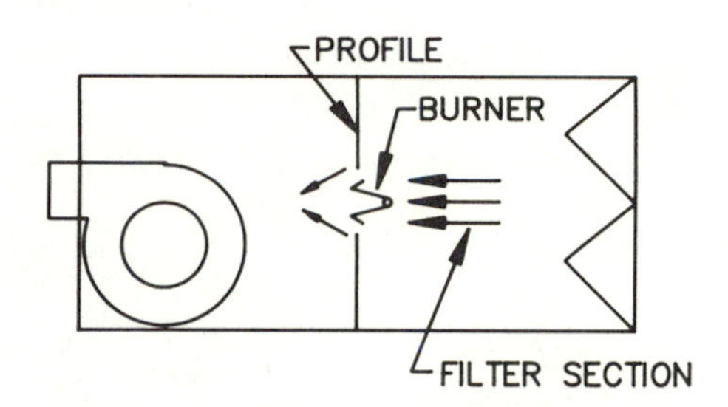

FIGURE 7–12 DIRECT FIRED UNIT

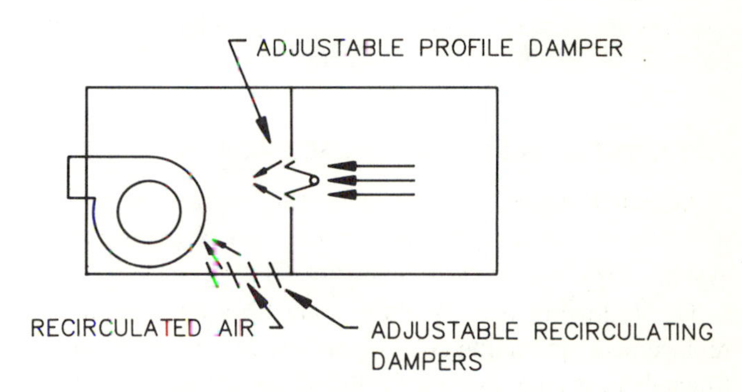

FIGURE 7–13 DIRECT FIRED BY-PASS UNIT

TABLE 7-4. Comparison of heater advantages and disadvantages

Advantages	Disadvantages
Direct-Fired Unvented:	
1. Good turndown ratio — 8:1 in small sizes; 25:1 in large sizes. Better control; lower operating costs.	1. Products of combustion in heater air stream (some CO_2, CO, oxides of nitrogen and water vapor present.)
2. No vent stack, flue or chimney necessary. Can be located in side walls of building.	2. First cost higher in small size units.
3. Higher efficiency (90%). Lower operating costs. (Efficiency based on available sensible heat.)	3. May be limited in application by governmental regulations. Consult local ordinances.
4. Can heat air over a wide temperature range.	4. Extreme care must be exercised to prevent minute quantities of chlorinated or other hydrocarbons from entering air intake or toxic products may be produced in heated air.
5. First cost lower in large size units.	5. Can be used only with natural gas or LPG.
	6. Burner must be tested to assure low CO and oxides of nitrogen content in air stream.
Indirect Exchanger:	
1. No products of combustion; outdoor air only is discharged into building.	1. First cost higher in large size units.
2. Allowable in all types of applications and buildings if provided with proper safety controls.	2. Turn down ratio is limited — 3:1 usual, maximum 5:1.
3. Small quantities of chlorinated hydrocarbons will not normally break down on exchanger to form toxic products in heated air.	3. Flue or chimney required. Can be located only where flue or chimney is available.
4. Can be used with oil, LPG and natural gas as fuel.	4. Low efficiency (80%). Higher operating cost.
5. First cost lower in small size units.	5. Can heat air over a limited range of temperatures.
6. Can be used for recirculation as well as replacement.	6. Heat exchanger subject to severe corrosion condition. Needs to be checked periodically for leaks after a period of use.
	7. Difficult to adapt to all combustion air from outdoors unless roof or outdoor mounted.

gained acceptance in larger plants where there is a desire to circulate large air flows at all times (see Figure 7-13). In this design controls are arranged to reduce the flow of outside air across the burner and permit the entry of room air into the fan compartment. In this way the fan air flow rate remains constant and circulation in the space is maintained. It is important to note that the bypass air does not cross the burner — 100% outside air only is allowed to pass through the combustion zone. Controls are arranged to regulate outside air flow and also to insure that burner profile velocity remains within the limits specified by the burner manufacturer, usually in the range of 2000 to 3000 fpm. This is accomplished by providing a variable profile which changes area as the damper positions change.

Inasmuch as there are advantages and disadvantages to both direct-fired and indirect-fired replacement air heaters, a careful consideration of characteristics of each heater should be made. A comparison of the heaters is given in Table 7-4.

7.10 COST OF HEATING REPLACEMENT AIR

As noted above, the cost of heating replacement air is probably the most significant annual cost of a ventilation system. This cost needs to be estimated at the design stage.

The following two equations may be used to estimate replacement air heating costs on an *hourly* and *yearly* basis. Since there is an allowance for the efficiency of the replacement air unit, these equations will tend to give a low result if air is allowed to enter by infiltration only.

$$C_1 = \text{Hourly cost} = 0.001 \frac{QN}{q} c \qquad \textbf{[7.1]}$$

$$C_2 = \text{Yearly cost} = \frac{0.154(Q)(dg)(T)(c)}{q} \qquad \textbf{[7.2]}$$

where:

Q = air flow rate, cfm

N = required heat, BTU/hr/1000 cfm (Figure 7-14 and Table 7-5)

T = operating time, hours/week

q = available heat per unit of fuel (Table 7-6)

dg = annual degree days (Table 7-7)

c = cost of fuel, $/unit

EXAMPLE PROBLEM 1

Find the hourly and yearly cost of tempering 10,000 cfm of replacement air to 70 F in St. Louis, Missouri, using oil at $1.35/gallon.

Average winter temperature = 31 F (Figure 7-14).

$$\text{Hourly cost} = \frac{0.001\ QN}{q} c$$

$$= 0.001 \times 10^4 \times \frac{42{,}000}{106{,}500} \times \$1.35$$

$$= \$5.32$$

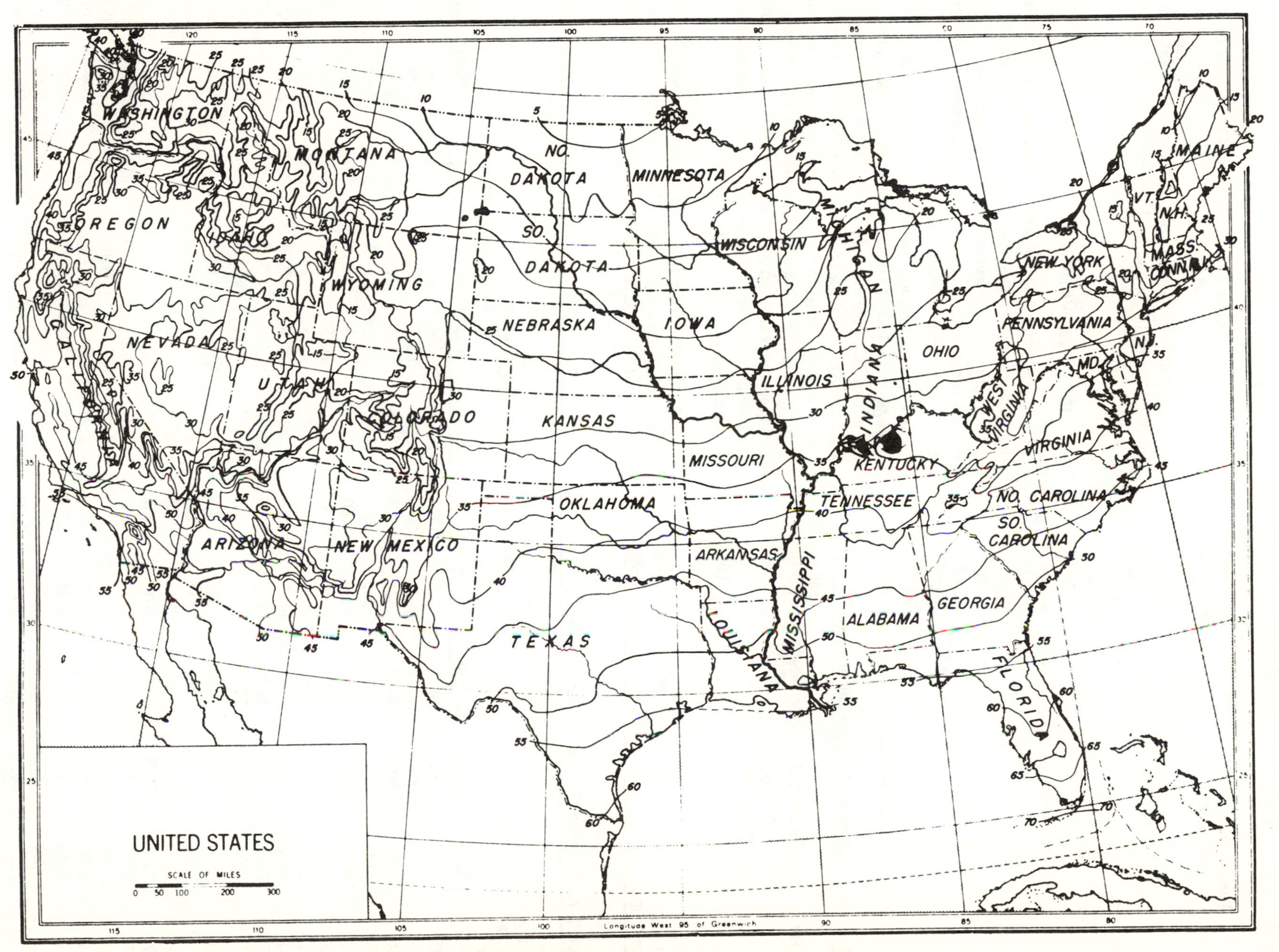

FIGURE 7–14. AVERAGE WINTER TEMPERATURES DECEMBER–FEBRUARY, INCLUSIVELY. (Courtesy U.S. Weather Bureau.)

TABLE 7-5. Required Heat for Outside Air Temperature

Average Outside Air Temperature, F	N, Required Heat Btu/hr/1000 cfm @ 70F
0	75,500
5	70,000
10	65,000
15	59,500
20	54,000
25	48,500
30	43,000
35	38,000
40	32,500
45	27,000
50	21,500
55	16,000
60	11,000
65	5,500

TABLE 7-6. Available Heat per Unit of Fuel

Fuel	Btu Per Unit	Efficiency %	Available Btu Per Unit q
Coal	12,000 Btu/lb	50	6,000
Oil	142,000 Btu/gal	75	106,500
Gas			
Heat Exchanger	1,000 Btu/ft³	80	800
Direct Fired		90	900

$$\text{Yearly cost} = \frac{(0.154)(10^4)(6023)(40)}{106,500} \times \$1.35$$

$$= \$4,700 \text{ (assuming 40 hr/week of operation).}$$

The yearly cost is more representative because both the length and severity of the heating season are taken into account.

7.11 AIR CONSERVATION

The supply and exhaust of air represent both a capital cost for equipment and an operating cost which is often sizable in northern climates. Concerned designers, recognizing these cost and energy conservation needs, are unanimous in their desire for reduced ventilation rates.

There are four methods by which the cost of heating and cooling a large flow of outside air can be reduced: 1) reduction in the total flow of air handled, 2) delivery of untempered outdoor air to the space, 3) recovery of energy from the exhaust air, and 4) recovery of warm, uncontaminated air from processes. The successful application of these engineering methods without reduction in health hazard control and without impairing the inplant environment requires careful consideration.

7.11.1 Reduced Flow Rate: A reduction of total air flow rate handled can be accomplished by conducting a careful inventory of all exhaust and supply systems in the plant.

TABLE 7-7. Heating Degree-Day Normals [79]

Air Discharge Temperature F (Base)	Albany	Boston	Chicago	Cleveland	Detroit	Minneapolis	N.Y.	Philadelphia	Pittsburgh	St. Louis	Wash., D.C.
	Annual Heating Degree-Day Normals										
80	11782	10409	10613	11343	10959	13176	9284	9652	10797	8943	8422
79	11425	10049	10277	10982	10605	12826	8937	9300	10436	8624	8089
78	11062	9690	9940	10621	10256	12478	8596	8954	10076	8310	7764
77	10709	9342	9610	10265	9914	12135	8265	8619	9723	8003	7446
76	10356	8994	9283	9915	9581	11797	7938	8285	9379	7702	7139
75	10009	8652	8972	9570	9247	11475	7620	7959	9036	7413	6835
74	9669	8317	8656	9229	8920	11142	7308	7641	8702	7121	6538
73	9333	7990	8349	8898	8599	10816	7004	7328	8373	6839	6250
72	9007	7668	8046	8567	8291	10496	6706	7028	8050	6560	5974
71	8682	7354	7750	8248	7981	10180	6421	6728	7740	6289	5703
70	8364	7046	7468	7928	7678	9870	6146	6438	7429	6023	5438
69	8056	6749	7183	7617	7383	9567	5871	6158	7127	5767	5179
68	7750	6458	6905	7313	7100	9269	5606	5886	6833	5523	4929
67	7452	6175	6635	7016	6816	8975	5349	5618	6546	5277	4690
66	7162	5903	6373	6722	6543	8687	5101	5360	6272	5053	4455
65	6881	5633	6122	6445	6278	8410	4858	5109	5997	4822	4229
64	6607	5370	5875	6165	6020	8131	4621	4864	5734	4595	4014
63	6340	5118	5638	5897	5772	7858	4394	4628	5483	4379	3798
62	6081	4873	5399	5636	5533	7590	4176	4397	5234	4168	3588
61	5829	4634	5164	5381	5290	7339	3957	4172	5006	3963	3383
60	5586	4399	4936	5140	5054	7086	3747	3952	4769	3761	3182

Determine which are necessary, which can be replaced with more efficient systems or hood designs and which systems may have been rendered obsolete by changes.

Numerous hood designs presented in Chapter 10 are intended specifically to provide for adequate contaminant capture at reduced air flow rates. For instance, the use of horizontal sliding sash in the laboratory hood can provide a 30% saving in exhaust air flow rate without impairing capture velocity. The use of a tailored hood design, such as the evaporation hood shown in VS-206, provides good contaminant capture with far lower exhaust flow rates than would be required for a typical laboratory bench hood. Low Volume-High Velocity hoods and systems such as those illustrated in VS-801 through VS-807 are used for many portable hand tool and fixed machining operations and can provide contaminant capture at far lower air handling requirements.

Throughout industry there are many applications of window exhaust fans and power roof exhausters to remove heat or nuisance contaminants which would be captured more readily at the source with lower air flow rates. Many roof exhausters, as noted earlier, have been installed initially to combat problems which were really caused by a lack of replacement air. When air supply and balanced ventilation conditions are established, their use no longer may be necessary.

Good design often can apply proven principles of local exhaust capture and control to reduce air flow rates with improved contaminant control.

7.11.2 Untempered Air Supply: In many industries utilizing hot processes, cold outside air is supplied untempered or moderately tempered to dissipate sensible heat loads on the workers and to provide effective temperature relief for workers exposed to radiant heat loads. The air required for large compressors, as well as for cooling tunnels in foundries, also can come directly from outside the plant and thus eliminate a load that is otherwise replaced with tempered air.

7.11.3 Energy Recovery: Energy recovery from exhaust air can be considered in two aspects: 1) the use of heat exchange equipment to extract heat from the air stream before it is exhausted to the outside and 2) the return (recirculation) of cleaned air from industrial exhaust systems. Heat exchanger application to industrial exhaust systems has been limited primarily by the ratio of installed cost to annual return.

Heat Exchangers — Air-to-air heat exchangers have been used to reduce energy consumption. This is achieved by transferring waste energy from the exhaust to replacement air streams of a building or process. The methods and equipment used will depend on the characteristics of the air streams. Major categories of equipment include heat wheels, fixed plate exchangers, heat pipes, and run-around coils.

A heat wheel is a revolving cylinder filled with an air permeable media. As the exhaust air passes through the media, heat is transferred to the media. Since the media rotates, the warm media transfers heat to the cooler replacement air. Special care is required to ensure that this transfer does not cause a transfer of contaminants.

A fixed plate exchanger consists of intertwined tunnels of exhaust and replacement air separated by plates (or sometimes a combination of plates and fins). The warm exhaust air heats the plates which in turn heat the cool replacement air on the other side of the plate. This exchanger uses no transfer media other than the plate forming wall of the unit.

A heat pipe, or thermo siphon, uses a pipe manifold with one end in the warm exhaust air stream and the other in the cool replacement air stream. The pipe contains a fluid which boils in the warm exhaust air stream extracting heat and condenses in the cool replacement air stream releasing heat. Thus the heat pipe operates in a closed loop evaporation/condensation cycle.

A run-around coil exchanger uses a pair of finned-tube coils. A circulating fluid is heated by warm exhaust air and in turn warms the cool replacement air. An advantage of the run-around coil is that the exhaust and supply duct systems can be separated by a significant distance which results in a reduced potential for re-entry; usually less duct in the systems and usually less roof area occupied by the units.

Several factors are important in the selection of the appropriate heat exchanger. A partial list is as follows:

1. The nature of the exhaust stream. A corrosive or dust laden stream may need to be precleaned.
2. The need to isolate the contaminated exhaust stream from the clean replacement air stream.
3. The temperature of the exhaust stream. Unless the hot air stream is well above the desired delivery temperature of the replacement air stream and the exhaust air stream is at elevated temperatures whenever heat is demanded by the replacement air stream, additional heating capacity will be required.
4. Space requirements. Space requirements for some heat exchangers can be very extensive, especially when the additional duct runs are considered.
5. The nature of the air stream. Many exhaust air streams are corrosive or dirty and special construction materials may be required.
6. The need for a by-pass. During failure mode or summer conditions, a by-pass will be required.

Recirculation of Air from Industrial Exhaust Systems: Where large amounts of air are exhausted from a room or building in order to remove particulates, gases, fumes or vapors, an equivalent amount of fresh tempered replacement air must be supplied to the room. If the amount of replacement air is large, the cost of energy to condition the air can be very high. Recirculation of the exhaust air after *thorough* cleaning is one method that can reduce the amount of energy consumed. Acceptance of such recirculating systems will depend on the degree of health hazard associated with the particular contaminant being exhausted as well as other safety, technical and economic factors. A logic diagram listing the factors that must be evaluated is provided in Figure 7-15.(7.6)

Essentially this diagram states that recirculation may be permitted if the following conditions are met:

1. The chemical, physical, and toxicological characteristics of the chemical agents in the air stream to be recirculated must be identified and evaluated. Exhaust air containing chemical agents whose toxicity is unknown or for which there is no established safe exposure level should not be recirculated.
2. All governmental regulations regarding recirculation must be reviewed to determine whether it is restricted or prohibited for the recirculation system under review.
3. The effect of a recirculation system malfunction must be considered. Recirculation should not be attempted if a malfunction could result in exposure levels that would cause worker health problems. Substances which can cause permanent damage or significant physiological harm from a short overexposure shall not be recirculated.
4. The availability of a suitable air cleaner must be determined. An air cleaning device capable of providing an effluent air stream contaminant concentration sufficiently low to achieve acceptable workplace concentrations must be available.
5. The effects of minor contaminants should be reviewed. For example, welding fumes can be effectively removed from an air stream with a fabric filter; however, if the welding process produces oxides of nitrogen, recirculation could cause a concentration of these gases to reach an unacceptable level.
6. Recirculation systems must incorporate a monitoring system that provides an accurate warning or signal capable of initiating corrective action or process shutdown before harmful concentrations of the recirculated chemical agents build up in the workplace. Monitoring can be accomplished by a number of methods and must be determined by the type and hazard of the substance. Examples include area monitoring for nuisance type substances and secondary high efficiency filter pressure drop and on-line monitors for more hazardous materials.

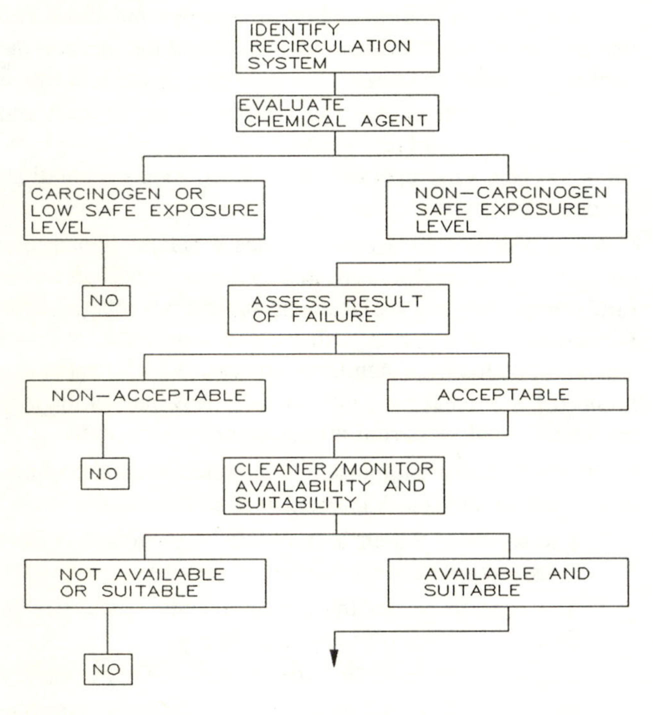

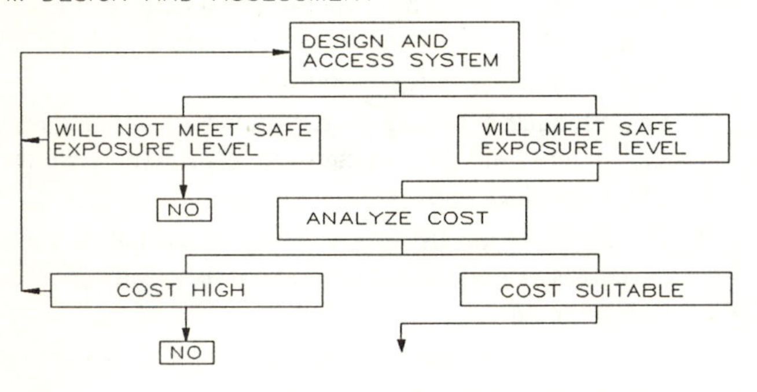

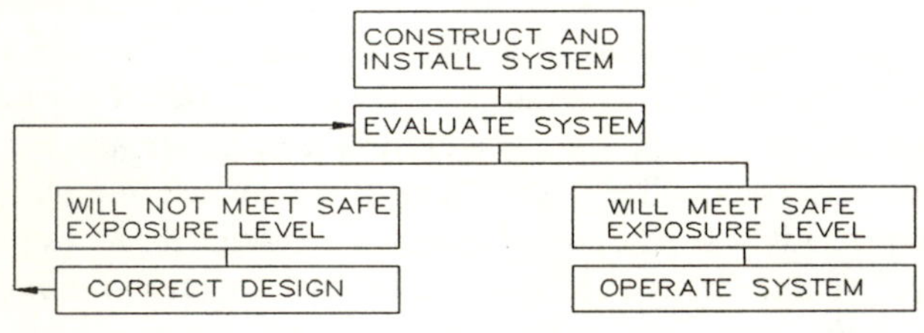

FIGURE 7-15 RECIRCULATION DECISION LOGIC

While all system components are important, special consideration should be given to the monitor. The prime requisites are that the monitor be capable of sensing a system malfunction or failure and of providing a signal which will initiate an appropriate sequence of actions to assure that overexposure does not occur. The sophistication of the monitoring system can vary widely. The type of monitor selected will depend on various parameters (i.e., location, nature of contaminant — including shape and size — and degree of automation).

7.11.4 Selection of Monitors: The safe operation of a recirculating system depends on the selection of the best monitor for a given system. Reference 7.7 describes four basic components of a complete monitoring system which includes signal transfer, detector/transducer, signal conditioner and information processor. Figure 7-16 shows a schematic diagram of the system incorporating these four

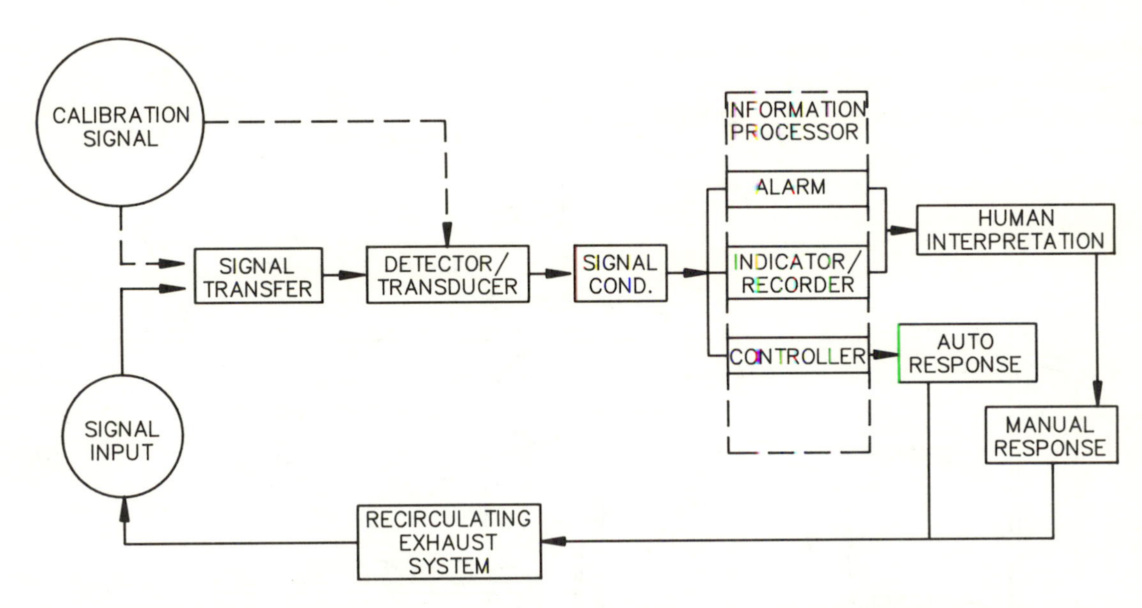

FIGURE 7–16. SCHEMATIC DIAGRAM OF RECIRCULATION MONTIORING SYSTEM

components. It is quite likely that commercially available monitors may not contain all of the above four components and may have to be custom engineered to the need.

In addition, the contaminant must be collected from the air stream either as an extracted sample or in toto. If a sample is taken it must be representative of the average conditions of the air stream. At normal duct velocities, turbulence assures perfect mixing so gas and vapor samples should be representative. For aerosols, however, the particle size discrimination produced by the probe may bias the estimated concentration unless isokinetic conditions are achieved.

The choice of detection methods depends on the measurable chemical and physical properties of the contaminants in the air stream. Quantifying the collected contaminants is generally much easier for particulate aerosols than for gases, vapors or liquid aerosols.

Particulates: Where the hazardous contaminant constitutes a large fraction of the total dust weights, filter samples may allow adequate estimation of concentration. Better, if the primary collector (e.g., bag filters, cartridge filters) allows very low penetration rates, it may be economical to use high efficiency filters as secondary filters. If the primary filter fails, the secondary filter not only will experience an easily measured increase in pressure drop, but will filter the penetrating dust as well — earning this design the sobriquet, "safety monitor" systems (see Figure 7-17).

Non-particulates: Continuously detecting and quantifying vapor and gas samples reliably and accurately is a complex subject beyond the scope of this manual.

Reference 7.8 (see Chapters U and V) describes and evaluates different air monitoring devices. The monitor in a recirculating system must be capable of reliably monitoring continuously and unattended for an extended period of time. It must also be able to quickly and accurately sense a change in system performance and provide an appropriate warning if a preselected safety level is reached. In order to function properly, monitors must be extremely reliable and properly maintained. Monitors should be designed so that potential malfunctions are limited in number and can be detected easily by following recommended procedures. Required maintenance should be simple, infrequent, and of short duration.

7.12 EVALUATION OF EMPLOYEE EXPOSURE LEVELS

Under equilibrium conditions, the following equations may be used to determine the concentration of a contaminant permitted in the recirculation return air stream:

$$C_R = \frac{(1-\eta)(C_E - K_R C_M)}{1 - [(K_R)(1-\eta)]} \quad [7.3]$$

where:

C_R = air cleaner discharge concentration after recirculation, mg/m³

η = fractional air cleaner efficiency

C_E = local exhaust duct concentration before recirculation, mg/m³

K_R = factor which represents a fraction of the recirculated exhaust stream that is composed of the recirculation return air (range 0 to 1.0)

C_M = replacement air concentration, mg/m³

$$C_B = \frac{Q_B}{Q_A}(C_G - C_M)(1-f) + (C_O - C_M)f + K_B C_R + (1 - K_B)(C_M) \quad [7.4]$$

where:

C_B = 8-hr TWA worker breathing zone concentration after recirculation, mg/m³

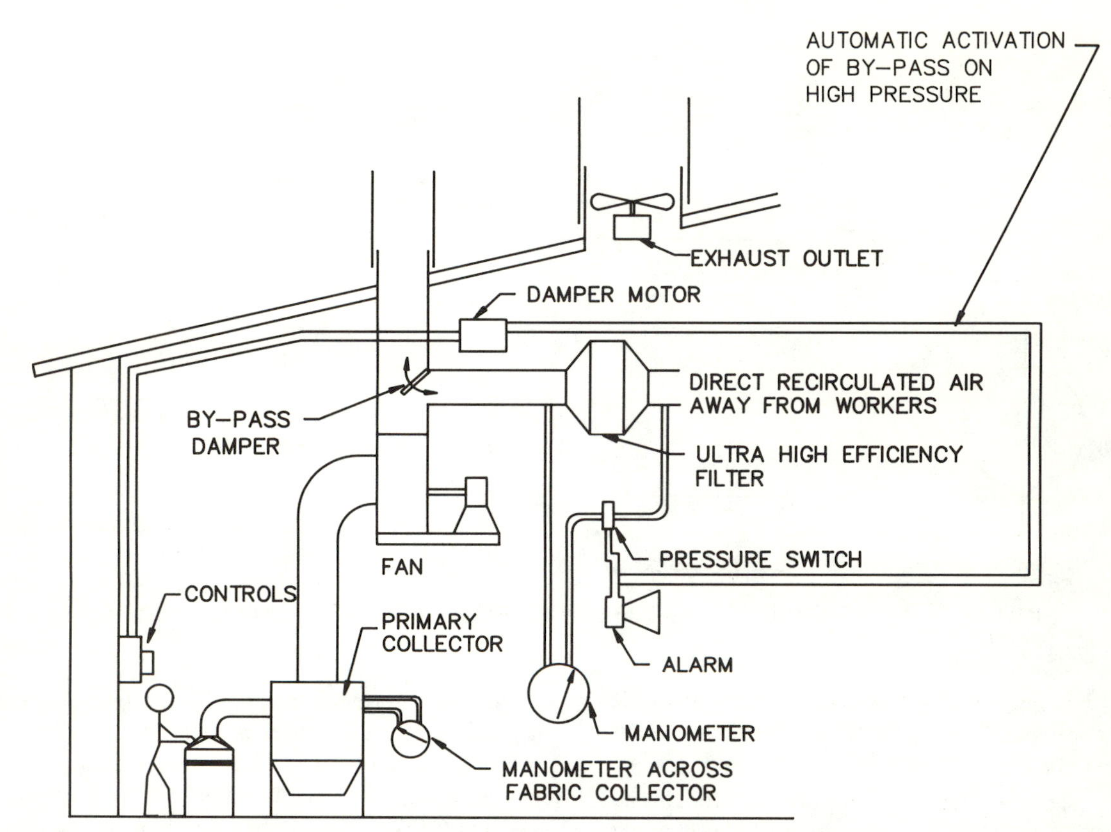

FIGURE 7–17. SCHEMATIC OF RECIRCULATION FROM AIR CLEANING DEVICES (PARTICULATES).

Q_B = total ventilation air flow before recirculation

Q_A = total ventilation air flow after recirculation

C_G = general room concentration before recirculation, mg/m^3

f = factor which represents the fraction of time the worker spends at the work station

C_O = 8 hr TWA breathing zone concentration at work station before recirculation

K_B = fraction of worker's breathing zone air that is composed of recirculation return air (range 0 to 1.0)

The factors K_R, K_B and f are dependent on work station and worker's position in relation to the source of the recirculation return air and the worker's position in relation to the exhaust hood. The value of K_R can range from 0 to 1.0 where 0 indicates no recirculation return air entering the hood and 1.0 indicates 100% recirculation air entering the hood. Similarly, the value of K_B can range from 0 to 1.0 where 0 indicates there is no recirculation return air in the breathing zone and 1.0 indicates that the breathing zone air is 100% recirculated return air. The factor f varies from 0 where the worker does not spend any time at the work station where the air is being recirculated to 1.0 where the worker spends 100% time at the work station.

In many cases it will be difficult to attempt quantification of the values required for solution of these equations for an operation not yet in existence. Estimates based on various published and other available data for the same or similar operations may be useful. The final system must be tested to demonstrate that it meets design specifications.

An example of use of Equations 7.3 and 7.4 and the effect of the various parameters is as follows:

Consider a system with 10,000 cfm total ventilation before recirculation (Q_B) consisting of 5,000 cfm of general exhaust and 5,000 cfm of local exhaust. The local exhaust is recirculated resulting in 10,000 cfm after recirculation air flow consisting of 5,000 cfm recirculated and 5,000 cfm fresh air flow.

Assume poor *placement* of the recirculation return (K_R and K_B = 1) and that the worker spends all his time at the work station (f = 1); the air cleaner efficiency (η) = 0.95; exhaust duct concentration (C_E) = 500 ppm; general room concentration (C_G) = 20 ppm; replacement air concentration (C_M) = 5 ppm; work station (breathing zone) concentration before recirculation (C_O) = 35 ppm; and a contaminant TLV of 50 ppm.

Equation 7.3 gives recirculation air return concentration:

$$C_R = \frac{(1 - 0.95)(500 - 1 \times 5)}{1 - [(1)(1 - 0.95)]} = 26.1 \text{ ppm}$$

Equation 7.4 gives the worker breathing zone concentration:

$$C_B = \frac{Q_B}{Q_A}(C_G - C_M)(1 - f) + (C_O - C_M)\,f + K_B C_R$$

$$+ (1 - K_B)(C_M)$$

$$= \left(\frac{10^4}{10^4}\right) (20 - 5)(1 - 1) + (35 - 5)(1) + (26.1) + (1 - 1)(5)$$

$$= 56.1 \text{ ppm}$$

Obviously, 56.1 ppm exceeds the TLV of 50 ppm and therefore is unacceptable.

In order to achieve lower concentrations (C_B), the system configuration must be redesigned so that only 50% of the recirculation return air reaches the work station. Thus, K_R and K_B are reduced to 0.5. Substituting these new data in Equation 7.4, the breathing zone concentration calculates as 45.3 ppm. This is lower than the TLV of 50 ppm and therefore acceptable.

Several potential problems may exist in the design of recirculated air systems. Factors to be considered are:

1. Recirculating systems should, whenever practicable, be designed to bypass to the outdoors, rather than to recirculate, when weather conditions permit. If a system is intended to conserve heat in winter months and if adequate window and door openings permit sufficient replacement air when open, the system can discharge outdoors in warm weather. In other situations where the work space is conditioned or where mechanically supplied replacement air is required at all times, such continuous bypass operation would not be attractive.
2. Wet collectors also act as humidifiers. Recirculation of humid air from such equipment can cause uncomfortably high humidity and require auxiliary ventilation or some means must be used to prevent excess humidity.
3. The exit concentration of typical collectors can vary with time. Design data and testing programs should consider all operational time periods.
4. The layout and design of the recirculation duct should provide adequate mixing with other supply air and avoid uncomfortable drafts on workers or air currents which would upset the capture velocity of local exhaust hoods.
5. A secondary air cleaning system, as described in the example on particulate recirculation, is preferable to a monitoring device because it is usually more reliable and requires a less sophisticated degree of maintenance.
6. Odors or nuisance value of contaminants should be considered as well as the official TLV values. In some areas, adequately cleaned recirculated air, provided by a system with safeguards, may be of better quality than the ambient outside air available for replacement air supply.
7. Routine testing, maintenance procedures and records should be developed for recirculating systems.
8. Periodic testing of the workroom air should be provided.
9. An appropriate sign shall be displayed in a prominent place reading as follows:

CAUTION

AIR CONTAINING HAZARDOUS SUBSTANCES IS BEING CLEANED TO A SAFE LEVEL IN THIS EQUIPMENT AND RETURNED TO THE BUILDING. SIGNALS OR ALARMS INDICATE MALFUNCTIONS AND MUST RECEIVE IMMEDIATE ATTENTION: STOP RECIRCULATION, DISCHARGE THE AIR OUTSIDE, OR STOP THE PROCESS IMMEDIATELY.

REFERENCES

7.1. American Industrial Hygiene Association: *Heating and Cooling Man and Industry*. Akron, OH (1969).

7.2. Hart and Cooley Manufacturing Co.: "Bulletin E-6." Holland, MI

7.3. Hama, G.: "How Safe Are Direct-Fired Makeup Units?" *Air Engineering*, p. 22 (September 1962)

7.4. National Fire Protection Association, 470 Atlantic Ave., Boston, MA 02210.

7.5. American Society of Heating, Refrigeration and Air Conditioning Engineers: *Heating Ventilating and Air Conditioning Guide*. Atlanta, GA (1963).

7.6. R.T. Hughes and A.A. Amendola: "Recirculating Exhaust Air: Guides, Design Parameters and Mathematical Modeling." *Plant Engineering* (March 18, 1982).

7.7. *The Recirculation of Industrial Exhaust Air — Symposium Proceedings*. Department of Health, Education and Welfare (NIOSH) Pub. No. 78-141 (1978).

7.8. American Conference of Governmental Industrial Hygienists: *Air Sampling Instruments for Evaluation of Atmospheric Contaminants*, 6th ed., Chap. U and V. Cincinnati, OH (1983).

Chapter 8

CONSTRUCTION GUIDELINES FOR LOCAL EXHAUST SYSTEMS

8.1 INTRODUCTION 8-2

8.2 GENERAL 8-2

8.3 MATERIALS 8-2

8.4 CONSTRUCTION 8-2

8.5 SYSTEM DETAILS 8-4

8.6 CODES 8-6

8.7 OTHER TYPES OF DUCT MATERIALS ... 8-6

8.8 TESTING 8-6

REFERENCES 8-6

8.1 INTRODUCTION

Ducts in industry may be used for many diverse applications. They are specified most often for use in the low static pressure range (−10 "wg to +10 "wg), but higher static pressures are occasionally encountered. The duct conveys air or gas which is sometimes at high temperatures and often contaminated with abrasive particulate or corrosive aerosols. Whether conditions are mild or severe, correct design and competent installation of ducts and hoods are necessary for proper functioning of any ventilation system. The following minimum specifications are recommended.

8.2 GENERAL

Exhaust systems should be constructed with materials suitable for the conditions of service and installed in a permanent and workmanlike manner. To minimize friction loss and turbulence, the interior of all ducts should be smooth and free from obstructions — especially at joints.

8.3 MATERIALS

Ducts usually are constructed of black iron which has been welded, flanged and gasketed, or of welded galvanized sheet steel unless the presence of corrosive gases, vapors and mists or other conditions make such material impractical. Arc welding of black iron lighter than 18 gauge is not recommended. Galvanized construction is not recommended for temperatures exceeding 400 F. The presence of corrosive gases, vapor and mist may require the selection of corrosive resistant metals, plastics or coatings. It is recommended that a specialist be consulted for the selection of materials best suited for applications when corrosive atmospheres are anticipated. Table 8-2 provides a guide for selection of materials for corrosive conditions.

8.4 CONSTRUCTION

1. There are four classifications for exhaust systems on non-corrosive applications:
   A. *Class 1 — Light Duty:* Includes nonabrasive applications, e.g., replacement air, general ventilation, gaseous emissions control.
   B. *Class 2 — Medium Duty:* Includes applications with moderately abrasive particulate in light concentrations, e.g., buffing and polishing, woodworking, grain dust.
   C. *Class 3 — Heavy Duty:* Includes applications with highly abrasive particulate in low concentrations, e.g., abrasive cleaning operations, dryers and kilns, boiler breeching, sand handling.
   D. *Class 4 — Extra Heavy Duty:* Includes applications with highly abrasive particles in high concentrations, e.g., materials conveying high concentrations of particulate in all examples listed under Class 3 (usually used in heavy industrial plants such as steel mills, foundries, mining, and smelting).
2. For most conditions, round duct is recommended for industrial ventilation, air pollution control, and dust collecting systems. Compared to non-round duct, it provides for lower friction loss and its higher structural integrity allows lighter gauge materials and fewer reinforcing members. Round duct should be constructed in accordance with Reference 8.1. Metal thickness required for round industrial duct varies with classification, static pressure, reinforcement, and span between supports. Metal thicknesses required for the four classes are based on design and use experience (see Table 8-1).

TABLE 8-1. Table of Duct Gauge Recommendations

Diameter of Straight Ducts	RANGE OF METAL THICKNESSES* U.S. Standard Gauge for Steel Duct			
	Class 1	Class 2	Class 3	Class 4
4" to 8"	22-20	22-18	16	14
Over 8" to 18"	22-12	22-12	16-11	14-11
Over 18" to 30"	18-7	16-7	16-6	14-6
Over 30"	14-2	14-2	12-2	12-2

*Thickness varies with classification, pressure, reinforcement, and span requirements. *NOTE:* 24 and 26 gauge metal can not be welded.

3. Rectangular ducts should only be used when space requirements preclude the use of round construction. Rectangular ducts should be as nearly square as possible to minimize resistance, and they should be constructed in accordance with Reference 8.2.
4. For many applications, spiral wound ductwork is adequate and less expensive than custom construction. However, spiral wound duct should not be used for Classes 3 and 4 because it does not withstand abrasion well. Elbows, branch entries, and similar fittings should be fabricated, if necessary, to achieve good design. Special considerations concerning use of spiral duct are as follows:
   A. Unless flanges are used for joints, the duct should be supported close to each joint, usually within 2 inches. Additional supports may be needed. See Reference 8.1.
   B. Joints should be sealed by methods shown to be adequate for the service.
   C. Systems should be leak tested after installation at the maximum expected static pressure. Leakage should be no more than 1% of the design volume.
5. The following formula, taken from Reference 8.1, can be used for specifying ducts to be constructed of metals other than steel.

TABLE 8-2. Typical Physical and Chemical Properties of Fabricated Plastics and Other Materials

				Resistance To							
Chemical Type	Trade Names	Max. Opr. Temp., F	Flammability	Gasoline	Mineral Oil	Strong Alk.	Weak Alk.	Strong Acid	Weak Acid	Salt Solution	Solvents
Urea Formaldehyde	Beetle Plaskon Sylplast	170	Self Ext.	Good	Good	Unac.	Fair	Poor	Poor	—	Good
Melamine Formaldehyde	Cymel Plaskon Resimene	210-300	Self Ext.	Good	Good	Poor	Good	Poor	Good	—	Good
Phenolic	Bakelite Durite Durez G.E. Resinox	250-450	Self Ext.	Fair	—	Poor	Fair	Poor	Fair	—	Fair
Alkyd	Plaskon	—	Self Ext.	Good	—	Unac.	Poor	—	Good	—	Good
Silicone	Bakelite G.E.	550	—	Good	Good	—	—	Good	Good	—	Unac.
Epoxy	Epiphen Araldite Maraset Renite Tool Plastik Epon Resin	50-200	Self Ext.	Good	—	Good	Good	Good	Good	—	Good
Cast Phenolic	Marblette	—	Self Ext.	—	—	Unac.	Fair	Good	Good	—	Good to Unac.
Allyl & Polyester	Laminac Bakelite Plaskon Glykon Paraplex	300-450	Self Ext.	—	—	Poor	Fair	Poor	Fair	—	Fair
Acrylic	Lucite Plexiglas Wascoline	140-200	0.5-2.0 in/min	—	—	—	Good	Unac.	Good	—	Good to Unac.
Polyethylene	Tenite Irrathene	140-200	Burns Slowly	—	—	—	—	—	—	—	Unac.
Tetrafluoroehtylene Chlorotrifluoro-ethylene	Teflon Kel F	500	Non-Fl.	Good	—	Good	Good	Good	Good	—	Good
Polyvinyl Formal & Butyral	Vinylite Butacite Saflex Butvar Formuar	—	Slow Burning	Good	Good	Good	Good	Unac.	Unac.	—	Unac.
Vinyl Chloride Polyner & Copolyner	Krene Bakelite Vinyl Dow pvc Vygen	130-175	Slow Burning	—	—	Good	Good	Good	Good	—	Unac.
Vinylidene Chloride	Saran	160-200	Self Ext.	Good	Good	Good	Good	Good	Good	—	Fair
Styrene	Bakelite Catalin Styron Dylene Luxtrex	150-165	0.5-2.0 in/min	Unac.	Fair	Good	Good	—	—	Good	Poor
Polystyrene Reinforced with Fibrous Glass				Unac.	Fair	Good	Good	—	—	Good	Poor
Cellulose Acetate	Celanese Acetate Tenite	Thermo Plastic	0.5-2.0 in/min	Good	Good	Unac.	Unac.	Unac.	Fair	—	Poor
Nylon	Plaskon Zytel Tynex	250	Self Ext.	Good	Good	Good	Good	Unac.	Good	—	Good
Glass	Pyrex	450	Non-Fl.	Good	Good	Good	Good	Good	Good	Good	Good

NOTE: Each situation must be thoroughly checked for compatability of materials during the design phase or if usage is changed.

For a duct of infinite length, the required thickness may also be determined from the following equation:[8.1]

$$\frac{t}{D} = \sqrt[3]{0.035714\ p \left(\frac{1 - \nu^2}{E}\right) (52 + D)}$$

where:

t = the thickness of the duct in inches

D = the diameter of the duct in inches

p = the intensity of the negative pressure on the duct in pounds per square inch

E = modulus of elasticity in pounds per square inch

ν = Poisson's ratio

The above equation for Class 1 ducts incorporates a safety factor which varies linearly with the diameter (D), beginning at 4 for small ducts and increasing to 8 for duct diameters of 60 inches. This safety factor has been adopted by the sheet metal industry to provide for lack of roundness, excesses in negative pressure due to particle accumulation in the duct and other manufacturing or assembly imperfections unaccounted for by quality control, and tolerances provided by design specifications.

Additional metal thickness must be considered for Classes 2, 3 and 4. The designer is urged to consult the SMACNA standards for complete engineering design procedures.

6. Hoods should be a minimum of two gauges heavier than straight sections of connecting branches, free of sharp edges or burrs, and reinforced to provide necessary stiffness.
7. Longitudinal joints or seams should be welded. All welding should conform to the standards established by the American Welding Society (AWS) structural code.[8.3] Double lock seams are limited to Class 1 applications.
8. Duct systems subject to wide temperature fluctuations should be provided with expansion joints. Flexible materials used in the construction of expansion joints should be selected with temperature and corrosion conditions considered.
9. Elbows and bends should be a minimum of 2 gauges heavier than straight lengths of equal diameter and have a centerline radius of at least 1.5 and preferably 2 times the pipe diameter (see Figure 8-1). Large centerline radius elbows are recommended where highly abrasive dusts are being conveyed.
10. Elbows of 90° should be of a five-piece construction for round ducts up to six inches and of a seven-piece construction for larger diameters. Bends less than 90° should have a proportional number of pieces. Prefabricated elbows of smooth construction may be used (see Figure 8-2 for heavy duty elbows).
11. Transitions in mains and sub mains should be tapered. The taper should be at least 5 units long for each 1 unit change in diameter or 30° included angle (see Figure 8-3).
12. All branches should enter the main at the center of the transition at an angle not to exceed 45° with 30° preferred. To minimize turbulence and possible particulate fall out, connections should be to the top or side of the main with no two branches entering at opposite sides (see Figure 8-3).
13. Where the air contaminant includes particulates that may settle in the ducts, clean-out doors should be provided in horizontal runs, near elbows, junctions, and vertical runs. The spacing of clean-out doors should not exceed 12 feet for ducts of 12 inches diameter and less but may be greater for larger duct sizes (see Figure 8-4). Removable caps should be installed at all terminal ends and the last branch connection should not be more than six inches from the capped end.
14. Where condensation may occur, the duct system should be liquid tight and provisions made for proper sloping and drainage.
15. A straight duct section of at least six equivalent duct diameters should be used when connecting to a fan (see Figure 8-5). Elbows or other fittings at the fan inlet will seriously reduce the flow rate (see Figures 6-16 through 6-21, and AMCA 201).[8.3] The diameter of the duct should be approximately equal to the fan inlet diameter.
16. Discharge stacks should be vertical and terminate at a point where height or air velocity limit re-entry into supply air inlets or other plant openings (see Figures 8-6 and 8-7).

8.5 SYSTEM DETAILS

1. Provide duct supports of sufficient capacity to carry the weight of the system plus the weight of the duct half filled with material with no load placed on the connecting equipment. (See SMACNA standards.[8.1,8.2])
2. Provide adequate clearance between ducts and ceilings, walls and floors for installation and maintenance.
3. Install fire dampers, explosion vents, etc., in accordance with the National Fire Protection Association Codes and other applicable codes and standards.
4. Avoid using blast gates or other dampers. However, if blast gates are used for system adjustment, place each near the connection of the branch to the main. To reduce tampering, provide a means of locking dampers in place after the adjustments have been made. (see Figure 8-8 for types.)
5. Allow for vibration and expansion. If no other considerations make it inadvisable, provide a flexible connection between the duct and the fan. The fan housing and drive motor should be mounted on a common base of sufficient

weight to dampen vibration, or on a properly designed vibration isolator.

6. Exhaust fans handling explosive or flammable atmospheres require special construction (see Section 6.3.9).
7. Do not allow hoods and duct to be added to an existing exhaust system unless specifically provided for in the original design or unless the system is modified.

8.6 CODES

Where laws conflict with the preceding, the more stringent requirement should be followed. Deviation from existing regulations may require approval.

8.7 OTHER TYPES OF DUCT MATERIALS

1. Avoid use of flexible ducts. Where required, use a non-collapsible type that is no longer than necessary. Refer to the manufacturer's data for friction and bend losses.
2. Commercially available seamless tubing for small duct sizes (e.g., up to 6 inches) may be more economical on an installed cost basis than other types.
3. Plastic pipe may be the best choice for some applications (e.g., corrosive conditions at low temperature). For higher temperatures consider fiberglass or a coated duct.
4. Friction losses for non-fabricated duct will probably be different than shown in Figures 5-18A and 5-18B. For specific information, see manufacturer's data or Figure 5-22.

8.8 TESTING

The exhaust system should be tested and evaluated (see Chapter 9). Openings for sampling should be provided in the discharge stack or duct to test for compliance with air pollution codes or ordinances.

REFERENCES

8.1. Sheet Metal and Air Conditioning Contractors' National Assoc., Inc.: *Round Industrial Duct Construction Standards.* 8224 Old Courthouse Rd., Tysons Corner, Vienna, VA 22180 (1982).

8.2. Sheet Metal and Air Conditioning Contractors' National Assoc., Inc., *Rectangular Industrial Duct Construction Standards.* 8224 Old Courthouse Rd., Tysons Corner, Vienna, VA 22180 (1980).

8.3. American Welding Society: (AWS D1.1-72). P.O. Box 351040, Miami, FL 33135.

8.4. Air Movement & Control Associations, Inc.: *AMCA Publiction 201.* 30 West University Drive, Arlington Heights, IL 60004.

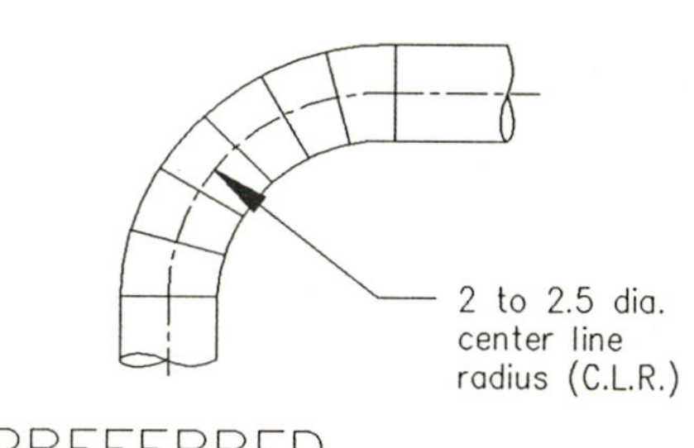

PREFERRED

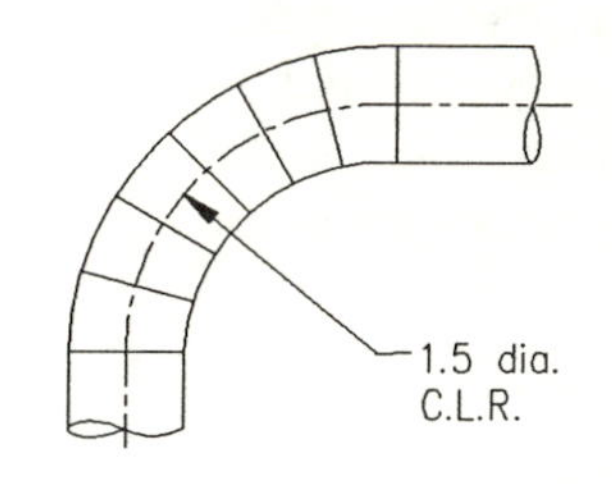

ACCEPTABLE

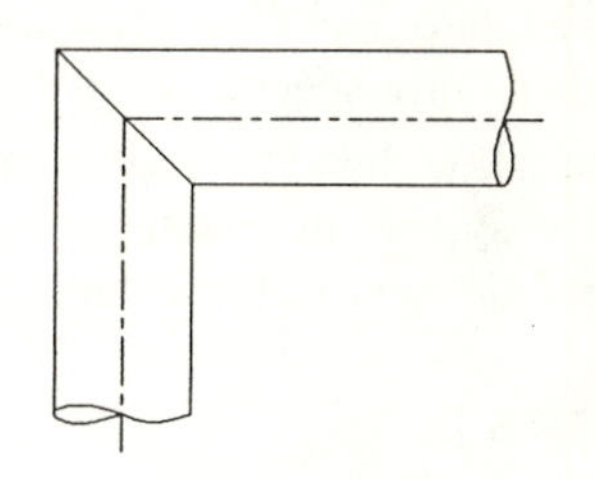

AVOID

ELBOW RADIUS

Elbows should be 2 to 2.5 diameter centerline radius except where space does not permit.

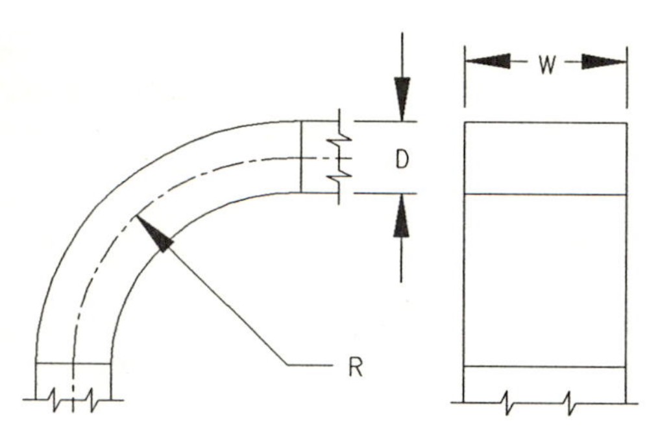

PREFERRED

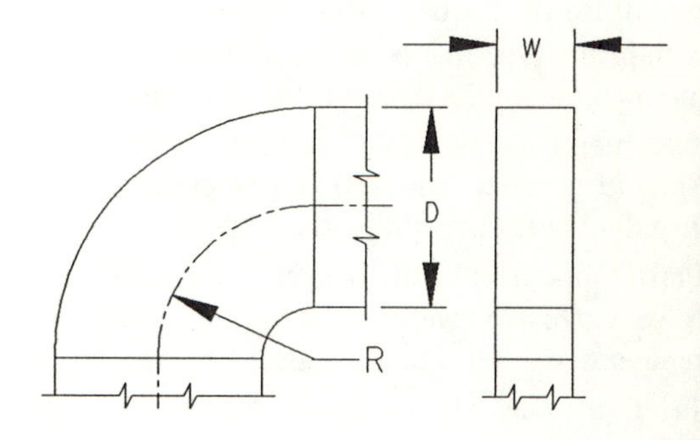

AVOID

ASPECT RATIO $\left(\frac{W}{D}\right)$

Elbows should have $\left(\frac{W}{D}\right)$ and $\left(\frac{R}{D}\right)$ equal to or greater than (1).

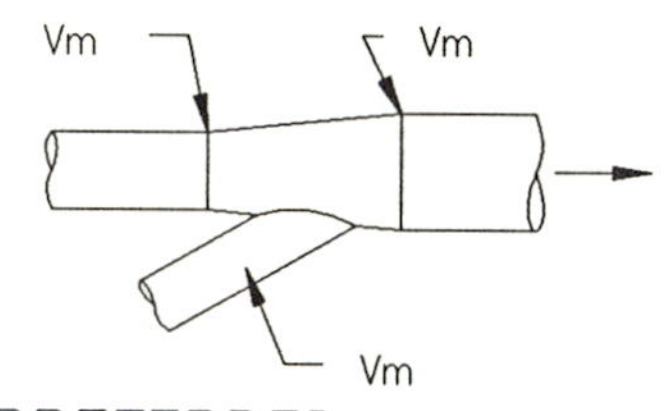

PREFERRED

A_1 $A_3 = A_1 + A_2 + 20\%$ A_2

AVOID

Vm = Minimum transport velocity

A = Cross section area

PROPER DUCT SIZE

Size the duct to hold the selected transport velocity or higher.

AMERICAN CONFERENCE OF GOVERNMENTAL INDUSTRIAL HYGIENISTS	*PRINCIPLES OF DUCT CONSTRUCTION*	
	DATE *1–88*	FIGURE *8–1*

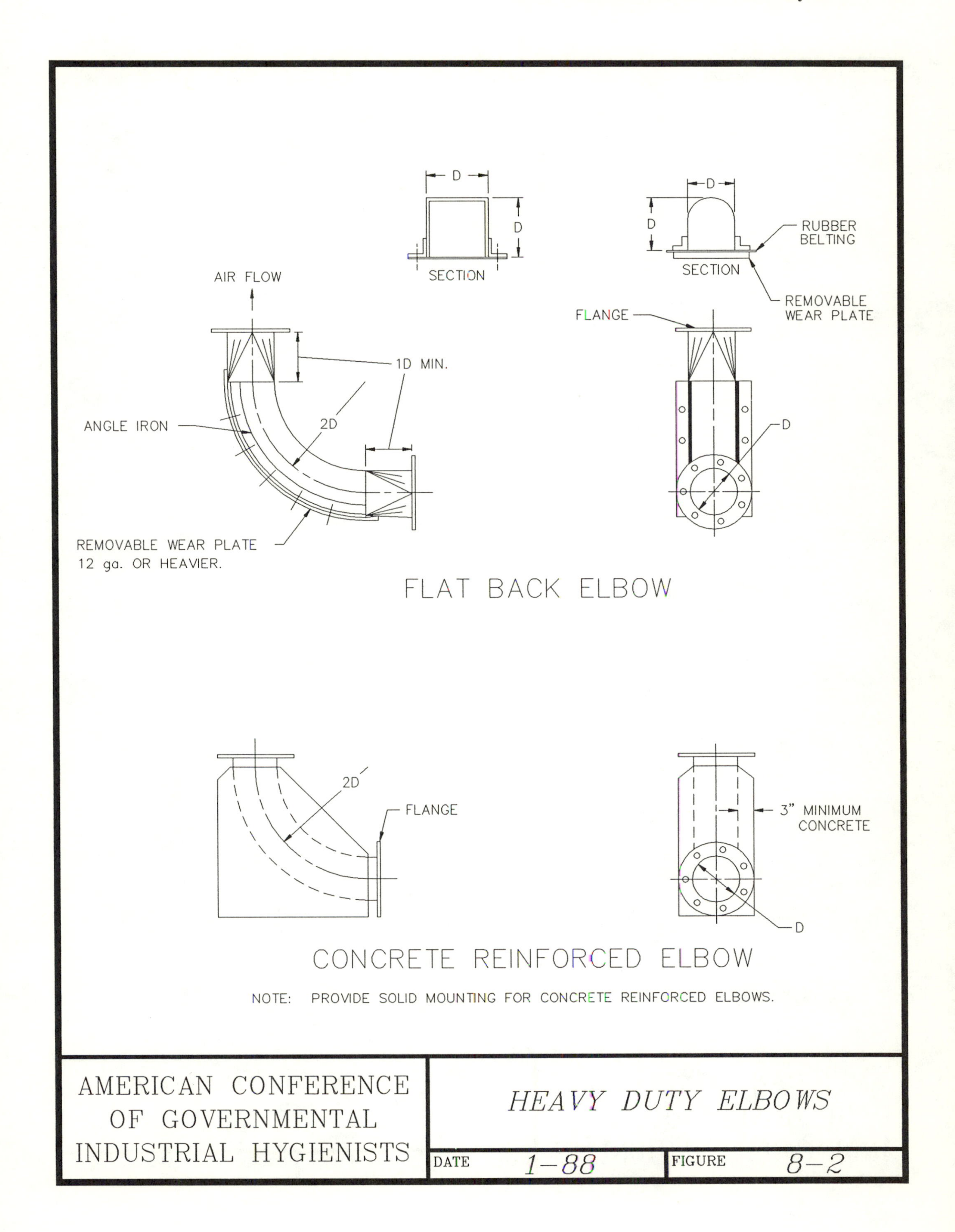
D
D
SECTION
D
D
RUBBER
BELTING
SECTION
REMOVABLE
WEAR PLATE
AIR FLOW
FLANGE
1D MIN.
2D
ANGLE IRON
D
REMOVABLE WEAR PLATE
12 ga. OR HEAVIER.
FLAT BACK ELBOW
2D
FLANGE
3" MINIMUM
CONCRETE
D
CONCRETE REINFORCED ELBOW
NOTE: PROVIDE SOLID MOUNTING FOR CONCRETE REINFORCED ELBOWS.
AMERICAN CONFERENCE
OF GOVERNMENTAL
INDUSTRIAL HYGIENISTS
HEAVY DUTY ELBOWS
DATE 1-88
FIGURE 8-2

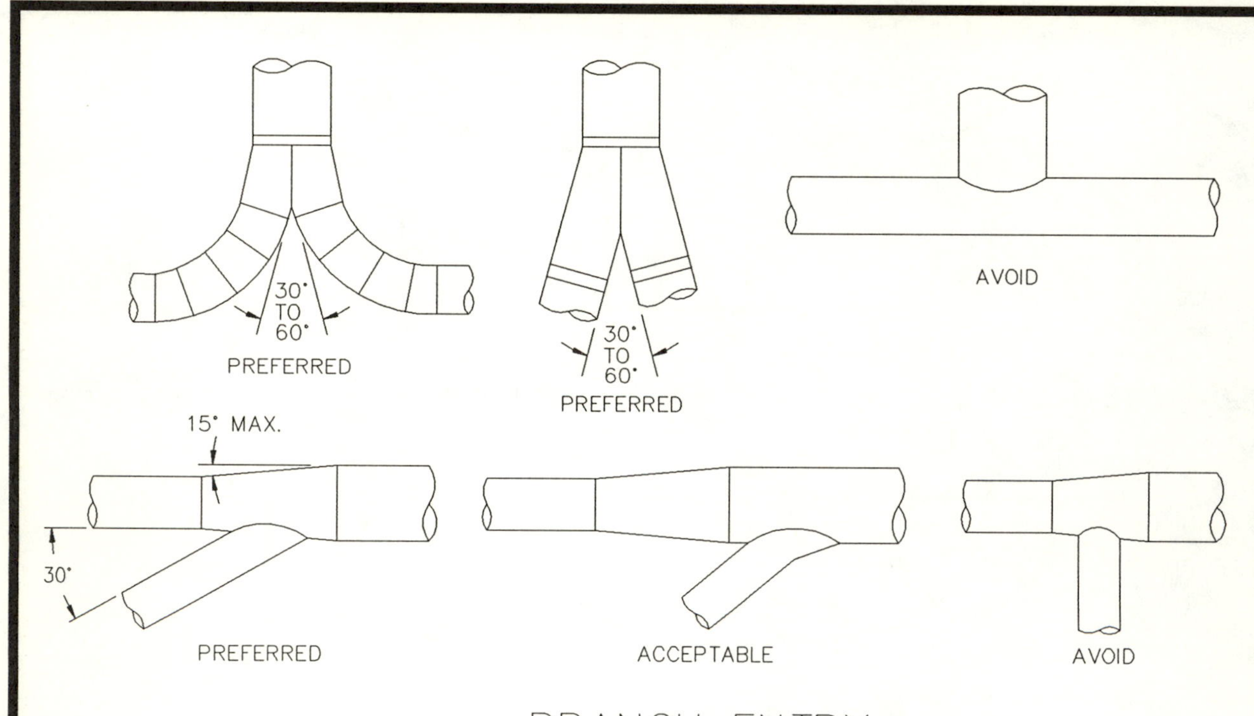

BRANCH ENTRY

BRANCHES SHOULD ENTER AT GRADUAL EXPANSIONS AND AT AN ANGLE OF 30° OR LESS (PREFERRED) TO 45° IF NECESSARY.

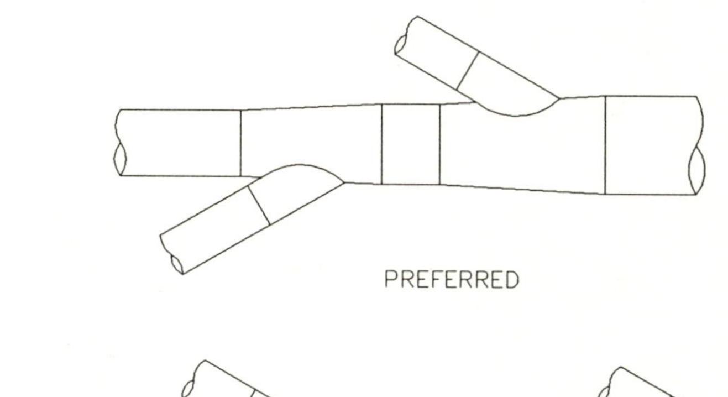

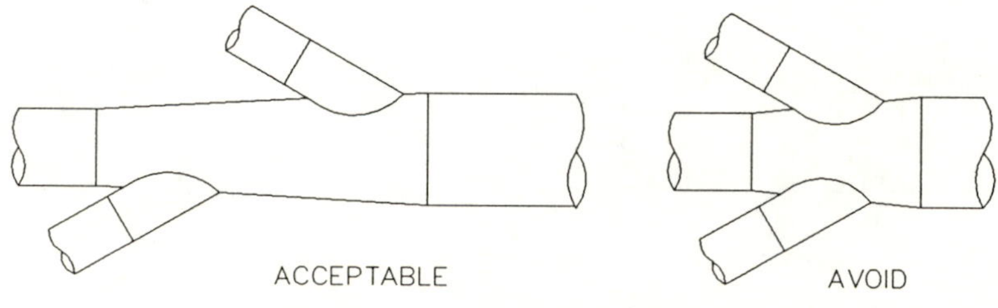

BRANCH ENTRY

BRANCHES SHOULD NOT ENTER DIRECTLY OPPOSITE EACH OTHER.

AMERICAN CONFERENCE OF GOVERNMENTAL INDUSTRIAL HYGIENISTS	*PRINCIPLES OF DUCT CONSTRUCTION*
	DATE *1–88* FIGURE *8–3*

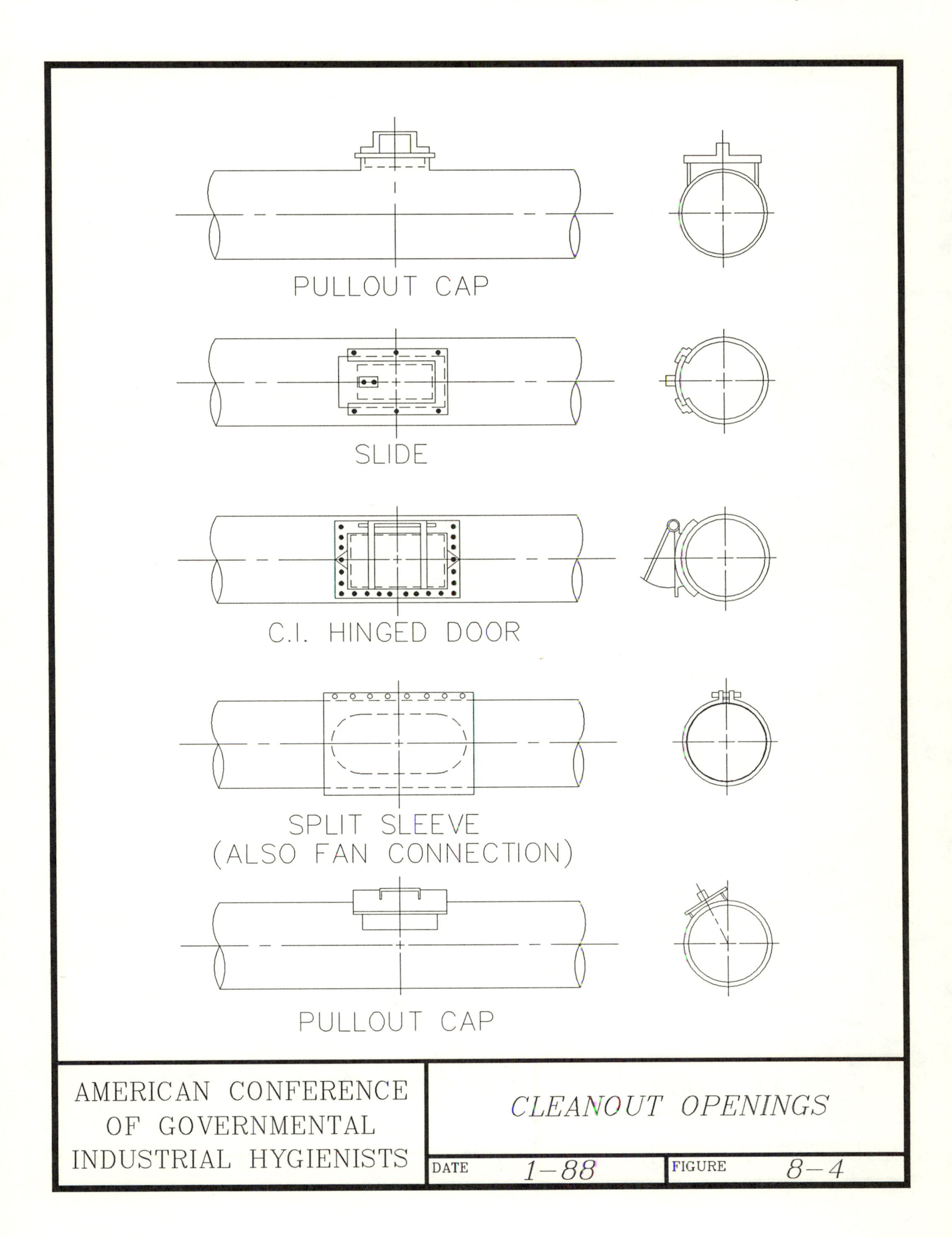
PULLOUT CAP
SLIDE
C.I. HINGED DOOR
SPLIT SLEEVE
(ALSO FAN CONNECTION)
PULLOUT CAP
AMERICAN CONFERENCE
OF GOVERNMENTAL
INDUSTRIAL HYGIENISTS
CLEANOUT OPENINGS
DATE 1-88
FIGURE 8-4

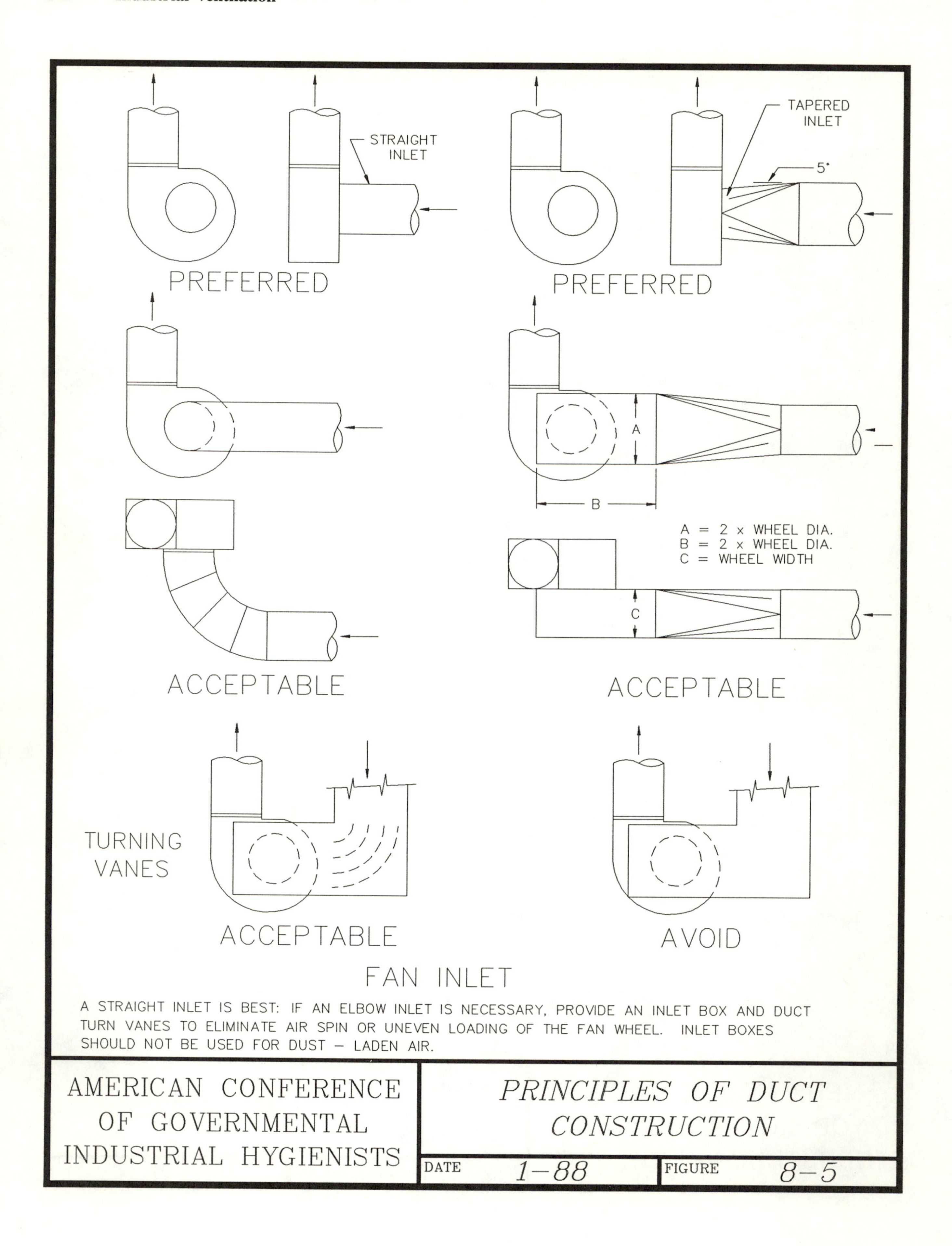
STRAIGHT INLET
TAPERED INLET
5°
PREFERRED
PREFERRED
A
B
A = 2 x WHEEL DIA.
B = 2 x WHEEL DIA.
C = WHEEL WIDTH
C
ACCEPTABLE
ACCEPTABLE
TURNING VANES
ACCEPTABLE
AVOID
FAN INLET
A STRAIGHT INLET IS BEST: IF AN ELBOW INLET IS NECESSARY, PROVIDE AN INLET BOX AND DUCT TURN VANES TO ELIMINATE AIR SPIN OR UNEVEN LOADING OF THE FAN WHEEL. INLET BOXES SHOULD NOT BE USED FOR DUST – LADEN AIR.
AMERICAN CONFERENCE OF GOVERNMENTAL INDUSTRIAL HYGIENISTS
PRINCIPLES OF DUCT CONSTRUCTION
DATE 1–88
FIGURE 8–5

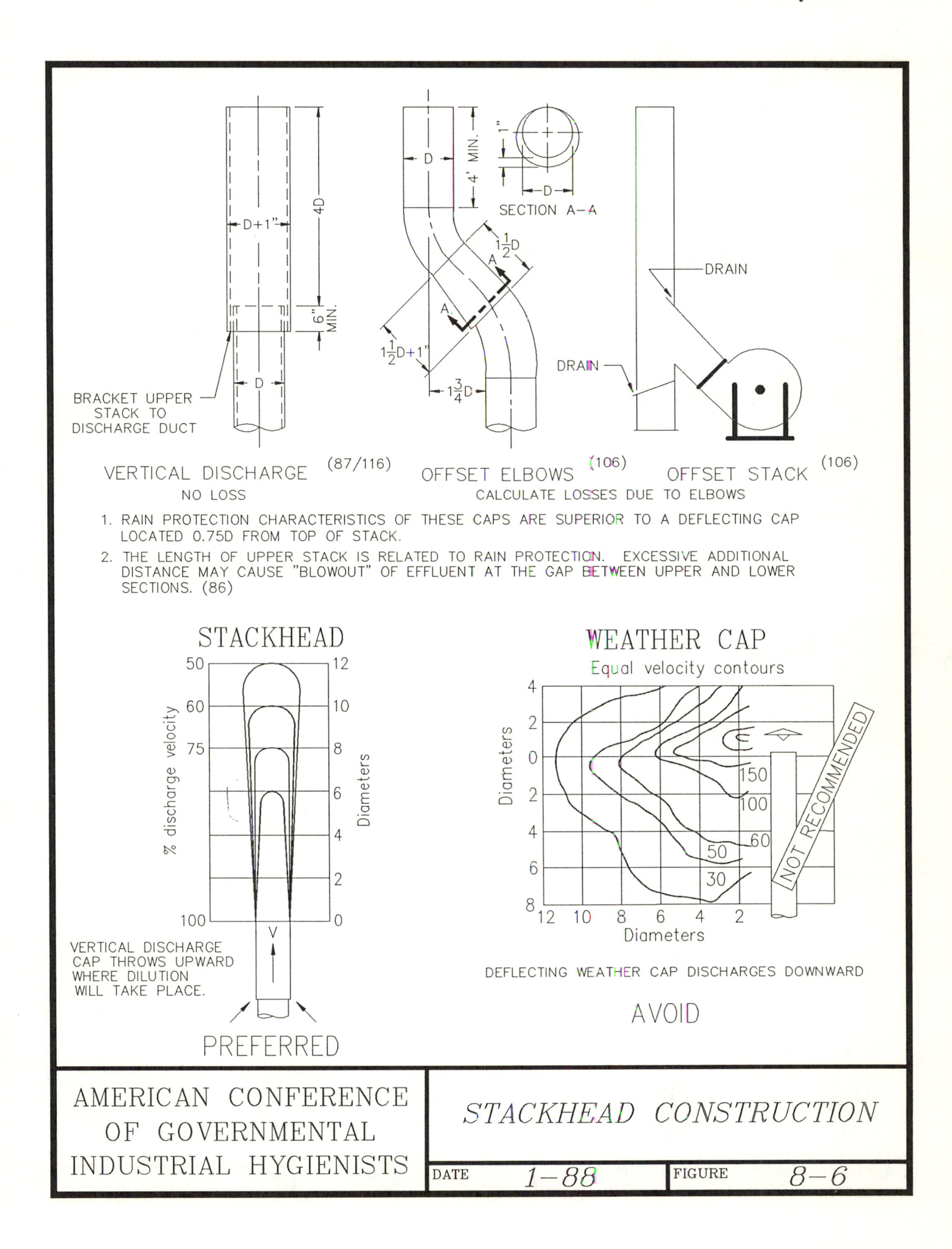
D+1"
4D
6" MIN.
D
BRACKET UPPER STACK TO DISCHARGE DUCT
VERTICAL DISCHARGE (87/116)
NO LOSS
D
4' MIN.
1"
D
SECTION A–A
1½D
A
A
1½D+1"
1¾D
OFFSET ELBOWS (106)
CALCULATE LOSSES DUE TO ELBOWS
DRAIN
DRAIN
OFFSET STACK (106)
1. RAIN PROTECTION CHARACTERISTICS OF THESE CAPS ARE SUPERIOR TO A DEFLECTING CAP LOCATED 0.75D FROM TOP OF STACK.
2. THE LENGTH OF UPPER STACK IS RELATED TO RAIN PROTECTION. EXCESSIVE ADDITIONAL DISTANCE MAY CAUSE "BLOWOUT" OF EFFLUENT AT THE GAP BETWEEN UPPER AND LOWER SECTIONS. (86)
STACKHEAD
% discharge velocity
50
60
75
100
12
10
8
6
4
2
0
Diameters
V
VERTICAL DISCHARGE CAP THROWS UPWARD WHERE DILUTION WILL TAKE PLACE.
PREFERRED
WEATHER CAP
Equal velocity contours
Diameters
4
2
0
2
4
6
8
12
10
8
6
4
2
Diameters
150
100
60
50
30
NOT RECOMMENDED
DEFLECTING WEATHER CAP DISCHARGES DOWNWARD
AVOID
AMERICAN CONFERENCE OF GOVERNMENTAL INDUSTRIAL HYGIENISTS
STACKHEAD CONSTRUCTION
DATE 1–88
FIGURE 8–6

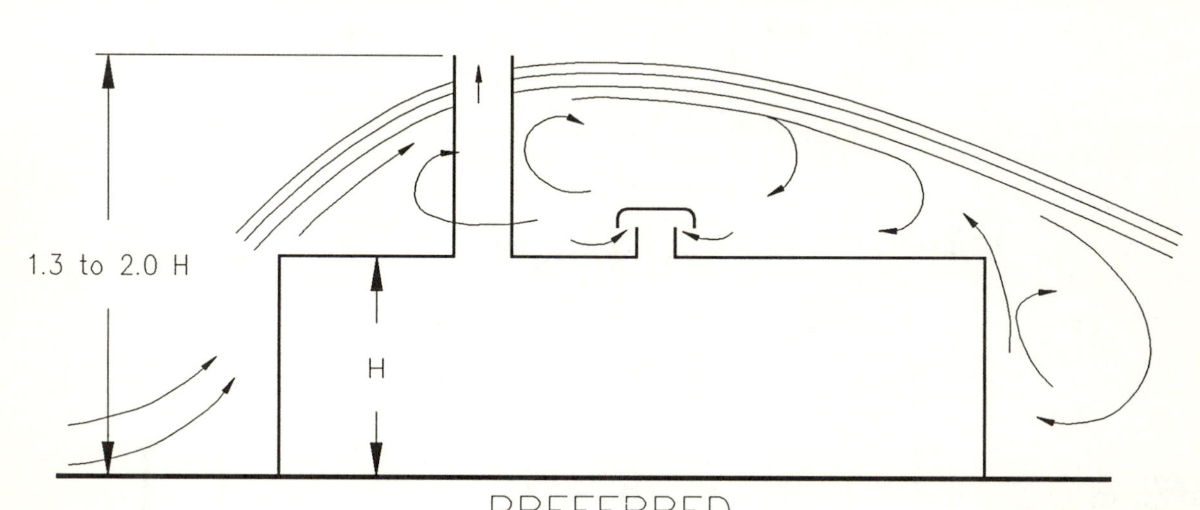

PREFERRED

High discharge stack relative to building height, air inlet on roof.

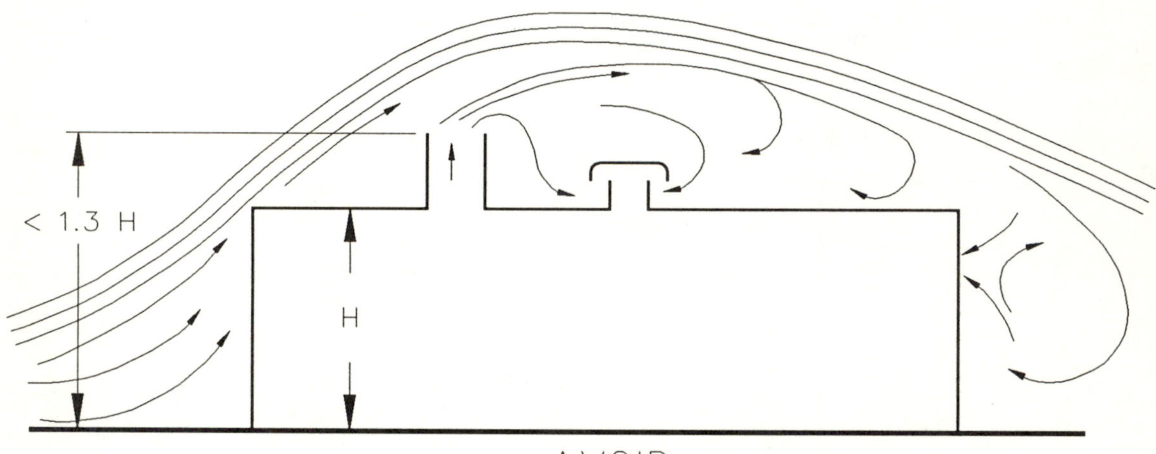

AVOID

Low discharge stack relative to building height and air inlets. (87)

These guidelines apply only to the simple case of a low building without surrounding obstructions on reasonably level terain.

Note: Low pressure on the lee of a building may cause return of contaminants into the building through openings.

AMERICAN CONFERENCE OF GOVERNMENTAL INDUSTRIAL HYGIENISTS	*BUILDING AIR INLETS AND OUTLETS*
	DATE *1-88* FIGURE *8-7*

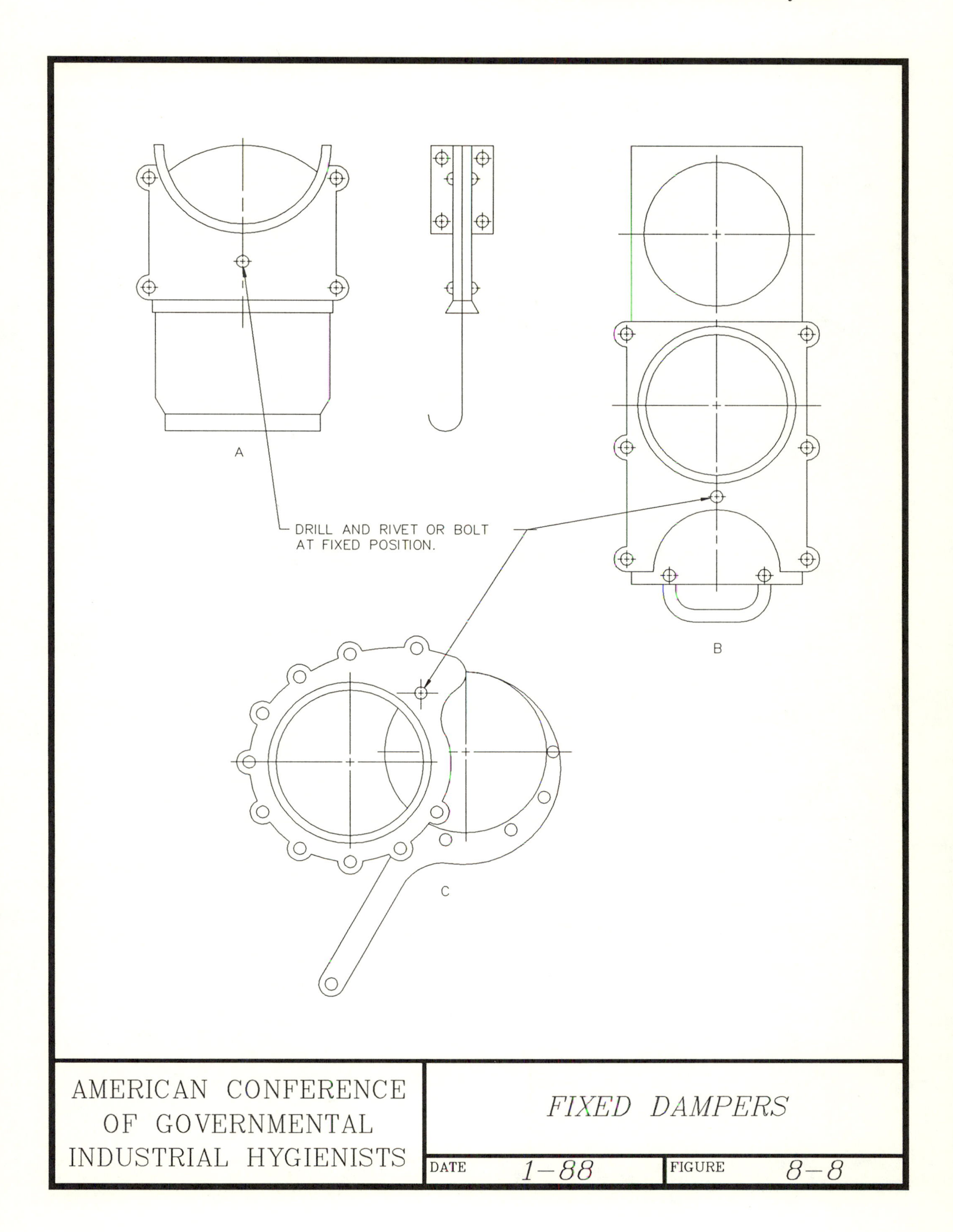
A
DRILL AND RIVET OR BOLT
AT FIXED POSITION.
B
C
AMERICAN CONFERENCE
OF GOVERNMENTAL
INDUSTRIAL HYGIENISTS
FIXED DAMPERS
DATE 1-88
FIGURE 8-8

Chapter 9
TESTING OF VENTILATION SYSTEMS

9.1 INTRODUCTION 9-2

9.2 PRESUREMENT MEASUREMENT 9-2
- 9.2.1 Static Pressure 9-2
- 9.2.2 Velocity Pressure 9-3

9.3 VOLUMETRIC FLOW MEASUREMENT . . . 9-4
- 9.3.1 Pitot Traverse Method 9-5
- 9.3.2 Hood Static Pressure 9-11
- 9.3.3 Hood Static Pressure Interpretation 9-11

9.4 AIR VELOCITY INSTRUMENTS 9-12
- 9.4.1 Rotating Vane Anemometer 9-12
- 9.4.2 Swinging Vane Anemometer (Velometer) 9-13
- 9.4.3 Thermal Anemometer 9-16
- 9.4.4 U-Tube Manometer 9-16
- 9.4.5 Inclined Manometer 9-17
- 9.4.6 Aneroid Gauges 9-17
- 9.4.7 Electronic Aneroid Gauges 9-17
- 9.4.8 Smoke Tubes 9-17
- 9.4.9 Tracer Gas 9-17

9.5 CALIBRATION OF AIR MEASURING INSTRUMENTS 9-19
- 9.5.1 Design of a Calibrating Wind Tunnel 9-19
- 9.5.2 Use of Calibrating Wind Tunnel . . . 9-19

9.6 EVALUATING EXHAUST SYSTEMS 9-22
- 9.6.1 New Installations 9-22
- 9.6.2 Periodic Testing 9-22
- 9.6.3 Check-out Procedure 9-22

9.7 DIFFICULTIES ENCOUNTERED IN FIELD MEASUREMENT 9-25
- 9.7.1 Selection of Instruments 9-25
- 9.7.2 Corrections for Non-Standard Conditions 9-26
- 9.7.3 Pitot Traverse Calculations 9-27

REFERENCES 9-28

9.1 INTRODUCTION

All ventilation systems should be tested at the time of initial installation to verify the volumetric flow rate(s) and to obtain other information which can be compared with the original design data. Testing is necessary to verify the setting of blast gates, fire dampers, and other air flow control devices which may be a part of the system. Initial testing will provide a baseline for periodic maintenance checks and isolation of system failures should a malfunction occur. Many governmental codes require initial and periodic testing of exhaust systems for certain types of processes. Exhaust system test data are useful also as a basis for design of future installations where satisfactory air contaminant control is currently being achieved.

Generally, the most important measurement obtained when testing a ventilation system is the measurement of volumetric flow rate. As described in Chapter 1, this flow at any point within a ventilation system can be determined by the following equation:

$$Q = V A \quad [9.1]$$

where:

Q = volumetric flow rate, cfm

V = average linear velocity, fpm

A = cross-sectional area of duct or hood at the measurement location, ft^2

Since most field meters measure the velocity of the air stream, it is necessary to determine not only the average air velocity but also the cross-sectional area of the duct or opening at the point of measurement in order to determine the volumetric flow rate. A number of direct reading air velocity instruments available for field use are described later in this chapter. A more frequent method of determining velocity is to measure one or more of the air pressures existing in a cross-sectional plane and to use these values in conjunction with the air density to determine velocity. Air pressure measurements are used also to determine fan static pressure and resistance or pressure drop through hoods, dust collectors, and other parts of an exhaust system. Pressure measurements can be helpful in locating obstructions in a duct and isolating points of excessive air leakage.

9.2 PRESSURE MEASUREMENT

At any point in an exhaust system, three air pressures exist which can be compared to the atmospheric pressure immediately surrounding the system. Typically, these pressures are measured in inches water gauge ("wg) and are related to each other as follows:

$$TP = SP + VP \quad [9.2]$$

where:

TP = total pressure, "wg

SP = static pressure, "wg

VP = velocity pressure, "wg

Static pressure is that pressure which tends to burst or collapse a duct and is positive when the pressure is above atmospheric and negative when below atmospheric. Velocity pressure is the pressure resulting from the movement of air and is always positive. Total pressure is the algebraic sum of the static pressure and velocity pressure and can be either positive or negative.

9.2.1 Static Pressure is measured by a pressure measuring device, usually a simple U-tube manometer filled with

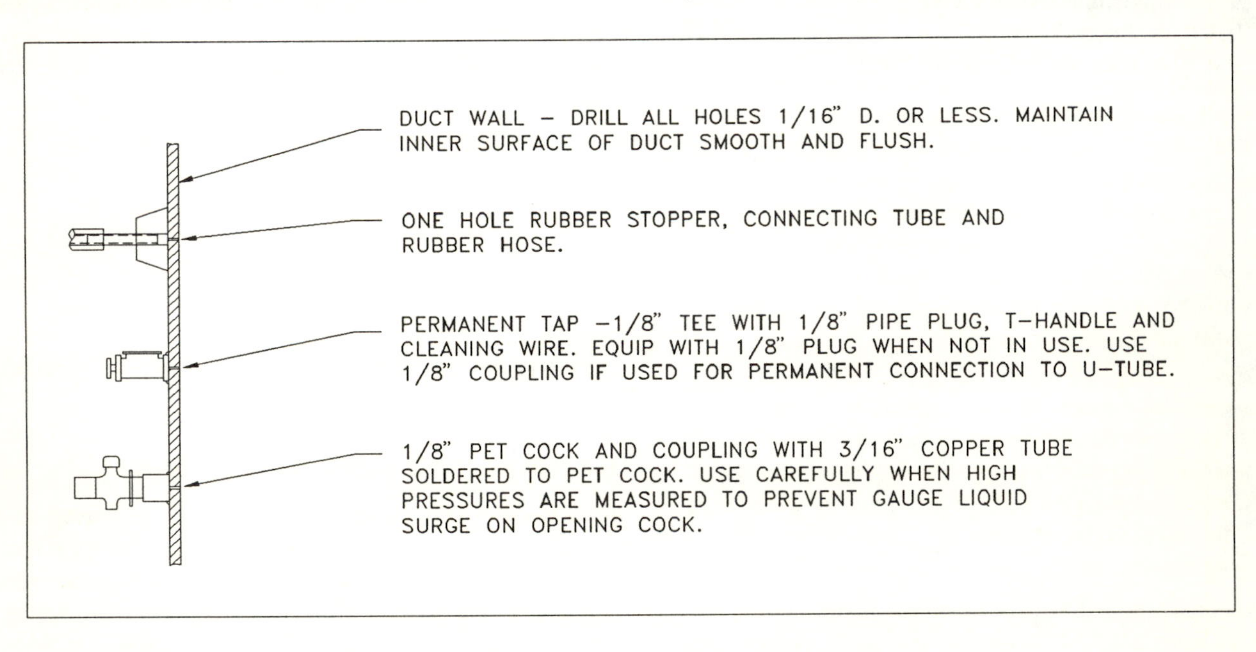

FIGURE 9-1. STATIC TAP CONNECTIONS.

oil, water or other appropriate liquid and graduated in inches water gauge or similar reading pressure gauge. A vertical manometer is suitable for most static pressure measurements. The use of an inclined manometer will give increased accuracy and permits reading of lower values. For field measurement, one leg of the manometer is open to the atmosphere and the other leg is connected with tubing held flush and tight against a small opening in the side of the pipe. Additional information concerning manometers and their construction can be found in References 9.1 and 9.2.

The location of the static pressure opening is usually not too important in obtaining a correct measurement except that one should avoid pressure measurement at the heel of an elbow or other location where static pressure may be incorrect because the direction of the velocity component is not parallel with the duct wall. It is usually advisable to drill 2 to 4 pressure holes at uniform distances around the duct in order to obtain an average and to detect any discrepancy in value.

The static pressure opening should be flush with the inner surface of the pipe wall and there should be no burrs or projections on the inner surface. The hole should be *drilled*, not punched. A 1⁄16″ to 1⁄8″ hole is usually satisfactory since the size is not too important except for some types of instruments where air actually flows through the device (see Figure 9-1). In such cases, the manufacturers' recommendations concerning the size of the static pressure opening should be followed. A second method less likely to involve error is to use the static pressure element of a Pitot tube as shown in Figure 9-2. A static tube of the same general design as the Pitot tube also may be used by omitting the center tube. In use, the instrument must be pointed upstream to avoid impact or eddies.

9.2.2 Velocity Pressure: For measuring velocity pressure to determine air velocity, a standard Pitot tube may be used. A large volume of research and many applications have been devoted to the subject of flow measurements by this instrument which was developed by Henri Pitot in 1734 while a student in Paris, France. A standard Pitot tube (see Figure 9-3) needs no calibration if carefully made and the velocity pressure readings obtained are considered to be ± 1.0% at velocities above 2000 fpm. For more details concerning specifications and application of the Pitot tube, see the *Standard Test Code* published by ASHRAE and the Air Moving and Conditioning Association.[9.1,9.2]

The device consists of two concentric tubes — one measures the total or impact pressure existing in the air stream, the other measures the static pressure only. When the annular space and the center tube are connected across a manometer, the difference between the total pressure and the static pressure is indicated on the manometer. This difference is the velocity pressure.

The velocity pressure can be used to compute the velocity of the air stream if the density of the air is known. The following equation, developed in Chapter 1, can be used:

$$V = 1096\sqrt{\frac{VP}{\rho}} \qquad [9.3]$$

where:

ρ = actual gas density, lbm/ft^3

Where air is at standard conditions (ρ = 0.075 lbm/ft^3), Equation 9.3 becomes:

$$V = 4005\sqrt{VP} \qquad [9.4]$$

If the temperature of the air stream varies more than 30 F from standard air (70 F and 29.92 ″Hg) or the altitude of the site is more than 1,000 feet above or below sea level and/or the moisture content of the air is 0.02 lb/lb of dry air or greater, the actual gas density (ρ) must be corrected as described in Chapter 5.

Tables 9-1A and 9-1B are velocity vs. velocity pressure tables for standard air. These tables can be used for air at densities other than standard conditions by correcting the measured velocity pressure as in the following equation:

$$VP_C = VP_M \times \frac{0.075}{\rho} \qquad [9.5]$$

where:

VP_C = corrected velocity pressure, ″wg

VP_M = measured velocity pressure, ″wg

The corrected VP then can be used in the velocity vs. velocity pressure tables (Table 9-1A & B) to give the actual velocity at duct conditions.

From Tables 9-1A & B, it can be seen that at low velocities (below 1000 fpm) the VP values are small (less than 0.06 ″wg). The accuracy of the Pitot tube is limited at these velocities as the manometer is not precise enough to accurately measure the small pressures. A calibrated inclined manometer can be read to approximately ± 0.005 ″wg. A standard Pitot tube with an inclined manometer can be used with the following degree of accuracy:

Velocity, fpm	% Error (±)
4000	0.25
3000	0.3
2000	1.0
1000	4.0
800	6.0
600	15.0

Note that the use of the Pitot tube in the field is limited at velocities lower than 600-800 fpm.

9.3 VOLUMETRIC FLOW MEASUREMENT

A number of techniques can be used to determine the volumetric flow rate at hood openings and at other points in an exhaust system using the fluid flow principles previously described. The method selected will depend on the degree of accuracy required, time available for testing and the type of test data required. It is extremely important that measure-

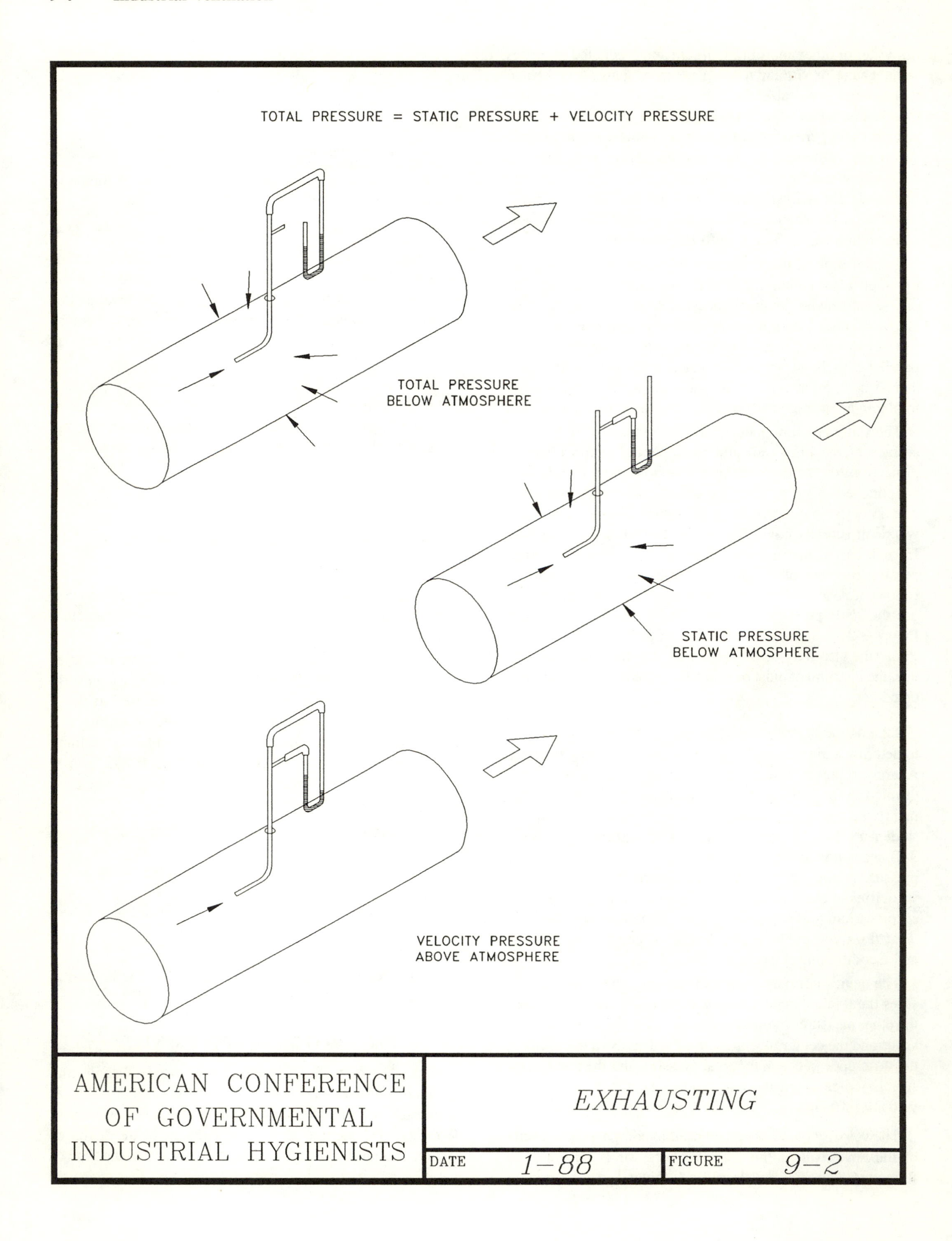
TOTAL PRESSURE = STATIC PRESSURE + VELOCITY PRESSURE
TOTAL PRESSURE
BELOW ATMOSPHERE
STATIC PRESSURE
BELOW ATMOSPHERE
VELOCITY PRESSURE
ABOVE ATMOSPHERE
AMERICAN CONFERENCE
OF GOVERNMENTAL
INDUSTRIAL HYGIENISTS
EXHAUSTING
DATE 1-88
FIGURE 9-2

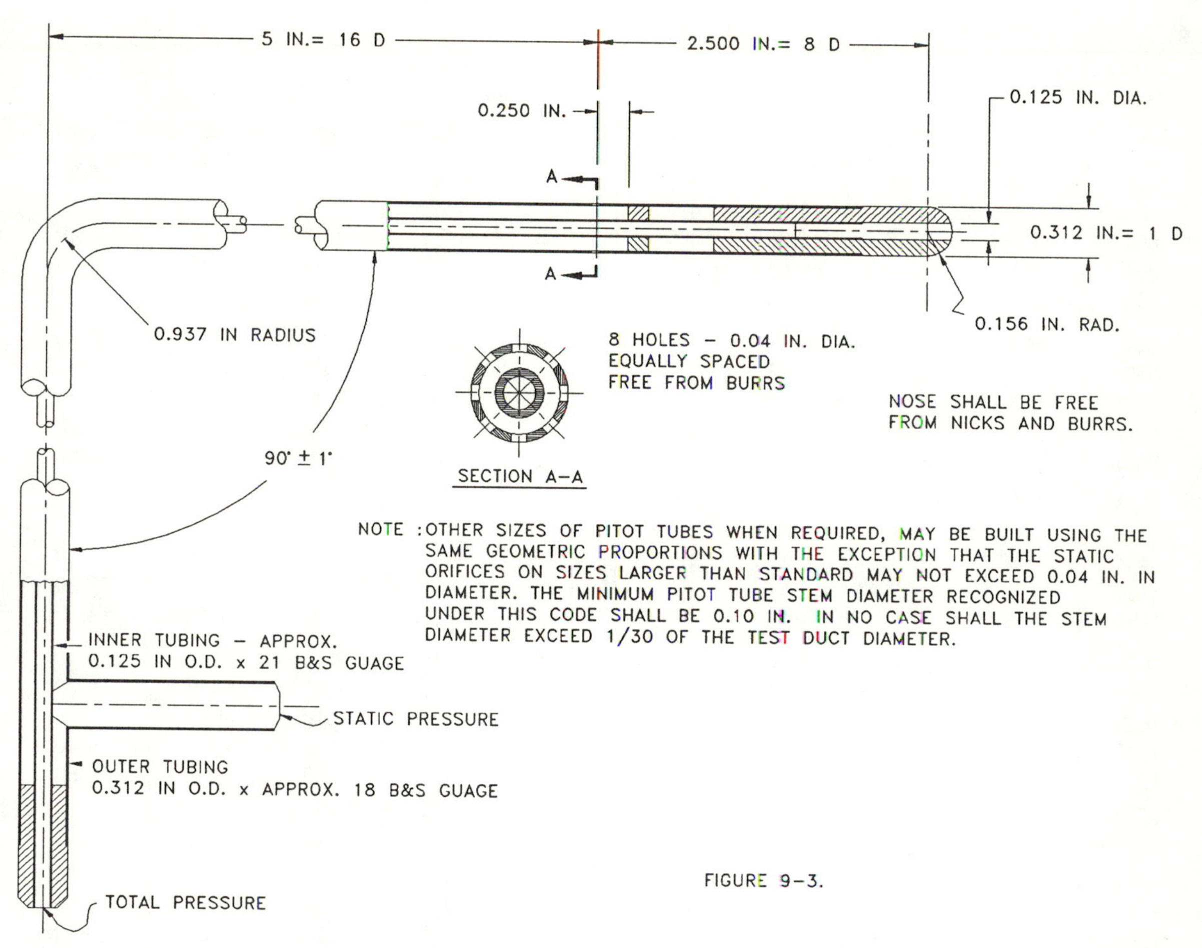

FIGURE 9-3.

ments taken at the time of the tests include all necessary information to determine the gas density to permit the calculation of the actual velocity and volumetric flow rate.

9.3.1 Pitot Traverse Method: Because the air flow in the cross-section of a duct is not uniform, it is necessary to obtain an average by measuring VP at points in a number of equal areas in the cross-section. The usual method is to make two traverses across the diameter of the duct at right angles to each other. Readings are taken at the center of annular rings of equal area (see Figure 9-4). Whenever possible, the traverse should be made 7.5 duct diameters or more downstream from any major air disturbance such as an elbow, hood, branch entry, etc. Where measurements are made closer to disturbances, the results must be considered subject to some doubt and checked against a second location. If agreement within 10% of the two traverses is obtained, reasonable accuracy can be assumed and the average of the two readings used. Where the variation exceeds 10%, a third location should be selected and the two air flows in the best agreement averaged and used (see Figure 9-5). The use of a single centerline reading for obtaining average velocity is a very coarse approximation and is NOT recommended.

For round ducts 6″ and smaller, at least 6 traverse points should be used. For round ducts larger than 6″ diameter, at least 10 traverse points should be employed. For very large

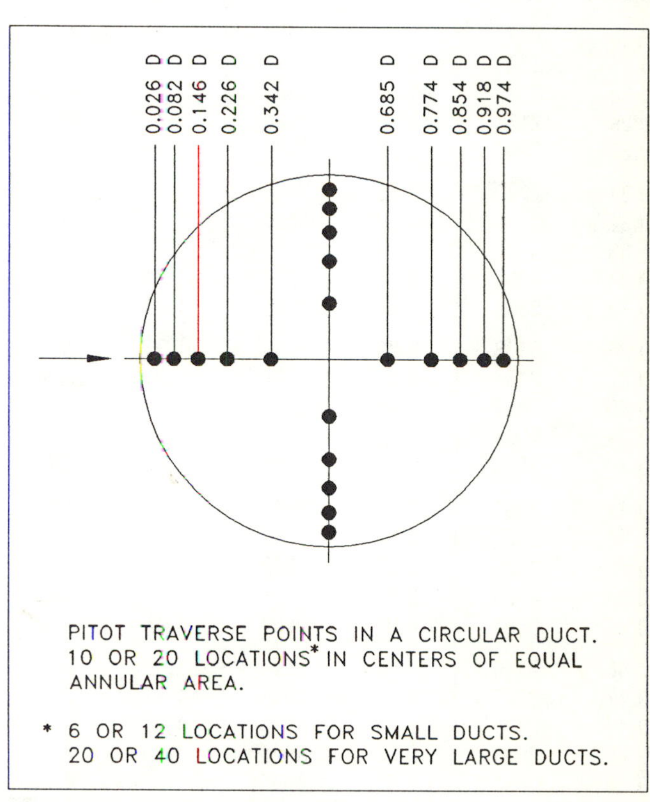

FIGURE 9-4.

TABLE 9-1A. Velocity Pressure to Velocity Conversion — Standard Air

FROM: $V = 4005 \sqrt{VP}$

V = Velocity, fpm
VP = Velocity Pressure, "wg

VP	V	VP	V	VP	V	VP	V	VP	V	VP	V
0.01	401	0.51	2860	1.01	4025	1.51	4921	2.01	5678	2.60	6458
0.02	566	0.52	2888	1.02	4045	1.52	4938	2.02	5692	2.70	6581
0.03	694	0.53	2916	1.03	4065	1.53	4954	2.03	5706	2.80	6702
0.04	801	0.54	2943	1.04	4084	1.54	4970	2.04	5720	2.90	6820
0.05	896	0.55	2970	1.05	4104	1.55	4986	2.05	5734	3.00	6937
0.06	981	0.56	2997	1.06	4123	1.56	5002	2.06	5748	3.10	7052
0.07	1060	0.57	3024	1.07	4143	1.57	5018	2.07	5762	3.20	7164
0.08	1133	0.58	3050	1.08	4162	1.58	5034	2.08	5776	3.30	7275
0.09	1201	0.59	3076	1.09	4181	1.59	5050	2.09	5790	3.40	7385
0.10	1266	0.60	3102	1.10	4200	1.60	5066	2.10	5804	3.50	7493
0.11	1328	0.61	3128	1.11	4220	1.61	5082	2.11	5818	3.60	7599
0.12	1387	0.62	3154	1.12	4238	1.62	5098	2.12	5831	3.70	7704
0.13	1444	0.63	3179	1.13	4257	1.63	5113	2.13	5845	3.80	7807
0.14	1499	0.64	3204	1.14	4276	1.64	5129	2.14	5859	3.90	7909
0.15	1551	0.65	3229	1.15	4295	1.65	5145	2.15	5872	4.00	8010
0.16	1602	0.66	3254	1.16	4314	1.66	5160	2.16	5886	4.10	8110
0.17	1651	0.67	3278	1.17	4332	1.67	5176	2.17	5900	4.20	8208
0.18	1699	0.68	3303	1.18	4351	1.68	5191	2.18	5913	4.30	8305
0.19	1746	0.69	3327	1.19	4369	1.69	5206	2.19	5927	4.40	8401
0.20	1791	0.70	3351	1.20	4387	1.70	5222	2.20	5940	4.50	8496
0.21	1835	0.71	3375	1.21	4405	1.71	5237	2.21	5954	4.60	8590
0.22	1879	0.72	3398	1.22	4424	1.72	5253	2.22	5967	4.70	8683
0.23	1921	0.73	3422	1.23	4442	1.73	5268	2.23	5981	4.80	8775
0.24	1962	0.74	3445	1.24	4460	1.74	5283	2.24	5994	4.90	8865
0.25	2003	0.75	3468	1.25	4478	1.75	5298	2.25	6007	5.00	8955
0.26	2042	0.76	3491	1.26	4496	1.76	5313	2.26	6021	5.50	9393
0.27	2081	0.77	3514	1.27	4513	1.77	5328	2.27	6034	6.00	9810
0.28	2119	0.78	3537	1.28	4531	1.78	5343	2.28	6047	6.50	10211
0.29	2157	0.79	3560	1.29	4549	1.79	5358	2.29	6061	7.00	10596
0.30	2194	0.80	3582	1.30	4566	1.80	5373	2.30	6074	7.50	10968
0.31	2230	0.81	3604	1.31	4584	1.81	5388	2.31	6087	8.00	11328
0.32	2266	0.82	3627	1.32	4601	1.82	5403	2.32	6100	8.50	11676
0.33	2301	0.83	3649	1.33	4619	1.83	5418	2.33	6113	9.00	12015
0.34	2335	0.84	3671	1.34	4636	1.84	5433	2.34	6126	9.50	12344
0.35	2369	0.85	3692	1.35	4653	1.85	5447	2.35	6140	10.00	12665
0.36	2403	0.86	3714	1.36	4671	1.86	5462	2.36	6153	10.50	12978
0.37	2436	0.87	3736	1.37	4688	1.87	5477	2.37	6166	11.00	13283
0.38	2469	0.88	3757	1.38	4705	1.88	5491	2.38	6179	11.50	13582
0.39	2501	0.89	3778	1.39	4722	1.89	5506	2.39	6192	12.00	13874
0.40	2533	0.90	3799	1.40	4739	1.90	5521	2.40	6205	12.50	14160
0.41	2564	0.91	3821	1.41	4756	1.91	5535	2.41	6217	13.00	14440
0.42	2596	0.92	3841	1.42	4773	1.92	5549	2.42	6230	13.50	14715
0.43	2626	0.93	3862	1.43	4789	1.93	5564	2.43	6243	14.00	14985
0.44	2657	0.94	3883	1.44	4806	1.94	5578	2.44	6256	14.50	15251
0.45	2687	0.95	3904	1.45	4823	1.95	5593	2.45	6269	15.00	15511
0.46	2716	0.96	3924	1.46	4839	1.96	5607	2.46	6282	15.50	15768
0.47	2746	0.97	3944	1.47	4856	1.97	5621	2.47	6294	16.00	16020
0.48	2775	0.98	3965	1.48	4872	1.98	5636	2.48	6307	16.50	16268
0.49	2803	0.99	3985	1.49	4889	1.99	5650	2.49	6320	17.00	16513
0.50	2832	1.00	4005	1.50	4905	2.00	5664	2.50	6332	17.50	16754

TABLE 9-1B. Velocity to Velocity Pressure Conversion — Standard Air

FROM: $V = 4005 \sqrt{VP}$

V = Velocity, fpm
VP = Velocity Pressure, "wg

V	VP	V	VP	V	VP	V	VP	V	VP	V	VP
400	0.01	2600	0.42	3850	0.92	4880	1.48	5690	2.02	6190	2.39
500	0.02	2625	0.43	3875	0.94	4900	1.50	5700	2.03	6200	2.40
600	0.02	2650	0.44	3900	0.95	4920	1.51	5710	2.03	6210	2.40
700	0.03	2675	0.45	3925	0.96	4940	1.52	5720	2.04	6220	2.41
800	0.04	2700	0.45	3950	0.97	4960	1.53	5730	2.05	6230	2.42
900	0.05	2725	0.46	3975	0.99	4980	1.55	5740	2.05	6240	2.43
1000	0.06	2750	0.47	4000	1.00	5000	1.56	5750	2.06	6250	2.44
1100	0.08	2775	0.48	4020	1.01	5020	1.57	5760	2.07	6260	2.44
1200	0.09	2800	0.49	4040	1.02	5040	1.58	5770	2.08	6270	2.45
1300	0.11	2825	0.50	4060	1.03	5060	1.60	5780	2.08	6280	2.46
1400	0.12	2850	0.51	4080	1.04	5080	1.61	5790	2.09	6290	2.47
1450	0.13	2875	0.52	4100	1.05	5100	1.62	5800	2.10	6300	2.47
1500	0.14	2900	0.52	4120	1.06	5120	1.63	5810	2.10	6310	2.48
1550	0.15	2925	0.53	4140	1.07	5140	1.65	5820	2.11	6320	2.49
1600	0.16	2950	0.54	4160	1.08	5160	1.66	5830	2.12	6330	2.50
1650	0.17	2975	0.55	4180	1.09	5180	1.67	5840	2.13	6340	2.51
1700	0.18	3000	0.56	4200	1.10	5200	1.69	5850	2.13	6350	2.51
1750	0.19	3025	0.57	4220	1.11	5220	1.70	5860	2.14	6360	2.52
1800	0.20	3050	0.58	4240	1.12	5240	1.71	5870	2.15	6370	2.53
1825	0.21	3075	0.59	4260	1.13	5260	1.72	5880	2.16	6380	2.54
1850	0.21	3100	0.60	4280	1.14	5280	1.74	5890	2.16	6390	2.55
1875	0.22	3125	0.61	4300	1.15	5300	1.75	5900	2.17	6400	2.55
1900	0.23	3150	0.62	4320	1.16	5320	1.76	5910	2.18	6410	2.56
1925	0.23	3175	0.63	4340	1.17	5340	1.78	5920	2.18	6420	2.57
1950	0.24	3200	0.64	4360	1.19	5360	1.79	5930	2.19	6430	2.58
1975	0.24	3225	0.65	4380	1.20	5380	1.80	5940	2.20	6440	2.59
2000	0.25	3250	0.66	4400	1.21	5400	1.82	5950	2.21	6450	2.59
2025	0.26	3275	0.67	4420	1.22	5420	1.83	5960	2.21	6460	2.60
2050	0.26	3300	0.68	4440	1.23	5440	1.84	5970	2.22	6470	2.61
2075	0.27	3325	0.69	4460	1.24	5460	1.86	5980	2.23	6480	2.62
2100	0.27	3350	0.70	4480	1.25	5480	1.87	5990	2.24	6490	2.63
2125	0.28	3375	0.71	4500	1.26	5500	1.89	6000	2.24	6500	2.63
2150	0.29	3400	0.72	4520	1.27	5510	1.89	6010	2.25	6550	2.67
2175	0.29	3425	0.73	4540	1.29	5520	1.90	6020	2.26	6600	2.72
2200	0.30	3450	0.74	4560	1.30	5530	1.91	6030	2.27	6650	2.76
2225	0.31	3475	0.75	4580	1.31	5540	1.91	6040	2.27	6700	2.80
2250	0.32	3500	0.76	4600	1.32	5550	1.92	6050	2.28	6750	2.84
2275	0.32	3525	0.77	4620	1.33	5560	1.93	6060	2.29	6800	2.88
2300	0.33	3550	0.79	4640	1.34	5570	1.93	6070	2.30	6900	2.97
2325	0.34	3575	0.80	4660	1.35	5580	1.94	6080	2.30	7000	3.05
2350	0.34	3600	0.81	4680	1.37	5590	1.95	6090	2.31	7100	3.14
2375	0.35	3625	0.82	4700	1.38	5600	1.96	6100	2.32	7200	3.23
2400	0.36	3650	0.83	4720	1.39	5610	1.96	6110	2.33	7300	3.32
2425	0.37	3675	0.84	4740	1.40	5620	1.97	6120	2.34	7400	3.41
2450	0.37	3700	0.85	4760	1.41	5630	1.98	6130	2.34	7500	3.51
2475	0.38	3725	0.87	4780	1.42	5640	1.98	6140	2.35	7600	3.60
2500	0.39	3750	0.88	4800	1.44	5650	1.99	6150	2.36	7700	3.70
2525	0.40	3775	0.89	4820	1.45	5660	2.00	6160	2.37	7800	3.79
2550	0.41	3800	0.90	4840	1.46	5670	2.00	6170	2.37	7900	3.89
2575	0.41	3825	0.91	4860	1.47	5680	2.01	6180	2.38	8000	3.99

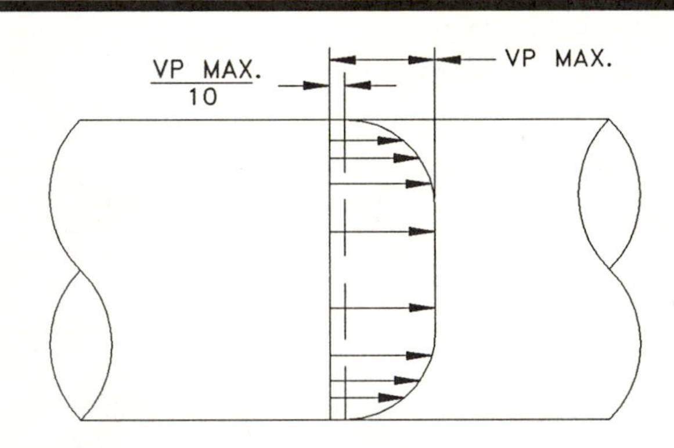

A: IDEAL VP DISTRIBUTION

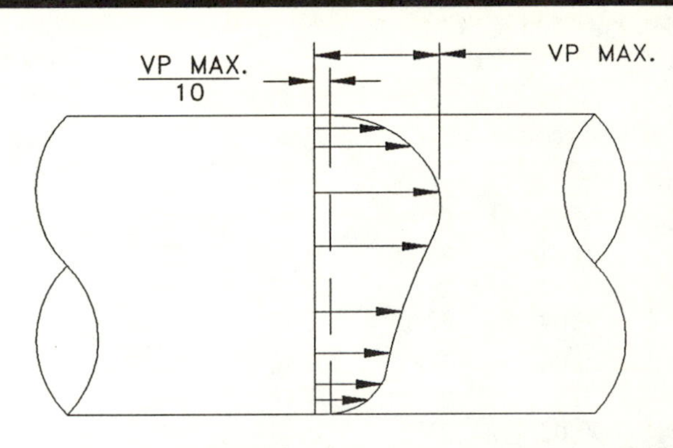

B: GOOD VP DISTRIBUTION. (ALSO SATISFACTORY FOR FLOW INTO FAN INLETS. BUT MAY BE UNSATISFACTORY FOR FLOW INTO INLET BOXES – MAY PRODUCE SWIRL IN BOXES.)

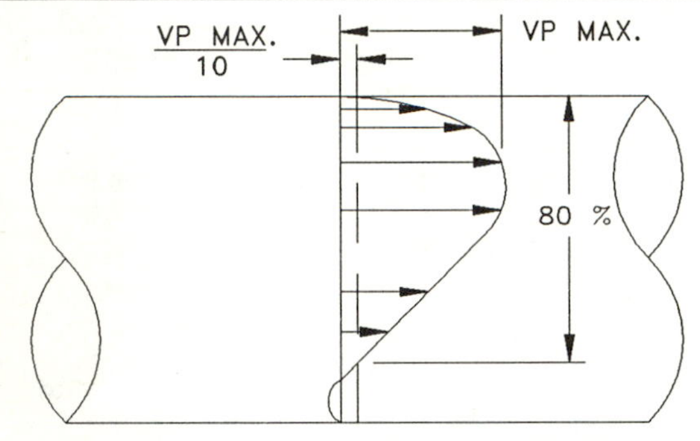

C: SATISFACTORY VP DISTRIBUTION – MORE THAN 75% OF VP READINGS GREATER THAN $\frac{\text{VP MAX.}}{10}$
(ALSO UNSATISFACTORY FOR FLOW INTO FAN INLETS OR INLET BOXES)

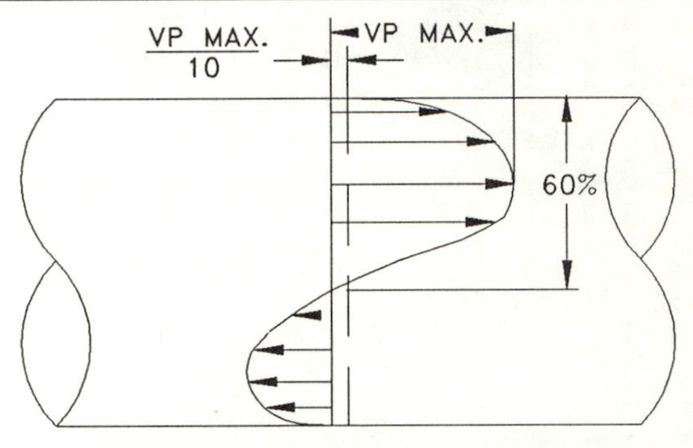

D: DO NOT USE! UNSATISFACTORY VP DISTRIBUTION – LESS THAN 75% OF VP READINGS GREATER THAN $\frac{\text{VP MAX.}}{10}$
(ALSO UNSATISFACTORY FOR FLOW INTO FAN INLETS OR INLET BOXES)

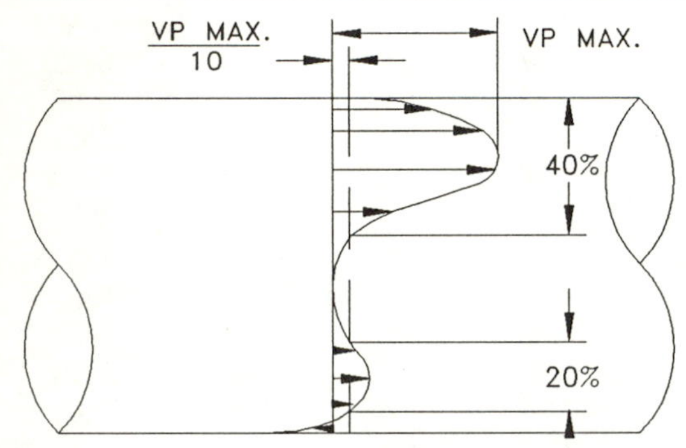

E: DO NOT USE! UNSATISFACTORY VP DISTRIBUTION – LESS THAN 75% OF VP READINGS GREATER THAN $\frac{\text{VP MAX.}}{10}$
(ALSO UNSATISFACTORY FOR FLOW INTO FAN INLETS OR INLET BOXES)

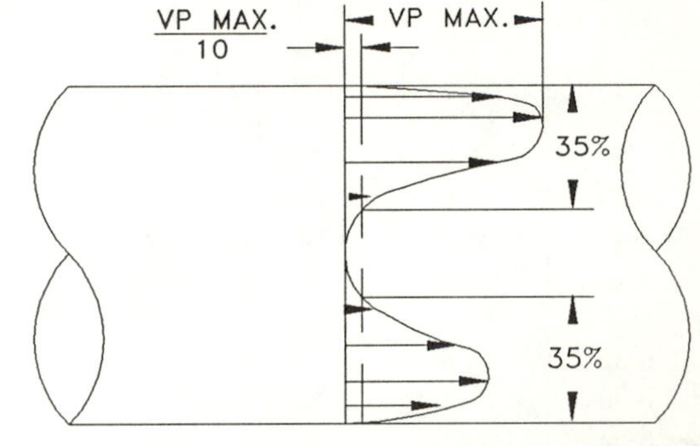

F: DO NOT USE! UNSATISFACTORY VP DISTRIBUTION – LESS THAN 75% OF VP READINGS GREATER THAN $\frac{\text{VP MAX.}}{10}$
(ALSO UNSATISFACTORY FOR FLOW INTO FAN INLETS OR INLET BOXES)

AMERICAN CONFERENCE OF GOVERNMENTAL INDUSTRIAL HYGIENISTS	*TYPICAL VELOCITY DISTRIBUTIONS*	
	DATE *7–89*	FIGURE *9–5*

TABLE 9-2. Distance from Wall of Round Pipe to Point Reading (nearest ⅛ inch) for 6 Point Traverse

DUCT DIA	R_1 .043 DIA	R_2 .146 DIA	R_3 .296 DIA	R_4 .704 DIA	R_5 .854 DIA	R_6 .957 DIA
3	⅛	½	⅞	2⅛	2½	2⅞
3½	⅛	½	1	2½	3	3⅜
4	⅛	⅝	1⅛	2⅞	3⅜	3⅞
4½	¼	⅝	1⅜	3⅛	3⅞	4¼
5	¼	¾	1½	3½	4¼	4¾
5½	¼	¾	1⅝	3⅞	4¾	5¼
6	¼	⅞	1¾	4¼	5⅛	5¾

TABLE 9-3. Distance from Wall of Round Pipe to Point of Reading (nearest ⅛ inch) for 10 Point Traverse

DUCT DIA	R_1 0.026 DIA	R_2 0.082 DIA	R_3 0.146 DIA	R_4 0.226 DIA	R_5 0.342 DIA	R_6 0.658 DIA	R_7 0.774 DIA	R_8 0.854 DIA	R_9 0.918 DIA	R_{10} 0.974 DIA
4	⅛	⅜	⅝	⅞	1⅜	2⅝	3⅛	3⅜	3⅝	3⅞
4½	⅛	⅜	⅝	1	1½	3	3½	3⅞	4⅛	4⅜
5	⅛	⅜	¾	1⅛	1¾	3¼	3⅞	4¼	4⅝	4⅞
5½	⅛	½	¾	1¼	1⅞	3⅝	4¼	4¾	5	5⅜
6	⅛	½	⅞	1⅜	2	4	4⅝	5⅛	5½	5⅞
7	⅛	⅝	1	1⅝	2⅜	4⅝	5⅜	6	6⅜	6⅞
8	¼	⅝	1⅛	1¾	2¾	5¼	6¼	6⅞	7⅜	7¾
9	¼	¾	1¼	2	3⅛	5⅞	7	7¾	8¼	8¾
10	¼	⅞	1½	2¼	3⅜	6⅝	7¾	8½	9⅛	9¾
11	¼	⅞	1⅝	2½	3¾	7¼	8½	9⅜	10⅛	10¾
12	⅜	1	1¾	2¾	4⅛	7⅞	9¼	10¼	11	11⅝
13	⅜	1	1⅞	2⅞	4½	8½	10⅛	11⅛	12	12⅝
14	⅜	1⅛	2	3⅛	4¾	9¼	10⅞	12	12⅞	13⅝
15	⅜	1¼	2¼	3⅜	5⅛	9⅞	11⅝	12¾	13¾	14⅝
16	⅜	1¼	2⅜	3⅝	5½	10½	12⅜	13⅝	14¾	15⅝
17	½	1⅜	2½	3⅞	5¾	11¼	13⅛	14½	15⅝	16½
18	½	1½	2⅝	4⅛	6⅛	11⅞	13⅞	15⅜	16½	17½
19	½	1½	2¾	4¼	6½	12½	14¾	16¼	17½	18½
20	½	1⅝	2⅞	4½	6⅞	13⅛	15½	17⅛	18⅜	19½
22	⅝	1¾	3¼	5	7½	14½	17	18¾	20¼	21⅜
24	⅝	2	3½	5½	8¼	15¾	18½	20½	22	23⅜
26	⅝	2⅛	3¾	5⅞	8⅞	17⅛	20⅛	22¼	23⅞	25⅜
28	¾	2¼	4⅛	6⅜	9⅝	18⅜	21⅝	23⅞	25¾	27¼
30	¾	2½	4⅜	6¾	10¼	19¾	23¼	25⅝	27½	29¼
32	⅞	2⅝	4⅝	7¼	11	21	24¾	27⅜	29⅜	31⅛
34	⅞	2¾	5	7¾	11⅝	22⅜	26¼	29	31¼	33⅛
36	1	3	5¼	8⅛	12⅜	23⅝	27⅞	30¾	33	35
38	1	3⅛	5½	8⅝	13	25	29⅜	32½	34⅞	37
40	1	3¼	5⅞	9	13⅝	26⅜	31	34⅛	36¾	39
42	1⅛	3⅜	6⅛	9½	14⅜	27⅝	32½	35⅞	38⅝	40⅞
44	1⅛	3⅝	6⅜	10	15	29	34	37⅝	40⅜	42⅞
46	1¼	3¾	6¾	10⅜	15¾	30¼	35⅝	39¼	42¼	44¾
48	1¼	4	7	10⅞	16⅜	31⅝	37⅛	41	44	46¾

ducts and discharge stacks with wide variation in velocity, 20 traverse points will increase the precision of the air flow measurement. Six, ten and twenty point traverse points for various duct diameters are given in Tables 9-2, 9-3, and 9-4. To minimize errors, a Pitot tube smaller than the standard ⁵⁄₁₆″ O.D. should be used in ducts less than 12″ in diameter.

For square or rectangular ducts, the procedure is to divide the cross-section into a number of equal rectangular areas and measure the velocity pressure at the center of each. The number of readings should not be less than 16. However, enough readings should be made so the greatest distance between centers is approximately six inches (see Figure 9-6).

The following data are essential and more detailed data may be taken if desired:

- The area of the duct at the traverse location.
- Velocity pressure at each point in the traverse.
- Temperature of the air stream at the time and location of the traverse.

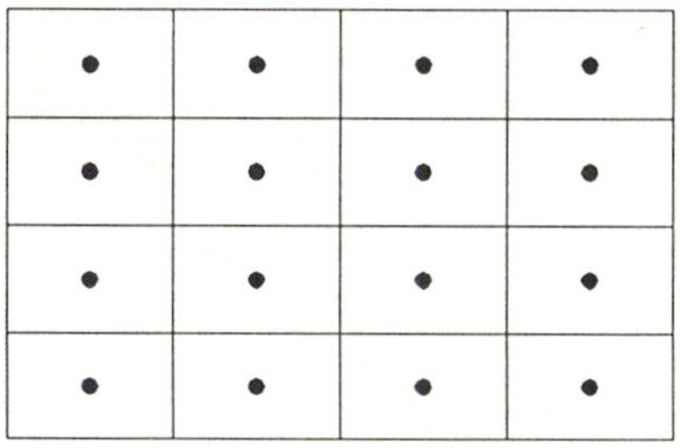

PITOT TRAVERSE POINTS IN A RECTANGULAR DUCT. CENTERS OF 16 TO 64 EQUAL AREAS. LOCATIONS NOT MORE THAN 6" APART.

FIGURE 9–6.

TABLE 9-4. Distance from Wall of Round Pipe to Point of Reading (nearest ⅛ inch) for 20 Point Traverse

Duct dia	R_1 0.013D	R_2 0.039D	R_3 0.067D	R_4 0.097D	R_5 0.129D	R_6 0.165D	R_7 0.204D	R_8 0.250D	R_9 0.306D	R_{10} 0.388D
	R_{11} 0.612D	R_{12} 0.694D	R_{13} 0.750D	R_{14} 0.796D	R_{15} 0.835D	R_{16} 0.871D	R_{17} 0.903D	R_{18} 0.933D	R_{19} 0.961D	R_{20} 0.987D
40	1/2	1 1/2	2 5/8	3 7/8	5 1/8	6 5/8	8 1/8	10	12 1/4	15 1/2
	24 1/2	27 3/4	30	31 7/8	33 3/8	34 7/8	36 1/8	37 3/8	38 1/2	39 1/2
42	1/2	1 5/8	2 7/8	4 1/8	5 3/8	6 7/8	8 5/8	10 1/2	12 7/8	16 1/4
	25 3/4	29 1/8	31 1/2	33 3/8	35 1/8	36 5/8	37 7/8	39 1/8	40 3/8	41 1/2
44	1/2	1 3/4	3	4 1/4	5 5/8	7 1/4	9	11	13 1/2	17 1/8
	26 7/8	30 1/2	33	35	36 3/4	38 3/8	39 3/4	41	42 1/4	43 1/2
46	5/8	1 3/4	3 1/8	4 1/2	6	7 5/8	9 3/8	11 1/2	14 1/8	17 7/8
	28 1/8	31 7/8	34 1/2	36 5/8	38 3/8	40	41 1/2	42 7/8	44 1/4	45 3/8
48	5/8	1 7/8	3 1/4	4 5/8	6 1/4	7 7/8	9 3/4	12	14 3/4	18 5/8
	29 3/8	33 1/4	36	38 1/4	40 1/8	41 3/4	43 3/8	44 3/4	46 1/8	47 3/8
50	5/8	2	3 3/8	4 7/8	6 1/2	8 1/4	10 1/4	12 1/2	15 3/8	19 3/8
	30 5/8	34 5/8	37 1/2	39 3/4	41 3/4	43 1/2	45 1/8	46 5/8	48	49 3/8
52	5/8	2	3 1/2	5	6 3/4	8 1/2	10 5/8	13	15 7/8	20 1/8
	31 7/8	36 1/8	39	41 3/8	43 1/2	45 1/4	47	48 1/2	50	51 3/8
54	5/8	2 1/8	3 5/8	5 1/4	7	8 7/8	11	13 1/2	16 1/2	21
	33	37 1/2	40 1/2	43	45 1/8	47	48 3/4	50 3/8	51 7/8	53 3/8
56	3/4	2 1/8	3 3/4	5 3/8	7 1/4	9 1/4	11 3/8	14	17 1/8	21 3/4
	34 1/4	38 7/8	42	44 5/8	46 3/4	48 3/4	50 5/8	52 1/4	53 7/8	55 1/4
58	3/4	2 1/4	3 7/8	5 5/8	7 1/2	9 1/2	11 7/8	14 1/2	17 3/4	22 1/2
	35 1/2	40 1/4	43 1/2	46 1/8	48 1/2	50 1/2	52 3/8	54 1/8	55 3/4	57 1/4
60	3/4	2 3/8	4	5 7/8	7 3/4	9 7/8	12 1/4	15	18 3/8	23 1/4
	36 3/4	41 5/8	45	47 3/4	50 1/8	52 1/4	54 1/8	56	57 5/8	59 1/4
62	3/4	2 3/8	4 1/8	6	8	10 1/4	12 5/8	15 1/2	19	24 1/8
	37 7/8	43	46 1/2	49 3/8	51 3/4	54	56	57 7/8	59 5/8	61 1/4
64	3/4	2 1/2	4 1/4	6 1/4	8 1/4	10 1/2	13 1/8	16	19 5/8	24 7/8
	39 1/8	44 3/8	48	50 7/8	53 1/2	55 3/4	57 3/4	59 3/4	61 1/2	63 1/4
66	7/8	2 5/8	4 3/8	6 3/8	8 1/2	10 7/8	13 1/2	16 1/2	20 1/4	25 5/8
	40 3/8	45 3/4	49 1/2	52 1/2	55 1/8	57 1/2	59 5/8	61 5/8	63 3/8	65 1/8
68	7/8	2 5/8	4 1/2	6 5/8	8 3/4	11 1/4	13 7/8	17	20 7/8	26 3/8
	41 5/8	47 1/8	51	54 1/8	56 3/4	59 1/4	61 3/8	63 1/2	65 3/8	67 1/8
70	7/8	2 3/4	4 3/4	6 3/4	9	11 1/2	14 1/4	17 1/2	21 1/2	27 1/8
	42 7/8	48 1/2	52 1/2	55 3/4	58 1/2	61	63 1/4	65 1/4	67 1/4	69 1/8
72	7/8	2 3/4	4 7/8	7	9 1/4	11 7/8	14 3/4	18	22	28
	44	50	54	57 1/4	60 1/8	62 3/4	65	67 1/8	69 1/4	71 1/8
74	7/8	2 7/8	5	7 1/8	9 1/2	12 1/8	15 1/8	18 1/2	22 5/8	28 3/4
	45 1/4	51 3/8	55 1/2	58 7/8	61 7/8	64 1/2	66 7/8	69	71 1/8	73 1/8
76	1	3	5 1/8	7 3/8	9 7/8	12 1/2	15 1/2	19	23 1/4	29 1/2
	46 1/2	52 3/4	57	60 1/2	63 1/2	66 1/8	68 5/8	70 7/8	73	75
78	1	3	5 1/4	7 1/2	10 1/8	12 7/8	15 7/8	19 1/2	23 7/8	30 1/4
	47 3/4	54 1/8	58 1/2	62 1/8	65 1/8	67 7/8	70 1/2	72 3/4	75	77
80	1	3 1/8	5 3/8	7 3/4	10 3/8	13 1/8	16 3/8	20	24 1/2	31
	49	55 1/2	60	63 5/8	66 7/8	69 5/8	72 1/4	74 5/8	76 7/8	79

The velocity pressure readings obtained are converted to velocities and the velocities, not the velocity pressures, are averaged. Where more convenient, *the square root of each of the velocity pressures may be averaged* and this value then converted to velocity (average). The measured air flow *at the temperature in the duct* is then the average velocity multiplied by the cross-sectional area of the duct (Q = VA). Where conditions are not standard, see Section 9.7.2.

The Pitot tube cannot be used for measuring low velocities in the field. It is not a direct-reading air meter. If used with a liquid manometer, a vibration free mounting is necessary. It is susceptible to plugging in air streams with heavy dust and/or moisture loadings.

Modified Pitot Tubes: Modifications of Pitot tubes have been made in an effort to reduce plugging difficulties encountered in heavy dust streams or to increase manometer differentials enabling the measurement of lower velocities in the field. These are referred to as "S" type (Staubscheid) tubes since they usually take the form of two relatively large impact openings, one facing upstream and the other opening facing downstream. Such tubes are also useful when thick walled ducts, such as boiler stacks, make it difficult or impossible to insert a conventional Pitot tube through any reasonably sized port opening.

Other modified forms of the Pitot tube are the air foil pitometer, the Pitot venturi and the air speed nozzle, to name a few. Some of these instruments are of considerable size. A limitation for field use is that "S" type Pitot tubes and other modifications require calibration under conditions similar to those in which they are to be used.

9.3.2 Hood Static Pressure: The hood static pressure method of estimating air flow into an exhaust hood or duct is based on the principle of the orifice; i.e., the inlet opening simulating an orifice. This method is quick, simple, and practical. It is a fairly accurate estimation of the volumetric flow in branch exhaust ducts if the static pressure or suction measurement can be made at a point one to three duct diameters of straight duct downstream from the throat of the exhaust inlet and if an accurate analysis of the hood entry loss can be made.

This technique involves the measuring of hood static pressure by means of a U-tube manometer at one or more holes (preferably four, spaced 90° apart), one duct diameter downstream from the throat for all hoods having tapers, and three duct diameters from the throat for flanged or plain duct ends. The holes should be drilled 1/16" to 1/8" in diameter or less; the holes should not be punched as inwardly projecting jagged edges or metal will disturb the air stream. The U-tube manometer is connected to each hole in turn by means of a thick walled soft rubber tube and the difference in the height of the water columns is read in inches.

If an elbow intervenes between the hood and the suction measurement location, the pressure loss caused by the elbow should be determined as described in Chapter 5 and subtracted from the reading to indicate the suction produced by the hood and throat alone.

The values for hood entry loss factor (F_h) for various hood shapes are listed in Chapter 5, Figure 5-15. When the hood static pressure (SP_h) is known, the volumetric flow rate can be determined by the following equation:

$$Q = 1096\,A\sqrt{\frac{SP_h}{(1 + F_h)\,\rho}} \quad [9.6]$$

where:

Q = volumetric flow rate, cfm

A = area of duct attached to hood, ft^2

SP_h = U-tube average manometer reading, "wg

F_h = hood entry loss factor

For standard air, Equation 9.6 becomes:

$$Q = 4005\,A\sqrt{\frac{SP_h}{1 + F_h}} \quad [9.7]$$

9.3.3 Hood Static Pressure Interpretation: If the hood static pressure is known while a system is functioning properly, its continued effectiveness can be assured so long as the original value is not reduced. Any change from the original measurement can only indicate a change in velocity in the branch and, consequently, a change in volumetric flow through the hood. This relationship will be true unless: 1) a hood design change has affected the entrance loss; 2) there are obstructions or accumulations in the hood or branch ahead of the point where the hood static pressure reading was taken; or 3) the system has been altered or added to. Depending on the location of the obstruction in the duct system, restrictions of the cross-sectional area will reduce the air flow although hood suction may increase or decrease.

Pressure readings vary as the square of the velocity or volumetric flow rate. To illustrate, an indicated reduction in static pressure readings of 30% would reflect a volumetric flow rate (or velocity) decrease of 14%.

A marked reduction in hood static pressure often can be traced to one or more of the following conditions:

1. Reduced performance of the exhaust fan caused by reduced shaft speed due to belt slippage, wear, or accumulation of debris on the rotor or casing that would obstruct air flow.
2. Reduced performance caused by defects in the exhaust piping such as accumulations in branch or main ducts due to insufficient conveying velocities, condensation of oil or water vapors on duct walls, adhesive characteristics of material exhausted, or leakage losses caused by loose clean-out doors, broken joints, holes worn in duct (most frequently in elbows), poor connection to exhauster inlet, or accumulations in ducts or on fan blades.
3. Losses in hood static pressure also can be charged to additional exhaust duct openings added to the system

(sometimes systems are designed for future connections and more air than required is handled by present branches until future connections are made) or change of setting of blast gates in branch lines. Blast gates adjust the air distribution between the various branches. Tampering with the blast gates can seriously affect such distribution and therefore gates should be locked in place immediately after the system has been installed and its effectiveness verified. Fan volume control dampers also should be checked.

4. Reduced volumetric flow may be caused by increased pressure loss through the dust collector due to lack of maintenance, improper operation, wear, etc. These effects will vary with the collector design. Refer to operation and maintenance instructions furnished with the collector or consult the equipment manufacturer. Also, see Chapter 4.

9.4 AIR VELOCITY INSTRUMENTS

The volumetric flow rate of an exhaust system can be determined by the use of various types of field instruments which measure air velocity directly. Typically these instruments are used at exhaust and discharge openings or, depending on size and accessibility, inside a duct. The field technique is based on measuring air velocities at a number of points in a plane and averaging the results. The average velocity is used in Equation 9.1 to determine the volumetric flow rate. Due to the difficulty of measuring the area of an irregularly shaped cross-section and the rapid change in velocity as air approaches an exhaust opening, measurements obtained should be considered an approximation of the true air flow. All instruments should be handled and used in strict compliance with the recommendations and directions of the manufacturers. Table 9-5 lists some characteristics of typical air velocity instruments designed for field use.

FIGURE 9-7.

9.4.1 Rotating Vane Anemometer: This instrument (Figure 9-7) is accurate and can be used to determine air flow through large supply and exhaust openings. Where possible, the cross-sectional area of the instrument should not exceed 5.0% of the cross-sectional area of the duct or hood opening. The standard instrument consists of a propeller or revolving vane connected through a gear train to a set of recording dials that read the linear feet of air passing in a measured length of time. It is made in various sizes; 3″, 4″, and 6″ are the most common. It gives average flow for the time of the test (usually one minute). The instrument requires frequent calibration and the use of a calibration card or curve to determine actual velocity. The instrument may be used for either pres-

TABLE 9-5. Characteristics of Flow Instruments

Instrument	Range (fpm)	Hole Size (for ducts)	Range, Temp.*	Dust, Fume Difficulty	Calibration Requirements	Ruggedness	General Usefullness and Comments
Pitot Tubes with Inclined manometer	600 - up	3/8″	Wide	Some	None	Good	
Standard	600 - up	3/8″	Wide	Some	None	Good	Good except at low velocities
Small Size	600 - up	3/16″	Wide	Yes	Once	Good	Good except at low velocities
Double	500 - up	3/4″	Wide	Small	Once	Good	Special
Swinging Vane Anemometers	25-10,000	1/2-1″	Medium	Some	Frequent	Fair	Good
Rotating Vane Anemometers							
Mechanical	30-10,000	Not for duct use	Narrow	Yes	Frequent	Poor	Special; limited use
Electronic	25-200 25-500 25-2,000 25-5,000	Not for duct use	Narrow	Yes	Frequent	Poor	Special; can record; direct reading

*Temperature range: Narrow, 20–150 F; Medium, 20–300 F; Wide, 0–800 F.

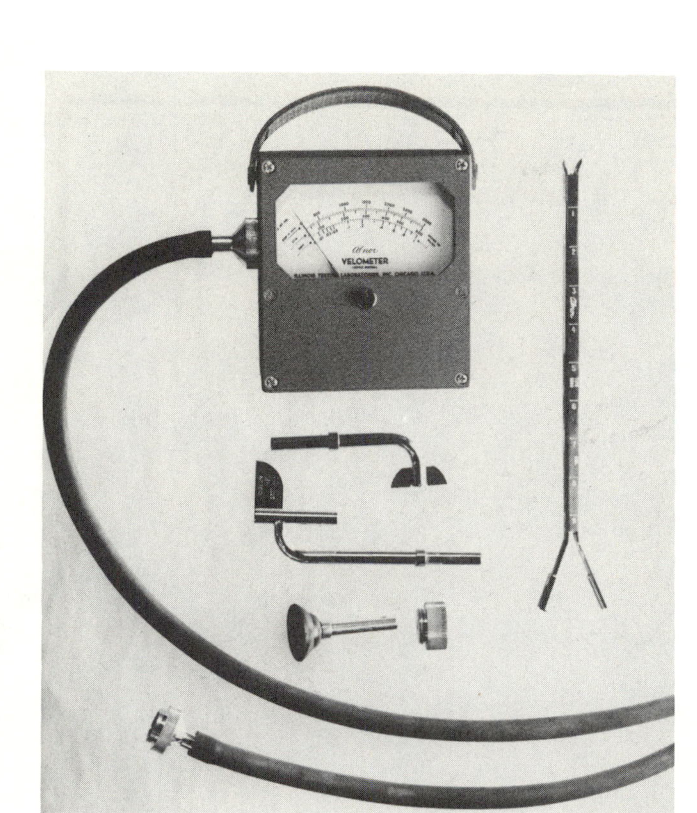

FIGURE 9-8.

sure or suction measurements using the correction factors listed by the manufacturer. The standard instrument has a useful range of 200–3000 fpm; specially built models will read lower velocities.

Direct recording and direct reading rotating vane anemometers are available. These instruments record and meter electrical pulses developed by a capacitance or inductive transducer. The impulses are fed to an indicator unit where they are integrated to operate a conventional meter dial. Readings as low as 25 fpm can be measured and recorded.

The rotating vane anemometer is unsuited for measurement in ducts less than 20 inches in diameter as it has too large a finite area and its equivalent cross-sectional area is difficult to compute. The conventional meter is not a direct reading velocity meter and must be timed. It is fragile and care must be used in dusty or corrosive atmospheres.

9.4.2 Swinging Vane Anemometer (Velometer): This instrument (Figure 9-8) is used extensively in field measurements because of its portability, wide scale range, and instantaneous reading features. Where accurate readings are desired, the correction factors in Table 9-6 should be applied. The instrument has wide application and, by a variety of fittings, can be used to check static pressures and a wide range of linear velocities. The minimum velocity is 50 fpm unless specially adapted for a lower range. The instrument is fairly rugged and accuracy is suitable for most field checks. Uses of the swinging vane anemometer and its various fittings have been illustrated in Figure 9-10.

TABLE 9-6. Correction Factors for the Swinging Vane and Thermal Anemometers

Grill Openings	Correction Factor (CF) (%)
Pressure	
More than 4 in wide and up to 600-in^2 area, free opening 70% or more of gross area, no directional vanes. Use free open area.	93
Suction	
Square punched grille (use free open area)	88
Bar grille (use gross area)	78
Strip grille (use gross area)	73
Free open, no grille	No correction

Before using, check the meter for zero setting by holding it horizontal and covering both ports so that no air can flow through. If the pointer does not come to rest at zero, the "Z" adjustment must be turned to make the necessary correction. Check the meter for balance. After setting to zero as above, the pointer should not deviate more than ⅛″ from zero when both ports are closed regardless of the position in which the meter is held. A velometer with its fittings was originally calibrated as a unit and the fittings cannot be interchanged with another instrument. The serial number on the fittings and on the meter must agree. If the meter was originally calibrated for a filter, it must always be used.

The meter should be used in an upright position and, when using fittings, it must be held *out of the air* stream so that the air flows freely into the opening. The length and inside diameter of the connecting tubing will affect the calibration of the meter. When replacement is required, use only connecting tubing of the same length and inside diameter as that originally supplied with the meter. Some manufacturers have developed instruments (see Figure 9-9) which do not require

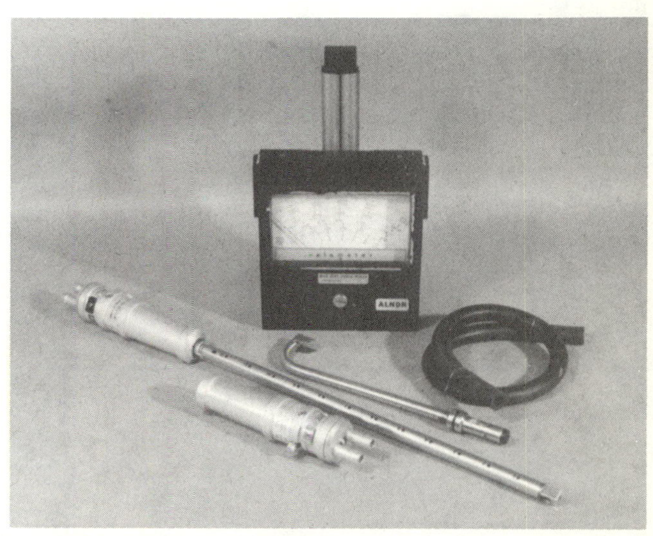

FIGURE 9-9.

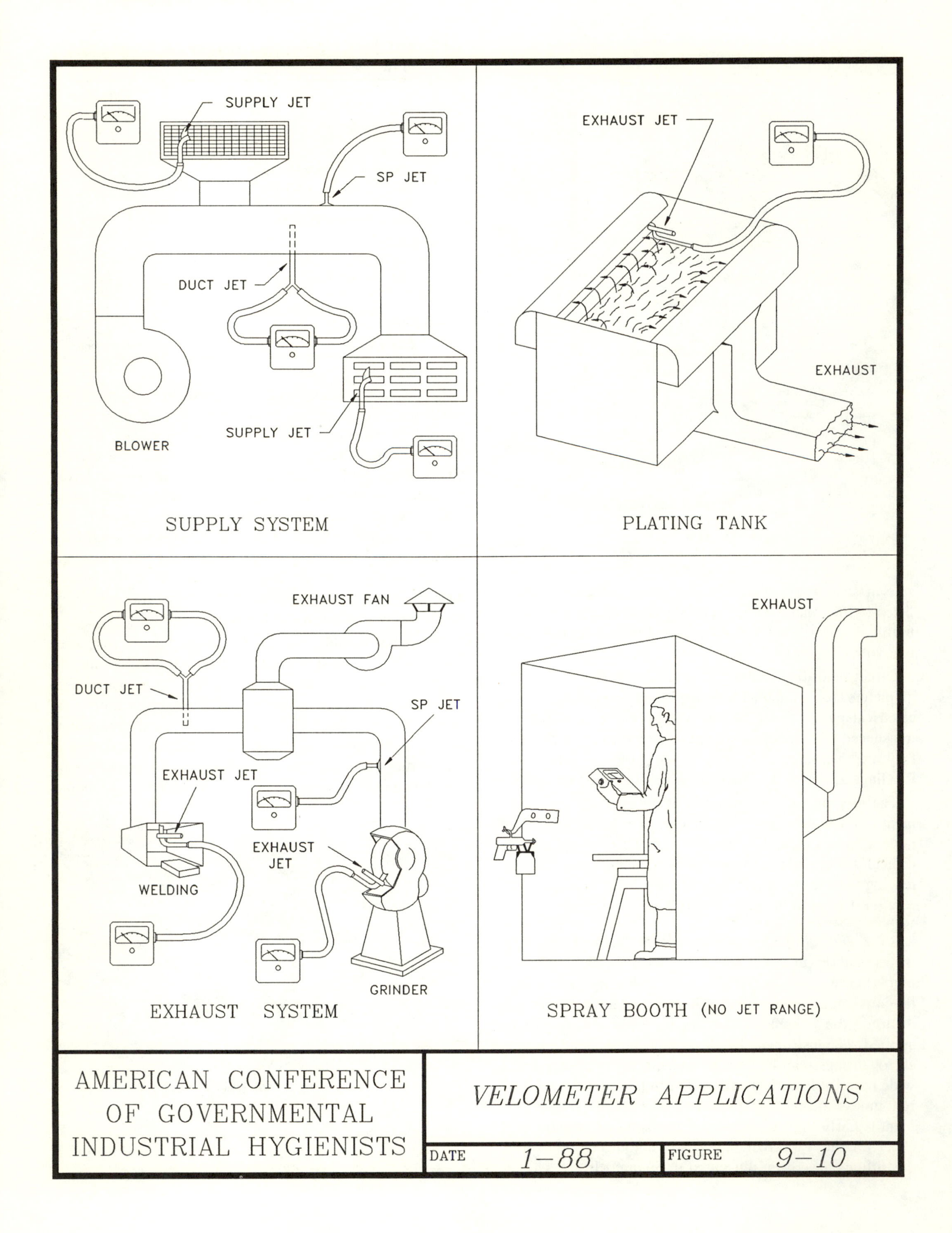
SUPPLY JET
SP JET
DUCT JET
SUPPLY JET
BLOWER
SUPPLY SYSTEM
EXHAUST JET
EXHAUST
PLATING TANK
EXHAUST FAN
DUCT JET
SP JET
EXHAUST JET
EXHAUST JET
WELDING
GRINDER
EXHAUST SYSTEM
EXHAUST
SPRAY BOOTH (NO JET RANGE)
AMERICAN CONFERENCE OF GOVERNMENTAL INDUSTRIAL HYGIENISTS
VELOMETER APPLICATIONS
DATE 1-88
FIGURE 9-10

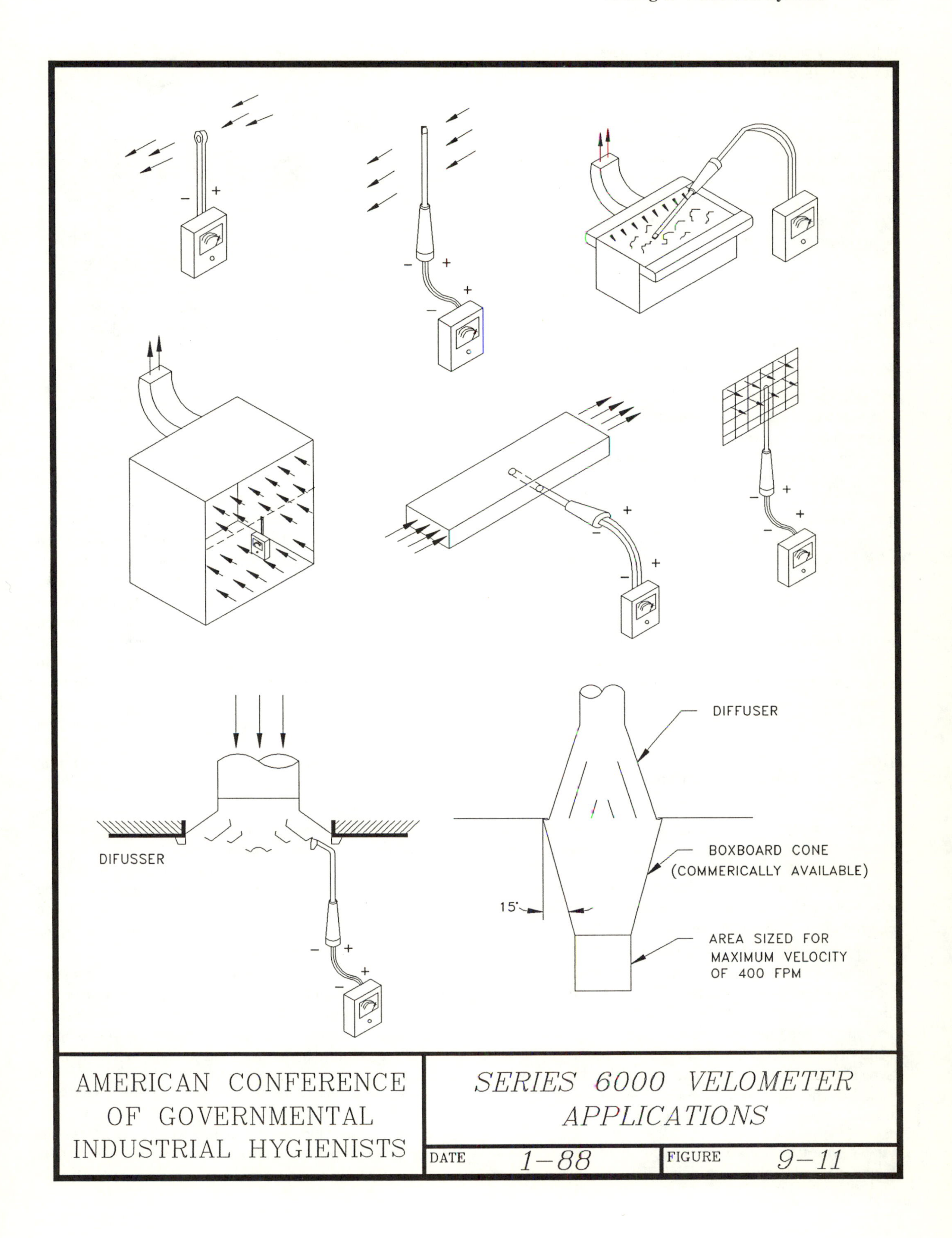
DIFFUSER
BOXBOARD CONE
(COMMERICALLY AVAILABLE)
15°
AREA SIZED FOR
MAXIMUM VELOCITY
OF 400 FPM
DIFUSSER
AMERICAN CONFERENCE
OF GOVERNMENTAL
INDUSTRIAL HYGIENISTS
SERIES 6000 VELOMETER
APPLICATIONS
DATE 1-88
FIGURE 9-11

individually calibrated fittings. Other operating characteristics are similar to those described above. Figure 9-11 illustrates some applications of this model.

Where temperatures of an air stream vary more than 30 F from the standard temperature of 70 F and/or if the altitude is greater than 1,000 feet, it is advisable to make a correction for temperature and pressure. Corrections for change in density from variations in altitude and temperature can be made by using the actual gas density (ρ) shown in Equation 9.3 in the following equation:

$$V_c = V_r \sqrt{\frac{0.075}{\rho}} \qquad [9.8]$$

where:

V_c = corrected velocity, fpm

V_r = velocity reading of instrument, fpm

Use at Supply Openings: On large (at least 3 ft^2) supply openings where the instrument itself will not block the opening seriously and where the velocities are low, the instrument itself may be held in the air stream with the air impinging directly in the left port. When the opening is smaller than 3 ft^2 and/or where the velocities are above the "No Jet" scale, appropriate fittings must be used as illustrated in Figure 9-9. On measurements at the face of exhaust openings of less than 3 ft^2, fittings are required which make readings below 100 fpm impossible.

Because the velocity and static gradient in front of an exhaust opening is steep, the finned opening of the fitting must be held flush with the exhaust opening. If the opening is covered by a grille, hold the fin directly against the grille and use the correction factors listed in Table 9-6 when computing exhaust volumes.

$$Q = C_f VA \qquad [9.9]$$

where:

C_f = correction factor in per cent of scale reading

While it can be used to measure air velocities, static pressure and total pressure in ducts, it has several disadvantages. Used in place of a Pitot tube for velocity or total pressure measurements, it necessitates a much larger hole in the duct, often difficult and impractical to provide. When the velocities are high there may be no appreciable errors at the high end of the scale and the instrument tends to read low on the discharge side of the fan and high on the inlet side.

The presence of dust, moisture, or corrosive material in the atmosphere presents a problem since the air passes through the instrument. In those instruments calibrated for use with a filter (the filter must be used always), the filter itself is a source of error since as the filter becomes plugged its resistance increases and thus alters the amount of air passed to the swinging vane. The instrument requires periodic calibration and adjustment.

9.4.3 Thermal Anemometer: This type of instrument (Figure 9-12) employs the principle that the amount of heat removed by an air stream passing a heated object is related to the velocity of the air stream. Since heat transfer to the air is a function of the number of molecules of air moving by a fixed monitoring point, the sensing element can be calibrated as a mass flow meter as well as a velocity recorder. Commercial instruments use a probe which consists of two integral sensors: a velocity sensor and a temperature sensor. The velocity sensor operates at a constant temperature — typically about 75 F above ambient conditions. Heating energy is supplied and controlled electrically by a battery powered amplifier in the electronics circuit. The electrical current required to maintain the probe temperature in conjunction with the temperature sensor will provide an electrical signal which is proportional to the air velocity and is displayed on either a digital or analog meter. Additional features often include time integration of fluctuating readings and air temperature at the probe. Displays are available in either English or SI units.

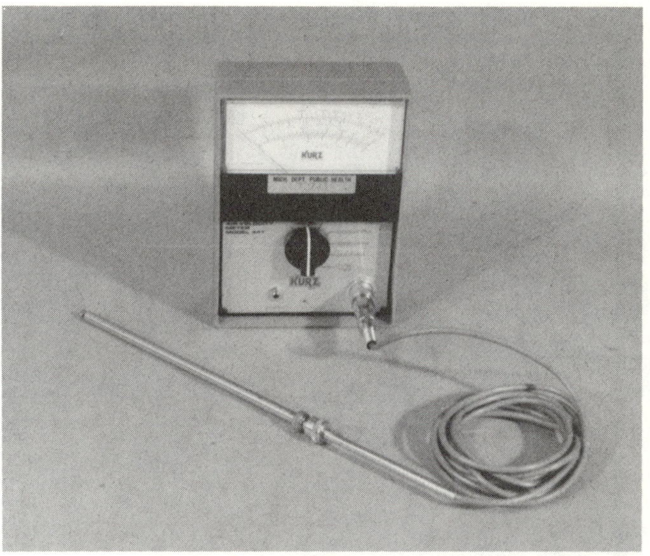

FIGURE 9-12.

The velocity sensor should be used with care in normal field use and is insensitive to mild particulate contamination. The probe can be used directly to measure air velocity in open spaces at air exhaust and supply air openings. Attachments are available to measure static pressure. Due to the small diameter of the probe, measurements can be made directly inside ducts using the measurement techniques described for Pitot traverses in Section 9.3.1.

Battery charging and maintenance is extremely important and the battery voltage must be checked prior to instrument use. The correction factors for this instrument are the same as a swinging vane anemometer (see Table 9-6). Instruments of this type require both initial and periodic calibration.

9.4.4 U-Tube Manometer: The vertical U-tube is the simplest type of pressure gauge. Usually calibrated in "wg, it is used with various fluid media such as alcohol, mercury, oil, water, kerosene and special manometer fluids. The U-tube may be used for either portable or stationary applications.

Available commercial units offer a wide latitude in range, number of columns and styles. Tubes are usually of all plastic construction to minimize breakage. One leg may be replaced by a reservoir or well (well-type manometer) with the advantage of easier manometer reading.

9.4.5 Inclined Manometer: Increased sensitivity and scale magnification is realized by tilting one leg of the U-tube to form an inclined manometer or draft gauge. The inclined manometer spreads the scale thus increasing the accuracy when reading small changes in pressure. In commercial versions, only one tube of the small bore is used and the other leg is replaced by a reservoir. The accuracy of the gauge is dependent on the slope of the tubes. Consequently, the base of the gauge must be leveled carefully and the mounting must be firm enough to permit accurate leveling. The better draft gauges are equipped with a built in level, leveling adjustment and, in addition, a means of adjusting the scale to zero. Some models include "over pressure" safety traps to prevent loss of fluid in event of pressure surges beyond the manometer range.

A modification of the inclined manometer is the inclined vertical gauge in which the indicator leg is bent or shaped to give both a vertical and inclined portion — the advantage is smaller physical size for a given range while retaining the refined measurement afforded by the inclined manometer. As with the U-tube and inclined gauges, available commercial units offer a wide choice in range, number of columns and calibration units.

9.4.6 Aneroid Gauges: This type of gauge is used as a field instrument in ventilation studies for measuring static, velocity or total pressure with a Pitot tube or for single tube static pressure measurements. A number of manufacturers offer gauges suitable for the measurement of the low pressures encountered in ventilation studies. Perhaps the best known of this type is the Magnehelic™ gauge (Figure 9-13). The principal advantages of this gauge can be listed as follows: easy to read, greater response than manometer types, very portable (small physical size and weight), absence of fluid means less maintenance, and mounting and use in any position is possible without loss of accuracy. Principal disadvantages are that the gauge is subject to mechanical failure, requires periodic calibration checks, and occasional recalibration.

9.4.7 Electronic Aneroid Gauges: Commercial instruments are now available which will measure and record static pressure as well as integrate velocity pressure directly to velocity using the pressure sensing principles of an aneroid gauge. This type of instrument can be connected directly to a standard Pitot tube and used in the same manner as a U-tube manometer. The instruments are light in weight, easily hand held, and can be equipped with an electronic digital display or print recorder with measurement data in either English or SI units. Because they are battery powered, periodic servicing and calibration are required.

FIGURE 9-13.

9.4.8 Smoke Tubes: Low velocity measurements may be made by timing the travel of smoke clouds through a known distance. Smoke trail observations are limited to velocities less than 150 fpm since high air velocities diffuse the smoke too rapidly. Commercially available, smoke tubes and candles are useful in the observation of flow patterns surrounding exhaust or supply openings. They also can be used for checking air movement and direction in plant space.

The visible plume is corrosive and should be used with care near sensitive processes or food preparation. Smoke candles are incendiary and thus cannot be used in flammable atmospheres. They should not be hand-held.

9.4.9 Tracer Gas: The principle of dilution sometimes is used to determine rate of air flow. A tracer gas is metered continuously into one or more intake ports (hood or duct openings) along with the entering air stream. After thorough mixing and system equilibrium has been established, air samples are collected at some point downstream — usually at or near the effluent point — and the concentration of the tracer gas in the exit stream is determined. The rate of air flow is readily calculated from the degree of dilution noted in the exit and feed gas concentrations (rate of air flow equals rate of feed divided by tracer gas concentration[9.5]).

The tracer gas usually is selected on the basis of the following: 1) ease of collection and analysis, 2) not present naturally in the process being studied, 3) not absorbed chemically or physically in the duct system, 4) non-reactive with other constituents of the gas stream, and 5) non-toxic or non-explosive.

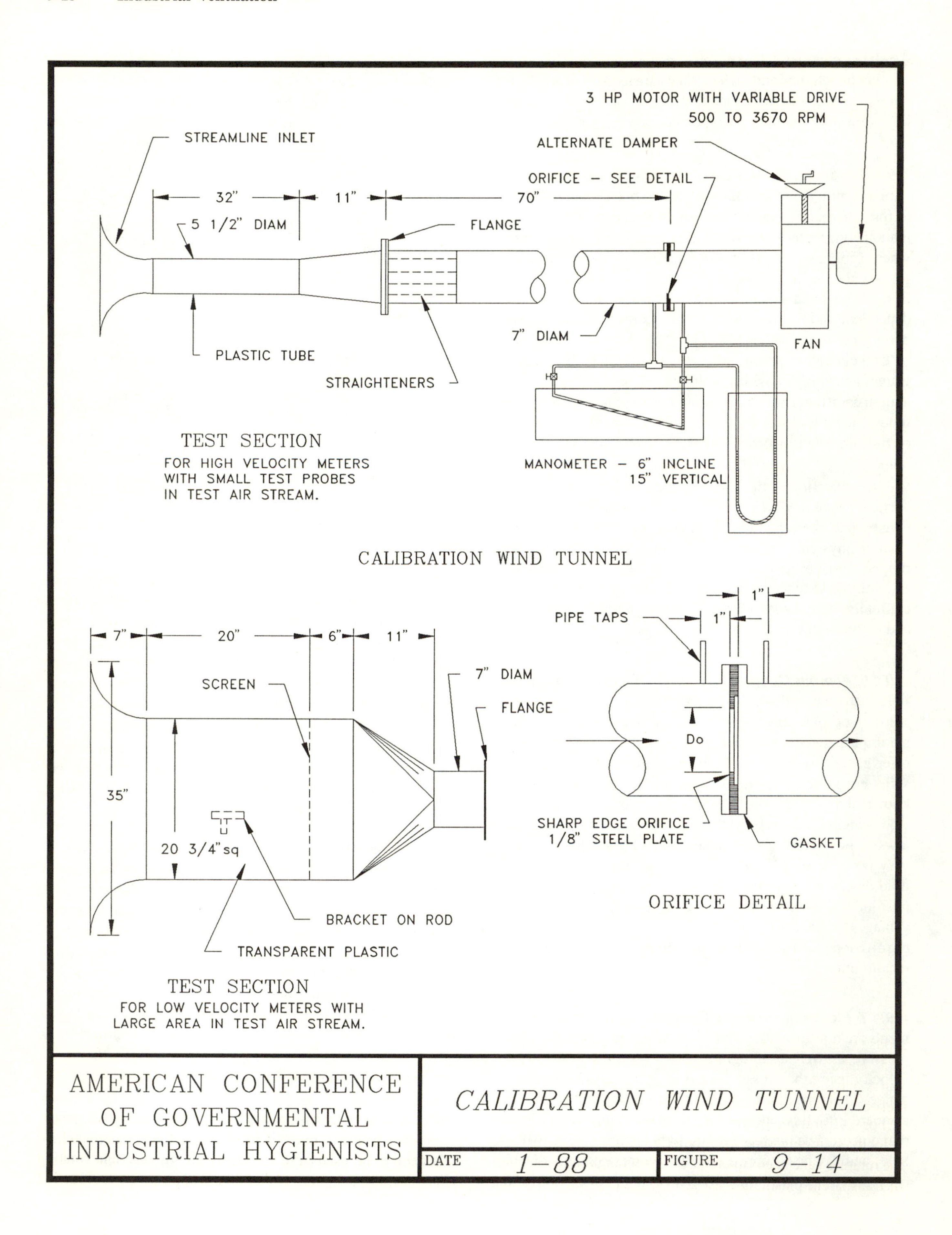
3 HP MOTOR WITH VARIABLE DRIVE
500 TO 3670 RPM
STREAMLINE INLET
ALTERNATE DAMPER
ORIFICE – SEE DETAIL
32"
11"
70"
5 1/2" DIAM
FLANGE
7" DIAM
FAN
PLASTIC TUBE
STRAIGHTENERS
TEST SECTION
FOR HIGH VELOCITY METERS
WITH SMALL TEST PROBES
IN TEST AIR STREAM.
MANOMETER – 6" INCLINE
15" VERTICAL
CALIBRATION WIND TUNNEL
1"
PIPE TAPS
1"
7"
20"
6"
11"
7" DIAM
SCREEN
FLANGE
Do
35"
20 3/4"sq
SHARP EDGE ORIFICE
1/8" STEEL PLATE
GASKET
ORIFICE DETAIL
BRACKET ON ROD
TRANSPARENT PLASTIC
TEST SECTION
FOR LOW VELOCITY METERS WITH
LARGE AREA IN TEST AIR STREAM.
AMERICAN CONFERENCE
OF GOVERNMENTAL
INDUSTRIAL HYGIENISTS
CALIBRATION WIND TUNNEL
DATE 1–88
FIGURE 9–14

9.5 CALIBRATION OF AIR MEASURING INSTRUMENTS

Direct reading meters need regular calibration because they can be easily impaired by shock (dropping, jarring), dust, high temperatures, and corrosive atmospheres. Meters should be calibrated regularly and must be calibrated if they will not adjust to zero properly or if they have been subjected to rough handling and adverse atmospheres.

9.5.1 Design of a Calibrating Wind Tunnel: A typical calibrating wind tunnel for testing air flow meters must have the following components:

1. *A satisfactory test section.* This is the section where the sensing probe or instrument is placed; it must be uniform in air flow both across the air stream and in line with the air flow. A section with a pronounced vena contracta and turbulence will not give satisfactory results.
2. *A satisfactory means of precisely metering the air flow.* The meter on this system must be accurate and with large enough scale graduations so that the volumetric flow rate is indicated within ± 1%. For convenience and time saving, a fixed single reading meter such as a venturi meter or orifice meter is preferable to a multipoint traverse type instrument such as a Pitot tube.
3. *A means of regulating and effecting air flow through the tunnel.* For usual calibrations of instruments used on heating, ventilating, and industrial exhaust systems, test velocities from approximately 50 to 8000 fpm are needed. Air flow regulation must be such that there is no disturbance in the test section. The regulating device must be easily and precisely set to the desired velocities. The fan must have sufficient capacity to develop the maximum velocity in the test section against the static pressure of the entire system.

To provide a satisfactory uniform flow in the test section, a bell shaped streamline entry is necessary (Figure 9-14). There are various designs for this entry. One type is the elliptical approach in which curvature is similar to a one quarter section of an ellipse in which the semi-major axis of the ellipse is equal to the duct diameter to which the entry is placed and the semi-minor axis is two-thirds of the semi-major axis. This type of entry can be made on a spinning lathe.

Actually, any type of smooth curved, bell shaped entry which directs the air into the duct over a 180° angle should be satisfactory. A readily available entry is a tuba or Sousaphone bell. This bell entry should be connected to a 5.5″ diameter smooth, seamless plastic tube. Ridges, small burrs or obstructions should be filed so a smooth connection between horn and tube results.

For calibrating larger instruments such as the lower velocity swinging vane anemometer (Alnor velometer) and the rotating vane anemometer, a large rectangular test section of transparent plastic at least 2.5 ft^3 in cross-sectional area can be constructed with curved air foil inlets as shown in Figure 9-15. A fine mesh screen placed deep in the enclosure will assist in providing a uniform air flow in the test section.

A sharp-edged orifice, venturi meter or a flow nozzle can be used as a metering device. Of these, the sharp-edged orifice has more resistance to flow but is more easily constructed, and it can be designed to be readily interchangeable for several orifice sizes. The orifice can be mounted between two flanged sections sealed with gaskets as shown in Figure 9-14. Each orifice should be calibrated using a standard Pitot tube and manometer prior to use. For velocity measurements below 2,000 fpm, a micromanometer should be used.(9.6)

Table 9-7 lists calculations for three sizes or orifices: 1.400″, 2.625″, and 4.900″ diameters. When the orifices are placed in a 7″ diameter duct and made to the precise dimensions given, no calibration is needed and the tabulated data in the Table will give volumetric flow rates within ± 5% over the range of values shown for standard air density.

A centrifugal fan with sufficient capacity to exhaust 1,100 cfm at 10 ″wg static pressure is needed for a wind tunnel with a 5.5″ diameter test section using an orifice meter. Radial and backward inclined blade centrifugal fans are available with the required characteristics. The air flow can be changed with an adjustable damper at the discharge, a variable speed motor or an adjustable drive on the fan.

The air flow for a sharp-edged orifice with pipe taps located 1″ on either side of the orifice can be computed from the following equation for 2″ to 14″ diameter ducts:

$$Q = 6KD^2 \sqrt{\frac{h}{\rho}} \qquad [9.10]$$

where:

Q = volumetric flow rate, cfm

K = coefficient of air flow

D = orifice diameter, inches

h = pressure drop across orifice, ″wg

ρ = density, lbm/ft^3

The coefficient, K, is affected by the Reynolds number — a dimensionless value expressing flow conditions in a duct. The following equation gives a simplified method of calculating Reynolds number for standard air:

$$R = 8.4\ DV \qquad [9.11]$$

where:

R = Reynolds number, dimensionless

V = velocity of air through orifice, fpm

The coefficient, K, can be selected from Table 9-8.(9.3)

9.5.2 Use of Calibrating Wind Tunnel: Air velocity measuring instruments must be calibrated in the manner in which they are to be used in the field. Swinging vane and rotating vane anemometers are placed in the appropriate test section

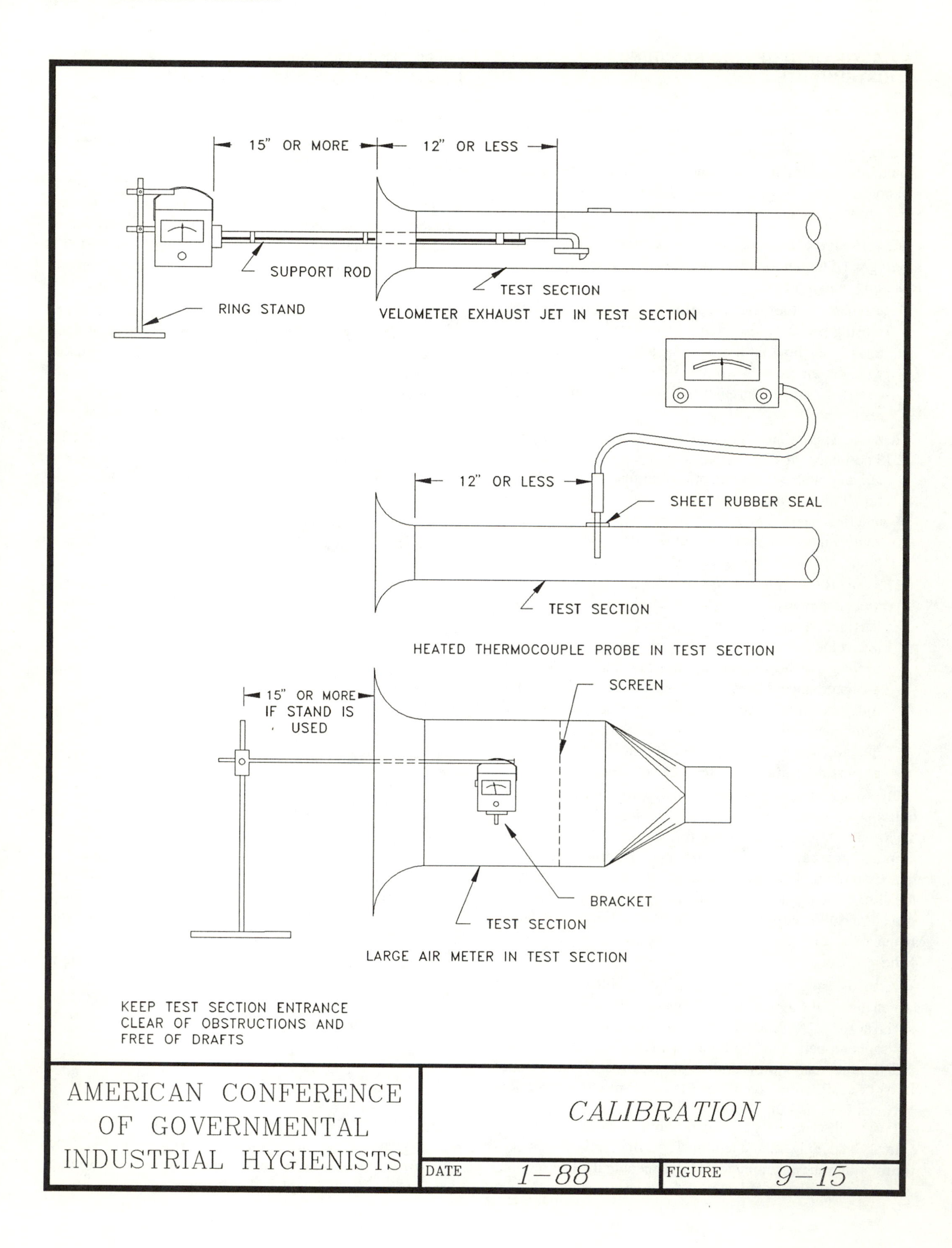

15" OR MORE
12" OR LESS
SUPPORT ROD
TEST SECTION
RING STAND
VELOMETER EXHAUST JET IN TEST SECTION
12" OR LESS
SHEET RUBBER SEAL
TEST SECTION
HEATED THERMOCOUPLE PROBE IN TEST SECTION
15" OR MORE IF STAND IS USED
SCREEN
BRACKET
TEST SECTION
LARGE AIR METER IN TEST SECTION
KEEP TEST SECTION ENTRANCE CLEAR OF OBSTRUCTIONS AND FREE OF DRAFTS
AMERICAN CONFERENCE OF GOVERNMENTAL INDUSTRIAL HYGIENISTS
CALIBRATION
DATE 1-88
FIGURE 9-15

TABLE 9-7. Orifice Flow Rate (scfm) Versus Pressure Differential (in. of water)

ΔP	ORIFICE SIZE			ΔP	ORIFICE SIZE			ΔP	ORIFICE SIZE		
"WC	1.4"	2.625"	4.90"	"WC	1.4"	2.625"	4.90"	"WC	1.4"	2.625"	4.90"
0.02			57.1	1.22	28.7	101.4	410.3	4.10	52.3	185.3	746
0.04		18.7	78.8	1.24	28.9	102.3	413.6	4.20	52.9	187.5	755
0.06		22.8	95.3	1.26	29.2	103.1	416.9	4.30	53.5	189.7	763
0.08		26.2	109.2	1.28	29.4	103.9	420.1	4.40	54.1	191.9	772
0.10		29.3	121.5	1.30	29.6	104.7	423.4	4.50	54.7	194.0	781
0.12		32.1	132.6	1.32	29.8	105.5	426.5	4.60	55.3	196.2	789
0.14		34.6	142.8	1.34	30.1	106.3	429.7	4.70	55.9	198.3	797
0.16		37.0	152.3	1.36	30.3	107.1	432.9	4.80	56.5	200.4	806
0.18		39.2	161.2	1.38	30.5	107.9	436.0	4.90	57.1	202.4	814
0.20		41.3	169.6	1.40	30.7	108.6	439.1	5.00	57.6	204.4	822
0.22		43.3	177.6	1.42	30.9	109.4	442.2	5.10	58.2	206.5	830
0.24		45.2	185.2	1.44	31.2	110.2	445.2	5.20	58.8	208.5	838
0.26		47.0	192.6	1.46	31.4	110.9	448.3	5.30	59.3	210.4	846
0.28		48.8	199.6	1.48	31.6	111.7	451.3	5.40	59.9	212.4	854
0.30		50.5	206.5	1.50	31.8	112.4	454.3	5.50	60.4	214.3	862
0.32		52.1	213.0	1.52	32.0	113.2	457.2	5.60	61.0	216.3	869
0.34		53.7	219.4	1.54	32.2	113.9	460.2	5.70	61.5	218.2	877
0.36		55.3	225.6	1.56	32.4	114.6	463.1	5.80	62.0	220.0	884
0.38		56.8	231.6	1.58	32.6	115.4	466.0	5.90	62.6	221.9	892
0.40		58.3	237.5	1.60	32.8	116.1	468.9	6.00	63.1	223.8	899
0.42		59.7	243.2	1.62	33.0	116.8	471.8	6.10	63.6	225.6	907
0.44		61.1	248.8	1.64	33.2	117.5	474.7	6.20	64.1	227.4	914
0.46		62.4	254.3	1.66	33.4	118.2	477.5	6.30	64.6	229.2	921
0.48		63.8	259.6	1.68	33.6	118.9	480.3	6.40	65.1	231.0	928
0.50	18.5	65.1	264.9	1.70	33.8	119.6	483.1	6.50	65.6	232.8	935
0.52	18.8	66.4	270.0	1.72	34.0	120.3	485.9	6.60	66.1	234.6	942
0.54	19.2	67.6	275.0	1.74	34.2	121.0	488.7	6.70	66.6	236.3	949
0.56	19.5	68.9	280.0	1.76	34.4	121.7	491.5	6.80	67.1	238.1	956
0.58	19.9	70.1	284.8	1.78	34.6	122.4	494.2	6.90	67.6	239.8	963
0.60	20.2	71.3	289.6	1.80	34.8	123.1	496.9	7.00	68.1	241.5	970
0.62	20.6	72.4	294.3	1.82	35.0	123.8	499.7	7.10	68.5	243.2	977
0.64	20.9	73.6	298.9	1.84	35.2	124.4	502.4	7.20	69.0	244.9	984
0.66	21.2	74.7	303.4	1.86	35.4	125.1	505.0	7.30	69.5	246.5	990
0.68	21.5	75.8	307.9	1.88	35.5	125.8	507.7	7.40	69.9	248.2	997
0.70	21.8	76.9	312.3	1.90	35.7	126.4	510.4	7.50	70.4	249.9	1003
0.72	22.1	78.0	316.7	1.92	35.9	127.1	513.0	7.60	70.9	251.5	1010
0.74	22.4	79.1	320.9	1.94	36.1	127.8	515.6	7.70	71.3	253.1	1017
0.76	22.7	80.2	325.2	1.96	36.3	128.4	518.2	7.80	71.8	254.7	1023
0.78	23.0	81.2	329.3	1.98	36.5	129.1	520.8	7.90	72.2	256.4	1029
0.80	23.3	82.2	333.5	2.00	36.6	129.7	523.4	8.00	72.7	257.9	1036
0.82	23.6	83.2	337.5	2.10	37.5	132.9	536.2	8.10	73.1	259.5	1042
0.84	23.9	84.2	341.6	2.20	38.4	136.0	548.6	8.20	73.6	261.1	1048
0.86	24.1	85.2	345.5	2.30	39.3	139.0	560.8	8.30	74.0	262.7	1055
0.88	24.4	86.2	349.4	2.40	40.1	142.0	572.6	8.40	74.5	264.2	1061
0.90	24.7	87.2	353.3	2.50	40.9	144.9	584.3	8.50	74.9	265.8	1067
0.92	25.0	88.1	357.2	2.60	41.7	147.8	595.7	8.60	75.3	267.3	1073
0.94	25.2	89.1	361.0	2.70	42.5	150.6	606.9	8.70	75.7	268.8	1079
0.96	25.5	90.0	364.7	2.80	43.3	153.3	617.9	8.80	76.2	270.4	1085
0.98	25.8	91.0	368.4	2.90	44.0	156.0	628.6	8.90	76.6	271.9	1091
1.00	26.0	91.9	372.1	3.00	44.8	158.7	639.2	9.00	77.0	273.4	1097
1.02	26.3	92.8	375.7	3.10	45.5	161.3	649.6	9.10	77.4	274.9	1103
1.04	26.5	93.7	379.3	3.20	46.2	163.8	659.9	9.20	77.9	276.4	1109
1.06	26.8	94.6	382.9	3.30	46.9	166.4	670.0	9.30	78.3	277.8	1115
1.08	27.0	95.5	386.4	3.40	47.6	168.8	679.9	9.40	78.7	279.3	1121
1.10	27.3	96.3	390.0	3.50	48.3	171.3	689.7	9.50	79.1	280.8	1127
1.12	27.5	97.2	393.4	3.60	49.0	173.7	699.3	9.60	79.5	282.2	1132
1.14	27.8	98.1	396.9	3.70	49.7	170.1	708.8	9.70	79.9	283.6	1138
1.16	28.0	98.9	400.3	3.80	50.3	178.4	718.2	9.80	80.3	285.1	1144
1.18	28.2	99.8	403.7	3.90	51.0	180.7	727.5	9.90	80.7	286.5	1150
1.20	28.5	100.6	407.0	4.00	51.6	183.0	736.6	10.00	81.1	287.9	1155

TABLE 9-8. Values of K in Equation 9.10 for Different Orifice Diameters to Duct-Diameter Ratios (d/D) and Different Reynolds Numbers*

	Reynolds Number in Thousands						
d/D	25	50	100	230	500	1000	10,000
0.100	0.605	0.601	0.598	0.597	0.596	0.595	0.595
0.200	0.607	0.603	0.600	0.599	0.598	0.597	0.597
0.300	0.611	0.606	0.603	0.603	0.601	0.600	0.600
0.400	0.621	0.615	0.611	0.610	0.609	0.608	0.608
0.450	0.631	0.624	0.619	0.617	0.615	0.615	0.615
0.500	0.644	0.634	0.628	0.626	0.624	0.623	0.623
0.550	0.663	0.649	0.641	0.637	0.635	0.634	0.634
0.600	0.686	0.668	0.658	0.653	0.650	0.649	0.649
0.650	0.717	0.695	0.680	0.674	0.670	0.668	0.667
0.700	0.755	0.723	0.707	0.699	0.694	0.692	0.691
0.750	0.826	0.773	0.747	0.734	0.726	0.723	0.721

*For duct diameters of 2 in to 14 in inclusive.

on a suitable support and the air velocity varied through the operating range of interest. Heated thermocouple instruments are calibrated in the same manner. Special Pitot tubes and duct probes of direct reading instruments are placed through a suitable port in the circular duct section and calibrated throughout the operating range (Figure 9-15).[9.9]

9.6 EVALUATING EXHAUST SYSTEMS

Knowledge of the air measurement equipment and methods previously discussed is essential before a ventilation system can be tested. Testing should be conducted after installation and at later intervals to verify system performance.

9.6.1 New Installations: Sufficient data should be taken on completion of every installation to verify that 1) volumetric flow rates, distribution and system balance are in agreement with design data and 2) contaminant control is effective. As a first step, a sketch of the system, not necessarily to scale but indicating size, length, and relative location of all ducts, fittings and associated system components, should be made. The sketch will serve as a guide in the selection of measuring points and very often will bring to light incorrect installation and poor design features. Physical system changes which may occur at a later date (addition of branches, alteration of hoods or ducts) are easily noted if such a sketch is available as a permanent record.

Initial air measurements should include Q, velocity and static pressure measurements in each branch and in the main; static pressure measurements (throat suction) at each exhaust opening; static and total pressure measurements at both fan inlet and outlet; and pressure measurements at collection equipment inlet and outlet (differential pressure). Measurement data will indicate any variation from design data, the need for balancing to obtain distribution and verify that transport velocities in branches and main are sufficient to convey the material handled. Where imbalance is found, it is advisable to remeasure pressure readings after the system has been balanced. All air measurements and location of measuring points should be recorded to serve as a basis for future tests designed to monitor possible variations in volumetric air flow from original values.

Hood design should be verified to be sure the contaminant source is hooded as completely as possible without interfering with the operation. Air analysis will vary with contaminant: use air sampling at operator's breathing zone where toxic materials are involved. Observation for escape may be a satisfactory alternative where visible non toxic nuisance materials are exhausted.

In Figure 9-16, an example of an exhaust system with possible testing points is illustrated. The selection of testing points will be dictated by the system under consideration. Rarely will it be possible to attain the ideal — as in laboratory test installations. However, with judicious selection of measuring stations to avoid excessively turbulent flow and with proper attention to calibration, alignment and positioning of instruments, accurate measurements are possible.

Figure 9-17 illustrates one type of survey form. Collection of data will serve little purpose if a permanent record is not maintained so periodic test data can be compared. Any type of form which suits the individual's needs will suffice.

9.6.2 Periodic Testing: For most existing installations, frequent, elaborate air sampling studies are seldom needed. Unless the process is changed, the hoods or enclosures altered or the method of materials handling revised, the hazard, if initially controlled, should remain controlled as long as the exhaust system functions *properly*.

The word "properly" can be underscored because in many cases little attention is given to the installation after the project has been completed. Mechanical exhaust equipment and dust collectors require the same attention that machine tools and other plant equipment usually receive.

9.6.3 Check-out Procedure: The following procedure may be used on systems which were designed to balance without the aid of blast gates. It is intended as an initial verification of the design computations and contractor's construction in new systems, but it may be used also for existing systems when design calculations are available or can be

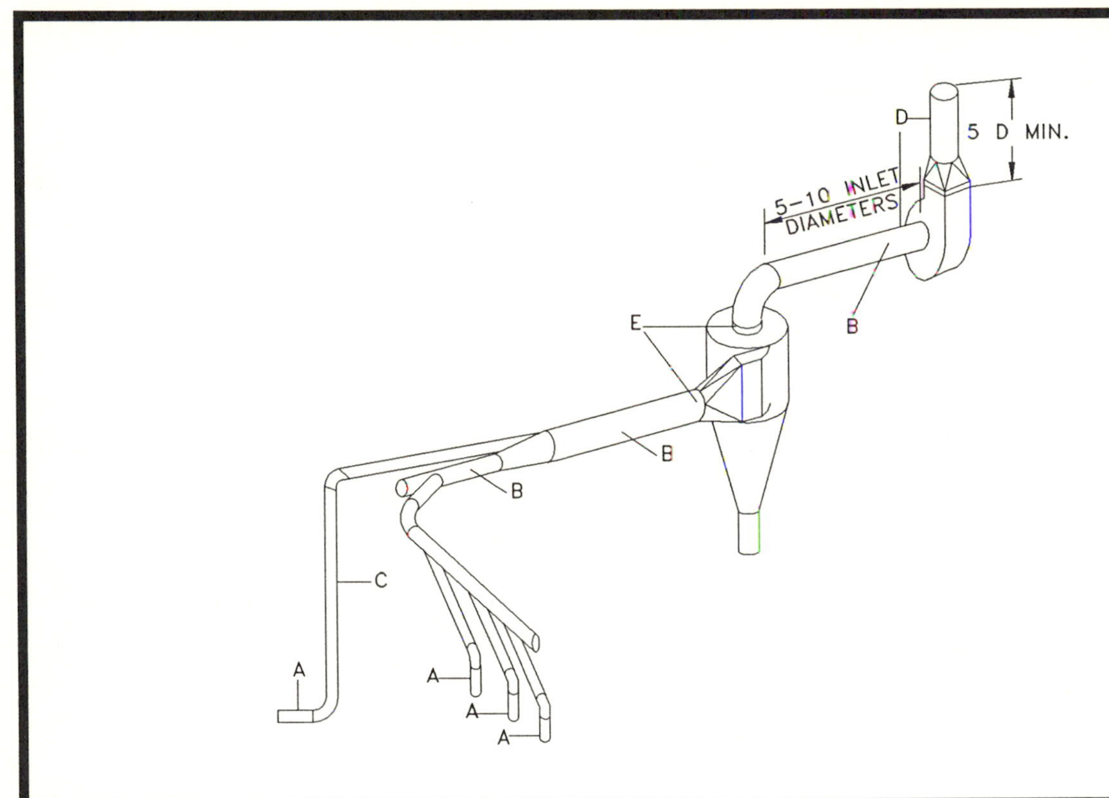

POINT	MEASUREMENT	LOCATION OF MEASUREMENT	MEASUREMENT USE
A	HOOD STATIC PRESSURE	DISTANCE FROM HOOD – 3 PIPE ø'S–FLANGED OR PLAIN HOOD 1 ø–TAPERED HOOD	1. ESTIMATE FLOW: $Q=4005C_eA\sqrt{SP_h}$ 2. CHECK POINT FOR HOOD AND SYSTEM PROFORMANCE.
B	VELOCITY AND STATIC PRESSURE	BRANCH AND MAINS–PREFERABLY 7.5 ø'S STRAIGHT RUN DOWNSTREAM FROM NEAREST AIR DISTURBANCE (EL,ENTRY, ETC..)	1. TRANSPORT VELOCITY 2. EXHAUST VOLUME: Q=VA 3. SP AS SYSTEM CHECK POINT
C	CENTERLINE VP	SMALL DUCTS LOCATION AS ABOVE. CENTERLINE VELOCITY READING ONLY.	ROUND DUCT ONLY. USE ON SMALL DUCTS WHERE TRANSVERSE IMPRACTICAL OR WHERE APPROXIMATE VOLUME WANTED.
D	STATIC,VELOCITY AND TOTAL PRESSURES	INLET AND OUTLET OF FAN–ANY TWO OF THREE READINGS AT EACH LOCATION	1. FAN STATIC AND TOTAL PRESSURES $FSP = SP_o - SP_i - VP_i$ $TP = SP_o - SP_i + VP_o - VP_i$ 2. MOTOR SIZE OR GFM ESTIMATE $BHP = \frac{CFM \times TP}{6356 \times ME\ OF\ FAN}$ 3. SP AS SYSTEM CHECK POINT
E	STATIC PRESSURE	INLET AND OUTLET OF COLLECTOR DIFFERENTIAL PRESSURE	1. COMPARE PRESSURE DROP WITH NORMAL OPERATING RANGE 2. CHECKPOINTS FOR MAINTAINENCE. READINGS ABOVE OR BELOW NORMAL INDICATE PLUGGING, WEAR OR DAMAGE TO COLLECTOR ELEMENTS,NEED OF CLEANING

IN ADDITION TO THE ABOVE, FACE VELOCITY (HOOD FACE) AND CAPTURE VELOCITY (POINT OF CONTAMINANT DISPERSION) MEASUREMENTS ARE USUALLY MADE TO DEFINE HOOD PERFORMANCE. OBSERVATION OF AIR FLOWS SURROUNDING EXHAUST OPENINGS MAY BE VISUALLY AUGMENTED BY USE OF SMOKE GENERATORS, TRAILS, AND STREAMERS.

AMERICAN CONFERENCE OF GOVERNMENTAL INDUSTRIAL HYGIENISTS

SAMPLE SYSTEM

DATE *1–88* FIGURE *9–16*

PLANT ________________ DEPT. ____________ DATE __________
OPERATION EXHAUSTED ______________________ BY __________

LINE SKETCH SHOWING POINTS OF MEASUREMENT

DATE SYSTEM INSTALLED __________

HOOD AND TRANSPORT VELOCITY

POINT	DUCT		VP	SP	FPM	CFM	REMARKS
	D	AREA (Tbl. 5-5)	IN. H_2O	IN. H_2O	(Tbl. 9-1)	Q =VA	

PITOT TRAVERSE
PITOT READINGS– SEE TABLES 9-1 TO 9-4

POINTS	VP	VEL.	VP	VEL.	VP	VEL.
1						
2						
3						
4						
5						
6						
7						
8						
9						
10						
TOTAL VEL.						
AVERAGE VEL.						
CFM						

FAN
TYPE __________
SIZE __________

POINT	DIA.	SP	VP	TP	CFM
INLET					
OUTLET					

FAN SP______ (SEE SECTION 6)
MOTOR
NAME __________ SIZE ______
HP______ E____ I____ W____
COLLECTOR
TYPE & SIZE __________

POINT	DIA.	SP	△ SP
INLET			
OUTLET			

NOTES ______________________________________

AMERICAN CONFERENCE OF GOVERNMENTAL INDUSTRIAL HYGIENISTS

SURVEY FORM

DATE 1-88 FIGURE 9-17

recomputed. It does not detect poor choices of design criteria, such as low conveying or capture velocities, and consequently will not reveal inadequate control due to this type of error. Agreement with design within ± 10% is considered acceptable.

1. Determine volumetric flow in duct with a pitot traverse. If volumetric flow matches design, go to Step 4; otherwise, continue with 1a.
   a. Check fan size against plan.
   b. Check fan speed and direction of rotation against design.
   c. Check fan inlet and outlet configuration against plan.
2. If a discrepancy is found and corrected, return to Step 1. If not, measure fan inlet and outlet static pressures and compute the fan static pressure. Using fan table, check flow, fan static pressure, and fan speed (rpm). If agreement is acceptable although at some other operating point than specified, the fan is satisfactory and trouble is elsewhere in the system. Go to Step 3.
3. If fan *inlet* static pressure is greater (more negative) than calculated in the design, proceed to Step 4. If fan *outlet* static pressure is greater (more positive) than design, proceed to Step 8.
4. Measure hood static pressure on each hood and check against design. If correct, go to Step 10; otherwise, continue with Step 4a.
   a. Check size and design of hoods and slots against plan.
   b. Examine each hood for obstructions.
5. After all hood construction errors and obstructions have been corrected, if hood static pressures are correct, return to Step 1; if too low, proceed to Step 6.
6. Measure static pressure at various junctions in ducts and compare with design calculations. If too high at a junction, proceed up stream until static pressures are too low and isolate the trouble. In an area where the loss exceeds design:
   a. Check angle of entries to junctions against plan.
   b. Check radii of elbows against plan.
   c. Check duct diameters against plan.
   d. Check duct for obstructions.
7. After correcting all construction details which deviate from specifications, return to Step 1.
8. Measure pressure differential across air cleaning device and check against manufacturer's data. If loss is excessive, make necessary corrections and return to Step 1. If loss is less than anticipated, proceed to Step 8a.
   a. Check ducts, elbows and entries as in Step 6a and 6d.
   b. Check system discharge type and dimensions against plans.
9. If errors are found, correct and return to Step 1. If no errors can be detected, recheck design against plan, recalculate and return to Step 1 with new expected design parameters.
10. Measure control velocities at all hoods where possible. If control is inadequate, redesign or modify hood.
11. The above process should be repeated until all defects are corrected and hood static pressures and control velocities are in reasonable agreement with design. The actual hood static pressures should then be recorded for use in periodic system checks. A file should be prepared containing the following documents:

    System plan
    Design calculations
    Fan rating table
    Hood static pressures after field measurement
    Maintenance schedule
    Periodic hood static pressure measurement log
    Periodic maintenance log

9.7 DIFFICULTIES ENCOUNTERED IN FIELD MEASUREMENT

The general procedures and instrumentation for the measurement of air flow have been previously discussed in this chapter. However, special problems connected with air flow necessitate a somewhat more detailed discussion.

Some of these special problems are as follows:

1. Measurement of air flows in highly contaminated air which may contain corrosive gases, dusts, fumes, or mists.
2. Measurement of air flows at high temperatures.
3. Measurement of air flow in high concentrations of water vapor and mist.
4. Measurement of air flow where the velocity is very low.
5. Measurement of air flow in locations of turbulence and non uniform air flow, e.g., discharge of cupolas, locations near bends, or enlargements.
6. Measurement of air flow in connection with isokinetic sampling when the velocity is constantly changing.

9.7.1 Selection of Instruments: The selection of the proper instrument will depend on the range of air flow to which the instrument is sensitive; its vulnerability to high temperatures, corrosive gases and contaminated atmospheres; its portability and ruggedness, and the size of the measuring probe relative to the available sampling port. A brief summary of the characteristics of a few of the instruments which have been used is given in Table 9-5.

In many cases, conditions for air flow measurement are so severe that it is difficult to select an instrument. Generally speaking, the Pitot tube is the most serviceable instrument, inasmuch as it has no moving parts, it is rugged and will stand high temperatures and corrosive atmospheres when it is made of stainless steel. It is subject to plugging, however, when it is used in a dusty atmosphere. It cannot be used for measurement of low velocities. A special design of Pitot tube can be used for dusty atmospheres. In many cases, it is difficult to set up an inclined manometer in the field because many readings are made from ladders, scaffolds, and difficult places. This greatly limits the lower range of the Pitot tube. A mechanical gauge can be used in place of a manometer. A mechanical gauge is estimated to be accurate to 0.02 "wg with proper calibration.

For lower velocities, the swinging vane anemometer previously described can be used if conditions are not too severe. The instrument can be purchased with a special dust filter which allows its use in light dust loadings. It can be used in temperatures up to 1000 F if the jet is exposed to the high temperature gases only for a very short period of time (30 seconds or less). It cannot be used in corrosive gases. If the very low velocity jet is used, a hole over 1" in diameter must be cut into the duct or stack.

For very low velocities, anemometers utilizing the heated thermocouple principle can be used under special conditions. In most cases, these anemometers cannot be used in temperatures above 300 F. Contact the manufacturer to determine to what degree the thermocouple probe will withstand corrosive gases.

In sampling work where a match of velocities in the sampling nozzle and air stream under changing velocities is required, the null method is sometimes used. This method uses two static tubes or inverted impact tubes, one located within the sampling nozzle and the other in the air stream. Each is connected to a leg of the manometer; the sampling rate is adjusted until the manometer reading is zero.

9.7.2 Corrections for Non-Standard Conditions: Air velocities are sometimes measured at conditions significantly different from standard. If these conditions are ignored, serious errors can be introduced in the determination of the actual duct velocity and the volumetric flow rate(s) in the system. Elevation, pressure, temperature, and moisture content all affect the density of the air stream. The actual density present in the system must be used in either Equation 9.3 or 9.5 to determine the actual velocity.

Correction for changes in elevation, duct pressure, and temperature can be made independently of each other with reasonable accuracy. The individual correction factors are multiplied together to determine the change from standard density. The actual density becomes:

$$\rho = 0.075\, C_e\, C_p\, C_t \qquad [9.12]$$

where:

C_e = correction for elevations outside the range of ± 1000 ft

C_p = correction for local duct pressures greater than ± 20 "wg

C_t = correction for temperatures outside the 40 to 100 F range

One exception to this general rule is when elevations significantly different from sea level are coupled with high moisture content. Where this occurs, a psychrometric chart based upon the barometric pressure existing at the elevation of concern should be used. See Chapter 5, Section 5.13, for an explanation of the determination of density using a psychrometric chart when moisture content and temperature are significantly different from standard.

The correction factor for elevation, C_e, can be given by

$$C_e = [1 - (6.73 \times 10^{-6})z]^{5.258} \qquad [9.13]$$

where:

z = elevation, ft

The correction factor for local duct pressure, C_p, can be given by

$$C_p = \frac{407 + SP}{407} \qquad [9.14]$$

where:

SP = static pressure, "wg. (Note that the sign of SP is important.)

The correction factor for temperature, C_t, can be given by

$$C_t = \frac{530}{t + 460} \qquad [9.15]$$

where:

t = dry bulb temperature, F

EXAMPLE

A velocity pressure reading of 1.0 "wg was taken with a Pitot tube in a duct where the dry bulb temperature is 300 F, the moisture content is negligible, and the static pressure is –23.5 "wg. The system is installed at an elevation of 5000 feet. What would the density and actual velocity be at that point?

As the moisture content is unimportant, Equation 9.12 can be used directly to determine the density.

The individual correction factors can be found from Equations 9.13 through 9.15 as

$$C_e = [1 - (6.73 \times 10^{-6})(5000)]^{5.258} = 0.84$$

$$C_p = \frac{407 - 23.5}{407} = 0.94$$

$$C_t = \frac{530}{300 + 460} = 0.70$$

Then the density at this condition would be

$$\rho = (0.075)(0.84)(0.94)(0.70) = 0.0415 \text{ lbm/ft}^3$$

and the velocity from Equation 9.3 would be

$$V = 1096\sqrt{\frac{1.0}{0.0415}} = 5380 \text{ fpm}$$

Note that an error of 26% would result if standard density had been assumed.

EXAMPLE

A velometer is used to determine the velocity in a duct at sea level where the dry bulb temperature is 250 F, the SP = – 10 "wg and moisture is negligible. What is the actual duct velocity if the velometer reading is 3150?

The temperature correction factor is 0.75 from Equation 9.15 and the density would be

$$(0.75)(0.075) = 0.0563 \text{ lbm/ft}^3$$

Therefore, the actual velocity in the duct would be

$$\frac{(3150)(0.075)}{0.0563} = 4196 \text{ fpm}$$

9.7.3 Pitot Traverse Calculations: Measurement of air velocity at non standard conditions requires calculation of the true air velocity, accounting for difference in air density due to air temperature, humidity, and barometric pressure. The following calculations illustrate the method of calculation and the effect of varying air density.

1. ***Standard Conditions:***
Air Temp. = 79 F; Wet Bulb Temp. = 50 F
Barometer = Std. (29.92 "Hg); 24" Duct Diameter

Pitot Traverse #1			Pitot Traverse #2 (⊥ to Traverse #1)		
Traverse Pt.	VP_M	V_s*	Traverse Pt.	VP_M	V_s*
1	0.22	1879	1	0.23	1921
2	0.28	2119	2	0.27	2081
3	0.32	2260	3	0.33	2301
4	0.33	2301	4	0.34	2335
5	0.34	2335	5	0.34	2335
6	0.35	2369	6	0.35	2369
7	0.33	2301	7	0.34	2335
8	0.31	2230	8	0.32	2260
9	0.30	2193	9	0.31	2230
10	0.24	1962	10	0.25	2003
		21949			22170

*Calculated from Equation 9.4 or Table 9-1.

$$\text{Average Velocity} = \frac{21949 + 22170}{20} = \frac{44119}{20}$$

$$= 2206 \text{ fpm}$$

$$Q_s = VA = 2206 \times 3.142 = 6931.2 = 6931 \text{ scfm}$$

2. ***Elevated Temperature:***
Air Temp. = 150 F; Wet Bulb Temp. = 80 F
Barometer = Std; 24" Outside Diameter Duct

To determine the air velocity at standard conditions (V_s) for each VP_M, the air density (ρ) can be calculated using Equation 9.15:

$$\rho = 0.075 \times (530 \div 610) = 0.065 \text{ lbm/ft}^3$$

Pitot Traverse #1			Pitot Traverse #2 (⊥ to Traverse #1)		
Traverse Pt.	VP_M	V_s*	Traverse Pt.	VP_M	V_s*
1	0.22	2015	1	0.23	2060
2	0.28	2275	2	0.27	2235
3	0.32	2430	3	0.33	2465
4	0.33	2470	4	0.34	2505
5	0.34	2505	5	0.34	2505
6	0.35	2540	6	0.35	2540
7	0.33	2470	7	0.34	2505
8	0.31	2395	8	0.32	2430
9	0.30	2355	9	0.31	2395
10	0.24	2105	10	0.25	2150
		23560			23790

*Calculated from Equation 9.4 or Table 9-1.

Using Equation 9.5, each VP_M is multiplied by the ratio 0.075 ÷ 0.065 and the resulting V_s values averaged.

$$\text{Average Velocity, } V_s = (23560 + 23790) \div 20$$
$$= 47350 \div 20 = 2368 \text{ scfm}$$

$$Q_s = VA = 2368 \times 3.142 = 7440 \text{ scfm}$$

Short Method:

Find: "standard velocity" average from measured VPs = 2206 fpm (from #1)

VP for 2206 fpm = 0.30 (Fig. 6-16)

$$\text{at 150 F, density} = \frac{0.075}{\rho} = 0.065 \text{ lbm/ft}^3$$

$$VP_s = VP_M \times \frac{0.075}{0.065} = 0.346 = 0.35 \text{ "wg}$$

$$V_s = 2369 \text{ fpm}$$

3. ***Elevated Temperature and Moisture:***
Air Temp. = 150 F; Wet Bulb Temp. = 140 F;
Barometer = Standard; 24" Outside Diameter Duct

To determine the air velocity at standard conditions (V_s) for each VP_M, the air density (ρ) can be calculated using the psychrometric charts found in Chapter 5.

Pitot Traverse #1			Pitot Traverse #2 (⊥ to Traverse #1)		
Traverse Pt.	VP_M	V_s*	Traverse Pt.	VP_M	V_s*
1	0.22	2100	1	0.23	2145
2	0.28	2370	2	0.27	2325
3	0.32	2530	3	0.33	2570
4	0.33	2570	4	0.34	2610
5	0.34	2610	5	0.34	2610
6	0.35	2645	6	0.35	2645
7	0.33	2570	7	0.34	2610
8	0.31	2490	8	0.32	2530
9	0.30	2450	9	0.31	2490
10	0.24	2190	10	0.25	2235
		24525			24770

*Calculated from Equation 9.4 or Table 9-1.

$$\rho = 0.075 \times 0.80 \text{ (density factor} - \text{mixture)}$$
$$= 0.06 \text{ lbm/ft}^3$$

Using Equation 9.5, each VP_M is multiplied by the ratio $0.075 \div 0.06$ and the resulting V_s values averaged.

$$\text{Average velocity, } V_s = \frac{24525 + 24770}{20}$$
$$= \frac{49295}{20} = 2464.7$$
$$= 2465 \text{ fpm}$$

(V_s may be found also by the Short Method found in #2.)

$$Q_s = VA = 2465 \times 3.142$$
$$= 7745 \text{ cfm of air and water mixture.}$$
$$\text{Weight of mixture} = Q_s \times 0.075 \times d$$
$$= 7745 \times 0.075 \times 0.80 = 465 \text{ lb.}$$

From Chapter 5, weight of water in mixture = 0.15 lb H_2O/lb dry air.

Weight of dry air = weight of meixture/weight of dry air and moisture = $465 \div 1.15 = 404$ lb.

Alternate Method:
From Figure 5-25, humid volume = 19.3 ft^3 of mixture/lb dry air (Interpolate).

$$\text{Weight of dry air} = \frac{Q}{19.3} = \frac{7745}{19.3} = 401 \text{ lb}$$
$$Q = \frac{404 \text{ lb}}{0.075} = 5387 \text{ scfm}$$

4. ***High or Low Altitudes:*** $Q_s = V_s \times A$ where V_s can be obtained from Equations 9.3 or 9.13 or Table 5-7, Chapter 5.

REFERENCES

9.1 American Society of Heating, Refrigerating and Air-Conditioning Engineers: *Fundamentals*, 1985.

9.2 Air Moving and Control Association, Inc.: AMCA Publication 203-81, *Field Performance Measurements,* 30 West University Dr., Arlington Heights, IL 60004.

9.3 A.D. Brandt: *Industrial Health Engineering*, John Wiley and Sons, New York, 1947.

9.4 Air Movement and Control Association, Inc.: *AMCA Standard 210-86: Test Code for Air Moving Devices*, 30 West University Dr., Arlington Heights, IL 60004.

9.5 J.P. Farant, D.L. McKinnon and T.A. McKenna: "Tracer Gases as a Ventilation Tool: Methods and Instrumentation," *Ventilation '85 — Proceedings of the first International Symposium on Ventilation for Contaminant Control,* pp. 263-274, October 1-3, 1985, Toronto, Canada.

9.6 *ASME Power Test Codes*, Chapter 4, Flow Measurement, P.T.C. 19.5: 4-1959.

9.7 M.W. First and L. Silverman: "Airfoil Pitometer," *Ind. Engr. Chem 42*: 301-308 (February 1950).

9.8 American Society of Mechanical Engineers: *Fluid Meters — Their Theory and Applications,* 1959.

9.9 G. Hama: "A Calibrating Wind Tunnel for Air Measuring Instruments," *Air Engr. 41*:18-20 (December 1967).

9.10 G., Hama: "Calibration of Alnor Velometers," *Am. Ind. Assoc. J. 19*:477 (December 1958).

9.11 G. Hama and L.S. Curley: "Instrumentation for the Measurement of Low Velocities with a Pitot Tube," *Air Engineering,* July 1967 and *Am. Ind. Hyg. Assoc. J. 28*:204 (May-June 1967).

Chapter 10
SPECIFIC OPERATIONS

The following illustrations of hoods for specific operations are intended as guides for design purposes and apply to usual or typical operations. In most cases they are taken from designs used in actual installations of successful local exhaust ventilation systems. All conditions of operation cannot be categorized and because of special conditions, (i.e., cross drafts, motion, differences in temperature or use of other means of contaminant suppression) modifications may be in order.

Unless it is specifically stated, the design data are not to be applied indiscriminately to materials of high toxicity, i.e., beryllium and radioactive materials. Thus the designer may require higher or lower air flow rates or velocities or other modifications because of the peculiarities of the process in order to adequately control the air contaminant.

Index of Prints

Group	Operation	Print No.	Page No.
1. Foundry	Abrasive Blasting	VS-101	10-4
		VS-101.1	10-5
	Core Grinder	VS-102	10-6
	Melting Furnace		
	Crucible, Non-Tilt	VS-103	10-7
	Electric Rocking	VS-104	10-8
	Electric, Top Electrode	VS-105	10-9
	Tilting	VS-106	10-10
	Mixer and Muller Hood	VS-107	10-11
	Mixer and Muller Ventilation	VS-108	10-12
	Pouring Station	VS-109	10-13
	Shakeout	VS-110	10-14
		VS-111	10-15
		VS-112	10-16
	Tumbling Mills	VS-113	10-17
	Shell Core Molding	VS-114	10-18
	Core Making Machine;		
	Small Roll-over Type	VS-115	10-19
2. High Toxicity Materials	Crucible Furnace	VS-201	10-20
	Dry Box	VS-202	10-21
	Laboratory Hood	VS-203	10-22
	Laboratory Data	VS-204 & 204.1	10-23 & 10-24
	General Use Laboratory Hoods	VS-205	10-25
	Perchloric Acid Hood Data	VS-205.1	10-26
	Work Practices for Laboratory Hoods	VS-205.2	10-27
	Specialized Laboratory Hood Designs	VS-206	10-28
	Lathe	VS-207	10-29
	Metal Shears	VS-208	10-30
	Milling Machine	VS-209	10-31
	Shaft Seal Enclosure	VS-210	10-32
	Sampling Box	VS-211	10-33

Index of Prints

Group	Operation	Print No.	Page No.
3. Material Handling	Bag Filling	VS-301	10-34
	Bag Tube Packer	VS-302	10-35
	Barrel Filling	VS-303	10-36
	Bin and Hopper	VS-304	10-37
	Bucket Elevator	VS-305	10-38
	Conveyor Belt	VS-306	10-39
	Screens	VS-307	10-40
4. Metal Working	Abrasive Cutoff Saw	VS-401	10-41
	Buffing and Polishing		
	Belts		
	Backstand idler	VS-402	10-42
	Metal Polishing	VS-403	10-43
	Wheels		
	Automatic Circular	VS-404	10-44
	Automatic Straight Line	VS-405	10-45
	Manual	VS-406	10-46
	Lathe	VS-407	10-47
	Grinding		
	Disc		
	Horizontal Double-Spindle	VS-408	10-48
	Horizontal Single-Spindle	VS-409	10-49
	Vertical Spindle	VS-410	10-50
	Wheel		
	Grinder Wheel Hood Speeds		
	Below 6500 sfm	VS-411	10-51
	Above 6500 sfm	VS-411.1	10-52
	Portable Hand Grinding	VS-412	10-53
	Portable Grinding Table	VS-413	10-54
	Swing Grinder	VS-414	10-55
	Metal Spraying	VS-415	10-56
	Welding Bench	VS-416	10-57
	Welding Bench	VS-416.1	10-58
	Surface Grinder	VS-417	10-59
	Metal Cutting Bandsaw	VS-418	10-60
5. Open Surface Tanks	Degreasing - Solvent	VS-501	10-61
	Solvent Vapor Degreasing	VS-501.1	10-62
	Dip Tank	VS-502	10-63
	Open Surface Tanks	VS-503	10-64
		VS-503.1	10-65
	Push-Pull Hood Design Data	VS-504	10-66
		VS-504.1	10-67
	Push Nozzle Plenum Pressure	VS-504.2	10-68
	Table Slot	VS-505	10-69
	Open Surface Tank Data		10-70 to 10-77
6. Painting	Auto Spray Booth	VS-601	10-78
	Drying Oven	VS-602	10-79
	Paint Spray Booth		
	Large	VS-603	10-80
	Small	VS-604	10-81
	Trailer Interior Spray Painting	VS-605	10-82
	Large Drive-through Spray Paint Booth	VS-606	10-83
7. Wood Working	Jointer	VS-701	10-84
	Sanders		
	Horizontal Belt	VS-702	10-85
	Vertical Belt	VS-702.1	10-86

Index of Prints

Group	Operation	Print No.	Page No.
	Disc	VS-703	10-87
	Drum-Multiple	VS-704	10-88
	Single	VS-705	10-89
	Saws		
	Band	VS-706	10-90
	Swing	VS-707	10-91
	Table	VS-708	10-92
	Radial	VS-709	10-93
	Miscellaneous Data	Table 10-7.1	10-94
8. Low Volume – High Velocity	Data		10-95 to 10-96
	Cone Wheels	VS-801	10-97
	Cup Wheels & Brushes	VS-802	10-98
	Pneumatic Chisel	VS-803	10-99
	Radial Grinders	VS-804	10-100
	Disc Sander	VS-805	10-101
	Vibratory Sander	VS-806	10-102
	Typical System for Low Volume-High Velocity	VS-807	10-103
9. Miscellaneous	Banbury Mixer	VS-901	10-104
	Calender Rolls	VS-902	10-105
	Rubber Mill Ventilation	VS-902.1	10-106
	Canopy Hood	VS-903	10-107
	Die Casting Hood	VS-904	10-108
	Die Casting Machine	VS-905	10-109
	Melting Pot	VS-906	10-110
	Service Garages		
	Overhead	VS-907	10-111
	Underfloor	VS-908	10-112
	Fuel Powered Lift Truck	VS-908.1	10-113
	Exhaust Requirements for Typical Diesel Engines Under Load	VS-908.2	10-114
	Granite Cutting & Finishing	VS-909	10-115
	Kitchen Range Hoods	VS-910	10-116
	Kitchen Range and Data	VS-911	10-117
	Dishwasher	VS-912	10-118
	Charcoal Broiler and Barbeque	VS-913	10-119
	Pistol Range (indoor)	VS-914	10-120
	Fluidized Beds	VS-915	10-121
	Torch Cutting	VS-916	10-122
	Clean Room Air Flow	VS-917	10-123
	Clean Room Air Flow Data	VS-917.1	10-124
	Clean Air Exhaust Hood	VS-918.1	10-125
	Clean Bench Work Station	VS-918.2	10-126
	Cold Heading Machine Ventilation	VS-919	10-127
	Outboard Motor Test	VS-920	10-128
	Fumigation Booth	VS-921	10-129
	Fumigation Booth Data	VS-921.1	10-130
	Asbestos Fiber Bag Opening	VS-1001	10-131
	Asbestos Fiber Belt Conveying	VS-1002	10-132
	Grain Industry Data	Table 10-9.1	10-133
	Miscellaneous Data	Table 10-9.2	10-134 to 10-135

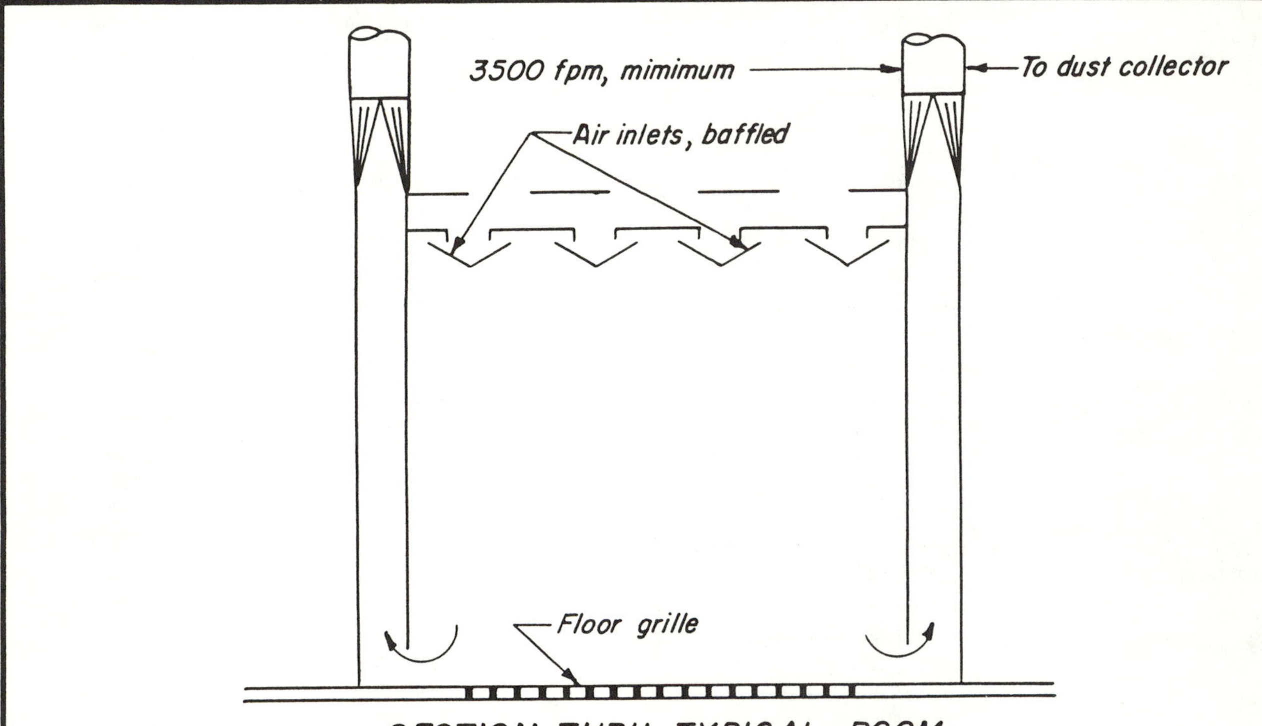

SECTION THRU TYPICAL ROOM

Rooms: 60-100 fpm downdraft; usual choice 80 fpm ; or 100 fpm cross-draft. 25 fpm for steel or iron abrasives. Operator in room requires NIOSH certified respiratory protective equipment.

Note: Above ventilation for operator visibility, not for control of dust hazards.

Rotary tables: 200 cfm/sq ft of total openings(taken without curtains).

Cabinets: 20 air changes per minute.
Atleast 500 fpm inward velocity at all operating openings.
Openings to be baffled.
For details see VS-101.1

Entry loss: 1 VP, or calculate from individual losses.

AMERICAN CONFERENCE OF GOVERNMENTAL INDUSTRIAL HYGIENISTS	
ABRASIVE BLASTING VENTILATION	
DATE 1-84	VS-101

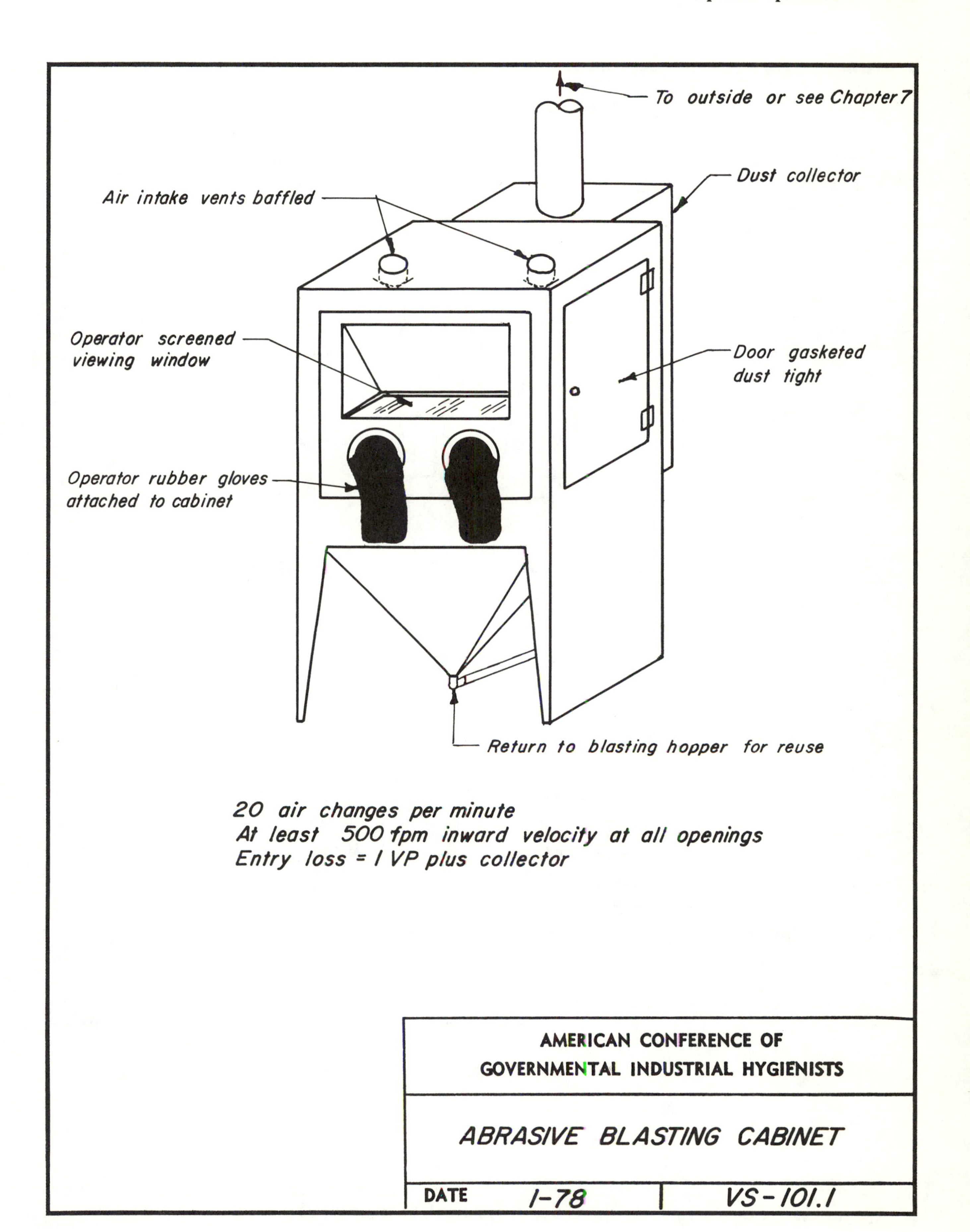
To outside or see Chapter 7
Dust collector
Air intake vents baffled
Operator screened viewing window
Door gasketed dust tight
Operator rubber gloves attached to cabinet
Return to blasting hopper for reuse
20 air changes per minute
At least 500 fpm inward velocity at all openings
Entry loss = 1 VP plus collector
AMERICAN CONFERENCE OF GOVERNMENTAL INDUSTRIAL HYGIENISTS
ABRASIVE BLASTING CABINET
DATE 1-78
VS-101.1

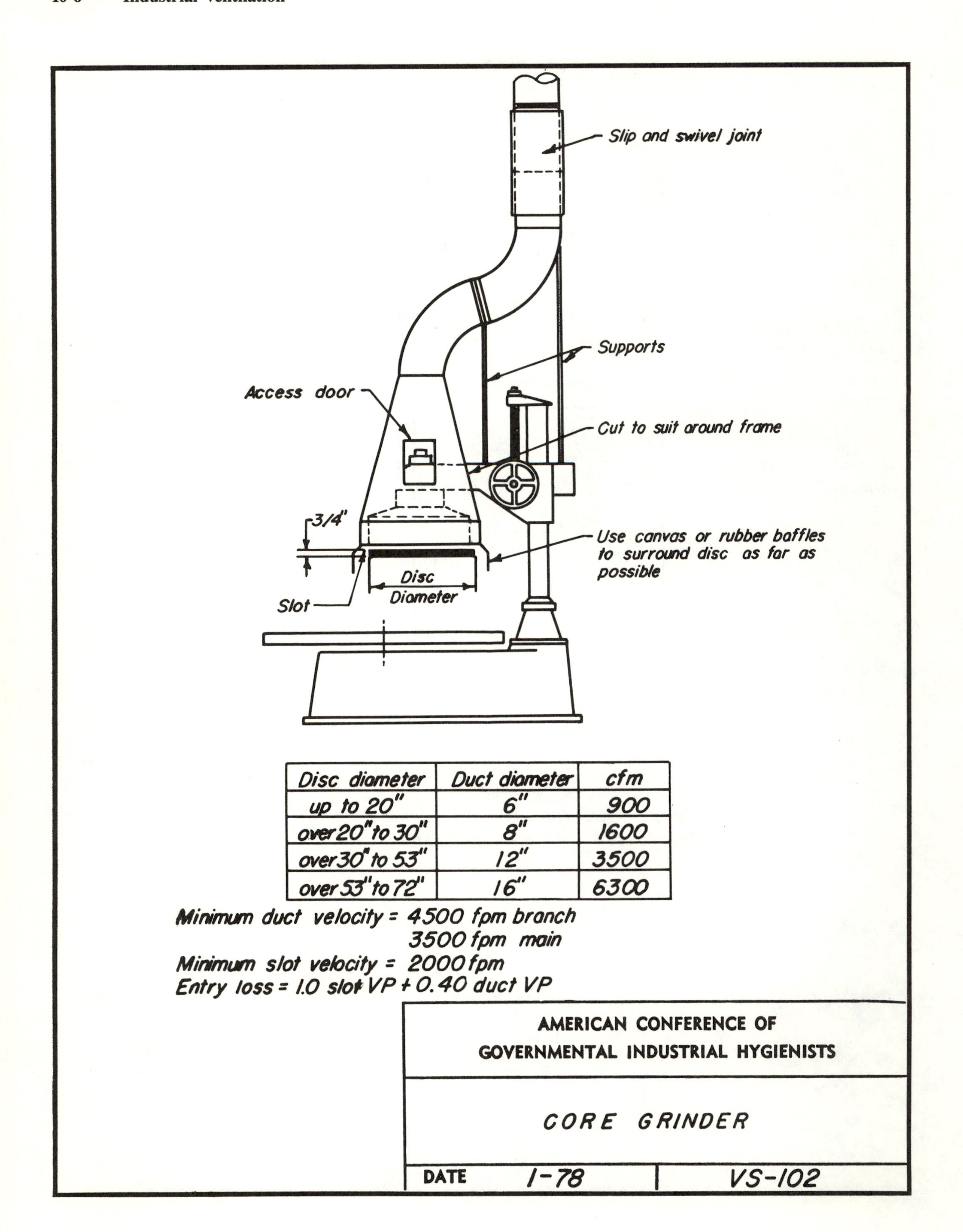

Disc diameter	Duct diameter	cfm
up to 20"	6"	900
over 20" to 30"	8"	1600
over 30" to 53"	12"	3500
over 53" to 72"	16"	6300

Minimum duct velocity = 4500 fpm branch
3500 fpm main
Minimum slot velocity = 2000 fpm
Entry loss = 1.0 slot VP + 0.40 duct VP

AMERICAN CONFERENCE OF GOVERNMENTAL INDUSTRIAL HYGIENISTS

CORE GRINDER

DATE 1-78 | VS-102

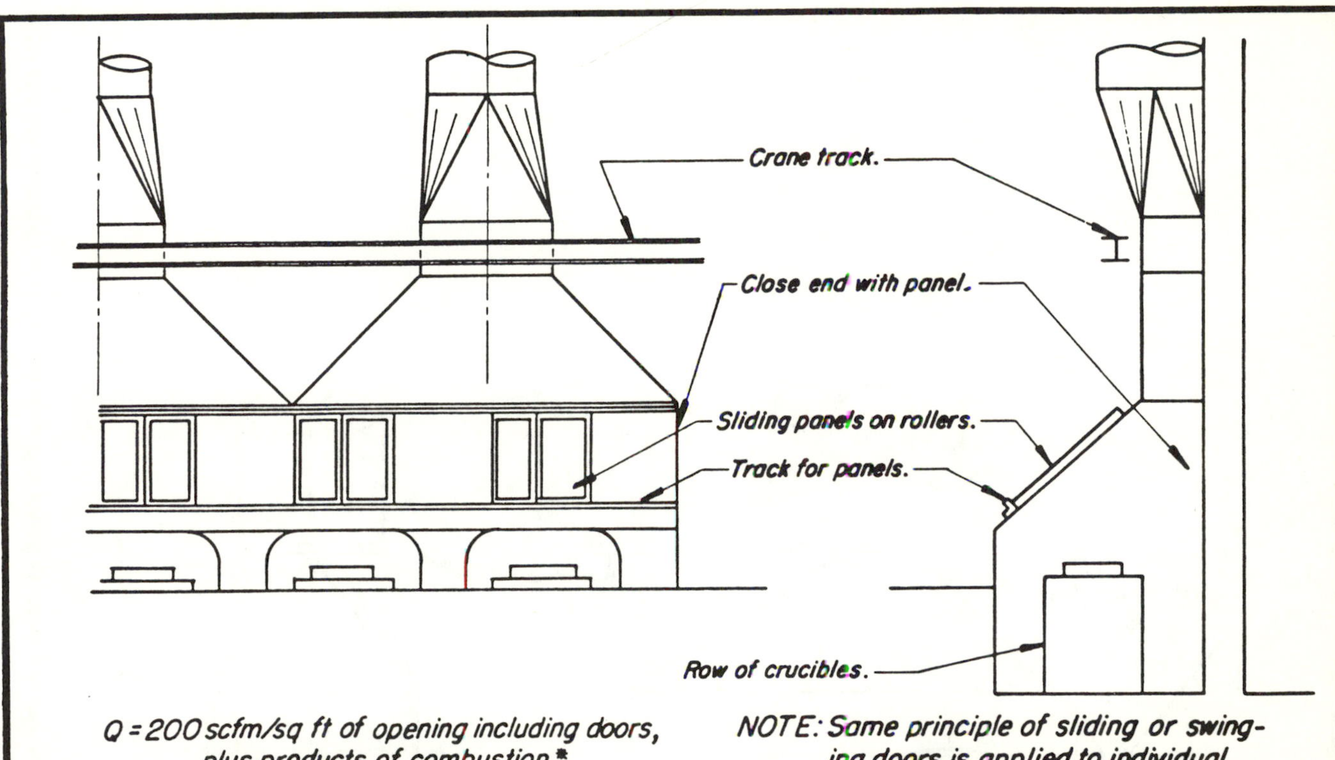

Q = 200 scfm/sq ft of opening including doors, plus products of combustion.*
Entry loss = 0.5 duct VP
Duct velocity = 1000 - 3500 fpm **
* Correct for temperature.
** For horizontal runs, transport velocity is necessary.

NOTE: Same principle of sliding or swinging doors is applied to individual furnace enclosures.

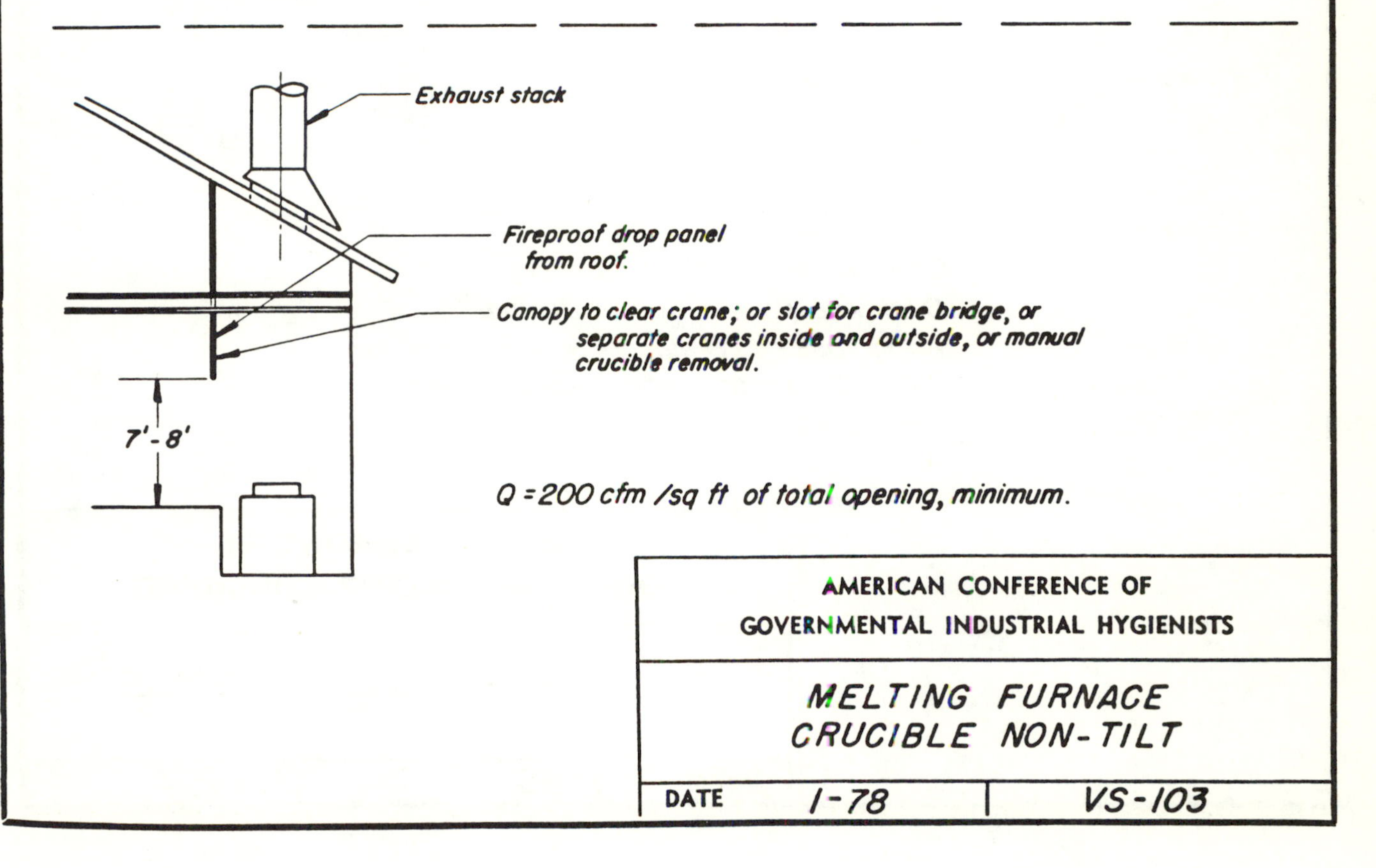

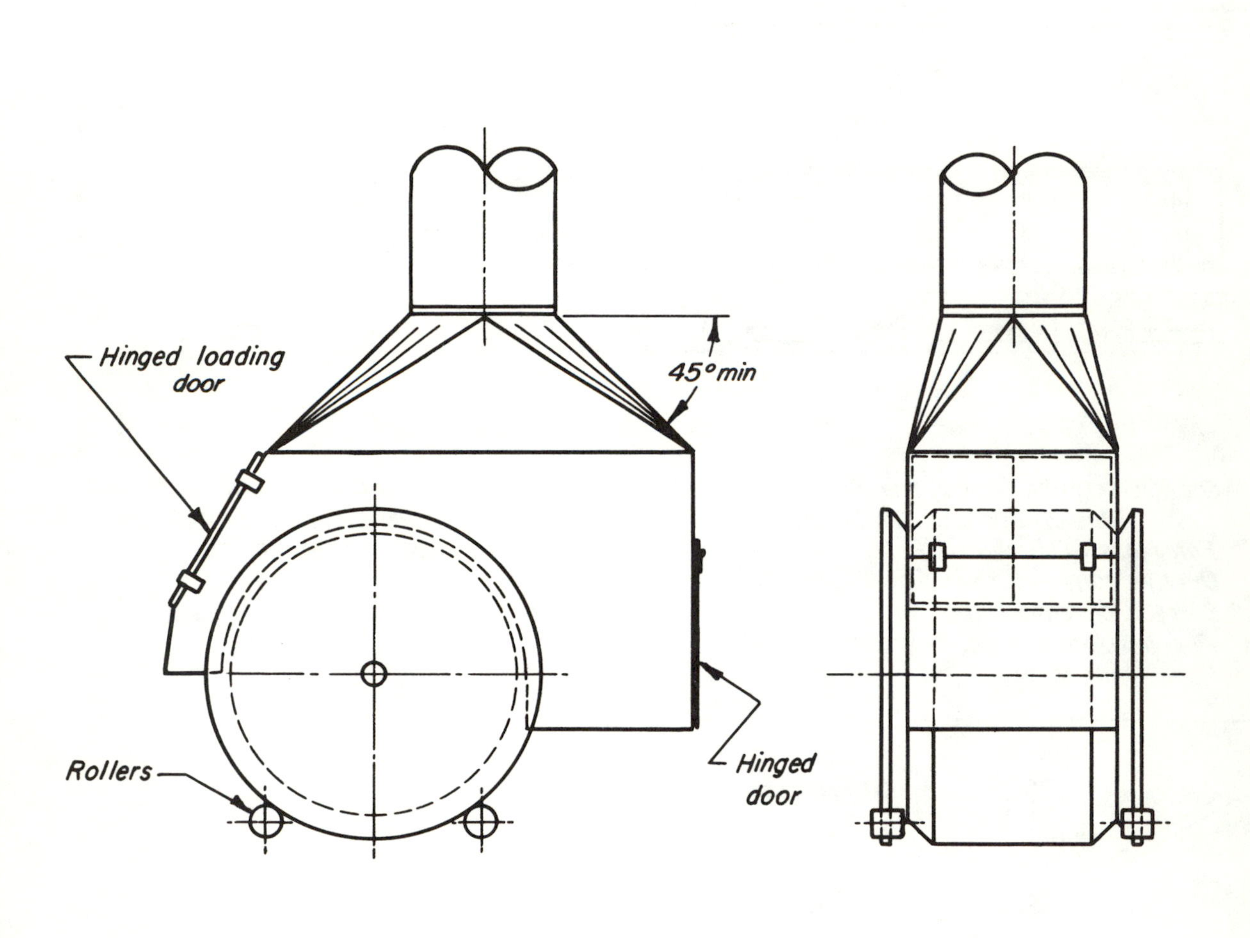

Q= 400 cfm/sq ft of opening
Duct velocity = 1000-3500 fpm*
Entry loss= 1.78 slot VP + 0.25 duct VP
* For horizontal runs, transport velocity is necessary

AMERICAN CONFERENCE OF GOVERNMENTAL INDUSTRIAL HYGIENISTS

ELECTRIC ROCKING FURNACE

DATE	1-80	VS-104

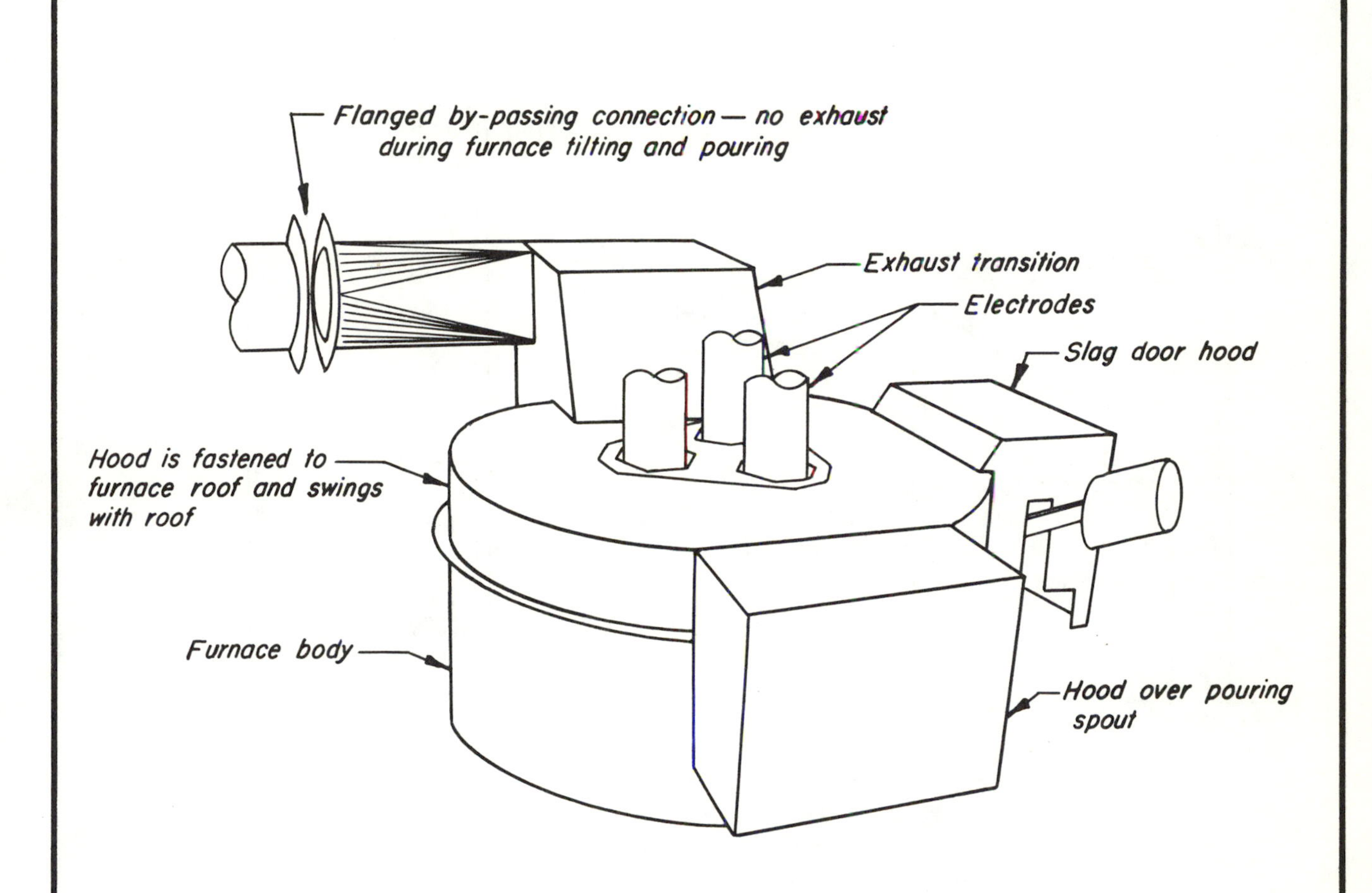

For Q, SP and operating temperature, consult manufacturers
Approximate exhaust volume = 2500 scfm/ton of charge[1][22][23]

Alternate designs:
1. *Other exhaust designs utilize direct furnace roof tap. For details consult manufacturers.*
2. *Canopy hood exhaust can be utilized but requires large exhaust air volumes; Q = 200 scfm/sq ft of open area between furnace and lower edge of canopy.*

AMERICAN CONFERENCE OF GOVERNMENTAL INDUSTRIAL HYGIENISTS

HOOD FOR TOP ELECTRODE MELTING FURNACE

DATE 1-78	VS-105

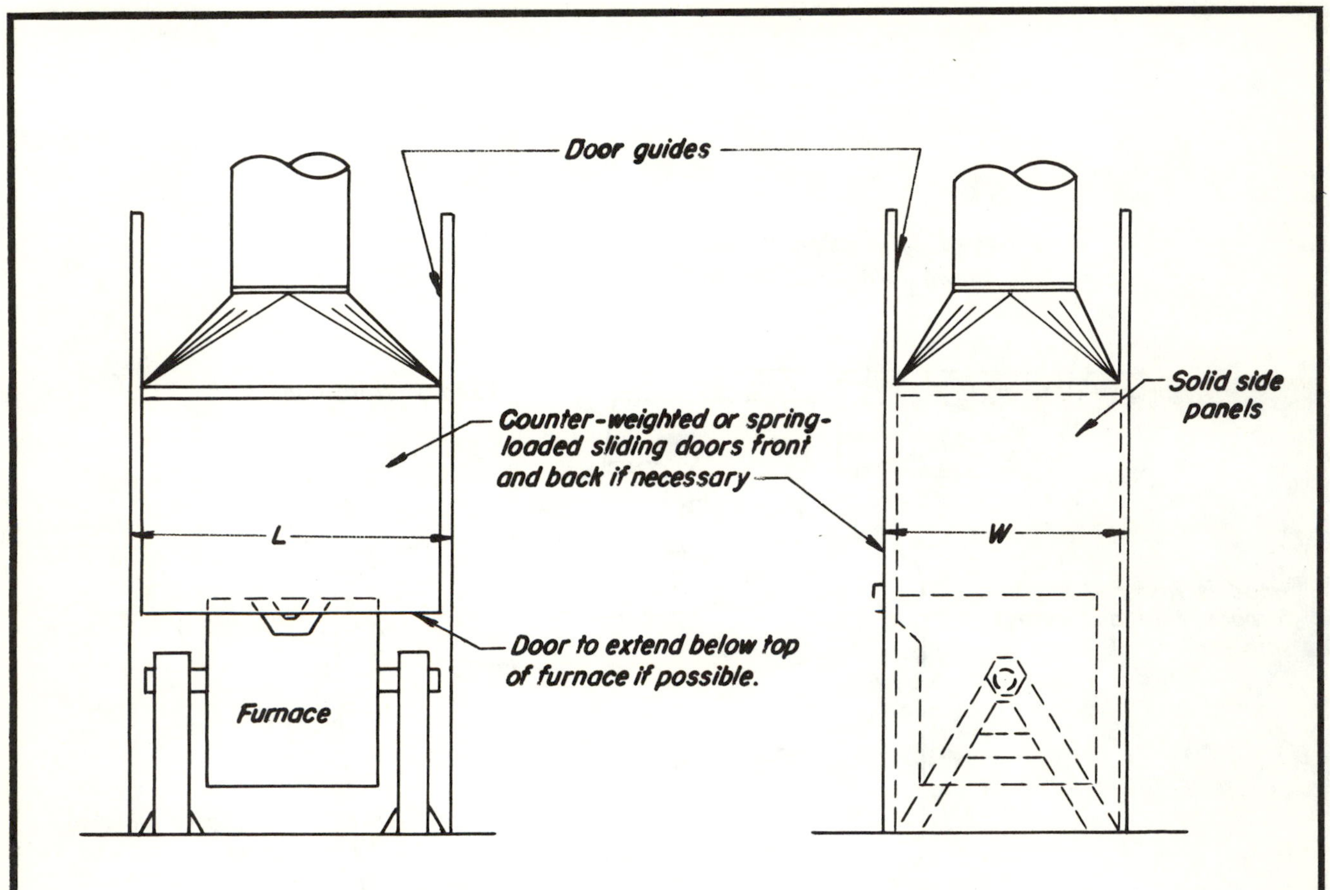

*Q = 200 LW; but not less than 200 scfm/sq ft of all openings with doors open.**

Entry loss = 0.25 VP

*Duct velocity = 1000 - 3500 fpm***

**Correct for temperature and combustion products.*

***For horizontal runs, transport velocity is necessary.*

AMERICAN CONFERENCE OF
GOVERNMENTAL INDUSTRIAL HYGIENISTS

MELTING FURNACE - TILTING

DATE 1-64	VS-106

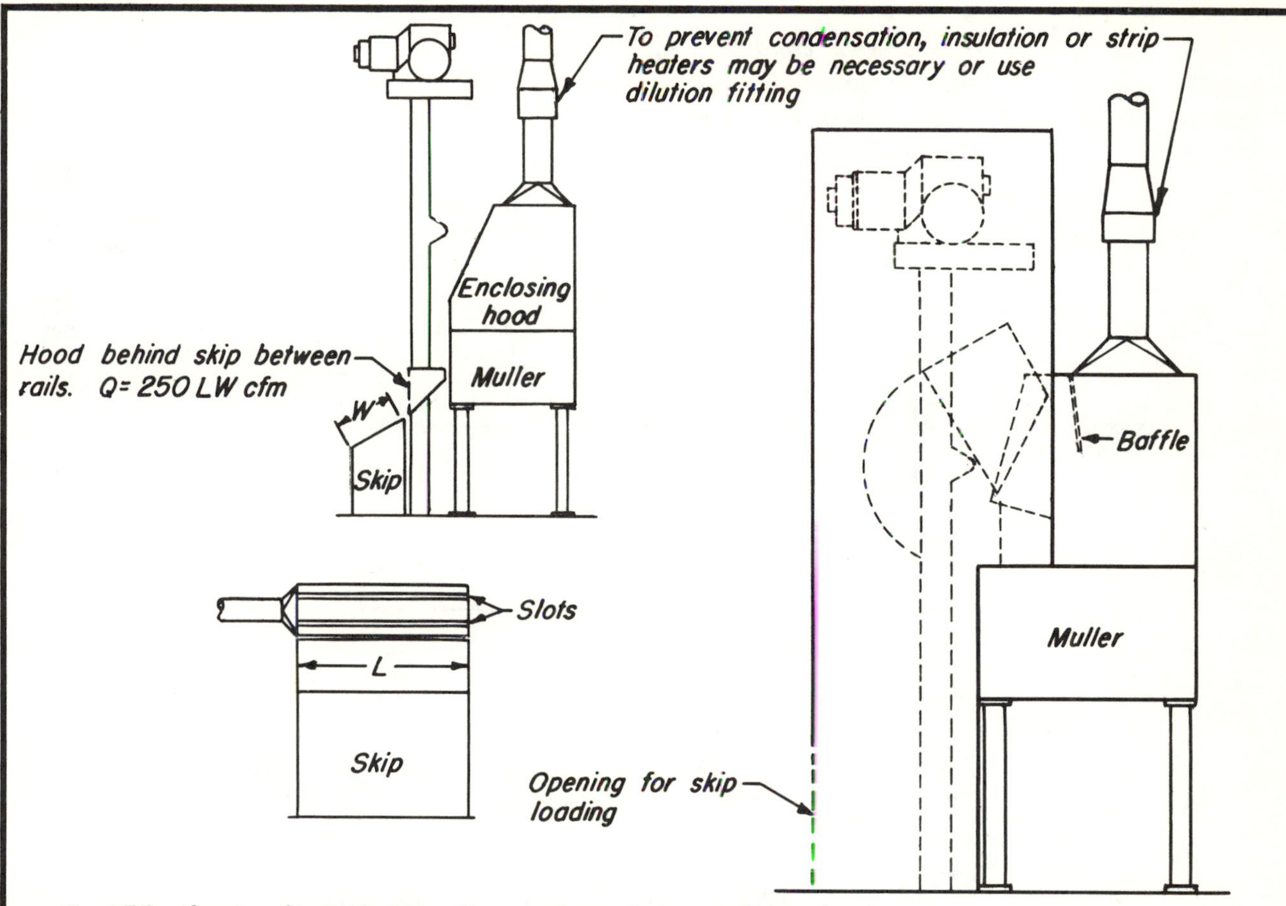

Q= 150 cfm/sq ft through all openings but not less than:

Mixer diam, feet	Exhaust, cfm
4	750
6	900
7	1050
8	1200
10	1575

For Cooling Mullers, See VS-108

Other types of mixers: enclose as much as possible and provide 150 cfm/sq ft of remaining openings

When flammable solvents are used in mixer, calculate minimum exhaust volume for dilution to 25% of the L E L See Chapter 2

Duct velocity = 3500 fpm, min

Entry loss = 0.25 VP

AMERICAN CONFERENCE OF GOVERNMENTAL INDUSTRIAL HYGIENISTS

MIXER AND MULLER HOOD

DATE	1-72	VS-107

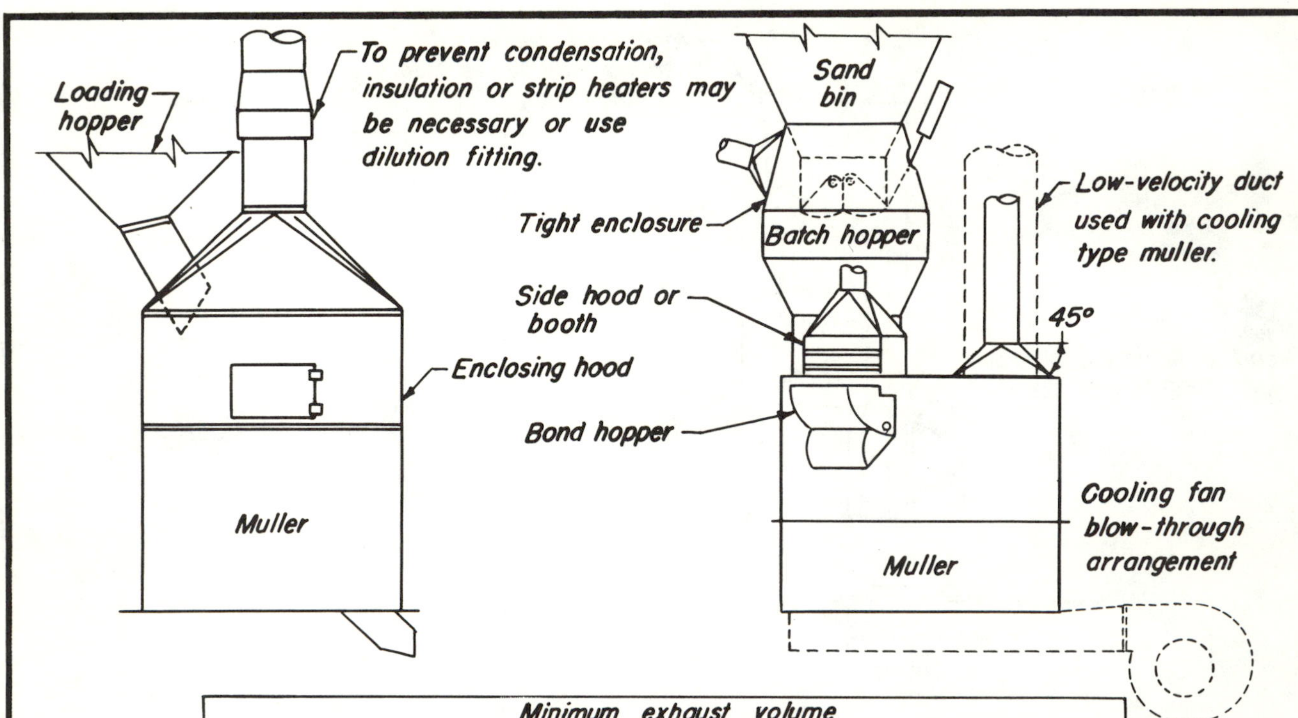

Location	Minimum exhaust volume		
	Muller type		
	No cooling	Blow-thru cooling	Draw-thru cooling
Batch hopper	Note 1	600	Note 1
Bond hopper	600	600	600
Muller:	Note 2	Note 3	Note 3
4' diameter	750	"	"
6' diameter	900	"	"
7' diameter	1050	"	"
8' diameter	1200	"	"
10' diameter	1575	"	"

Duct velocity = 4500 fpm minimum
Entry loss = 0.25 VP

Notes:

1. Batch hopper requires separate exhaust with blow-thru cooling. With other fan arrangement, (muller under suction) separate exhaust may not be required. (If skip hoist is used, see VS-107)
2. Maintain 150 fpm velocity through all openings in muller hood. Exhaust volume shown are the minimum to be used.
3. Cooling mullers do not require exhaust if maintained in dust tight condition. Blow-thru fan must be off during loading. If muller is not dust tight, exhaust as in note 2 plus cooling air volume.
4. When flammable solvents are used in mixer, calculate minimum exhaust volume for dilution to 25% of the LEL See Chapter 2

AMERICAN CONFERENCE OF GOVERNMENTAL INDUSTRIAL HYGIENISTS

MIXER AND MULLER VENTILATION

DATE 1-66 | VS-108

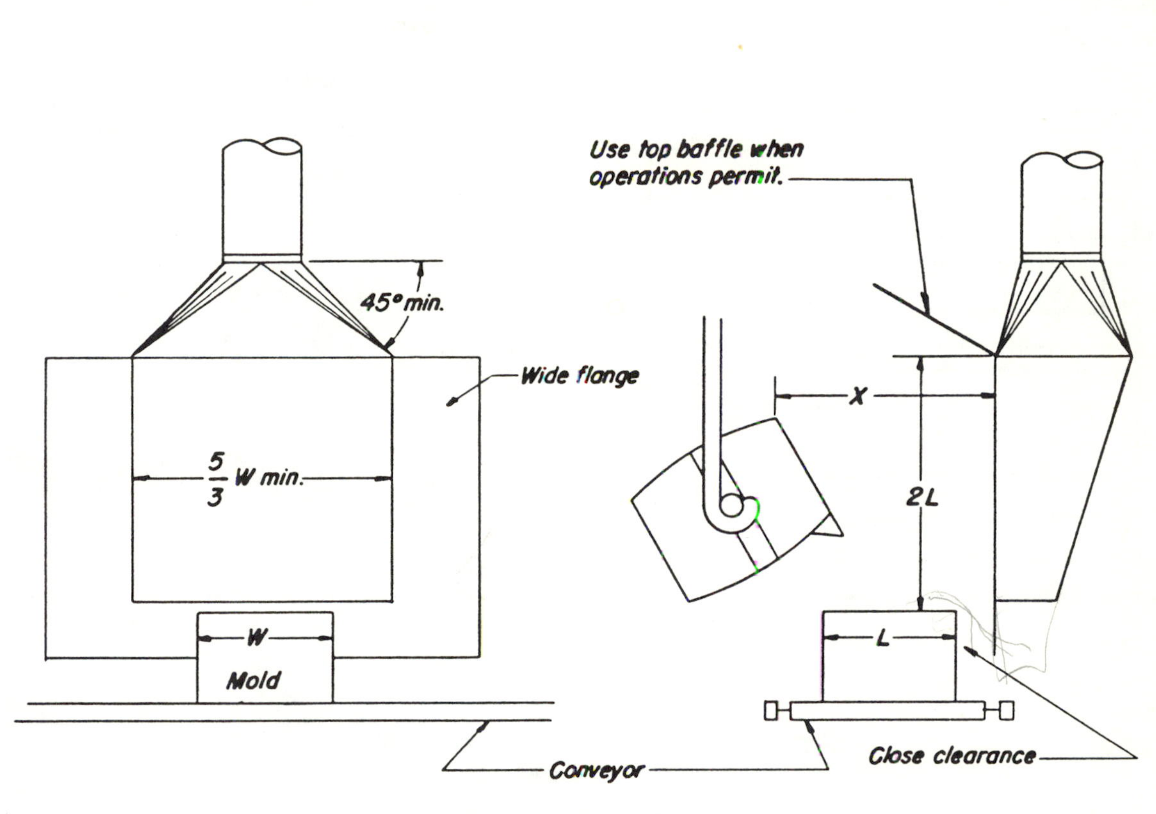

SMALL MOLDS

Unflanged hood: $Q = 200\,(10X^2 + \text{hood area})$.

Flanged hood, reduce Q 25%

Duct velocity = 2000 fpm

Entry loss = 0.25 VP (For slots, 1.78 slot VP + 0.25 duct VP).

PARTIAL SIDE ENCLOSURE

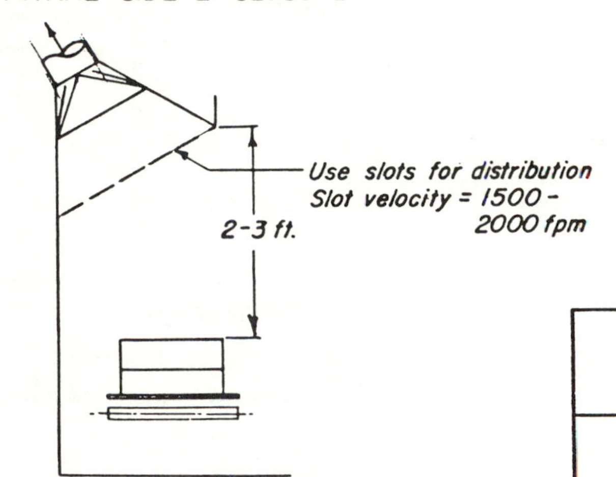

NOTE:

For large molds and ladles provide large side-draft hood similar to shakeout.

Q = 400 cfm/sq ft working area.

Q = 200 - 300 cfm/lin ft of hood.

AMERICAN CONFERENCE OF GOVERNMENTAL INDUSTRIAL HYGIENISTS

POURING STATION

DATE	1-64	VS-109

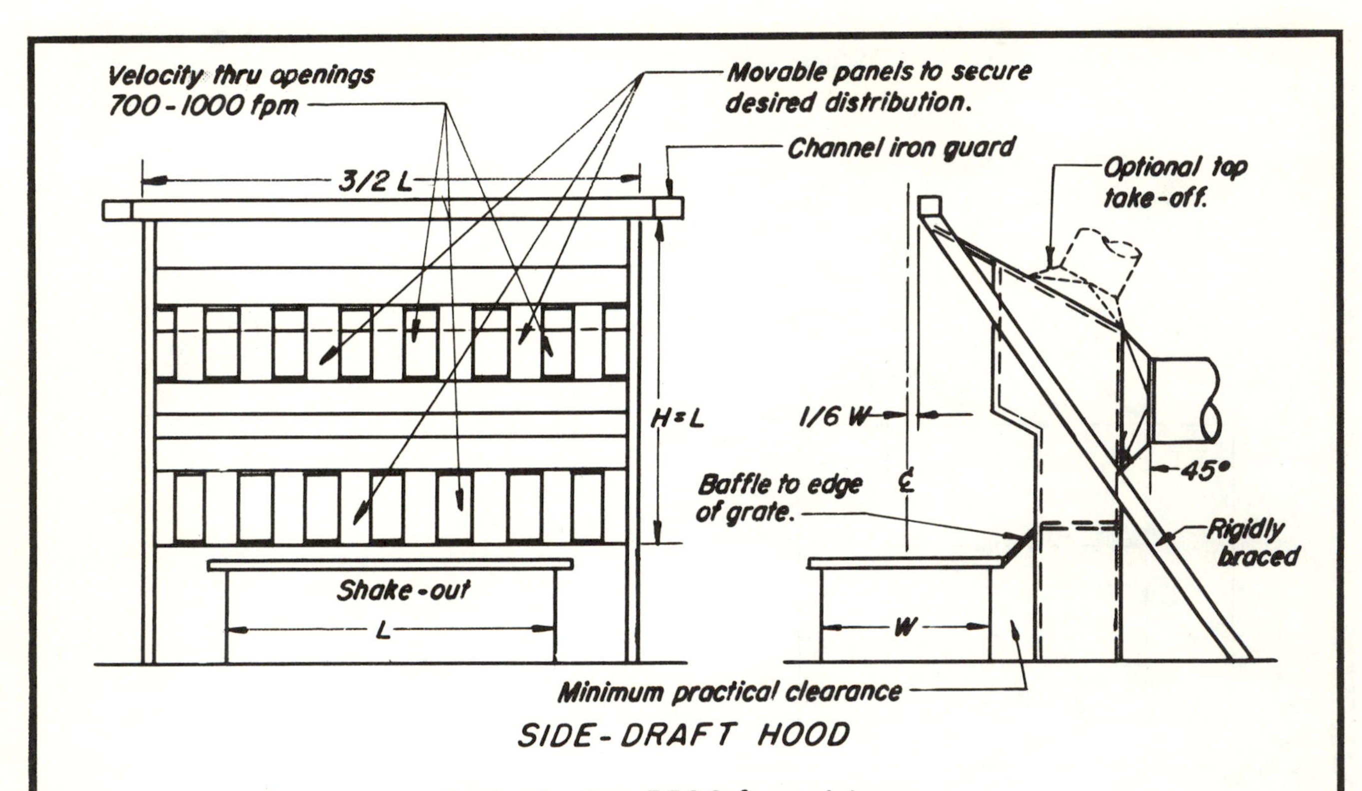

SIDE-DRAFT HOOD

Duct velocity = 3500 fpm minimum.
Entry loss = 1.78 slot VP + 0.25 duct VP

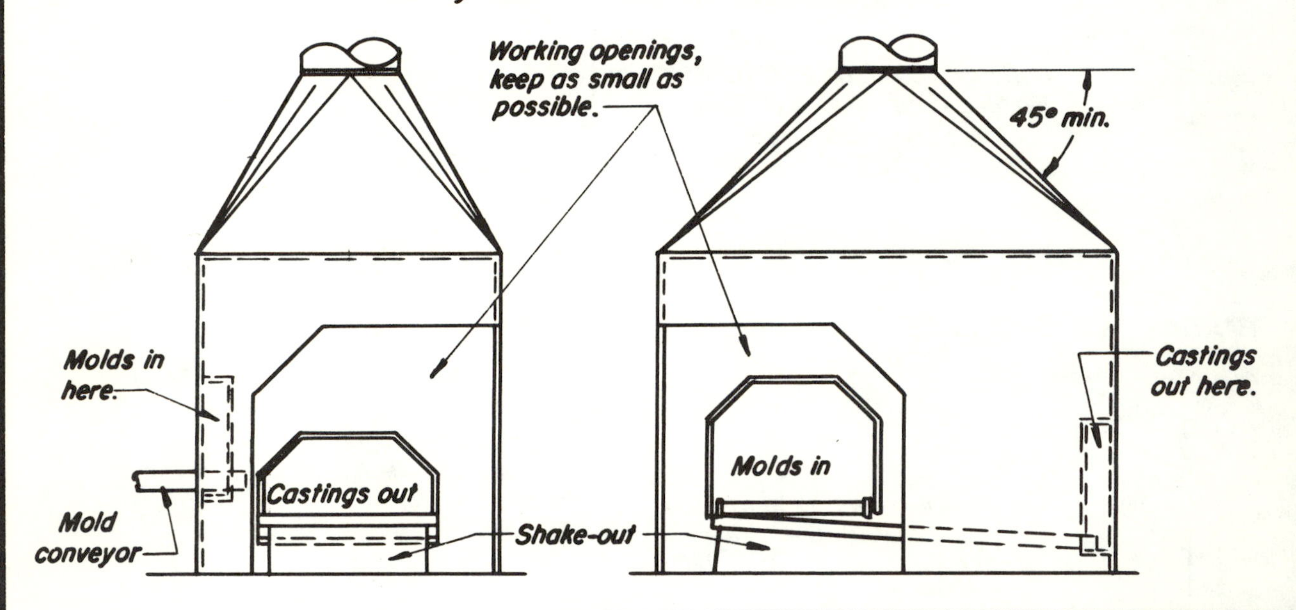

ENCLOSING HOOD

Provides best control with least volume.
Duct velocity = 3500 fpm minimum.
Entry loss = 0.25 VP

See VS-112

AMERICAN CONFERENCE OF
GOVERNMENTAL INDUSTRIAL HYGIENISTS

FOUNDRY SHAKEOUT

DATE 1-64 | VS-110

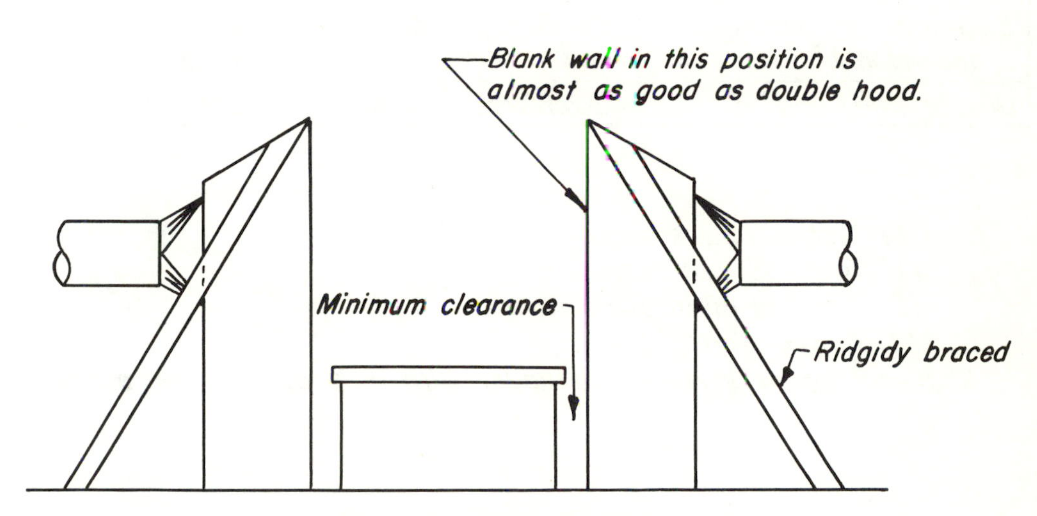

DOUBLE SIDE-DRAFT

Proportions same as single side-draft hood except for overhang.

Slot sized for 1500-2000 fpm

Duct velocity = 4000 fpm minimum

Entry loss = 1.78 slot VP plus fittings

See VS-112

AMERICAN CONFERENCE OF GOVERNMENTAL INDUSTRIAL HYGIENISTS

FOUNDRY SHAKEOUT

DATE 1-82 | VS-111

Shakeout exhaust, minimum*

Type of hood	Hot castings	Cool castings
Enclosing**	200 cfm/sq ft opening At least 200 cfm/sq ft grate area	200 cfm/sq ft opening At least 150 cfm/sq ft grate area
Enclosed two sides and 1/3 top area**	300 cfm/sq ft grate area	275 cfm/sq ft grate area
Side hood (as shown or equivalent)**	400-500 cfm/sq ft grate area	350-400 cfm/sq ft grate area
Double side hood**	400 cfm/sq ft grate area	300 cfm/sq ft grate area

*Choose higher values when
(1) Castings are quite hot
(2) Sand to metal ratio is low
(3) Cross-drafts are high
**Shakeout hoppers require exhaust with 10% of the total exhaust volume.

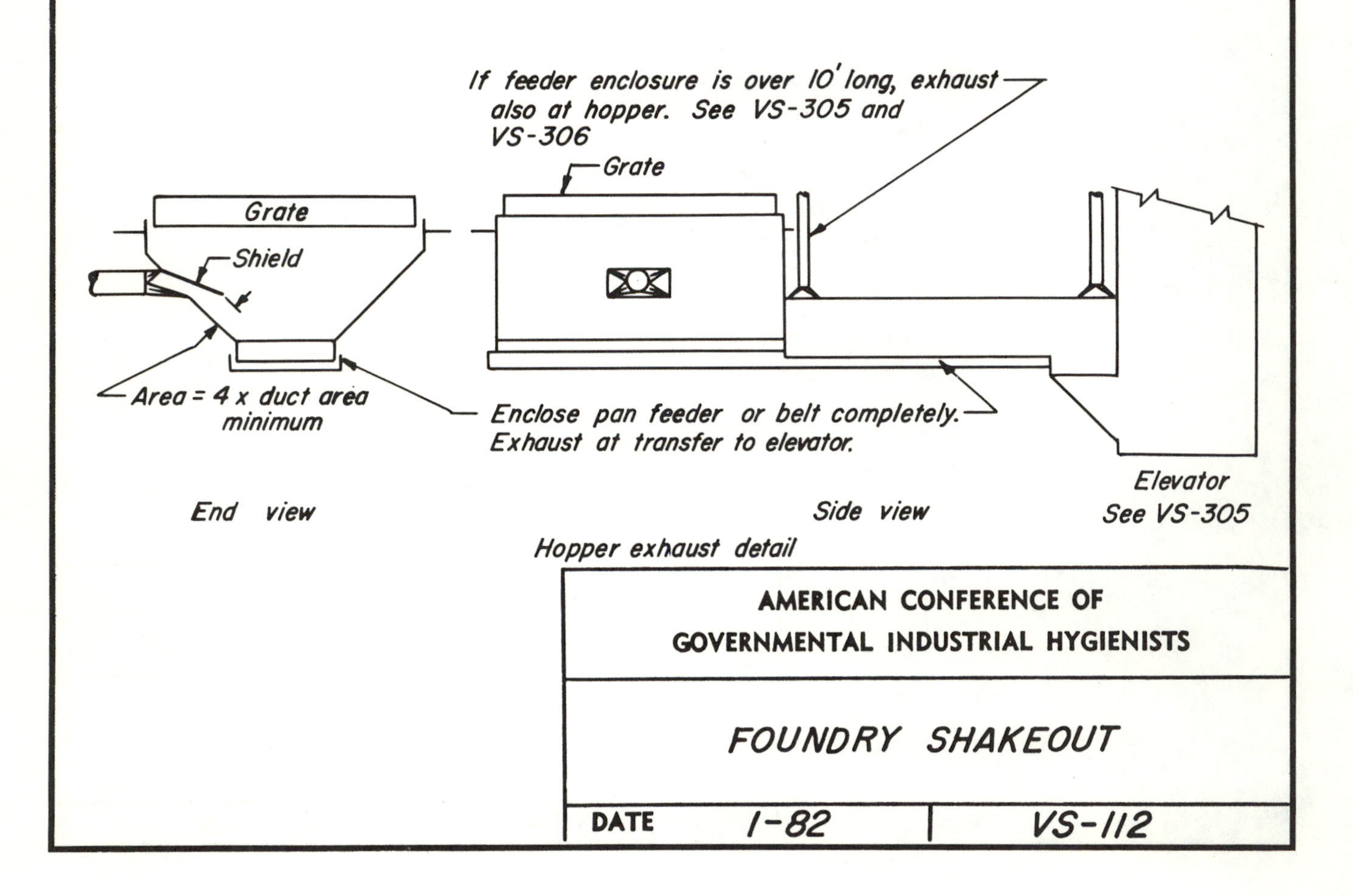

AMERICAN CONFERENCE OF GOVERNMENTAL INDUSTRIAL HYGIENISTS

FOUNDRY SHAKEOUT

DATE 1-82 | VS-112

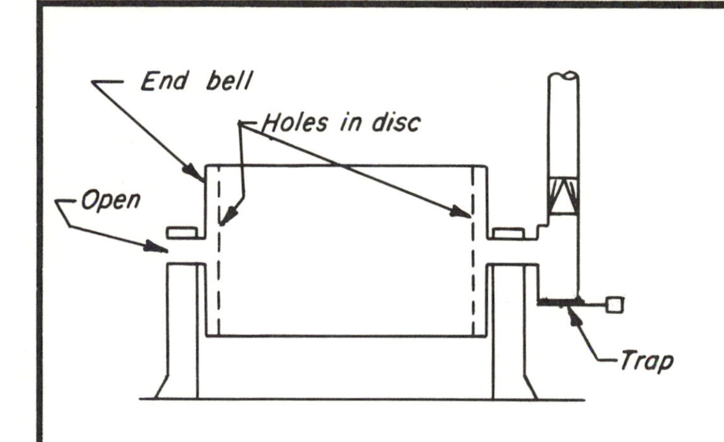

SECTION THRU HOLLOW TRUNNION TUMBLER

Duct velocity = 5000 fpm
Entry loss = 3.25"- 8.25" H_2O (depends on design*)

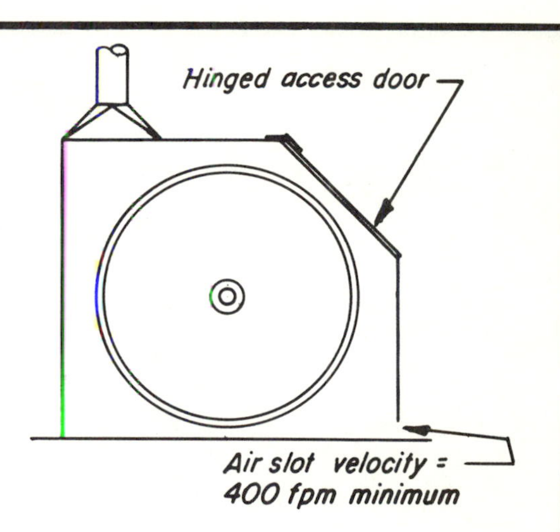

STAVE MILL (END SECTION)

Duct velocity = 3500 fpm minimum
Entry loss varies with take-off 0.25 - 0.50 VP

EXHAUST VOLUMES

Square mill side diam in.	Round mill I. D. in inches	cfm **	
		Trunnion	Stave
	Up to 24 incl.	430	800
Up to 24 incl.	24 - 30	680	900
25 - 30	31 - 36	980	980
31 - 36	37 - 42	1330	1330
37 - 42	43 - 48	1750	1750
43 - 48	49 - 54	2200	2200
49 - 54	55 - 60	2730	2730
55 - 60	61 - 66	3300	3300
61 - 66	67 - 72	3920	3920
67 - 72		4600	4600

* Low-loss designs have large air inlet openings in end bell. Holes in end discs are sized for velocities of 1250-1800 fpm
** For lengths over 70", increase cfm proportionately

AMERICAN CONFERENCE OF GOVERNMENTAL INDUSTRIAL HYGIENISTS

TUMBLING MILLS

DATE 1-64 | VS-113

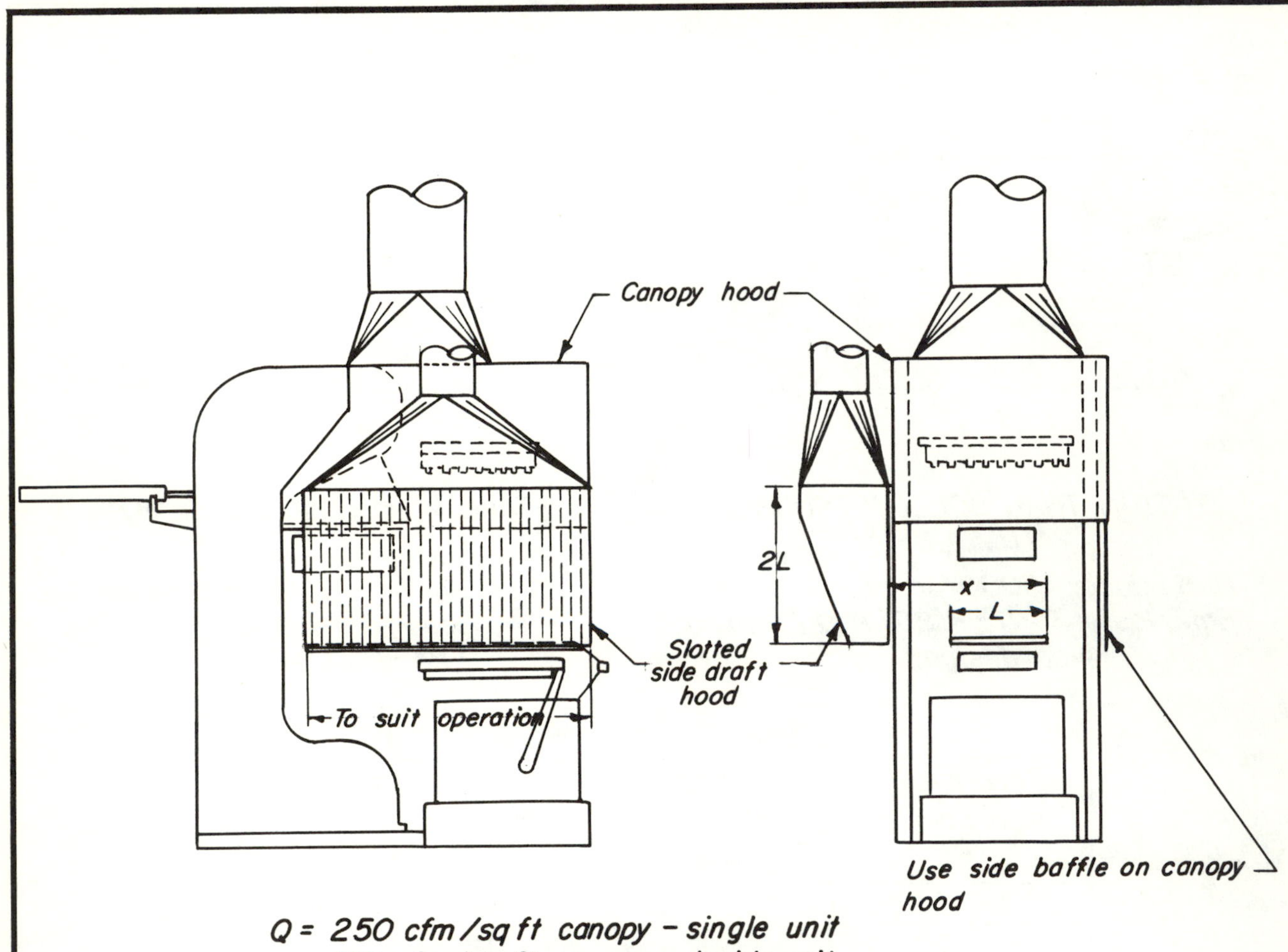

Q = 250 cfm/sq ft canopy - single unit
150 cfm/sq ft canopy - double unit
Entry loss = 0.25 VP for tapered take-off

Slotted side draft hoods required to remove smoke as hot cores emerge from machine.
Capture velocity = 75 fpm minimum
$Q = 75(10x^2 + \text{hood area})$
Entry loss = 1.78 slot VP + 0.25 duct VP

Conveyor or cooling area require ventilation for large cores. Scrap conveyor or tote boxes may require ventilation also.

AMERICAN CONFERENCE OF GOVERNMENTAL INDUSTRIAL HYGIENISTS

SHELL CORE MOLDING

DATE 1-72 | VS-114

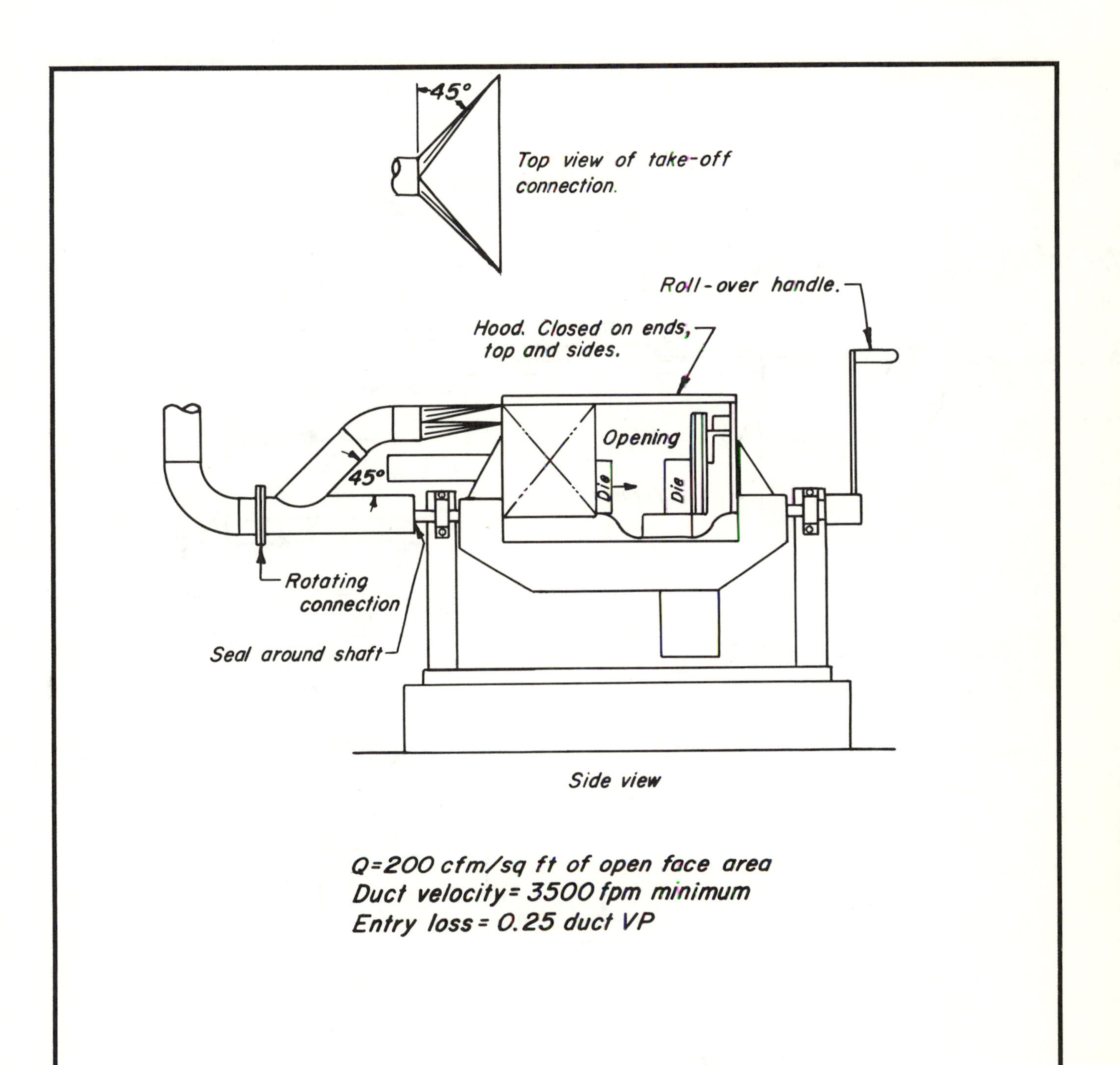

Q=200 cfm/sq ft of open face area
Duct velocity=3500 fpm minimum
Entry loss=0.25 duct VP

AMERICAN CONFERENCE OF GOVERNMENTAL INDUSTRIAL HYGIENISTS

CORE MAKING MACHINE SMALL ROLL OVER TYPE

DATE 1-70 | VS-115

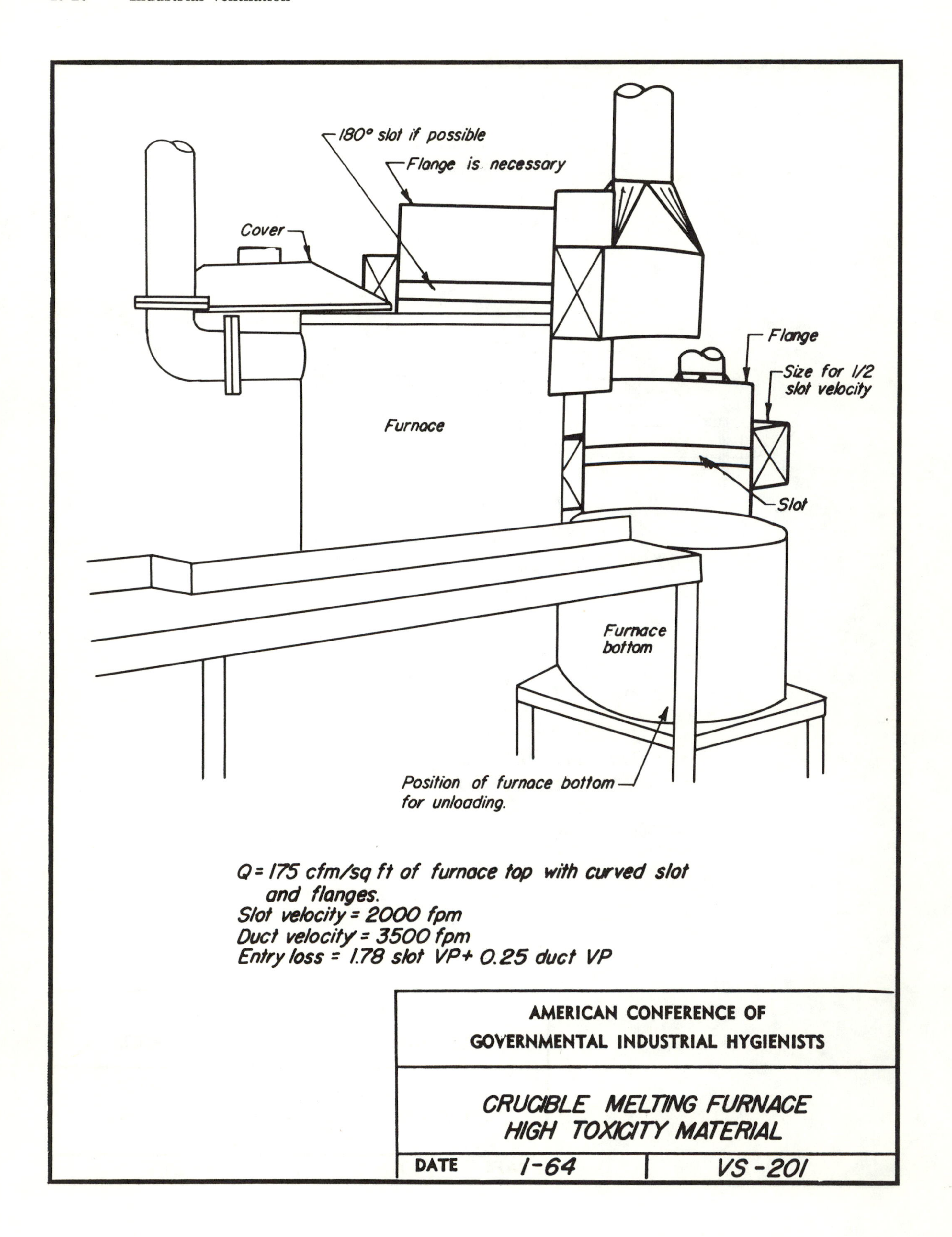
180° slot if possible
Flange is necessary
Cover
Furnace
Flange
Size for 1/2 slot velocity
Slot
Furnace bottom
Position of furnace bottom for unloading.
Q = 175 cfm/sq ft of furnace top with curved slot and flanges.
Slot velocity = 2000 fpm
Duct velocity = 3500 fpm
Entry loss = 1.78 slot VP + 0.25 duct VP
AMERICAN CONFERENCE OF GOVERNMENTAL INDUSTRIAL HYGIENISTS
CRUCIBLE MELTING FURNACE HIGH TOXICITY MATERIAL
DATE 1-64
VS-201

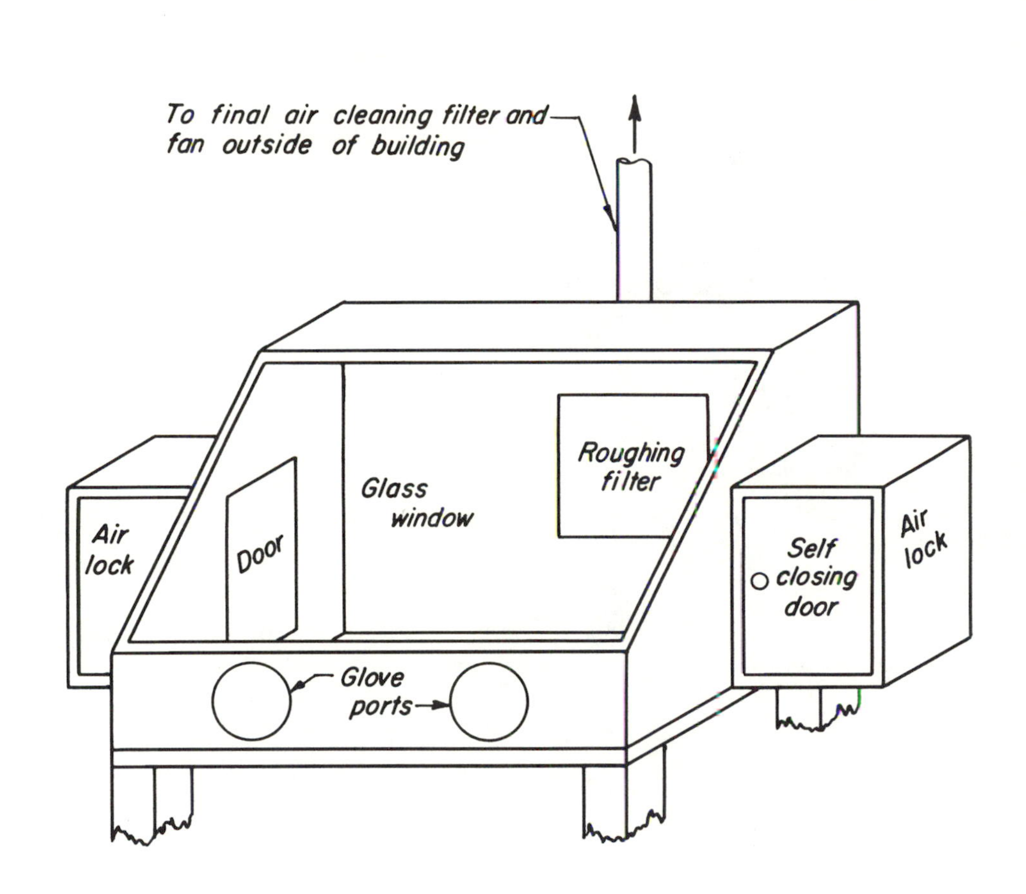

Q = 50 cfm/sq ft of open door area and 0.25" SP
on a closed system.
Entry loss = 0.50 VP
Duct velocity = 2000 - 4000 fpm
Filters: 1. Inlet air filters in doors.
2. Roughing filter at exhaust connection to hood.
3. Final air cleaning filter.
All facilities totally enclosed in hood. Exterior controls may be advisable.
Arm length rubber gloves are sealed to glove port rings.
Strippable plastic on interior and air cleaner on exhaust outlet may be used to facilitate decontamination of the system.
Filter units may be installed in the doors to allow the air flow necessary for burners etc.
For filters, see Chapter 4

AMERICAN CONFERENCE OF GOVERNMENTAL INDUSTRIAL HYGIENISTS	
DRY BOX OR GLOVE HOOD FOR HIGH TOXICITY & RADIOACTIVE MATERIALS	
DATE 1-66	VS-202

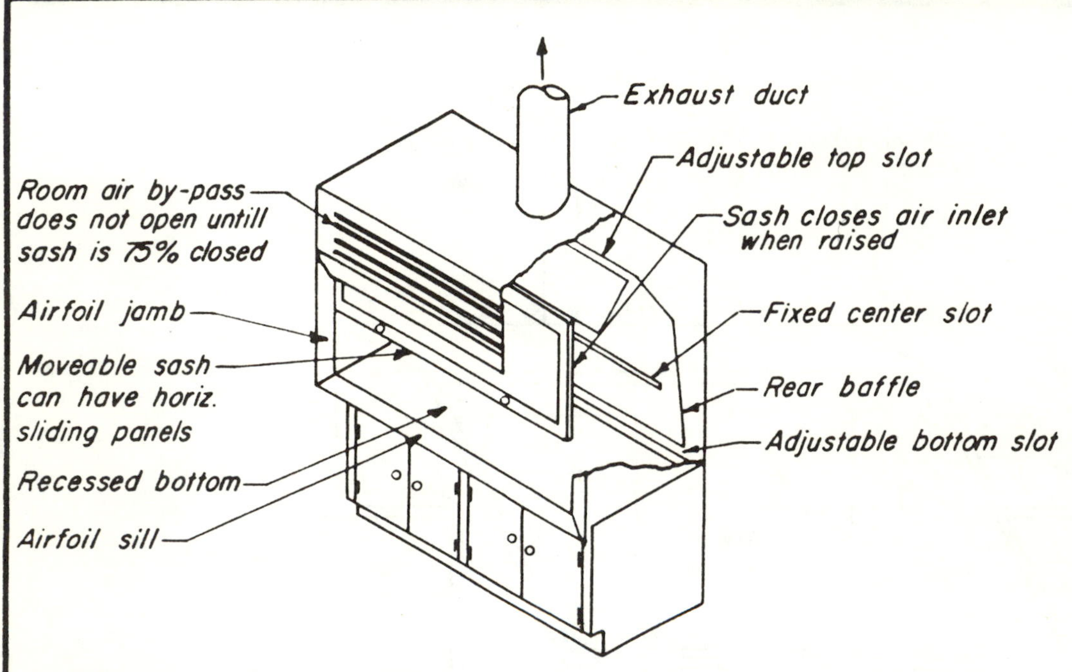

VERTICAL SASH AIRFOIL HOOD

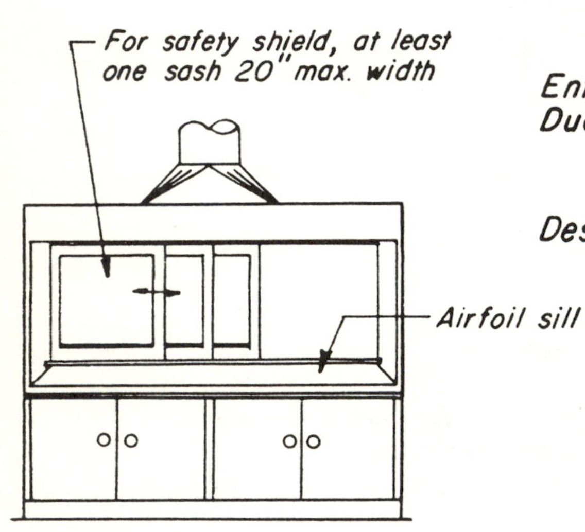

HORIZONTAL SASH AIRFOIL HOOD

Q = 60 - 150 cfm/sq ft full open area depending on quality of supply air distribution
Entry loss = 0.5 VP
Duct velocity = 1000-2000 fpm to suit conditions

Design specifications:

General use laboratory hoods — See VS-205

Perchloric acid - See VS-205.1

"Auxiliary Air" or "Compensating" hoods furnish some make-up air at hood face, design varies with vendor - See VS-204.1

AMERICAN CONFERENCE OF GOVERNMENTAL INDUSTRIAL HYGIENISTS

LABORATORY HOOD

DATE 1-84	VS-203

SUPPLY AIR DISTRIBUTION:

For typical operations at a laboratory fume hood, the worker stands at the face of the hood and manipulates the apparatus in the hood. The indraft at the hood face creates eddy currents around the worker's body which can drag contaminants in the hood back to the body and up to the breathing zone. The higher the face velocity, the greater the eddy currents. For this reason, higher face velocities do not result in as much greater protection as might be supposed.

Room air currents have a large effect on the performance of the hood. Thus the design of the room air supply distribution system is as important in securing good hood performance as is the face velocity of the hood. ASHRAE research project RP-70 results, reported by Caplan and Knutson (Ref 136), concludes in part:

1. Lower breathing zone concentrations can be attained at 50 cfm/sq.ft. face velocities with good air supply distribution than at 150 cfm/sq.ft. with poor air distribution. With a good air supply system, and tracer gas released at 8 liters per minute inside the hood, breathing zone concentrations can be kept below 0.1 ppm and usually below 0.01 ppm.

2. The terminal throw velocity of supply air jets should be no more than 1/2 to 2/3 the hood face velocity; such terminal throw velocities are far less than conventional practice.

3. Perforated ceiling panels provide a better supply system than grilles or ceiling diffusers in that the system design criteria are simplier and easier to apply, and precise adjustment of the fixtures is not required.

For the reasons described, an increased hood face velocity may be self-defeating because the increased air volume handled through the room makes the low-velocity distribution of supply air more difficult.

SELECTION OF HOOD FACE VELOCITY:

The interaction of supply air distribution and hood face velocity makes any blanket specification of hood face velocity inappropriate. Higher hood face velocities will be wasteful of energy and may provide no better or even poorer worker protection. The performance test developed by Caplan and Knutson may be used as a specification. The specified performance should be required of both the hood manufacturer and the designer of the room air supply system.

The specification takes the form xx AU YYY

where:

xx = tracer release rate in hood using the specified diffuser apparatus. Rates are as follows:

- 1 liters/minute approximates pouring volatile solvents back and forth from one beaker to another.
- 4 liters/minute is an intermediate rate between 1 lpm and 8 lpm.
- 8 liters/minute approximates violently boiling water on a 500 watt hotplate.
- (other release rates can be specified for special cases).

YYY = control level, ppm, at the breathing zone of the worker.

AU = "as used" in the laboratory. "AM" would indicate "as manufactured" presumably tested in the manufacturer's test room.

AMERICAN CONFERENCE OF GOVERNMENTAL INDUSTRIAL HYGIENISTS

LABORATORY HOOD DATA

DATE 1-82 | VS-204

Any well-designed airfoil hood, properly balanced, can achieve < 0.10 ppm control level when the supply air distribution is good. Therefore, it would seem appropriate that the "AM" requirements would be < 0.10 ppm. The "AU" requirement involves the design of the room supply system and the toxicity of the materials handled in the hood. The AU specification would be tailored to suit the needs of the laboratory room location.

For projected new buildings, it is frequently necessary to estimate the cost of air conditioning early, before the detailed design and equipment specifications are available. For that early estimating, the following guidelines can be used.

	Condition	cfm/ft^2 Open Hood Face
1.	Ceiling panels properly located with average panel face velocity < 40 fpm.(137) Horizontal-sliding sash hoods. No equipment in hood closer than 12 inches to face of hood. Hoods located away from doors and trafficways.*	60
2.	Same as 1 above, some traffic past hoods. No equipment in hoods closer than 6 inches to face of hood. Hoods located away from doors and trafficways.*	80
3.	Ceiling panels properly located with average panel face velocity < 60 fpm(137) or ceiling diffusers properly located; no diffuser immediately in front of hoods, quadrant facing hood blocked, terminal throw velocity < 60 fpm. No equipment in hood closer than 6 inches to face of hood. Hoods located way from doors or trafficways.*	80
4.	Same as 3 above; some traffic past hoods. No equipment in hoos closer than 6 inches to face of hood.	100
	Wall grilles. Possible but not recommended for advance planning of new facilities.	

* Hoods near doors are acceptable if 1) there is a second safe egress from the room, 2) traffic past hood is low, and 3) door is normally open.

AUXILIARY AIR HOODS

Auxiliary air hoods are of proprietary design and a quantitative analysis cannot be provided here. Some designs blow contaminants out of the hood into the room; others are quite effective. The referenced performance test can and has been used to demonstrate the control level achieved by any specific design. Well-designed auxiliary air hoods perform as well as any other hoods in this regard.

Some auxiliary airhoods, introducing untreated or partially treated air at low velocity, may degrade the room air conditioning if the auxiliary air is as much as 20 F warmer than the room air. This behavior may be observed with a smoke test, but it is difficult to quantify and there is not a valid, demonstrated, quantifying test.

If the laboratory room air is to be maintained at some specified condition of temperature or humidity (and perhaps cleanliness), use of auxiliary hoods may not be economic or energy-conserving as compared to regular airfoil hoods with well-designed room air supply.

AMERICAN CONFERENCE OF GOVERNMENTAL INDUSTRIAL HYGIENISTS

LABORATORY HOOD DATA

DATE	VS-204.1

GENERAL USE LABORATORY HOODS:

A. Provide uniform exhaust air distribution in hood. Adjust baffles and air flow for less than ±10% variation in point-to-point face velocity with sash in maximum open position.

B. Locate hood away from heavy traffic aisles and doorways. Hoods near doors are acceptable if, (1) There is a second safe means of egress from room, (2) Traffic past hood is low and (3) Door is normally open.

C. Use corrosion resisting materials suitable for expected use.

D. Provide air cleaning on exhaust air if necessary and adequate stack height to minimize re-entry of contaminants to comply with air pollution regulations.

E. Avoid sharp corners at jambs and sill. Tapered or round hood inlets are desirable; an air foil shroud at sill is important.

F. Provide filters for radioactive materials in greater than "exempt" quantities.

G. By-pass opening in hood is desirable to avoid excessive indraft under partially-closed sash condition. Opening to be baffled to prevent splash from eruption in hood as shown in VS-203.

H. Provide tempered or conditioned make-up air to laboratory. Make-up air volume to be selected for desired air balance with adjoining spaces. See VS-204.

I. In order to reduce exhaust volumes, local exhaust hood should be considered instead of laboratory bench hoods for fixed set-ups.

J. For air conservation use horizontal sliding sash; sill airflow required.

K. All bench hoods should have a recessed work surface and airfoil sill.

AMERICAN CONFERENCE OF GOVERNMENTAL INDUSTRIAL HYGIENISTS

GENERAL USE LABORATORY HOODS

DATE 1-86 | VS-205

PERCHLORIC ACID HOODS

Perchloric acid is extremely dangerous because it is a very strong oxidizer. When the acid reacts with organic material, an explosive reaction product may be formed.

1. Do not use perchloric acid in a hood designed for other purposes. Identify Perchloric Acid Hoods with large warning signs.
2. Provide exhaust ventilation and room supply air in accordance with VS-204.
3. Utilize local exhaust ventilation within the hood to minimize condensation of vapors inside the hood.
4. Locate all utility controls outside the hood.
5. Materials of construction for this type of hood and ductwork must be non-reactive, acid resistant, and relatively impervious. AVOID ORGANIC MATERIALS unless known to be safe. Stainless steel type 316 with welded joints is preferred. Unplasticized polyvinyl chloride or an inorganic ceramic coating such as porcelain are acceptable.
6. Ease of cleanliness is paramount. Use stainless steel with accessible rounded corners and all-welded construction.
7. The work surface should be water tight with a minimum of 0.5-inch dished front and sides and an integral trough at the rear to collect the washdown water.
8. Design washdown facilities into the hood and ductwork. Use daily or more often to thoroughly clean perchloric acid from the exhaust system surfaces.
9. Each perchloric acid hood should have an individual exhaust system. Slope horizontal runs to drain. Avoid sharp turns.
10. Construct the hood and ductwork to allow easy visual inspection.
11. Where required, use a high efficiency (greater than 80%) wet collector constructed for perchloric acid service. Locate as close to the hood as possible to minimize the accumulation of perchloric acid in the exhaust duct.
12. Use only an acid-resistant metallic fan, a metallic fan protected by an inorganic coating, or an air injector.
13. Lubricate the fan with a fluorocarbon-type grease.
14. Locate the fan outside the building.
15. The exhauset discharge must terminate out of doors, preferably using a vertical discharge cap which extends well above the roof eddy zone. See Figures 5-25 and 5-26.

AMERICAN CONFERENCE OF
GOVERNMENTAL INDUSTRIAL HYGIENISTS

PERCHLORIC ACID HOOD DATA

DATE 1-86 | VS-205.1

WORK PRACTICES FOR LABORATORY HOODS

No large open face hood with a low face velocity and a work standing at the face can provide complete safety against all events which may occur in the hood, nor for volatile or otherwise airborne contaminants with a TLV in the low part per billion range. For more ordinary exposures, a properly designed hood in a properly ventilated room can provide adequate protection. However, certain work practices are necessary in order for the hood to perform capably. The following work practices are generally required; more stringent practices may be necessary in some circumstances.

1. Conduct all operations which may generate air contaminants at or above the appropriate TLV inside a hood.
2. Keep all apparatus at least 6 inches back from the face of the hood. A stripe on the bench surface is a good reminder.
3. Do not put your head in the hood when contaminants are being generated.
4. Do not use the hood as a waste disposal mechanism except for very small quantities of volatile materials.
5. Do not store chemicals or apparatus in the hood. Store hazardous chemicals in an approved safety cabinet.
6. Keep the hood sash closed as much as possible.
7. Keep the slots in the hood baffle free of obstruction by apparatus or containers.
8. Minimize foot traffic past the face of the hood.
9. Keep laboratory doors closed (exception: some of the laboratory design requires the lab doors to be open).
10. Do not remove hood sash or panels except when necessary for apparatus set-up; replace sash or panels before operating.
11. Do not place electrical receptacles or other spark sources inside the hood when flammable liquids or gases are present. No permanent electrical receptacles are permitted in the hood.
12. Use an appropriate barricade if there is a chance of explosion or eruption.
13. Provide adequate maintenance for the hood exhaust system and the building supply system. Use static pressure gauges on the hood throat, across any filters in the exhaust system, or other appropriate indicators to insure that exhaust flow is appropriate.
14. If hood sash is supposed to be partially closed for operation, the hood should be so labeled and the appropriate closure point clearly indicated.

AMERICAN CONFERENCE OF GOVERNMENTAL INDUSTRIAL HYGIENISTS	
WORK PRACTICES FOR LABORATORY HOODS	
DATE 1-86	VS-205.2

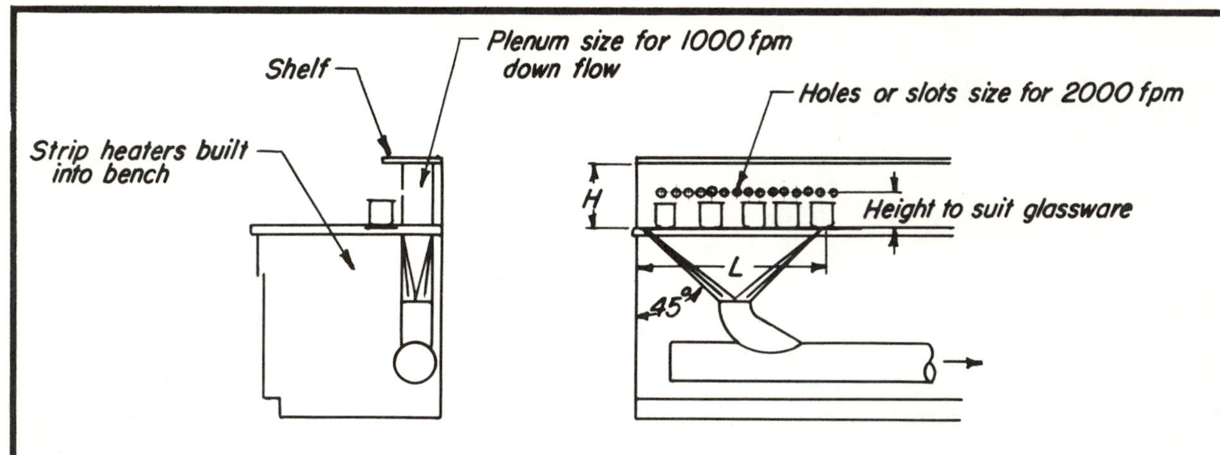

EVAPORATION BENCH

Q= 20 cfm/lineal foot of hood or 50 HL
Duct velocity = 2000 fpm
Entry loss = 1.78 slot VP + 0.25 duct VP

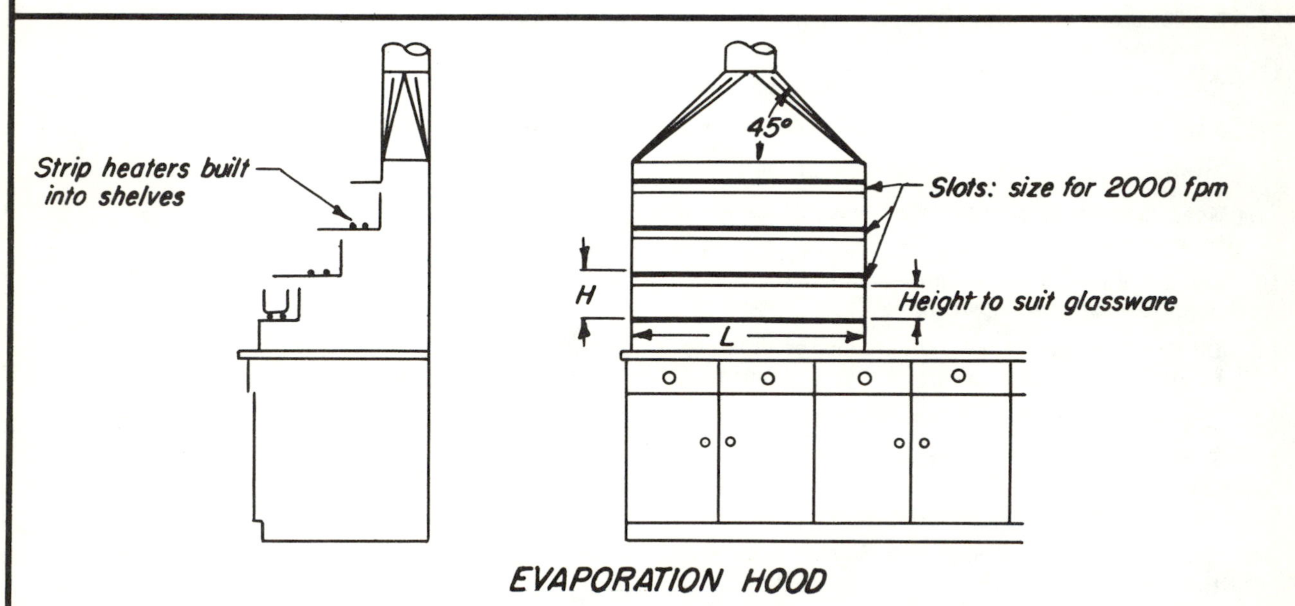

EVAPORATION HOOD

Q= 20 cfm/lineal foot of shelf or 50 HL for each shelf
Duct velocity = 2000 fpm
Entry loss = 1.78 slot VP + 0.25 duct VP

AMERICAN CONFERENCE OF GOVERNMENTAL INDUSTRIAL HYGIENISTS

SPECIALIZED LABORATORY HOOD DESIGNS

DATE 1-68 | VS-206

Reference 95

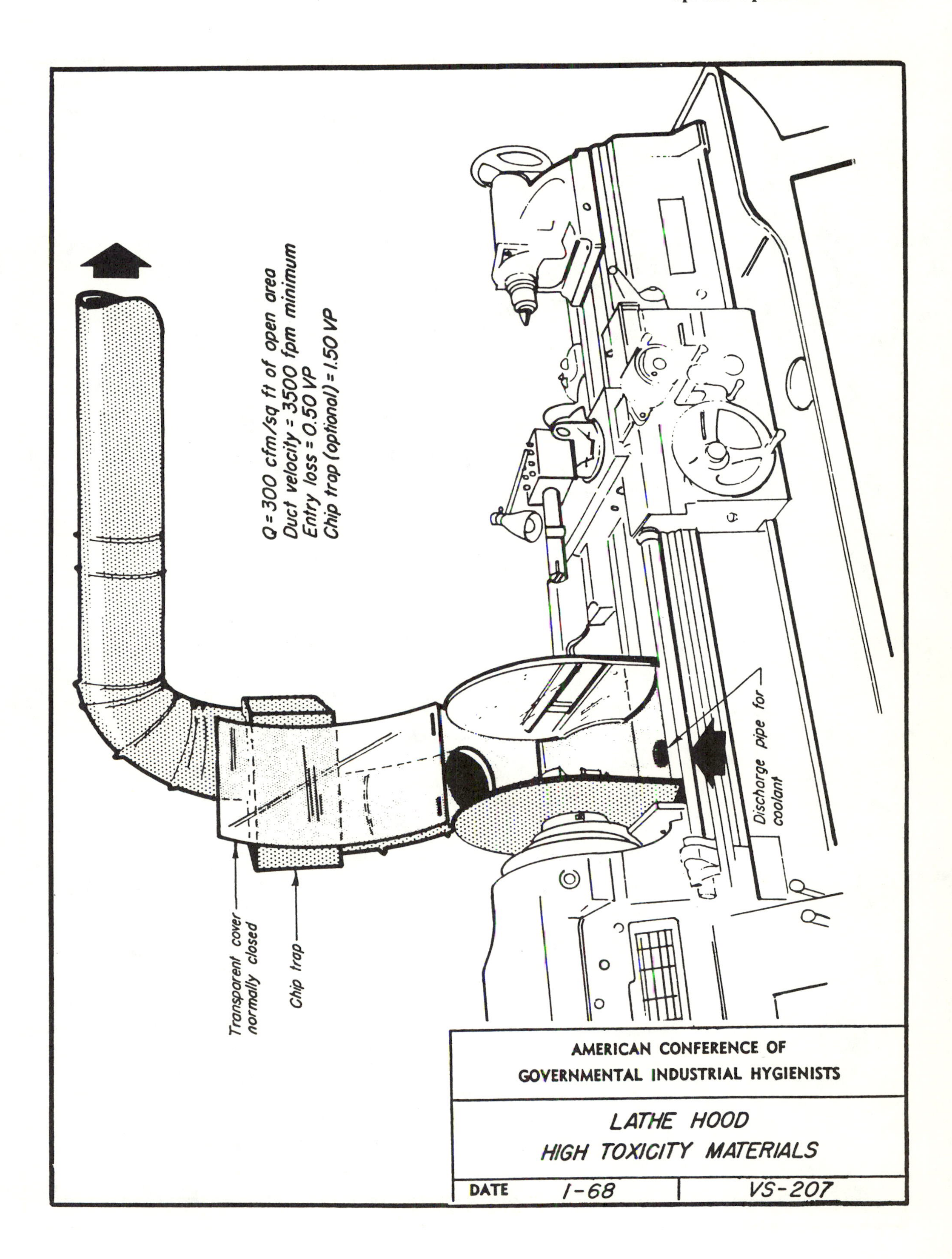
Q = 300 cfm/sq ft of open area
Duct velocity = 3500 fpm minimum
Entry loss = 0.50 VP
Chip trap (optional) = 1.50 VP
Transparent cover normally closed
Chip trap
Discharge pipe for coolant
AMERICAN CONFERENCE OF GOVERNMENTAL INDUSTRIAL HYGIENISTS
LATHE HOOD
HIGH TOXICITY MATERIALS
DATE 1-68
VS-207

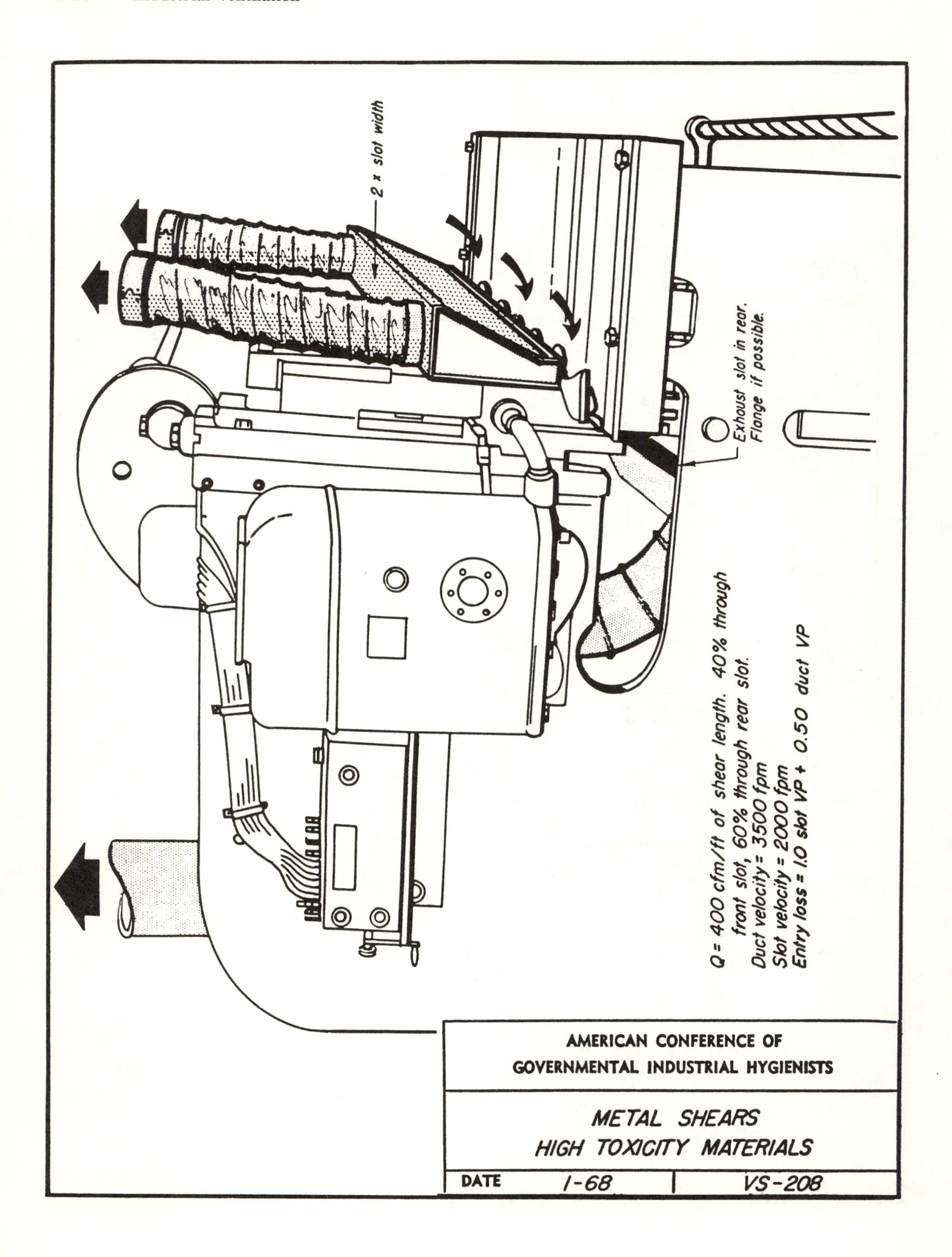
2 x slot width
Exhaust slot in rear.
Flange if possible.
Q = 400 cfm/ft of shear length. 40% through
front slot, 60% through rear slot.
Duct velocity = 3500 fpm
Slot velocity = 2000 fpm
Entry loss = 1.0 slot VP + 0.50 duct VP
AMERICAN CONFERENCE OF
GOVERNMENTAL INDUSTRIAL HYGIENISTS
METAL SHEARS
HIGH TOXICITY MATERIALS
DATE
1-68
VS-208

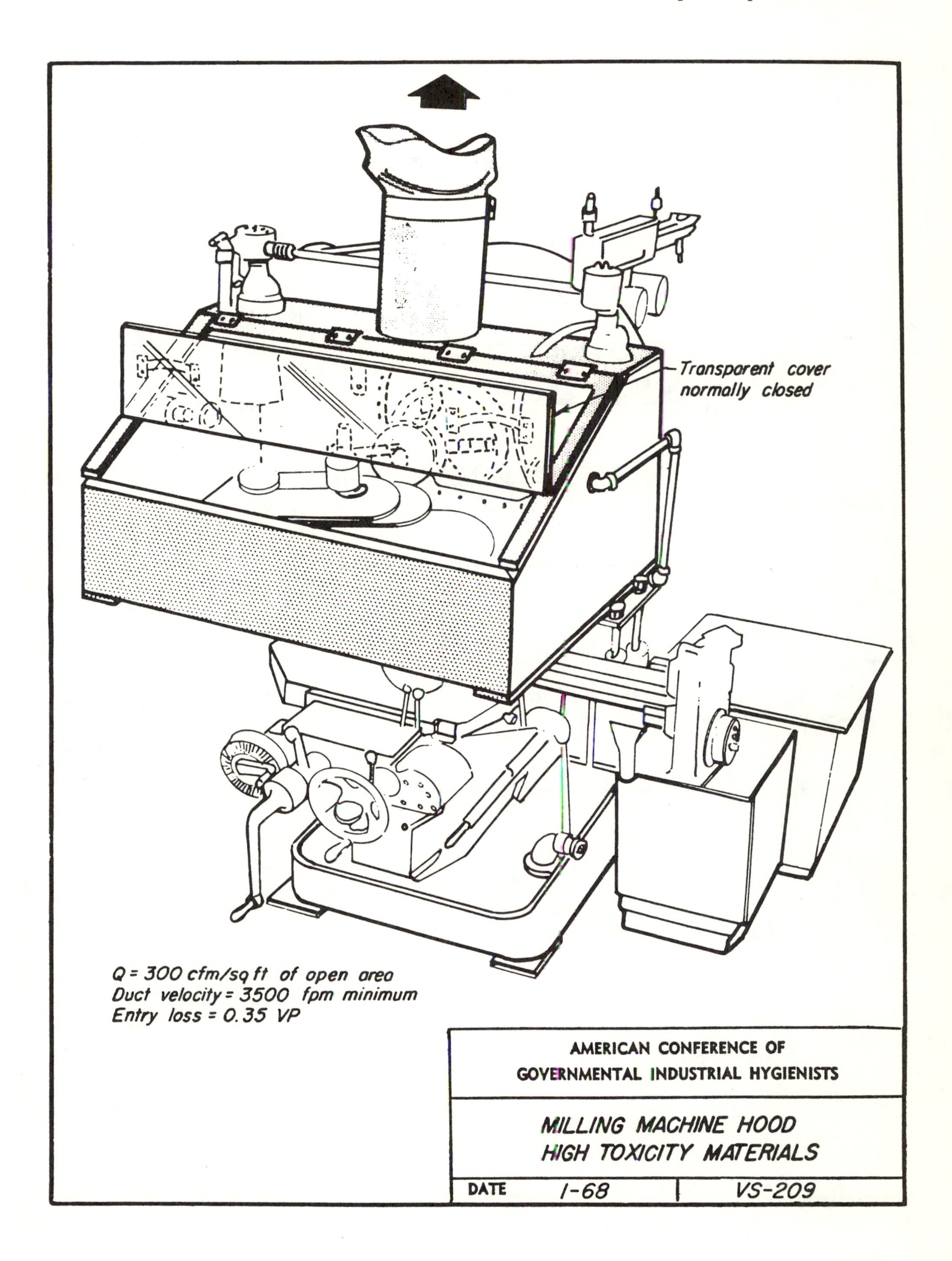
Transparent cover normally closed
Q = 300 cfm/sq ft of open area
Duct velocity = 3500 fpm minimum
Entry loss = 0.35 VP
AMERICAN CONFERENCE OF GOVERNMENTAL INDUSTRIAL HYGIENISTS
MILLING MACHINE HOOD HIGH TOXICITY MATERIALS
DATE 1-68
VS-209

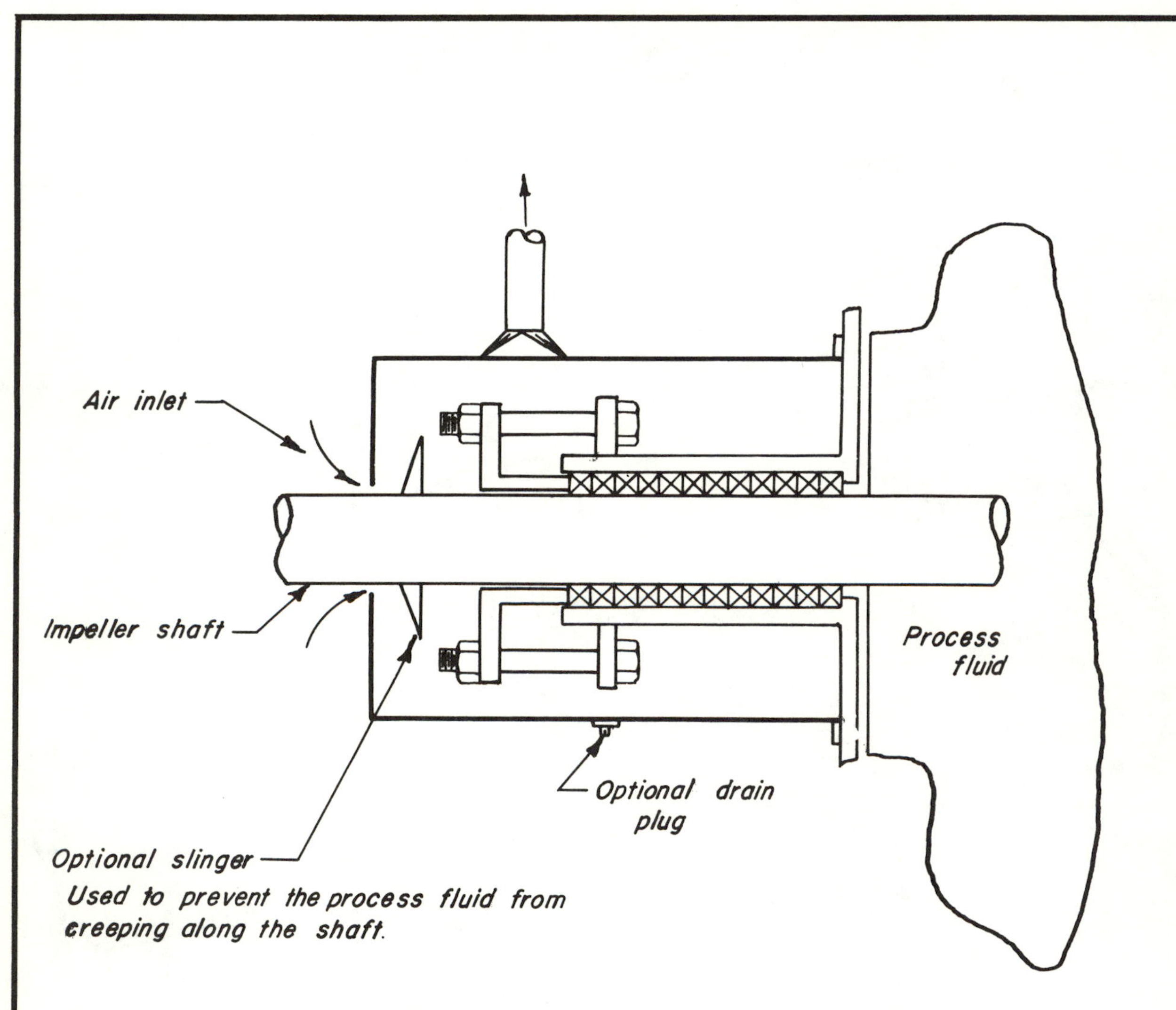

Q = 500 cfm/sq ft of open area minimum
(typically 10-40 cfm)
Note: Sufficient air must be provided to dilute flammable gases and/or vapors to 25% of LEL See Section 2
Duct velocity = 2000 fpm
Entry loss = 1.78 VP slot plus 0.25 duct VP

AMERICAN CONFERENCE OF GOVERNMENTAL INDUSTRIAL HYGIENISTS

SHAFT SEAL ENCLOSURE

DATE 1-78 | VS-210

Reference 129

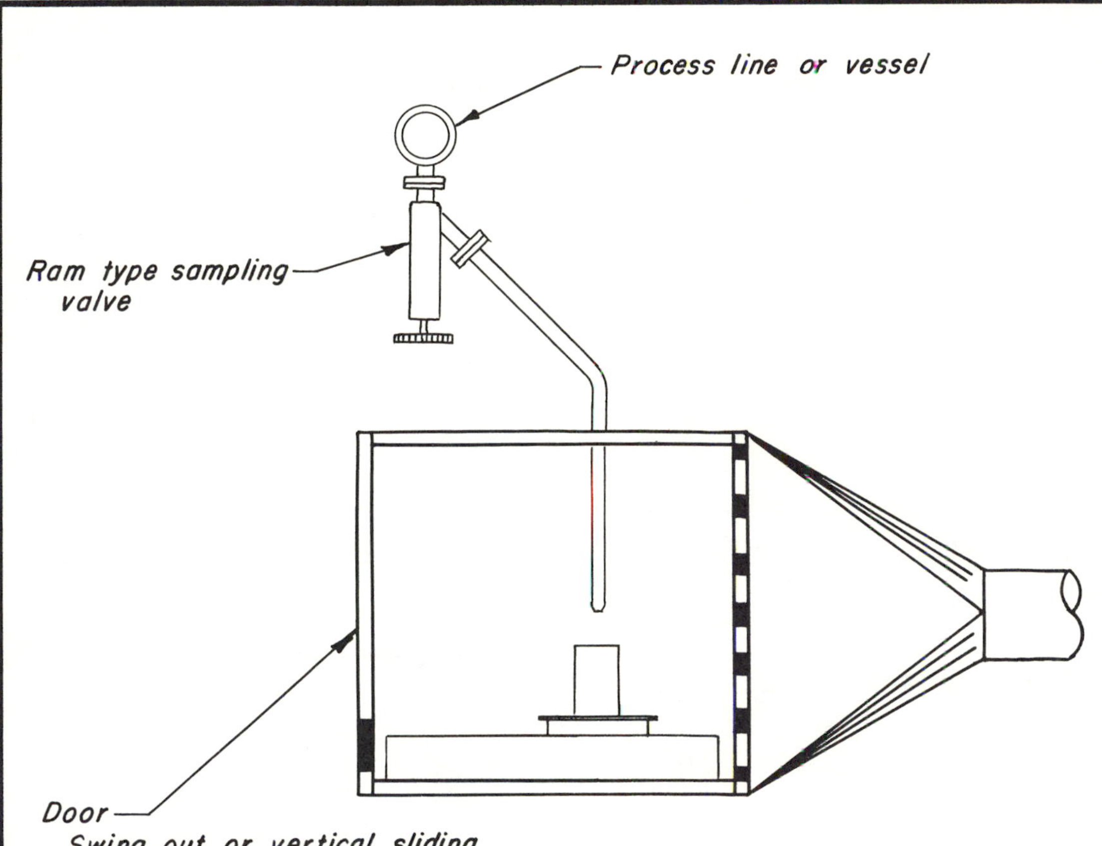

Q = 125 cfm/sq ft of open area (door area) minimum

Note: Sufficient air must provided when door closed to dilute flammable gases and/or vapors to 25% of LEL. See Chapter 2

Duct velocity = 2000 fpm

Entry loss = 1.78 VP + 0.50 duct VP

Reference 129

AMERICAN CONFERENCE OF GOVERNMENTAL INDUSTRIAL HYGIENISTS	
SAMPLING BOX	
DATE 1-80	VS-211

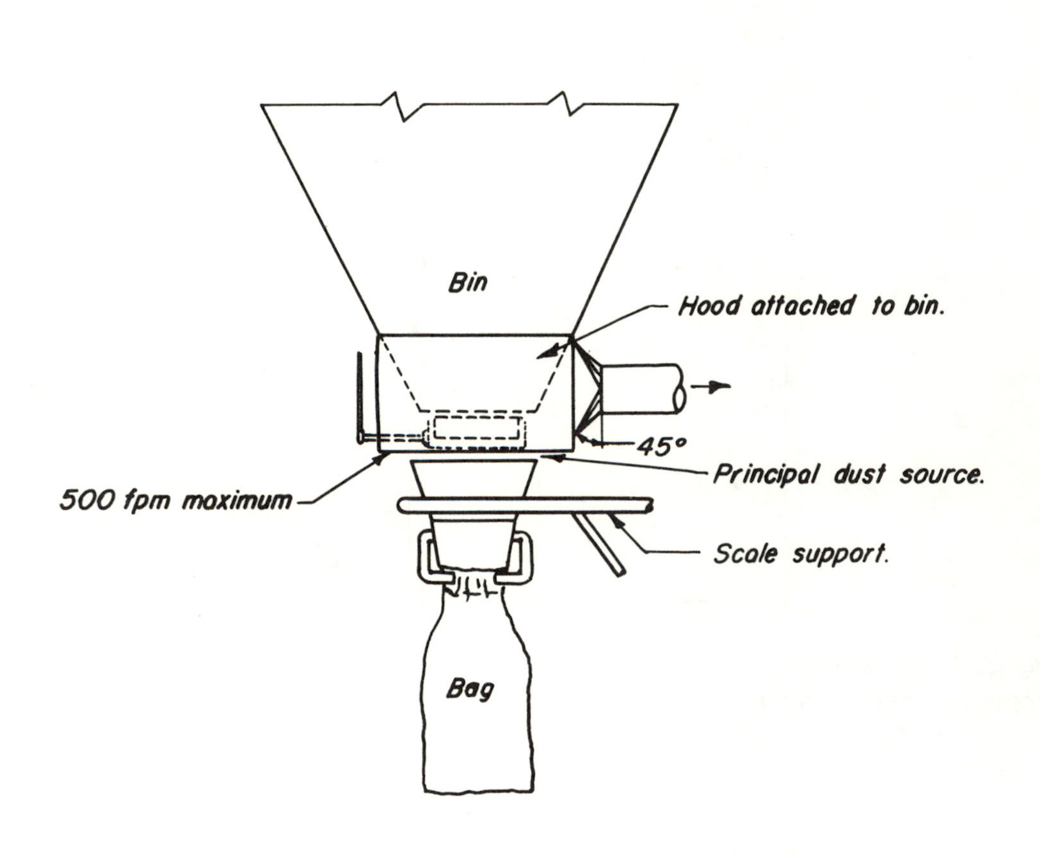

Q = 400 - 500 cfm - non-toxic dust
1000 - 1500 cfm - toxic dust
Duct velocity = 3500 fpm minimum
Entry loss = 0.25 VP

AMERICAN CONFERENCE OF
GOVERNMENTAL INDUSTRIAL HYGIENISTS

BAG FILLING

DATE 1-64	VS-301

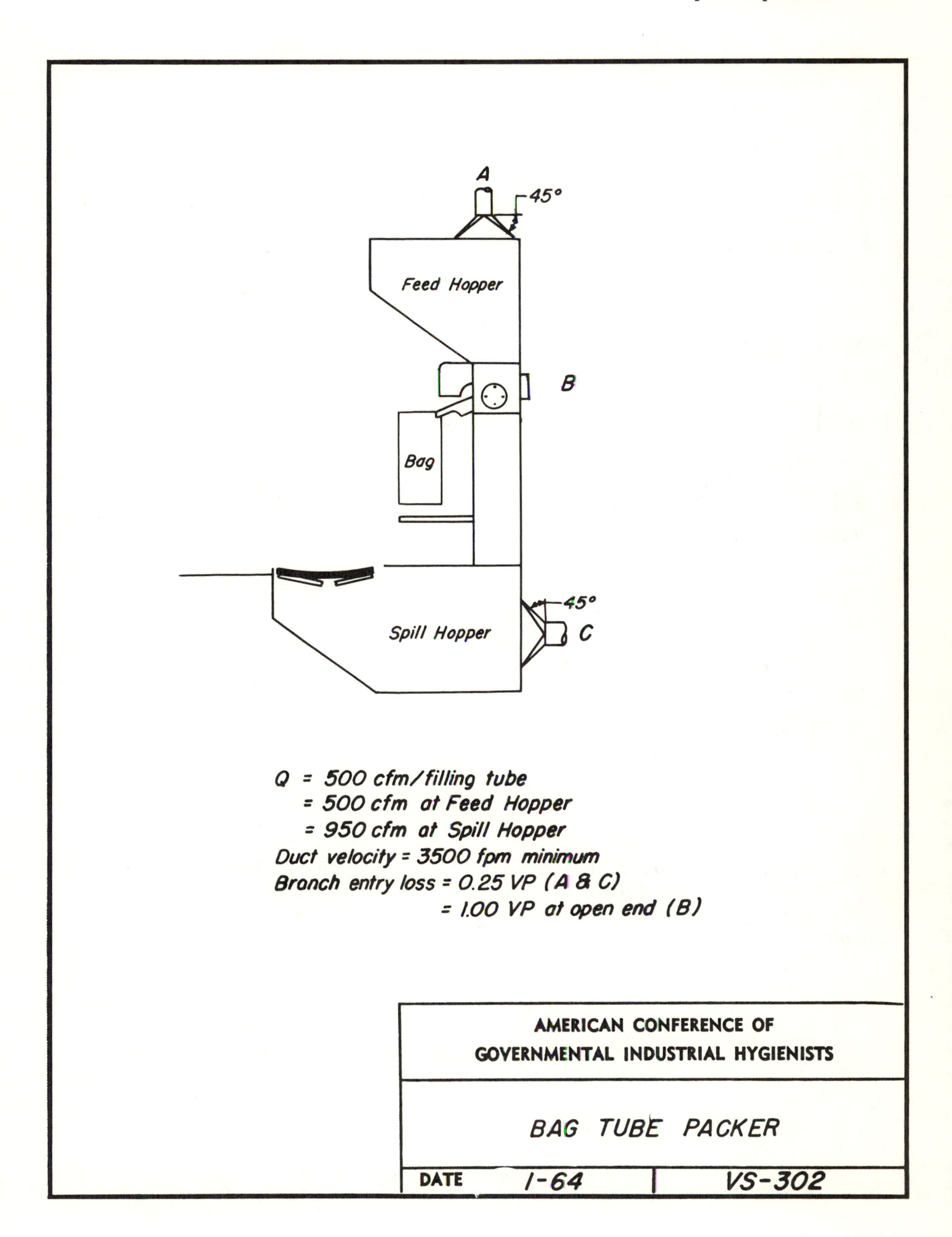
A
45°
Feed Hopper
B
Bag
45°
Spill Hopper
C
Q = 500 cfm/filling tube
= 500 cfm at Feed Hopper
= 950 cfm at Spill Hopper
Duct velocity = 3500 fpm minimum
Branch entry loss = 0.25 VP (A & C)
= 1.00 VP at open end (B)
AMERICAN CONFERENCE OF
GOVERNMENTAL INDUSTRIAL HYGIENISTS
BAG TUBE PACKER
DATE 1-64
VS-302

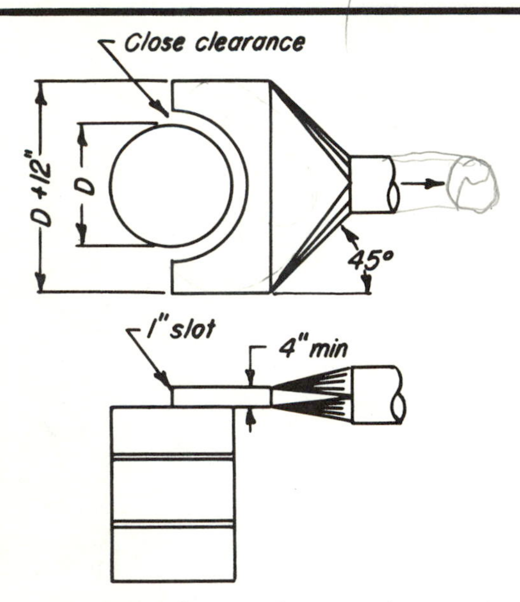

Q = 100 cfm/sq ft barrel top min
Duct velocity = 3500 minimum
Entry loss = 0.25 VP + 1.78 slot VP
Manual loading.

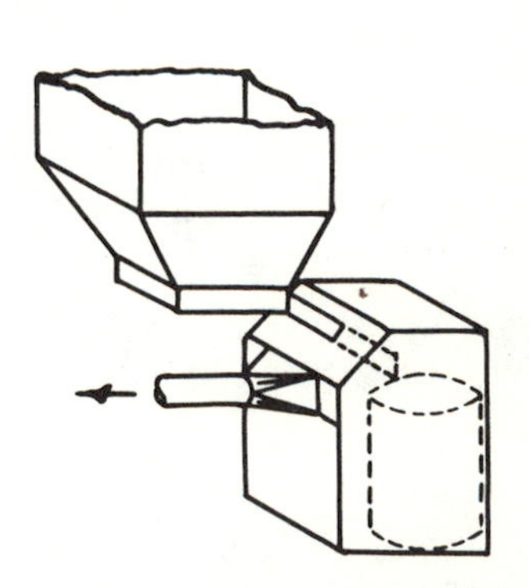

Q = 150 cfm/sq ft open face area
Duct velocity = 3500 fpm minimum
Entry loss = 0.25 VP for 45° taper

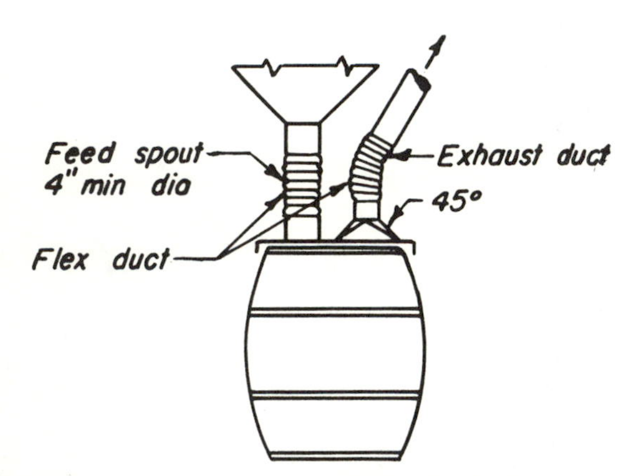

Q = 50 cfm × drum dia (ft) for weighted lid
150 cfm × drum dia (ft) for loose lid
Duct velocity = 3500 fpm minimum
Entry loss = 0.25 VP

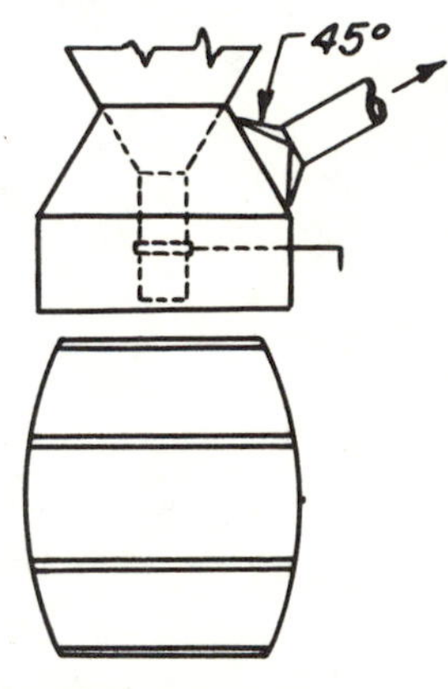

Q = 300-400 cfm
Duct velocity = 3500 fpm min
Entry loss = 0.25 VP

AMERICAN CONFERENCE OF
GOVERNMENTAL INDUSTRIAL HYGIENISTS

BARREL FILLING

DATE 1-64 | VS-303

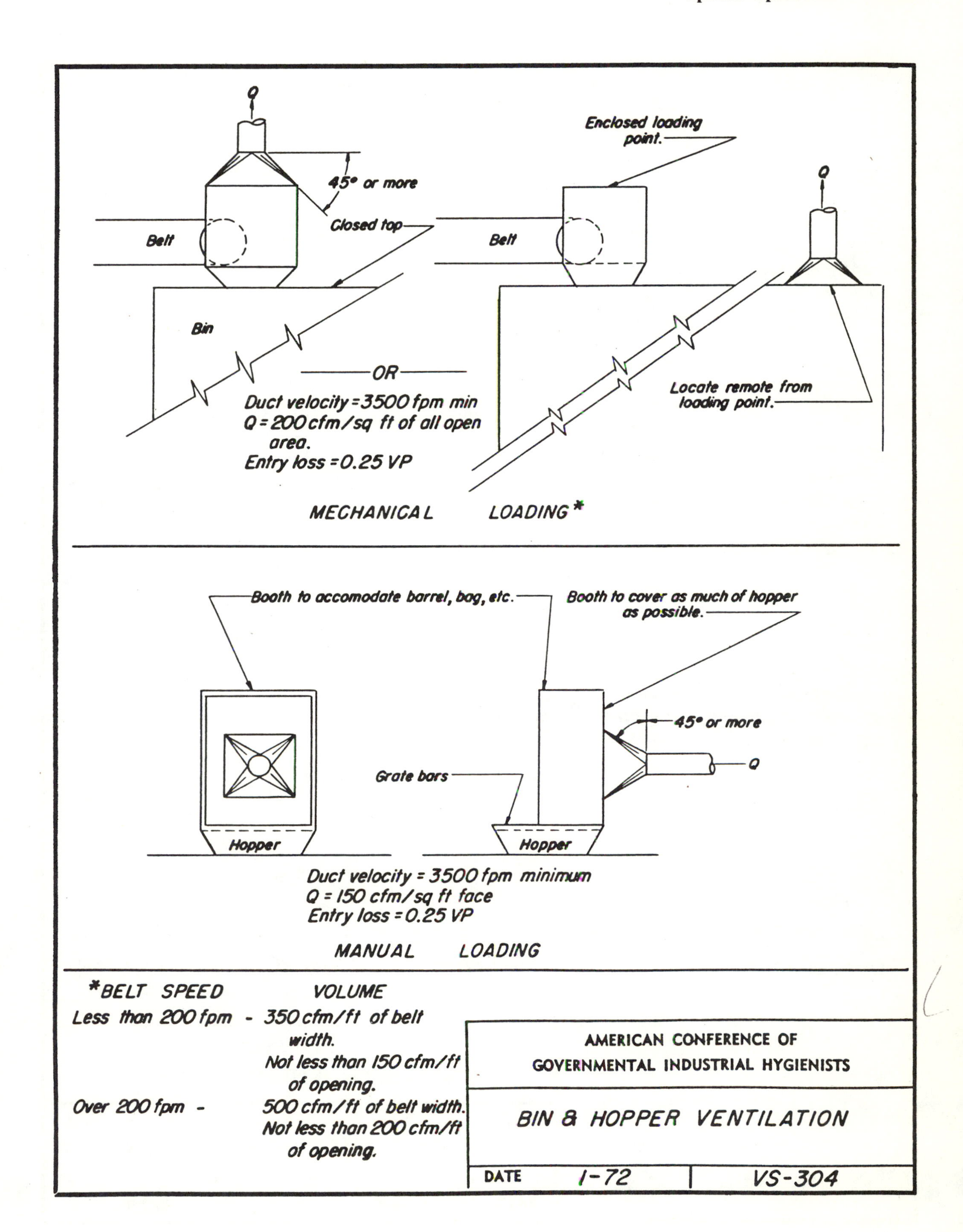
Q
45° or more
Closed top
Belt
Bin
Enclosed loading point.
Belt
Q
OR
Duct velocity = 3500 fpm min
Q = 200 cfm/sq ft of all open area.
Entry loss = 0.25 VP
Locate remote from loading point.
MECHANICAL LOADING*
Booth to accomodate barrel, bag, etc.
Booth to cover as much of hopper as possible.
45° or more
Grate bars
Q
Hopper
Hopper
Duct velocity = 3500 fpm minimum
Q = 150 cfm/sq ft face
Entry loss = 0.25 VP
MANUAL LOADING
*BELT SPEED
VOLUME
Less than 200 fpm - 350 cfm/ft of belt width.
Not less than 150 cfm/ft of opening.
Over 200 fpm - 500 cfm/ft of belt width.
Not less than 200 cfm/ft of opening.
AMERICAN CONFERENCE OF GOVERNMENTAL INDUSTRIAL HYGIENISTS
BIN & HOPPER VENTILATION
DATE 1-72
VS-304

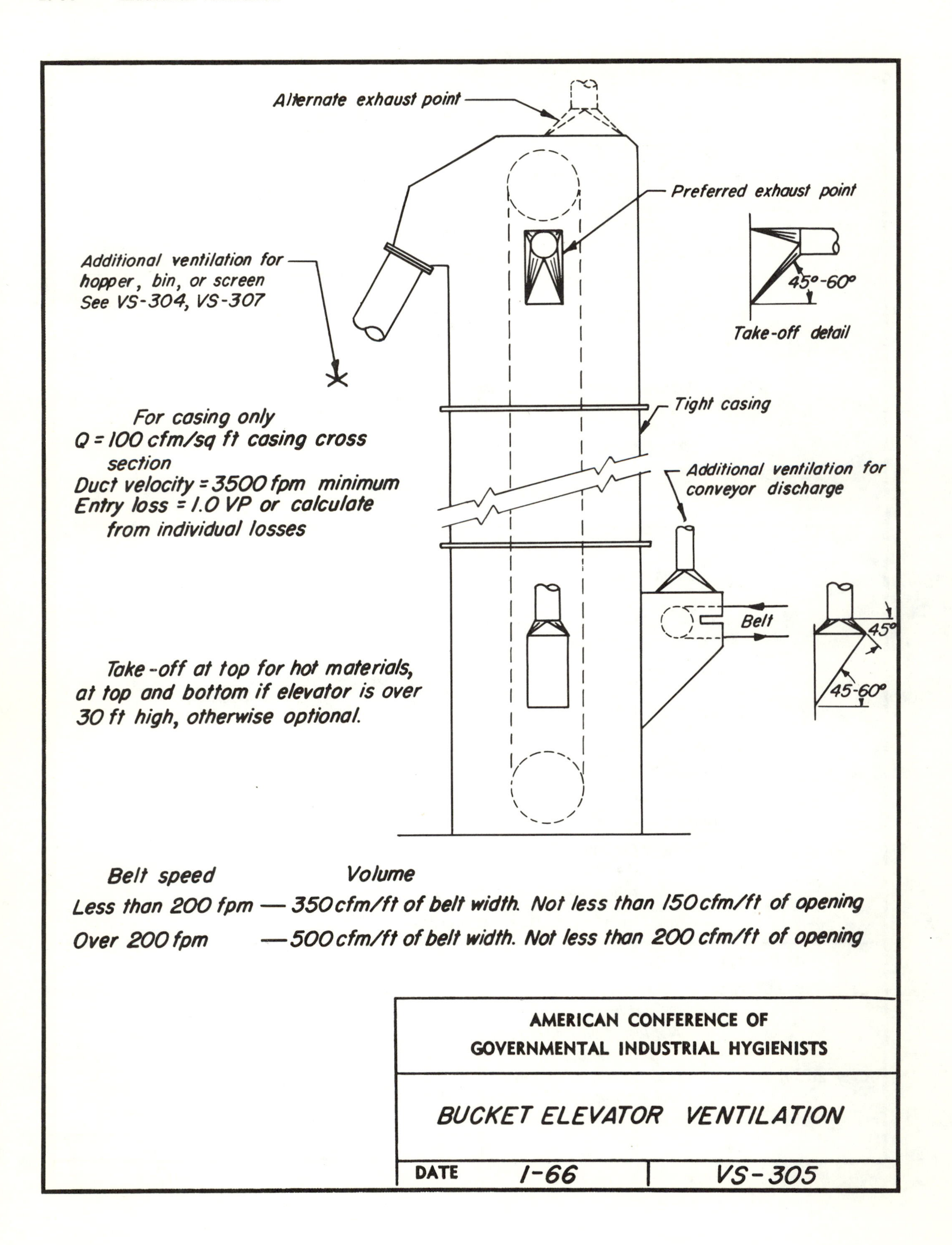

Alternate exhaust point
Preferred exhaust point
Additional ventilation for hopper, bin, or screen See VS-304, VS-307
45°-60°
Take-off detail
For casing only
Q = 100 cfm/sq ft casing cross section
Duct velocity = 3500 fpm minimum
Entry loss = 1.0 VP or calculate from individual losses
Tight casing
Additional ventilation for conveyor discharge
Belt
45°
45-60°
Take-off at top for hot materials, at top and bottom if elevator is over 30 ft high, otherwise optional.
Belt speed
Volume
Less than 200 fpm — 350 cfm/ft of belt width. Not less than 150 cfm/ft of opening
Over 200 fpm — 500 cfm/ft of belt width. Not less than 200 cfm/ft of opening
AMERICAN CONFERENCE OF GOVERNMENTAL INDUSTRIAL HYGIENISTS
BUCKET ELEVATOR VENTILATION
DATE 1-66
VS-305

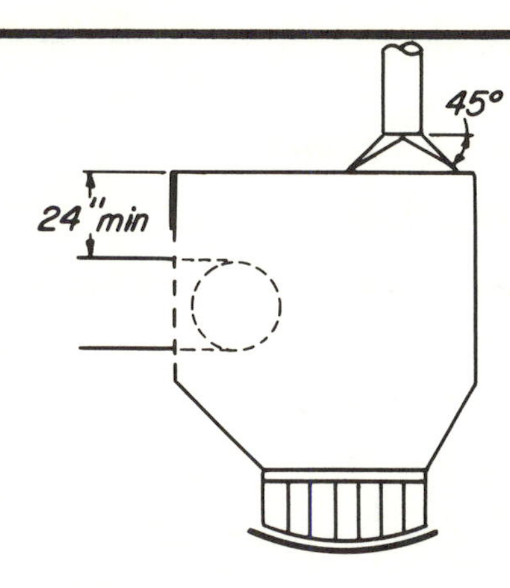

1. Conveyor transfer less than 3' fall. For greater fall provide additional exhaust at lower belt. See 3 below.

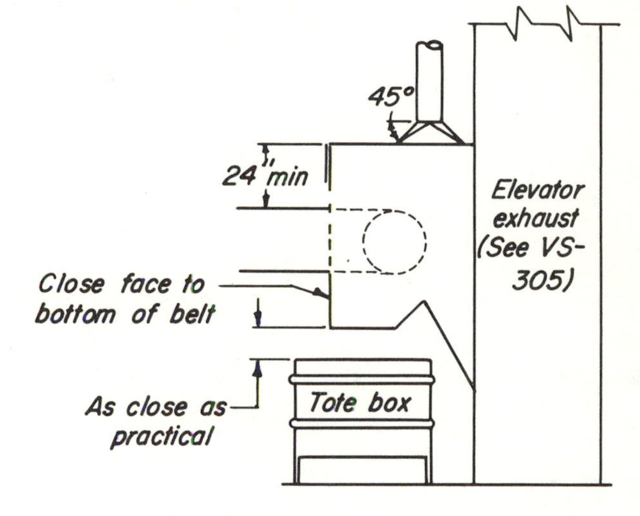

2. Conveyor to elevator with magnetic separator.

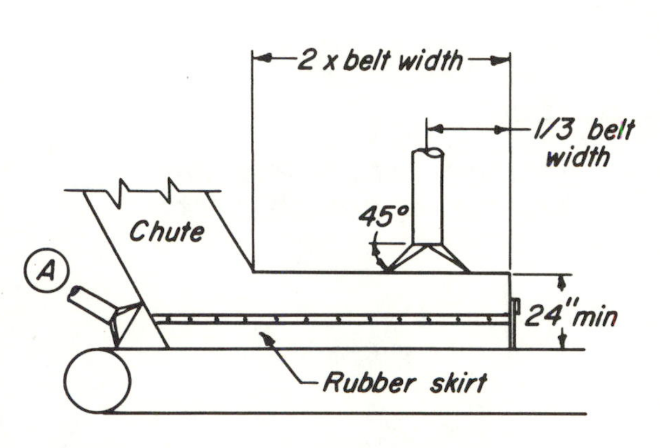

3. Chute to belt transfer and conveyor transfer, greater than 3' fall. Use additional exhaust at (A) for dusty material as follows:
Belt width 12"-36", Q=700 cfm
above 36", Q=1000 cfm

Detail of belt opening

DESIGN DATA

Transferpoints:
Enclose to provide 150-200 fpm indraft at all openings.
Minimum Q=350 cfm/ft belt width for belt speeds under 200 fpm
= 500 cfm/ft belt width for belt speeds over 200 fpm and for magnetic separators
Duct velocity = 3500 fpm minimum
Entry loss = 0.25VP

Conveyor belts:
Cover belt between transfer points
Exhaust at transfer points
Exhaust additional 350 cfm/ft of belt width at 30' intervals. Use 45° tapered connections.
Entry loss = 0.25 VP

Note:
Dry, very dusty materials may require exhaust volumes 1.5 to 2.0 times stated values.

AMERICAN CONFERENCE OF GOVERNMENTAL INDUSTRIAL HYGIENISTS

CONVEYOR BELT VENTILATION

DATE 1-72 | VS-306

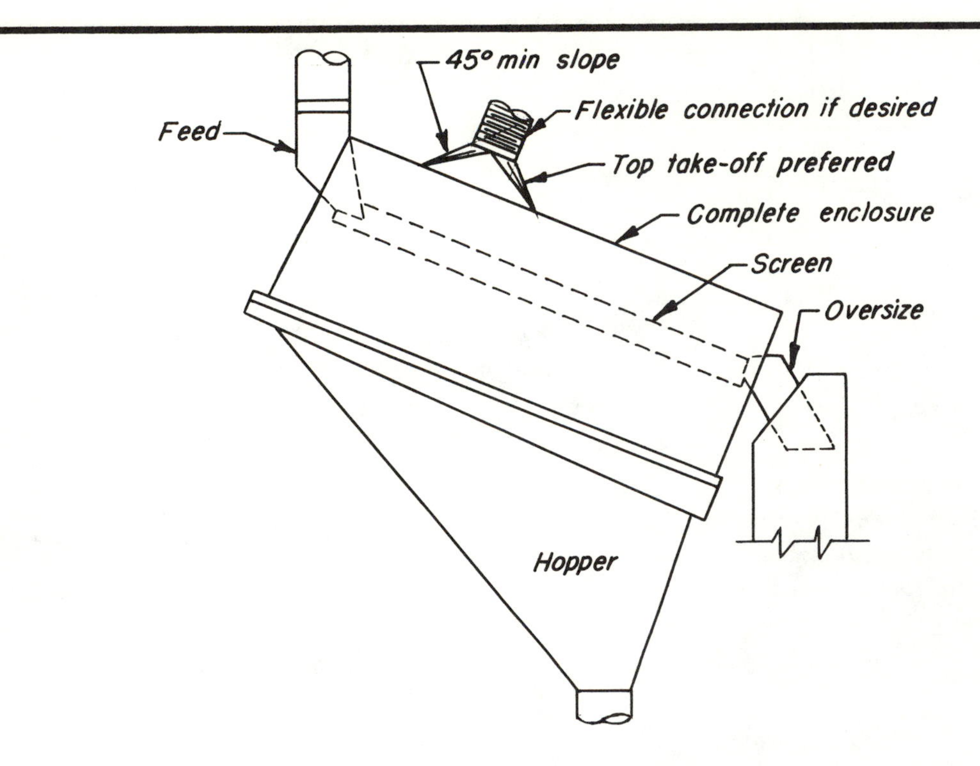

FLAT DECK SCREEN

Q = 200 cfm/sq ft through hood openings, but not less than 50 cfm/sq ft screen area. No increase for multiple decks
Duct velocity = 3500 fpm minimum
Entry loss = 0.50 VP

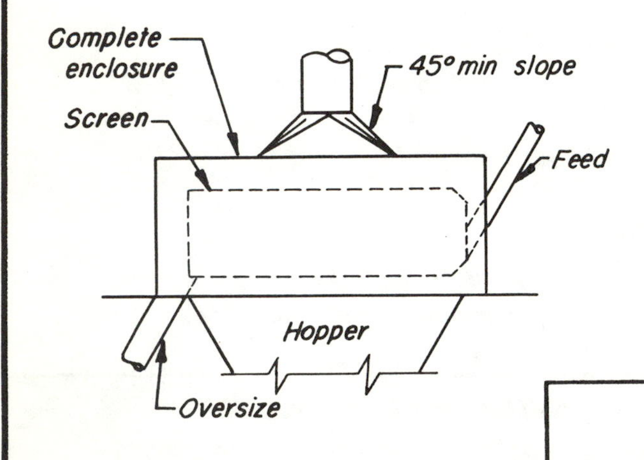

CYLINDRICAL SCREEN

Q = 100 cfm/sq ft circular cross section of screen; at least 400 cfm/sq ft of enclosure opening
Duct velocity = 3500 fpm minimum
Entry loss = 0.50 VP

AMERICAN CONFERENCE OF GOVERNMENTAL INDUSTRIAL HYGIENISTS

SCREENS

DATE 1-64	VS-307

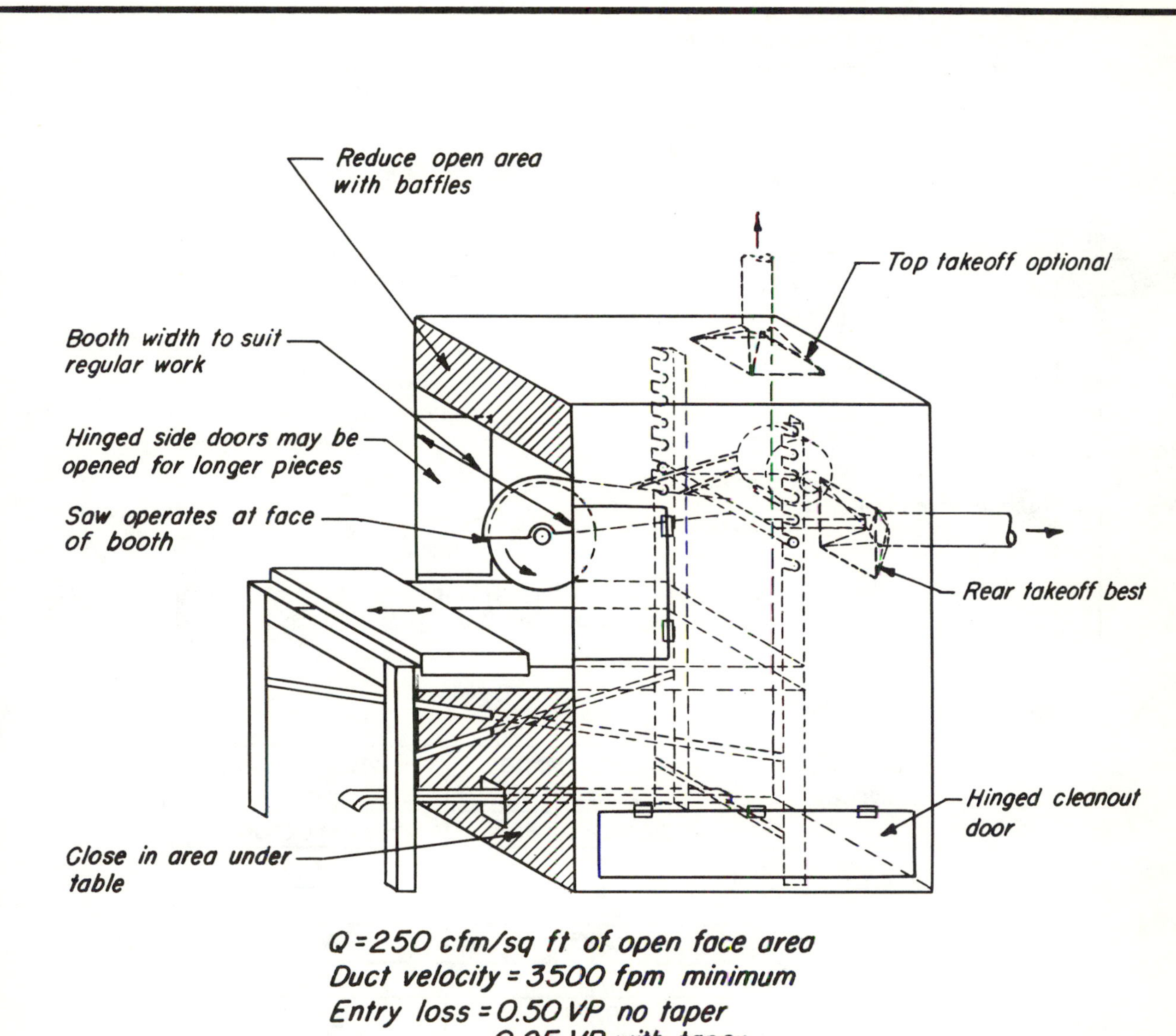

AMERICAN CONFERENCE OF GOVERNMENTAL INDUSTRIAL HYGIENISTS

ABRASIVE CUT-OFF SAW VENTILATION

DATE 1-78	VS-401

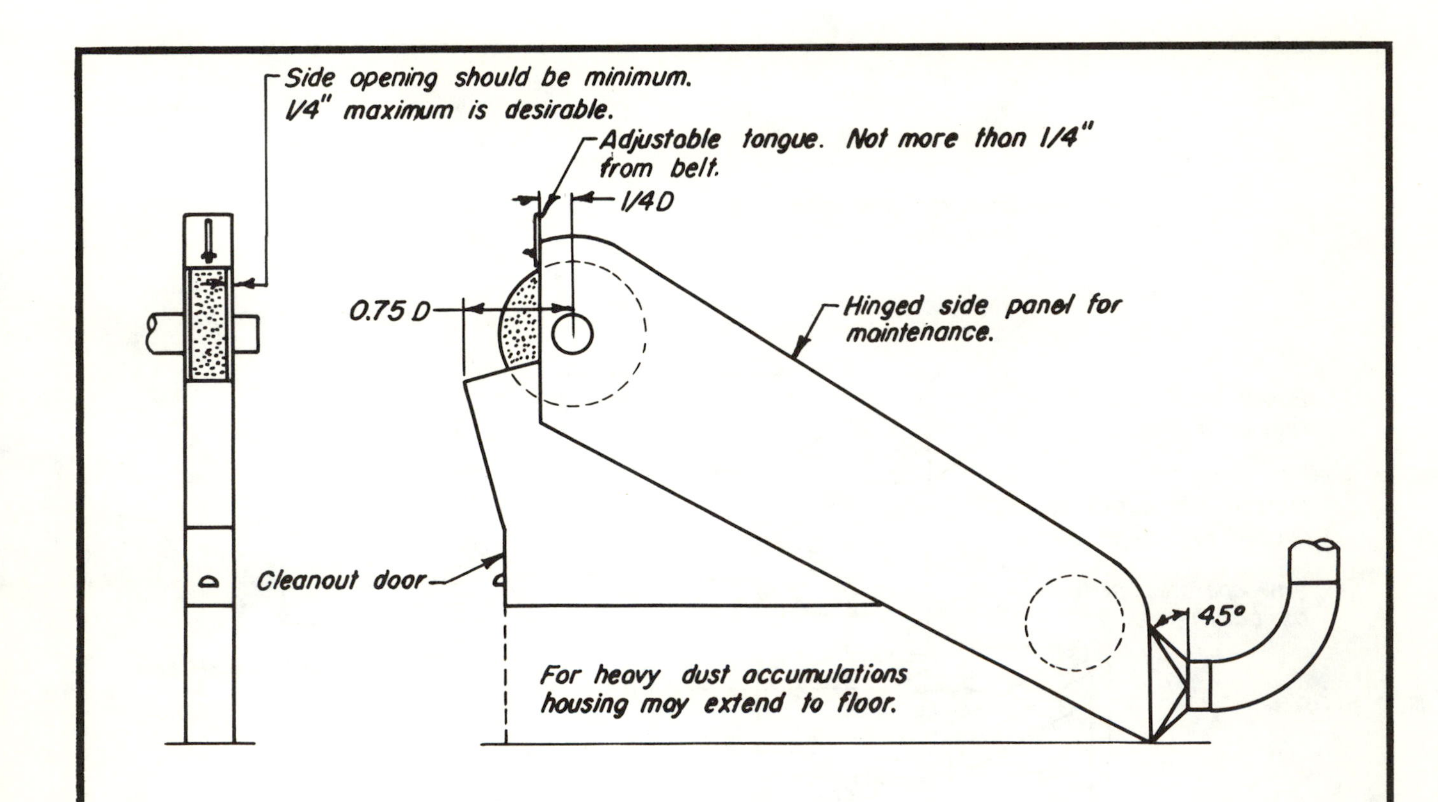

Belt width inches	Exhaust volume cfm Good enclosure *	Exhaust volume cfm Poor enclosure
1 1/2	220	300
2	390	610
3	500	740
4	610	880
5	880	1200
6	1200	1570

* Hood as shown. No more than 25% of wheel exposed.

Entry loss = 0.40 VP

Duct velocity = 3500 fpm mimimum

Note:

For titanium and magnesium eliminate hopper and use 5000 fpm through hood cross section.

AMERICAN CONFERENCE OF GOVERNMENTAL INDUSTRIAL HYGIENISTS

BACKSTAND IDLER POLISHING MACHINE

DATE 1-74 | VS-402

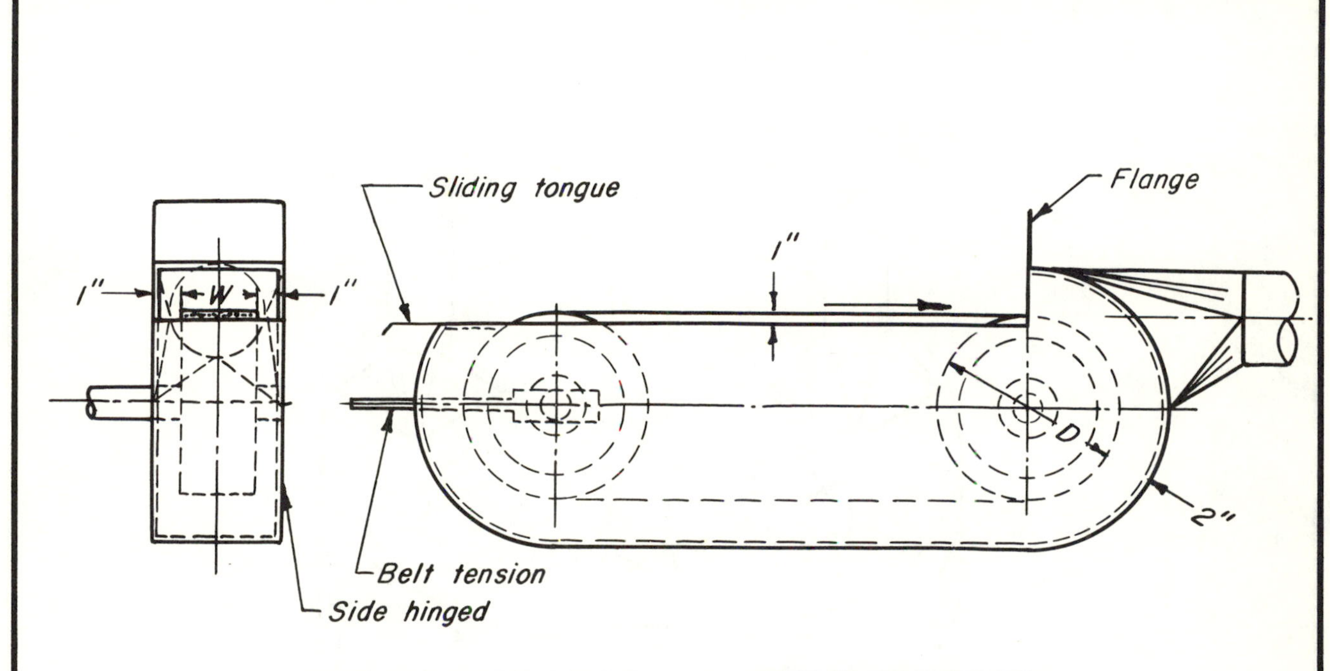

Belt width, inches	Exhust volume, cfm
up to 3	220
3 to 5	300
5 to 7	390
7 to 9	500
9 to 11	610
11 to 13	740

Minimum duct velocity = 3500 fpm
Entry loss = 0.65 VP for straight take-off
0.45 VP for tapered take-off

AMERICAN CONFERENCE OF
GOVERNMENTAL INDUSTRIAL HYGIENISTS

METAL POLISHING BELT

DATE 1-82 | VS-403

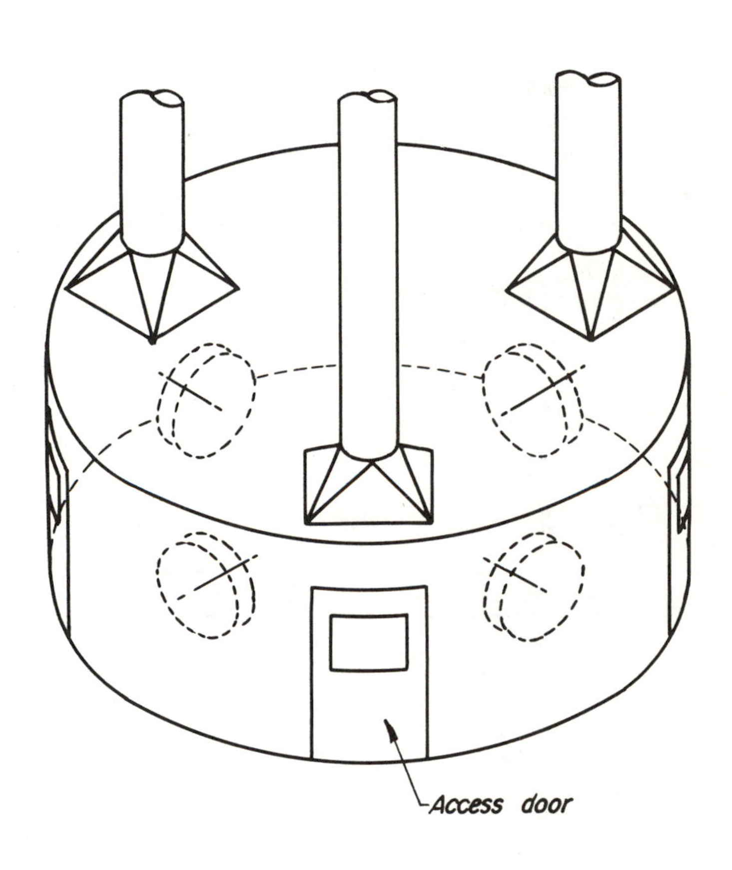

Q = 500 cfm/wheel, minimum
Not less than 250 cfm/sq ft total open area
Duct velocity = 3500 fpm minimum
Entry loss = 1.78 slot VP plus 0.25 duct VP
Use ammeters to gage wheel pressures

On small, 2 or 3 spindle machines, one take-off may be used
Multiple take-offs desirable
Provide automatic sprinklers or other fire protection. Consult Fire and Insurance Codes

AMERICAN CONFERENCE OF GOVERNMENTAL INDUSTRIAL HYGIENISTS

CIRCULAR AUTOMATIC BUFFING

DATE 1-64	VS-404

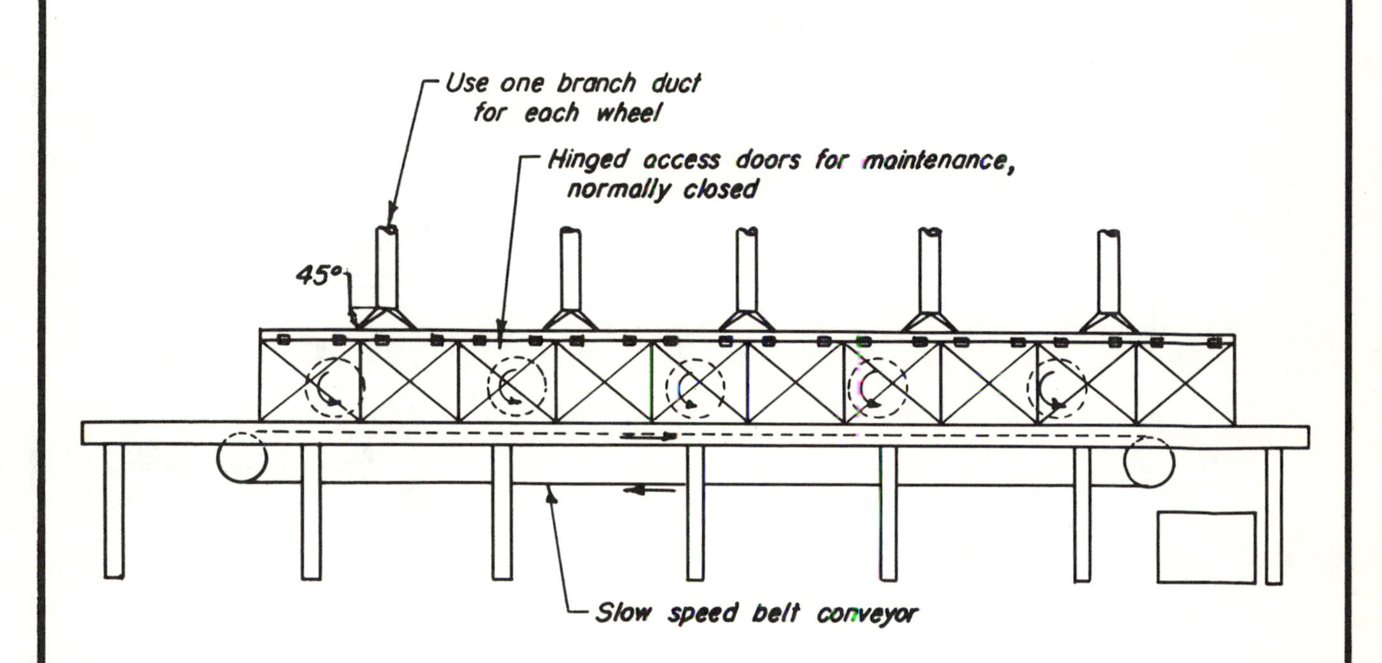

Q = 500 cfm/wheel, minimum
Not less than 250 cfm/sq ft total open area
Duct velocity = 4500 fpm minimum
Entry loss = 1.78 slot VP plus 0.25 duct VP
Use ammeters to gage wheel pressures
Wheel adjustments on outside of enclosure at the rear

AMERICAN CONFERENCE OF GOVERNMENTAL INDUSTRIAL HYGIENISTS	
STRAIGHT LINE AUTOMATIC BUFFING	
DATE 1-68	VS-405

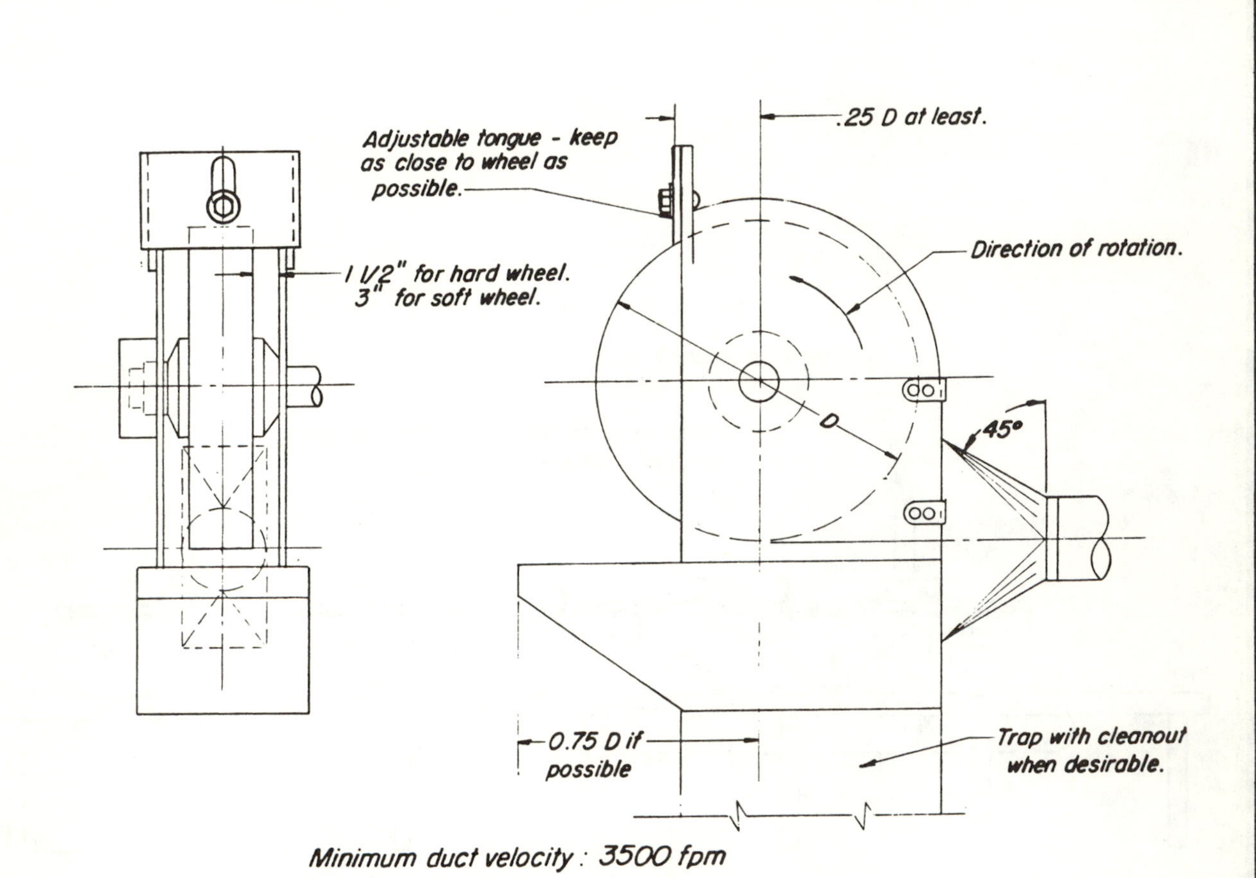

Minimum duct velocity : 3500 fpm

Entry loss : 0.65 VP for straight take-off.
0.40 VP for tapered take-off.

Wheel diam. inches	Wheel width * inches	Exhaust volume cfm	Exhaust volume cfm
		Good enclosure	Poor enclosure
to 9	2	300	400
over 9 to 16	3	500	610
over 16 to 19	4	610	740
over 19 to 24	5	740	1200
over 24 to 30	6	1040	1500
over 30 to 36	6	1200	1990

* In cases of extra wide wheels, use wheel width to determine exhaust volume.

AMERICAN CONFERENCE OF GOVERNMENTAL INDUSTRIAL HYGIENISTS

BUFFING AND POLISHING

DATE 1-82 | VS-406

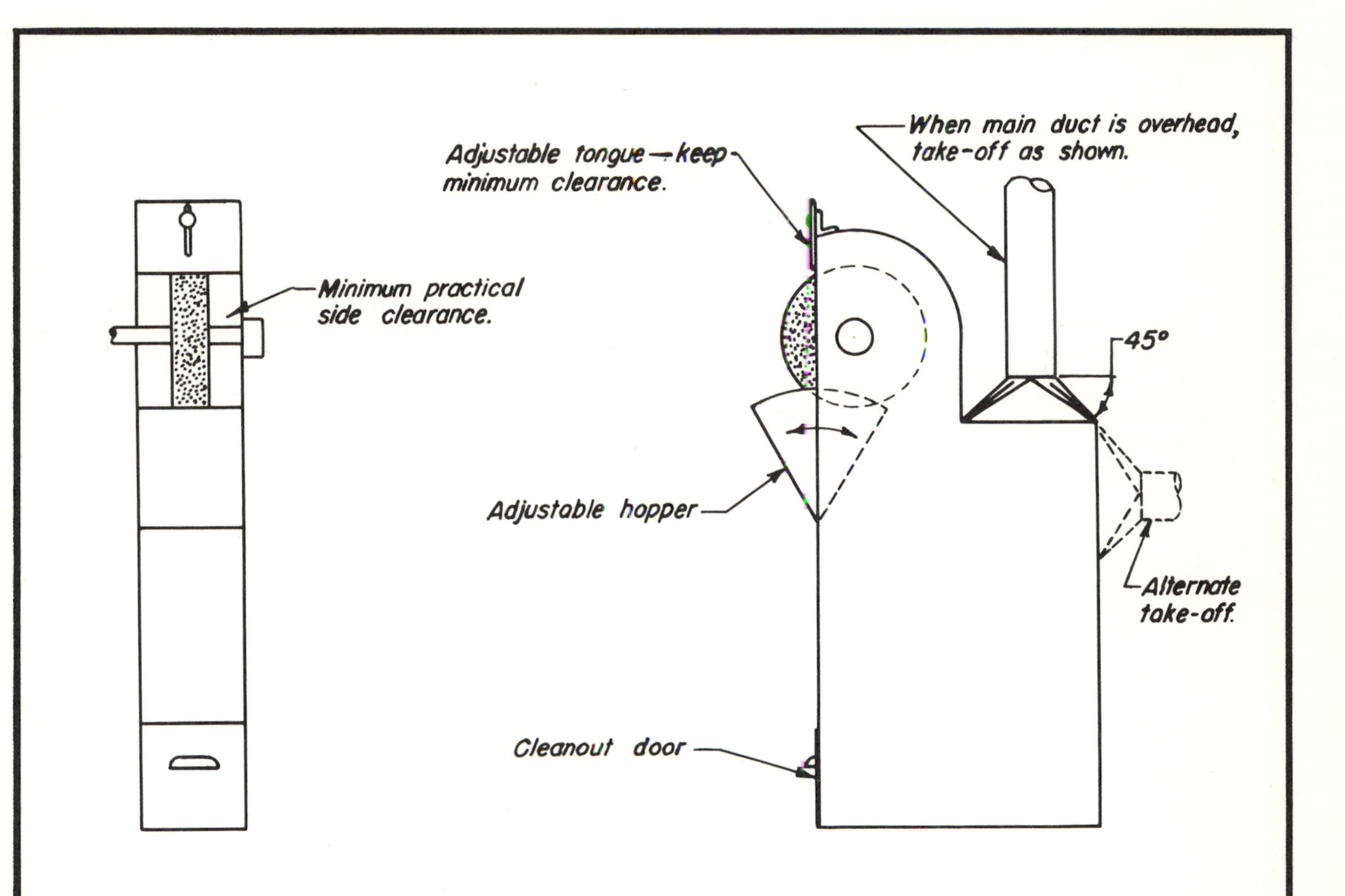

Wheel diam inches	Wheel width inches	Exhaust volume cfm
to 9	2	400
over 9 to 16	3	610
over 16 to 19	4	740
over 19 to 24	5	1200
over 24 to 30	6	1500
over 30 to 36	6	1900

Note: For wider wheels than listed, increase cfm with width
Duct velocity = 4500 fpm minimum.
Entry loss = 0.40 VP

AMERICAN CONFERENCE OF GOVERNMENTAL INDUSTRIAL HYGIENISTS

BUFFING LATHE

DATE 1-74 | VS-407

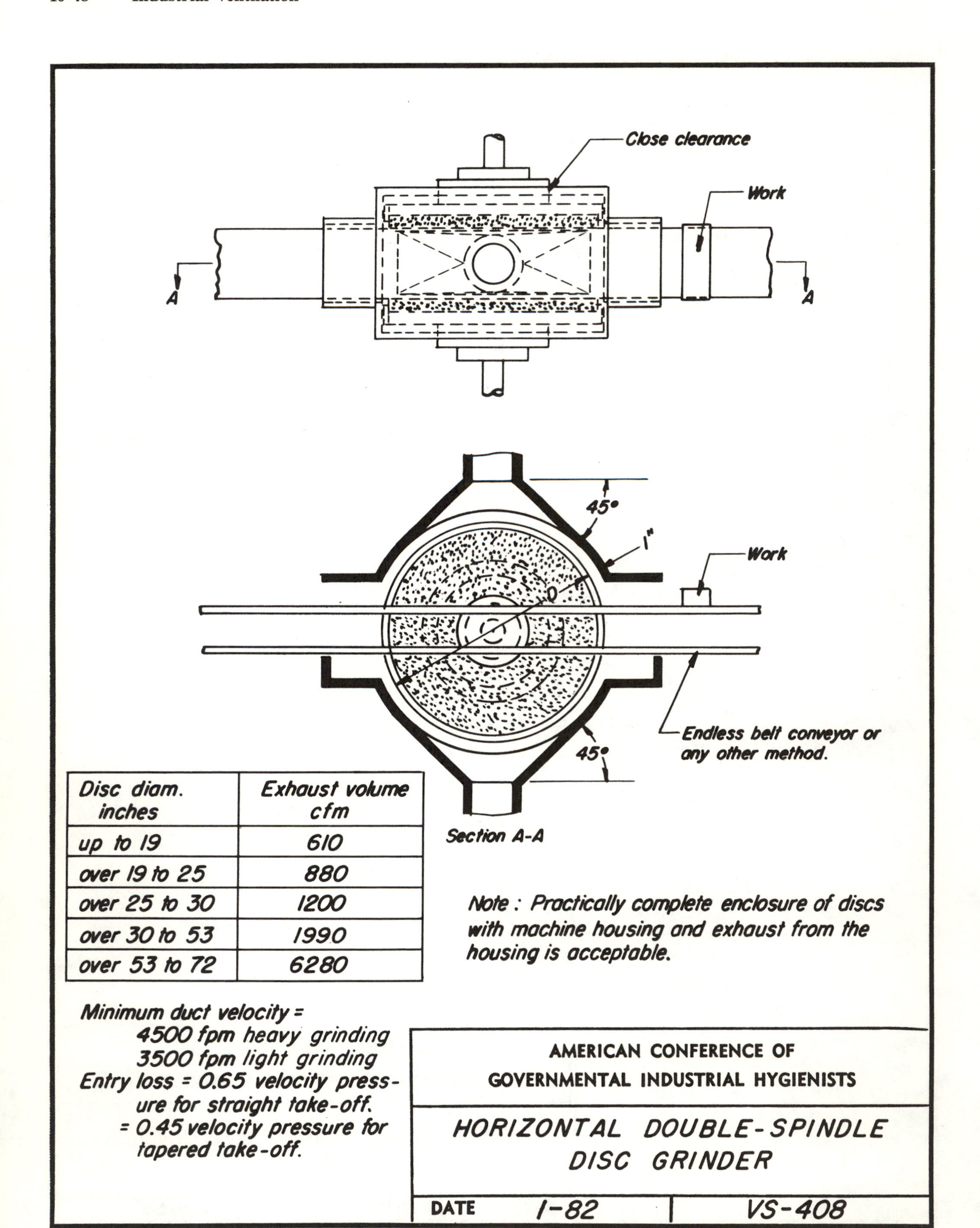

Disc diam. inches	Exhaust volume cfm
up to 19	610
over 19 to 25	880
over 25 to 30	1200
over 30 to 53	1990
over 53 to 72	6280

Note : Practically complete enclosure of discs with machine housing and exhaust from the housing is acceptable.

Minimum duct velocity =
4500 fpm heavy grinding
3500 fpm light grinding
Entry loss = 0.65 velocity pressure for straight take-off.
= 0.45 velocity pressure for tapered take-off.

AMERICAN CONFERENCE OF GOVERNMENTAL INDUSTRIAL HYGIENISTS

HORIZONTAL DOUBLE-SPINDLE DISC GRINDER

DATE 1-82	VS-408

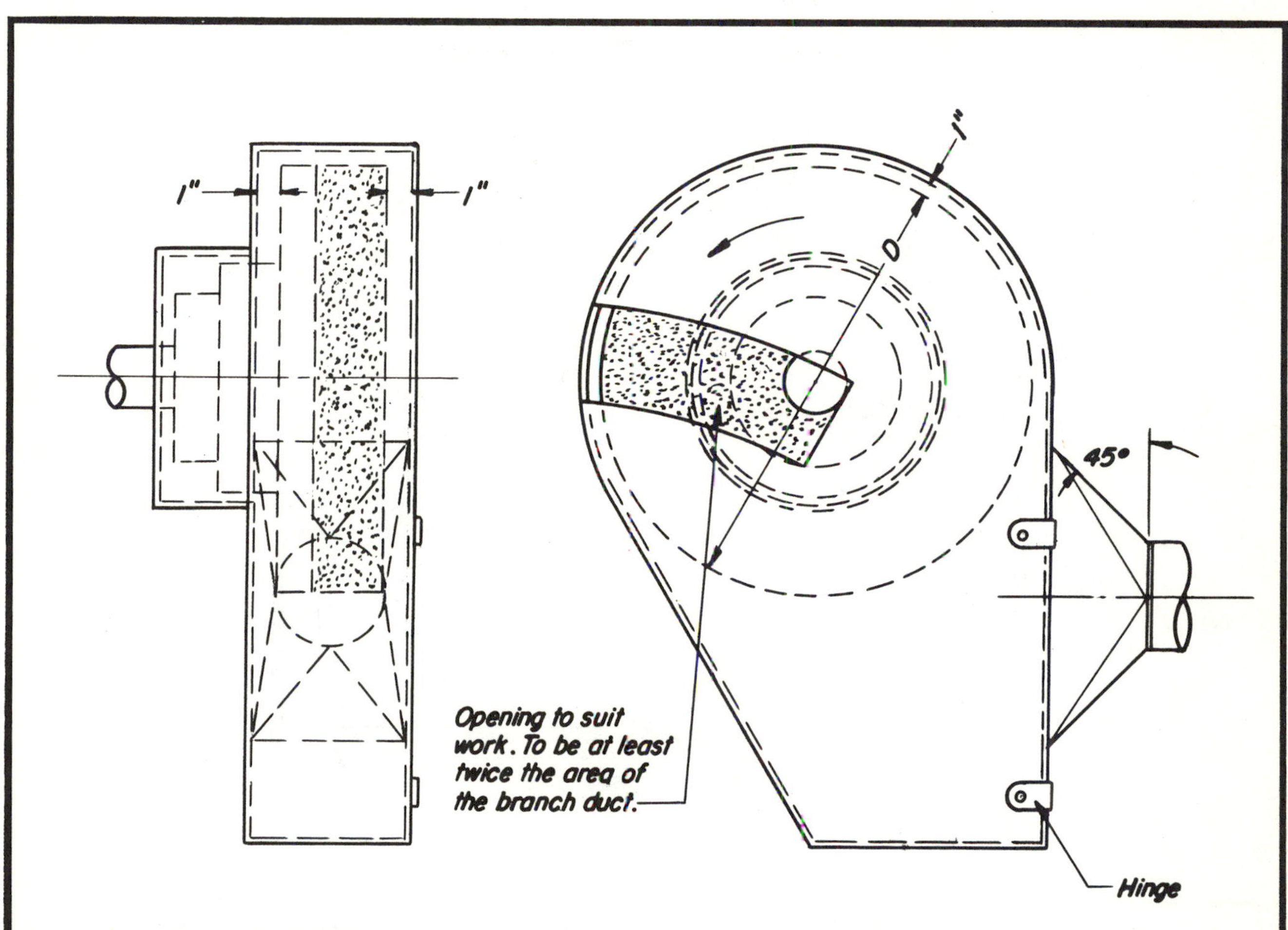

Disc diam., inches	Exhaust volume, cfm
Up to 12	220
over 12 to 19	390
over 19 to 30	610
over 30 to 36	880

Minimum duct velocity = 4500 fpm heavy grinding, 3500 fpm light grinding
Entry loss = 0.65 VP for straight take-off.
= 0.45 VP for tapered take-off.

Note: If best practical hood is a poor enclosure, increase exhaust volume accordingly.

AMERICAN CONFERENCE OF GOVERNMENTAL INDUSTRIAL HYGIENISTS

HORIZONTAL SINGLE-SPINDLE DISC GRINDER

DATE 1-82 | VS-409

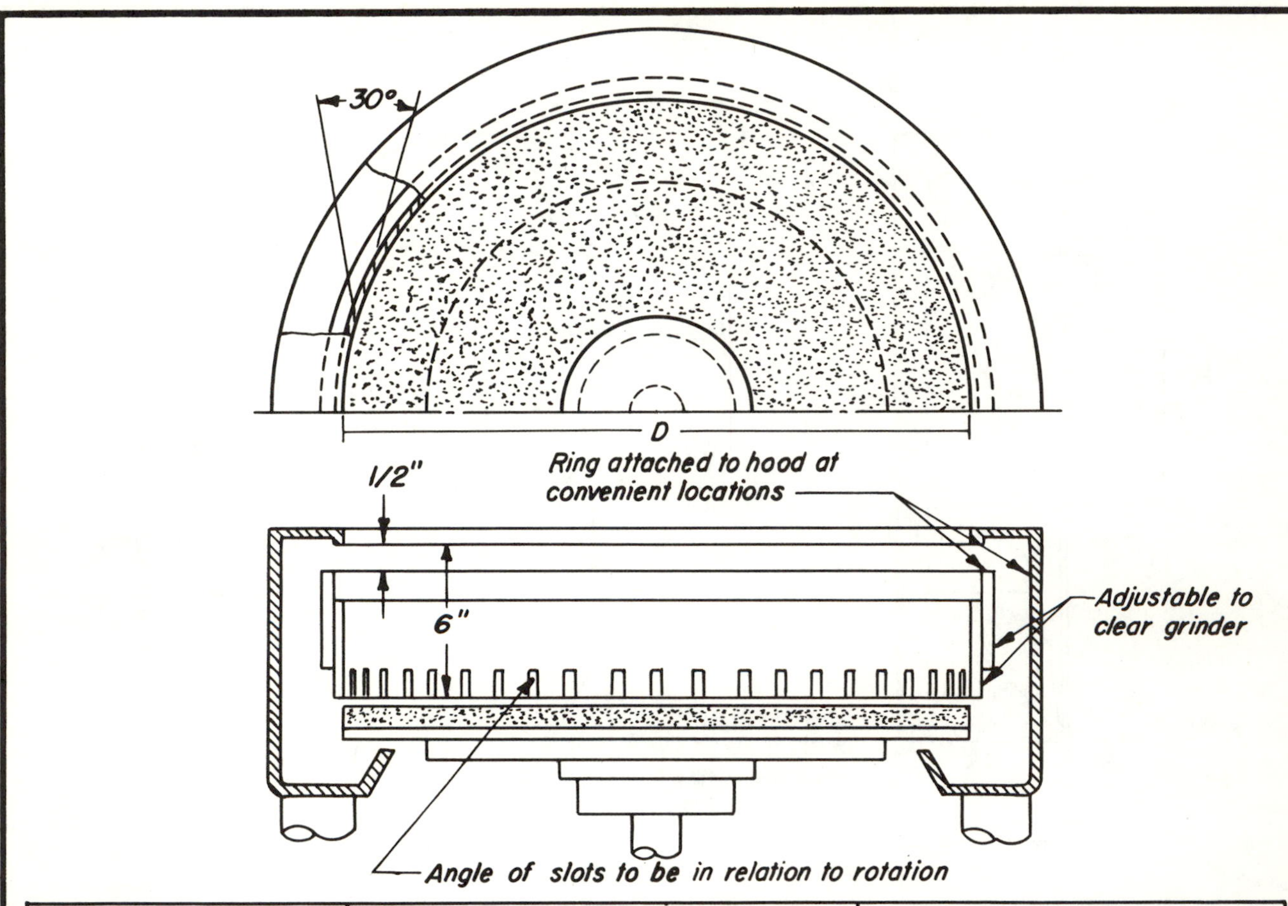

Disc diam, inches	1/2 or more of disc covered		Disc not covered	
	No.*	Exhaust, cfm	No.*	Exhaust, cfm
up to 20	1	500	2	780
over 20 to 30	2	780	2	1480
over 30 to 53	2	1770	4	3530
over 53 to 72	2	3140	5	6010

* Number of exhaust outlets around periphery of hood; or equal distribution provided by other means.

Slot velocity = 2000 fpm
Minimum duct velocity = 4500 fpm heavy grinding
3500 fpm light grinding
Entry loss = 1.0 slot VP + 0.5 branch duct VP

AMERICAN CONFERENCE OF GOVERNMENTAL INDUSTRIAL HYGIENISTS

VERTICAL SPINDLE DISC GRINDER

DATE 1-82 | VS-410

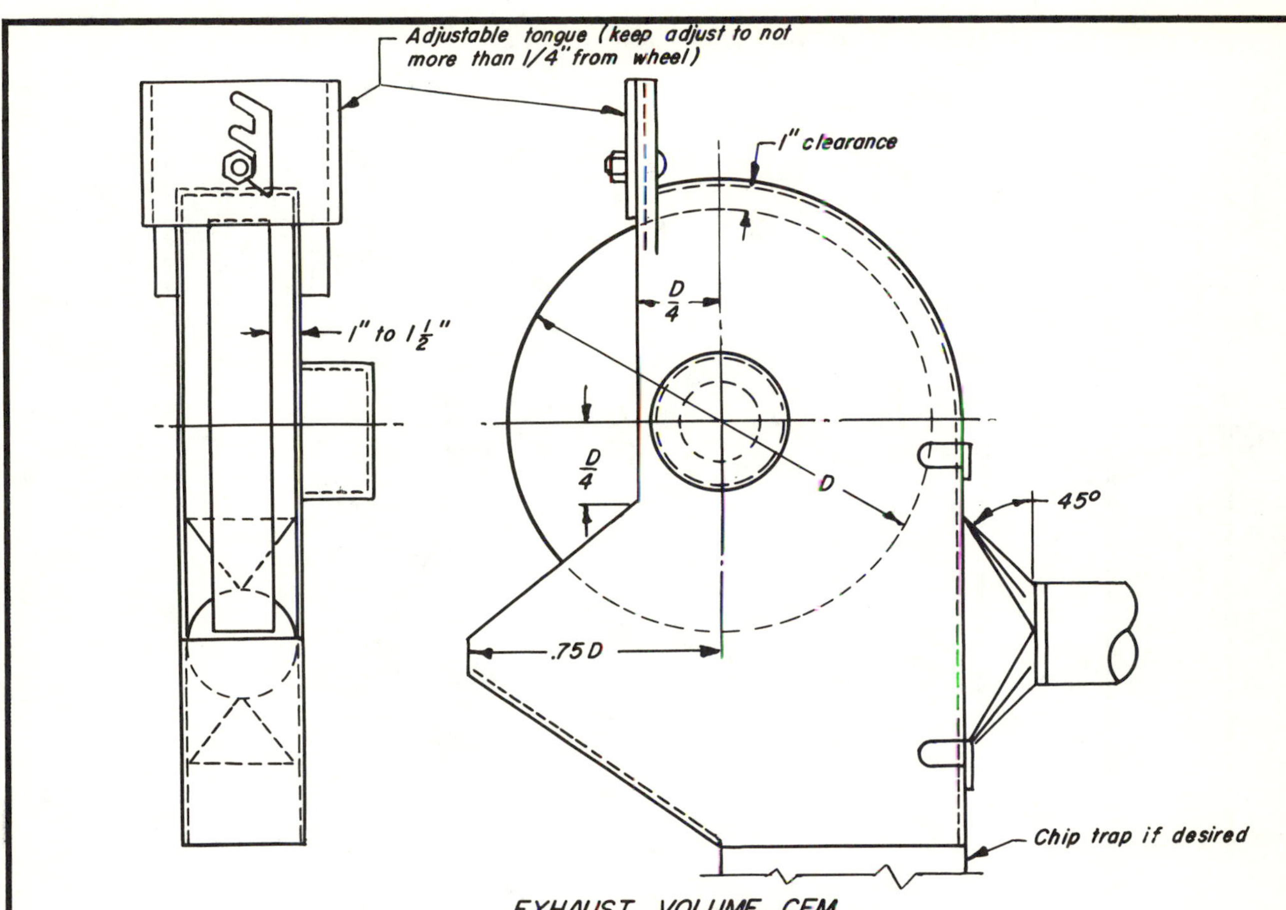

EXHAUST VOLUME, CFM

Wheel diam inches	Wheel width inches	Good enclosure*	Poor enclosure
to 5	1	220	220
over 5 to 10	11/2	220	300
over 10 to 14	2	300	500
over 14 to 16	2	390	610
over 16 to 20	3	500	740
over 20 to 24	4	610	880
over 24 to 30	5	880	1200
over 30 to 36	6	1200	1570

* No more than 25% of wheel exposed.

Minimum duct velocity = 4500 fpm heavy grinding
3500 fpm light grinding

Entry loss = 0.65 VP for straight takeoff
0.40 VP for tapered takeoff

AMERICAN CONFERENCE OF GOVERNMENTAL INDUSTRIAL HYGIENISTS

GRINDER WHEEL HOOD
SPEEDS BELOW 6500 sfm

DATE 1-82 | VS-411

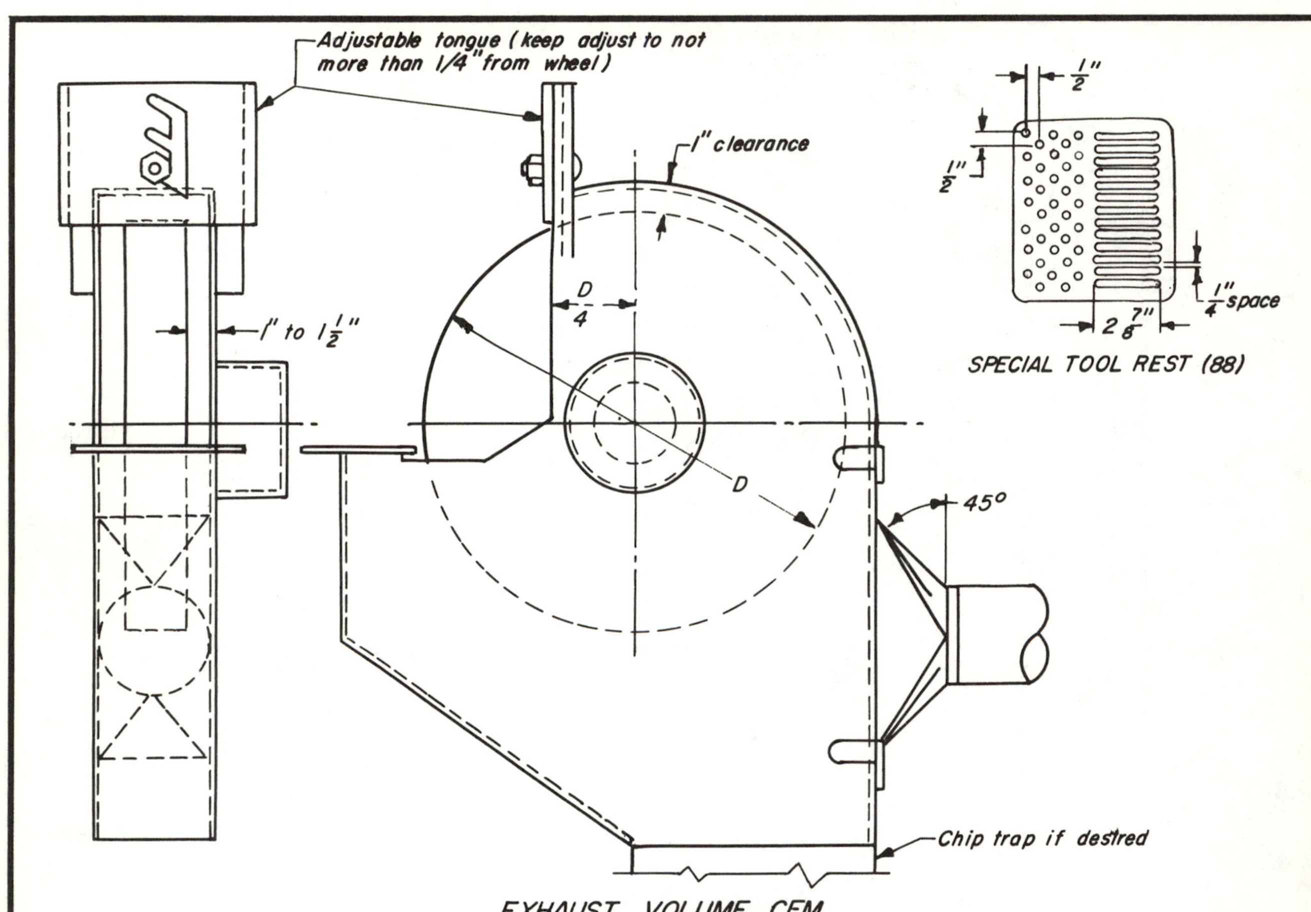

EXHAUST VOLUME, CFM

Wheel diam inches	Wheel width inches	Good enclosure *	Poor enclosure
to 5	1	220	390
over 5 to 10	11/2	390	610
over 10 to 14	2	500	740
over 14 to 16	2	610	880
over 16 to 20	3	740	1040
over 20 to 24	4	880	1200
over 24 to 30	5	1200	1570
over 30 to 36	6	1570	1990

* Special hood and tool rest as shown.

Minimum duct velocity = 4500 fpm heavy grinding
3500 fpm light grinding

Entry loss = 0.65 VP for straight takeoff
0.40 VP for tapered takeoff

AMERICAN CONFERENCE OF GOVERNMENTAL INDUSTRIAL HYGIENISTS

GRINDER WHEEL HOOD
SPEEDS ABOVE 6500 sfm

DATE 1-82 | VS-411.1

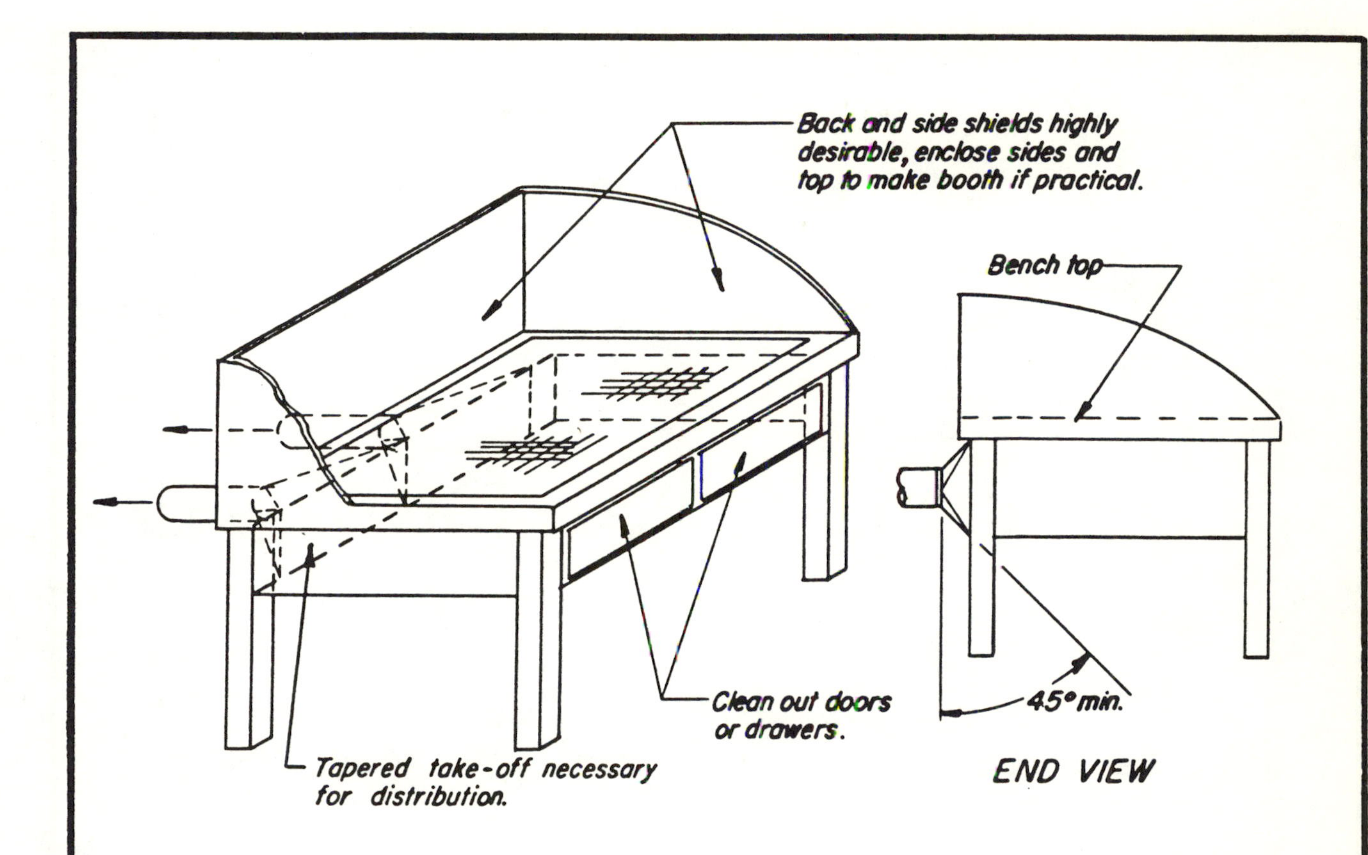

Q = 150 - 250 cfm / sq ft of bench area.
Minimum duct velocity = 3500 fpm
Entry loss = 0.25 VP for tapered take-off.

Grinding in booth, 100 fpm face velocity also suitable.

For downdraft grilles in floor: Q = 100 cfm / sq ft of working area.

Provide equal distribution. Provide for cleanout.

AMERICAN CONFERENCE OF GOVERNMENTAL INDUSTRIAL HYGIENISTS	
PORTABLE HAND GRINDING	
DATE 1-64	VS-412

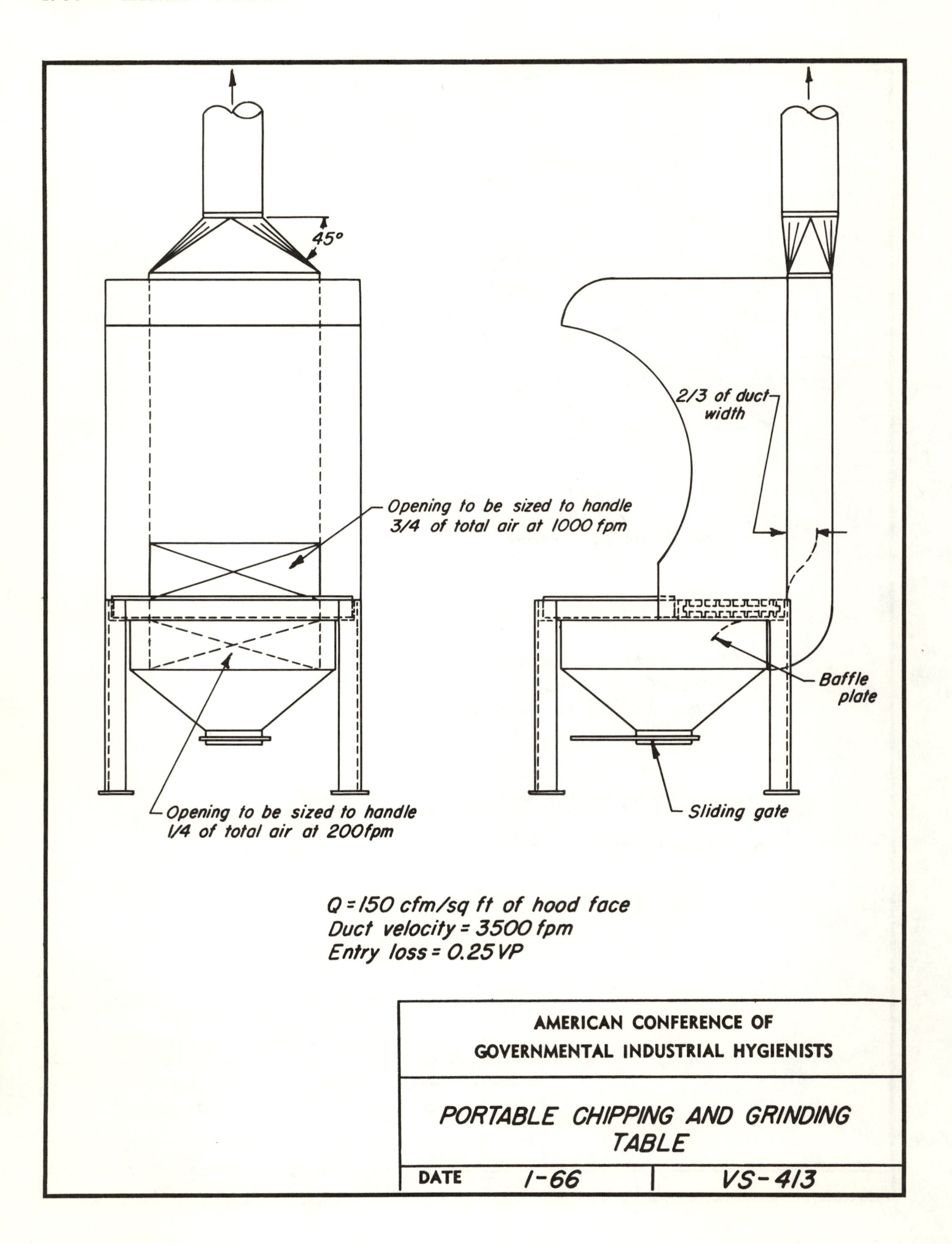
45°
2/3 of duct width
Opening to be sized to handle 3/4 of total air at 1000 fpm
Baffle plate
Opening to be sized to handle 1/4 of total air at 200fpm
Sliding gate
Q = 150 cfm/sq ft of hood face
Duct velocity = 3500 fpm
Entry loss = 0.25 VP
AMERICAN CONFERENCE OF
GOVERNMENTAL INDUSTRIAL HYGIENISTS
PORTABLE CHIPPING AND GRINDING TABLE
DATE 1-66
VS-413

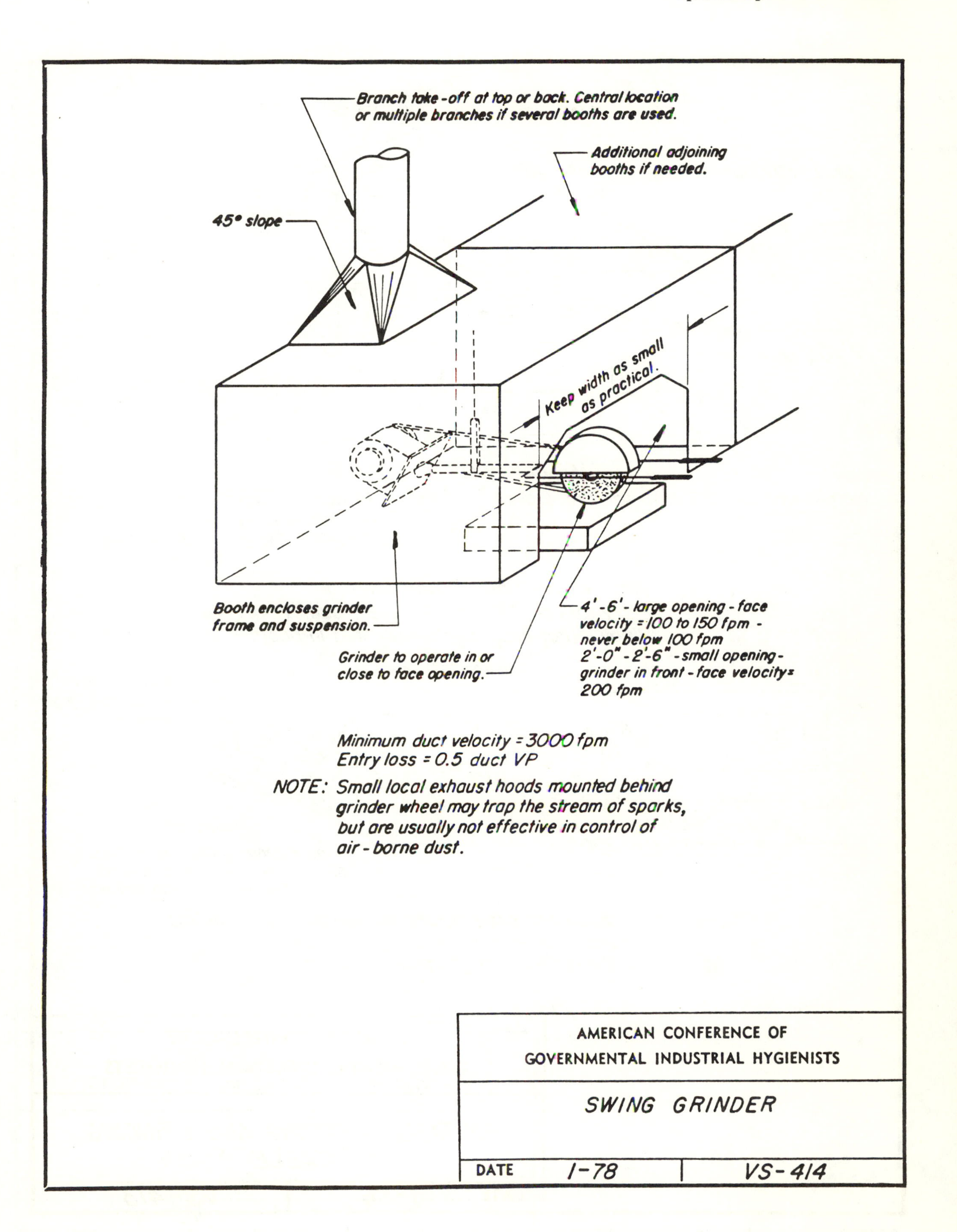
Branch take-off at top or back. Central location or multiple branches if several booths are used.
Additional adjoining booths if needed.
45° slope
Keep width as small as practical.
Booth encloses grinder frame and suspension.
Grinder to operate in or close to face opening.
4'-6' - large opening - face velocity = 100 to 150 fpm - never below 100 fpm
2'-0" - 2'-6" - small opening - grinder in front - face velocity = 200 fpm
Minimum duct velocity = 3000 fpm
Entry loss = 0.5 duct VP
NOTE: Small local exhaust hoods mounted behind grinder wheel may trap the stream of sparks, but are usually not effective in control of air-borne dust.
AMERICAN CONFERENCE OF GOVERNMENTAL INDUSTRIAL HYGIENISTS
SWING GRINDER
DATE 1-78
VS-414

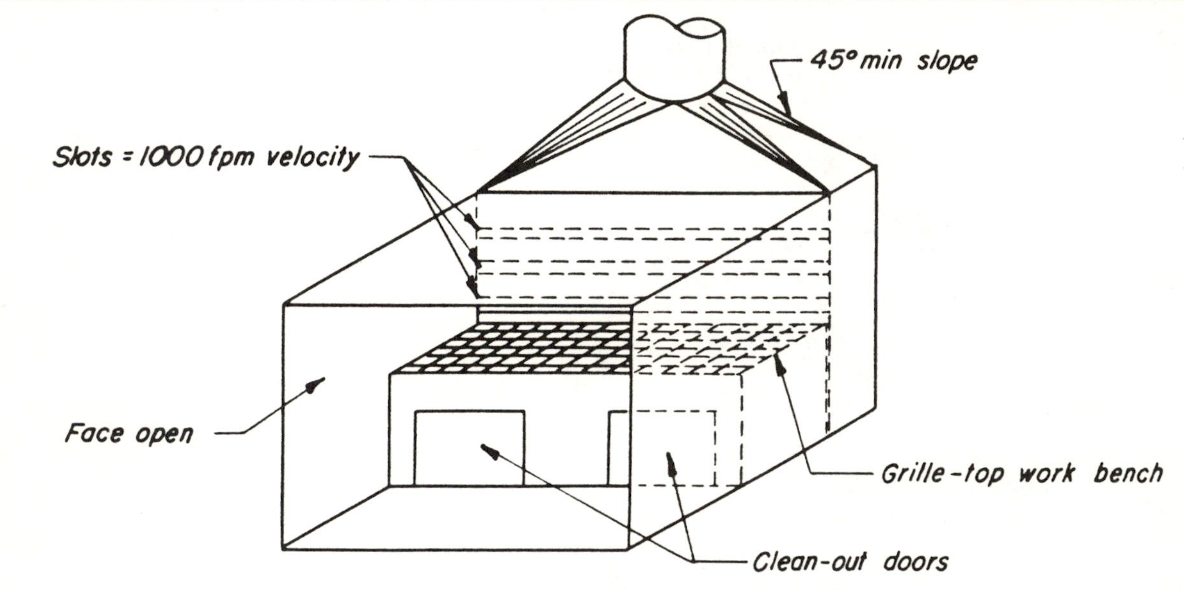

METALLIZING BOOTH

Non-toxic: Q = 125 cfm/sq ft face area

Toxic: Provide NIOSH certified air-supplied respirator. Q = 200 cfm/sq ft face area

Duct velocity = 3000 fpm minimum
Entry loss = 1.78 slot VP + 0.25 duct VP
Small lathe, etc., may be mounted in booth

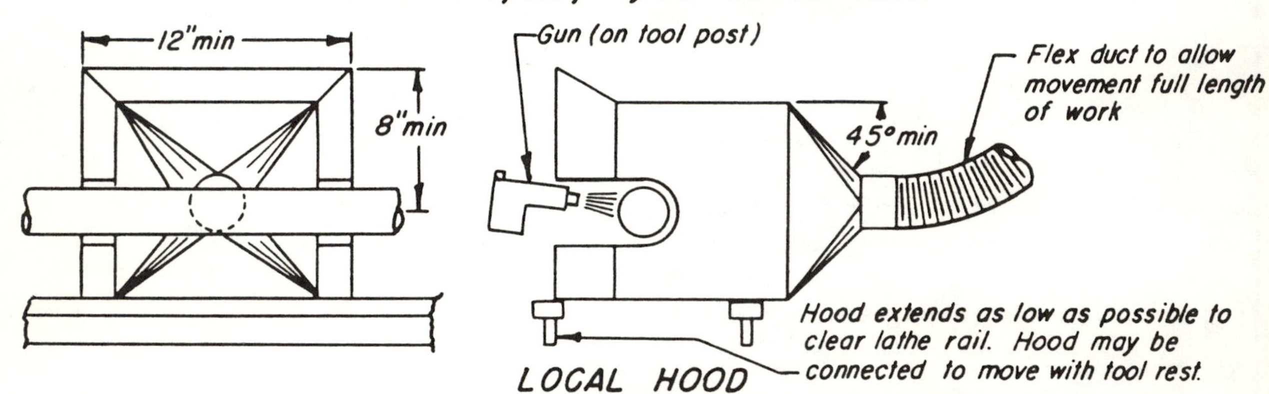

LOCAL HOOD

Note: Local hood not satisfactory for spraying toxic metals.

Q = 200 cfm/sq ft face openings
Duct velocity = 3500 fpm minimum
Entry loss = 0.25 VP

AMERICAN CONFERENCE OF GOVERNMENTAL INDUSTRIAL HYGIENISTS

METAL SPRAYING

DATE 1-84 | VS-415

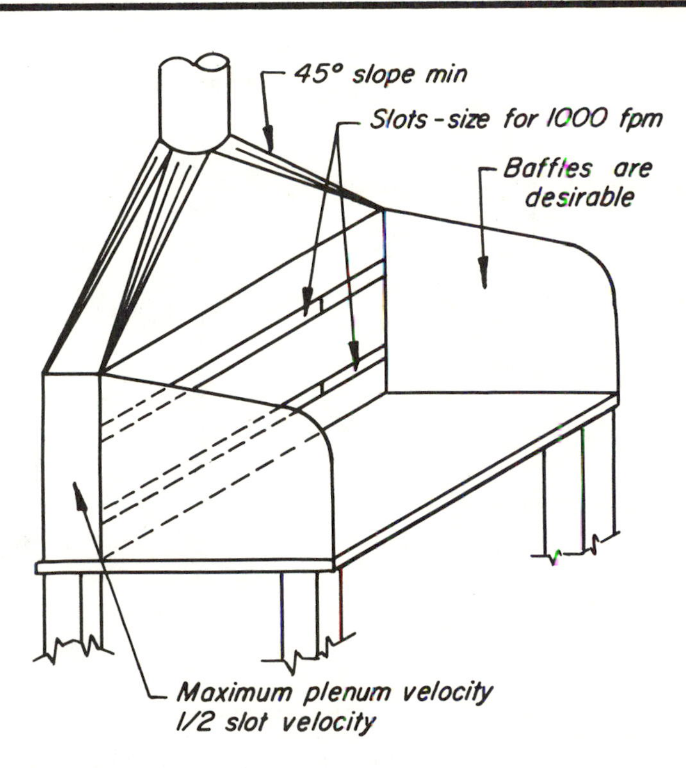

Q = 350 cfm/lineal ft of hood
Hood length = required working space
Bench width = 24" maximum
Duct velocity = 1000 - 3000 fpm
Entry loss = 1.78 slot VP + 0.25 duct VP

GENERAL VENTILATION, where local exhaust cannot be used:

Rod, diam	cfm/welder*
5/32	1000
3/16	1500
1/4	3500
3/8	4500

OR

A. For open areas, where welding fume can rise away from the breathing zone:
cfm required = 800 × lb/hour rod used

B. For enclosed areas or positions where fume does not readily escape breathing zone:
cfm required = 1600 × lb/hour rod used

*For toxic materials higher airflows are necessary and operator may require respiratory protection equipment.

OTHER TYPES OF HOODS

Local exhaust: See VS-416.1
Booth: For design See VS-415, VS-604
Q = 100 cfm/sq ft of face opening

AMERICAN CONFERENCE OF GOVERNMENTAL INDUSTRIAL HYGIENISTS

WELDING BENCH

DATE 1-76 | VS-416

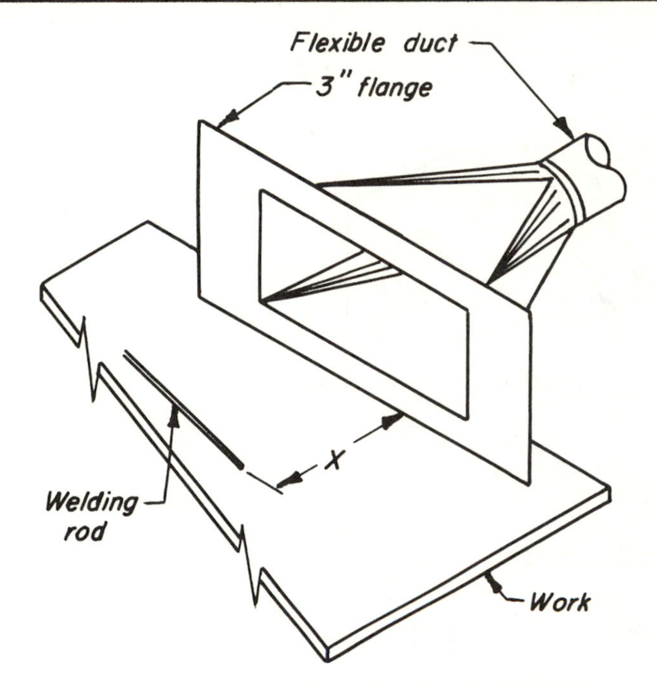

PORTABLE EXHAUST

X, inches	Plain duct cfm	Flange or cone cfm
up to 6	335	250
6 - 9	755	560
9 -12	1335	1000

Face velocity = 1500 fpm
Duct velocity = 3000 fpm minimum
Plain duct entry loss = 0.93 duct VP
Flange or cone entry loss = 0.25 duct VP

GENERAL VENTILATION, where local exhaust cannot be used:

Rod, diam	cfm/welder
5/32	1000
3/16	1500
1/4	3500
3/8	4500

OR

A. For open areas, where welding fume can rise away from the breathing zone:
cfm required = 800 x lb/hour rod used

B. For enclosed areas or positions where fume does not readily escape breathing zone:
cfm required = 1600 x lb/hour rod used

For toxic materials higher airflows are necessary and operator may require respiratory protection equipment.

OTHER TYPES OF HOODS

Bench: See VS-416
Booth: For design See VS-415, VS-604
Q=100 cfm/sq ft of face opening
"Granite Cutting" VS-909

AMERICAN CONFERENCE OF GOVERNMENTAL INDUSTRIAL HYGIENISTS

WELDING BENCH

DATE 1-78 | VS-416.1

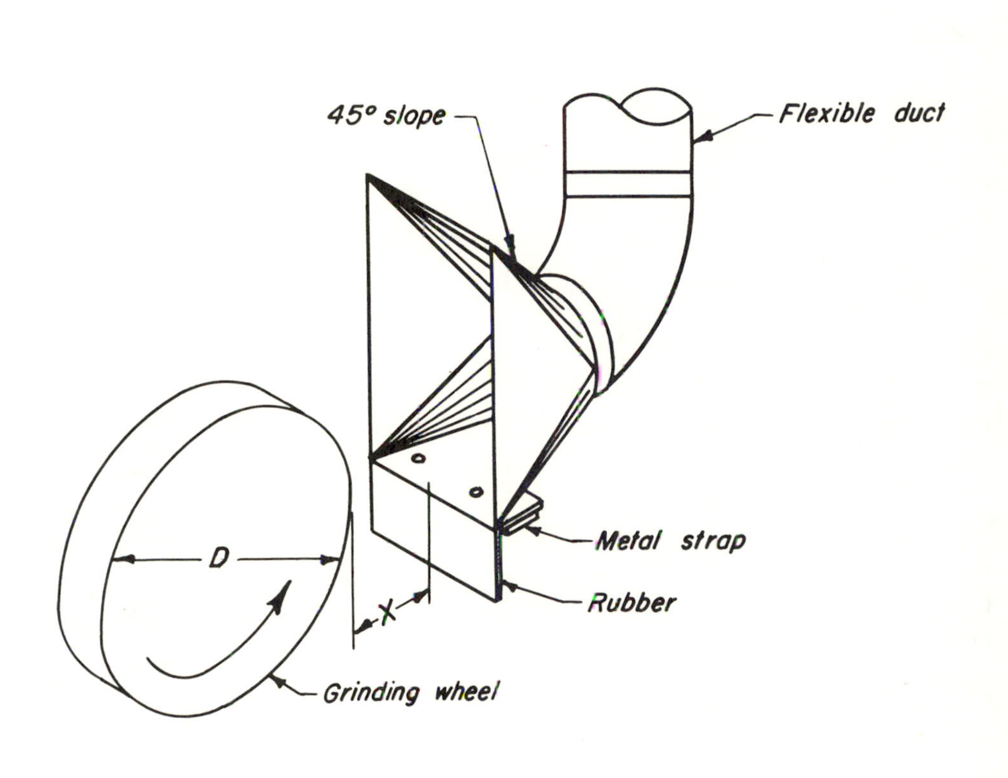

$Q = 0.0003\ Vs(10X^2 + A)$

Entry loss = 0.25 VP

Duct velocity = 3500 fpm minimum

X = Hood wheel distance in inches (measured from center of hood face to nearest point on wheel surface)

A = Hood face area (sq in)

Vs = Wheel surface speed (sfm)
$= 3.1416 \times rpm \times \frac{D}{12}$

D = Diam. in inches

Example for
X = 4"
A = 3" x 4 1/2"

Vs	Q, cfm
1 000	52
2 000	104
3 000	156
4 000	208
5 000	260
6 000	312
7 000	364
8 000	416
9 000	468
10 000	520

(Ref 123)

AMERICAN CONFERENCE OF GOVERNMENTAL INDUSTRIAL HYGIENISTS

SURFACE GRINDER

DATE 1-76 | VS-417

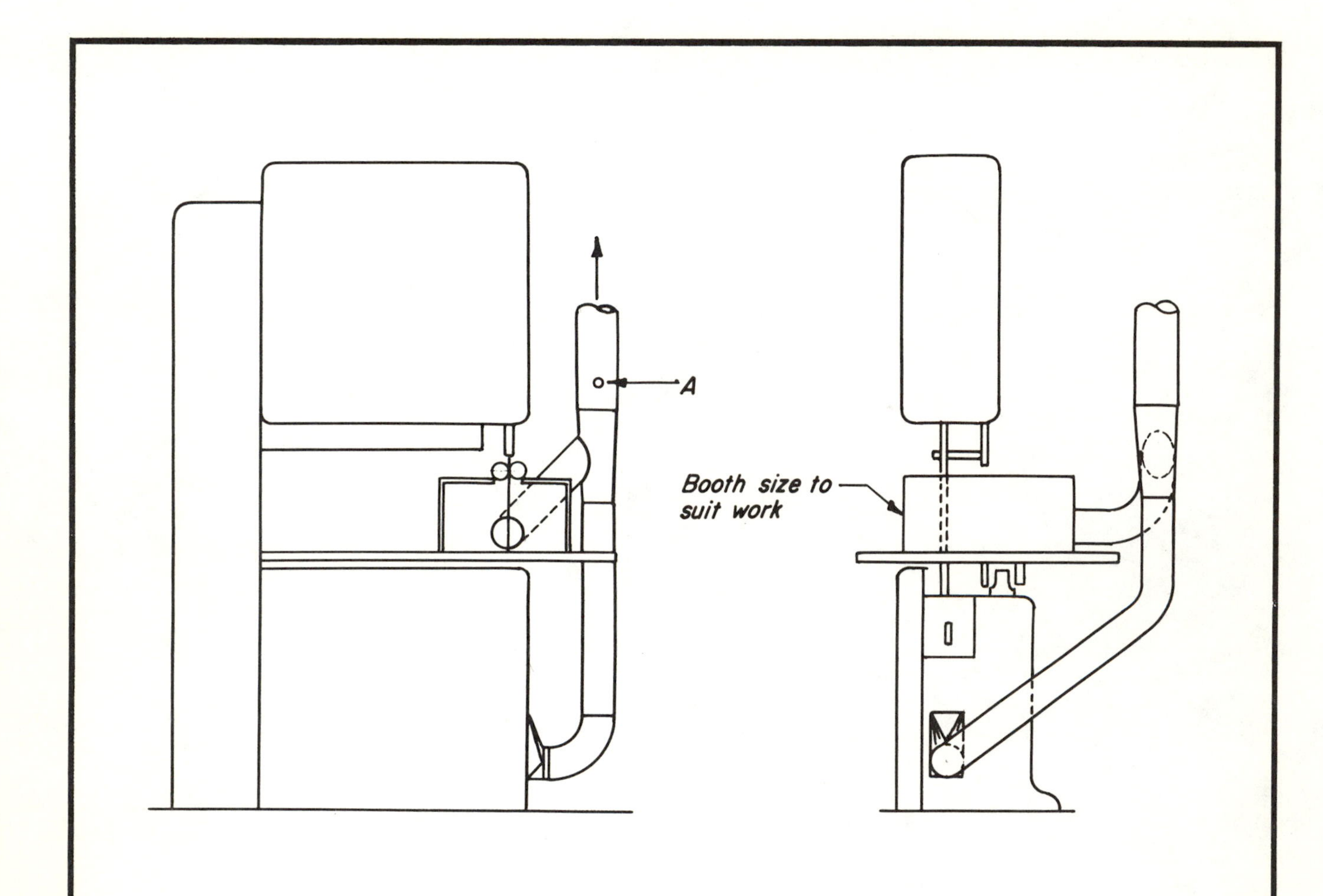

Q, booth = 225 cfm/sq ft open area
Q, bottom = 350 cfm
Duct velocity = 4000 fpm
Entry loss = 1.75 VP in riser (point A)

AMERICAN CONFERENCE OF GOVERNMENTAL INDUSTRIAL HYGIENISTS	
METAL CUTTING BANDSAW	
DATE *1-70*	*VS-418*

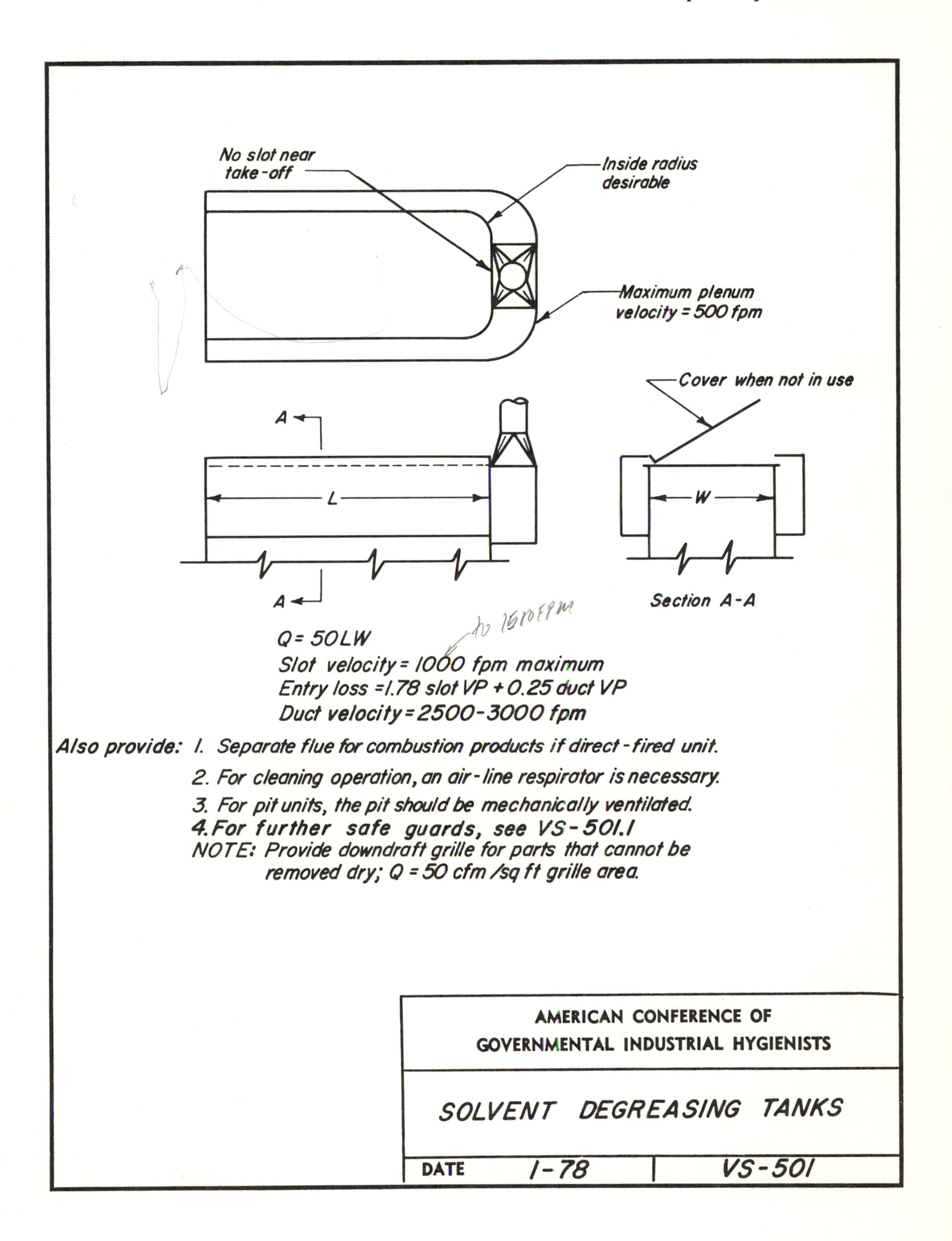
No slot near take-off
Inside radius desirable
Maximum plenum velocity = 500 fpm
Cover when not in use
A
L
A
W
Section A-A
Q = 50LW
Slot velocity = 1000 fpm maximum
Entry loss = 1.78 slot VP + 0.25 duct VP
Duct velocity = 2500-3000 fpm
Also provide: 1. Separate flue for combustion products if direct-fired unit.
2. For cleaning operation, an air-line respirator is necessary.
3. For pit units, the pit should be mechanically ventilated.
4. For further safe guards, see VS-501.1
NOTE: Provide downdraft grille for parts that cannot be removed dry; Q = 50 cfm/sq ft grille area.
AMERICAN CONFERENCE OF GOVERNMENTAL INDUSTRIAL HYGIENISTS
SOLVENT DEGREASING TANKS
DATE 1-78
VS-501

Solvent vapor degreasing refers to boiling liquid cleaning systems utilizing trichloroethylene, perchloroethylene, methylene chloride, Freons®, or other halogenated hydrocarbons. Cleaning action is accomplished by the condensation of the solvent vapors in contact with the work surface producing a continuous liquid rinsing action. Cleaning ceases when the temperature of the work reaches the temperature of the surrounding solvent vapors. Since halogenated hydrocarbons are somewhat similar in their physical, chemical, and toxic characteristics, the following safeguards should be provided to prevent the creation of a health or life hazard:

1. Vapor degreasing tanks should be equipped with a condenser or vapor level thermostat to keep the vapor level below the top edge of the tank by a distance equal to one-half the tank width or 36 inches, whichever is shorter.

2. Where water-type condensers are used, inlet water temperatures should not be less than 80 F (27 C) and the outlet temperature should not exceed 110 F (43 C).

3. Degreasers should be equipped with a boiling liquid thermostat to regulate the rate of vapor generation, and with a safety control at an appropriate height above the vapor line to prevent the escape of solvent in case of a malfunction.

4. Tanks or machines of more than 4 square feet of vapor area should be equipped with suitable gasketed cleanout or sludge doors, located near the bottom, to facilitate cleaning.

5. Work should be placed in and removed slowly from the degreaser, at a rate no greater than 11 ft/min (0.055 m/s), to prevent sudden disturbances of the vapor level.

6. CARE MUST BE TAKEN TO PREVENT DIRECT SOLVENT CARRYOUT DUE TO THE SHAPE OF THE PART. Maximum rated workloads as determined by the rate of heat transfer (surface area and specific heat) should not be exceeded.

7. Special precautions should be taken where natural gas or other open flames are used to heat the solvent to prevent vapors* from entering the combustion air supply.

8. Heating elements should be designed and maintained so that their surface temperature will not cause the solvent or mixture to breakdown* or produce excessive vapors.

9. Degreasers should be located in such a manner that vapors* will not reach or be drawn into atmospheres used for gas or electric arc welding, high temperature heat treating, combustion air, or open electric motors.

10. Whenever spray or other mechanical means are used to disperse solvent liquids, sufficient enclosure or baffling should be provided to prevent direct release of airborne vapor above the top of the tank.

11. An emergency quick-drenching facility should be located in near proximity to the degreaser for use in the event of accidental eye contact with the degreasing liquid.

---

* Electric arcs, open flames, and hot surfaces will thermally decompose halogenated hydrocarbons to toxic and corrosive substances (such as hydrochloric and/or hydrofluoric acid). Under some circumstances, phosgene may be formed.

AMERICAN CONFERENCE OF GOVERNMENTAL INDUSTRIAL HYGIENISTS

SOLVENT VAPOR DEGREASING

DATE 1-78 | VS-501.1

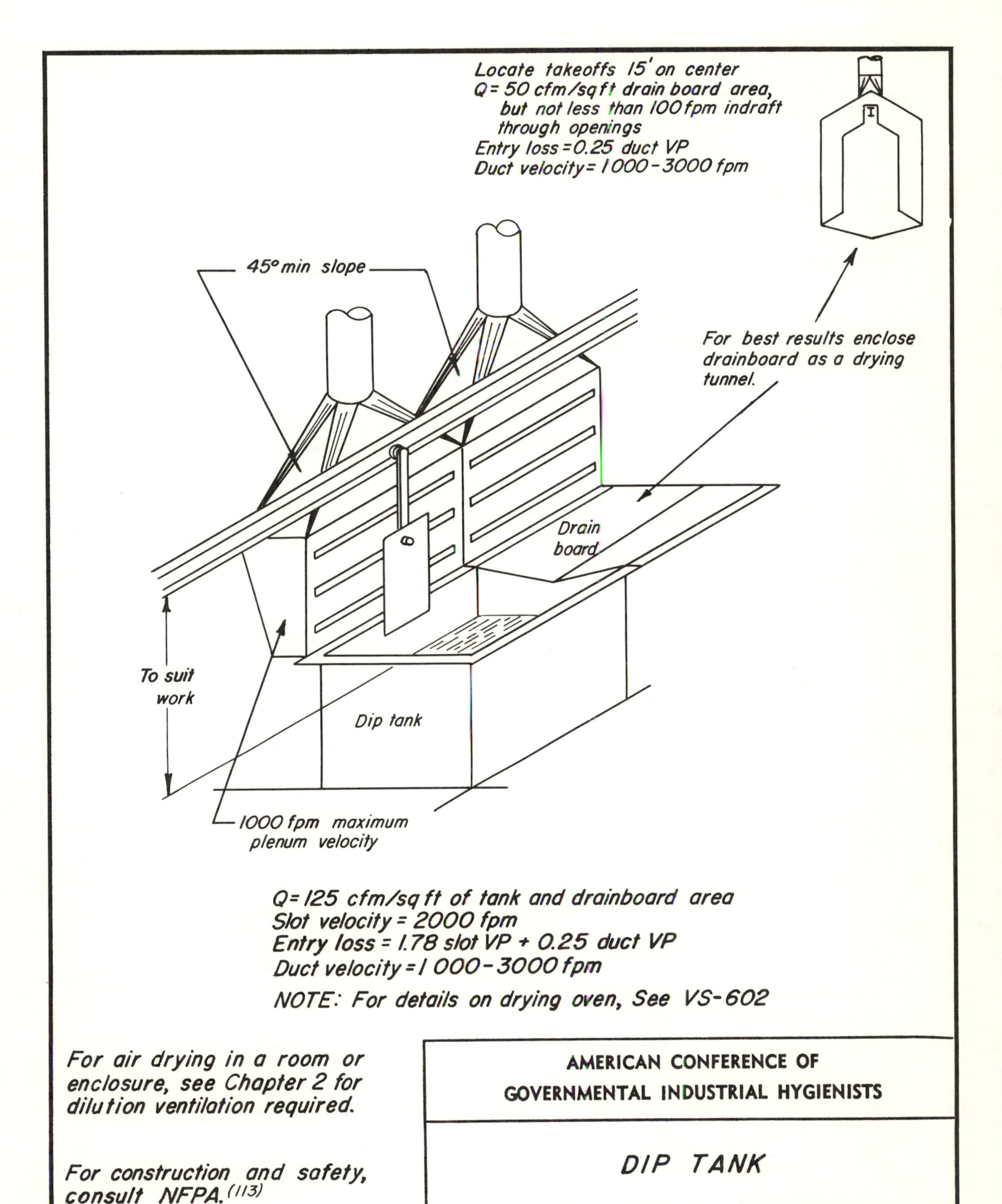

For air drying in a room or enclosure, see Chapter 2 for dilution ventilation required.

For construction and safety, consult NFPA.[113]

AMERICAN CONFERENCE OF GOVERNMENTAL INDUSTRIAL HYGIENISTS

DIP TANK

DATE 1-78 | VS-502

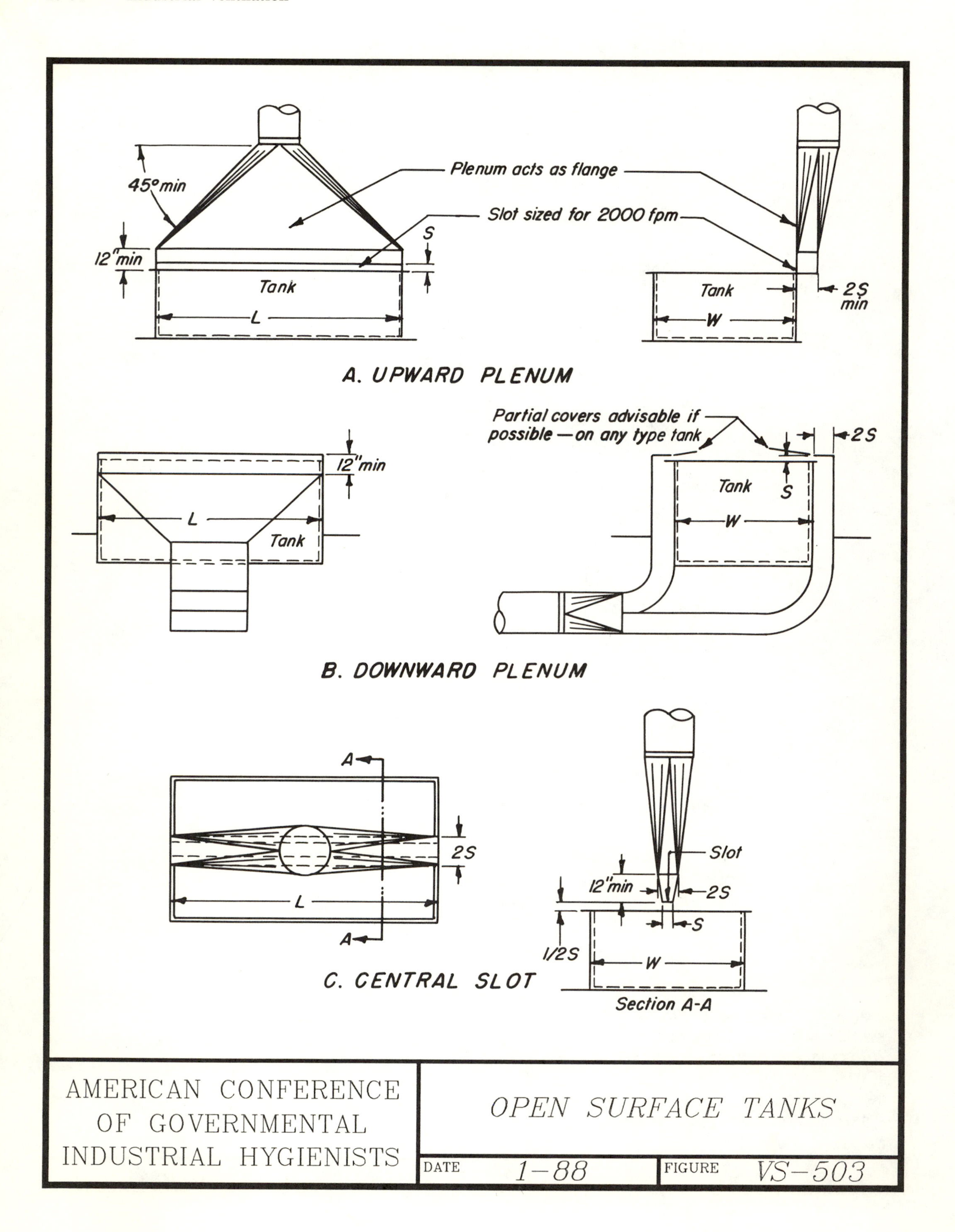

Plenum acts as flange
45° min
Slot sized for 2000 fpm
12" min
S
Tank
L
Tank
W
2S min
A. UPWARD PLENUM
Partial covers advisable if possible — on any type tank
2S
12" min
L
Tank
Tank
S
W
B. DOWNWARD PLENUM
A
2S
L
A
Slot
12" min
2S
S
1/2S
W
C. CENTRAL SLOT
Section A-A
AMERICAN CONFERENCE OF GOVERNMENTAL INDUSTRIAL HYGIENISTS
OPEN SURFACE TANKS
DATE 1-88
FIGURE VS-503

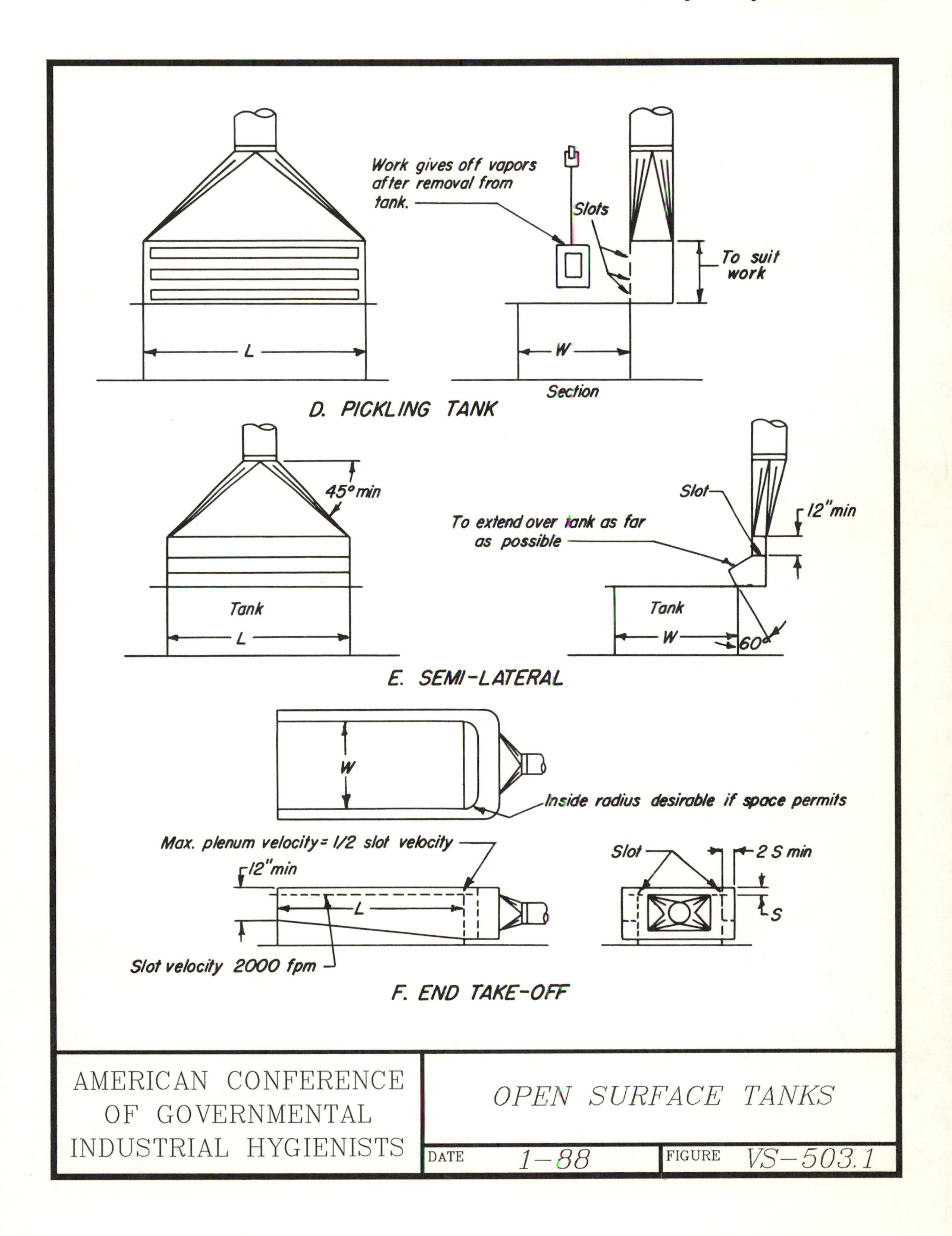
Work gives off vapors after removal from tank.
Slots
To suit work
L
W
Section
D. PICKLING TANK
45° min
Slot
12" min
To extend over tank as far as possible
Tank
L
Tank
W
60°
E. SEMI-LATERAL
W
Inside radius desirable if space permits
Max. plenum velocity = 1/2 slot velocity
12" min
L
Slot velocity 2000 fpm
Slot
2 S min
S
F. END TAKE-OFF
AMERICAN CONFERENCE OF GOVERNMENTAL INDUSTRIAL HYGIENISTS
OPEN SURFACE TANKS
DATE 1-88
FIGURE VS-503.1

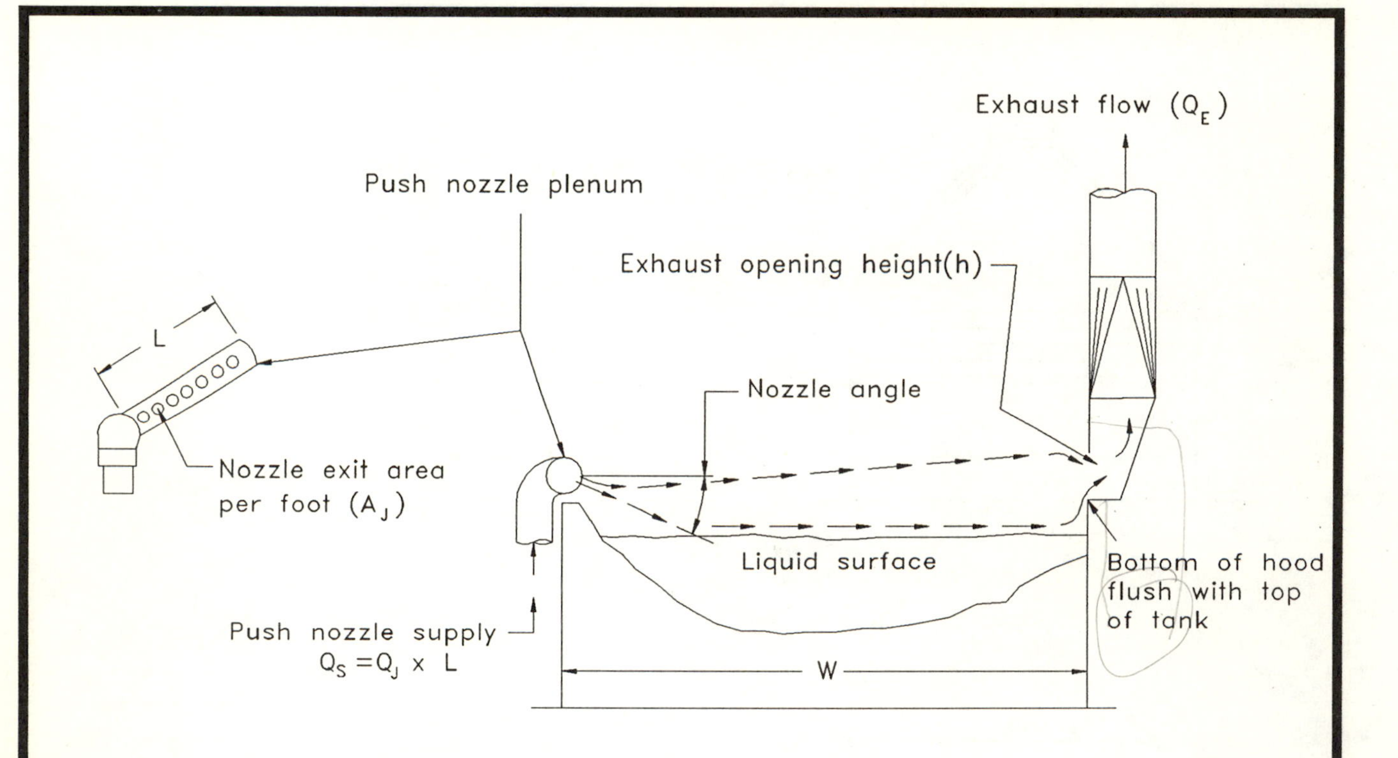

Nozzle openings — 1/8" to 1/4" slot. or 1/4" diameter holes with 3/4" to 2" spacing.

Push nozzle plenum — Circular, rectangular or square. plenum cross-sectional area should be at least 3 times the total nozzle area.

Nozzle angle — 0° to 20° down.

Push nozzle supply — $Q_J = 243\sqrt{A_J}$ cfm/ft of nozzle plenum length

Exhaust flow — $Q_E = 75$ cfm/ft^2 of tank surface area

Exhaust opening height — h=0.14W

Tank surface dimensions — L=length of tank
W=width of tank

Design procedure: Select nozzle opening within above limits and calculate pushnozzle supply

Reference 141,142

AMERICAN CONFERENCE OF GOVERNMENTAL INDUSTRIAL HYGIENISTS	*PUSH–PULL HOOD DESIGN DATA FOR WIDTHS UP TO 10'*	
	DATE *1–88*	FIGURE *VS–504*

In push-pull ventilation, a nozzle pushes a jet of air across the vessel surface into an exhaust hood. Effectiveness of a push jet is a function of its momentum which can be related to the product of the nozzle supply air flow (Q_J) and the nozzle exit velocity (V_J). For a jet used for plating tanks or other open surface vessels, a push supply flow can be determined from:

$$Q_J = 243 \sqrt{A_J}$$

Where: Q_J = push nozzle supply, cfm per foot of push nozzle plenum length
A_J = push nozzle exit area, ft^2 per foot of push nozzle plenum length

Using this approach, a push nozzle design is first selected and the nozzle exit area (A_J) determined.

The push nozzle plenum may be round, rectangular, or square in cross-section. The push nozzle may be a 1/8-inch to 1/4-inch horizontal slot or 1/4-inch diameter drilled holes on 3 and 8 diameter spacing.

It is important that the air flow from the nozzle be evenly distributed along the length of the supply plenum. To achieve this, the total nozzle exit area should not exceed 33% of the plenum cross-sectional area. Multiple supply plenum inlets should be used where practical.

The push nozzle should be located as near the vessel edge as possible to minimize the height above the liquid surface. The nozzle axis can be angled down a maximum of 20° to permit the jet to clear obstructions and to maintain the jet at the vessel surface. It is essential any opening between the nozzle and tank be sealed.

An exhaust flow of 75 cfm/ft^2 of vessel area will adequately capture and remove the push jet.[141,142] The exhaust hood opening (h) should be 0.14 times the distance from the push nozzle to the hood (0.14 W). If multiple slots are used, they should be located within the 0.14 W height. A flanged hood design is to be used where ever practical. The exhaust hood should be located at the vessel edge so as not to leave a gap between the hood and vessel.

Design and location of an open surface vessel encompasses a number of variables. In some cases, vessel shape, room location, cross drafts, etc., may create conditions requiring adjustment of the push and/or pull flow rates in order to achieve effective control. Cross-draft velocities over 75 ft/min or very wide vessels (8 feet or more) may require increased push and/or pull flows. To account for the effects of these variables, a $\pm$ 20% flow adjustment should be designed into both the push and the pull flow systems where ever practical. Once designed and installed, push-pull systems can be initially evaluated by use of a visual tracer technique.

AMERICAN CONFERENCE OF GOVERNMENTAL INDUSTRIAL HYGIENISTS	PUSH–PULL HOOD DESIGN DATA	
	DATE 1–88	FIGURE VS–504.1

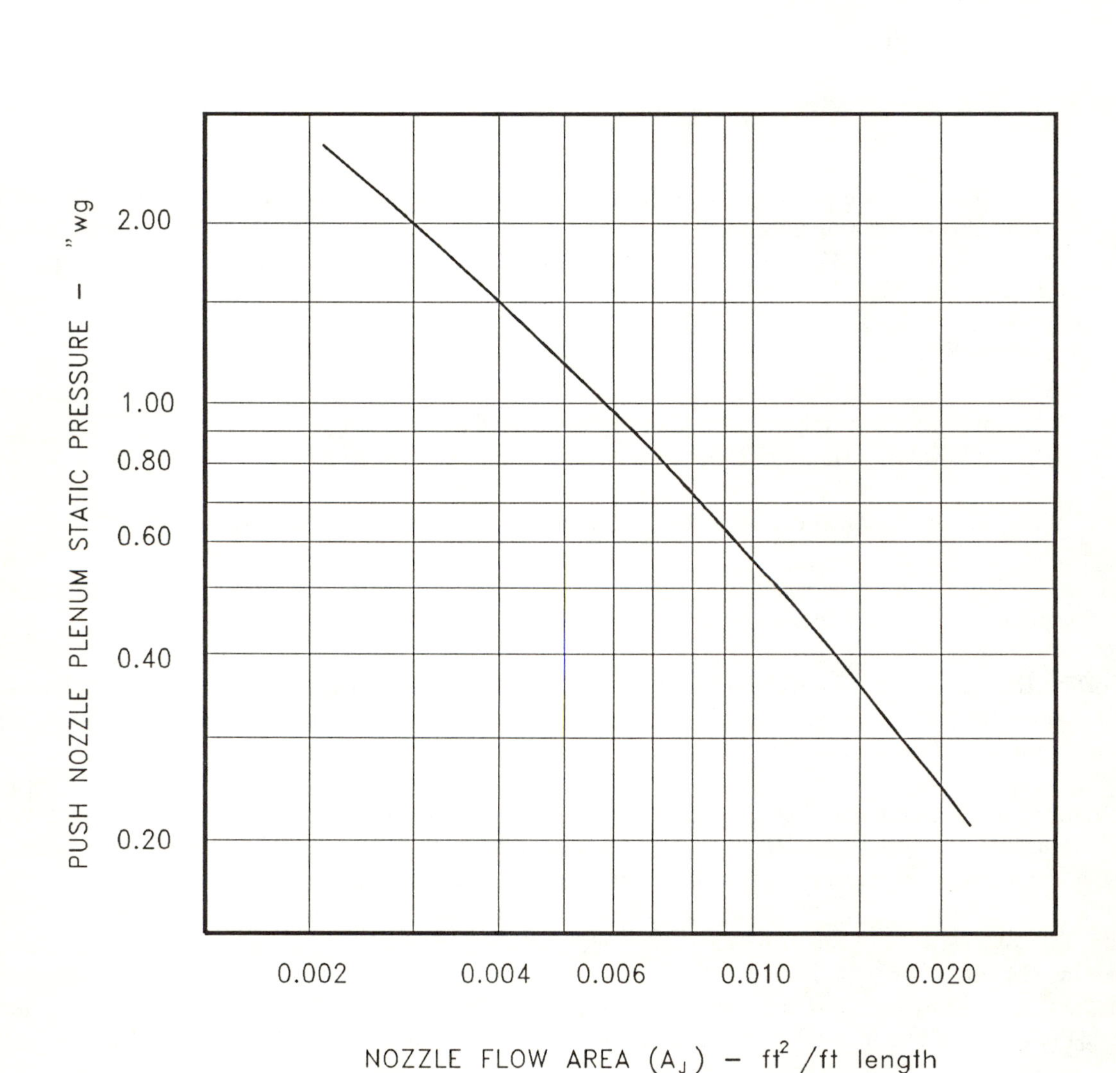

PUSH NOZZLE SUPPLY = $250\sqrt{A_J}$ cfm/ft of length FOR NOZZLES WITH 1/8 TO 1/4 INCH WIDE SLOTS OR 1/4 INCH DIAMETER HOLES ON 3 TO 8 DIAMETER SPACING.

AMERICAN CONFERENCE OF GOVERNMENTAL INDUSTRIAL HYGIENISTS	*PUSH NOZZLE PLENUM PRESSURE*	
	DATE *1–88*	FIGURE *VS–504.2*

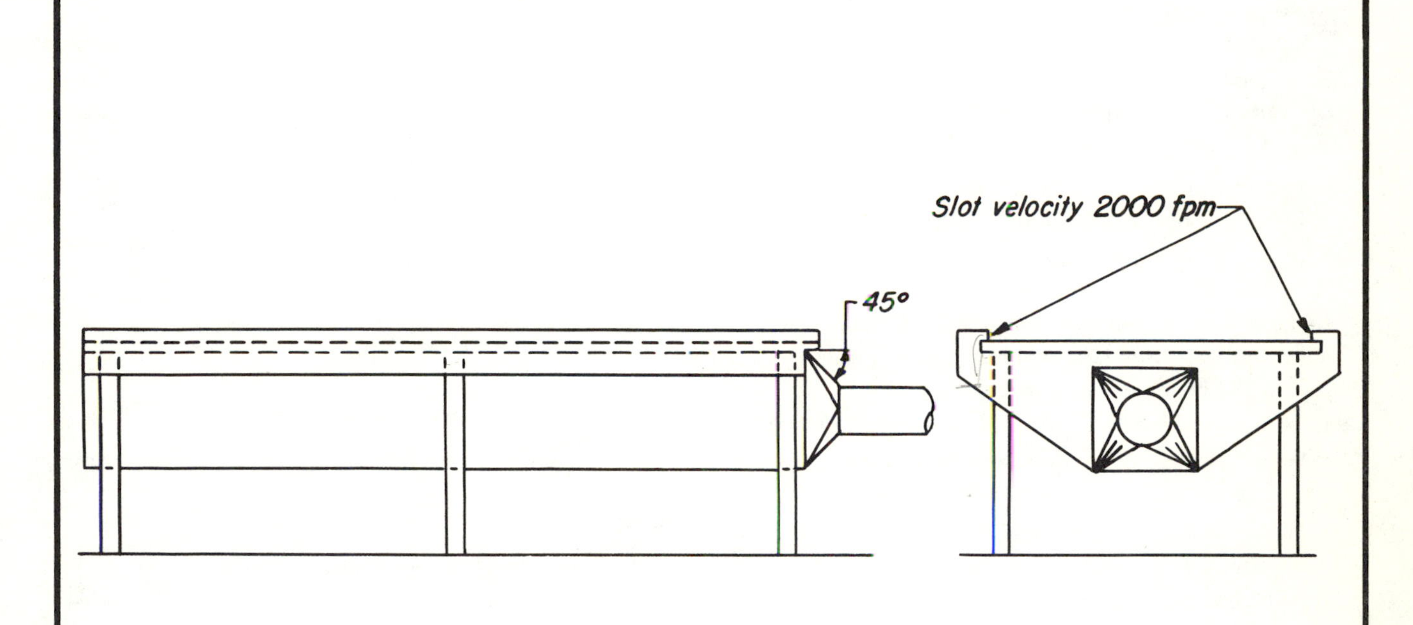

Q = 50-100 cfm/sq ft of table top.
Duct velocity = 2500-3000 fpm
Entry loss = 1.78 slot VP + 0.25 duct VP

Note: See "Open Surface Tanks", VS-503 and VS-504 for other suitable slot types. Air quantities may be calculated on dilution basis if data is available. Maximum plenum velocity = 1/2 slot velocity. Large plenum essential for good distribution.

AMERICAN CONFERENCE OF GOVERNMENTAL INDUSTRIAL HYGIENISTS	
TABLE SLOT	
DATE 1-70	VS-505

OPEN SURFACE TANK DESIGN DATA[108]

A. Duct velocity = any desired velocity. See Chapter 3.

B. Entry loss = 1.78 VP plus duct entry loss.

C. Maximum plenum velocity = 1/2 slot velocity. See Chapter 3.

D. Slot velocity = 2000 fpm unless distribution provided by well-designed, tapered takeoff.

E. Provide ample area at small end of plenum.

F. If L = 6 ft or greater, multiple takeoffs are desirable. If L = 10 ft or greater, multiple takeoffs are necessary.

G. Tank width (W) means the effective width over which the hood must pull air to operate (i.e., where the hood face is set back from the edge of the tank, this set back must be added in measuring tank width.

If W = 20 inches, slot on one side suitable.

If W = 20 to 36 inches, slots on both sides are desirable.

If W = 36 to 48 inches, slots on both sides are necessary unless all other conditions are optimum.

If W = 48 inches or greater, local exhaust is usually not practical. Enclosure is best.

It is not practicable to ventilate across the long dimension of a tank whose ratio W ÷ L exceeds 2.0. It is undesirable to do so when W ÷ L exceeds 1.0.

H. Liquid level to be at least 6 inches below top of tank.

I. Hood types A, C, D and E are preferred — plenum acts as baffle to room air currents.

J. Provide enclosures or removable covers on tank if possible.

K. Provide duct with cleanouts and drains and corrosion-resistant coating if necessary. Use flexible connection at fan inlet.

L. Install baffles to reduce cross-drafts. If impossible, increase control velocity by vector analysis. Baffle is a vertical plate the same length as tank and with top of plate as high as tank is wide. If exhaust hood is on side of tank against a building wall or close to it, it is perfectly baffled.

Flow Rate Calculation for Good Conditions (No cross-drafts, adequate and well-distributed replacement air.)

1. Determine hazard potention from Table 10.5-1 using information from Threshold Limit Values, Solvent Flash Point, Solvent Drying Time Tables in Appendices A and B and Table 10.5-6.

TABLE 10.5-1. Determination of Hazard Potential

Hazard Potential	HYGIENIC STANDARDS		Flash Point (see Appendix B)
	Gas and Vapor (see Appendix A)	Mist (see Appendix A)	
A	0 – 10 ppm	0 – 0.1 mg/m^3	—
B	11 – 100 ppm	0.11 – 1.0 mg/m^3	Under 100 F
C	101 – 500 ppm	1.1 – 10 mg/m^3	100 – 200 F
D	Over 500 ppm	Over 10 mg/m^3	Over 200 F

TABLE 10.5-2. Determination of Rate of Gas, Vapor, or Mist Evolution

Rate	Liquid Temperature (F)	Degrees Below Boiling Point (F)	Relative Evaporation* (Time for 100% Evaporation)	Gassing**
1	Over 200	0-20	Fast (0-3 hours)	High
2	150–200	21–50	Medium (3-12 hours)	Medium
3	94–149	51–100	Slow (12-50 hours)	Low
4	Under 94	Over 100	Nil (Over 50 hours)	Nil

*Dry Time Relation (see Appendix B). Below 5 — Fast; 5-15 — Medium, 15-75 — Slow; 75-over — Nil.

**Rate of gassing depends on rate of chemical or electrochemical action and therefore depends on the material treated and the solution used in the tank and tends to increase with, 1) amount of work in the tank at any one time, 2) strength of the solution in the tank, 3) temperature of the solution in the tank, and 4) current density applied to the work in electrochemical tanks.

TABLE 10.5-3. Minimum Control Velocity (FPM) for Undisturbed Locations

Class (See Tables 10.5-1 & 10.5-2)	Enclosing Hood		Lateral Exhaust (see VS-503/504) (Note 1)	Canopy Hoods see Fig. 3-8 & VS-903	
	One Open Side	Two Open Sides		Three Open Sides	Four Open Sides
A-1, and A-2 (Note 2)	100	150	150	Do not use	Do not use
A-3 (Note 2), B-1, B-2, and C-1	75	100	100	125	175
B-3, C-2, and D-1 (Note 3)	65	90	75	100	150
A-4 (Note 2) C-3, and D-2 (Note 3)	50	75	50	75	125
B-4, C-4, D-3 (Note 3), and D-4—Adequate General Room Ventilation Required (see Chap. 2).					

Notes: 1. Use aspect ratio to determine air volume; see Table 10.5-4 form computation.
2. Do not use canopy hood for Hazard Potential A processes.
3. Where complete control of hot water is desired, design as next highest class.

2. Determine contaminant evolution rate from Table 10.5-2 employing number denoting highest range (see Table 10.5-6).

3. From Table 10-5.3, choose minimum control velocity according to hazard potential, evolution rate and hood design (see Table 10.5-5 for typical processes).

4. From Table 10.5-4, select the cfm/ft^2 for tank dimensions and tank location.

5. Multiply tank area by value obtained from Table 10.5-4 to calculate required air volume.

EXAMPLE

Given: Chrome Plating Tank 6′ × 2.5′.
High production decorative chrome.
Free standing in room.
No cross-drafts.

a: Tank Hood. See Vs-503. Use hood "A" long 6′ side. Hood acts as baffle.

W = 2.5′; L = 6.0′; W/L = 0.42

b: Component — Chromic Acid.

Hazard potential: A (from Table 10.5-1; from Appendix A: TLV = 0.05 mg/m^3 from Appendix A: Flash point = Negligible)

Rate of Evolution: 1 (from Table 10.5-2; from Table 10.5-6: Gassing rate = *high*)

Class: A-1

Control Velocity = 150 fpm (from Table 10.5-3)

Minimum Exhaust Rate = 225 cfm/ft^2 (from Table 10.5-4; Baffled tank, W/L = 0.42)

Minimum Exhaust Flow Rate = 225 × 15 = 3375 cfm

c: Hood Design

Design slot velocity = 2000 fpm

Slot Area = Q/V = 3375 cfm/2000 fpm = 1.69 ft^2

Slot Width = A/L = 1.69 ft^2/6 ft = 0.281 ft = 3.375 in

Plenum depth = (2) (slot width) = (2)(3.375) = 6.75″

Duct area = Q/V = 3375 cfm/2500 fpm = 1.35 ft^2. Use 16″ duct, area = 1.396 ft^2

Final duct velocity = Q/A = 3375/1.396 = 2420 fpm

Hood SP = Entry loss + Acceleration

$= 1.78\ VP_s + 0.25\ VP_d + 1.0\ VP_d$ (see Chapter 3)

$= (1.78 \times 0.25'') + (0.25 \times 0.37'') + 0.37''$

$= 0.45 + 0.09 + 0.37$

Hood SP = 0.91″

TABLE 10.5-4. Minimum Rate, cfm/ft^2 of Tank Area for Lateral Exhaust

Required Minimum Control Velocity, fpm (from Table 10.5-3)	cfm/ft^2 to maintain required minimum control velocities at following $\frac{\text{tank width}}{\text{tank length}}\left(\frac{W}{L}\right)$ ratios.				
	0.0 – 0.09	0.1 – 0.24	0.25 – 0.49	0.5 – 0.99	1.0 – 2.0 Note 2
Hood against wall or baffle (see Note 1 below and Note L, page 10-70.) See VS-503 and VS-503.1 D and E.					
50	50	60	75	90	100
75	75	90	110	130	150
100	100	125	150	175	200
150	150	190	225	[250] Note 3	[250] Note 3
Hood on free standing tank (see Note 1). See VS-503 B and VS-503.1 F.					
50	75	90	100	110	125
75	110	130	150	170	190
100	150	175	200	225	250
150	225	[250] Note 3	[250] Note 3	[250] Note 3	[250] Note 3

Notes: 1. Use W/2 as tank width in computing W/L ratio for hood along centerline or two parallel sides of tank. See VS-503 B and C and VS-503.1 F.
2. See Notes F and G, page 10-70.
3. While bracketed values may not produce 150 fpm control velocity at all aspect ratios, the 250 cfm/ft^2 is considered adequate for control.

TABLE 10.5-5. Typical Processes Minimum Control Velocity (fpm) for Undisturbed locations

Operation	Contaminant	Hazard	Contaminant Evolution	Lateral Exhaust Control Velocity (See VS-503/504)	Collector Recommended
Anodizing Alum.	Chromic-Sulfuric Acids	A	1	150	X
Alum. Bright Dip	Nitric + Sulf. Acids	A	1	150	X
	Nitric + Phosphoric Acids	A	1	150	X
Plating — Chromium	Chromic Acid	A	1	150	X
Copper Strike	Cyanide Mist	C	2	75	X
Metal Cleaning (Boiling)	Alkaline Mist	C	1	100	X
Hot Water (If vent desired)					
Not Boiling	Water Vapor	D	2	50*	
Boiling		D	1	75*	
Stripping — Copper	Alkaline-Cyanide Mists	C	2	75	X
Nickel	Nitrogen Oxide Gases	A	1	150	X
Pickling — Steel	Hydrochloric Acid	A	2	150	X
	Sulfuric Acid	B	1	100	X
Salt Solution					
(Bonderizing & Parkerizing)	Water Vapor	D	2	50*	
Not Boiling	Water Vapor	D	2	50*	
Boiling		D	1	75*	
Salt Baths (Molten)	Alkaline Mist	C	1	100	X

*Where complete control of water vapor is desired, design as next highest class.

TABLE 10.5-6. Airborne Contaminants Released by Metallic Surfaced Treatment, Etching, Pickling, Acid Dipping and Metal Cleaning Operations

Process	Type	Notes	Component of Bath Which May be Released to Atmosphere (13)	Physical and Chemical Nature of Major Atmospheric Contaminant	Class (12)	Usual Temp. Range F
Surface Treatment	Anodizing Aluminum		Chromic-Sulfuric Acids	Chromic Acid Mist	A-1	95
	Anodizing Aluminum		Sulfuric Acid	Sulfuric Acid Mist	B-1	60-80
	Black Magic		Conc. Sol. Alkaline Oxidizing Agents	Alkaline Mist, Steam	C-1	260-350
	Bonderizing	1	Boiling Water	Steam	D-2,1 (14, 15)	140-212
	Chemical Coloring		None	None	D-4	70-90
	Descaling	2	Nitric-Sulfuric, Hydrofluoric Acids	Acid Mist, Hydrogen Fluoride Gas, Steam	B-2,1 (15)	70-150
	Ebonol		Conc. Sol. Alkaline Oxidizing Agents	Alkaline Mist, Steam	C-1	260-350
	Galvanic-Anodize	3	Ammonium Hydroxide	Ammonia Gas, Steam	B-3	140
	Hard-Coating Aluminum		Chromic-Sulfuric Acids	Chromic Acid Mist	A-1	120-180
	Hard Coating Aluminum		Sulfuric Acid	Sulfuric Acid Mist	B-1	120-180
	Jetal		Conc. Sol. Alkaline Oxidizing Agents	Alkaline Mist, Steam	C-1	260-350
	Magcote	4	Sodium Hydroxide	Alkaline Mist, Steam	C-3,2 (15)	105-212
	Magnesium Pre-Dye Dip		Ammonium Hydroxide-Ammonium Acetate	Ammonia Gas, Steam	B-3	90-180
	Parkerizing	1	Boiling Water	Steam	D-2,1 (14,15)	140-212
	Zincete Immersion	5	None	None	D-4	70-90
Etching	Aluminum		Sodium Hydroxide-Soda Ash-Trisodium Phosphate	Alkaline Mist, Steam	C-1	160-180
	Copper	6	Hydrochloric Acid	Hydrogen Chloride Gas	A-2	70-90
	Copper	7	None	None	D-4	70
Pickling	Aluminum		Nitric Acid	Nitrogen Oxide Gases	A-2	70-90
	Aluminum		Chromic, Sulfuric Acids	Acid Mists	A-3	140
	Aluminum		Sodium Hydroxide	Alkaline Mist	C-1	140
	Cast Iron		Hydrofluoric-Nitric Acids	Hydrogen Fluoride-Nitrogen Oxide Gases	A-2,1 (15)	70-90
	Copper		Sulfuric Acid	Acid Mist, Steam	B-3,2 (15)	125-175
	Copper	8	None	None	D-4	70-175
	Duralumin		Sodium Flouride, Sulfuric Acid	Hydrogen Fluoride Gas, Acid Mist	A-3	70
	Inconel		Nitric, Hydrofluoric Acids	Nitrogen Oxide, HF Gases, Steam	A-1	150-165
	Inconel		Sulfuric Acid	Sulfuric Acid Mist, Steam	B-2	160-180
	Iron and Steel		Hydrochloric Acid	Hydrogen Chloride Gas	A-2	70
	Iron and Steel		Sulfuric Acid	Sulfuric Acid Mist, Steam	B-1	70-175
	Magnesium		Chromic-Sulfuric, Nitric Acids	Nitrogen Oxide Gases, Acid Mist, Steam	A-2	70-160
	Monel and Nickel		Hydrochloric Acid	Hydrogen Chloride Gas, Steam	A-2	180
	Monel and Nickel		Sulfuric Acid	Sulfuric Acid Mist, Steam	B-1	160-190
	Nickel Silver		Sulfuric Acid	Acid Mist, Steam	B-3,2 (15)	70-140
	Silver		Sodium Cyanide	Cyanide Mist, Steam	C-3	70-210
	Stainless Steel	9	Nitric, Hydrofluoric Acids	Nitrogen Oxide, Hydrogen Fluoride Gases	A-2	125-180
	Stainless Steel	9,10	Hydrochloric Acid	Hydrogen Chloride Gas	A-2	130-140
	Stainless Steel	9,10	Sulfuric Acid	Sulfuric Acid Mist, Steam	B-1	180
	Stainless Steel Immunization		Nitric Acid	Nitrogen Oxide Gases	A-2	70-120
	Stainless Steel Passivation		Nitric Acid	Nitrogen Oxide Gases	A-2	70-120

TABLE 10.5-6. Airborne Contaminants Released by Metallic Surfaced Treatment, Etching, Pickling, Acid Dipping and Metal Cleaning Operations (con't)

Process	Type	Notes	Component of Bath Which May be Released to Atmosphere (13)	Physical and Chemical Nature of Major Atmospheric Contaminant	Class (12)	Usual Temp. Range F
Acid Dipping	Aluminum Bright Dip		Phosphoric, Nitric Acids	Nitrogen Oxide Gases	A-1	200
	Aluminum Bright Dip		Nitric, Sulfuric Acids	Nitrogen Oxide Gases, Acid Mist	A-2,1 (15)	70-90
	Cadmium Bright Dip		None	None	D-4	70
	Copper Bright Dip		Nitric, Sulfuric Acids	Nitrogen Oxide Gases, Acid Mist	A-2,1 (15)	70-90
	Copper Semi-Bright Dip		Sulfuric Acid	Acid Mist	B-2	70
	Copper Alloys Bright Dip		Nitric, Sulfuric Acids	Nitrogen Oxide Gases, Acid Mist	A-2,1 (15)	70-90
	Copper Matte Dip		Nitric, Sulfuric Acids	Nitrogen Oxide Gases, Acid Mist	A-2,1 (15)	70-90
	Magnesium Dip		Chromic Acid	Acid Mist, Steam	A-2	190-212
	Magnesium Dip		Nitric, Sulfuric Acids	Nitrogen Oxide Gases, Acid Mist	A-2,1 (15)	70-90
	Monel Dip		Nitric, Sulfuric Acids	Nitrogen Oxide Gases, Acid Mist	A-2,1 (15)	70-90
	Nickel and Nickel Alloys Dip		Nitric, Sulfuric Acids	Nitrogen Oxide Gases, Acid Mist	A-2,1 (15)	70-90
	Silver Dip		Nitric Acid	Nitrogen Oxide Gases	A-1	70-90
	Silver Dip		Sulfuric Acid	Sulfuric Acid Mist	B-2	70-90
	Zinc and Zinc Alloys Dip		Chromic, Hydrochloric Acids	Hydrogen Chloride Gas (If HCl attacks Zn)	A-4,3 (15)	70-90
Metal Cleaning	Alkaline Cleaning	11	Alkaline Sodium Salts	Alkaline Mist, Steam	C-2,1 (15)	160-210
	Degreasing		Tricloroethylene-Perchloroethylene	Trichloroethylene-Perchloroethylene Vapors	B (16)	188-250
	Emulsion Cleaning		Petroleum-Coal Tar Solvents	Petroleum-Coal Tar Vapors	B-3,2 (15)	70-140
					(17)	70-140
	Emulsion Cleaning		Chlorinated Hydrocarbons	Chlorinated Hydrocarbon Vapors	(17)	70-140

Notes:
1 Also Aluminum Seal, Magnesium Seal, Magnesium Dye Set, Dyeing Anodized Magnesium, Magnesium Alkaline Dichromate Soak, Coloring Anodized Aluminum.
2 Stainless Steel before Electropolishing.
3 On Magnesium.
4 Also Manodyz, Dow-12.
5 On Aluminum.
6 Dull Finish.
7 Ferric Chloride Bath.
8 Sodium Dichromate, Sulfuric Acid Bath and Ferrous Sulfate, Sulfuric Acid Bath.
9 Scale Removal.
10 Scale Loosening.
11 Soak and Electrocleaning.
12 Class as described in Chapter 2 for use in Table 10.5-3 based on hazard potential (Table 10.5-1) and rate of evolution (Table 10.5-2) for usual operating conditions. Higher temperatures, agitation or other conditions may result in a higher rate of evolution.
13 Hydrogen gas also released by many of these operations.
14 Rate where essentially complete control of steam is required. Otherwise, adequate dilution ventilation may be sufficient.
15 The higher rate is associated with the higher value in the temperature range.
16 For vapor degreasers, rate is determined by operating procedure. See VS-501.
17 Class of operation is determined by nature of the hydrocarbon. Refer to Appendix A.

TABLE 10.5-7. Airborne Contaminants Released by Electropolishing, Electroplating and Electroless Plating Operations

Process	Type	Notes	Component of Bath Which May be Released to Atmosphere (19)	Physical and Chemical Nature of Major Atmospheric Contaminant	Class (18)	Usual Temp. Range F
Electropolishing	Aluminum	1	Sulfuric, Hydrofluoric Acids	Acid Mist, Hydrogen Flouride Gas, Steam	A-2	140-200
	Brass, Bronze	1	Phosphoric Acid	Acid Mist	B-3	68
	Copper	1	Phosphoric Acid	Acid Mist	B-3	68
	Iron	1	Sulfuric, Hydrochloric, Perchloric Acids	Acid Mist, Hydrogen Chloride Gas, Steam	A-2	68-175
	Monel	1	Sulfuric Acid	Acid Mist, Steam	B-2	86-160
	Nickel	1	Sulfuric Acid	Acid Mist, Steam	B-2	86-160
	Stainless Steel	1	Sulfuric, Hydrofluoric, Chromic Acids	Acid Mist, Hydrogen Flouride Gas, Steam	A-2,1 (20)	70-300
	Steel	1	Sulfuric, Hydrochloric, Perchloric Acids	Acid Mist, Hydrogen Chloride Gas, Steam	A-2	68-175
Strike Solutions	Copper		Cyanide Salts	Cyanide Mist	C-2	70-90
	Silver		Cyanide Salts	Cyanide Mists	C-2	70-90
	Wood's Nickel		Nickel Chloride, Hydrochloric Acid	Hydrogen Chloride Gas, Chloride Mist	A-2	70-90
Electroless Plating	Copper		Formaldehyde	Formaldehyde Gas	A-1	75
	Nickel	2	Ammonium Hydroxide	Ammonia Gas	B-1	190
Electroplating Alkaline	Platium		Ammonium Phosphate, Ammonia Gas	Ammonia Gas	B-2	158-203
	Tin		Sodium Stannate	Tin Salt Mist, Steam	C-3	140-170
	Zinc	3	None	None	D-4	170-180
Electroplating Fluoborate	Cadmium		Fluoborate Salts	Fluoborate Mist, Steam	C-3,2 (20)	70-170
	Copper		Copper Fluoborate	Fluoborate Mist, Steam	C-3,2 (20)	70-170
	Indium		Fluoborate Salts	Fluoborate Mist, Steam	C-3,2 (20)	70-170
	Lead		Lead Fluoborate-Fluoboric Acid	Fluoborate Mist, Hydrogen Fluoride Gas	A-3	70-90
	Lead-Tin Alloy		Lead Fluoborate-Fluoboric Acid	Fluoborate Mist	C-3,2 (20)	70-100
	Nickel		Nickel Fluoborate	Fluoborate Mist	C-3,2 (20)	100-170
	Tin		Stannous Fluoborate, Fluoboric Acid	Fluoborate Mist	C-3,2 (20)	70-100
	Zinc		Fluoborate Salts	Fluoborate Mist, Steam	C-3,2 (20)	70-170
Electroplating Cyanide	Brass, Bronze	4,5	Cyanide Salts, Ammonium Hydroxide	Cyanide Mist, Ammonia Gas	B-4,3 (20)	60-100
	Bright Zinc	5	Cyanide Salts, Sodium Hydroxide	Cyanide, Akaline Mists	C-3	70-120
	Cadmium	5	None	None	D-4	70-100
	Copper	5,6	None	None	D-4	70-160
	Copper	5,7	Cyanide Salts, Sodium Hydroxide	Cyanide, Alkaline Mists, Steam	C-2	110-160
	Indium	5	Cyanide Salts, Sodium Hydroxide	Cyanide, Alkaline Mists	C-3	70-120
	Silver	5	None	None	D-4	72-120
	Tin-Zinc Alloy	5	Cyanide Salts, Potassium Hydroxide	Cyanide, Alkaline Mists, Steam	C-3,2 (20)	120-140
	White Alloy	5,8	Cyanide Salts, Sodium Stannate	Cyanide, Alkaline Mists	C-3	120-150
	Zinc	5,9	Cyanide Salts, Sodium Hydroxide	Cyanide, Alkaline Mists	C-3,2 (7)	70-120

TABLE 10.5-7. Airborne Contaminants Released by Electropolishing, Electroplating and Electroless Plating Operations (Con't)

Process	Type	Notes	Component of Bath Which May be Released to Atmosphere (19)	Physical and Chemical Nature of Major Atmospheric Contaminant	Class (18)	Usual Temp. Range F
Electroplating Acid	Chromium		Chromic Acid	Chromic Acid Mists	A-1	90-140
	Copper	10	Copper Sulfate, Sulfuric Acid	Sulfuric Acid Mist	B-4,3 (20,21)	75-120
	Indium	12	None	None	D-4	70-120
	Indium	13,14	Sulfamic Acid, Sulfamate Salts	Sulfamate Mist	C-3	70-90
	Iron		Chloride Salts, Hydrochloric Acid	Hydrochloric Acid Mist, Steam	A-2	190-210
	Iron	12	None	None	D-4	70-120
	Nickel	3	Ammonium Fluoride, Hydrofluoric Acid	Hydrofluoric Acid Mist	A-3	102
	Nickel and Black Nickel	12,15	None	None	C-4 (22)	70-150
	Nickel	9,12	Nickel Sulfate	Nickel Sulfate Mist	B-2	70-90
	Nickel	13,14	Nickel Sulfamate	Sulfamate Mist	C-3	75-160
	Palladium	15	None	None	D-4	70-120
	Rhodium	12,17	None	None	D-4	70-120
	Tin		Tin Halide	Halide Mist	C-2	70-90
	Tin	12	None	None	D-4	70-120
	Zinc		Zinc Chloride	Zinc Chloride Mist	B-3	75-120
	Zinc	12	None	None	D-4	70-120

Notes:
1 Arsine may be produced due to the presence of arsenic in the metal or polishing bath.
2 Alkaline Bath
3 On Magnesium
4 Also Copper-Cadmium Bronze
5 HCN gas may be evolved due to the acidic action of CO_2 in the air at the surface of the bath
6 Conventional Cyanide Bath
7 Except Conventional Cyanide Bath
8 Albaloy, Spekwhite, Bonwhite (Alloys of Copper, Tin, Zinc)
9 Using Insoluble Anodes
10 Over 90 F
11 Mild Organic Acid Bath
12 Sulfate Bath
13 Sulfamate Bath
14 Air Agitated
15 Chloride Bath
16 Nitrite Bath
17 Phosphate Bath
18 Class as described in Chapter 2 for use in Table 3 based on hazard potential (Table 1) and rate of evolution (Table 2) for usual operating conditions. Higher temperatures, agitation, high current density or other conditions may result in a higher rate of evolution.
19 Hydrogen gas also released by many of these operations.
20 The higher rate is associated with the higher value in the temperature range.
21 Baths operated at a temperature of over 140 F with a current density of over 45 amps/ft^2 and with air agitation will have a higher rate of evolution.
22 Local exhaust ventilation may be desired to control steam and water vapor.

TABLE 10.5-8. Airborne Contaminants Released by Stripping Operations

Coating to be Stripped	Base Metal (Footnote)	Component of Batch Which May be Released to Atmosphere (f)	Physical and Chemical Nature of Major Atmospheric Contaminant	Class (e)	Usual Temp. Range F
Anodized Coatings	1,7	Chromic Acid	Acid Mist, Steam	A-2	120-200
Black Oxide Coatings	14	Hydrochloric Acid	Hydrogen Chloride Gas	A-3,2 (g)	70-125
Brass and Bronze	8,14 (a)	Sodium Hydroxide, Sodium Cyanide	Alkaline, Cyanide Mists	C-3,2 (g)	70-90
Cadmium	8,14 (a)	Sodium Hydroxide, Sodium Cyanide	Alkaline, Cyanide Mists	C-3,2 (g)	70-90
	2,4,14	Hydrochloric Acid	Acid Mist, Hydrogen Chloride Gas	A-3,2 (g)	70-90
Chromium	7,8,14 (a)	Sodium Hydroxide	Alkaline Mist, Steam	C-3	70-150
	2,4,8,14	Hydrochloric Acid	Hydrogen Chloride Gas	A-2	70-125
	2,4,8,18 (a)	Sulfuric Acid	Acid Mist	B-2	70-90
Copper	8,14	Sodium Hydroxide, Sodium Cyanide	Alkaline, Cyanide Mists	C-3,2 (g)	70-90
	7,12,14 (b)	None	None	D-4	70-90
	14 (a)	Alkaline Cyanide	Cyanide Mist	C-3,2 (g)	70-160
	1	Nitric Acid	Nitrogen Oxide Gases	A-1	70-120
	18 (a)	Sodium Hydroxide-Sodium Sulfide	Alkaline Mist, Steam	C-2	185-195
Gold	4,5,6,8,9,14 (a)	Sodium Hydroxide, Sodium Cyanide	Alkaline, Cyanide Mists	C-3,2 (g)	70-90
	4,5,18 (a)	Sulfuric Acid	Acid Mist	B-3,2 (g)	70-100
Lead	13 (c)	Acetic Acid, Hydrogen Peroxide	Oxygen Mist	D-3	70-90
	14 (a),(c)	Sodium Hydroxide	Alakaline Mist, Steam	C-3,2 (g)	70-140
Nickel	2,4	Sulfuric, Nitric Acids	Nitrogen Oxide Gases	A-2,1 (g)	70-90
	2,4 (a)	Hydrochloric Acid	Hydrogen Chloride Gas	A-3	70-90
	2,4,14 (a)	Sulfuric Acid	Acid Mist	B-3	70-90
	7	Hydrofluoric Acid	Hydrogen Fluoride Gas	A-3,2 (g)	70-90
	14	Fuming Nitric Acid	Nitrogen Oxide Gases	A-1	70-90
	(a),(d)	Hot Water	Steam	D-2 (h)	200
	1,18,19 (a)	Sulfuric Acid	Acid Mist, Steam	B-3,2 (g)	70-150
Phosphate Coatings	15	Chromic Acid	Acid Mist, Steam	A-3	165
	16	Ammonium Hydroxide	Ammonia Gas	B-3,2 (g)	70-90
Rhodium	10	Sulfuric, Hydrochloric Acids	Acid Mist, Hydrogen Chloride Gas	A-3,2 (g)	70-100
Silver	1	Nitric Acid	Nitrogen Oxide Gases	A-1	70-90
	2,11	Sulfuric, Nitric Acids	Nitrogen Oxide Gases, Steam	A-1	180
	8,14 (a)	Sodium Hydroxide, Sodium Cyanide	Alkaline, Cyanide Mists	C-3	70-90
	17 (a)	Sodium Cyanide	Cyanide Mist	C-3	70-90
Tin	2,3,4	Ferric Chloride, Copper Sulfate Acetic Acid	Acid Mist	B-4,3 (g)	70-90
	(a)	Sodium Hydroxide	Alkaline Mist	C-3	70-90
	2,14,14	Hydrochloric Acid	Hydrogen Chloride Gas	A-3,2 (g)	70-90
	14 (a)	Sodium Hydroxide	Alkaline Mist, Steam	C-2	70-200

TABLE 10.5-8. Airborne Contaminants Released by Stripping Operations (Con't)

Coating to be Stripped	Base Metal (Footnote)	Component of Batch Which May be Released to Atmosphere (f)	Physical and Chemical Nature of Major Atmospheric Contaminant	Class (e)	Usual Temp. Range F
Zinc	1	Nitric Acid	Nitrogen Oxide Gases	A-1	70-90
	8,14	Sodium Hydroxide, Sodium	Alkaline, Cyanide Mists	C-3	70-90

Base Metal:
1. Aluminum
2. Brass
3. Bronze
4. Copper
5. Copper Alloys
6. Ferrous Metals
7. Magnesium
8. Nickel
9. Nickel Alloys
10. Nickel Plated Brass
11. Nickel Silver
12. Non-Ferrous Metals
13. Silver
14. Steel
15. Steel (Manganese Type Coatings)
16. Steel (Zinc Type Coatings)
17. White Metal
18. Zinc
19. Zinc Base Die Castings

Notes:
(a) Electrolytic Process
(b) Refers only to steel (14) when Chromic, Sulfuric Acids Bath is used.
(c) Also Lead Alloys
(d) Sodium Nitrate Bath
(e) Class as described in Chapter 2 for use in Table 3 based on hazard potential (Table 1) and rate of evolution (Table 2) for usual operating conditions. Higher temperatures, agitation or other conditions may result in a higher rate of evolution.
(f) Hydrogen gas also released by some of these operations.
(g) The higher rate is associated with the higher value in the temperature range.
(h) Rate where essentially complete control of steam is required. Otherwise, adequate dilution ventilation may be sufficient.

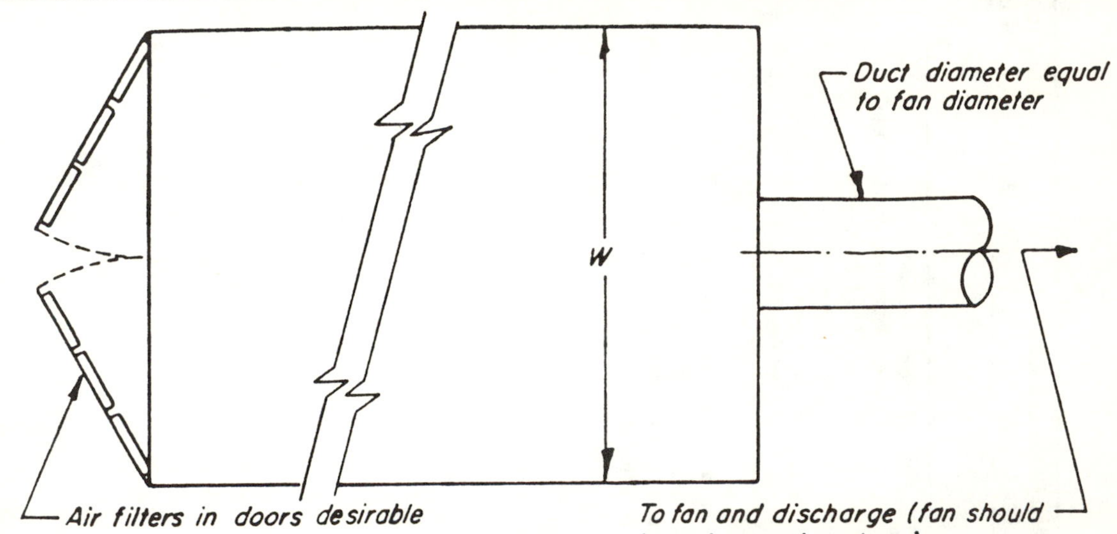

PLAN VIEW

Q = 100 cfm/sq ft of cross-sectional area *
(When WxH is greater than 150 sq ft, Q= 50 cfm/sq ft)
Entry loss =0.50 VP plus resistance of each filter bank when dirty
Duct velocity = 1000 - 3000 fpm
Paint arresting filters to be sized for 100 - 500cfm/sq ft of filter. Consult manufacturer for specfic details.

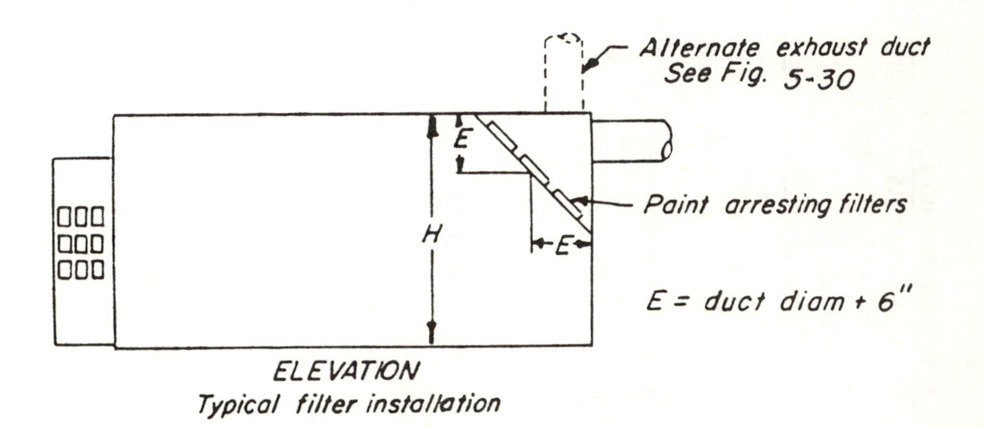

ELEVATION
Typical filter installation

* Airless spray painting
Q= 60cfm/sq ft of cross-section area

For construction and safety, consult NFPA(113)

AMERICAN CONFERENCE OF GOVERNMENTAL INDUSTRIAL HYGIENISTS

AUTO SPRAY PAINT BOOTH

DATE 1-86	VS-601

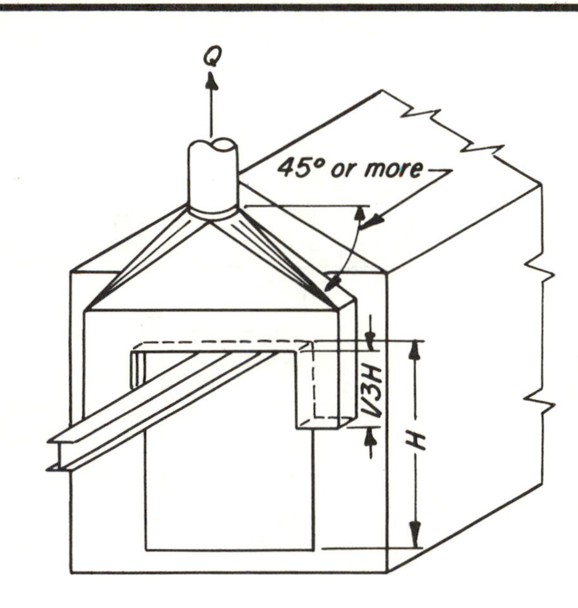

SLOT TYPE

Q = 100 cfm/sq ft door plus 1/2 products of combustion
Entry loss = 1.0 slot VP plus 0.25 duct VP
Duct velocity = 1000-3000 fpm

Size plenum for 500 fpm maximum

Slot on three sides size for 1000 fpm
Locate on inside or outside of door.

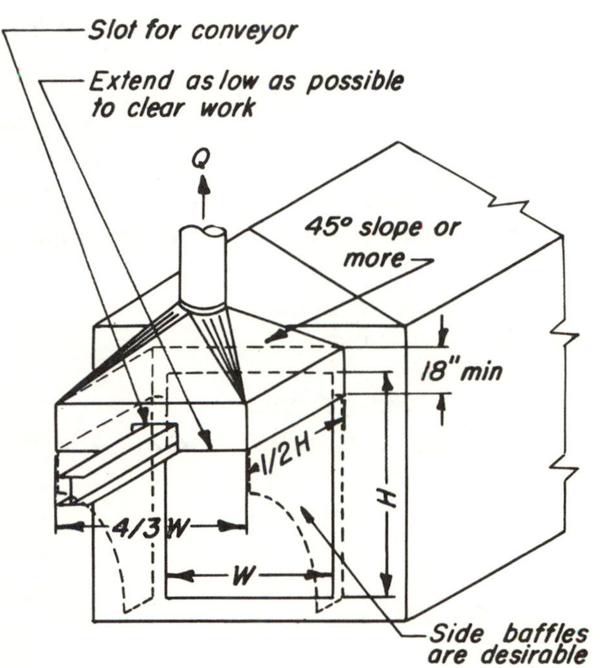

CANOPY TYPE

Q = 200 cfm/sq ft of hood face plus 1/2 products of combustion
Entry loss = 0.25 VP
Duct velocity = 1000-3000 fpm

Note:

For dryers, include volume of water vapor liberated.

For flammable solvent drying refer to Chapter 2, "Dilution Ventilation for Fire and Explosions".

Note:

Hoods at each end of oven. Reduce size of doors as much as possible. Separate vent must be added for products of combustion.

For construction and safety, consult NFPA(113)

AMERICAN CONFERENCE OF GOVERNMENTAL INDUSTRIAL HYGIENISTS	
DRYING OVEN VENTILATION	
DATE 1-78	VS-602

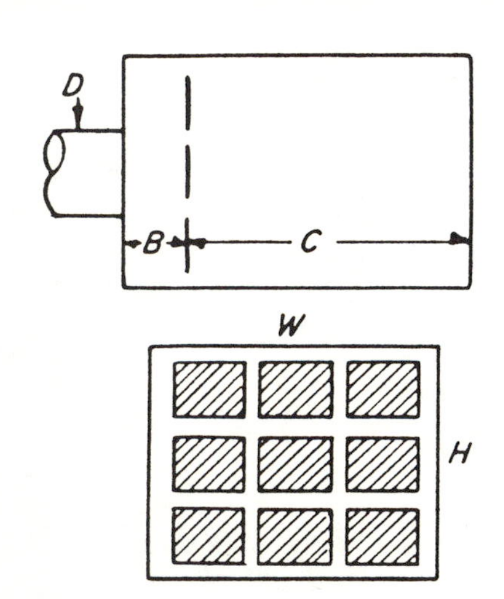

1. Split Baffle or Filters
B = 0.75 D
Baffle area = 0.75 WH
For filter area, See Note 2

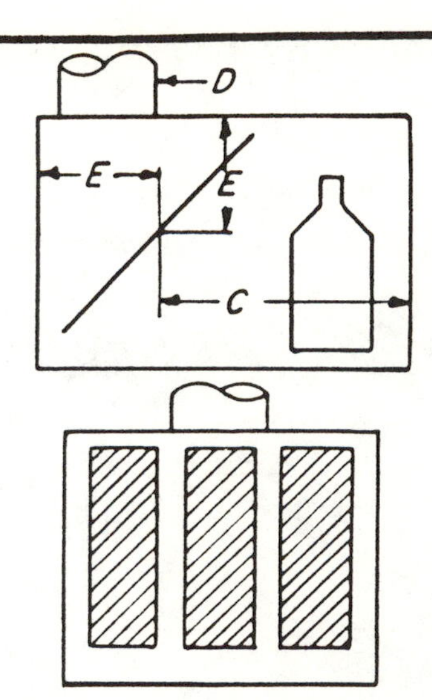

2. Angular Baffle
E = D + 6"
Baffle area = 0.40WH
For filter area, See Note 2

Air spray paint design data

Any combination of duct connections and baffles may be used. Large, deep booths do not require baffles. Consult manufactures for water-curtain designs. Use explosion proof fixtures and non-sparking fan. Electrostatic spray booth requires automatic high-voltage disconnects for conveyor failure, fan failure or grounding.

Walk-in booth
W = work size + 6'
H = work size + 3' (minimum = 7')
C = work size + 6'
Q = 100 cfm/sq ft booth cross section
May be 75 cfm/sq ft for very large, deep, booth. Operator may require a NIOSH certified respirator.
Entry loss = Baffles: 1.78 slot VP + 0.50 duct VP
= Filters: Dirty filter resistance + 0.50 duct VP
Duct velocity = 1000-2000 fpm

Operator outside booth
W = work size + 2'
H = work size + 2'
C = 0.75 x larger front dimension
Q = 100-150 cfm/sq ft of open area, includling conveyor openings.

Airless spray paint design
Q = 60 cfm/sq ft booth cross section, walk-in booth
= 60-100 cfm/sq ft of total open area, operator outside of booth

Notes:
1. Baffle arrangements shown are for air distribution only.
2. Paint arresting filters usually selected for 100-500 fpm, consult manufacturer for specific details.
3. For construction and safety, consult NFPA[113]

AMERICAN CONFERENCE OF GOVERNMENTAL INDUSTRIAL HYGIENISTS

LARGE PAINT BOOTH

DATE 1-86 | VS-603

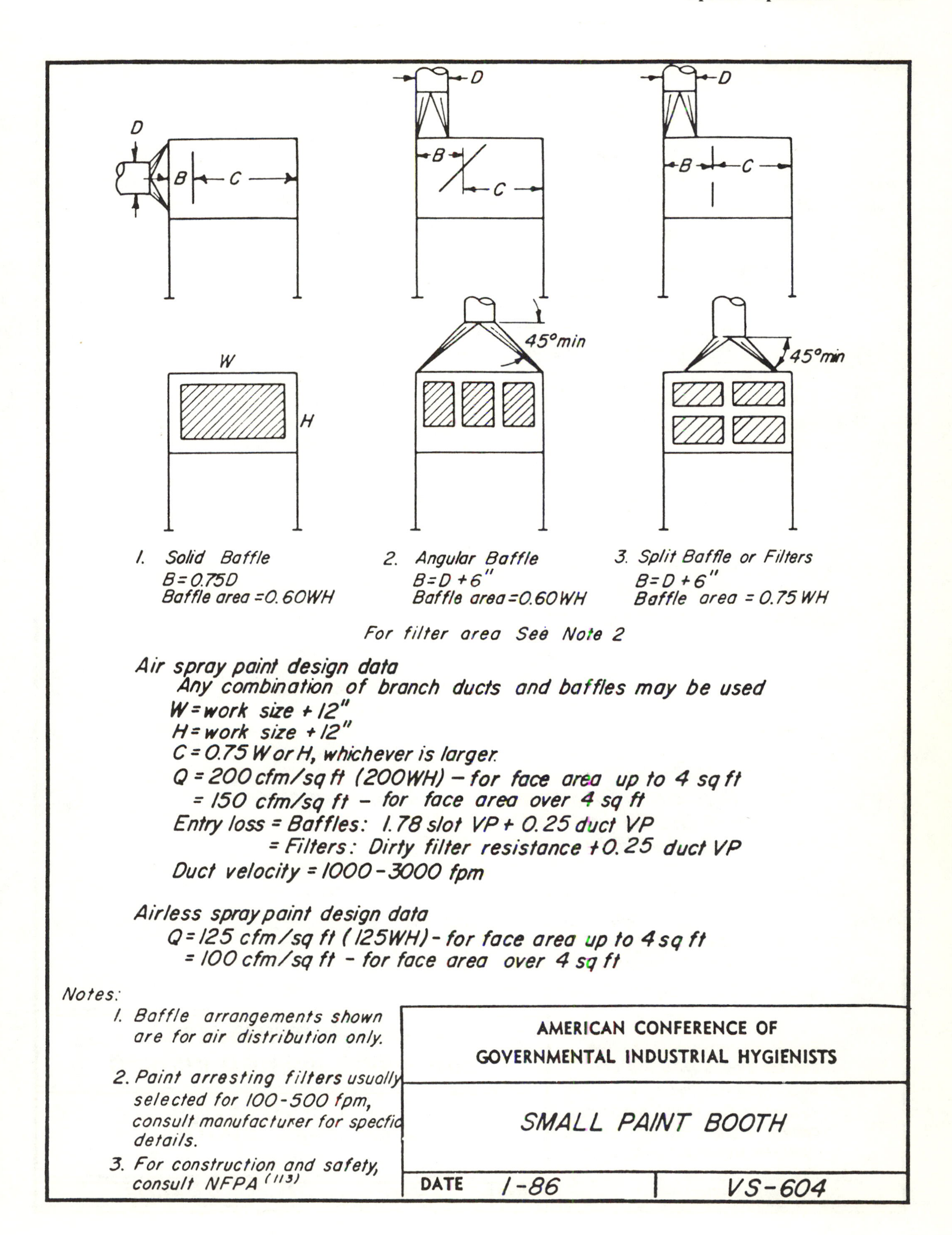
D
B
C
D
B
C
D
B
C
W
H
45°min
45°min
1. Solid Baffle
B=0.75D
Baffle area=0.60WH
2. Angular Baffle
B=D+6"
Baffle area=0.60WH
3. Split Baffle or Filters
B=D+6"
Baffle area = 0.75 WH
For filter area See Note 2
Air spray paint design data
Any combination of branch ducts and baffles may be used
W=work size + 12"
H=work size + 12"
C=0.75 W or H, whichever is larger.
Q=200 cfm/sq ft (200WH) – for face area up to 4 sq ft
= 150 cfm/sq ft – for face area over 4 sq ft
Entry loss = Baffles: 1.78 slot VP + 0.25 duct VP
= Filters: Dirty filter resistance + 0.25 duct VP
Duct velocity = 1000-3000 fpm
Airless spray paint design data
Q=125 cfm/sq ft (125WH) - for face area up to 4 sq ft
= 100 cfm/sq ft – for face area over 4 sq ft
Notes:
1. Baffle arrangements shown are for air distribution only.
2. Paint arresting filters usually selected for 100-500 fpm, consult manufacturer for specific details.
3. For construction and safety, consult NFPA (113)
AMERICAN CONFERENCE OF GOVERNMENTAL INDUSTRIAL HYGIENISTS
SMALL PAINT BOOTH
DATE 1-86
VS-604

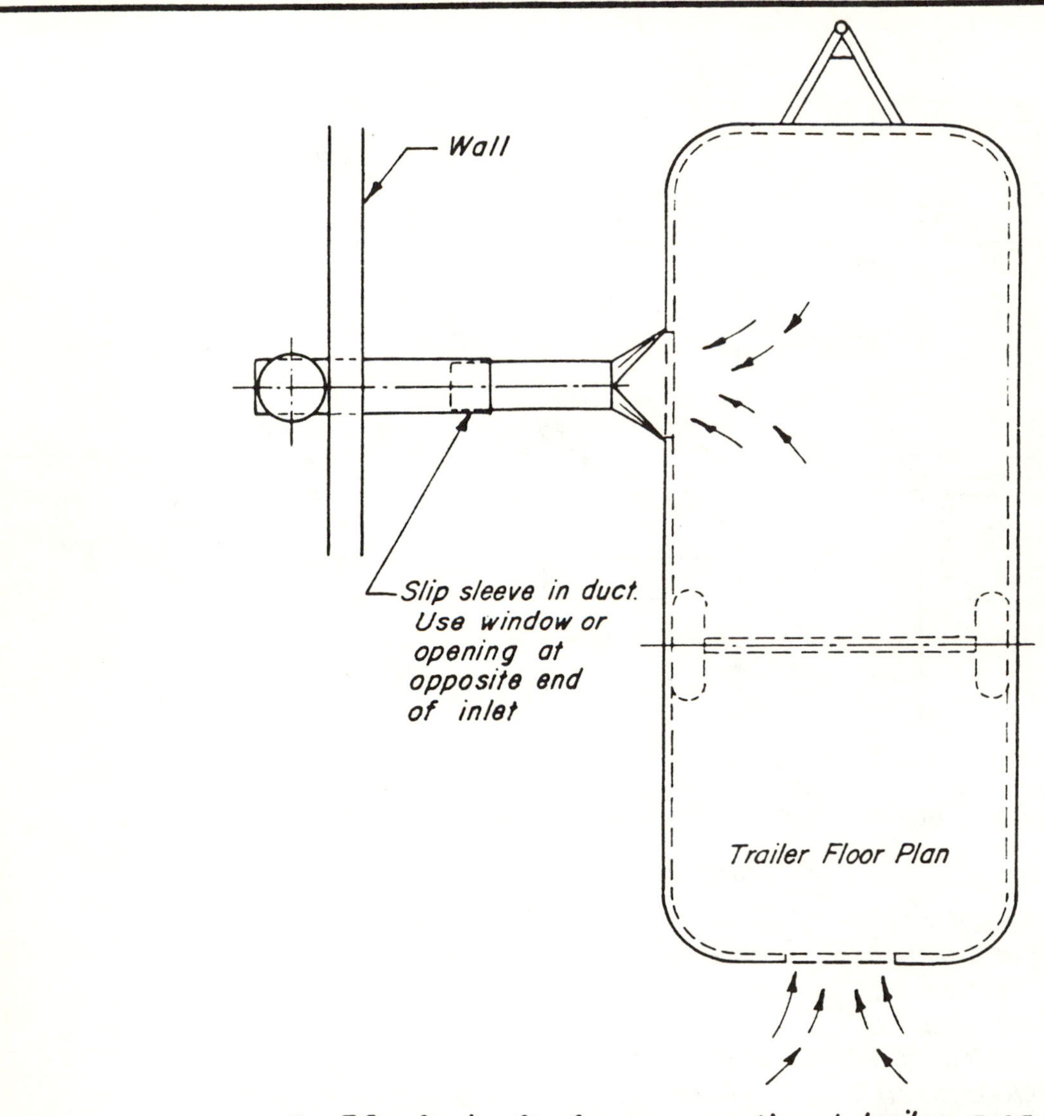

Q = 50 cfm/sq ft of cross-sectional trailer area
Entry loss = 0.25 VP
Duct velocity = 1000 - 3000 fpm

NOTE: Operator must wear a NIOSH certified air-supplied respirator

Notes:

1. Paint arresting filters usually selected for 100-500 fpm, consult manufacturer for specfic details.
2. For construction and safety, consult NFPA [113]

AMERICAN CONFERENCE OF GOVERNMENTAL INDUSTRIAL HYGIENISTS

TRAILER INTERIOR SPRAY PAINTING

DATE	1-86	VS-605

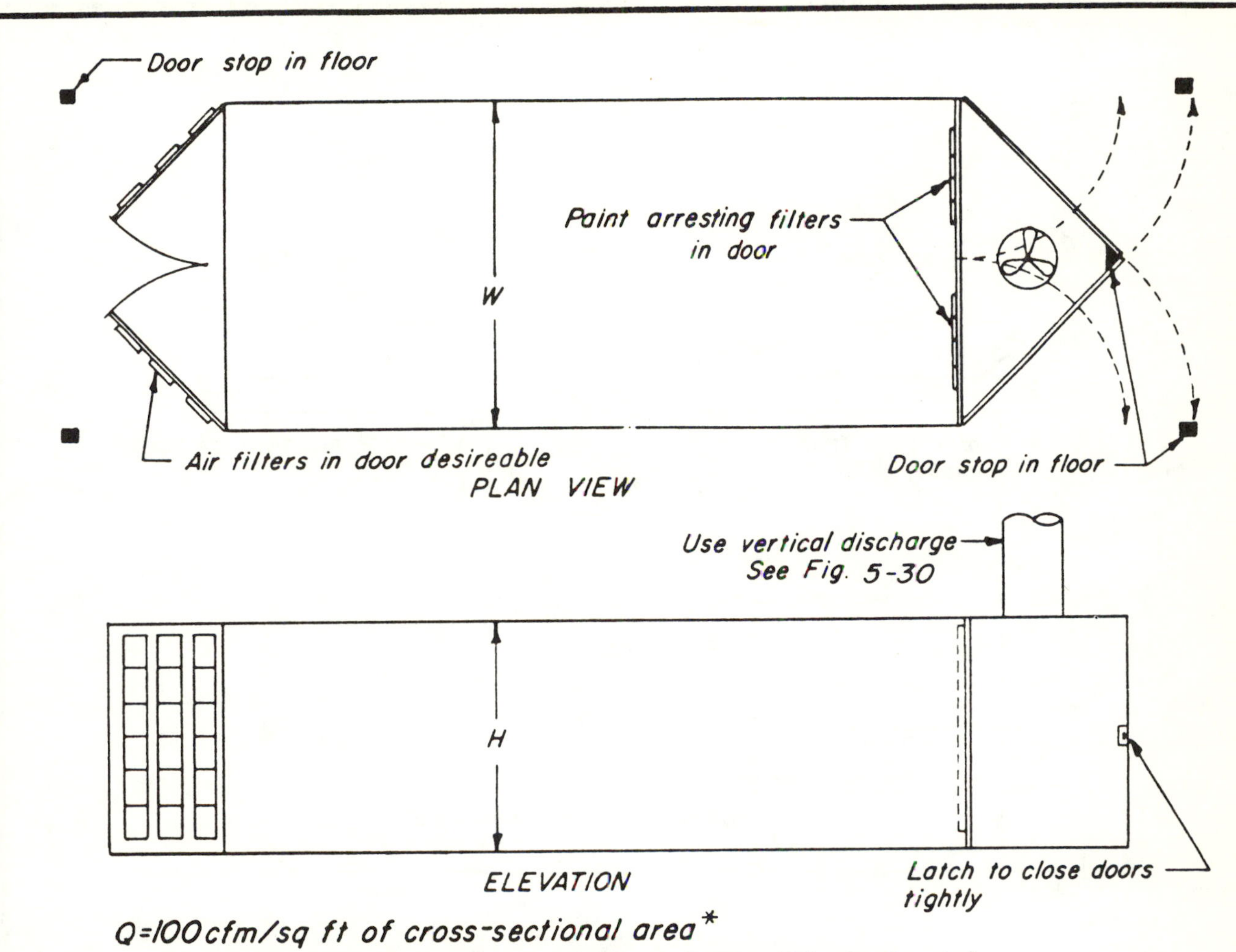

*Q=100 cfm/sq ft of cross-sectional area**
(when WxH is greater than 150 sq ft, Q= 50 cfm/sq ft)
Entry loss = 0.50 VP plus resistance of each filter bank when dirty
Duct velocity = 1000 - 3000 fpm

Notes:

1. *Exhaust fan interlock with make-up air supply and compressed air to spray gun is desirable.*
2. *Paint arresting filters usually selected for 100-500 fpm. Consult manufacture for specific details.*
3. *For construction and safety, consult NFPA.*[113]

**Airless spray painting*
Q=60 cfm/sq ft of cross-section area

AMERICAN CONFERENCE OF GOVERNMENTAL INDUSTRIAL HYGIENISTS	
LARGE DRIVE-THROUGH SPRAY PAINT BOOTH	
DATE *1-86*	*VS-606*

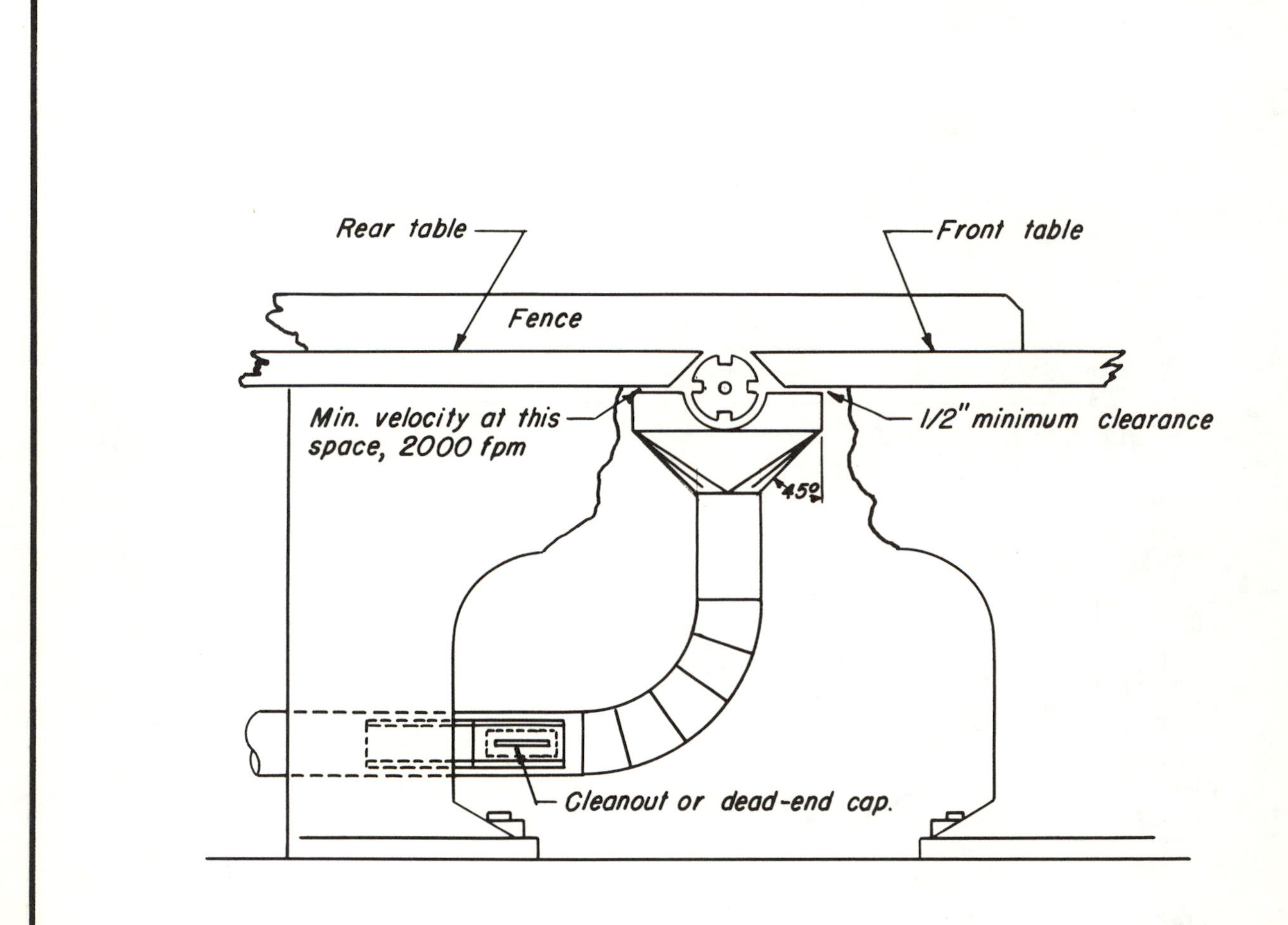

Knife length, inches	Exhaust volume, cfm
Up to 6 incl.	350
over 6 to 12 incl.	440
over 12 to 20 incl.	550
over 20	800

Duct velocity = 3500 fpm
Entry loss = 1.0 slot VP + 0.25 duct VP

AMERICAN CONFERENCE OF GOVERNMENTAL INDUSTRIAL HYGIENISTS

JOINTERS

DATE 1-64 | VS-701

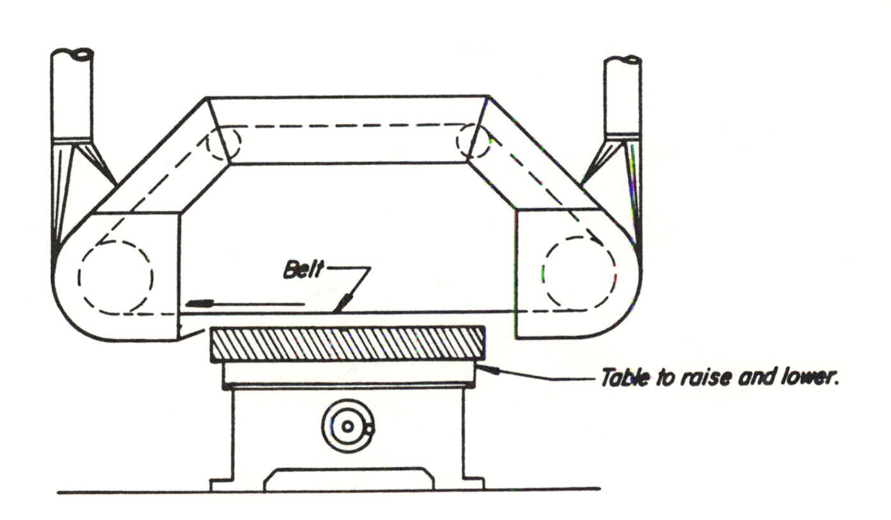

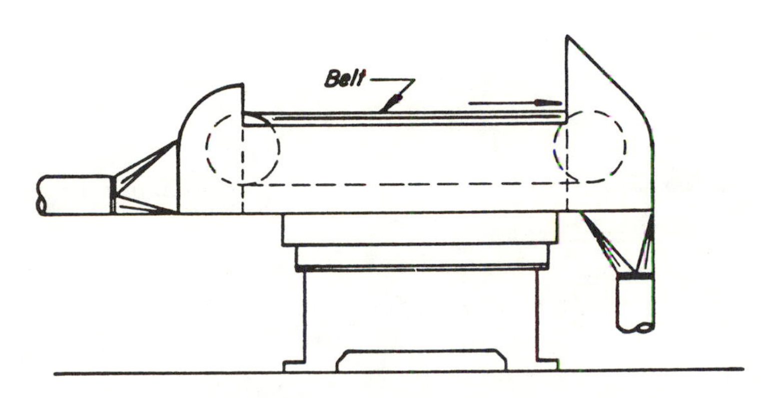

HORIZONTAL BELT SANDERS

Belt width, inches	Exhaust volume, cfm		
	Head end	Tail end	Total
Up to 6 incl.	440	350	790
over 6 to 9 incl.	550	350	900
over 9 to 14 incl	800	440	1240
over 14	1100	550	1650

Duct velocity = 3500 fpm
Entry losses = 0.40 VP for tapered take-off

AMERICAN CONFERENCE OF GOVERNMENTAL INDUSTRIAL HYGIENISTS

HORIZONTAL BELT SANDERS

DATE 1-64 | VS-702

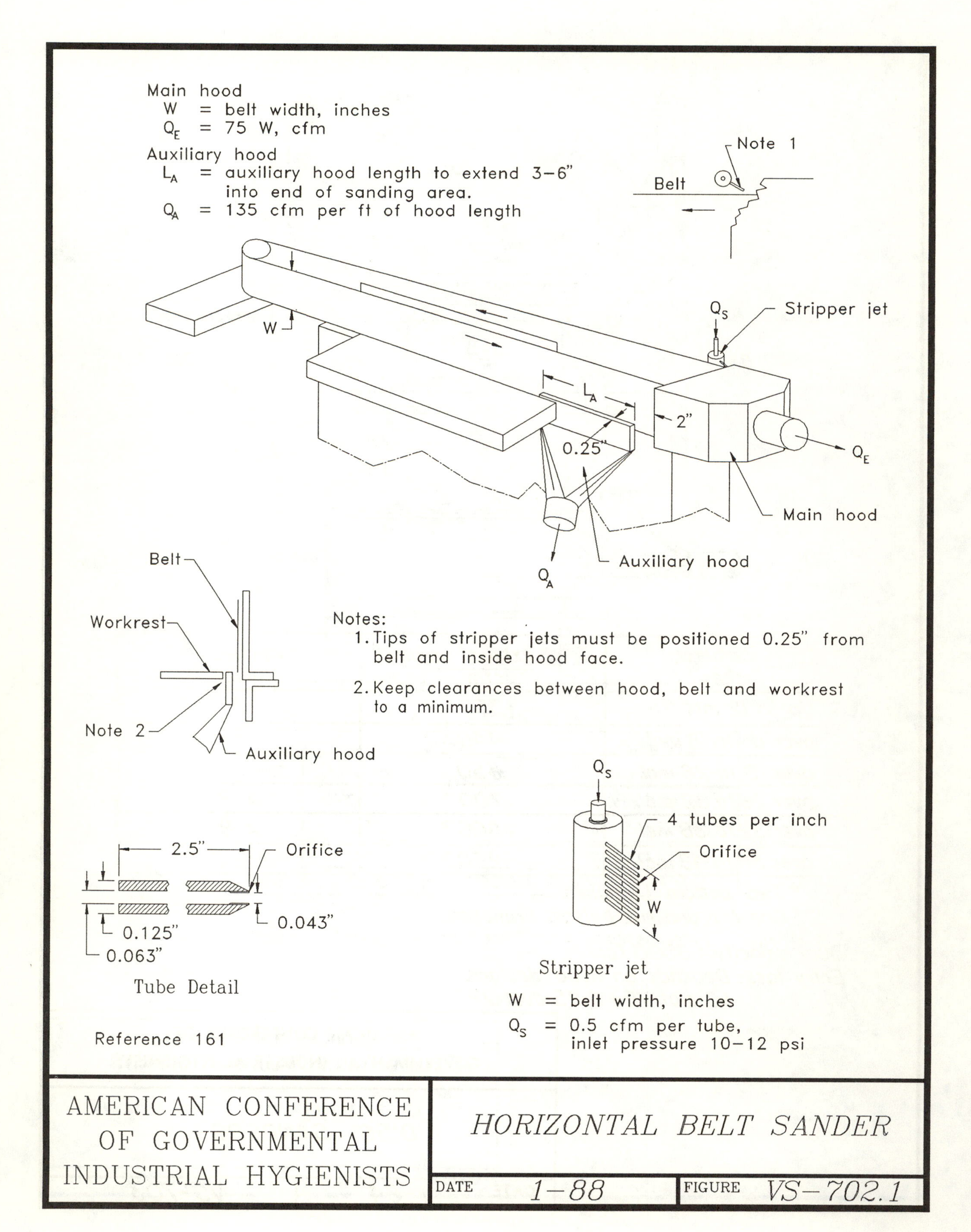
Main hood
W = belt width, inches
Q_E = 75 W, cfm
Auxiliary hood
L_A = auxiliary hood length to extend 3–6" into end of sanding area.
Q_A = 135 cfm per ft of hood length
Note 1
Belt
Q_S
Stripper jet
W
L_A
2"
0.25"
Q_E
Main hood
Auxiliary hood
Q_A
Belt
Workrest
Note 2
Auxiliary hood
Notes:
1. Tips of stripper jets must be positioned 0.25" from belt and inside hood face.
2. Keep clearances between hood, belt and workrest to a minimum.
Q_S
4 tubes per inch
Orifice
W
2.5"
Orifice
0.043"
0.125"
0.063"
Tube Detail
Stripper jet
W = belt width, inches
Q_S = 0.5 cfm per tube, inlet pressure 10–12 psi
Reference 161
AMERICAN CONFERENCE OF GOVERNMENTAL INDUSTRIAL HYGIENISTS
HORIZONTAL BELT SANDER
DATE 1–88
FIGURE VS–702.1

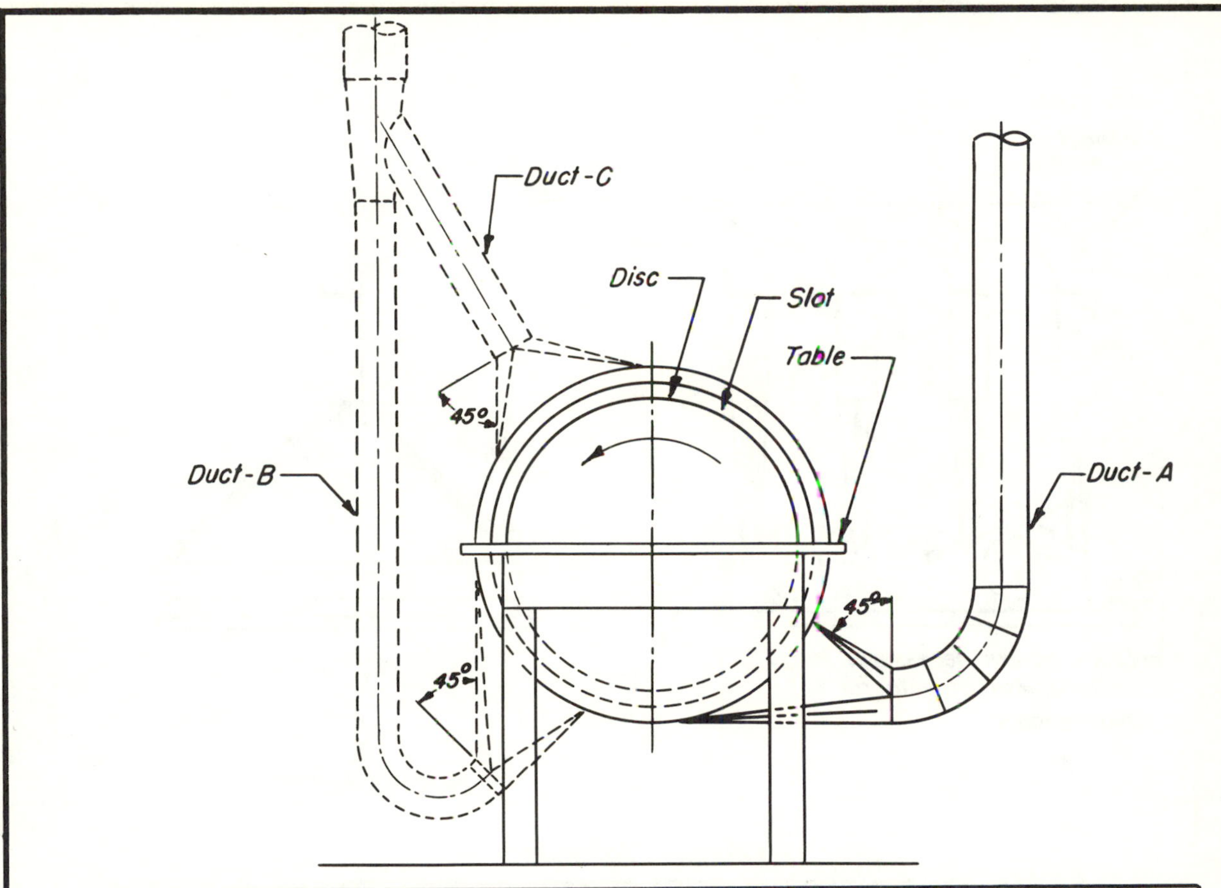

Disc diameter, inches	Total exhaust volume cfm	Applies to duct
Up to 12 incl.	350	A
over 12 to 18 incl.	440	A
over 18 to 26 incl.	550	A
over 26 to 32 incl.	700*	A-B
over 32 to 38 incl.	900*	A-B
over 38 to 48 incl.	1250**	A-B-C

* Two bottom branches.
** One top and two bottom branches.

Duct velocity = 3500 fpm
Entry loss: Depends on hood design.
1.0 slot VP + 0.25 duct VP

AMERICAN CONFERENCE OF GOVERNMENTAL INDUSTRIAL HYGIENISTS

DISC SANDERS

DATE 1-64	VS-703

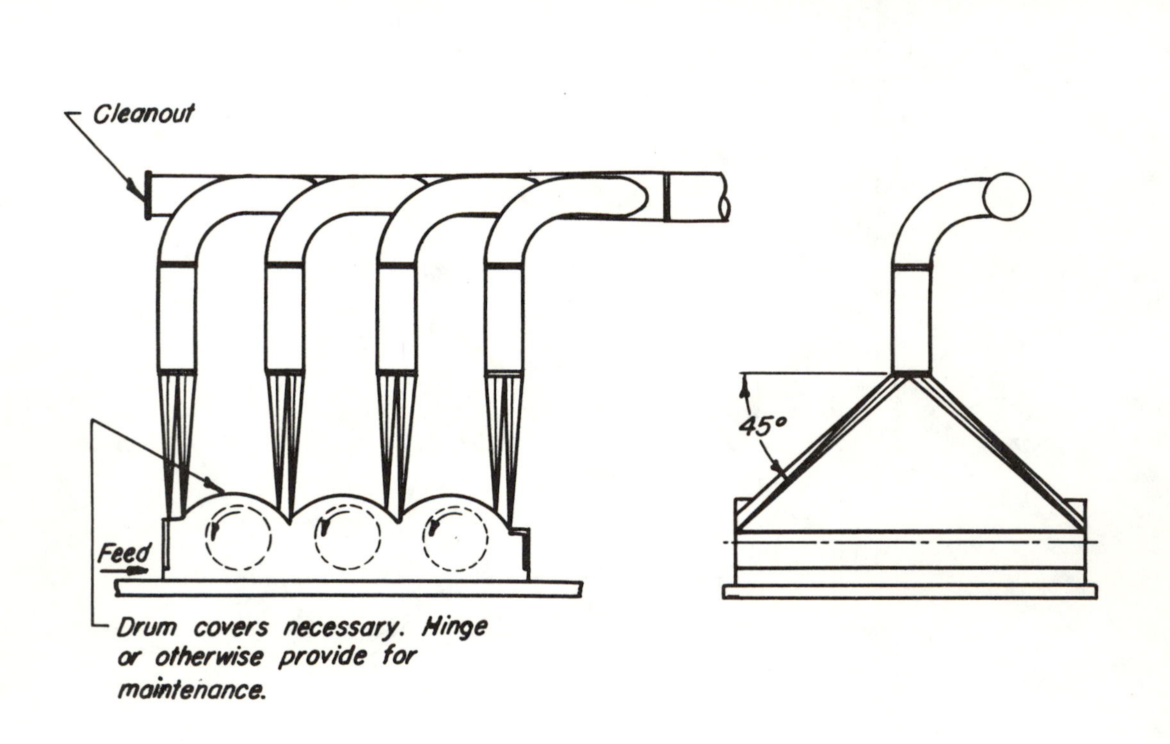

Exhaust Volumes

Drum length, inches	Total exhaust for machine cfm/drum*
Up to 31"	550
31" to 49"	785
49" to 67"	1100
over 67"	1400
Brush rolls	350 cfm at brush

*One hood per drum is minimum
Additional hood at feed side is desirable

Duct velocity = 3500 fpm
Entry loss = 0.25 duct VP

AMERICAN CONFERENCE OF GOVERNMENTAL INDUSTRIAL HYGIENISTS

MULTIPLE DRUM SANDER

DATE 1-64 | VS-704

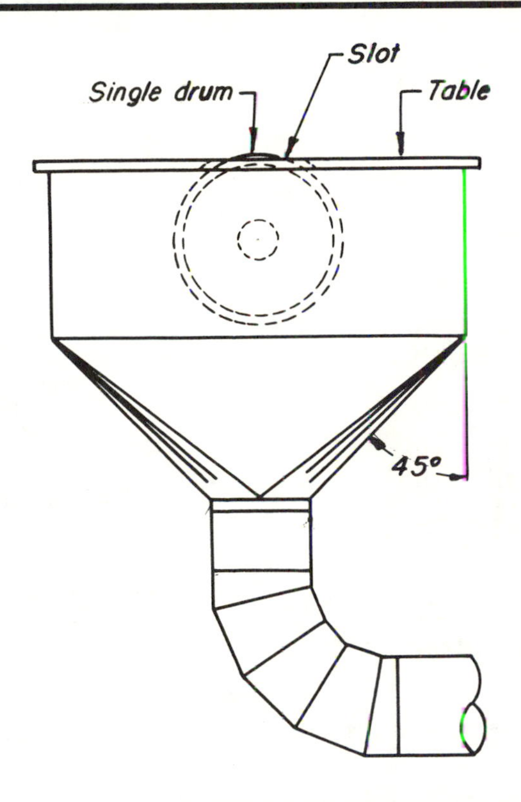

Drum surface, sq inches	Exhaust volume, cfm
Up to 200 incl. (and less than 10" dia.)	350
over 200 to 400 incl.	550
over 400 to 700 incl.	785
over 700 to 1400 incl.	1100
over 1400 to 2400 incl.	1400

Duct velocity = 3500 fpm
Entry loss: Depends on hood design.
1.78 slot VP plus 0.25 duct VP

AMERICAN CONFERENCE OF GOVERNMENTAL INDUSTRIAL HYGIENISTS

SINGLE DRUM SANDER

DATE 1-64 | VS-705

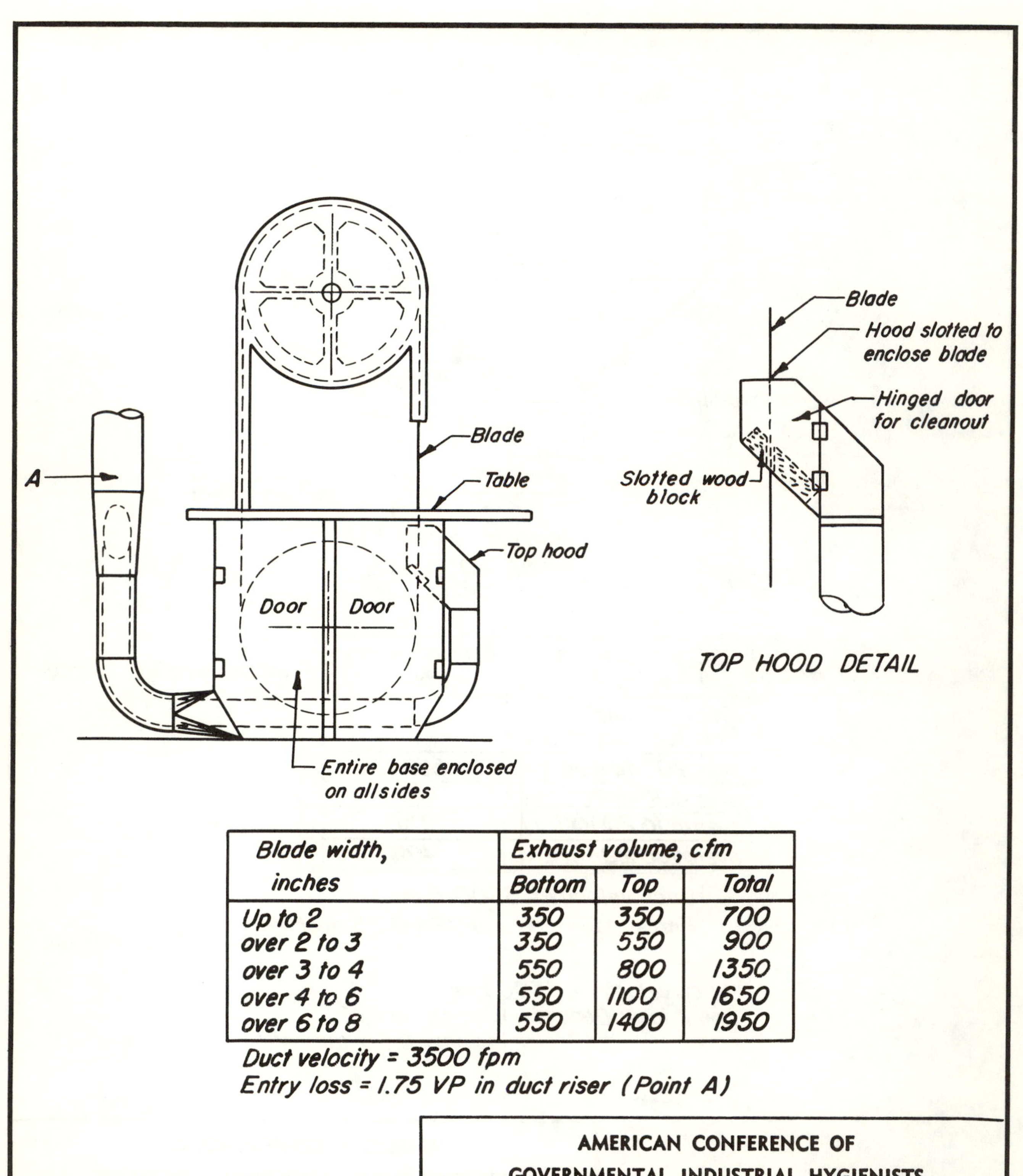

Blade width, inches	Exhaust volume, cfm		
	Bottom	Top	Total
Up to 2	350	350	700
over 2 to 3	350	550	900
over 3 to 4	550	800	1350
over 4 to 6	550	1100	1650
over 6 to 8	550	1400	1950

Duct velocity = 3500 fpm

Entry loss = 1.75 VP in duct riser (Point A)

AMERICAN CONFERENCE OF GOVERNMENTAL INDUSTRIAL HYGIENISTS

BAND SAW

DATE 1-66 | VS-706

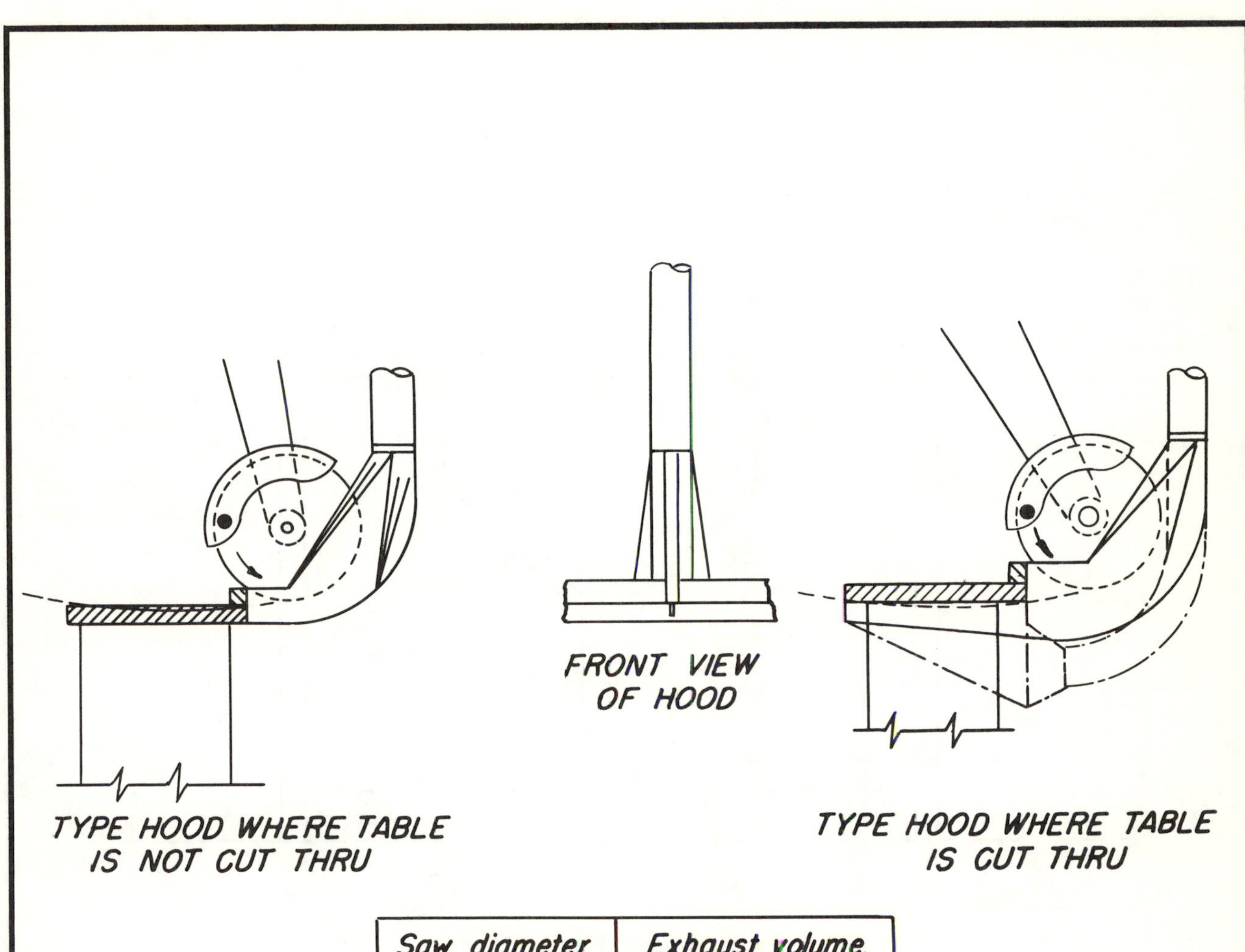

Saw diameter, inches	Exhaust volume, cfm
Up to 20 incl.	350
over 20	440

Duct velocity = 3500 fpm

Entry loss = 1.78 slot VP + 0.25 duct VP

AMERICAN CONFERENCE OF
GOVERNMENTAL INDUSTRIAL HYGIENISTS

SWING SAWS

DATE 1-64 | VS-707

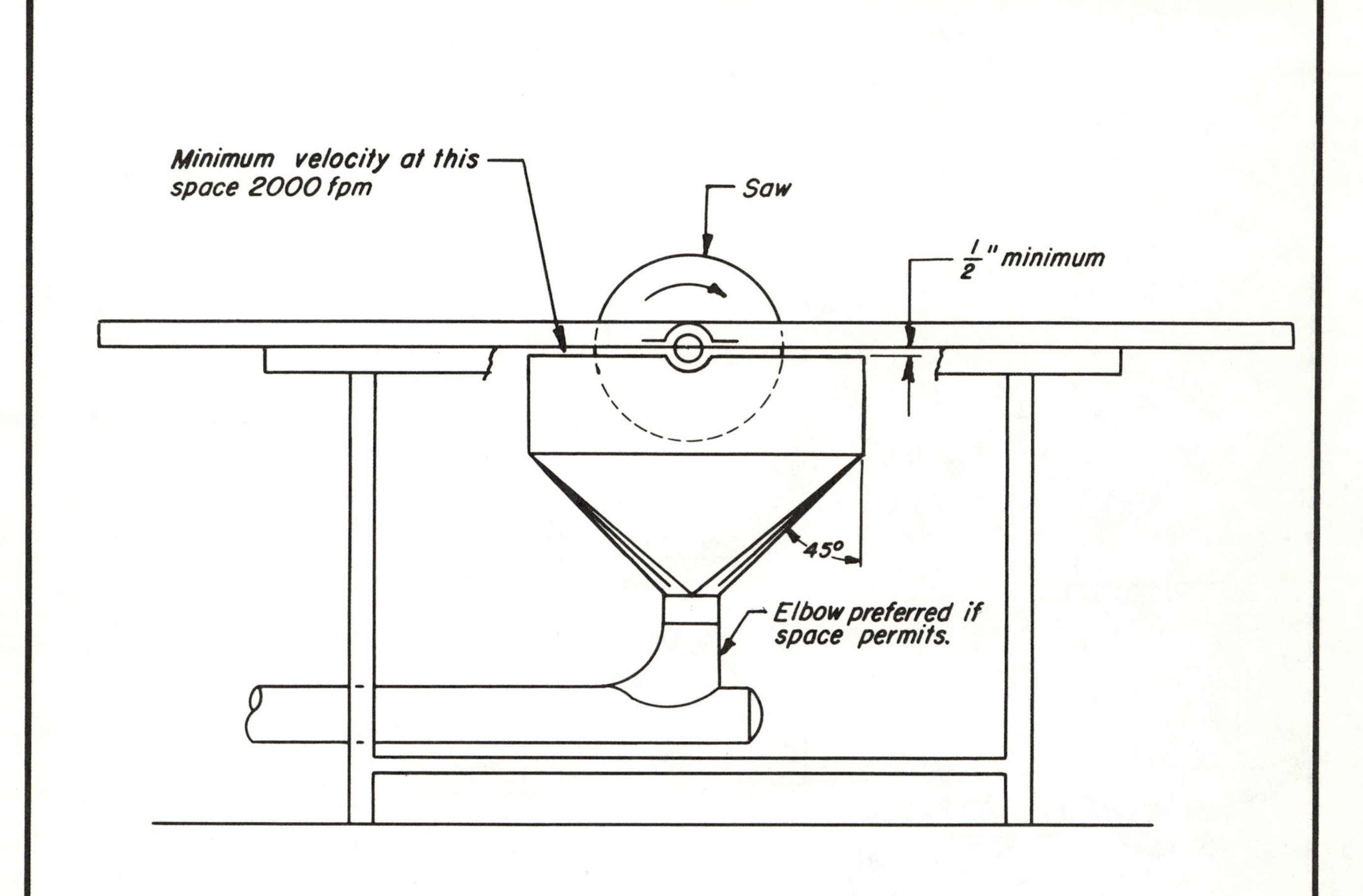

Table, rip, mitre and variety saws.

Saw diameter, inches	Exhaust volume, cfm
Up to 16 incl.	350
over 16 to 24 incl.	440
over 24	550
variety with dado	550

Duct velocity = 3500 fpm
Entry loss = 1.0 slot VP + 0.25 duct VP

AMERICAN CONFERENCE OF GOVERNMENTAL INDUSTRIAL HYGIENISTS

TABLE SAW

DATE 1-68 | VS-708

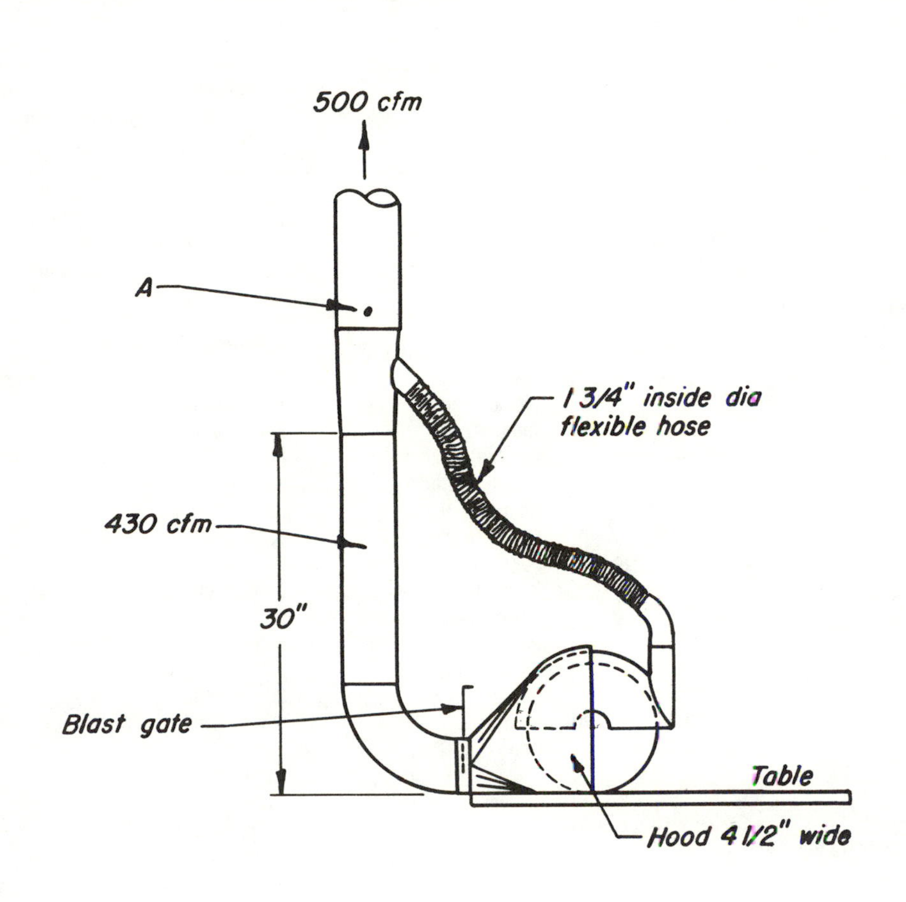

Duct velocity = 3500 fpm
Entry loss = 3.5 VP in duct riser (Point A)

For booth enclosure, see VS-401

AMERICAN CONFERENCE OF GOVERNMENTAL INDUSTRIAL HYGIENISTS

RADIAL SAW

DATE 1-68 | VS-709

TABLE 10-7.1. Miscellaneous Woodworking Machinery not Given in VS Prints

The following list of recommended exhaust volumes are for average-sized woodworking machines and are based on many years of experience. It must be noted that some modern high speed or extra large machines will produce such a large volume of waste that greater exhaust volumes must be used. Similarly, some small machines of the home workshop or bench type may use less exhaust air than listed.

Self-feed Table Rip Saw

Saw Diameter, inches	Exhaust Flow Rate, cfm		
	Bottom	Top	Total
Up to 16 inclusive	440	350	790
Over 16	550	350	900
Self-feed, not on table	800	550	1350

Gang Rip Saws

Saw Diameter, inches	Exhaust Flow Rate, cfm		
	Bottom	Top	Total
Up to 24, inclusive	550	350	900
Over 24 to 36, incl.	800	440	1240
Over 36 to 48, incl.	1100	550	1650
Over 48	1400	550	2060

Vertical Belt Sanders
(rear belt and both pulleys enclosed)
and
Top Run Horizontal Belt Sanders

Belt Width, inches	Exhaust Flow Rate, cfm
Up to 6, incl.	440
Over 6 to 9, incl.	550
Over 9 to 14, incl.	800
Over 14	1100

Swing Arm Sander: 440 cfm

Single Planers or Surfacers

	Exhaust Flow Rate, cfm
Up to 20″ knives	785
Over 20″ to 26″ knives	1100
Over 26″ to 32″ knives	1400
Over 32″ to 38″ knives	1765
Over 38″ knives	2200

Double Planers or Surfacers

	Exhaust Flow Rate, cfm		
	Bottom	Top	Total
Up to 20″ knives	550	785	1355
Over 20″ to 26″ knives	785	1100	1885
Over 26″ to 32″ knives	1100	1400	2500
Over 32″ to 38″ knives	1400	1800	3200
Over 38″ knives	1400	2200	3600

Molders, Matchers, & Sizers

Size, inches	Exhaust Flow Rate, cfm			
	Bottom	Top	Right	Left
Up to 7, incl.	440	550	350	350
Over 7 to 12, incl.	550	800	440	440
Over 12 to 18, incl.	800	1100	550	550
Over 18 to 24, incl.	1100	1400	800	800
Over 24	1400	1770	1100	1100

	Exhaust Flow Rate, cfm
Sash stickers	550
Woodshapers	440 to 1400
Tenoner	Same as moulder
Automatic lathe	800 to 5000
Forming lathe	350 to 1400
Chain mortise	350
Dowel machine	350 to 800
Panel raiser	550
Dove-tail and lock corner	550 to 800
Pulley pockets	550
Pulley stile	550
Glue jointer	800
Gainer	350 to 1400
Router	350 to 800
Hogs	
Up to 12″ wide	1400
Over 12 ″ wide	3100
Floorsweep	
6″ to 8″ diameter	800 to 1400

LOW VOLUME-HIGH VELOCITY EXHAUST SYSTEMS

The low volume-high velocity exhaust system is the unique application of exhaust which uses small volumes of air at relatively high velocities to control dust from portable hand tools and machining operations. Control is achieved by exhausting the air directly at the point of dust generation using close-fitting, custom-made hoods. Capture velocities are relatively high but the exhaust volume is low due to the small distance required. For flexibility, small diameter, lightweight plastic hoses are used with portable tools which results in very high duct velocities. This method allows the application of local exhaust ventilation to portable tools which otherwise would require relatively large air flow rates and large duct when controlled by conventional exhaust methods.

This technique has found a variety of applications although its use is not common. Rock drilling dust has been controlled by using hollow core drill steel with suitable exhaust holes in the drill bits. Air is exhausted either by a multi-stage turbine of the size generally used in industrial vacuum cleaners or, in the case of one manufacturer,(71) by the exhaust air from the pneumatic tool which operates a Venturi to withdraw air from the drill. Application has been made with flexible connections to a central vacuum system to aid in the control of graphite dust at conventional machining operations. One- to two-inch diameter flexible hose was used with simple exhaust hoods mounted directly at the cutting tool. In a similar application for the machining of beryllium,(72) a central vacuum system using 1.5-inch I.D. flexible hoses was employed. The exhaust hoods were made of lucite or transparent material and were tailor-made to surround the cutting tools and much of the work. Exhaust flow rates vary from 120 to 150 cfm with inlet velocities of 11,000 to 14,000 fpm. In another application,(73) a portable orbital sanding machine has been fitted with a small exhaust duct surrounding the edge of the plate. A fitting has been provided to connect this to the flexible hose of a standard domestic vacuum cleaner.

VS-801 to VS-806, 802, 803, and 804 illustrate a custom-made line of exhaust hoods available.(74) The required air flow rates range from 60 cfm for pneumatic chisels to 380 cfm for swing grinders (see Table 10.8-1). Due to the high entering velocities involved, static pressures are in the range of 7″ to 14″ of mercury (95″ to 290″ wg). This high pressure is necessary to create the high capture velocities at the dust source to control the dust.

The dust is conveyed at high velocities in small diameter flexible hoses ranging from 3/8″ to 2″ I.D. Exhaust is provided by a multi-stage centrifugal turbine capable of producing static pressures of approximately 12″ of mercury (163″ wg). A single stage positive displacement axial flow exhauster has the advantage of generating a vacuum in the range of 22″ of mercury (299″ wg) below atmospheric pressure. The fabric collector can be cleaned by a simple, manual valve which admits air into the clean side of the fabric bringing this side of the fabric to atmospheric pressure. Since the dirty side of the fabric is at a pressure far below atmospheric, this causes rapid air flow through the fabric and provides reverse cleaning.

TABLE 10.8-1. Exhaust Flow Rates Required for Low Volume-High Velocity System

	cfm	I.D. Plastic Hose Size (inches)
Disc sanders, 3–9-inch diameter	60–175	1–1.5
Vibratory pad sander — 4 × 9	100	1.25
Router, 1/8″ — 1″	80–100	1–1.25
Belt sander 3″ — 4,000 fpm	70	1
Pneumatic chisel	60	1
Radial wheel grinder	70	1
Surface die grinder, 1/4″	60	1
Cone wheel grinder	90	1.25
Cup stone grinder, 4″	100	1.25
Cup type brush, 6″	150	1.5
Radial wire brush, 6″	90	1.25
Hand wire brush 3 × 7	60	1
Rip out knife	175	1.5
Rip out cast cutter	150	1.5
Saber saw	120	1.5
Swing frame grinder 2 × 18	380	2.5
Saw abrasive 3″	100	1.25

Design—Calculations

With the exception of the proprietary system mentioned which can be purchased as a "package," the design calculations for these systems are largely empirical. In normal ventilation practice, air is considered to be incompressible since static pressures vary only slightly from atmospheric pressure. However, in these systems the extreme pressures required introduce problems of air density, compressibility, and viscosity which are not easily solved. Also, pressure drop data for small diameter pipe, especially flexible tubing, is not commonly available. For practical purposes, the turbine exhauster should be selected for the maximum simultaneous cfm exhaust required. Resistance in the pipe should be kept as low as possible; flexible tubing of less than 1- to 1.5-inch diameter should be limited to 10 feet or less. In most applications this is not a severe problem.

The main consideration in piping for such systems is to provide smooth internal configuration so as to reduce pressure loss at the high velocities involved and to minimize abrasion. Ordinary screwed-fitting piping is to be avoided because the lip of the pipe or male fitting, being of smaller diameter than the female thread, presents a discontinuity which increases pressure loss and may be a point of rapid abrasion.

If a screwed-fitting system is to be used, cast-iron drainage fittings and Schedule 40 pipe should be provided. Fittings of this type are recessed so that the I>D> of the pipe and fitting is the same. Tubing systems of 60 gauge wall thickness up to and including 4-inch diameter, and 14 gauge for 5-inch and above, provide for 10 percent less pressure loss than a Schedule 40 piping system and offer a lower installed cost in most cases. Commercially available tube fittings and clamps may be used, or a slip-on flange system, so that internal discontinuities are eliminated. In all cases, long-sweep elbows and bends should be provided.

For dust exhaust systems, a good collector should be mounted ahead of the exhauster to minimize erosion of the precision blades and subsequent loss in performance. Final balance of the system can be achieved by varying the length and diameters of the small flexible hoses.

It must be emphasized that although data are empirical, these systems require the same careful design as the more conventional ones. Abrupt changes of direction, expansions and contractions must be avoided and care must always be taken to minimize pressure losses.

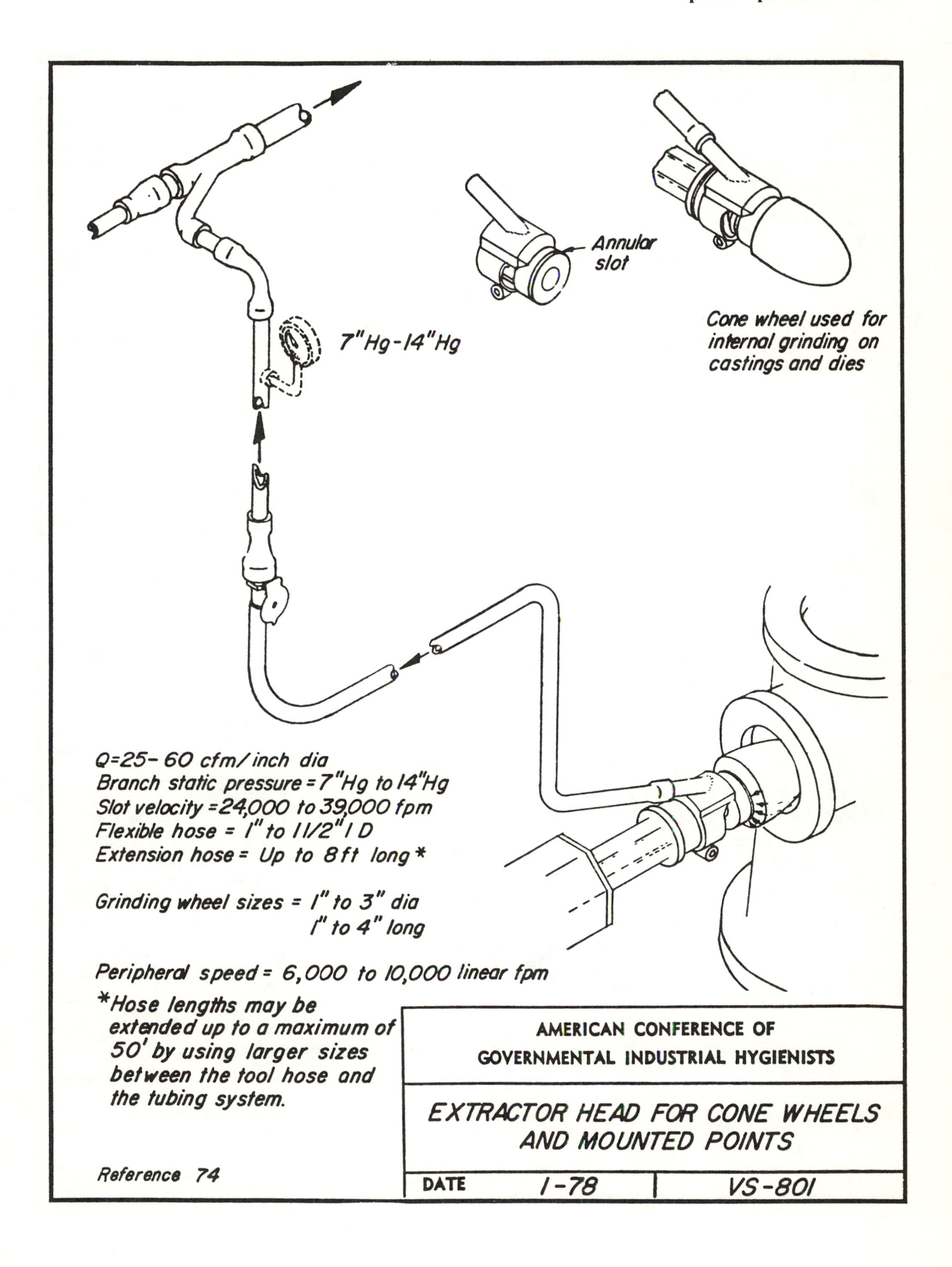
Annular slot
7"Hg-14"Hg
Cone wheel used for internal grinding on castings and dies
Q=25- 60 cfm/inch dia
Branch static pressure=7"Hg to 14"Hg
Slot velocity=24,000 to 39,000 fpm
Flexible hose = 1" to 1 1/2" I D
Extension hose= Up to 8 ft long *
Grinding wheel sizes = 1" to 3" dia
1" to 4" long
Peripheral speed= 6,000 to 10,000 linear fpm
*Hose lengths may be extended up to a maximum of 50' by using larger sizes between the tool hose and the tubing system.
Reference 74
AMERICAN CONFERENCE OF GOVERNMENTAL INDUSTRIAL HYGIENISTS
EXTRACTOR HEAD FOR CONE WHEELS AND MOUNTED POINTS
DATE 1-78
VS-801

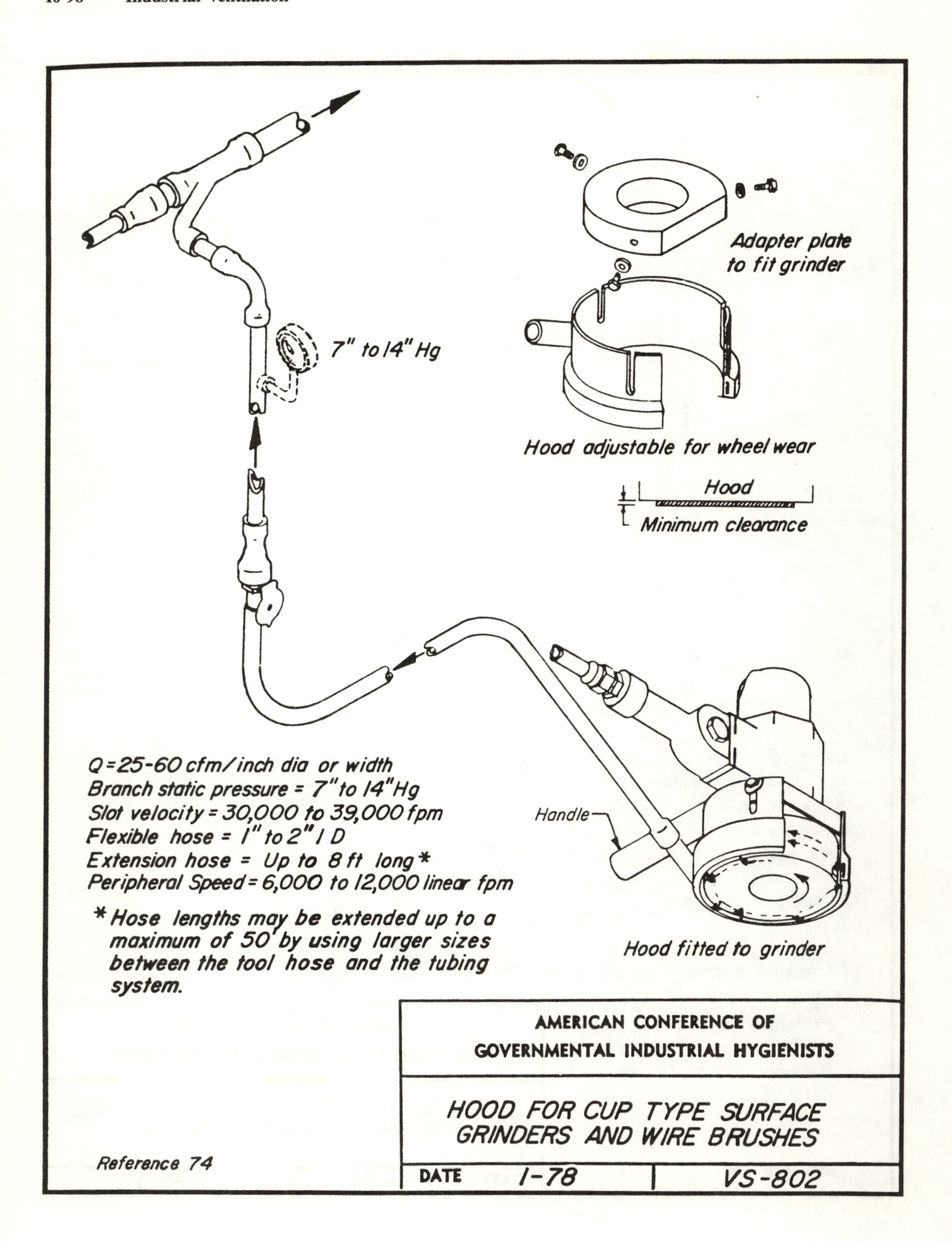
7" to 14" Hg
Adapter plate to fit grinder
Hood adjustable for wheel wear
Hood
Minimum clearance
Q = 25-60 cfm/inch dia or width
Branch static pressure = 7" to 14" Hg
Slot velocity = 30,000 to 39,000 fpm
Flexible hose = 1" to 2" I D
Extension hose = Up to 8 ft long*
Peripheral Speed = 6,000 to 12,000 linear fpm
* Hose lengths may be extended up to a maximum of 50' by using larger sizes between the tool hose and the tubing system.
Handle
Hood fitted to grinder
AMERICAN CONFERENCE OF GOVERNMENTAL INDUSTRIAL HYGIENISTS
HOOD FOR CUP TYPE SURFACE GRINDERS AND WIRE BRUSHES
Reference 74
DATE 1-78
VS-802

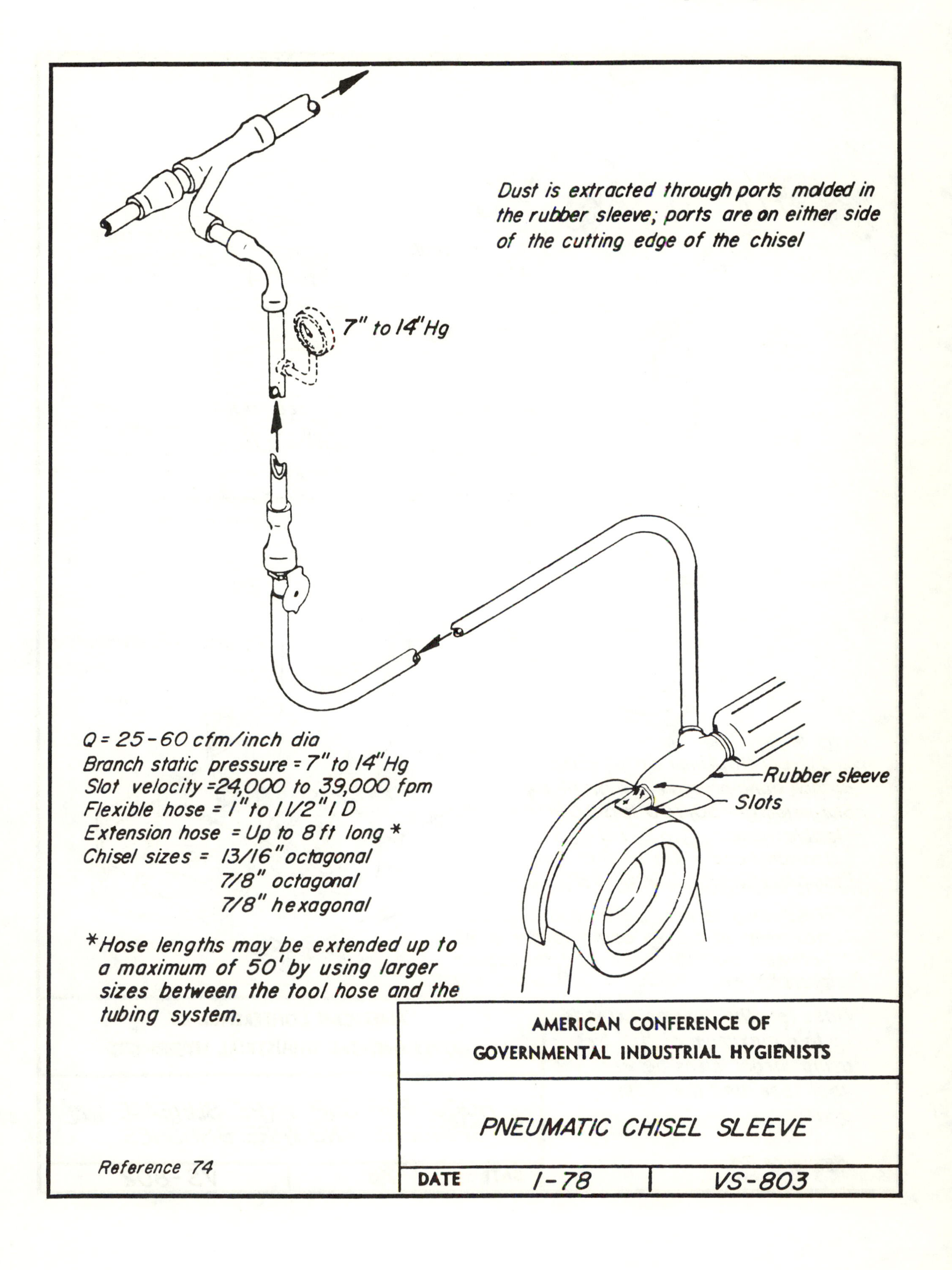

Dust is extracted through ports molded in the rubber sleeve; ports are on either side of the cutting edge of the chisel
7" to 14" Hg
Q = 25-60 cfm/inch dia
Branch static pressure = 7" to 14" Hg
Slot velocity = 24,000 to 39,000 fpm
Flexible hose = 1" to 1 1/2" I D
Extension hose = Up to 8 ft long *
Chisel sizes = 13/16" octagonal
7/8" octagonal
7/8" hexagonal
*Hose lengths may be extended up to a maximum of 50' by using larger sizes between the tool hose and the tubing system.
Rubber sleeve
Slots
Reference 74
AMERICAN CONFERENCE OF GOVERNMENTAL INDUSTRIAL HYGIENISTS
PNEUMATIC CHISEL SLEEVE
DATE 1-78
VS-803

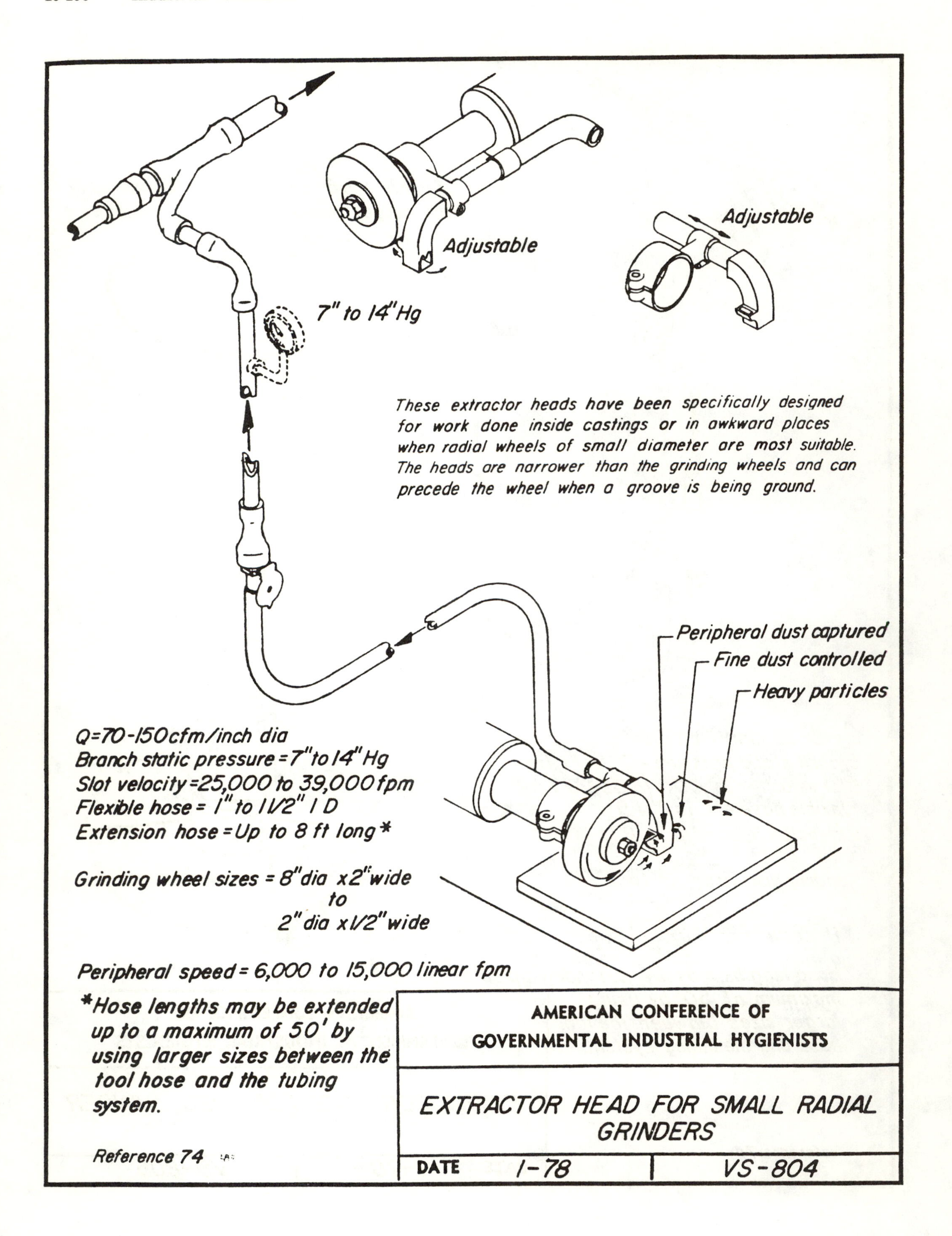

Adjustable
Adjustable
7" to 14" Hg
These extractor heads have been specifically designed for work done inside castings or in awkward places when radial wheels of small diameter are most suitable. The heads are narrower than the grinding wheels and can precede the wheel when a groove is being ground.
Peripheral dust captured
Fine dust controlled
Heavy particles
Q=70-150 cfm/inch dia
Branch static pressure = 7" to 14" Hg
Slot velocity = 25,000 to 39,000 fpm
Flexible hose = 1" to 1 1/2" I D
Extension hose = Up to 8 ft long*
Grinding wheel sizes = 8" dia x 2" wide
to
2" dia x 1/2" wide
Peripheral speed = 6,000 to 15,000 linear fpm
*Hose lengths may be extended up to a maximum of 50' by using larger sizes between the tool hose and the tubing system.
Reference 74
AMERICAN CONFERENCE OF GOVERNMENTAL INDUSTRIAL HYGIENISTS
EXTRACTOR HEAD FOR SMALL RADIAL GRINDERS
DATE 1-78
VS-804

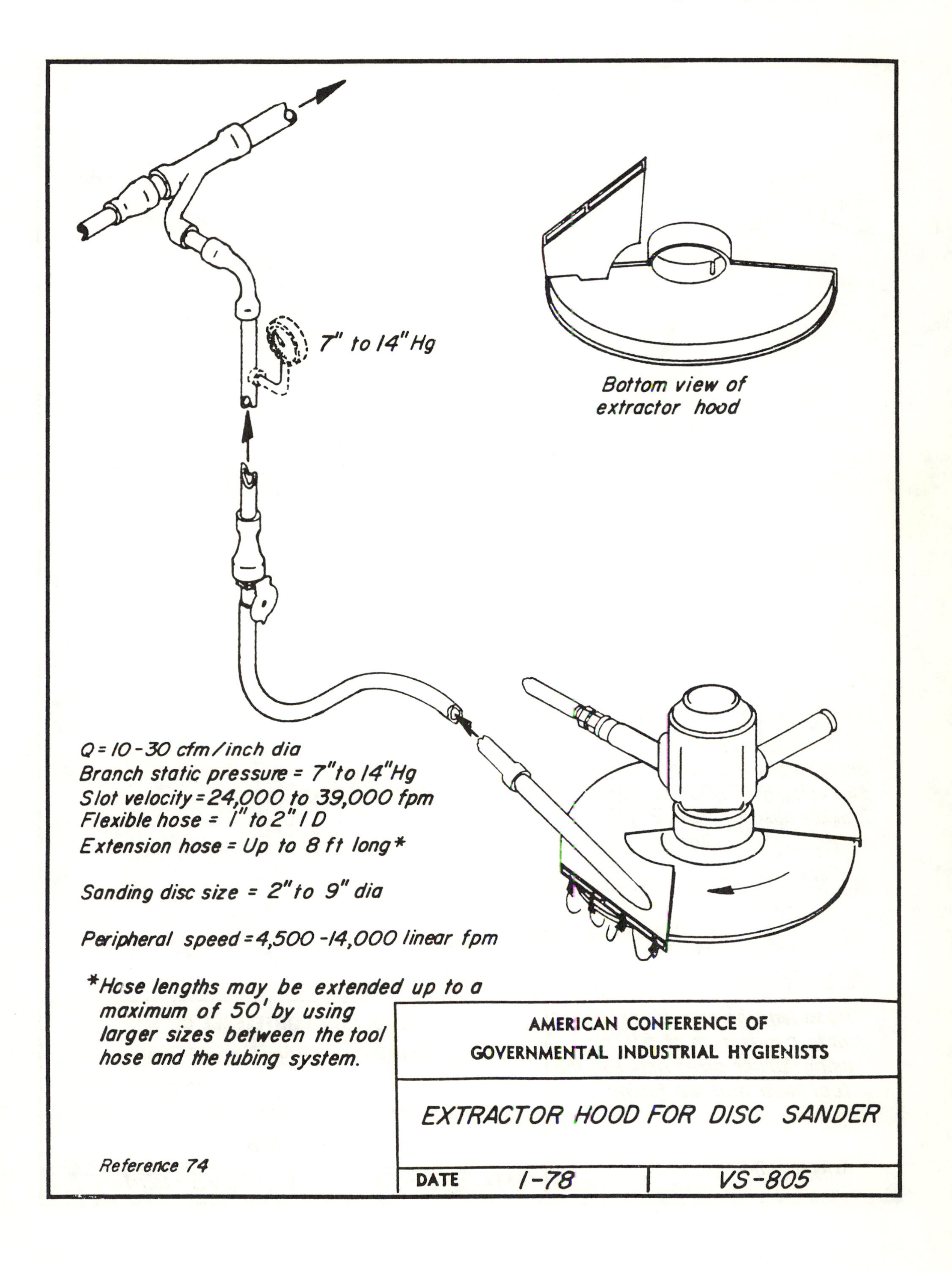
7" to 14" Hg
Bottom view of extractor hood
Q = 10-30 cfm/inch dia
Branch static pressure = 7" to 14" Hg
Slot velocity = 24,000 to 39,000 fpm
Flexible hose = 1" to 2" I D
Extension hose = Up to 8 ft long*
Sanding disc size = 2" to 9" dia
Peripheral speed = 4,500 - 14,000 linear fpm
*Hose lengths may be extended up to a maximum of 50' by using larger sizes between the tool hose and the tubing system.
Reference 74
AMERICAN CONFERENCE OF GOVERNMENTAL INDUSTRIAL HYGIENISTS
EXTRACTOR HOOD FOR DISC SANDER
DATE 1-78
VS-805

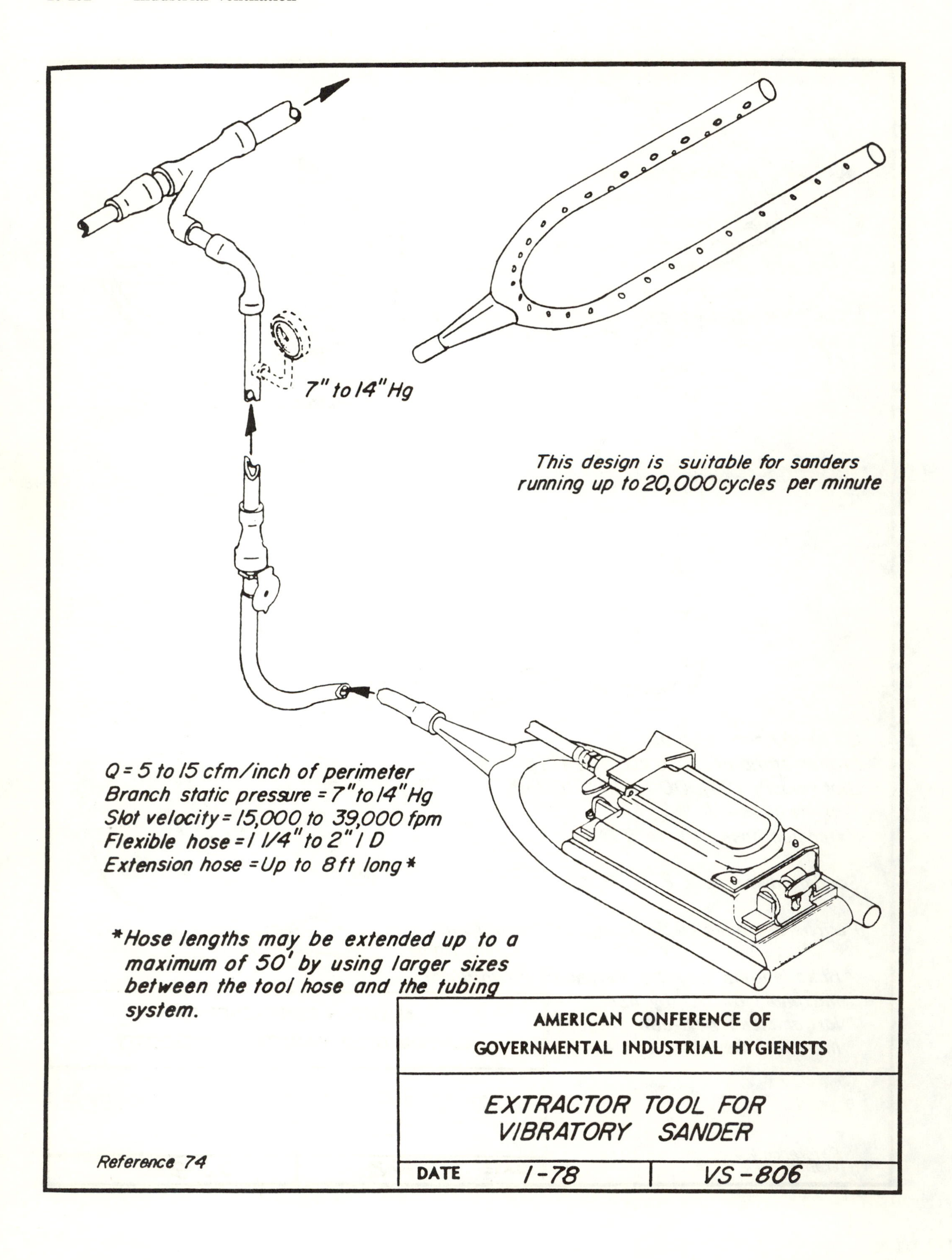
7" to 14" Hg
This design is suitable for sanders running up to 20,000 cycles per minute
Q = 5 to 15 cfm/inch of perimeter
Branch static pressure = 7" to 14" Hg
Slot velocity = 15,000 to 39,000 fpm
Flexible hose = 1 1/4" to 2" I D
Extension hose = Up to 8 ft long *
*Hose lengths may be extended up to a maximum of 50' by using larger sizes between the tool hose and the tubing system.
AMERICAN CONFERENCE OF GOVERNMENTAL INDUSTRIAL HYGIENISTS
EXTRACTOR TOOL FOR VIBRATORY SANDER
DATE 1-78
VS-806
Reference 74

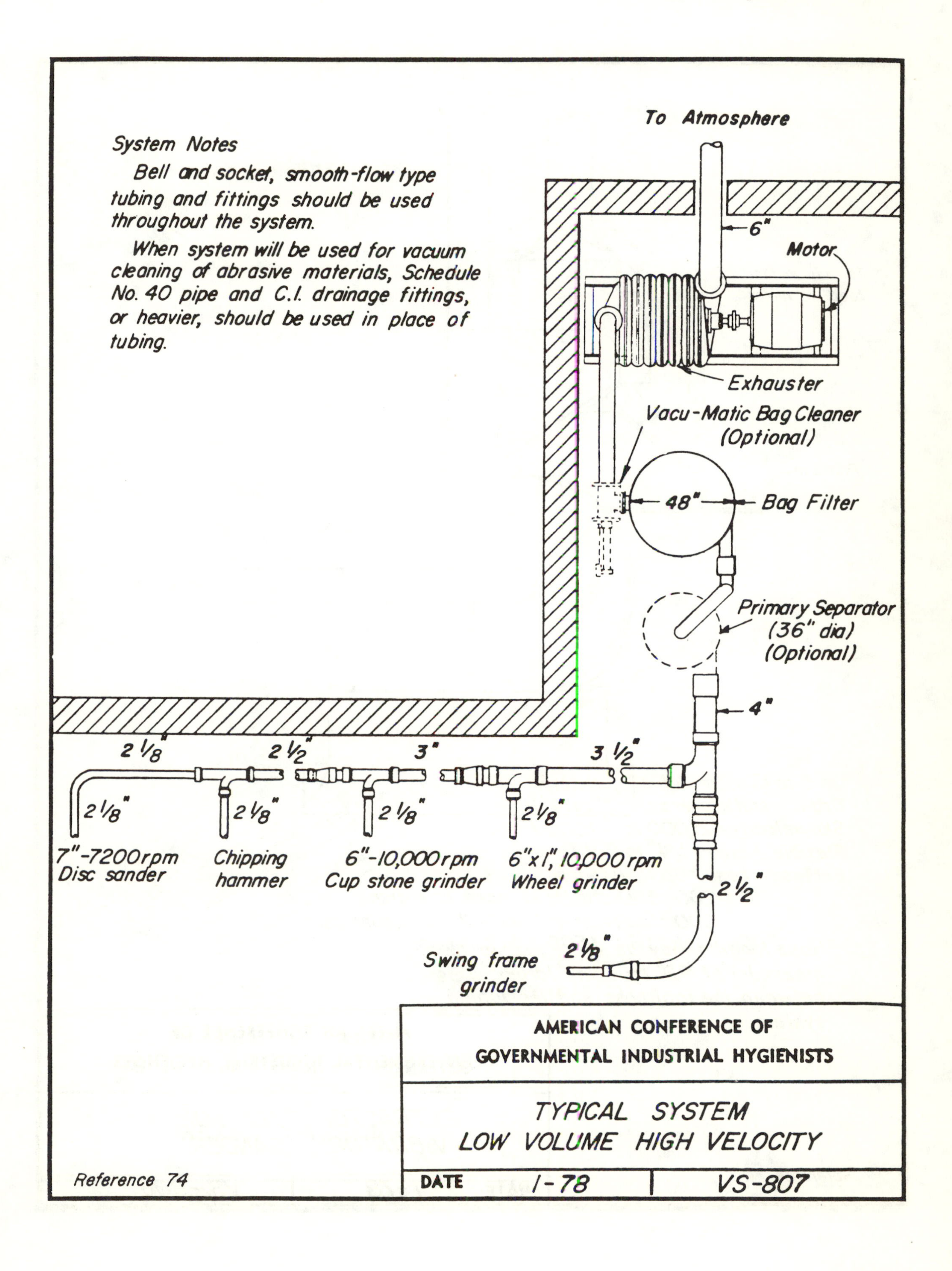
System Notes
Bell and socket, smooth-flow type tubing and fittings should be used throughout the system.
When system will be used for vacuum cleaning of abrasive materials, Schedule No. 40 pipe and C.I. drainage fittings, or heavier, should be used in place of tubing.
To Atmosphere
6"
Motor
Exhauster
Vacu-Matic Bag Cleaner (Optional)
48"
Bag Filter
Primary Separator (36" dia) (Optional)
4"
2 1/8"
2 1/2"
3"
3 1/2"
2 1/8"
2 1/8"
2 1/8"
2 1/8"
7"-7200 rpm Disc sander
Chipping hammer
6"-10,000 rpm Cup stone grinder
6"x 1", 10,000 rpm Wheel grinder
2 1/2"
2 1/8"
Swing frame grinder
Reference 74
AMERICAN CONFERENCE OF GOVERNMENTAL INDUSTRIAL HYGIENISTS
TYPICAL SYSTEM
LOW VOLUME HIGH VELOCITY
DATE 1-78
VS-807

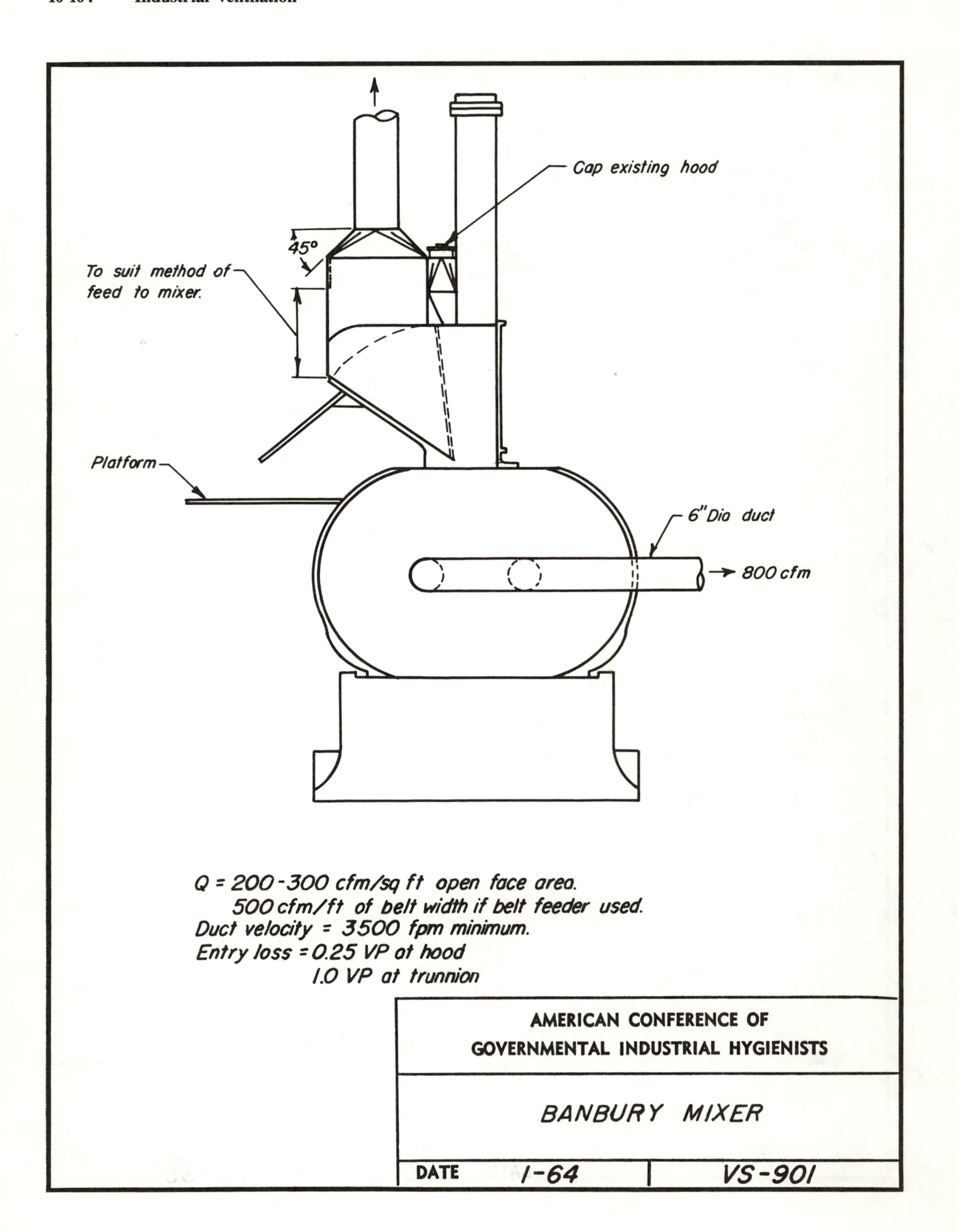
Cap existing hood
45°
To suit method of feed to mixer.
Platform
6" Dia duct
800 cfm
Q = 200-300 cfm/sq ft open face area.
500 cfm/ft of belt width if belt feeder used.
Duct velocity = 3500 fpm minimum.
Entry loss = 0.25 VP at hood
1.0 VP at trunnion
AMERICAN CONFERENCE OF GOVERNMENTAL INDUSTRIAL HYGIENISTS
BANBURY MIXER
DATE 1-64
VS-901

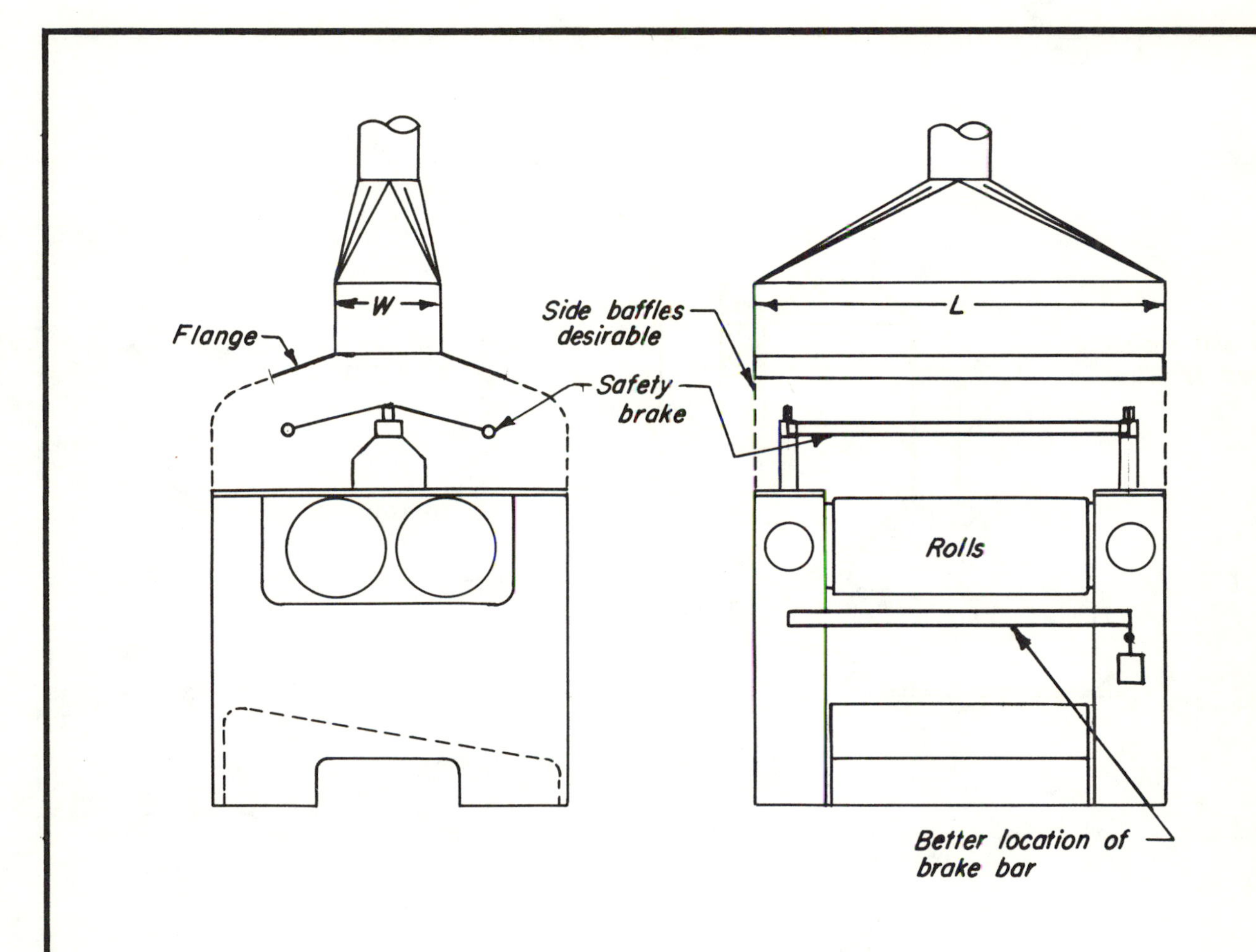

Q = 125 cfm/sq ft hood area (125 WL)
Duct velocity = 1000 - 3000 fpm
Entry loss = 0.25 duct VP

AMERICAN CONFERENCE OF GOVERNMENTAL INDUSTRIAL HYGIENISTS

RUBBER CALENDER ROLLS

DATE 1-70	VS-902

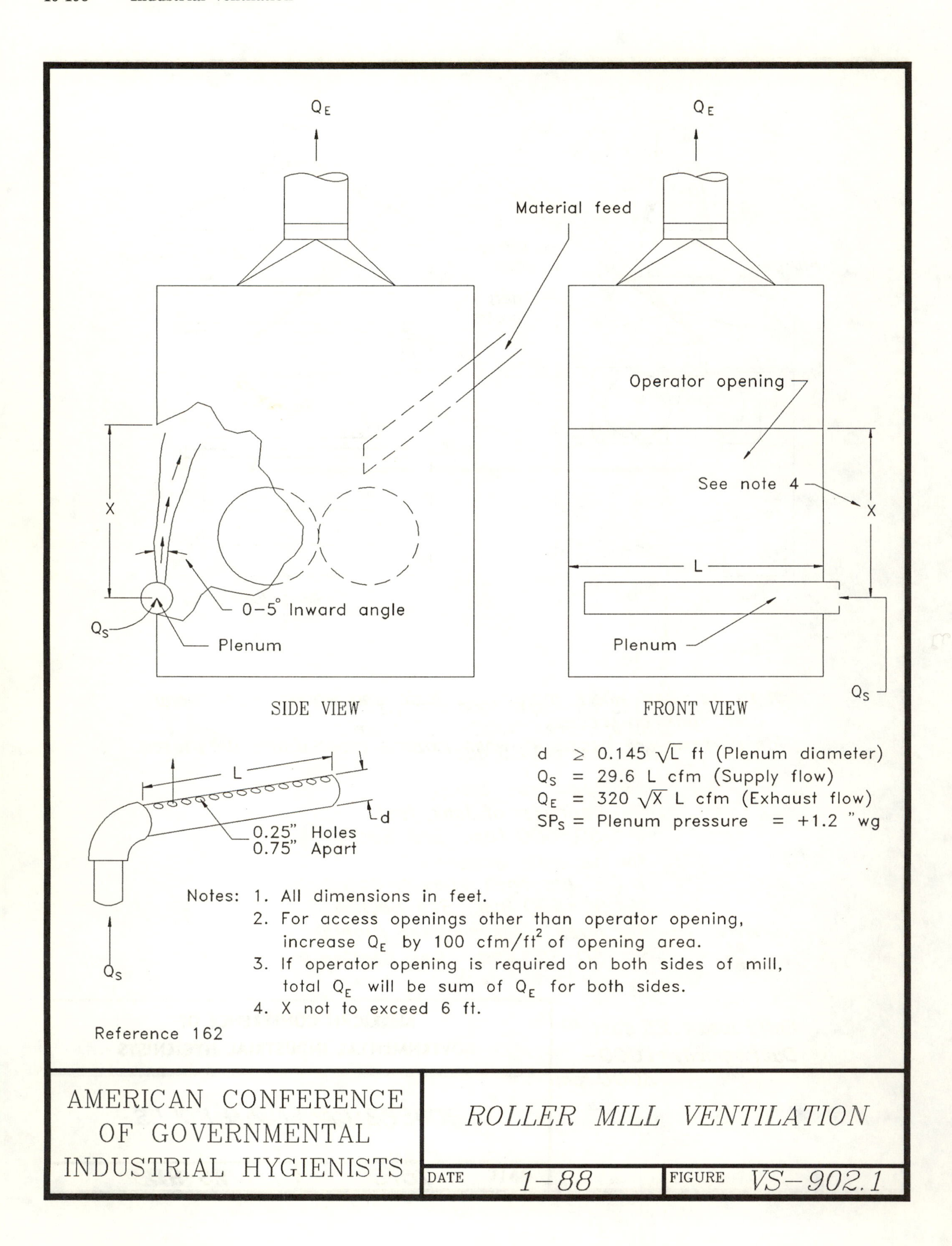

QE
QE
Material feed
Operator opening
See note 4
X
X
L
0–5° Inward angle
QS
Plenum
Plenum
QS
SIDE VIEW
FRONT VIEW
L
d
0.25" Holes
0.75" Apart
QS
d ≥ 0.145 √L ft (Plenum diameter)
QS = 29.6 L cfm (Supply flow)
QE = 320 √X L cfm (Exhaust flow)
SPS = Plenum pressure = +1.2 "wg
Notes: 1. All dimensions in feet.
2. For access openings other than operator opening, increase QE by 100 cfm/ft² of opening area.
3. If operator opening is required on both sides of mill, total QE will be sum of QE for both sides.
4. X not to exceed 6 ft.
Reference 162
AMERICAN CONFERENCE OF GOVERNMENTAL INDUSTRIAL HYGIENISTS
ROLLER MILL VENTILATION
DATE 1-88
FIGURE VS-902.1

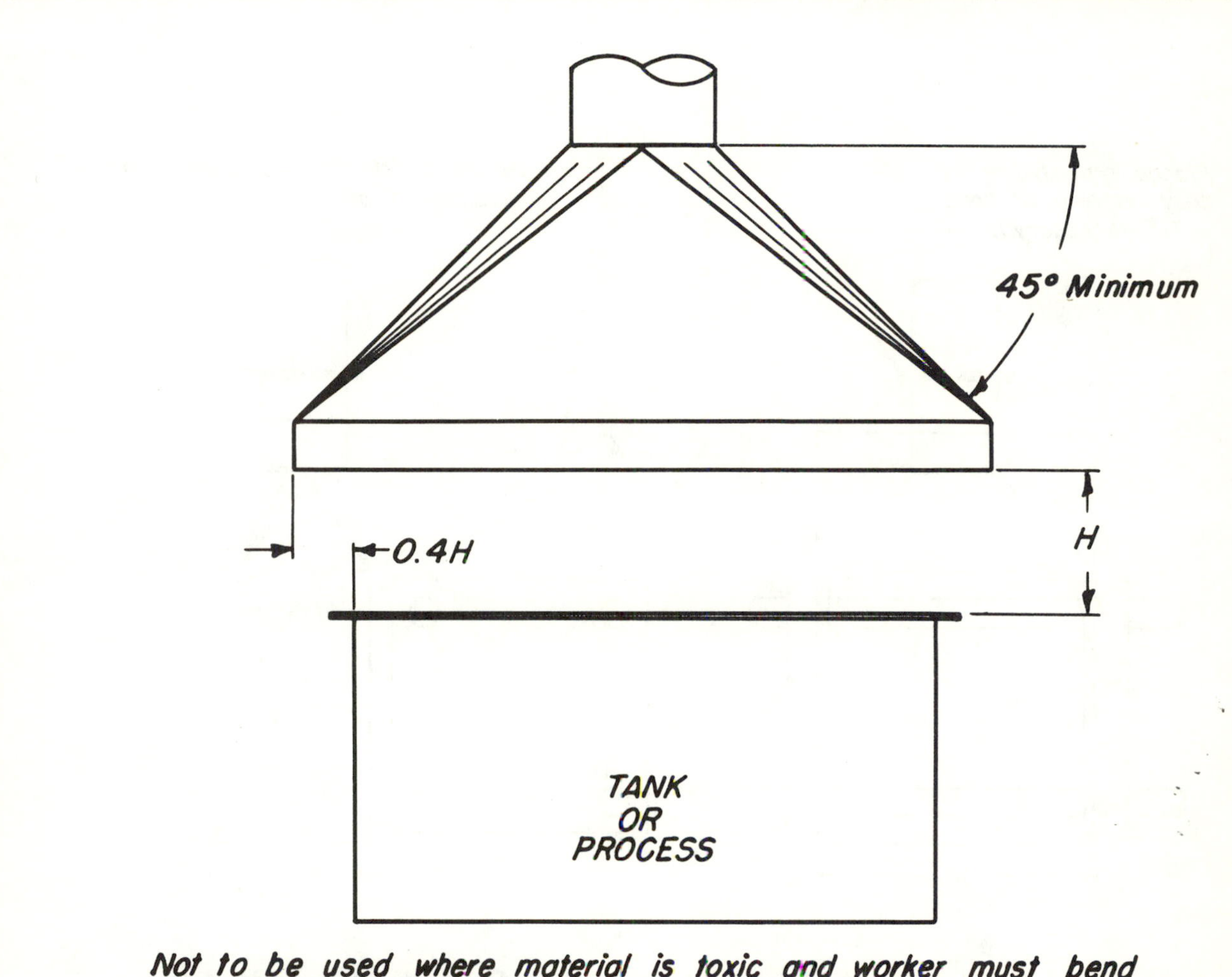

Not to be used where material is toxic and worker must bend over tank or process.
Side curtains are necessary when extreme cross-drafts are present.

$Q = 1.4PHV$	for open type canopy.
	P= perimeter of tank, feet.
	V= 50-500 fpm. See Section 4
$Q = (W+L)HV$	for two sides enclosed.
	W & L are open sides of hood.
	V = 50-500 fpm. See Section 4
$Q = WHV$ or LHV	for three sides enclosed. (Booth)
	V= 50-500 fpm. See Section 4

Entry loss= .25 duct VP
Duct velocity=1000-3000 fpm

AMERICAN CONFERENCE OF GOVERNMENTAL INDUSTRIAL HYGIENISTS

CANOPY HOOD

DATE 1-70 | VS-903

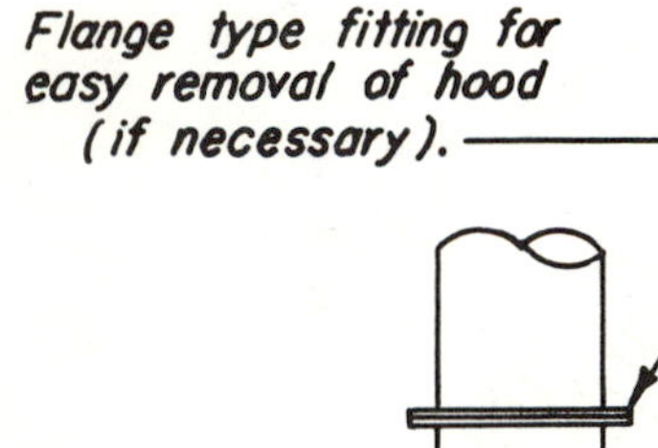

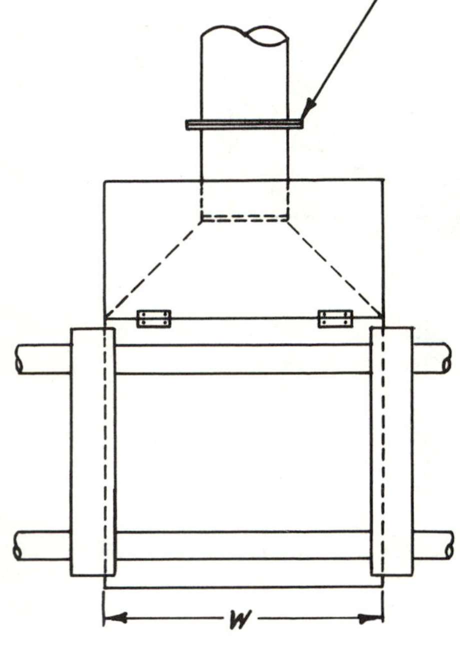

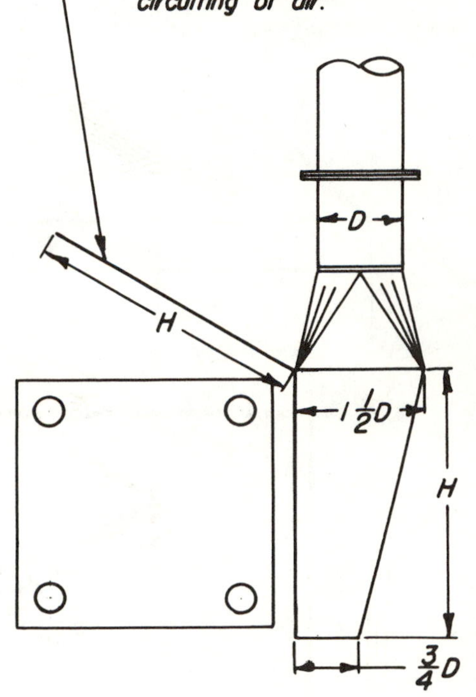

Note: Place hood as close to machine as possible. If more than 4 inches from back of machine, hinged side baffles should be used.

Note: Products of combustion require separate flue or may be vented into hood.

$Q = 300WH$
Entry loss = 0.25 duct VP
Duct velocity = 2500 – 3000 fpm

AMERICAN CONFERENCE OF
GOVERNMENTAL INDUSTRIAL HYGIENISTS

DIE CASTING HOOD

DATE 1-70	VS-904

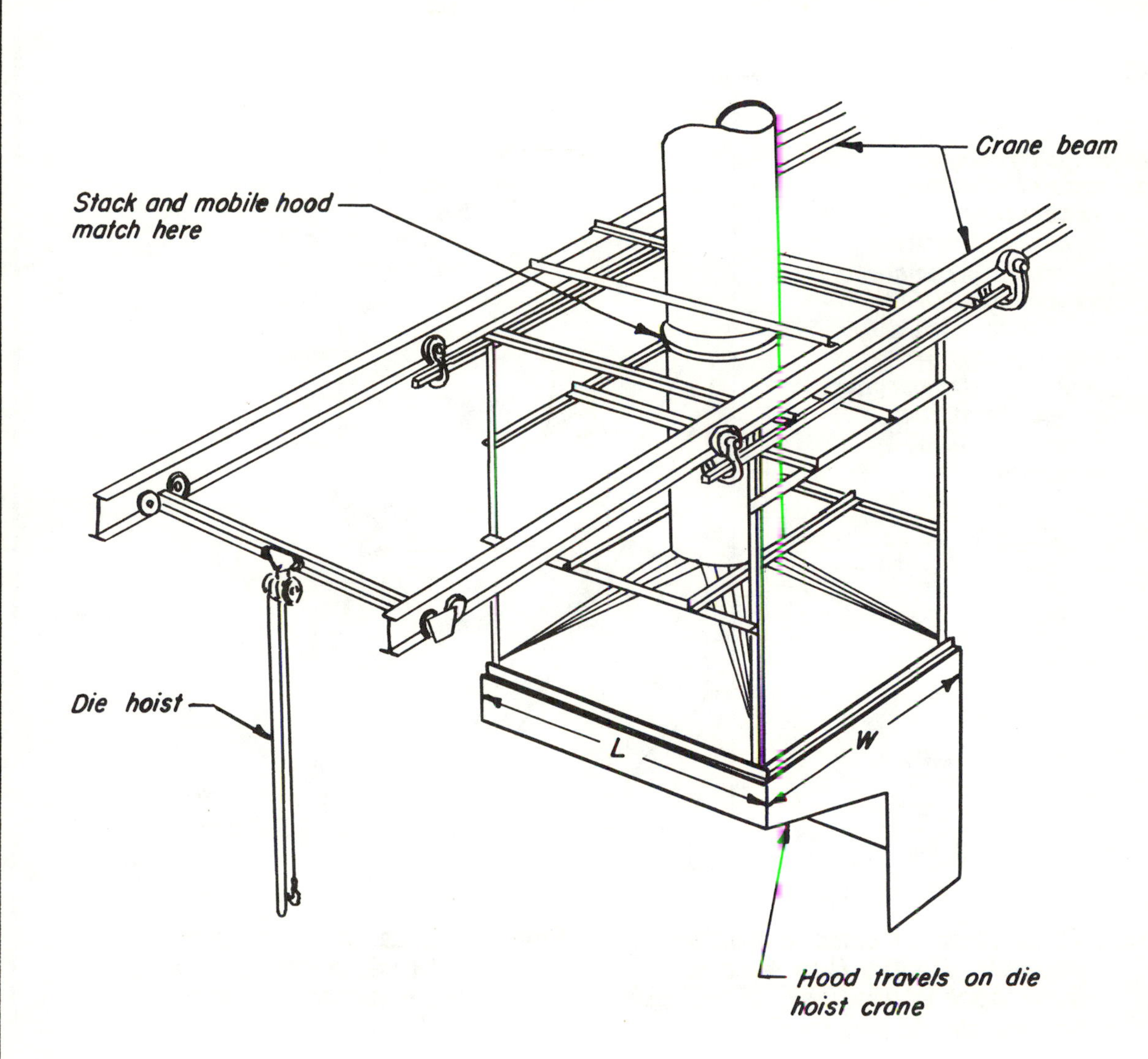

Q = 300WL
Duct velocity = 1000 - 3000 fpm
Entry loss = 0.25 duct VP

AMERICAN CONFERENCE OF GOVERNMENTAL INDUSTRIAL HYGIENISTS

DIE CASTING MACHINE OR MELTING FURNACE

DATE 1-70 | VS-905

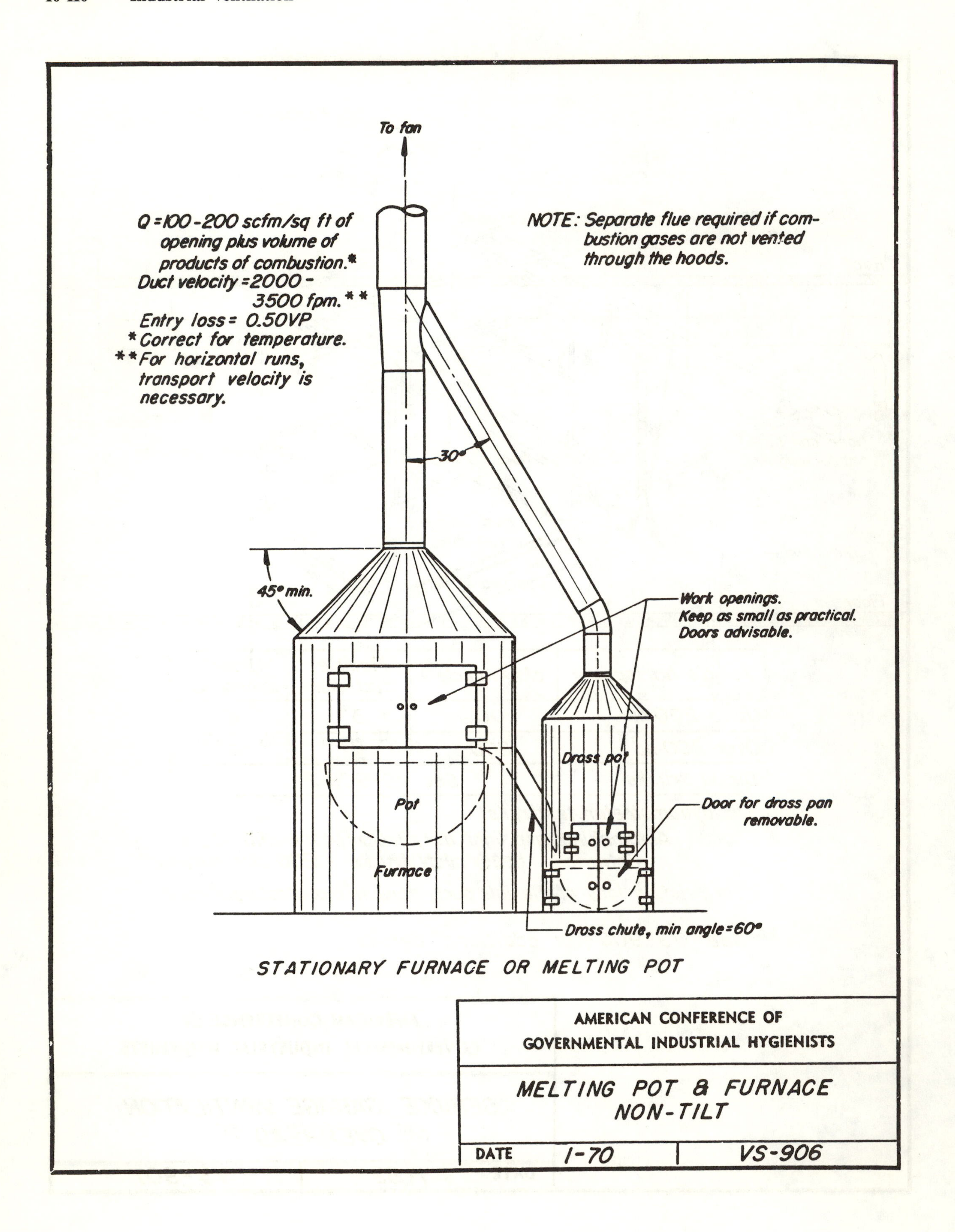

To fan
Q = 100-200 scfm/sq ft of opening plus volume of products of combustion.*
Duct velocity = 2000 - 3500 fpm.**
Entry loss = 0.50VP
* Correct for temperature.
** For horizontal runs, transport velocity is necessary.
NOTE: Separate flue required if combustion gases are not vented through the hoods.
30°
45° min.
Work openings. Keep as small as practical. Doors advisable.
Dross pot
Pot
Furnace
Door for dross pan removable.
Dross chute, min angle = 60°
STATIONARY FURNACE OR MELTING POT
AMERICAN CONFERENCE OF GOVERNMENTAL INDUSTRIAL HYGIENISTS
MELTING POT & FURNACE NON-TILT
DATE 1-70
VS-906

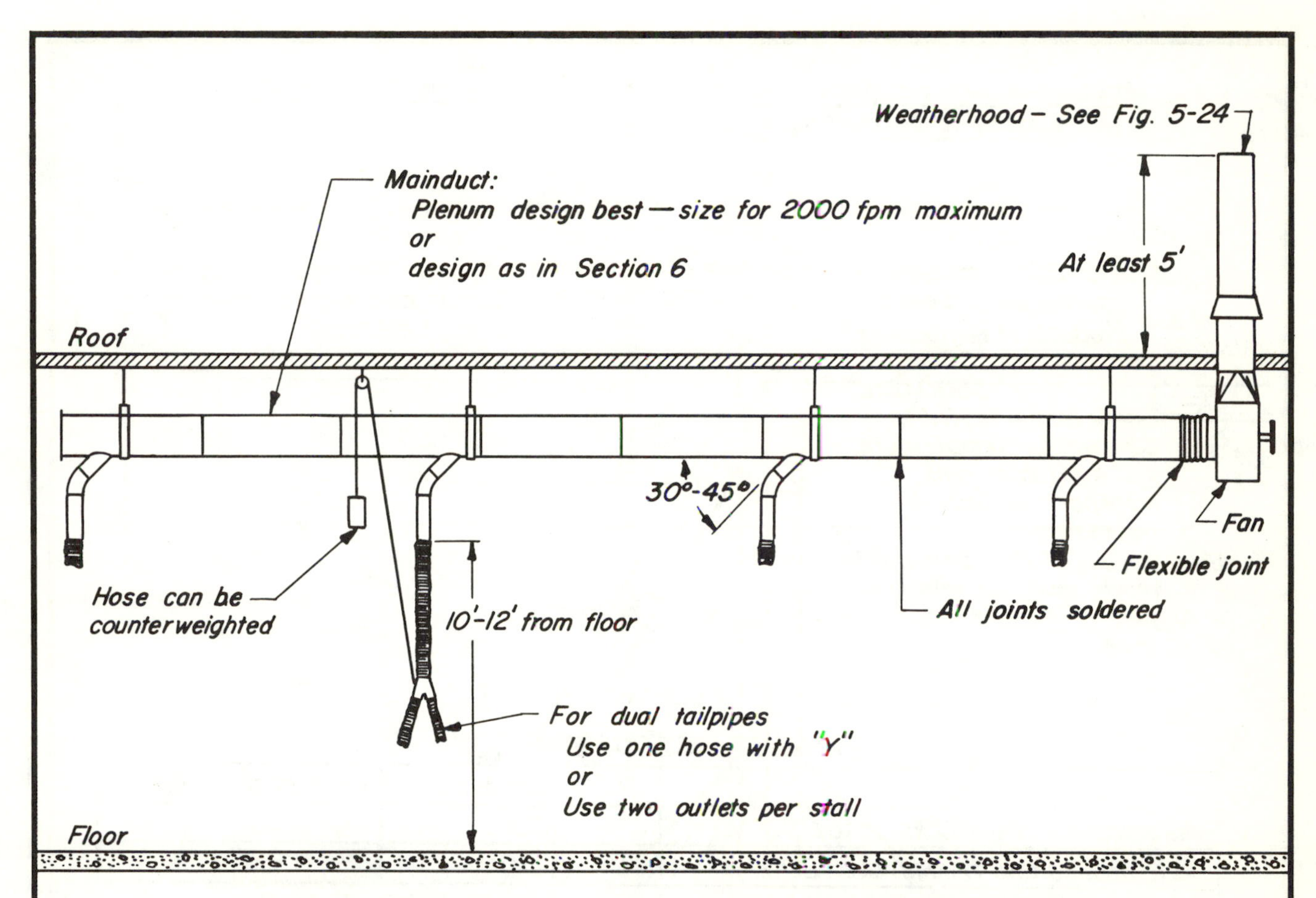

Vehicle horsepower	cfm/vehicle	Flexible duct diam	Branch connection
Up to 200 hp	100	3"	4"
Over 200 hp	200	4"	4"
Diesel trucks	See VS-908.2		

On dynamometer test rolls

Automobiles and light duty trucks = 2 x cfm above

Heavy duty trucks = 1200 cfm minimum

For friction loss of flexible duct; consult manufacturers' data

See VS-908 for additional details

AMERICAN CONFERENCE OF GOVERNMENTAL INDUSTRIAL HYGIENISTS

SERVICE GARAGE VENTILATION OVERHEAD

DATE 1-82	VS-907

Note: In ventilating a garage use either the overhead or under floor system. Exhaust to be discharged above roof

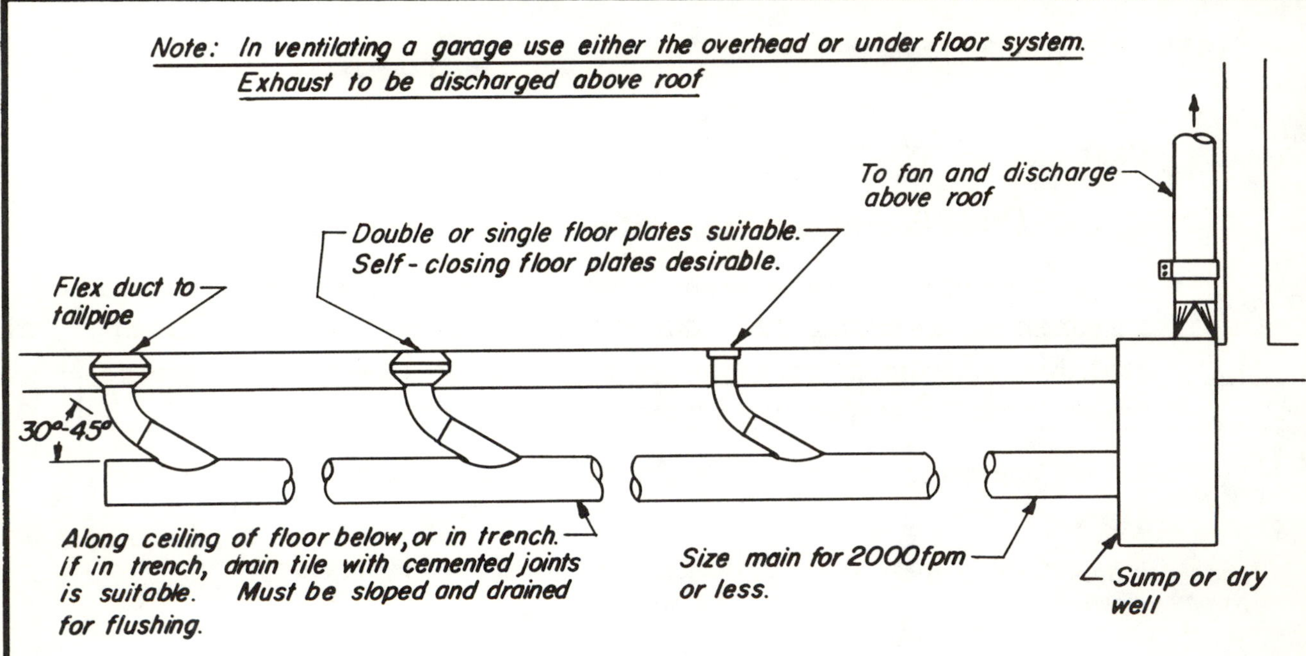

UNDER FLOOR SYSTEM

EXHAUST REQUIREMENTS *

Type	cfm per vehicle	Flex duct ID (min)
Automobiles and trucks up to 200 hp	100	3"
Automobiles and trucks over 200 hp	200	4" * *
Diesel	See VS-908.2	

* On dynamometer test rolls
Automobiles and light duty trucks = 2 x cfm above
Heavy duty trucks = 1200 cfm minimum.

** 3" dia permissible for short runs with proper fan.
For friction loss of flexible duct; consult manufacturers' data.

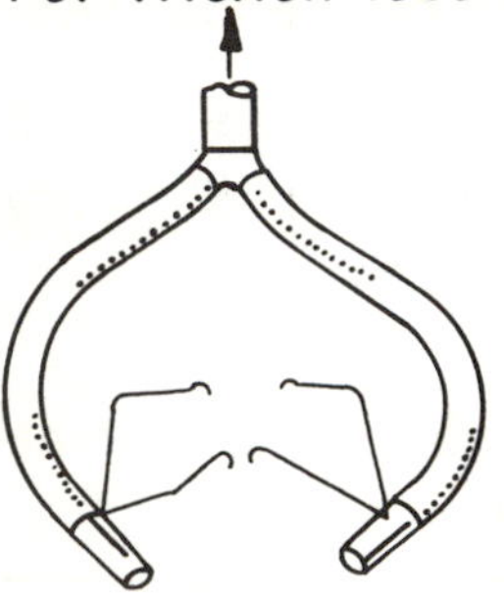

Dilution ventilation is necessary for cars in motion or idling outside of stalls.

DILUTION RATES:

5000 cfm/running automobile
10,000 cfm (or more)/truck.
100 cfm/horsepower for diesel.

For parking garages, see Table 10-9-2

Use adapters on dual exhausts and special tailpipes.

AMERICAN CONFERENCE OF GOVERNMENTAL INDUSTRIAL HYGIENISTS

SERVICE GARAGE VENTILATION UNDERFLOOR

DATE 1-82 | VS-908

BASIC DESIGN VENTILATION RATES

- 5000 cfm per propane fueled lift truck
- 8000 cfm per gasoline fueled lift truck

CONDITIONS UNDER WHICH BASIC DESIGN RATES APPLY

- A regular maintenance program incorporating final engine tuning through carbon monoxide analysis of exhaust gases must be provided. CO concentration of exhaust gases should be limited to 1 percent for propane fueled trucks, 2 percent for gasoline fueled trucks.
- Actual operating time of lift trucks must be 50 percent or less of total exposure time.
- A reasonably good distribution of air flow must be provided.
- The volume of space must amount to 150,000 cu. ft. per lift truck or more.

CORRECTIONS FOR CONDITIONS OTHER THAN THOSE ABOVE

- No regular maintenance program – multiply the basic design ventilation rate by three.
- Operating time greater than 50 percent – multiply the basic design ventilation rate by the actual operating time in percent divided by 50.
- Poor distribution of air flow – lift truck operation not recommended.
- Volume of space less than 150,000 cu. ft. per lift truck – multiply the basic design ventilation rate by a suitable factor based on the following:

  1.5 times design rate for 75,000 cu. ft.; 2.0 times design rate for 30,000 cu. ft. Lift truck operation in spaces of less than 25,000 cu. ft. is not recommended.
- Lift truck engine horsepower greater than 60 – multiply the basic design ventilation rate by the actual horsepower divided by 60.

AMERICAN CONFERENCE OF GOVERNMENTAL INDUSTRIAL HYGIENISTS	
FUEL POWERED LIFT TRUCK VENTILATION DESIGN	
DATE *1-74*	*VS-908.1*

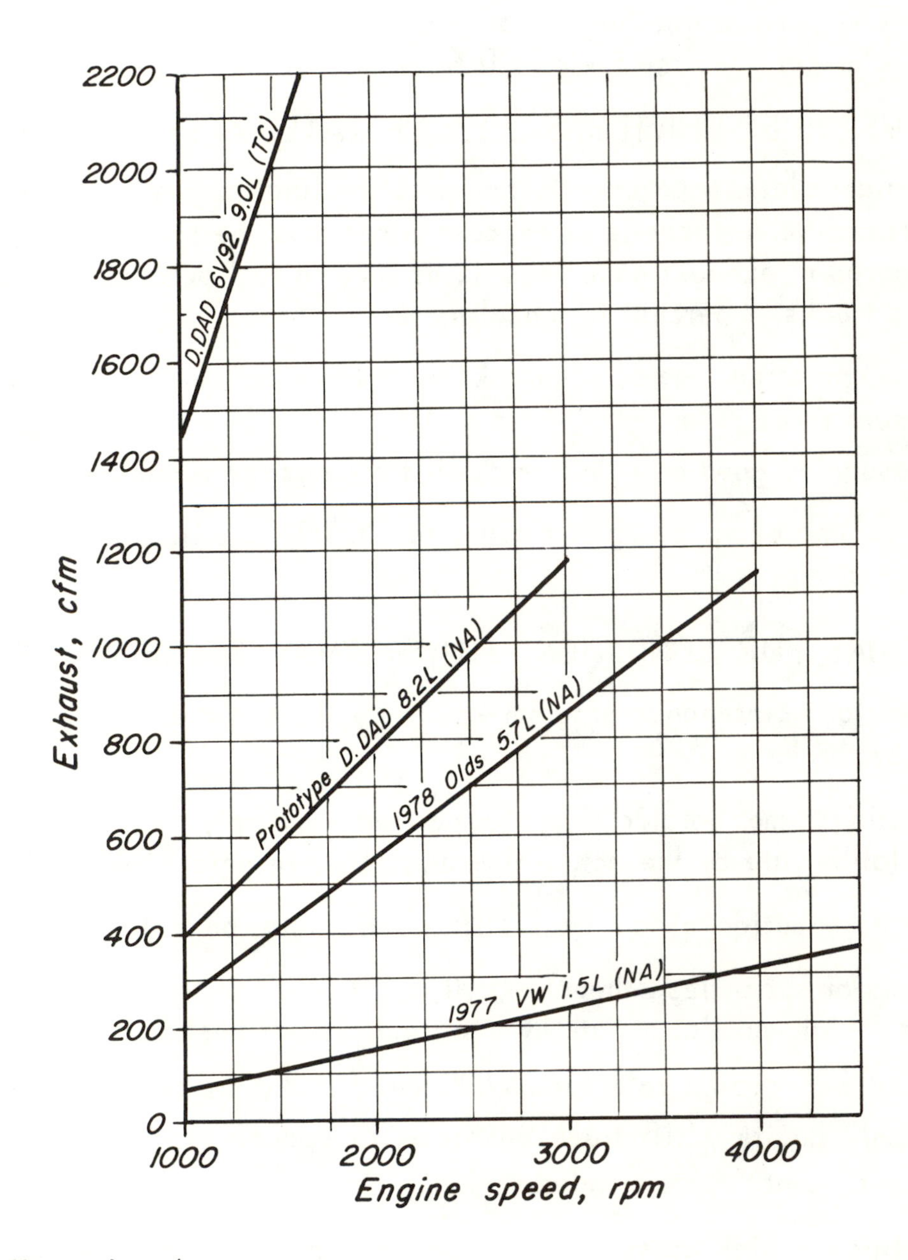

Note:
NA - normally aspirated
TC - turbo charged

Exhaust, cfm = acfm + 20% excess

For specific design information request manufactures 13 mode EPA engine bench test.

AMERICAN CONFERENCE OF GOVERNMENTAL INDUSTRIAL HYGIENISTS

EXHAUST SYSTEM REQUIREMENTS FOR TYPICAL DIESEL ENGINES UNDER LOAD

DATE 1-82	VS-908.2

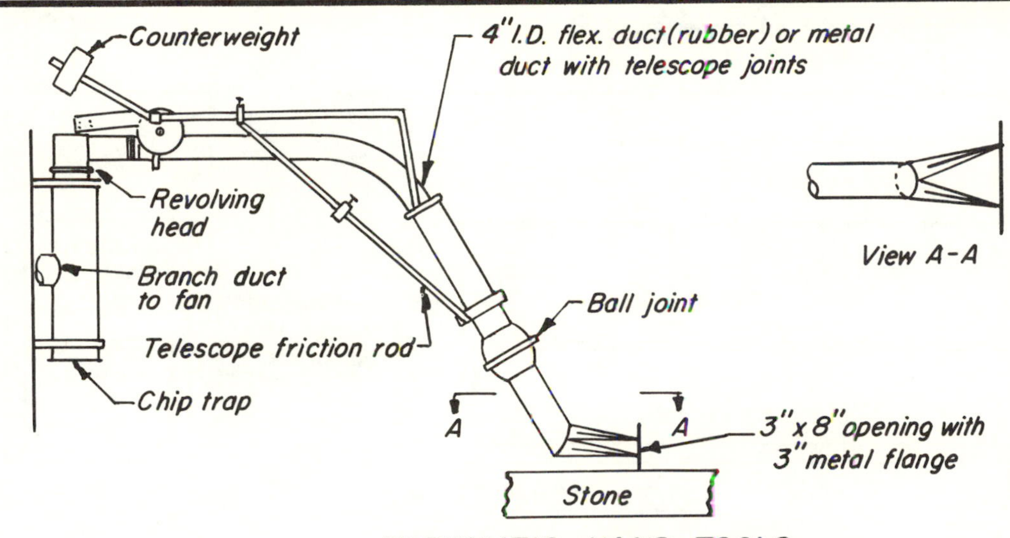

PNEUMATIC HAND TOOLS

Q = 400 cfm minimum, tool 10" max distance from hood
Minimum duct velocity = 3500-4000 fpm

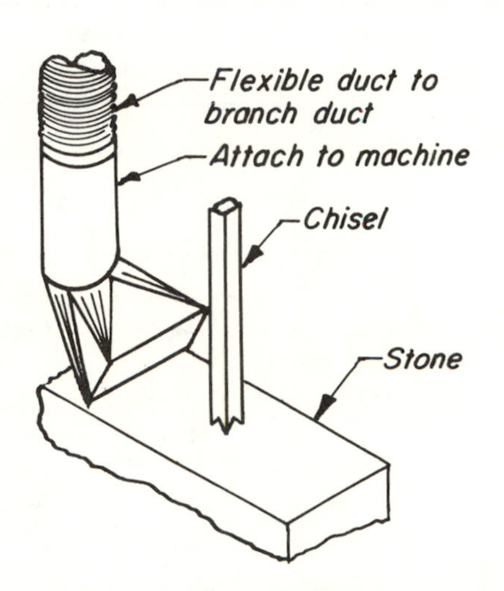

Abrasive blasting to be done in a room or cabinet; 500 fpm at all openings. See "Abrasive Blasting," VS-101

SURFACING MACHINE HOODS

Hood	cfm	Branch diam
Baby surfacer	400	4"
Medium surfacer	600	5"

Entry loss = 1.0 VP

AMERICAN CONFERENCE OF GOVERNMENTAL INDUSTRIAL HYGIENISTS

GRANITE CUTTING AND FINISHING

DATE 1-74	VS-909

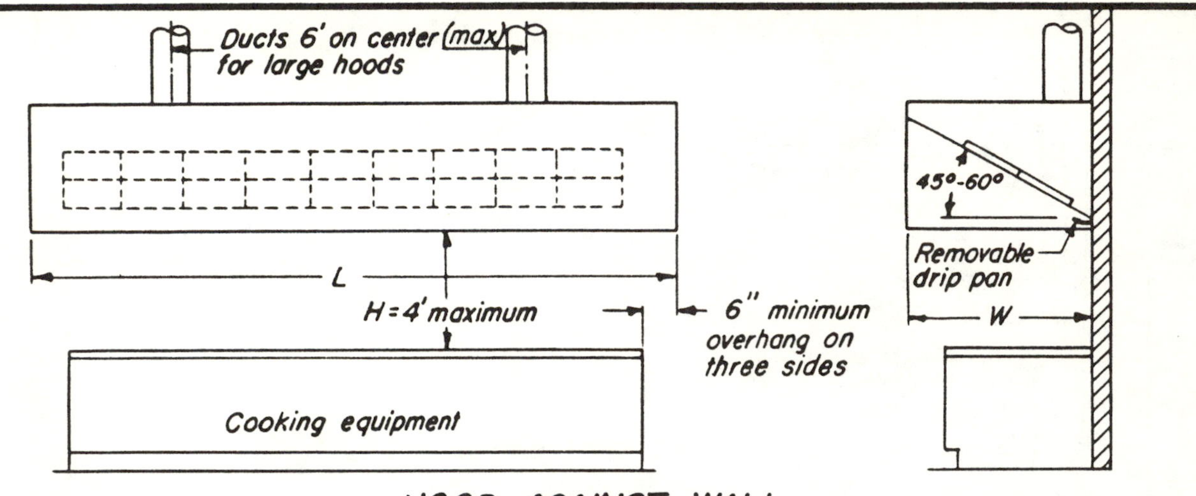

HOOD AGAINST WALL

Q= 80 cfm/sq ft of hood area (80 WL)
Not less than 50 cfm/sq ft of face area (50PH) P=perimeter of hood = 2W+L
Duct velocity = 1000-4000 fpm, to suit conditions
Entry loss =(filter resistance +0.1")+0.50VP(straight take off)
Entry loss =(filter resistance +0.1")+0.25VP(tapered take off)

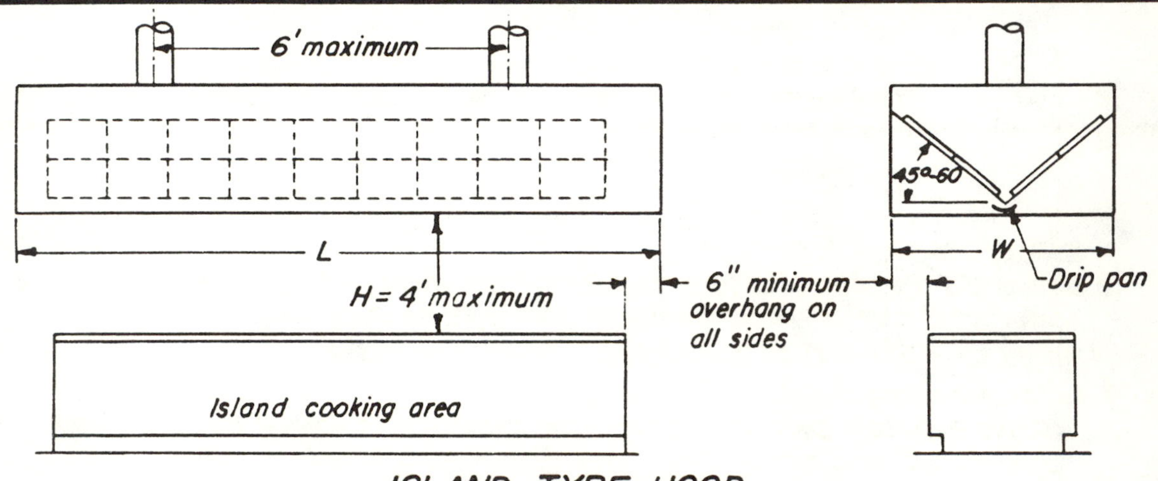

ISLAND TYPE HOOD

Q=125 cfm/sq ft of hood area (125 WL)
Not less than 50 cfm/sq ft of face area (50 PH) P=perimeter of hood = 2W+2L
Duct velocity= 1000-4000 fpm, to suit conditions
Entry loss =(filter resistance+0.1")+0.50VP (straight take off)
Entry loss=(filter resistance+ 0.1")+ 0.25 VP(tapered take off)

Note:
See VS-911 for information about filters and fans

AMERICAN CONFERENCE OF GOVERNMENTAL INDUSTRIAL HYGIENISTS

KITCHEN RANGE HOODS

DATE 1-84	VS-910

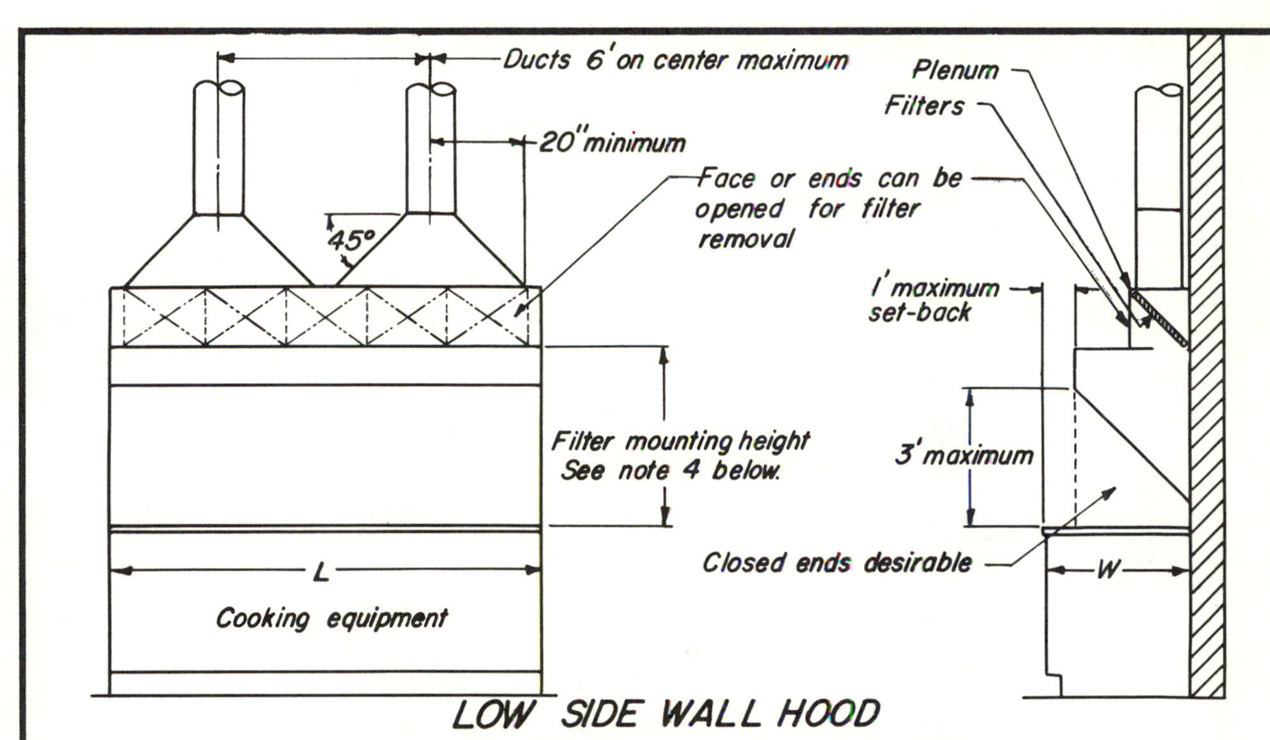

LOW SIDE WALL HOOD

Q=200 cfm/lineal ft of cooking surface (200 L)
Duct velocity = 1000 - 4000 fpm, to suit conditions
Entry loss = (filter resistance + 0.1") + 0.5 VP (straight take off)
Entry loss = (filter resistance + 0.1") + 0.25 VP (tapered take off)

NOTES FOR KITCHEN HOODS

Filters:

1. Select practical filter size.
2. Determine number of filters required from manufacturer's data. (Usually: 2 cfm maximum exhaust for each sq in of filter area)
3. Install at 45°- 60° to horizontal. <u>Never</u> horizontal.
4. Filter mounting height (Reference 66)
   a. No exposed cooking flame — $1\frac{1}{2}$' minimum to lowest edge of filter.
   b. Charcoal and similar fires — 4' minimum to lowest edge of filter.
5. Shield filters from direct radiant heat.
6. Provide removable grease drip pan.
7. Clean pan and filters regularly.

Fan:

1. Use upblast discharge fan. Downblast is not recommended.
2. Select fan for design Q and SP resistance of filters and ductwork.
3. Adjust fan specification for expected exhaust air temperature.

AMERICAN CONFERENCE OF GOVERNMENTAL INDUSTRIAL HYGIENISTS	
KITCHEN RANGE HOOD	
DATE 1-76	VS-911

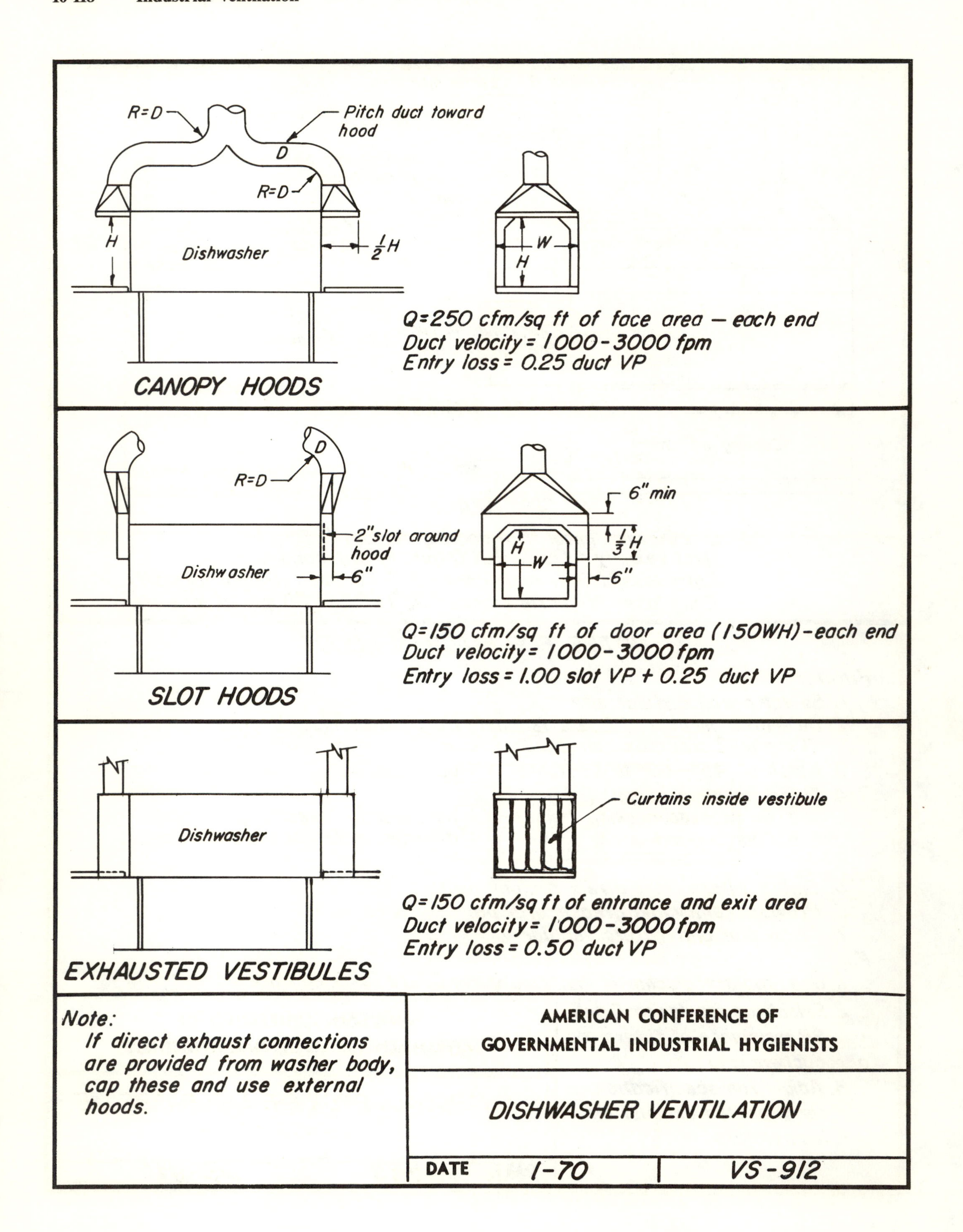
R=D
Pitch duct toward hood
D
R=D
H
Dishwasher
1/2 H
W
H
Q=250 cfm/sq ft of face area — each end
Duct velocity= 1000-3000 fpm
Entry loss= 0.25 duct VP
CANOPY HOODS
D
R=D
2" slot around hood
Dishwasher
6"
6" min
1/3 H
H
W
6"
Q=150 cfm/sq ft of door area (150WH)-each end
Duct velocity= 1000-3000 fpm
Entry loss= 1.00 slot VP + 0.25 duct VP
SLOT HOODS
Dishwasher
Curtains inside vestibule
Q=150 cfm/sq ft of entrance and exit area
Duct velocity= 1000-3000 fpm
Entry loss= 0.50 duct VP
EXHAUSTED VESTIBULES
Note:
If direct exhaust connections are provided from washer body, cap these and use external hoods.
AMERICAN CONFERENCE OF
GOVERNMENTAL INDUSTRIAL HYGIENISTS
DISHWASHER VENTILATION
DATE 1-70
VS-912

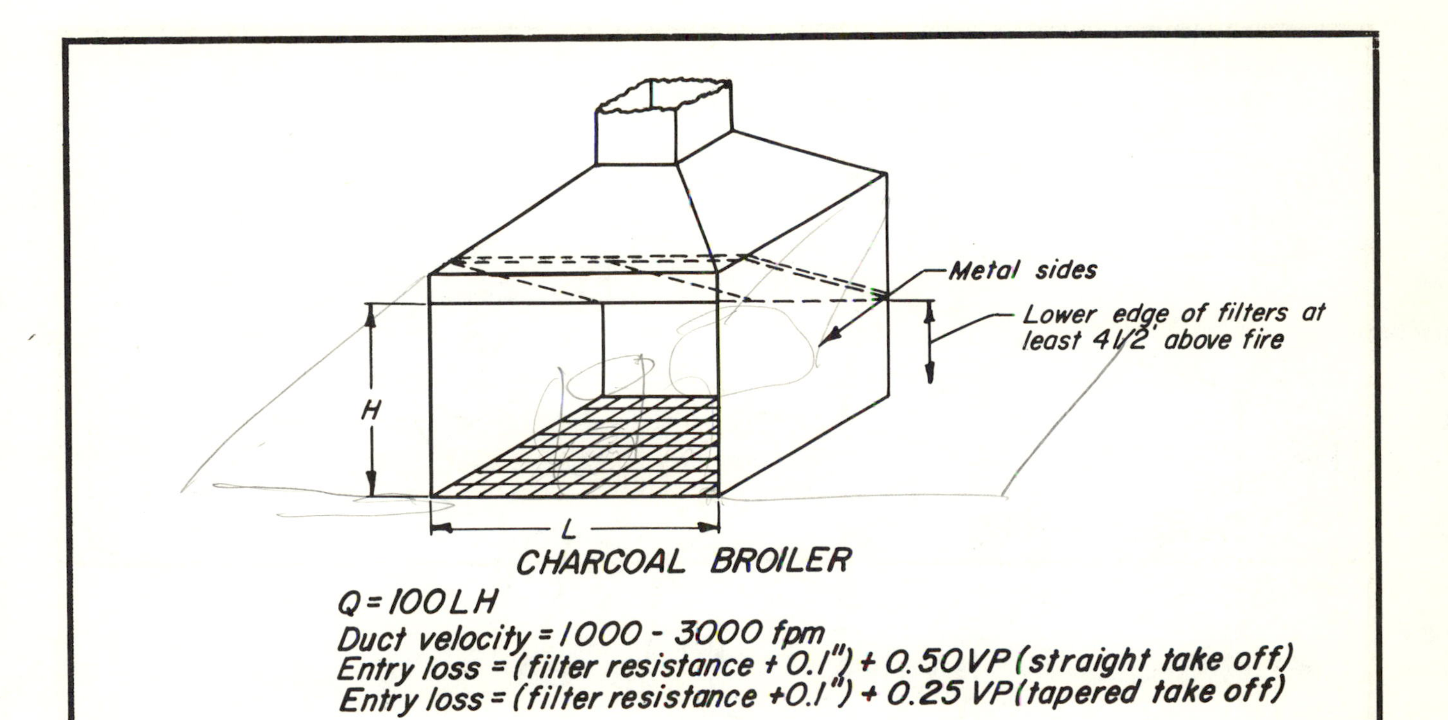
Metal sides
Lower edge of filters at least 41/2' above fire
H
L
CHARCOAL BROILER
Q=100LH
Duct velocity=1000 - 3000 fpm
Entry loss =(filter resistance + 0.1") + 0.50VP(straight take off)
Entry loss=(filter resistance +0.1") + 0.25 VP(tapered take off)

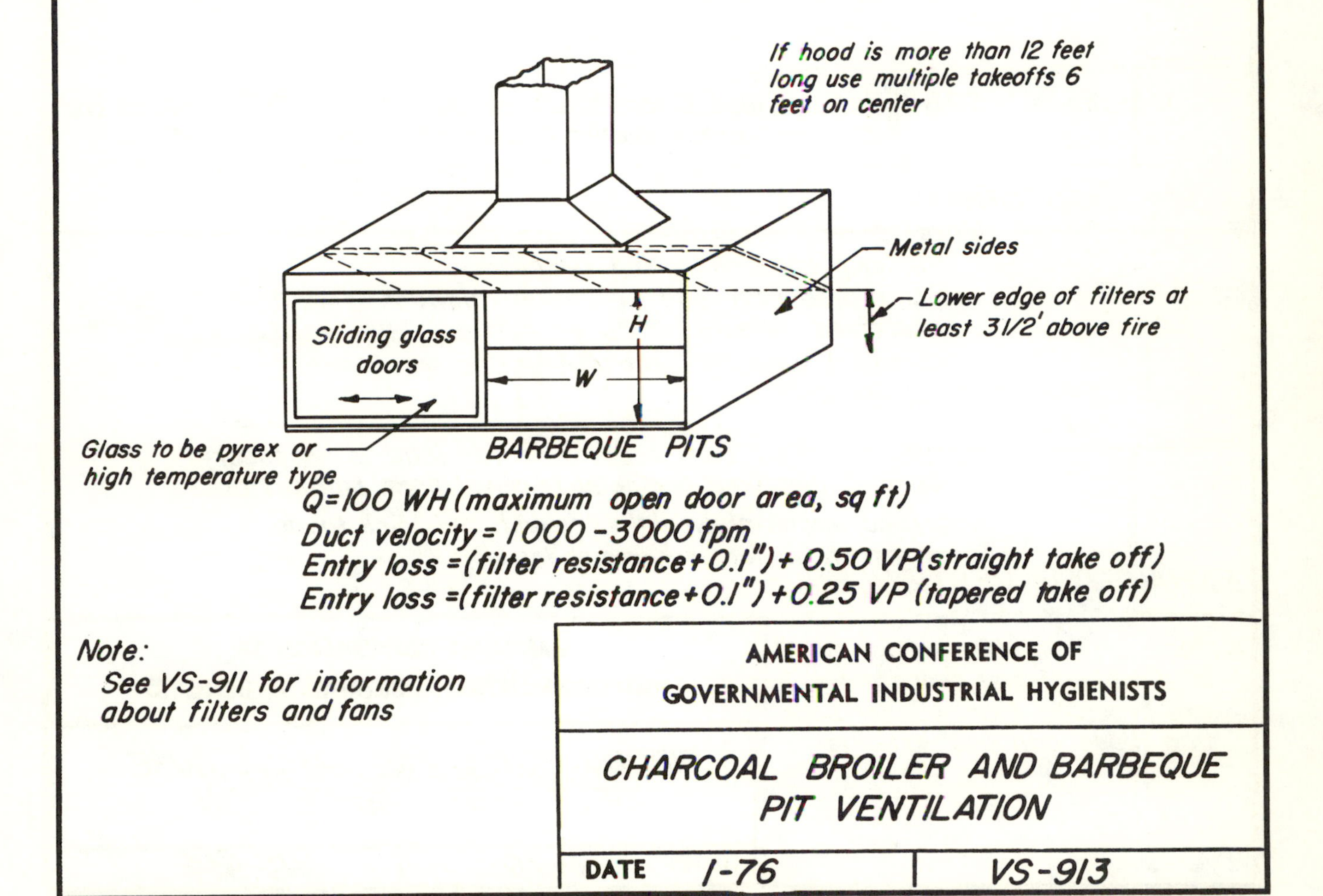
If hood is more than 12 feet long use multiple takeoffs 6 feet on center
Metal sides
Lower edge of filters at least 31/2' above fire
Sliding glass doors
H
W
Glass to be pyrex or high temperature type
BARBEQUE PITS
Q=100 WH(maximum open door area, sq ft)
Duct velocity= 1000 -3000 fpm
Entry loss =(filter resistance+0.1")+ 0.50 VP(straight take off)
Entry loss =(filter resistance+0.1") +0.25 VP (tapered take off)
Note:
See VS-911 for information about filters and fans
AMERICAN CONFERENCE OF GOVERNMENTAL INDUSTRIAL HYGIENISTS
CHARCOAL BROILER AND BARBEQUE PIT VENTILATION
DATE 1-76
VS-913

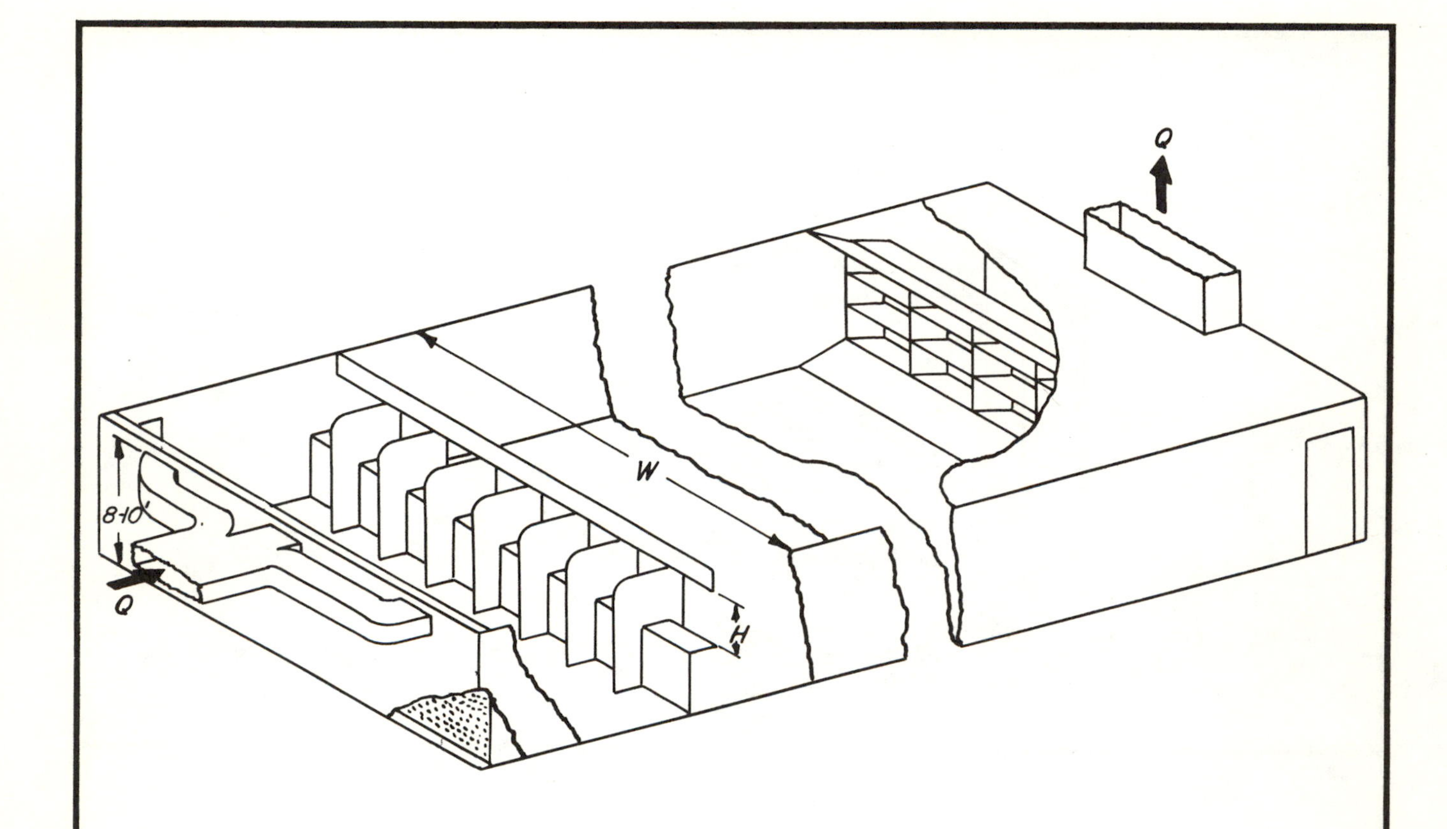

Q minimun = 50 HW, but not less than 20 cfm/sq ft of room cross sectional area

Notes:

Make-up air distribution:

Uniform air distribution necessary.
Perforated rear wall or ceiling plenum system preferred. Where grills or diffuser are used, terminal velocity near firing line must not exceed 50 fpm

NIOSH certified dust respirator for lead is necessary during clean-up and lead removal from bullet trap.

Acoustical material on walls, ceiling and thick fabric on bench top are recommended

AMERICAN CONFERENCE OF GOVERNMENTAL INDUSTRIAL HYGIENISTS	
INDOOR PISTOL AND SMALL BORE RIFLE RANGE VENTILATION	
DATE *1-78*	*VS-914*

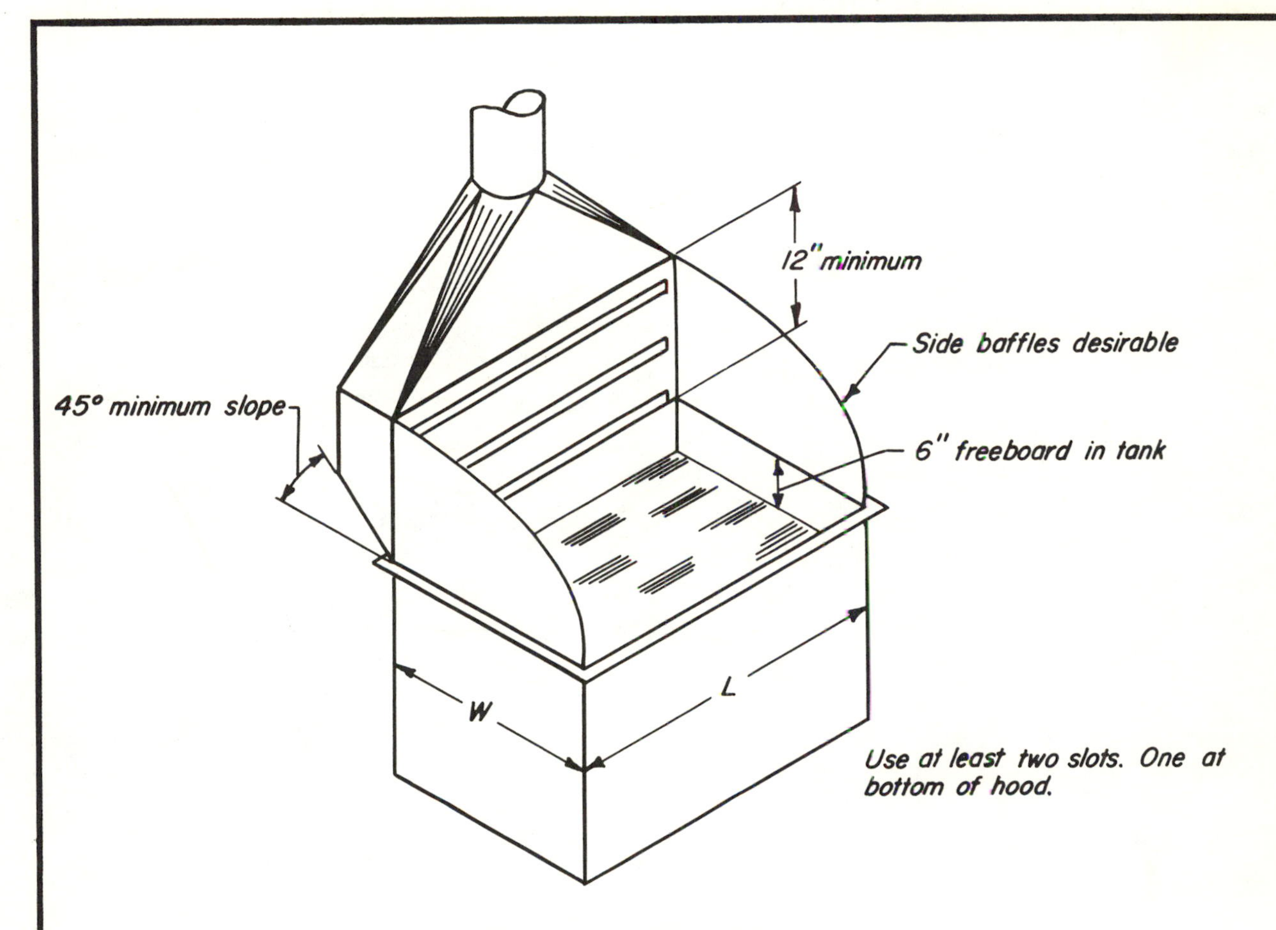

Q = 150 cfm/sq ft of bed (150LW)
Slot velocity = 2000 fpm
Entry loss = 1.78 slot VP + 0.25 duct VP
Duct velocity = 2500 - 3000 fpm
W not to exceed 36"

For circular beds or other hood designs, see VS-303, VS-504
Free board must be maintained to prevent material carryout.

AMERICAN CONFERENCE OF GOVERNMENTAL INDUSTRIAL HYGIENISTS	
FLUIDIZED BEDS	
DATE 1-70	VS-915

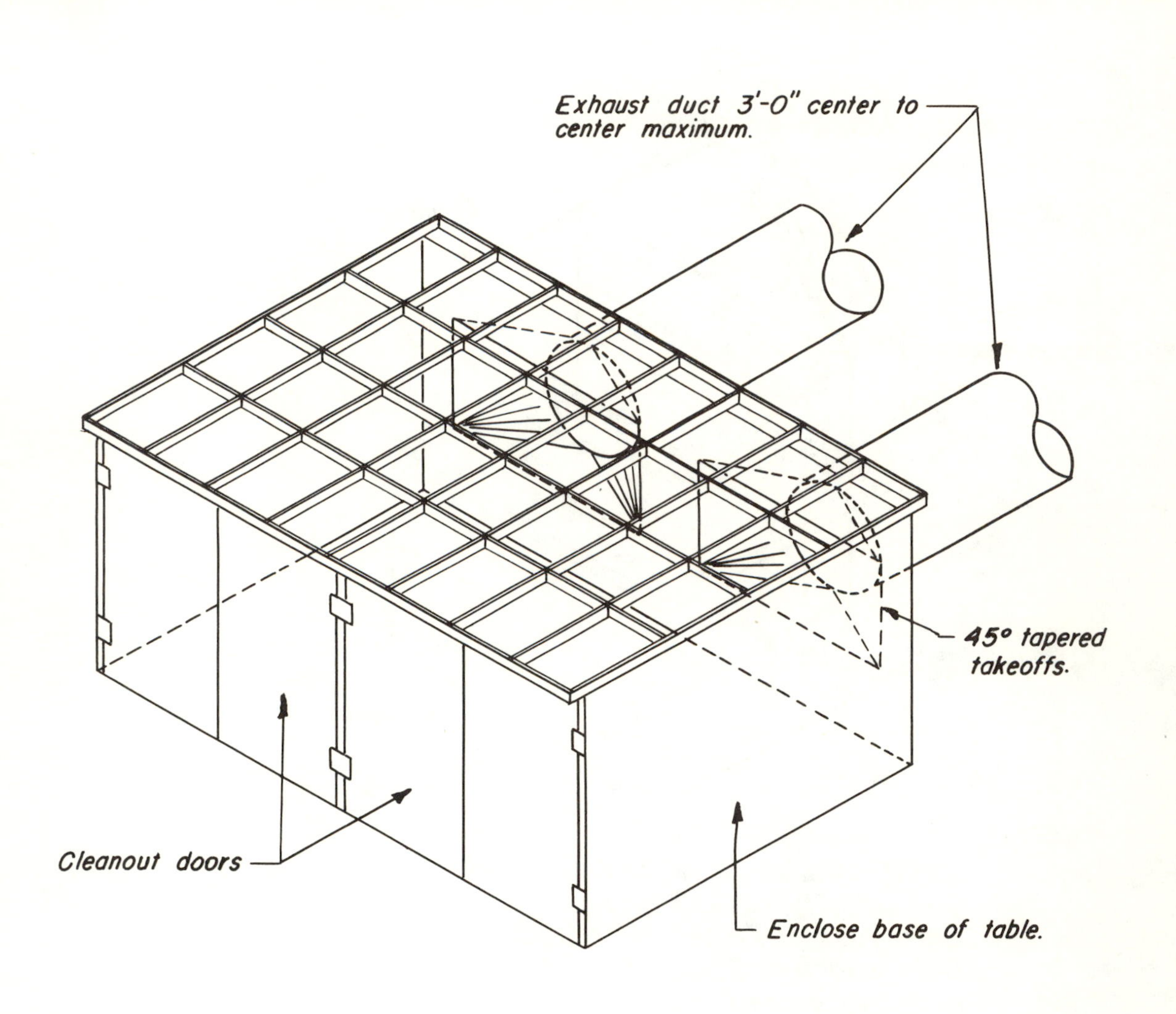

Q = 150 cfm/sq ft of gross table area
Duct velocity = 2000-4000 fpm *
Entry loss = 1.0 VP through grating
0.25 duct VP – tapered takeoff
* For horizontal runs, transport velocity is necessary

AMERICAN CONFERENCE OF GOVERNMENTAL INDUSTRIAL HYGIENISTS	
TORCH CUTTING VENTILATION	
DATE 1-68	VS-916

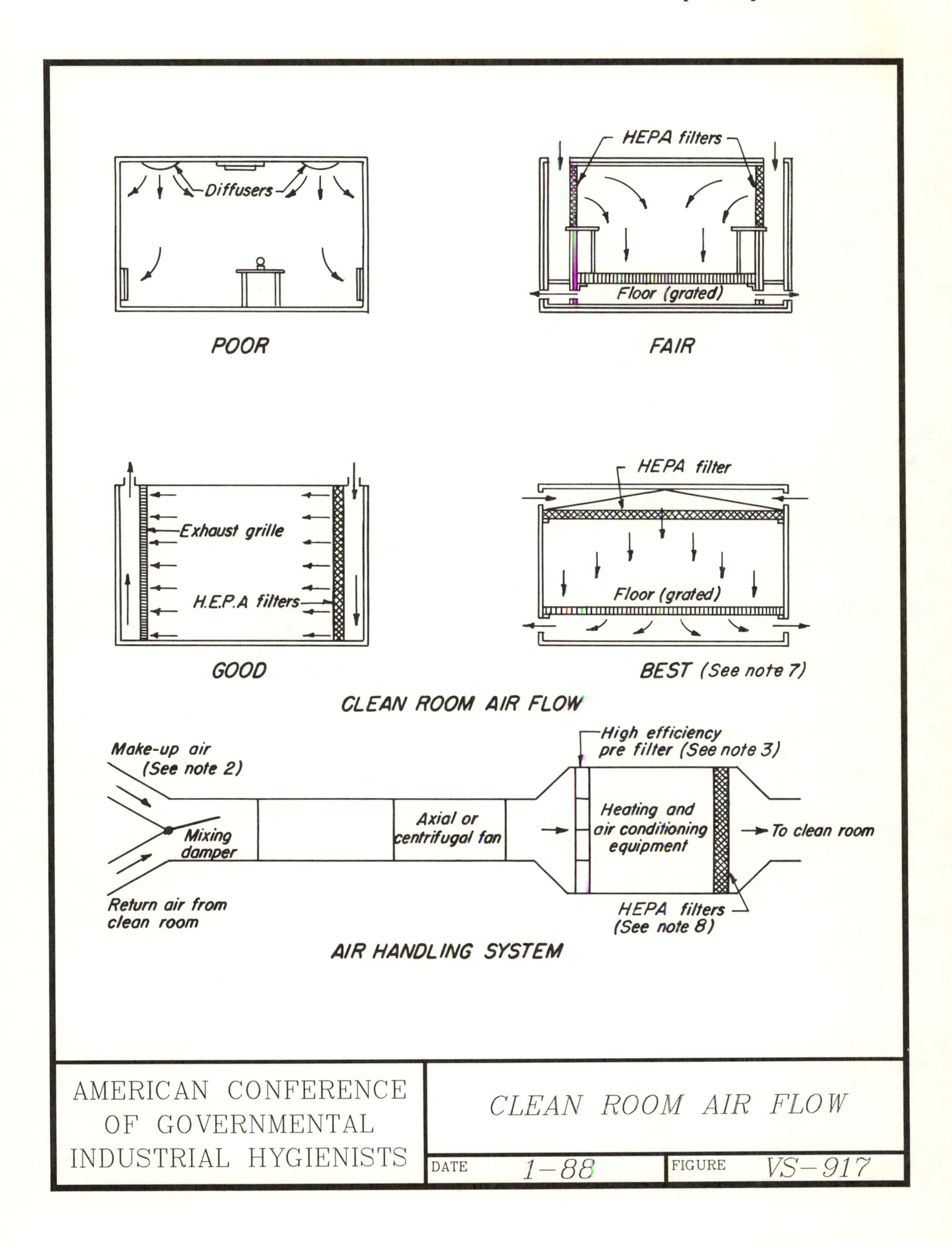
HEPA filters
Diffusers
Floor (grated)
POOR
FAIR
Exhaust grille
H.E.P.A filters
HEPA filter
Floor (grated)
GOOD
BEST (See note 7)
CLEAN ROOM AIR FLOW
Make-up air
(See note 2)
High efficiency
pre filter (See note 3)
Axial or
centrifugal fan
Mixing
damper
Heating and
air conditioning
equipment
To clean room
Return air from
clean room
HEPA filters
(See note 8)
AIR HANDLING SYSTEM
AMERICAN CONFERENCE
OF GOVERNMENTAL
INDUSTRIAL HYGIENISTS
CLEAN ROOM AIR FLOW
DATE 1-88
FIGURE VS-917

NOTES

1. A clean room is an enclosed area employing control over the particulate matter in air; with temperature, humidity and pressure control as required. Clean rooms are divided into classes as follows:[83]

   A. *Class 100*. Particle count not to exceed 100 particles per cubic foot of a size 0.5 micron and larger.

   B. *Class 10,000*. Particle count not to exceed 10,000 particles per cubic foot of a size 0.5 micron and larger, or 65 particles per cubic foot of a size 5.0 microns and larger.

   C. *Class 100,000*. Particle count not to exceed 100,000 particles per cubic foot of a size 0.5 micron and larger, or 700 particles per cubic foot of a size 5.0 microns and larger.

2. High Efficiency Particulate Air (HEPA) filters are 99.97% efficient by DOP test on 0.3 micron particles. Initial HEPA filter resistance is nominally 1.0 "wg. Final resistance is generally not more than 2.5 "wg to avoid unacceptable reductions in air flow.

3. Pre-filters greatly increase the life of HEPA filters by reducing the dust load. Filters with an ASHRAE efficiency of 90% are commonly used for this purpose.

4. HEPA filters should be installed in the clean room ceiling or walls. A room air velocity of 90 fpm is nominal for Class 100 clean rooms; however, 50–150 fpm may be used depending on the room usage.

5. Clean room temperatures are usually in the 68–72 F range, maintained ±3 F. Relative humidity is usually in the 33–40% range and maintained ± 5%. Specific applications may require a substantial deviation from these norms. For critical requirements, temperature can be maintained at ± 0.5 F and relative humidity at ± 1.0%. The control of electrostatic charges and vibration is essential for good clean room performance.

6. Non-laminar flow rooms (higher than Class 100) require 20–100 air changes per hour, depending on the requirements and operation of the individual room. A replacement air volume of 0.1 to 1.0 cfm/ft^2 is generally used. Additional air may be required to replace local exhaust for the control of toxic contaminants.

7. For sanitary reasons, specific industries, such as pharmaceutical, may not allow raised floors. In those cases, room width should not exceed 14 feet and the air should be taken from the room by wall grilles located close to the floor. Velocity through the grilles should not exceed 500 fpm.

8. Remote HEPA filter locations are not recommended and should be used only where ceiling or wall locations are not possible and for less stringent requirements.

AMERICAN CONFERENCE OF GOVERNMENTAL INDUSTRIAL HYGIENISTS

CLEAN ROOM AIR FLOW DATA

DATE 1–88 FIGURE VS–917.1

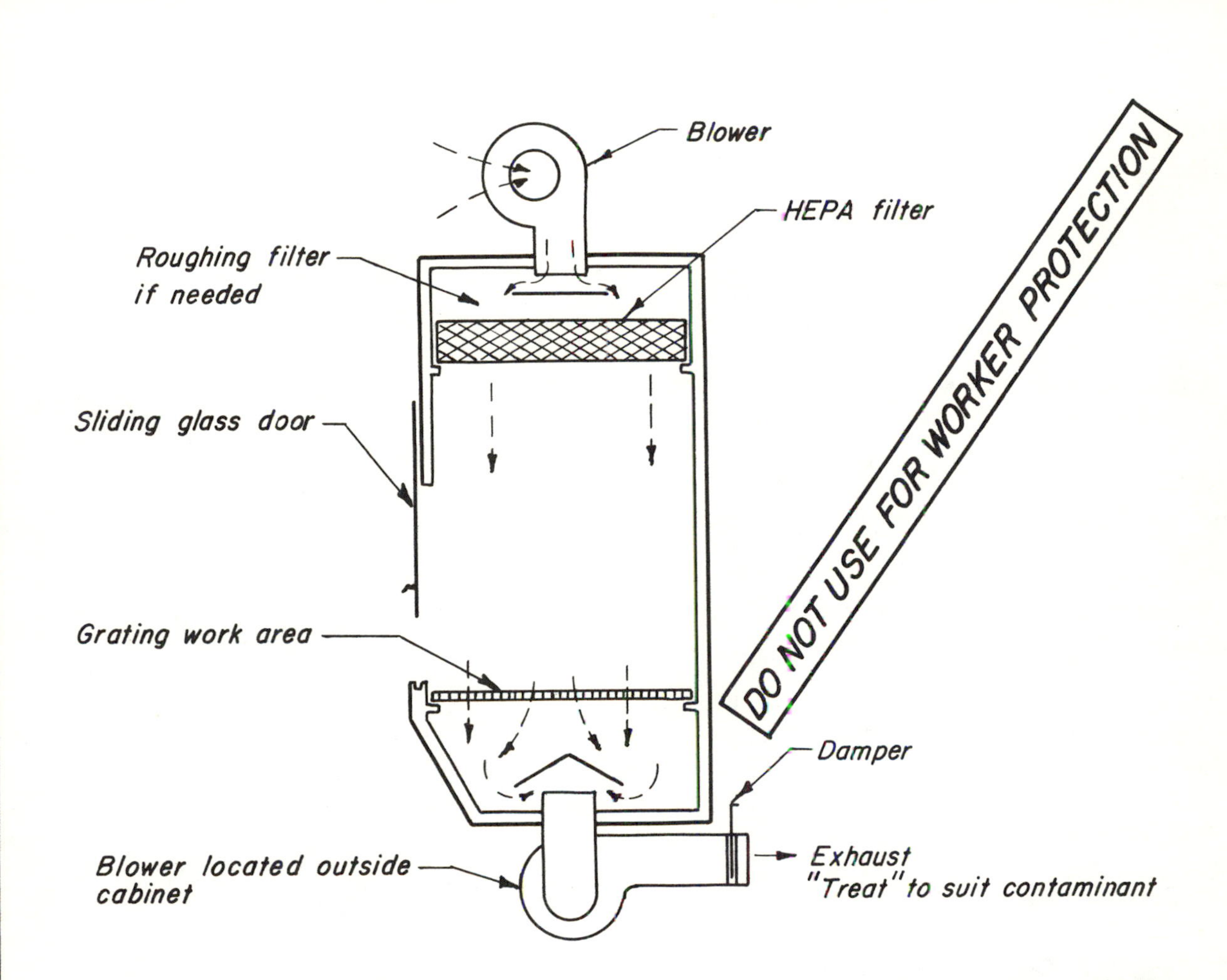

Velocity = 90 fpm with average minimum uniformity ±20 fpm
Duct velocity = 2000–4000 fpm to suit conditions

Clean station for control of air particulates[82,83,84]

Notes: Workers arms or other objects protruding into the hood opening may cause contaminant spillage.
Personal protective equipment or general ventilation to be provided as needed.
Supply and exhaust should be maintained equal by flow meter control techniques.

AMERICAN CONFERENCE OF GOVERNMENTAL INDUSTRIAL HYGIENISTS	CLEAN AIR EXHAUST HOOD (PRODUCT PROTECTION ONLY)	
	DATE 1–88	FIGURE VS–918.1

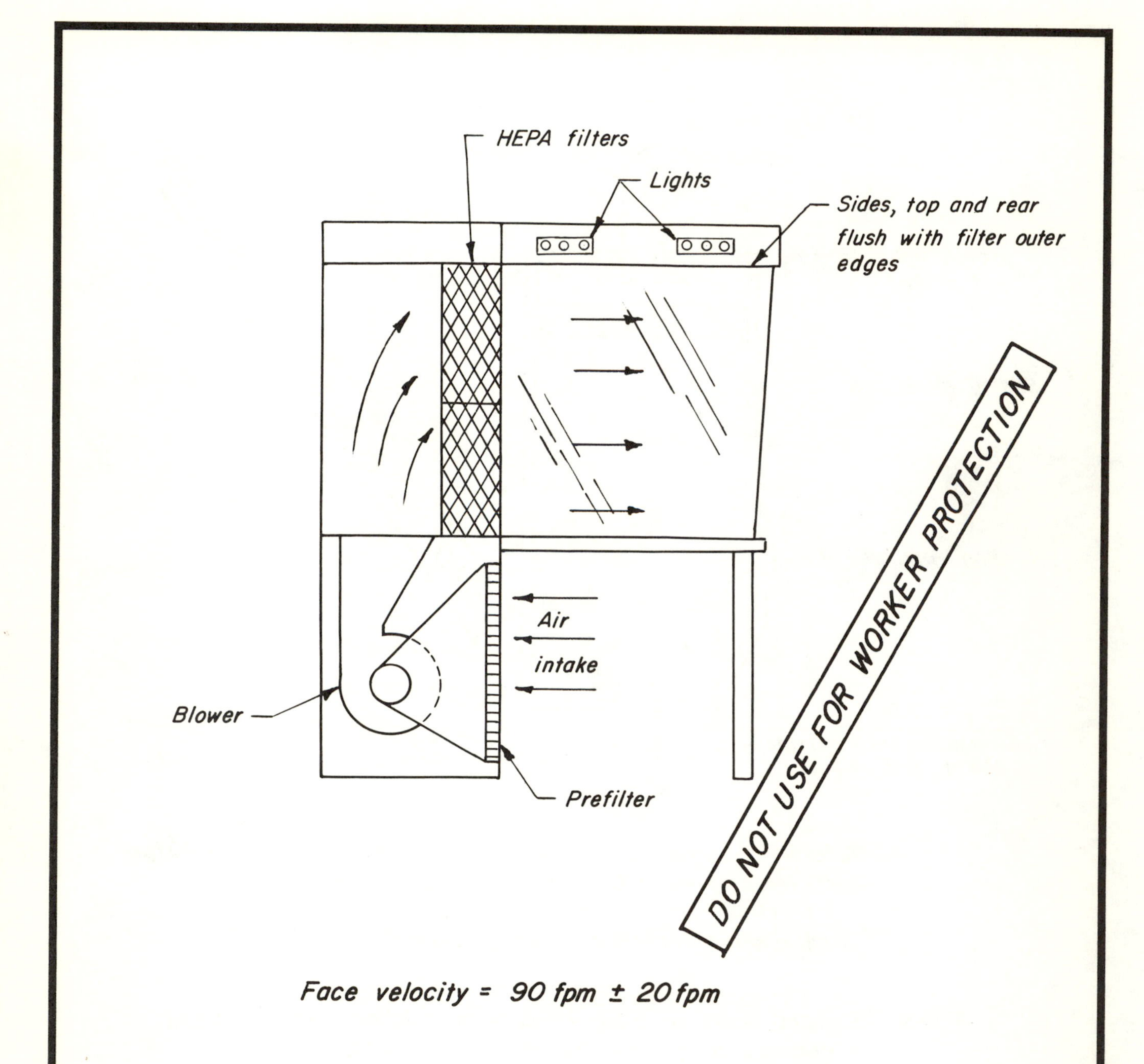

Note: Total power input must be considered as part of air conditioning load

Reference 82, 83, 84

AMERICAN CONFERENCE OF GOVERNMENTAL INDUSTRIAL HYGIENISTS	CLEAN BENCH WORK STATION (PRODUCT PROTECTION ONLY)
	DATE 1-88 FIGURE VS-918.2

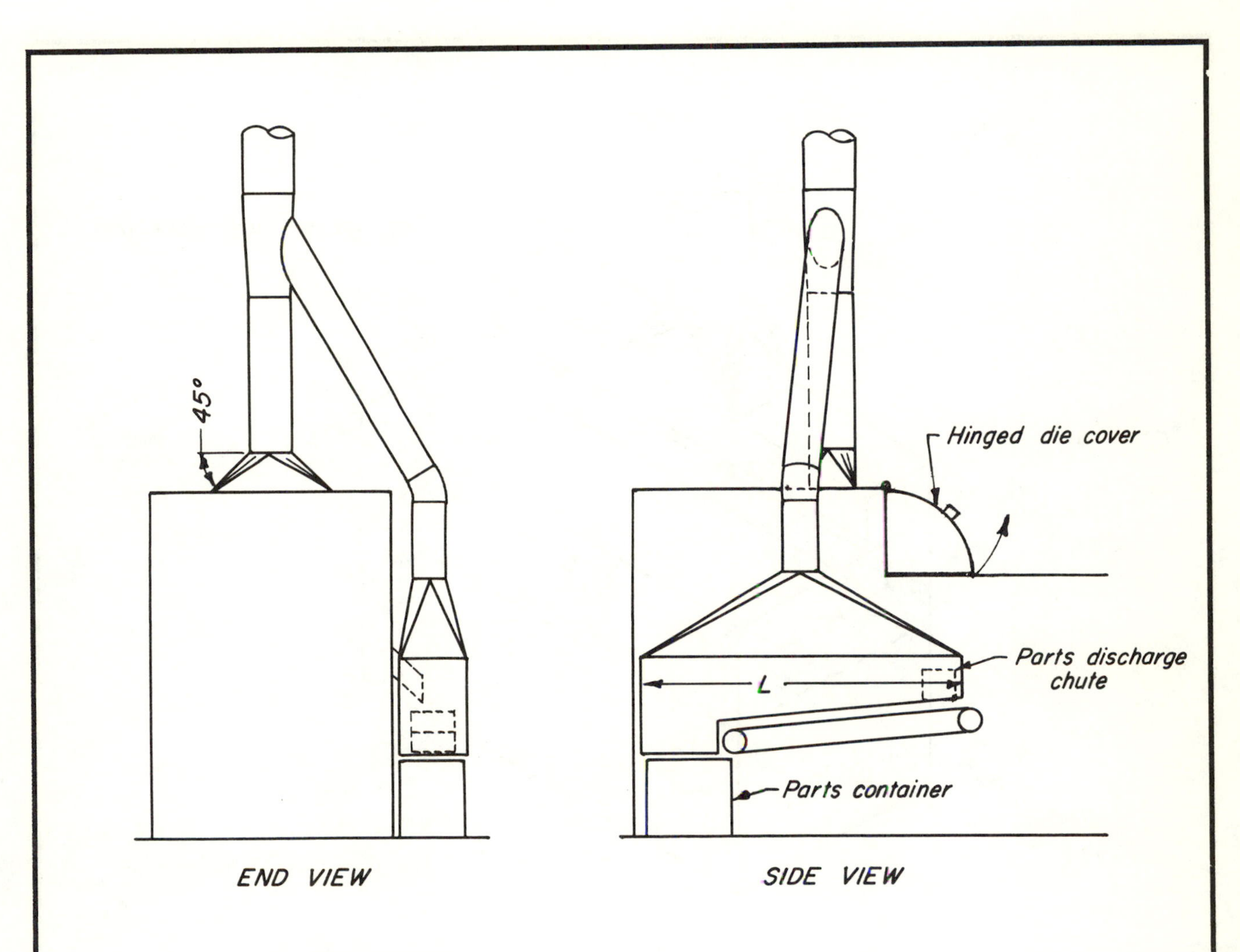

Cold header:
Q=750 cfm/sq ft of die opening
Duct velocity = 3500 fpm
Entry loss = 1.0 slot VP + 0.25 duct VP

Parts discharge and container:
Q = 100 cfm/ft of hood length
Duct velocity = 3500 fpm
Entry loss = 0.25 VP

AMERICAN CONFERENCE OF GOVERNMENTAL INDUSTRIAL HYGIENISTS	
COLD HEADING MACHINE VENTILATION	
DATE 1-72	VS-919

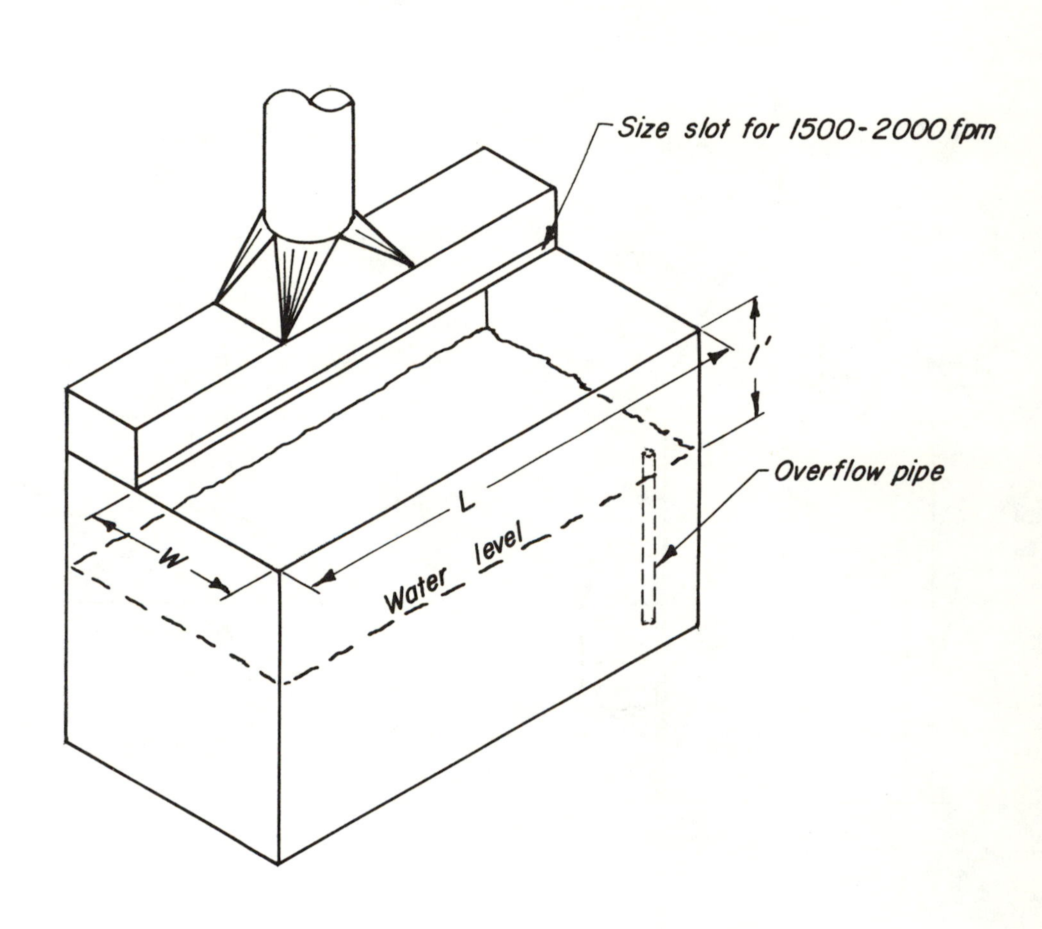

Q = 200L×W
Entry loss = 1.78 slot VP + 0.25 duct VP
Duct velocity = 2500–3000 fpm

AMERICAN CONFERENCE OF
GOVERNMENTAL INDUSTRIAL HYGIENISTS

OUTBOARD MOTOR TEST

DATE 1-78	VS-920

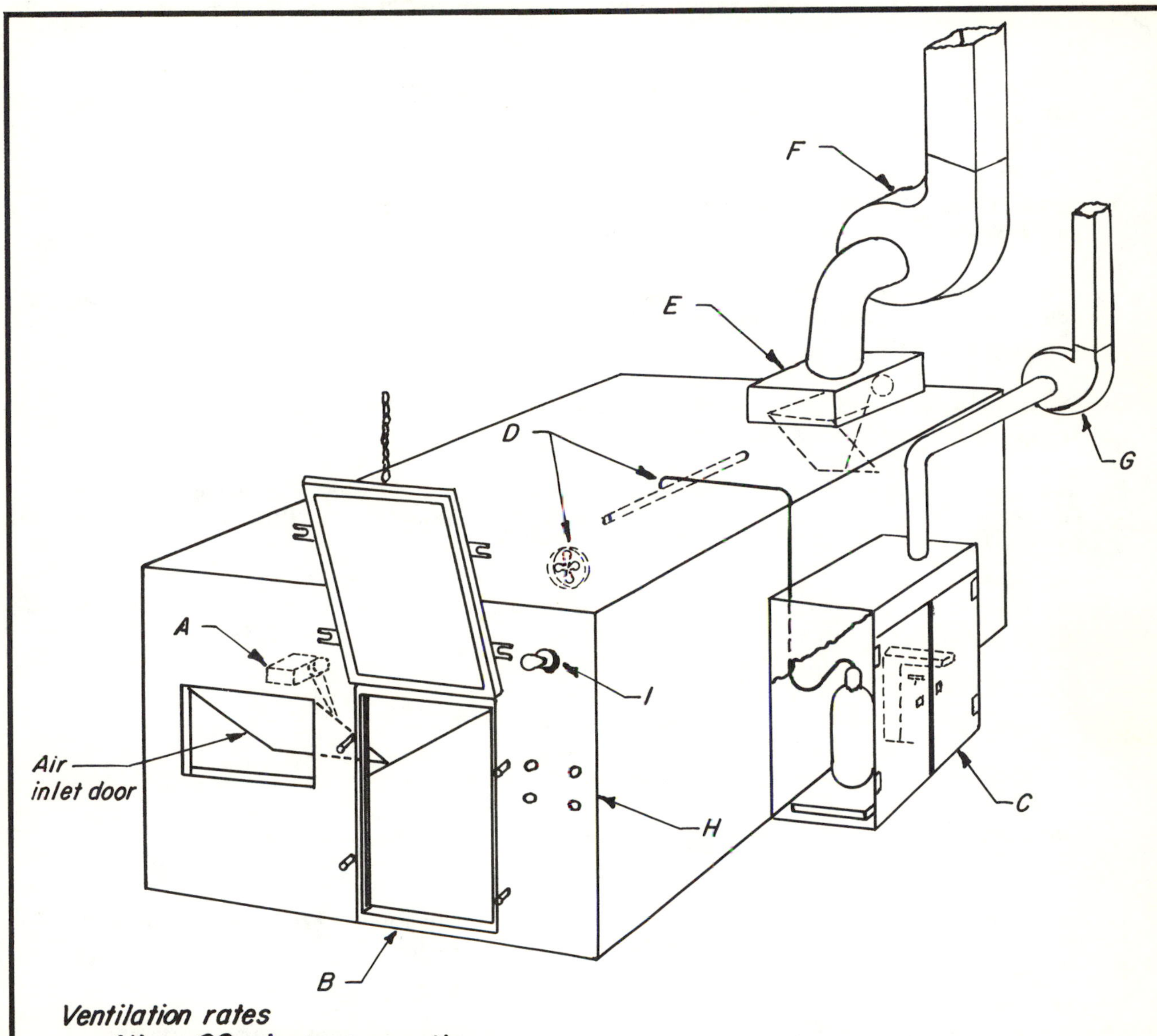

Ventilation rates

Allow 60 minute purge time

Ventilation rate must be 20 air changes per hour or greater

Air flow must provide:

(a) 500 fpm velocity or greator thru air inlet door if large access door is closed

and

(b) at least 100 fpm through loading door if booth are open

AMERICAN CONFERENCE OF
GOVERNMENTAL INDUSTRIAL HYGIENISTS

FUMIGATION BOOTH

DATE 1-78	VS-921

A. Provide an air inlet with automatic damper closure: damper must be interlocked with fan circuit to open only when fan is turned on. See Fig. 2 Size opening for a minimum velocity of 500 fpm. Air inlet must be located so ventilation air sweeps entire booth.

B. Loading door must be opened only when booth has been completely ventilated. Provide gaskets snd screw clamps and brackets for applying uniform pressure for a gas tight fit.

C. Provide ventilated cabinet for gas cylinders in use and being stored. Fan must be on continuously. and exhaust approximately 500 cfm to produce negative pressure in booth with doors closed.

D. Provide nozzle openings for introducing fumigant gas. A circulating fan should also be provided for obtaining good mixture of fumigant gas.

E. Mechanical fan damper must be provided which closes tightly when fan is shut off.during fumigation and opens when fan is turned on. Damper controls should be interlocked with fan controls.

F. Fan for ventilating fumigation booth. Fan must be sized to dilute air to safe limit in required time. Use vertical outside discharge stack away from windows, doors, and air intakes.

G. Fumigant gas cylinder cabinet fan must run continuously.

H. Control switches for fan and lights, air flow switch actuated pilot lights are recommended.

I. Red warning light to indicate booth is under fumigation as a protection against careless entry.

To facilitate penetration of fumigant gas and subsequent airing out, mattresses should be loaded with separators to allow free air space around each mattress.

Fumigants with no odor warning properties should be used together with odor indicating chemical.

Where toxic fumigants are used, a leak test should be made on the booth. The booth should be first tested by lighting several large smoke candles in it, with doors and dampers closed. Leaks can be noted by the presence of smoke at the point of escape. Where highly diffusable toxic gases are used, an additional test should be made with the booth under charge, at doors and dampers, with a sensitive detecting meter or sampling device.

AMERICAN CONFERENCE OF GOVERNMENTAL INDUSTRIAL HYGIENISTS

FUMIGATION BOOTH DATA

DATE 1-78 | VS-921.1

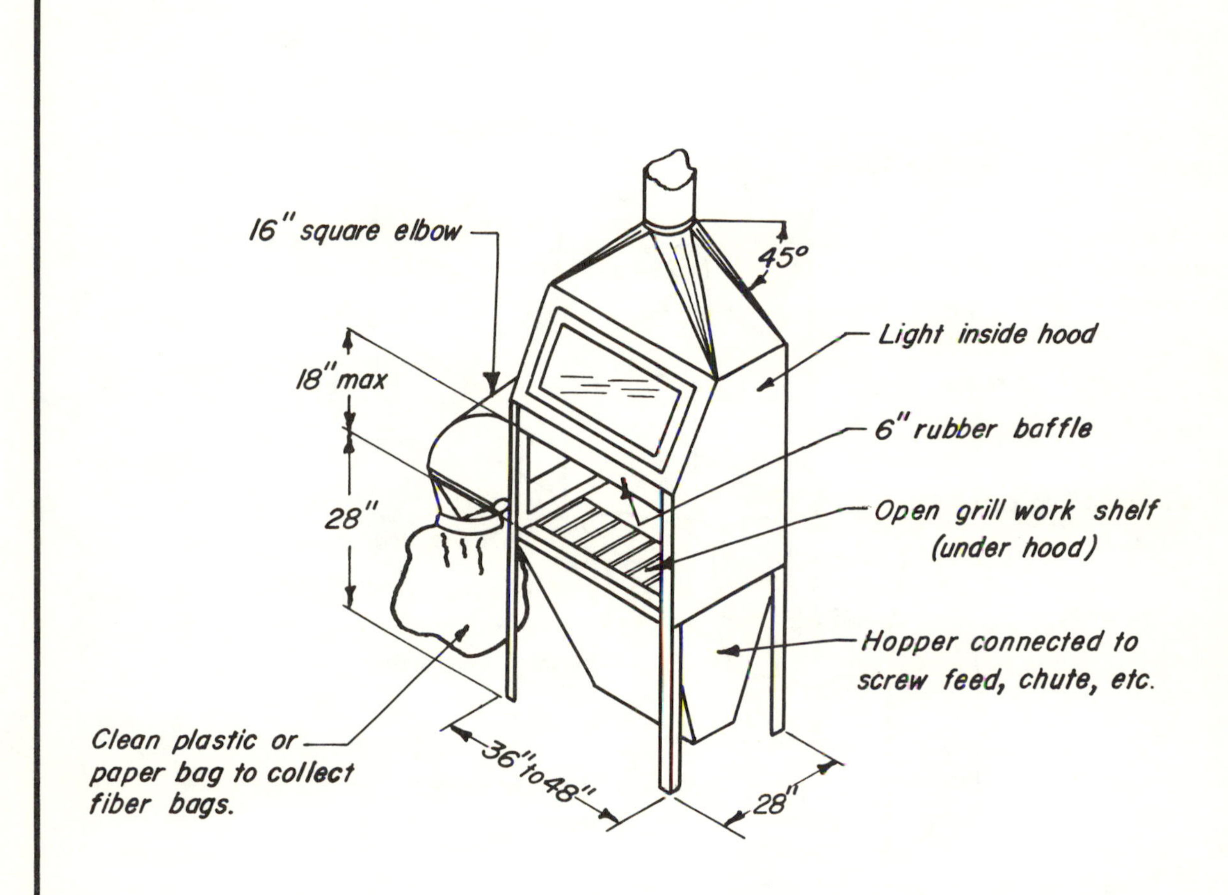

Q = minimum 250 cfm/sq ft of open area
Duct velocity = 3500 fpm minimum
Entry loss = 0.25 duct VP

Reference 125

AMERICAN CONFERENCE OF GOVERNMENTAL INDUSTRIAL HYGIENISTS

ASBESTOS FIBER BAG-OPENING

DATE 1-78 | VS-1001

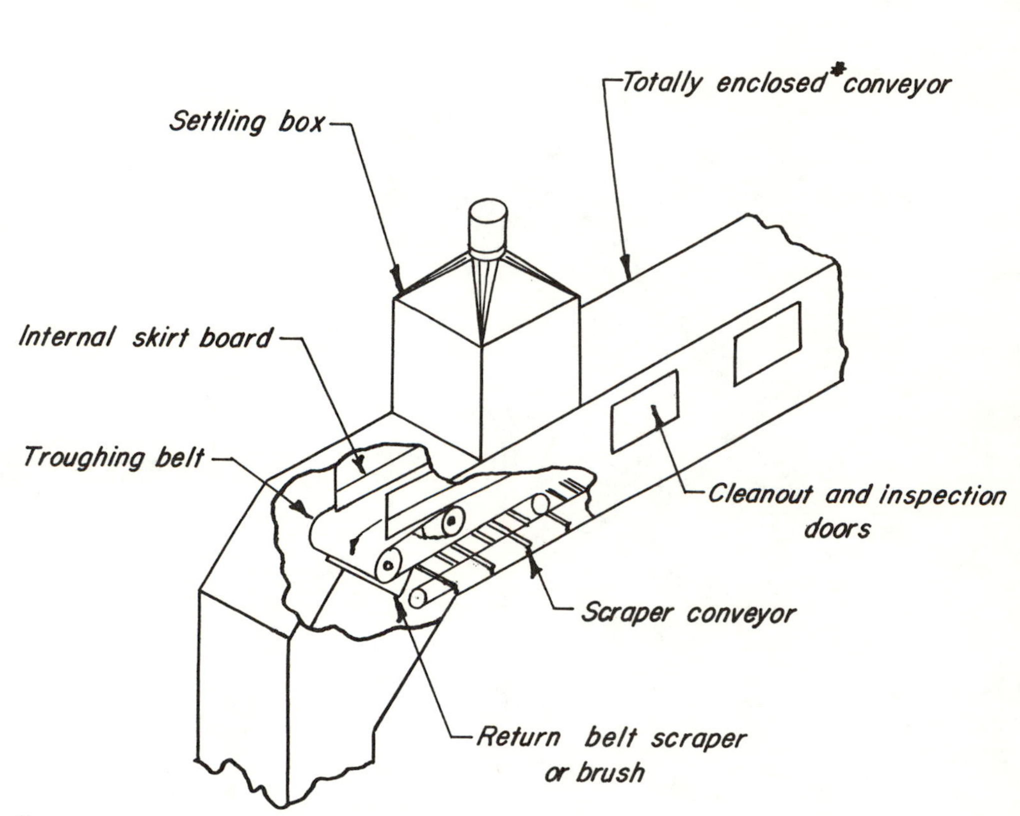

*Leakage factor depends on type of construction

Q = 50 cfm/sq ft of open area
Duct velocity = 3500 fpm minimum
Entry loss = 0.4 duct VP

(Ref 135)

AMERICAN CONFERENCE OF GOVERNMENTAL INDUSTRIAL HYGIENISTS

ASBESTOS FIBER BELT CONVEYING

DATE 1-80 | VS-1002

TABLE 10.9-1. Grain Elevators, Feed Mills, Flour Mills[66]

The following date are offered as guides. Air flow rates can vary considerably depending on degree of enclosure, flow rate of material, and dustiness of the grain. Minimum duct velocity = 3500 fpm. Ventilation control is desirable for these operations to minimize the explosive characteristics of grain dusts and to preserve plant housekeeping standards.

Operation	Hood Design	Air Volume
Bag Loading	VS-301, 302 Booth VS-303	As shown 100 cfm/ft^2 open face area
Belt Discharge	To belt — VS-306 To bin — VS-304 To elevator — VS-305, 306	150 cfm/ft of belt width up to 200 fpm belt speed 250 cfm/ft of belt width over 200 fpm belt speed Increase 1/3 if material drop is over 10 ft
Bins	Direct exhaust. Use taper.	500 cfm/bin
Bucket Elevator	VS-305	100 cfm/ft^2 cross-section
Cleaning Machines	Consult manufacturer	
Distributors	Enclose discharge 200 fpm in-draft through enclosure openings.	No. of Spouts / Diameter of Spouts (Exhaust cfm): 6″, 7″, 8″, 9″ 0 – 6: 550, 675, 950, 1250 6 – 12: 950, 1250, 1500, 1900 12 – 24: 1500, 1900, 2250, 2750
Feed Grinders	Consult manufacturer	
Floor Dump	Booth	200 cfm/ft^2 open face area
Floor Sweep		950 cfm in 4″ × 8″ opening
Garner Bin	Direct exhaust. Use taper.	cfm = 1.25 × bushels/min.
Mixers	Ventilated cover	Mixer Capacity — Exhaust, cfm Up to 0.5 ton — 300 0.5/1.5 tons — 675 Over 1.5 tons — 950
Percentage Feeders	Enclosed conveyor	200 cfm at each feeder
Purifiers	Enclosure	30 – 40 cfm/ft^2 screen area
Roll Stands	Enclosure	60 cfm/lineal ft
Scales	Enclosure	Scale Capacity, Bushels — Exhaust, cfm Up to 5 — 250 6 to 10 — 400 Over 11 — 600
Scale Hopper	Direct exhaust. Use taper.	cfm = 1.25 × bushels/min.
Screw Conveyor	Direct exhaust. Use taper.	200 cfm — ducts on 30 ft centers
Sifters	Enclosure	200 cfm/compartment
Track Sink	Direct exhaust from hopper. Use taper.	100 cfm/ft^2 grate area
Tripper Car	Belt discharge. VS-304, 305, 306. Spout ends — tapered connection. Spillage — exhaust under head pulley.	See "Belt Discharge" above. 200 cfm/ft^2 spout cross-section 90 cfm/ft belt width

TABLE 10.9-2. Miscellaneous Specific Operation Standards

Operation or Industry	Ventilation: Type of Hood	Ventilation: Air Flow or Capture Velocity	Minimum Design Duct Velocity (fpm)	Reference No. and Remarks
Abrasive Wheel Mfg.				33
Grading screen	Enclosure — booth	50 fpm at face	3000	
Barrels	Close canopy	400 fpm at face	3000	Bbls. receive dust from cyclone
Grinding wheel dressing	Enclosure — booth	400 fpm at face	3000	
Aluminum Furnaces	Enclosure	150-200 fpm through opening	2000	14
Asbestos				
Bagging	Enclosure — booth	250 fpm through all openings	3500	125,127
Carding	Enclosure	1600 cfm/card	3500	
Crushing	Enclosure	150 fpm through all openings	3500	124
Drilling of panels containing asbestos	Moveable hood	400 fpm capture velocity	4500	126
Dumping	Booth	250 fpm face velocity	3500	126
Grinding of brake shoes	Enclosure	400 fpm minimum capture at the tool rest	3500	126
Hot press for brake shoes	Enclosure	250 fpm through all openings	3500	126
Mixing	Booth	250 fpm face velocity	3500	126
Preform Press	Enclosure	250 fpm through all openings	3500	126
Screening	Enclosure	200 fpm through all openings but not less than 25 cfm/ft^2 screen areas	3000	125
Spool winding	Local Hoods	50 cfm/spool	3500	
Spinning and twisting	Partial	50 cfm/spool	3500	Hinged front panels and skirt, wet twisting preferred
Weaving	Canopy with baffles	50 fpm through openings	3500	Wet weaving preferred
Auto Parking Garage	2 Level	500 cfm/parking space		80
Ceramic	Enclosure	200 fpm through all openings	3500	24, 31, 39
Dry pan	Local at die	500 cfm	3500	Automatic feed
Dry Press	Local at die	500 cfm	3500	Manual feed
	At supply bin	500 cfm	3500	Manual feed
Aerographing	Booth	100 fpm (face)	—	
Spraying (lead glaze)	Booth	400 fpm (face)	2000	
Coating pans (pharmaceutical)	Air flow into opening of pan	100–150 fpm through opening	3000	7, 24 If heated air supplied to pan, add volume of heated air to exhaust
Cooling tunnels (foundry)	Enclosure	75–100 cfm per running foot of enclosure	—	22, 25
Core knockout (Manual)	Large side-draft or semi-booth—exhaust near floor	200–250 cfm/ft^2 dust producing working area	3500	13, 22, 25
Core sanding (on lathe)	Downdraft under work	100 fpm at source	3500	34
Crushers and grinders	Enclosure	200 fpm through openings	3500	30
Drilling (rocks)	Special trap (see references)	60 cfm — vertical (downward) work 200 cfm—horizontal work		9, 17, 18 May vary with size and speed of drill
Forge (hand)	Booth	200 fpm at face	1500	14
Outboard Motor Test Tank	Side draft	200 cfm/ft^2 of tank opening	—	
Packaging Machines	Booth Downdraft Complete Enclosure	50 – 100 fpm at face 95 – 150 fpm down 100 – 400 fpm opening	3000 to 4000	
Paper Machine	Canopy	200 – 300 fpm at face	1500	35, 36

TABLE 10.9-2. Miscellaneous Specific Operation Standards (con't)

Operation or Industry	Ventilation: Type of Hood	Ventilation: Air Flow or Capture Velocity	Minimum Design Duct Velocity (fpm)	Reference No. and Remarks
Quartz Fusing	Booth on bench	150 – 200 fpm at face	—	35, 36
Rotary Blasting Table	Enclosure	500 fpm through all openings when in operation	3500	
Silver Soldering	Free Hanging	100 fpm at source	2000	
Steam Kettles	Canopy	150 fpm at face	2000	
Varnish Kettles	Canopy	200–250 fpm at face	1500	14, 39
Wire Impregnating	Covered Tanks	200 cfm/ft^2 of opening	—	46 Chlorinated naphthalenes & diphenyls

BIBLIOGRAPHY

1. American Air Filter Co., "Rotoclone Dust Control," January 1946.
2. American Society of Heating, Air-Conditioning and Refrigerating Engineers, *Heating, Ventilating and Air Conditioning Guide,* 1963.
3. American Society of Mechanical Engineers, "Power Test Code 21," Test Code for Dust-Separating Apparatus, 1941.
4. American Society of Mechanical Engineers, "Power Test Code 19.2.5," Liquid Column Gauges, 1942.
5. Anemostat Corporation, "Anemotherm Air Meter," 10 East 29th St., New York, NY 10016.
6. Bloomfield, J.J., and DallaValle, J.M., "The Application of Engineering Surveys to the Hatters Fur Cutting Industry," U.S.P.H.S.
7. Brandt, A.D., *Industrial Health Engineering,* John Wiley and Sons, New York, 1947.
8. Brandt, Allen D., "Should Air Be Recirculated from Industrial Exhaust Systems?" *Heating, Piping and Air Conditioning, 19,* 69, August 1947.
9. DallaValle, J.M., *Exhaust Hoods,* Industrial Press, New York, 1946.
10. Dreesen, W.C., et al., "A Study of Asbestosis in the Asbestos Textile Industry," Public Health Bulletin 241, August 1938.
11. Drinker, P., and Hatch, T., *Industrial Dust,* McGraw-Hill, New York, 1936.
12. Drinker, P., and Snell, J.R., "Ventilation of Motion Picture Booths," *Journal of Industrial Hygiene and Toxicology, 20,* 321, April 1938.
13. Fen, O.E., "The Collection and Control of Dust and Fumes from Magnesium Alloy Processing," Peters-Dalton, Inc., January 1945.
14. Hartzell Propeller Fan Company, Bulletin 1001.
15. Hastings Instrument Company, "Air Meter," Box 1275, Hampton, Virginia.
16. Hatch, T., "Design of Exhaust Hoods for Dust Control Systems," *Journal of Industrial Hygiene and Toxicology, 18,* 595, 1936.
17. Hatch, T., et al, "Control of the Silicosis Hazard in the Hard Rock Industries. II. An Investigation of the Kelley Dust Trap for Use with Pneumatic Rock Drills of the Jackhammer Type," *Journal of Industrial Hygiene, 14,* 69, February 1932.
18. Hay, P.S., Capt., "Modified Design of Hay Dust Trap," *Journal of Industrial Hygiene, 12,* 28, January 1930.
19. Hemeon, W.C.L., "Air Dilution in Industrial Ventilation," *Heating and Ventilating, 38,* 41, February 1941.
20. Huebscher, R.G., "Friction Equivalents for Round, Square and Rectangular Ducts," *Heating, Piping and Air Conditioning, 19,* 127, December 1947.
21. Illinois Testing Laboratory, "Alnor Thermo-Anemometer," Chicago, IL 60010.
22. Kane, J.M., "Foundry Ventilation," *The Foundry,* February and March 1946.
23. Kane, J.M., "The Application of Local Exhaust Ventilation to Electric Melting Furnaces," *Trans. Am. Foundrymen's Assoc., 52,* 1351, 1945.
24. Kane, J.M., "Design of Exhaust Systems," *Heating and Ventilating, 42,* 68, November 1945.
25. Kane, J.M., "Foundry Ventilation," University of Michigan Inservice Training Course, October 1945.
26. Madison, R.D., and Elliot, W.R., "Friction Charts for Gases Including Correction for Temperature, Viscosity and Pipe Roughness," *Heating, Piping and Air Conditioning, 18,* 107, October 1946.
27. Moucher, S.C., "Principles of Air Flow," Sheet Metal Workers, September 1947.
28. Air Movement and Control Association, Inc., 30 West University Dr., Arlington Heights, IL 60004, AMCA Standard 210-74.
29. Neal, P.A., et al., "Mercurialism and Its Control in the Felt-Hat Industry," Public Health Bulletin 263, 1941.
30. New York Department of Labor, "Rules Relating to the Control of Silica Dust in Stone Crushing Operations," Industrial Code Rule No. 34, July 1942.
31. Oddie, W.M., "Pottery Dusts: Their Collection and Removal," *Pottery Gazette, 53,* 1280, 1928.
32. Page, R.T., and Bloomfield, J.J., "A Study of Dust Control Methods in an Asbestos Fabricating Plant," Reprint No. 1883, Public Health Reports, November 26, 1937.

33. Pennsylvania Department of Labor and Industry, "Abrasive Wheel Manufacture," Safe Practice Bulletin No. 13.
34. Postman, B.F., "Practical Application of Industrial Exhaust Ventilation for the Control of Occupational Exposures," *American Journal of Public Health, 30,* 149, 1940.
35. Riley, E.C., et al, "How to Design Exhaust Hoods for Quartz-Fusing Operations," *Heating and Ventilating, 37,* 23, April 1940.
36. Riley, E.C., and DallaValle, J.M., "A Study of Quartz-Fusing Operatings with Reference to Measurement and Control of Silica Fumes," *Public Health Reports, 54,* 532, 1939.
37. Rothmann, S.C., "Economic Recovery of Pottery Glazes with Reduction of Dust," *American Journal of Public Health, 29,* 511, 1939.
38. Silverman, Leslie, "Velocity Characteristics of Narrow Exhaust Slots," *Journal of Industrial Hygiene and Toxicology, 24,* 267, November 1942.
39. B. F. Sturtevant Company, "What We Make," Catalog No. 500.
40. Tuve, G. L., and Wright, D. K., "Air Flow Measurements at Intake and Discharge Openings and Grilles," *Heating, Piping and Air Conditioing, 12,* 501, August 1940.
41. Tuve, G. L., "Measuring Air Flow at Intake or Exhaust Grilles," *Heating, Piping and Air Conditioning, 13,* 740, December 1941.
42. Underwriters Laboratories, Inc., "Control of Floating Dust in Grain Elevators," Underwriters Labortories Bulletin of Research No. 1, December 1937.
43. Whalen, F.G., "The Whalen Gage," Engineering Experimental Station Bulletin 10, University of Illinois, March 1921.
44. Witheridge, W.N., "Principles of Industrial Process Ventilation," University of Michigan Inservice Training Course, October 1945.
46. Yaglou, C.P., "Ventilation of Wire Impregnating Tanks Using Chlorinated Hydrocarbons," *Journal of Industrial Hygiene and Toxicology, 20,* 401, June 1938.
47. Yaglou, C.P., Committee on Atmospheric Comport, A.P.H.A., "Report Presented at the 77th Annual Meeting, A.P.H.A., New York City, October 27, 1949."
48. Adolph, E.E., "Tolerance of Man Toward Hot Atmospheres," Supplement #192, Public Health Reports, 1946.
49. "Your Place in the 'Smart Man's War'," *Heating, Piping and Air Conditioning, 14,* 463, August 1942.
50. Factory Mutual Insurance Company, "Properties of Flammable Liquids, Gases and Solids," Factory Mutual Solvent Data Sheet 36.10, January 1945.
51. Malin, Benjamin S., "Practical Pointers on Industrial Exhaust Systems," *Heating and Ventilating, 42,* 75, February 1945.
52. National Board of Fire Underwriters, "Standard for Class A Ovens and Furnaces," Pamphlet #86.
53. United States Bureau of Mines, "Limits of Flammability of Gases and Vapors," Bulletin #503.
54. Silverman, Leslie, "Centerline Velocity Characteristics of Round Openings Under Suction," *Journal of Industrial Hygiene and Toxicology, 24,* 259, November 1942.
55. Schulte, H.F., Hyatt, E.C., and Smith, Jr., F.S., "Exhaust Ventilation for Machine Tools Used on Materials of High Toxicity," *A.M.A. Archives of Industrial Hygiene and Occupational Medicine, 5,* 21, January 1952.
56. Mitchell, R.N., and Hyatt, E.C., "Beryllium — Hazard Evaluation and Control Covering a Five-Year Study," *American Industrial Hygiene Quarterly, 18,* No. 3, September 1957.
57. Hemeon, W.C.L., *Plant and Process Ventilation,* Industrial Press.
58. Manufacturing Chemists' Association, "Technical Data on Plastics," February 1957.
59. First, M.W., and Silverman, L., "Airfoil Pitometer," *Industrial Engineering and Chemistry, 42,* 301-308, February 1950.
60. Stoll, H.W., "The Pitot-Venturi Flow Element," *Transactions ASME,* pp. 963-969, October 1951.
61. Republic Flow Meters Co., "Air Speed Nozzle," Bulletin ME-186-A, Chicago, IL, April 1948.
62. University of Michigan, "Encyclopedia of Instrumentation for Industrial Hygiene," Institute of Industrial Health, Ann Arbor, MI.
63. Dwyer Manufacturing Company, "Magnehelic Gage," P.O. Box 373, Michigan City, IN.
64. Burton, J.R., "Friction Chart," Quaker Oats Company, Chicago, IL.
65. U.S. Dept. of Health and Welfare, Public Health Service, Syllabus, "Short Course for Industrial Hygiene Engineers," p. B-25,7.
66. National Fire Protection Association, "Ventilation of Cooking Equipment 1971," Bulletin 96.
67. American Air Filter Co., Inc., "Usual Exhaust Requirements (for) Grain Elevators, Feed and Flour Mills," April 1956.
68. Air Movement and Control Association, Inc., 30 West University Drive, Arlington Heights, IL 60004.
69. Print, Robert T., "Dust Control in Large-Scale Ore-Concentrating Operations," American Institute of Mining and Metallurgical Engineering, Tech. Publication #1225, February 1940.
70. Langley, M.Y., Harris, Jr., R.L., Lee, D.H.K., "Calculation of Complex Radiant Heat Load from Surrounding Radiator Surface Temperatures," *American Industrial Hygiene Association Journal,* 24, 103-112, March 1963.
71. Thor Power Tool Company, Aurora, IL.
72. Chamberlin, Richard I., "The Control of Beryllium Machining Operations," *A.M.A. Archives of Industrial*

Health, 19, No. 2, February 1959.

73. The Black and Decker Tool Company, Townson, MD.
74. Hoffman Air and Filtration Div., Clarkson Industries, Inc., New York, NY.
75. Alexander, J.M., Croley, Jr., J.J., and Messick, R.R., "Use of Vortex Tube for Cooling Wearers of Industrial Protective Clothing," U.S. Atomic Energy Commission Report DP-861, Office of Technical Services, U.S. Dept. of Commerce, Washington, DC, October 1963.
76. Trickler, C.J., Engineering Letter E-4R, New York Blower Co., LaPorte, IN.
78. New York State Department of Labor, Division of Industrial Hygiene.
79. *Air Conditioning, Heating and Ventilating,* Vol. 60, No. 3, March 1963.
80. Hama, G., Frederick, W., and Monteith, H., "Air Flow Requirements for Underground Parking Garages," *American Industrial Hygiene Association Journal,* Vol. 22, No. 6, December 1961.
81. Feiner, B., and Kingsley, I., "Ventilation of Industrial Ovens," *Air Conditioning, Heating and Ventilating,* pp. 82-89, December 1956.
82. U.S. Air Force Technical Order 00-25-203, "Standards and Guidelines for the Design and Operation of Clean Rooms and Clean Work Stations," Office of Technical Services, Department of Commerce, Washington, DC, July 1963.
83. Federal Standard No. 290B, "Clean Room and Work Station Requirements, Controlled Environment," General Services Administration, Specifications Activity, Printed Material Supply Div., Bldg. 197, Naval Weapons Plant, Washington, DC 20407.
84. Austin, Philip R., and Timmerman, Stuart W., *Design and Operation of Clean Rooms,* Business News Publishing Co., Detroit, MI, 1965.
85. Constance, J.A., "Estimating Air Friction in Triangular Ducts," *Air Conditioning, Heating and Ventilating,* Vol. 60, No. 6, pp. 85-86, June 1963.
86. McKarns, J.S., Confer, R.G., and Brief, R.S., "Estimating Length Limits for Drain Type Stacks," *Heating, Piping and Air Conditiong,* Vol. 37, No. 7, July 1965.
87. Clarke, J.H., "Air Flow Around Buildings," *Heating, Piping and Air Conditioning,* Vol. 39, No. 5, pp. 145-154, May 1967.
88. British Steel Castings Reearch Association, "Dust Control on Stand Grinding Machines," *Conditions in Steel Foundries,* First Report of Joint Standing Committee, London, 1961.
89. The Kirk and Blum Mfg. Co., "Woodworking Plants," pp. W-9, Cincinnati, OH.
90. American Foundrymen's Society, *Engineering Manual for Control of In-Plant Environment in Foundries,* Des Plaines, IL, 1956.
91. Alden, John L., *Design of Industrial Exhaust Systems,* Industrial Press, 200 Madison Ave., New York, NY 10016, 1939.
92. The Quickdraft Corporation, P.O. Box 1353, Canton, OH.
93. Boles, Robert B., "Air Eductors Used to Handle Noxious and Corrosive Fumes," *Air Engineering,* Vol. 7, No. 6, June 1965.
94. Private Communication, E.A. Carsey, The Kirk and Blum Mfg. Co., Cincinnati, OH 45209.
95. Harris, W.B., Christofano, E.E., and Lippman, M., "Combination Hot Plate and Hood for Multiple Beaker Evaporation," *American Industrial Hygiene Assoc. Journal,* Vol. 22, No. 4, August 1961.
96. Dieter, W.E., Cohen, L., and Kundick, M.E., *A Stainless Steel Fume Hood for Safety in Use of Perchloric Acid,* U.S. Dept. of Interior, 1964.
97. Lynch, Jeremiah R., "Computer Design of Industrial Exhaust Systems," *Heating, Piping and Air Conditioning,* September 1968.
98. Hama, George, "A Calibrating Wind Tunnel for Air Measuring Instruments," *Air Engineering,* pp. 18-20, 41, December 1967.
99. Hama, George, "Calibration of Alnor Velometers," *American Industrial Hygiene Assoc. Journal,* December 1958.
100. Hama, George, and Curley, L.S., "Instrumentation for the Measurement of Low Velocities with a Pitot Tube," *Air Engineering,* July 1967, and *American Industrial Hygiene Assoc. Journal,* May, June 1967.
101. Yaffe, C.D., Byer, D.H., and Hosey, A.D., *Encyclopedia of Instrumentation for Industrial Hygiene,* pp. 703-709, University of Michigan, Ann Arbor, MI, 1956.
102. Airflow Developments Ltd., Lancaster Rd., High Wycombe, Bucks., England.
103. *Heating and Cooling for Man and Industry,* American Industrial Hygiene Association, 1969.
104. *ASHRAE Guide & Data Book,* p. 243, American Society of Heating, Refrigeration and Air Conditioning Engineers, 1961.
105. F.W. Dwyer Company, Michigan City, IN.
106. HPAC Data Sheet, "How to Design Drain Type Stacks," *Heating, Piping and Air Conditioning,* p. 143, June 1964.
107. Air Movement and Control Association, Inc., 30 W. University Dr., Arlington Heights, IL 60064, AMCA Standard 99-2408-69.
108. Adapted from U.S. Dept. of Labor, Occupational Safety and Health Administration, Washington, DC, *Federal Register,* Vol. 36, No. 105, May 29, 1971, "Occupational Safety and Health Standards; National Concensus Standards and Established Federal Standards."
109. Hama, George, "How Safe Are Direct-fired Makeup Units," *Air Engineering,* p. 22, September 1962.
110. Hama, George M., and Butler, Jr., Kerrel E., "Ventila-

tion Requirements for Lift Truck Operation," *Heating, Piping and Air Conditioning,* January 1970.

111. Hama, George M., "Is Makeup Air Necessary?", *Air Conditioning, Heating and Ventilating,* November 1959.
112. Hama, George, and Bonkowski, K.J., "Ventilation Requirements for Airless Spray Painting," *Heating, Piping and Air Conditioning,* pp. 80-82, October 1970.
113. National Fire Protection Association, 470 Atlantic Ave., Boston, MA 02210.
114. L.J. Wing Mfg. Co., Bulletin IFB-61, p. 4, Linden, NJ.
115. Hart & Cooley Mfg. Co., Bulletin E-6, Holland, MI.
116. Hama, George, "The Characteristics of Weather Caps," *Air Engineering,* December 1973.
117. U.S. Dept. of Health, Education and Welfare, Health Services and Mental Health Administration, NIOSH, Report HSM 72-10269, *Criteria for a Recommended Standard....Occupational Exposure to Hot Environments.*
118. Rajhans, G.S., and Thompkins, R.W., "Critical Velocities of Mineral Dusts," *Canadian Mining Journal,* pp. 85-88, October 1967.
119. Djamgowz, O.T., and Ghoneim, S.A.A., "Determining the Pick-up Air Velocity of Mineral Dusts," *Canadian Mining Journal,* pp. 25-28, July 1974.
120. Baliff, J., Greenburg, L., and Stern, A.C., "Transport Velocities for Industrial Dusts — An Experimental Study," *Industrial Hygiene Quarterly,* Vol. 9, No. 4, pp. 85-88, December 1948.
121. DallaValle, J.M., "Determining Minimum Air Velocities for Exhaust Systems," *Heating, Piping and Air Conditioning,* 1932.
122. Hatch, T.F., "Economy in the Design of Exhaust Systems."
123. NIOSH Research Report #75-107, *Ventilation Requirements for Grinding, Buffing and Polishing Operations.*
124. Hutcheson, J.R.M., "Environmental Control in the Asbestos Industry of Quebec," C 1 MM Bulletin, Vol. 64, No. 712, pp. 83-89, August 1971.
125. Goldfield, J., and Brandt, F.E., "Dust Control Techniques in the Asbestos Industry," a paper presented at the A.I.H. Conference, Miami Beach, FL, May 12-17, 1974.
126. Private Communications, Occupational Health Protection Branch, Ontario Ministry of Labour, October 1976.
127. Hama, G.M., "Ventilation Control of Dust from Bagging Operations," *Heating and Ventilating,* p. 91, April 1948.
128. Ventilation and Air Contracting Contractors Association of Chicago, *Testing and Balancing Manual for Ventilating and Air Conditioning Systems,* 228 N. LaSalle St., Chicago, IL, 1963.
129. Langner, Ralph R., "How to Control Carcinogens in Chemical Production," *Occupational Health and Safety, March-April 1977.*
130. Wright, Jr., D.K., "A New Friction Chart for Round Ducts," ASHVE Research Report No. 1280, *ASHVE Transactions,* Vol. 51, p. 303, 1945.
131. Leith, David, First, Melvin K.W., and Feldman, Henry, "Performance of a Pulse-Jet at High Filtration Velocity II, Filter Cake Redeposition," *Journal of the Air Pollution Control Assoc.,* Vol. 27, p. 636, 1977.
132. Beake, E., "Optimizing Filtration Parameters," *Journal of the Air Pollution Control Assoc.,* Vol. 24, p. 1150, 1974.
133. Leith, David, Gibson, Dwight D., and First, Melvin W., "Performance of Top and Bottom Inlet Pulse-Jet Fabric Filters," *Journal of the Air Pollution Control Association,* Vol. 28, p. 696, July 1978.
134. National Council on Radiation Protection and Measurements, *Basic Radiation Protection Criteria,* NCRP Report No. 39, January 15, 1971, 4201 Connecticut Ave., N.W., Washington, DC 20008.
135. Rajhans, G.S., and Bragg, G.M., *Engineering Aspects of Asbestos Dust Control,* Ann Arbor Science Publications, Inc., Ann Arbor, MI, 1978.
136. Caplan, K.J., and Knutson, G.W., "Laboratory Fume Hoods: A Performance Test," *ASHRAE Transactions,* Vol. 84, Part 1, 1978.
137. Caplan, K.J., and Knutson, G.W., "Laboratory Fume Hoods: Influence of Room Air Supply," *ASHRAE Transactions,* Vol. 84, Part 2, 1978.
138. Sheet Metal and Air Conditioning Contractors' National Assoc., Inc., *Round Industrial Duct Construction Standards,* 1977, 8224 Old Courthouse Rd., Tysons Corner, Vienna, VA 22180.
139. Sheet Metal and Air Conditioning Contractors' National Assoc., Inc., *Rectangular Industrial Duct Construction Standards,* 8224 Old Courthouse Rd., Tysons Corner, Vienna, VA 22180.
140. Air Movement and Control Association, Inc., 30 W. University Dr., Arlington Heights, IL 60004, AMCA Publication 201.
141. Huebener, D.J., and Hughes, R.T.: "Development of Push-Pull Ventilation," *American Industrial Hygiene Assoc. Journal,* Vol. 46, pp. 262-267, 1985.
142. Hughes, R.T., "Design Criteria for Plating Tank Push-Pull Ventilation," *Ventilation '85,* Elsiever Press, Amsterdam, 1986.
143. Hughes, R.T., Unpublished data.
144. Loeffler, J.J., "Simplified Equations for HVAC Duct Friction Factors," *ASHRAE Journal,* January 1980, pp. 76-79.
145. American Society of Heating, Refrigerating and Air-Conditioning Engineers, *Guide and Data Book — Fundamentals and Equipment,* 1985.
146. Air Moving and Control Association, Inc., AMCA Publication 203-81, *Field Performance Measurements,* 30 W. University Dr., Arlington Heights, IL 60004.

147. American Society of Mechanical Engineers, *Fluid Meters — Their Theory and Applications,* 1959.

148. Farant, J.P., McKinnon, D.L., and McKenna, T.A., "Tracer Gases as a Ventilation Tool: Methods and Instrumentation," *Ventilation '85 — Proceedings of the First International Symposium on Ventilation for Contaminant Control,* pp. 263-274, October 1-3, 1985, Toronto, Ont., Canada.

149. ASME Power Test Codes, Chapter 4, "Flow Measurement," P.T.C., 19.5:4-1959.

150. U.S. Dept. of Health, Education and Welfare, PHS, CDC, NIOSH, *The Industrial Environment — Its Evaluation and Control,* 1973.

151. U.S. Air Force, AFOSH Standard 161.2.

152. U.S. Dept. of Health and Human Services, PHS, CDC, NIOSH, *Occupational Exposure to Hot Environments,* Revised Criteria, 1986.

153. American Welding Society, (AWS D1.1-72), P.O. Box 351040, Miami, FL 33135.

154. Gibson, N., Lloyd, F.D., and Perry, G.R., *Fire Hazards in Chemical Plants from Friction Sparks Involving the Thermite Reaction,* Symposium Series No. 25, Insn. Chemical Engineers, London, 1968.

155. Air Movement and Control Association, Inc., *AMCA Standard 210-74,* 30 W. University Dr., Arlington Heights, IL 60004.

156. Hughes, R.T., and Amendola, A.A., "Recirculating Exhaust Air: Guides, Design Parameters and Mathematical Modeling," *Plant Engineering,* March 18, 1982.

157. U.S. Dept. of Health, Education and Welfare (NIOSH), *The Recirculation of Industrial Exhaust Air — Symposium Proceedings,* Pub. No. 78-141, 1978.

158. American Conference of Governmental Industrial Hygienists, *Air Sampling Instruments for Evaluation of Atmospheric Contaminants,* 6th ed., Chapters U and V, Cincinnati, OH, 1983.

159. Baturin, V.V., *Fundamentals Industrial Ventilation,* Pergamon Press, NY, 1972.

160. U.S. Public Health Service, *Air Pollution Engineering Manual,* Publication No. 999-AP-40, 1973.

161. Hampl, V., and Johnson, O.E.: "Control of Wood Dust from Horizontal Belt Sanding," *American Industrial Hygiene Assoc. Journal,* Vol. 46, No. 10, pp. 567-577, 1985.

162. Hampl, V., Johnston, O.E., and Murdock, D.M.: "Application of an Air Curtain Exhaust System at a Milling Process," *American Industrial Hygiene Assoc. Journal,* Vol. 49, No. 4, pp. 167-175, 1988.

163. Air Movement and Control Association, Inc., AMCA Publication 99-83, *Standards Handbook,* 30 W. University Dr., Arlington Heights, IL 60004.

APPENDICES

APPENDIX A

Threshold Limit Values for Chemical Substances in the Work Environment with Intended Changes for 1988-1989 . . . . **12-3**

APPENDIX B

Physical Constants **12-18**

Conversion Factors **12-22**

APPENDIX C

Metric Supplement **12-25**

APPENDIX A

Threshold Limit Values for Chemical Substances in the Work Environment Adopted by ACGIH

with Intended Changes for 1988-89

1987-88 CHEMICAL SUBSTANCES TLV COMMITTEE

Ernest Mastromatteo, M.D., University of Toronto—*Chair*
James R. Crawl, CIH, US Navy
D. Dwight Culver, M.D., University of California-Irvine
John L.S. Hickey, Ph.D., CIH, Retired
William S. Lainhart, M.D., Retired
Trent R. Lewis, Ph.D., U.S. Environmental Protection Agency
Jesse Lieberman, PE, CIH, Retired
Leonard D. Pagnotto, CIH, Massachusetts Dept. of Labor and Industry
Robert F. Phalen, Ph.D., University of California—Irvine, Air Sampling Procedures Liaison
Ronald S. Ratney, Ph.D., CIH, OSHA
Meier Schneider, PE, CIH, Retired
Raghubir Sharma, Ph.D., CIH, Utah State University
Judith A. Sparer, CIH, Yale University
Robert Spirtas, Dr. PH, National Cancer Institute
Vera F. Thomas, Ph.D., University of Miami, BEI Liaison
William D. Wagner, NIOSH
Elizabeth K. Weisburger, Ph.D., National Cancer Institute, Corresponding Liason
Margie E. Zalesak, CIH, USDL/MSHA

CONSULTANTS

Gerald L. Kennedy, Jr.
Georg Kimmerle, M.D., German MAK Commission Liaison
Theodore R. Torkelson, Sc.D.
Richard A. Youngstrom, CIH, CSP

POLICY STATEMENT ON THE USES OF TLVs AND BEIs

The Threshold Limit Values (TLVs) and Biological Exposure Indices (BEIs) are developed as guidelines to assist in the control of health hazards. These recommendations or guidelines are intended for use in the practice of industrial hygiene, to be interpreted and applied only by a person trained in this discipline. They are not developed for use as legal standards, and the American Conference of Governmental Industrial Hygienists (ACGIH) does not advocate their use as such. However, it is recognized that in certain circumstances individuals or organizations may wish to make use of these recommendations or guidelines as a supplement to their occupational safety and health program. The ACGIH will not oppose their use in this manner, if the use of TLVs and BEIs in these instances will contribute to the overall improvement in worker protection. However, the user must recognize the constraints and limitations subject to their proper use and bear the responsibility for such use.

The introductions to the TLV/BEI booklet and the TLV/BEI Documentation provide the philosophical and practical bases for the uses and limitations of the TLVs and BEIs. To extend those uses of the TLVs and BEIs to include other applications, such as use without the judgment of an industrial hygienist, application to a different population, development of new exposure/recovery time models, or new effect endpoints, stretch the reliability and even viability of the data base for the TLV or BEI as evidenced by the individual documentations.

It is not appropriate for individuals or organizations to impose on the TLVs or the BEIs their concepts of what the TLVs or BEIs should be or how they should be applied or to transfer regulatory standards requirements to the TLVs or BEIs.

INTRODUCTION TO THE CHEMICAL SUBSTANCES

Threshold limit values refer to airborne concentrations of substances and represent conditions under which it is believed that nearly all workers may be repeatedly exposed day after day without adverse effect. Because of wide variation in individual susceptibility, however, a small percentage of workers may experience discomfort from some substances at concentrations at or below the threshold limit; a smaller percentage may be affected more seriously by aggravation of a pre-existing condition or by development of an occupational illness. Smoking of tobacco may act synergistically with airborne chemicals encountered in the workplace, e.g., asbestos.

Individuals may also be hypersusceptible or otherwise unusually responsive to some industrial chemicals because of genetic factors, age, personal habits (smoking, other drugs), medication, or previous exposures. Such workers may not be adequately protected from adverse health effects from certain chemicals at concentrations at or below the threshold limits. An occupational physician should evaluate the extent to which such workers require additional protection.

Threshold limits are based on the best available information from industrial experience, from experimental human and animal studies, and, when possible, from a combination of the three. The basis on which the values are established may differ from substance to substance; protection against impairment of health may be a

---

The Policy Statement on the Uses of TLVs/BEIs was approved by the Board of Directors of ACGIH on March 1, 1988.

guiding factor for some, whereas reasonable freedom from irritation, narcosis, nuisance or other forms of stress may form the basis for others.

The amount and nature of the information available for establishing a TLV varies from substance to substance; consequently, the precision of the estimated TLV is also subject to variation and the latest *Documentation* should be consulted in order to assess the extent of the data available for a given substance.

These limits are intended for use in the practice of industrial hygiene as guidelines or recommendations in the control of potential health hazards and for no other use, e.g., in the evaluation or control of community air pollution nuisances, in estimating the toxic potential of continuous, uninterrupted exposures or other extended work periods, as proof or disproof of an existing disease or physical condition, or adoption by countries whose working conditions differ from those in the United States of America and where substances and processes differ. These limits *are not* fine lines between safe and dangerous concentration nor are they a relative index of toxicity, and *should not* be used by anyone untrained in the discipline of industrial hygiene.

The Threshold Limit Values, as issued by ACGIH, are recommendations and should be used as guidelines for good practices. In spite of the fact that serious injury is not believed likely as a result of exposure to the threshold limit concentrations, the best practice is to maintain concentrations of all atmospheric contaminants as low as is practical.

The ACGIH disclaims liability with respect to the use of TLVs.

Notice of Intent. At the beginning of each year, proposed actions of the Committee for the forthcoming year are issued in the form of a "Notice of Intended Changes." This Notice provides not only an opportunity for comment, *but solicits suggestions of substances to be added to the list. The suggestions should be accompanied by substantiating evidence.* The list of Intended Changes follows the Adopted Values in the TLV booklet. Values listed in parenthesis in the "Adopted" list are to be used during the period in which a proposed change for that Value is listed in the Notice of Intended Changes.

Definitions. Three categories of Threshold Limit Values (TLVs) are specified herein, as follows:

a) *The Threshold Limit Value-Time Weighted Average (TLV-TWA)*—the time-weighted average concentration for a normal 8-hour workday and a 40-hour workweek, to which nearly all workers may be repeatedly exposed, day after day, without adverse effect.

b) *Threshold Limit Value-Short Term Exposure Limit (TLV-STEL)*—the concentration to which workers can be exposed continuously for a short period of time without suffering from 1) irritation, 2) chronic or irreversible tissue damage, or 3) narcosis of sufficient degree to increase the likelihood of accidental injury, impair self-rescue or materially reduce work efficiency, and provided that the daily TLV-TWA is not exceeded. It is not a separate independent exposure limit, rather it supplements the time-weighted average (TWA) limit where there are recognized acute effects from a substance whose toxic effects are primarily of a chronic nature. STELs are recommended only where toxic effects have been reported from high short-term exposures in either humans or animals.

A STEL is defined as a 15-minute time-weighted average exposure which should not be exceeded at any time during a work day even if the eight-hour time-weighted average is within the TLV. Exposures at the STEL should not be longer than 15 minutes and should not be repeated more than four times per day. There should be at least 60 minutes between successive exposures at the STEL. An averaging period other than 15 minutes may be recommended when this is warranted by observed biological effects.

c) *Threshold Limit Value-Ceiling (TLV-C)*—the concentration that should not be exceeded during any part of the working exposure.

In conventional industrial hygiene practice if instantaneous monitoring is not feasible, then the TLV-C can be assessed by sampling over a 15-minute period except for those substances which may cause immediate irritation with exceedingly short exposures.

For some substances, e.g., irritant gases, only one category, the TLV-Ceiling, may be relevant. For other substances, either two or three categories may be relevant, depending upon their physiologic action. It is important to observe that if any one of these three TLVs is exceeded, a potential hazard from that substance is presumed to exist.

The Committee holds to the opinion that limits based on physical irritation should be considered no less binding than those based on physical impairment. There is increasing evidence that physical irritation may initiate, promote or accelerate physical impairment through interaction with other chemical or biologic agents.

Time-Weighted Average vs Ceiling Limits. Time-weighted averages permit excursions above the limit provided they are compensated by equivalent excursions below the limit during the workday. In some instances it may be permissible to calculate the average concentration for a workweek rather than for a workday. The relationship between threshold limit and permissible excursion is a rule of thumb and in certain cases may not apply. The amount by which threshold limits may be exceeded for short periods without injury to health depends upon a number of factors such as the nature of the contaminant, whether very high concentrations—even for short periods—produce acute poisoning, whether the effects are cumulative, the frequency with which high concentrations occur, and the duration of such periods. All factors must be taken into consideration in arriving at a decision as to whether a hazardous condition exists.

Although the time-weighted average concentration provides the most satisfactory, practical way of monitoring airborne agents for compliance with the limits, there are certain substances for which it is inappropriate. In the latter group are substances which are predominantly fast acting and whose threshold limit is more appropriately based on this particular response. Substances with this type of response are best controlled by a ceiling "C" limit that should not be exceeded. It is implicit in these definitions that the manner of sampling to determine noncompliance with the limits for each group must differ; a single brief sample, that is applicable to a "C" limit, is not appropriate to the time-weighted limit; here, a sufficient number of samples are needed to permit a time-weighted average concentration throughout a complete cycle of operations or throughout the work shift.

Whereas the ceiling limit places a definite boundary which concentrations should not be permitted to exceed, the time-weighted average limit requires an explicit limit to the excursions that are permissible above the listed values. It should be noted that the same factors are used by the Committee in determining the magnitude of the value of the STELs, or whether to include or exclude a substance for a "C" listing.

Excursion Limits. For the vast majority of substances with a TLV-TWA, there is not enough toxicological data available to warrant a STEL. Nevertheless, excursions above the TLV-TWA should be controlled even where the eight-hour TWA is within recommended limits. Earlier editions of the TLV list included such limits whose values depended on the TLV-TWAs of the substance in question.

While no rigorous rationale was provided for these particular values, the basic concept was intuitive: in a well controlled process exposure, excursions should be held within some reasonable limits. Unfortunately, neither toxicology nor collective industrial hygiene experience provide a solid basis for quantifying what those limits should be. The approach here is that the maximum recommended excursion should be related to variability generally observed in actual industrial processes. Leidel, Busch and Crouse,* in reviewing large numbers of industrial hygiene surveys conducted by NIOSH, found that short-term exposure measurements were generally log normally distributed with geometric standard deviation mostly in the range of 1.5 to 2.0.

While a complete discussion of the theory and properties of the log normal distribution is beyond the scope of this section, a brief description of some important terms is presented. The measure of central tendency in a log normal description is the antilog of the mean logarithm of the sample values. The distribution is skewed and the geometric mean is always smaller than the arithmetic mean by an amount which depends on the geometric standard deviation. In the log normal distribution, the geometric standard devia-

---

* Leidel, N.A., K.A. Busch and W.E. Crouse: *Exposure Measurement, Action Level and Occupational Environmental Variability.* NIOSH Pub. No. 76-131 (December 1975).

tion (sd_g) is the antilog of the standard deviation of the sample value logarithms and 68.26% of all values lie between m_g/sd_g and $m_g \times sd_g$.

If the short-term exposure values in a given situation have a geometric standard deviation of 2.0, 5% of all values exceed 3.13 times the geometric mean. If a process displays a variability greater than this, it is not under good control and efforts should be made to restore control. This concept is the basis for the new excursion limit recommendations which are as follows:

> *Short-term exposures should exceed three times the TLV-TWA for no more than a total of 30 minutes during a work day and under no circumstances should they exceed five times the TLV-TWA, provided that the TLV-TWA is not exceeded.*

The approach is a considerable simplification of the idea of the log normal concentration distribution but is considered more convenient to use by the practicing industrial hygienist. If exposure excursions are maintained within the recommended limits, the geometric standard deviation of the concentration measurements will be near 2.0 and the goal of the recommendations will be accomplished.

When the toxicological data for a specific substance are available to establish a STEL, this value takes precedence over the excursion limit regardless of whether it is more or less stringent.

"Skin" Notation. Listed substances followed by the designation "Skin" refer to the potential contribution to the overall exposure by the cutaneous route including mucous membranes and eye, either by airborne, or more particularly, by direct contact with the substance. Vehicles can alter skin absorption.

Little quantitative data are available describing absorption of vapors and gases through the skin. The rate of absorption is a function of the concentration to which the skin is exposed.

Substances having a skin notation and a low TLV may present a problem at high airborne concentrations, particularly if a significant area of the skin is exposed for a long period of time. Protection of the respiratory tract, while the rest of the body surface is exposed to a high concentration, may present such a situation.

Biological monitoring should be considered to determine the relative contribution of dermal exposure to the total dose.

This attention-calling designation is intended to suggest appropriate measures for the prevention of cutaneous absorption so that the threshold limit is not invalidated.

Mixtures. Special consideration should be given also to the application of the TLVs in assessing the health hazards which may be associated with exposure to mixtures of two or more substances. A brief discussion of basic considerations involved in developing threshold limit values for mixtures, and methods for their development, amplified by specific examples, are given in Appendix C.

Respirable and Total Dust. For solid substances and liquified mists, TLVs are expressed in terms of total dust except where the term "respirable dust" is used. See Appendix E, Particle Size-Selective Sampling Criteria for Airborne Particulate Matter, for the definition of respirable dust.

Nuisance Particulates. In contrast to fibrogenic dusts which cause scar tissue to be formed in lungs when inhaled in excessive amounts, so-called "nuisance" dusts have a long history of little adverse effect on lungs and do not produce significant organic disease or toxic effect when exposures are kept under reasonable control. The nuisance dusts have also been called (biologically) "inert" dusts, but the latter term is inappropriate to the extent that there is no dust which does not evoke some cellular response in the lung when inhaled in sufficient amount. However, the lung-tissue reaction caused by inhalation of nuisance dusts has the following characteristics: 1) the architecture of the air spaces remains intact; 2) collagen (scar tissue) is not formed to a significant extent; and 3) the tissue reaction is potentially reversible.

Excessive concentrations of nuisance dusts in the workroom air may seriously reduce visibility, may cause unpleasant deposits in the eyes, ears and nasal passages (Portland cement dust), or cause injury to the skin or mucous membranes by chemical or mechanical action *per se* or by the rigorous skin cleansing procedures necessary for their removal.

A threshold limit of 10 mg/m^3 of total dust containing no asbestos and < 1% free silica is recommended for substances in these categories and for which no specific threshold limits have been assigned. This limit, for a normal workday, does not apply to brief exposures at higher concentrations. Neither does it apply to those substances which may cause physiologic impairment at lower concentrations but for which a threshold limit has not yet been adopted. The separate appendix has been deleted in favor of individual entries in the adopted listing for those substances formerly carried in this appendix.

Simple Asphyxiants — 'Inert" Gases or Vapors. A number of gases and vapors, when present in high concentrations in air, act primarily as simple asphyxiants without other significant physiologic effects. A TLV may not be recommended for each simple asphyxiant because the limiting factor is the available oxygen. The minimal oxygen content should be 18 percent by volume under normal atmospheric pressure (equivalent to a partial pressure, pO_2 of 135 mm Hg). Atmospheres deficient in O_2 do not provide adequate warning and most simple asphyxiants are odorless. Several simple asphyxiants present an explosion hazard. Account should be taken of this factor in limiting the concentration of the asphyxiant. Specific examples are listed in Appendix D. This list is not meant to be all inclusive; the substances serve only as examples.

Physical Factors. It is recognized that such physical factors as heat, ultraviolet and ionizing radiation, humidity, abnormal pressure (altitude), and the like may place added stress on the body so that the effects from exposure at a threshold limit may be altered. Most of these stresses act adversely to increase the toxic response of a substance. *Although most threshold limits have built-in safety factors to guard against adverse effects to moderate deviations* from normal environments, the safety factors of most substances are not of such a magnitude as to take care of gross deviations. For example, continuous work at temperatures above 90°F, or overtime extending the workweek more than 25%, might be considered gross deviations. In such instances judgment must be exercised in the proper adjustments of the Threshold Limit Values.

Unlisted Substances. Many substances present or handled in industrial processes do not appear on the TLV list. In a number of instances, the material is rarely present as a particulate, vapor, or other airborne contaminant, and a TLV is not necessary. In other cases, sufficient information to warrant development of a TLV, even on a tentative basis, is not available to the Committee. Other substances, of low toxicity, could be classified as nuisance particulates.

In addition, there are some substances of not inconsiderable toxicity, which have been omitted primarily because only a limited number of workers (e.g., employees of a single plant) are known to have potential exposure to possibly harmful concentrations.

Operational Guidelines. The ACGIH Board of Directors has adopted operational guidelines for the Chemical Substances TLV Committee. These guidelines prescribe: charge, authority, policies, membership, organization, and operating procedures. The policies include the appeals procedures. Copies of the guidelines document are available from the Publications Office at a cost of $5 per copy. The guidelines were also published in the March 1988 issue of *Applied Industrial Hygiene.*

Substance [CAS #]	ADOPTED VALUES TWA ppm[a]	TWA mg/m3[b]	STEL ppm[a]	STEL mg/m3[b]
■Acetaldehyde [75-07-0]	100	180	150	270
Acetic acid [64-19-7]	10	25	15	37
Acetic anhydride [108-24-7]	C 5	C 20	—	—
•Acetone [67-64-1]	750	1,780	1,000	2,375
•Acetonitrile [75-05-8]—Skin	40	70	60	105
Acetylene [74-86-2]	D	—	—	—
Acetylene dichloride, *see* 1,2-Dichloroethylene				
Acetylene tetrabromide [79-27-6]	1	15	—	—
Acetylsalicylic acid (Aspirin) [50-78-2]	—	5	—	—
Acrolein [107-02-8]	0.1	0.25	0.3	0.8
■Acrylamide [79-06-1]—Skin	—	0.03,A2	—	—
‡Acrylic acid [79-10-7]	(10)	(30)	—	—
•■Acrylonitrile [107-13-1]—Skin	2,A2	4.5,A2	—	—
•■Aldrin [309-00-2] — Skin	—	0.25	—	—
Allyl alcohol [107-18-6]—Skin	2	5	4	10
■Allyl chloride [107-5-1]	1	3	2	6
Allyl glycidyl ether (AGE) [106-92-3]—Skin	5	22	10	44
Allyl propyl disulfide [2179-59-1]	2	12	3	18
α-Alumina, *see* Alumnium oxide				
Aluminum [7429-90-5], as Al				
Metal dust	—	10	—	—
Pyro powders	—	5	—	—
Welding fumes	—	5	—	—
Soluble salts	—	2	—	—
Alkyls (NOC†)	—	2	—	—
Aluminum oxide [1344-28-1], as Al	—	10[e]	—	—
4-Aminodiphenyl [92-67-1]—Skin	—	A1	—	—
2-Aminoethanol, *see* Ethanolamine				
2-Aminopyridine [504-29-0]	0.5	2	—	—
3-Amino 1,2,4-triazole, *see* Amitrole				
■Amitrole [61-82-5]	—	0.2	—	—
Ammonia [7664-41-7]	25	18	35	27
Ammonium chloride fume [12125-02-9]	—	10	—	20
*Ammonium perfluorooctanoate [3825-26-1]	—	0.1	—	—
Ammonium sulfamate [7773-06-0]	—	10	—	—
Amosite, *see* Asbestos				

(a) Parts of vapor or gas per million parts of contaminated air by volume at 25°C and 760 torr.

(b) Milligrams of substance per cubic meter of air. When entry is in this column only, the value is exact; when listed with a ppm entry, it is approximate.

(d) Fibers longer than 5 μm and with an aspect ratio equal to or greater than 3:1 as determined by the membrane filter method at 400-450X magnification (4 mm objective) phase contrast illumination.

(e) The value is for total dust containing no asbestos and < 1% free silica.

Capital letters, A, B, & D refer to Appendices; C denotes ceiling limit.

□ Identifies substances for which there are also BEIs (*see* BEI section). Substances identified in the BEI documentations for methemoglobin inducers and organophosphorus cholinesterase inhibitors are part of this notation.

■ Substance identified by other sources as a suspected or confirmed human carcinogen. See the compliation in the Appendix to the Documentation of TLVs, pp. A-5(86)–A-9(86).

• Substance for which OSHA and/or NIOSH has a Permissible Exposure Limit (PEL) or a Recommended Exposure Limit (REL) lower than the TLV.

‡ See Notice of Intended Changes.

* 1988-1989 Adoption.

† NOC = not otherwise classified.

Substance [CAS #]	ADOPTED VALUES TWA ppm[a]	TWA mg/m3[b]	STEL ppm[a]	STEL mg/m3[b]
n-Amyl acetate [628-63-7]	100	530	—	—
sec-Amyl acetate [626-38-0]	125	665	—	—
□■Aniline [62-53-3] & homologues—Skin	2	10	—	—
□■Anisidine [29191-52-4] (o-, p-isomers)—Skin	0.1	0.5	—	—
Antimony [7440-36-0] & compounds, as Sb	—	0.5	—	—
■Antimony trioxide [1309-64-4]				
Handling and use, as Sb	—	0.5	—	—
Production	—	A2	—	—
ANTU [86-88-4]	—	0.3	—	—
Argon [7440-37-1]	D	—	—	—
•■Arsenic [7440-38-2] & soluble compounds, as As	—	0.2	—	—
Arsenic trioxide production [1327-53-3]	—	A2	—	—
•Arsine [7784-42-1]	0.05	0.2	—	—
•■Asbestos[d]				
• Amosite [12172-73-5]		0.5 fiber/cc, A1		
• Chrysotile [12001-29-5]		2 fibers/cc, A1		
• Crocidolite [12001-28-4]		0.2 fiber/cc, A1		
• Other forms		2 fibers/cc, A1		
Asphalt (petroleum) fumes [8052-42-4]	—	5	—	—
Atrazine [1912-24-9]	—	5	—	—
□Azinphos-methyl [86-50-0]—Skin	—	0.2	—	—
Barium [7440-39-3], soluble compounds, as Ba	—	0.5	—	—
Barium sulfate [7727-43-7]	—	10[e]	—	—
Benomyl [17804-35-2]	0.8	10	—	—
□■•Benzene [71-43-2]	10,A2	30,A2	—	—
■Benzidine [92-87-5]—Skin	—	A1	—	—
p-Benzoquinone, *see* Quinone				
Benzoyl peroxide [94-36-0]	—	5	—	—
■Benzo(a)pyrene [50-32-8]	—	A2	—	—
■Benzyl chloride [100-44-7]	1	5	—	—
•■Beryllium and compounds [7440-41-7]	—	0.002,A2	—	—
Biphenyl [92-52-4]	0.2	1.5	—	—
Bismuth telluride [1304-82-1]	—	10	—	—
Se-doped	—	5	—	—
Borates, tetra, sodium salts [1303-96-4]				
Anhydrous	—	1	—	—
Decahydrate	—	5	—	—
Pentahydrate	—	1	—	—
Boron oxide [1303-86-2]	—	10	—	—
Boron tribromide [10294-33-4]	C 1	C 10	—	—
Boron trifluoride [7637-07-2]	C 1	C 3	—	—
Bromacil [314-40-9]	1	10	—	—
Bromine [7726-95-6]	0.1	0.7	0.3	2
Bromine pentafluoride [7789-30-2]	0.1	0.7	—	—
Bromochloromethane, *see* Chlorobromomethane				
Bromoform [75-25-2]—Skin	0.5	5	—	—
•■1,3-Butadiene [106-99-0]	10,A2	22,A2	—	—
Butane [106-97-8]	800	1,900	—	—
Butanethiol, *see* Butyl mercaptan				
2-Butanone, *see* Methyl ethyl ketone (MEK)				
2-Butoxyethanol [111-76-2]—Skin	25	120	—	—
n-Butyl acetate [123-86-4]	150	710	200	950
sec-Butyl acetate [105-46-4]	200	950	—	—
tert-Butyl acetate [540-88-5]	200	950	—	—

Substance [CAS #]	ADOPTED VALUES: TWA ppm[a]	TWA mg/m^3[b]	STEL ppm[a]	STEL mg/m^3[b]
Butyl acrylate [141-32-2]	10	55	—	—
n-Butyl alcohol [71-36-3]—Skin	C 50	C 150	—	—
‡sec-Butyl alcohol [78-92-2]	100	305	(150)	(455)
tert-Butyl alcohol [75-65-0]	100	300	150	450
Butylamine [109-73-9]—Skin	C 5	C 15	—	—
•tert-Butyl chromate, as CrO_3 [1189-85-1]—Skin	—	C 0.1	—	—
•n-Butyl glycidyl ether (BGE) [2426-08-6]	25	135	—	—
n-Butyl lactate [138-22-7]	5	25	—	—
Butyl mercaptan [109-79-5]	0.5	1.5	—	—
o-sec-Butylphenol [89-72-5]—Skin	5	30	—	—
p-tert-Butyltoluene [98-51-1]	10	60	20	120
▪•□‡Cadmium [7440-43-9] dusts & salts, as Cd	—	(0.05)	—	—
‡Cadmium oxide [1306-19-0]				
Fume, as Cd	—	(C 0.05)	—	—
Production	—	(0.05)	—	—
Calcium carbonate [1317-65-3]	—	10[e]	—	—
Calcium cyanamide [156-62-7]	—	0.5	—	—
Calcium hydroxide [1305-62-0]	—	5	—	—
Calcium oxide [1305-78-8]	—	2	—	—
Calcium silicate [1344-95-2]	—	10[e]	—	—
Calcium sulfate [7778-18-9]	—	10[e]	—	—
Camphor, synthetic [76-22-2]	2	12	3	18
‡Caprolactam [105-60-2]				
Dust	—	(1)	—	(3)
Vapor	(5)	(20)	(10)	(40)
Captafol [2425-06-1]—Skin	—	0.1	—	—
Captan [133-06-2]	—	5	—	—
Carbaryl [63-25-2]	—	5	—	—
Carbofuran [1563-66-2]	—	0.1	—	—
▪Carbon black [1333-86-4]	—	3.5	—	—
Carbon dioxide [124-38-9]	5,000	9,000	30,000	54,000
•□Carbon disulfide [75-15-0]—Skin	10	30	—	—
•□Carbon monoxide [630-08-0]	50	55	400	440
•▪Carbon tetrabromide [558-13-4]	0.1	1.4	0.3	4
Carbon tetrachloride [56-23-5]—Skin	5,A2	30,A2	—	—
Carbonyl chloride, *see* Phosgene				
Carbonyl fluoride [353-50-4]	2	5	5	15
Catechol [120-80-9]	5	20	—	—
Cellulose (paper fiber) [9004-34-6]	—	10[e]	—	—
Cesium hydroxide [21351-79-1]	—	2	—	—
‡▪Chlordane [57-74-9]—Skin	—	0.5	—	(2)
▪Chlorinated camphene [8001-35-2]—Skin	—	0.5	—	1
‡Chlorinated diphenyl oxide [57321-63-8]	—	0.5	—	(2)
‡Chlorine [7782-50-5]	(1)	(3)	(3)	(9)
Chlorine dioxide [10049-04-4]	0.1	0.3	0.3	0.9
Chlorine trifluoride [7790-91-2]	C 0.1	C 0.4	—	—
Chloroacetaldehyde [107-20-0]	C 1	C 3	—	—
α-Chloroacetophenone [532-27-4]	0.05	0.3	—	—
Chloroacetyl chloride [79-04-9]	0.05	0.2	—	—
Chlorobenzene [108-90-7]	75	350	—	—
o-Chlorobenzylidene malononitrile [2698-41-1]—Skin	C 0.05	C 0.4	—	—
‡Chlorobromomethane [74-97-5]	200	1,050	(250)	(1,300)
2-Chloro-1,3-butadiene, *see* β-Chloroprene				
‡▪Chlorodifluoromethane [75-45-6]	1,000	3,500	(1,250)	(4,375)
‡•Chlorodiphenyl (42% Chlorine) [53469-21-9]—Skin	—	1	—	(2)
‡•▪Chlorodiphenyl (54% Chlorine) [11097-69-1]—Skin	—	0.5	—	(1)
1-Chloro,2,3-epoxy-propane, *see* Epichlorohydrin				
2-Chloroethanol, *see* Ethylene chlorohydrin				
Chloroethylene, *see* Vinyl chloride				
•▪Chloroform [67-66-3]	10,A2	50,A2	—	—
•▪bis(Chloromethyl) ether [542-88-1]	0.001, A1	0.005, A1	—	—
▪Chloromethyl methyl ether [107-30-2]	A2	A2	—	—
1-Chloro-1-nitropropane [600-25-9]	2	10	—	—
Chloropentafluoroethane [76-15-3]	1,000	6,320	—	—
‡Chloropicrin [76-06-2]	0.1	0.7	(0.3)	(2)
•▪β-Chloroprene [126-99-8]—Skin	10	35	—	—
o-Chlorostyrene [2039-87-4]	50	285	75	430
‡o-Chlorotoluene [95-49-8]	50	250	(75)	(375)
2-Chloro-6-(trichloromethyl) pyridine, *see* Nitrapyrin				
‡Chlorpyrifos [2921-88-2]—Skin	—	0.2	—	(0.6)
•▪Chromite ore processing (Chromate), as Cr	—	0.05,A1	—	—
Chromium [7440-47-3] Metal	—	0.5	—	—
Chromium (II) compounds, as Cr	—	0.5	—	—
Chromium (III) compounds, as Cr	—	0.5	—	—
•▪□Chromium (VI) compounds, as Cr				
•□Water soluble	—	0.05	—	—
• Certain water insoluble	—	0.05,A1	—	—
•▪Chromyl chloride [14977-61-8]	0.025	0.15	—	—
▪Chrysene [218-01-9]	A2	A2	—	—
Chrysotile, *see* Asbestos				
‡Clopidol [2971-90-6]	—	10	—	(20)

(e) The value is for total dust containing no asbestos and < 1% free silica.
Capital letters, A, B, & D refer to Appendices; C denotes ceiling limit.

□ Identifies substances for which there are also BEIs (*see* BEI section). Substances identified in the BEI documentations for methemoglobin inducers and organophosphorus cholinesterase inhibitors are part of this notation.

▪ Substance identified by other sources as a suspected or confirmed human carcinogen. See the compliation in the Appendix to the Documentation of TLVs, pp. A-5(86)–A-9(86).

• Substance for which OSHA and/or NIOSH has a Permissible Exposure Limit (PEL) or a Recommended Exposure Limit (REL) lower than the TLV.

‡ See Notice of Intended Changes.

Substance [CAS #]	ADOPTED VALUES TWA ppm[a]	TWA mg/m^3 [b]	STEL ppm[a]	STEL mg/m^3 [b]
Coal dust	—	2,[f] Respirable fraction		
•■Coal tar pitch volatiles [65996-93-2], as benzene solubles	—	0.2,A1	—	—
■Cobalt [7440-48-4], as Co				
Metal dust & fume	—	0.05	—	—
Cobalt carbonyl [10210-68-1], as Co	—	0.1	—	—
Cobalt hydrocarbonyl [16842-03-8], as Co	—	0.1	—	—
Copper [7440-50-8]				
Fume	—	0.2	—	—
Dusts & mists, as Cu	—	1	—	—
Cotton dust, raw	—	0.2[g]	—	—
•Cresol [1319-77-3], all isomers—Skin	5	22	—	—
Cristobalite, *see* Silica—Crystaline				
Crocidolite, *see* Asbestos				
■Crotonaldehyde [4170-30-3]	2	6	—	—
‡Crufomate [299-86-5]	—	5	—	(20)
Cumene [98-82-8]—Skin	50	245	—	—
Cyanamide [420-04-2]	—	2	—	—
Cyanides [151-50-8; 143-33-9], as CN—Skin	—	5	—	—
Cyanogen [460-19-5]	10	20	—	—
Cyanogen chloride [506-77-4]	C 0.3	C 0.6	—	—
Cyclohexane [110-82-7]	300	1,050	—	—
Cyclohexanol [108-93-0]—Skin	50	200	—	—
Cyclohexanone [108-94-1]—Skin	25	100	—	—
Cyclohexene [110-83-8]	300	1,015	—	—
□Cyclohexylamine [108-91-8]	10	40	—	—
‡Cyclonite [121-82-4]—Skin	—	1.5	—	(3)
Cyclopentadiene [542-92-7]	75	200	—	—
Cyclopentane [287-92-3]	600	1,720	—	—
Cyhexatin [13121-70-5]	—	5	—	—
2,4-D [94-75-7]	—	10	—	—
■•DDT (Dichlorodiphenyl-trichloroethane) [50-29-3]	—	1	—	—
Decaborane [17702-41-9]—Skin	0.05	0.3	0.15	0.9
□Demeton [8065-48-3]—Skin	0.01	0.1	—	—
Diacetone alcohol [123-42-2]	50	240	—	—
1,2-Diaminoethane, *see* Ethylenediamine				
Diatomaceous earth, *see* Silica—Amorphous				
□Diazinon [333-41-5]—Skin	—	0.1	—	—
■Diazomethane [334-88-3]	0.2	0.4	—	—

Substance [CAS #]	ADOPTED VALUES TWA ppm[a]	TWA mg/m^3 [b]	STEL ppm[a]	STEL mg/m^3 [b]
Diborane [19287-45-7]	0.1	0.1	—	—
1,2-Dibromoethane, *see* Ethylene dibromide				
2-N-Dibutylaminoethanol [102-81-8]—Skin	2	14	—	—
Dibutyl phosphate [107-66-4]	1	5	2	10
Dibutyl phthalate [84-74-2]	—	5	—	—
■Dichloroacetylene [7572-29-4]	C 0.1	C 0.4	—	—
o-Dichlorobenzene [95-50-1]	C 50	C 300	—	—
p-Dichlorobenzene [106-46-7]	75	450	110	675
■3,3'-Dichlorobenzidine [91-94-1]—Skin	—	A2	—	—
Dichlorodifluoromethane [75-71-8]	1,000	4,950	—	—
1,3-Dichloro-5,5-dimethyl hydantoin [118-52-5]	—	0.2	—	0.4
•1,1-Dichloroethane [75-34-3]	200	810	250	1,010
1,2-Dichloroethane, *see* Ethylene dichloride				
1,1-Dichloroethylene, *see* Vinylidene chloride				
1,2-Dichloroethylene [540-59-0]	200	790	—	—
Dichloroethyl ether [111-44-4]—Skin	5	30	10	60
Dichlorofluoromethane [75-43-4]	10	40	—	—
Dichloromethane, *see* Methylene chloride				
1,1-Dichloro-1-nitroethane [594-72-9]	2	10	—	—
1,2-Dichloropropane, *see* Propylene dichloride				
■Dichloropropene [542-75-6]—Skin	1	5	—	—
2,2-Dichloropropionic acid [75-99-0]	1	6	—	—
Dichlorotetrafluoroethane [76-14-2]	1,000	7,000	—	—
□Dichlorvos [62-73-7]—Skin	0.1	1	—	—
Dicrotophos [141-66-2]—Skin	—	0.25	—	—
Dicyclopentadiene [77-73-6]	5	30	—	—
Dicyclopentadienyl iron [102-54-5]	—	10	—	—
•Dieldrin [60-57-1]—Skin	—	0.25	—	—
Diethanolamine [111-42-2]	3	15	—	—
Diethylamine [109-89-7]	10	30	25	75
2-Diethylaminoethanol [100-37-8]—Skin	10	50	—	—
Diethylene triamine [111-40-0]—Skin	1	4	—	—
Diethyl ether, *see* Ethyl ether				
Di(2-ethylhexyl)phthalate, *see* Di-sec-octyl phthalate				
Diethyl ketone [96-22-0]	200	705	—	—
Diethyl phthalate [84-66-2]	—	5	—	—
Difluorodibromomethane [75-61-6]	100	860	—	—
■Diglycidyl ether (DGE) [2238-07-5]	0.1	0.5	—	—
Dihydroxybenzene, *see* Hydroquinone				
Diisobutyl ketone [108-83-8]	25	150	—	—
Diisopropylamine [108-18-9]—Skin	5	20	—	—
Dimethoxymethane, *see* Methylal				
Dimethyl acetamide [127-19-5]—Skin	10	35	—	—
Dimethylamine [124-40-3]	10	18	—	—
Dimethylaminobenzene, *see* Xylidene				
□Dimethylaniline [121-69-7] (N,N-Dimethylaniline)—Skin	5	25	10	50

(f) The value is for dust containing < 5% free silica. For dust containing more than this percentage of free silica, the environment should be elevated against the TLV-TWA of 0.1 mg/m^3 for respirable quartz. The concentration of respirable dust for the application of this limit is to be determined from the fraction passing a size-selector with the characteristics defined in the "c." paragraphs of Appendix E.

(g) Lint-free dust as measured by the vertical elutriator cotton-dust sampler described in the *Transactions of the National Conference on Cotton Dust,* p. 33, by J.R. Lynch (May 2, 1970).

Capital letters, A, B, & D refer to Appendices; C denotes ceiling limit.

□ Identifies substances for which there are also BEIs (*see* BEI section). Substances identified in the BEI documentations for methemoglobin inducers and organophosphorus cholinesterase inhibitors are part of this notation.

■ Substance identified by other sources as a suspected or confirmed human carcinogen. See the compliation in the Appendix to the Documentation of TLVs, pp. A-5(86)–A-9(86).

• Substance for which OSHA and/or NIOSH has a Permissible Exposure Limit (PEL) or a Recommended Exposure Limit (REL) lower than the TLV.

‡ See Notice of Intended Changes.

Substance [CAS #]	ADOPTED VALUES TWA		STEL	
	ppm[a]	mg/m^3[b]	ppm[a]	mg/m^3[b]
Dimethylbenzene, *see* Xylene				
■Dimethyl carbamoyl chloride [79-44-7]	A2	A2	—	—
Dimethyl-1,2-dibromo-2-dichloroethyl phospate, *see* Naled				
□Dimethylformamide [68-12-2]—Skin	10	30	—	—
2,6-Dimethyl-4-heptanone, *see* Diisobutyl ketone				
•■1,1-Dimethylhydrazine [57-14-7]—Skin	0.5,A2	1,A2	—	—
Dimethylnitrosoamine, *see* N-Nitrosodimethylamine				
Dimethylphthalate [131-11-3]	—	5	—	—
■Dimethyl sulfate [77-78-1]—Skin	0.1,A2	0.5,A2	—	—
Dinitolmide [148-01-6]	—	5	—	—
□Dinitrobenzene [528-29-0; 99-65-0; 100-25-4] (all isomers)— Skin	0.15	1	—	—
Dinitro-o-cresol [534-52-1]—Skin	—	0.2	—	—
3,5-Dinitro-o-toluamide, *see* Dinitolmide				
•■□Dinitrotoluene [121-14-2]—Skin	—	1.5	—	—
•■Dioxane [123-91-1]—Skin	25	90	—	—
□Dioxathion [78-34-2]—Skin	—	0.2	—	—
Diphenyl, *see* Biphenyl				
Diphenylamine [122-39-4]	—	10	—	—
Diphenylmethane diisocyanate, *see* Methylene bisphenyl isocyanate				
Dipropylene glycol methyl ether [34590-94-8]—Skin	100	600	150	900
Dipropyl ketone [123-19-3]	50	235	—	—
Diquat [85-00-7]	—	0.5	—	—
■•Di-sec-octyl phthalate [117-81-7]	—	5	—	10
Disulfiram [97-77-8]	—	2	—	—
Disulfoton [298-04-4]	—	0.1	—	—
2,6-Di-tert-butyl-p-cresol [128-37-0]	—	10	—	—
Diuron [330-54-1]	—	10	—	—
Divinyl benzene [1321-74-0]	10	50	—	—
Emery [112-62-9]	—	10(e)	—	—
Endosulfan [115-29-7]—Skin	—	0.1	—	—
Endrin [72-20-8]—Skin	—	0.1	—	—
*Enflurane [13838-16-9]	75	575	—	—
Enzymes, *see* Subtilisins				
•■Epichlorohydrin [106-89-8]—Skin	2	10	—	—
□EPN [2104-64-5]—Skin	—	0.5	—	—
1,2-Epoxypropane, *see* Propylene oxide				
2,3-Epoxy-1-propanol, *see* Glycidol				
Ethane [74-84-0]	D	—	—	—
Ethanethiol, *see* Ethyl mercaptan				
Ethanol, *see* Ethyl alcohol				
Ethanolamine [141-43-5]	3	8	6	15
□Ethion [563-12-2]—Skin	—	0.4	—	—
•2-Ethoxyethanol [110-80-5]—Skin	5	19	—	—
2-Ethoxyethyl acetate [111-15-9]—Skin	5	27	—	—
Ethyl acetate [141-78-6]	400	1,400	—	—
‡Ethyl acrylate [140-88-5]	5	20	(25)	(100)
Ethyl alcohol [64-17-5]	1,000	1,900	—	—
Ethylamine [75-04-7]	10	18	—	—
Ethyl amyl ketone [541-85-5]	25	130	—	—
□Ethyl benzene [100-41-4]	100	435	125	545
Ethyl bromide [74-96-4]	200	890	250	1,110
Ethyl butyl ketone [106-35-4]	50	230	—	—
Ethyl chloride [75-00-3]	1,000	2,600	—	—
Ethylene [74-85-1]	D	—	—	—
Ethylene chlorohydrin [107-07-3]—Skin	C 1	C 3	—	—
Ethylenediamine [107-15-3]	10	25	—	—
•■Ethylene dibromide [106-93-4]—Skin	A2	A2	—	—
•■Ethylene dichloride [107-06-2]	10	40	—	—
Ethylene glycol [107-21-1] Vapor	C 50	C 125	—	—
•Ethylene glycol dinitrate [628-96-6]—Skin	0.05	0.3	—	—
Ethylene glycol methyl ether acetate, *see* 2-Methoxyethyl acetate				
■Ethylene oxide [75-21-8]	1,A2	2,A2	—	—
•■Ethylenimine [151-56-4]—Skin	0.5	1	—	—
Ethyl ether [60-29-7]	400	1,200	500	1,500
Ethyl formate [109-94-4]	100	300	—	—
Ethylidene chloride, *see* 1,1-Dichloroethane				
Ethylidene norbornene [16219-75-3]	C 5	C 25	—	—
Ethyl mercaptan [75-08-1]	0.5	1	—	—
N-Ethylmorpholine [100-74-3]—Skin	5	23	—	—
Ethyl silicate [78-10-4]	10	85	—	—
□Fenamiphos [22224-92-6]—Skin	—	0.1	—	—
□Fensulfothion [115-90-2]	—	0.1	—	—
□Fenthion [55-38-9]—Skin	—	0.2	—	—
Ferbam [14484-64-1]	—	10	—	—
Ferrovanadium dust [12604-58-9]	—	1	—	3
•Fibrous glass dust	—	10	—	—
□Fluorides, as F	—	2.5	—	—
•Fluorine [7782-41-4]	1	2	2	4
Fluorotrichloromethane, *see* Trichlorofluoromethane				
□Fonofos [944-22-9]—Skin	—	0.1	—	—
•■Formaldehyde [50-00-0]	1,A2	1.5,A2	2,A2	3,A2
*Formamide — Skin [75-12-7]	10	15	—	—
‡Formic acid [64-18-6]	5	9	(—)	(—)
Furfural [98-01-1]—Skin	2	8	—	—
Furfuryl alcohol [98-00-0]—Skin	10	40	15	60
Gasoline [8006-61-9]	300	900	500	1,500
Germanium tetrahydride [7782-65-2]	0.2	0.6	—	—
Glass, fibrous or dust, *see* Fibrous glass dust				
Glutaraldehyde [111-30-8]	C 0.2	C 0.7	—	—
Glycerin mist [56-81-5]	—	10(e)	—	—
Glycidol [556-52-5]	25	75	—	—
Glycol monoethyl ether, *see* 2-Ethoxyethanol				

(e) The value is for total dust containing no asbestos and < 1% free silica.
Capital letters, A, B, & D refer to Appendices; C denotes ceiling limit.

□ Identifies substances for which there are also BEIs (*see* BEI section). Substances identified in the BEI documentations for methemoglobin inducers and organophosphorus cholinesterase inhibitors are part of this notation.

■ Substance identified by other sources as a suspected or confirmed human carcinogen. See the compliation in the Appendix to the Documentation of TLVs, pp. A-5(86)—A-9(86).

• Substance for which OSHA and/or NIOSH has a Permissible Exposure Limit (PEL) or a Recommended Exposure Limit (REL) lower than the TLV.

‡ See Notice of Intended Changes.

* 1988-1989 Adoption.

Substance [CAS #]	ADOPTED VALUES TWA ppm[a]	TWA mg/m^3[b]	STEL ppm[a]	STEL mg/m^3[b]
Grain dust (oat, wheat, barley)	—	4[h]	—	—
‡Graphite (natural) [7782-42-5]	—	(2.5[i], Respirable dust)		
‡Graphite (synthetic)	—	(10[e])	—	—
Gypsum, *see* Calcium sulfate				
Hafnium [7440-58-6]	—	0.5	—	—
*Halothane [151-67-7]	50	400	—	—
Helium [7440-59-7]	D	—	—	—
■Heptachlor [76-44-8]—Skin	—	0.5	—	—
Heptane [142-82-5] (n-Heptane)	400	1,600	500	2,000
2-Heptanone, *see* Methyl n-amyl ketone				
3-Heptanone, *see* Ethyl butyl ketone				
■Hexachlorobutadiene [87-68-3]—Skin	0.02,A2	0.24,A2	—	—
Hexachlorocyclopentadiene [77-47-4]	0.01	0.1	—	—
•‡Hexachloroethane [67-72-1]	(10)	(100)	—	—
Hexachloronaphthalene [1335-87-1]—Skin	—	0.2	—	—
Hexafluoroacetone [684-16-2]—Skin	0.1	0.7	—	—
*Hexamethylene diisocyanate [822-06-0]	0.005	0.035	—	—
■Hexamethyl phosphoramide [680-31-9]—Skin	A2	A2	—	—
□Hexane (n-Hexane) [110-54-3]	50	180	—	—
• Other isomers	500	1,800	1,000	3,600
2-Hexanone, *see* Methyl n-butyl ketone				
Hexone, *see* Methyl isobutyl ketone				
sec-Hexyl acetate [108-84-9]	50	300	—	—
Hexylene glycol [107-41-5]	C 25	C 125	—	—
•■Hydrazine [302-01-2]—Skin	0.1,A2	0.1,A2	—	—
Hydrogen [1333-74-0]	D	—	—	—
Hydrogenated terphenyls [61788-32-7]	0.5	5	—	—
Hydrogen bromide [10035-10-6]	C 3	C 10	—	—
Hydrogen chloride [7647-01-0]	C 5	C 7	—	—
•Hydrogen cyanide [74-90-8]—Skin	C 10	C 10	—	—
Hydrogen fluoride [7664-39-3], as F	C 3	C 2.5	—	—
Hydrogen peroxide [7722-84-1]	1	1.5	—	—
Hydrogen selenide [7783-07-5], as Se	0.05	0.2	—	—

(e) The value is for total dust containing no asbestos and < 1% free silica.
(h) Total particulate.
(i) These TLVs are for the respirable fraction of dust for the substance listed. The concentration of respirable dust for the application of this limit is to be determined from the fraction passing a size-selector with the characteristics defined in the "c." paragraphs of Appendix E.

Capital letters, A, B, & D refer to Appendices; C denotes ceiling limit.

□ Identifies substances for which there are also BEIs (*see* BEI section). Substances identified in the BEI documentations for methemoglobin inducers and organophosphorus cholinesterase inhibitors are part of this notation.

■ Substance identified by other sources as a suspected or confirmed human carcinogen. See the compliation in the Appendix to the Documentation of TLVs, pp. A-5(86)–A-9(86).

• Substance for which OSHA and/or NIOSH has a Permissible Exposure Limit (PEL) or a Recommended Exposure Limit (REL) lower than the TLV.

‡ See Notice of Intended Changes.

* 1988-1989 Adoption.

Substance [CAS #]	ADOPTED VALUES TWA ppm[a]	TWA mg/m^3[b]	STEL ppm[a]	STEL mg/m^3[b]
•Hydrogen sulfide [7783-06-4]	10	14	15	21
Hydroquinone [123-31-9]	—	2	—	—
4-Hydroxy-4-methyl-2-pentanone, *see* Diacetone alcohol				
2-Hydroxypropyl acrylate [999-61-1] — Skin	0.5	3	—	—
Indene [95-13-6]	10	45	—	—
Indium [7440-74-6] & compounds, as In	—	0.1	—	—
Iodine [7553-56-2]	C 0.1	C 1	—	—
Iodoform [75-47-8]	0.6	10	—	—
Iron oxide fume (Fe_2O_3) [1309-37-1], as Fe	B2	5	—	—
Iron pentacarbonyl [13463-40-6], as Fe	0.1	0.8	0.2	1.6
Iron salts, soluble, as Fe	—	1	—	—
Isoamyl acetate [123-92-2]	100	525	—	—
Isoamyl alcohol [123-51-3]	100	360	125	450
‡Isobutyl acetate [110-19-0]	150	700	(187)	(875)
Isobutyl alcohol [78-83-1]	50	150	—	—
Isooctyl alcohol [26952-21-6]—Skin	50	270	—	—
Isophorone [78-59-1]	C 5	C 25	—	—
*Isophorone diisocyanate [4098-71-9]—Skin	0.005	0.045	—	—
Isopropoxyethanol [109-59-1]	25	105	—	—
Isopropyl acetate [108-21-4]	250	950	310	1,185
Isopropyl alcohol [67-63-0]	400	980	500	1,225
Isopropylamine [75-31-0]	5	12	10	24
N-Isopropylaniline [768-52-5]—Skin	2	10	—	—
Isopropyl ether [108-20-3]	250	1,050	310	1,320
•Isopropyl glycidyl ether (IGE) [4016-14-2]	50	240	75	360
Kaolin	—	10[e]	—	—
Ketene [463-51-4]	0.5	0.9	1.5	3
•□Lead [7439-92-1], inorg. dusts & fumes, as Pb	—	0.15	—	—
■Lead arsenate [3687-31-8], as $Pb_3(AsO_4)_2$	—	0.15	—	—
•■Lead chromate [7758-97-6], as Cr	—	0.05,A2	—	—
Limestone, *see* Calcium carbonate				
■Lindane [58-89-9]—Skin	—	0.5	—	—
Lithium hydride [7580-67-8]	—	0.025	—	—
L.P.G. (Liquified petroleum gas) [68476-85-7]	1,000	1,800	—	—
Magnesite [546-93-0]	—	10[e]	—	—
Magnesium oxide fume [1309-48-4]	—	10	—	—
□Malathion [121-75-5]—Skin	—	10	—	—
Maleic anhydride [108-31-6]	0.25	1	—	—
Manganese [7439-96-5], as Mn				
* Dust & compounds	—	5	—	—
Fume	—	1	—	3
Manganese cyclopentadienyl tricarbonyl [12079-65-1], as Mn—Skin	—	0.1	—	—
Marble, *see* Calcium carbonate				
Mercury [7439-97-6], as Hg—Skin				
Alkyl compounds	—	0.01	—	0.03
All forms except alkyl				
Vapor	—	0.05	—	—
• Aryl & inorganic compounds	—	0.1	—	—
•Mesityl oxide [141-79-7]	15	60	25	100

	ADOPTED VALUES			
	TWA		STEL	
Substance [CAS #]	ppm[a)]	mg/m3[b)]	ppm[a)]	mg/m3[b)]
Methacrylic acid [79-41-4]	20	70	—	—
Methane [74-82-8]	D	—	—	—
Methanethiol, *see* Methyl mercaptan				
Methanol, *see* Methyl alcohol				
□Methomyl [16752-77-5]	—	2.5	—	—
Methoxychlor [72-43-5]	—	10	—	—
•2-Methoxyethanol [109-86-4] — Skin	5	16	—	—
2-Methoxyethyl acetate [110-49-6] — Skin	5	24	—	—
4-Methoxyphenol [150-76-5]	—	5	—	—
Methyl acetate [79-20-9]	200	610	250	760
‡Methyl acetylene [74-99-7]	1,000	1,650	(1,250)	(2,040)
Methyl acetylene-propadiene mixture (MAPP)	1,000	1,800	1,250	2,250
Methyl acrylate [96-33-3] — Skin	10	35	—	—
Methylacrylonitrile [126-98-7] — Skin	1	3	—	—
Methylal [109-87-5]	1,000	3,100	—	—
Methyl alcohol [67-56-1] — Skin	200	260	250	310
Methylamine [74-89-5]	10	12	—	—
Methyl amyl alcohol, *see* Methyl isobutyl carbinol				
Methyl n-amyl ketone [110-43-0]	50	235	—	—
□N-Methyl aniline [100-61-8] — Skin	0.5	2	—	—
•■Methyl bromide [74-83-9] — Skin	5	20	—	—
•Methyl n-butyl ketone [591-78-6]	5	20	—	—
•■Methyl chloride [74-87-3]	50	105	100	205
•□Methyl chloroform [71-55-6]	350	1,900	450	2,450
Methyl 2-cyanoacrylate [137-05-3]	2	8	4	16
Methylcyclohexane [108-87-2]	400	1,600	—	—
Methylcyclohexanol [25639-42-3]	50	235	—	—
o-Methylcyclohexanone [583-60-8] — Skin	50	230	75	345
2-Methylcyclopentadienyl manganese tricarbonyl [12108-13-3], as Mn — Skin	—	0.2	—	—
□Methyl demeton [8022-00-2] — Skin	—	0.5	—	—
*Methylene bisphenyl isocyanate (MDI) [101-68-8]	0.005	0.055	—	—
*•Methylene chloride [75-09-2]	50,A2	175,A2	—	—
■□4,4'-Methylene bis(2-chloroaniline) [101-14-4] — Skin	0.02,A2	0.22,A2	—	—
*Methylene bis(4-cyclo-hexylisocyanate) [5124-30-1]	0.005	0.055	—	—
4,4'-Methylene dianiline [101-77-9] — Skin	0.1,A2	0.8,A2	—	—
Methyl ethyl ketone (MEK) [78-93-3]	200	590	300	885
□Methyl ethyl ketone peroxide [1338-23-4]	C 0.2	C 1.5	—	—
Methyl formate [107-31-3]	100	250	150	375
5-Methyl-3-heptanone, *see* Ethyl amyl ketone				
•■Methyl hydrazine [60-34-4] — Skin	C 0.2,A2	C 0.35,A2	—	—
•■Methyl iodide [74-88-4] — Skin	2,A2	10,A2	—	—
Methyl isoamyl ketone [110-12-3]	50	240	—	—
Methyl isobutyl carbinol [108-11-2] — Skin	25	100	40	165
Methyl isobutyl ketone [108-10-1]	50	205	75	300
Methyl isocyanate [624-83-9] — Skin	0.02	0.05	—	—
Methyl isopropyl ketone [563-80-4]	200	705	—	—
•Methyl mercaptan [74-93-1]	0.5	1	—	—
Methyl methacrylate [80-62-6]	100	410	—	—
□Methyl parathion [298-00-0] — Skin	—	0.2	—	—
•Methyl propyl ketone [107-87-9]	200	700	250	875
Methyl silicate [681-84-5]	1	6	—	—
α-Methyl styrene [98-83-9]	50	240	100	485
Metribuzin [21087-64-9]	—	5	—	—
□Mevinphos [7786-34-7] — Skin	0.01	0.1	0.03	0.3
Mica [12001-25-2]	—	3,[i] Respirable dust		
Mineral wool fiber	—	10[e]	—	—
Molybdenum [7439-98-7], as Mo				
Soluble compounds	—	5	—	—
Insoluble compounds	—	10	—	—
Monochlorobenzene, *see* Chlorobenzene				
Monocrotophos [6923-22-4]	—	0.25	—	—
Morpholine [110-91-8] — Skin	20	70	30	105
□Naled [300-76-5] — Skin	—	3	—	—
Naphthalene [91-20-3]	10	50	15	75
■□β-Naphthylamine [91-59-8]	—	A1	—	—
Neon [7440-01-9]	D	—	—	—
■Nickel [7440-02-0]				
Metal	—	1	—	—
Insoluable compounds, as Ni	—	1	—	—
• Soluble compounds, as Ni	—	0.1	—	—
•■Nickel carbonyl [13463-39-3], as Ni	0.05	0.1	—	—
•■Nickel sulfide roasting, fume & dust, as Ni	—	1,A1	—	—
Nicotine [54-11-5] — Skin	—	0.5	—	—
Nitrapyrin [1929-82-4]	—	10	—	20
Nitric acid [7697-37-2]	2	5	4	10
□Nitric oxide [10102-43-9]	25	30	—	—
□p-Nitroaniline [100-01-6] — Skin	—	3	—	—

(e) The value is for total dust containing no asbestos and < 1% free silica.

(i) These TLVs are for the respirable fraction of dust for the substance listed. The concentration of respirable dust for the application of this limit is to be determined from the fraction passing a size-selector with the characteristics defined in the "c." paragraphs of Appendix E.

Capital letters, A, B, & D refer to Appendices; C denotes ceiling limit.

□ Identifies substances for which there are also BEIs (*see* BEI section). Substances identified in the BEI documentations for methemoglobin inducers and organophosphorus cholinesterase inhibitors are part of this notation.

■ Substance identified by other sources as a suspected or confirmed human carcinogen. See the compliation in the Appendix to the Documentation of TLVs, pp. A-5(86)–A-9(86).

• Substance for which OSHA and/or NIOSH has a Permissible Exposure Limit (PEL) or a Recommended Exposure Limit (REL) lower than the TLV.

‡ See Notice of Intended Changes.

* 1988-1989 Adoption.

Substance [CAS #]	ADOPTED VALUES TWA ppm[a)]	mg/m³[b)]	STEL ppm[a)]	mg/m³[b)]
□Nitrobenzene [98-95-3] — Skin	1	5	—	—
*□p-Nitrochlorobenzene [100-00-5]—Skin	0.1	0.6	—	—
■4-Nitrodiphenyl [92-93-3]	—	A1	—	—
Nitroethane [79-24-3]	100	310	—	—
•Nitrogen dioxide [10102-44-0]	3	6	5	10
□Nitrogen trifluoride [7783-54-2]	10	30	—	—
Nitroglycerin (NG) [55-63-0] — Skin	0.05	0.5	—	—
Nitromethane [75-52-5]	100	250	—	—
1-Nitropropane [108-03-2]	25	90	—	—
•■□2-Nitropropane [79-46-9]	10, A2	35, A2	—	—
■N-Nitrosodimethylamine [62-75-9] — Skin	—	A2	—	—
Nitrotoluene [88-72-2; 99-08-1; 99-99-0] — Skin	2	11	—	—
Nitrotrichloromethane, *see* Chloropicrin				
Nonane [111-84-2]	200	1,050	—	—
Nuisance particulates	—	10(e)	—	—
Octachloronaphthalene [2234-13-1] — Skin	—	0.1	—	0.3
Octane [111-65-9]	300	1,450	375	1,800
Oil mist, mineral [8012-95-1]	—	5(j)	—	10
Osmium tetroxide [20816-12-0], as Os	0.0002	0.002	0.0006	0.006
Oxalic acid [144-62-7]	—	1	—	2
Oxygen difluoride [7783-41-7]	C 0.05	C 0.1	—	—
‡Ozone [10028-15-6]	(0.1)	(0.2)	(0.3)	(0.6)
Paraffin wax fume [8002-74-2]	—	2	—	—
Paraquat [4685-14-7], respirable sizes	—	0.1	—	—
•□Parathion [56-38-2] — Skin	—	0.1	—	—
Particulate polycyclic aromatic hydrocarbons (PPAH), *see* Coal tar pitch volatiles				
Pentaborane [19624-22-7]	0.005	0.01	0.015	0.03
Pentachloronaphthalene [1321-64-8]	—	0.5	—	—
□Pentachlorophenol [87-86-5] — Skin	—	0.5	—	—
Pentaerythritol [115-77-5]	—	10(e)	—	—
•Pentane [109-66-0]	600	1,800	750	2,250
2-Pentanone, *see* Methyl propyl ketone				
•□Perchloroethylene [127-18-4]	50	335	200	1,340
Perchloromethyl mercaptan [594-42-3]	0.1	0.8	—	—

(e) The value is for total dust containing no asbestos and < 1% free silica.
(j) As sampled by method that does not collect vapor.
Capital letters, A, B, & D refer to Appendices; C denotes ceiling limit.
□ Identifies substances for which there are also BEIs (*see* BEI section). Substances identified in the BEI documentations for methemoglobin inducers and organophosphorus cholinesterase inhibitors are part of this notation.
■ Substance identified by other sources as a suspected or confirmed human carcinogen. See the compliation in the Appendix to the Documentation of TLVs, pp. A-5(86)–A-9(86).
• Substance for which OSHA and/or NIOSH has a Permissible Exposure Limit (PEL) or a Recommended Exposure Limit (REL) lower than the TLV.
‡ See Notice of Intended Changes.
* 1988-1989 Adoption.

Substance [CAS #]	ADOPTED VALUES TWA ppm[a)]	mg/m³[b)]	STEL ppm[a)]	mg/m³[b)]
□Perchloryl fluoride [7616-94-6]	3	14	6	28
Precipitated silica, *see* Silica — Amorphous				
Perlite	—	10(e)	—	—
Petroleum distillates, *see* Gasoline; Stoddard solvent; VM&P naphtha				
Phenacyl chloride, *see* α-Chloroacetophenone				
□Phenol [108-95-2] — Skin	5	19	—	—
Phenothiazine [92-84-2] — Skin	—	5	—	—
■N-Phenyl-beta-naphthylamine [135-88-6]	A2	A2	—	—
p-Phenylene diamine [106-50-3] — Skin	—	0.1	—	—
Phenyl ether [101-84-8], vapor	1	7	2	14
Phenylethylene, *see* Styrene, monomer				
•■Phenyl glycidyl ether (PGE) [122-60-1]	1	6	—	—
•■Phenylhydrazine [100-63-0] — Skin	5,A2	20,A2	10,A2	45,A2
•Phenyl mercaptan [108-98-5]	0.5	2	—	—
Phenylphosphine [638-21-1]	C 0.05	C 0.25	—	—
Phorate [298-02-2] — Skin	—	0.05	—	0.2
Phosdrin, *see* Mevinphos				
Phosgene [75-44-5]	0.1	0.4	—	—
Phosphine [7803-51-2]	0.3	0.4	1	1
Phosphoric acid [7664-38-2]	—	1	—	3
Phosphorus (yellow) [7723-14-0]	—	0.1	—	—
‡Phosphorus oxychloride [10025-87-3]	0.1	0.6	(0.5)	(3)
Phosphorus pentachloride [10026-13-8]	0.1	1	—	—
Phosphorus pentasulfide [1314-80-3]	—	1	—	3
Phosphorus trichloride [7719-12-2]	0.2	1.5	0.5	3
Phthalic anhydride [85-44-9]	1	6	—	—
m-Phthalodinitrile [626-17-5]	—	5	—	—
‡Picloram [1918-02-1]	—	10	—	(20)
‡Picric acid [88-89-1] — Skin	—	0.1	—	(0.3)
Pindone [83-26-1]	—	0.1	—	—
Piperazine dihydrochloride [142-64-3]	—	5	—	—
2-Pivalyl-1,3-indandione, *see* Pindone				
Plaster of Paris, *see* Calcium sulfate				
Platinum [7440-06-4]				
Metal	—	1	—	—
Soluble salts, as Pt	—	0.002	—	—
Polychlorobiphenyls, *see* Chlorodiphenyls				
Polytetrafluoroethylene decomposition products	—	B1	—	—
Portland cement	—	10(e)	—	—
Potassium hydroxide [1310-58-3]	—	C 2	—	—
Propane [74-98-6]	D	—	—	—
■Propane sultone [1120-71-4]	A2	A2	—	—
Propargyl alcohol [107-19-7] — Skin	1	2	—	—
■•β-Propiolactone [57-57-8]	0.5,A2	1.5,A2	—	—
‡Propionic acid [79-09-4]	10	30	(15)	(45)
Propoxur [114-26-1]	—	0.5	—	—
n-Propyl acetate [109-60-4]	200	840	250	1,050
Propyl alcohol [71-23-8] — Skin	200	500	250	625
Propylene [115-07-1]	D	—	—	—

Substance [CAS #]	TWA ppm[a)]	TWA mg/m^3[b)]	STEL ppm[a)]	STEL mg/m^3[b)]
Propylene dichloride [78-87-5]	75	350	110	510
□Propylene glycol dinitrate [6423-43-4]—Skin	0.05	0.3	—	—
Propylene glycol mono-methyl ether [107-98-2]	100	360	150	540
■Propylene imine [75-55-8]—Skin	2,A2	5,A2	—	—
■Propylene oxide [75-56-9]	20	50	—	—
□n-Propyl nitrate [627-13-4]	25	105	40	170
Propyne, *see* Methyl acetylene				
Pyrethrum [8003-34-7]	—	5	—	—
Pyridine [110-86-1]	5	15	—	—
Pyrocatechol, *see* Catechol				
Quartz, *see* Silica—Crystaline				
Quinone [106-51-4]	0.1	0.4	—	—
RDX, *see* Cyclonite				
Resorcinol [108-46-3]	10	45	20	90
•Rhodium [7440-16-6]				
•Metal	—	1	—	—
•Insoluble compounds, as Rh	—	1	—	—
•Soluble compounds, as Rh	—	0.01	—	—
Ronnel [299-84-3]	—	10	—	—
Rosin core solder pyrolysis products, as formaldehyde	—	0.1	—	—
Rotenone (commercial) [83-79-4]	—	5	—	—
Rouge	—	10[(e)]	—	—
•Rubber solvent (Naphtha)	400	1,600	—	—
Selenium compounds [7782-49-2], as Se	—	0.2	—	—
Selenium hexafluoride [7783-79-1], as Se	0.05	0.2	—	—
Sesone [136-78-7]	—	10	—	—
Silane, *see* Silicon tetrahydride				
Silica—Amorphous				
Diatomaceous earth (uncalcined) [61790-53-2]	—	10[(e)]	—	—
Precipitated silica	—	10[(e)]	—	—
Silica gel	—	10[(e)]	—	—
Silica—Crystalline				
Cristobalite [14464-46-1]	—	0.05,[(i)] Respirable dust		
•Quartz [14808-60-7]	—	0.1,[(i)] Respirable dust		
•Silica, fused [60676-86-0]	—	0.1,[(i)] Respirable dust		
Tridymite [15468-32-3]	—	0.05,[(i)] Respirable dust		
•Tripoli [1317-95-9]	—	0.1,[(i)] of contained respirable quartz		
Silica, fused, *see* Silica—Crystaline				
Silica, gel, *see* Silica—Amorphous				
Silicon [7440-21-3]	—	10[(e)]	—	—
Silicon carbide [409-21-2]	—	10[(e)]	—	—
Silicon tetrahydride [7803-62-5]	5	7	—	—
Silver [7440-22-4]				
Metal	—	0.1	—	—
Soluble compounds, as Ag	—	0.01	—	—
Soapstone				
Respirable dust	—	3[(i)]	—	—
Total dust	—	6[(e)]	—	—
Sodium azide [26628-22-8]	C 0.1	C 0.3	—	—
Sodium bisulfite [7631-90-5]	—	5	—	—
Sodium 2,4-dichloro-phenoxyethyl sulfate, *see* Sesone				
Sodium fluoroacetate [62-74-8] — Skin	—	0.05	—	0.15

Substance [CAS #]	TWA ppm[a)]	TWA mg/m^3[b)]	STEL ppm[a)]	STEL mg/m^3[b)]
Sodium hydroxide [1310-73-2]	—	C 2	—	—
Sodium metabisulfite [7681-57-4]	—	5	—	—
Starch [9005-25-8]	—	10[(e)]	—	—
*Stearates[(n)]	—	10[(e)]	—	—
Stibine [7803-52-3]	0.1	0.5	—	—
Stoddard solvent [8052-41-3]	100	525	—	—
Strychnine [57-24-9]	—	0.15	—	—
□Styrene, monomer [100-42-5] — Skin	50	215	100	425
Subtilisins [1395-21-7] (Proteolytic enzymes as 100% pure crystalline enzyme)	—	C0.00006[(k)]	—	—
Sucrose [57-50-1]	—	10[(e)]	—	—
□Sulfotep [3689-24-5]—Skin	—	0.2	—	—
•Sulfur dioxide [7446-09-5]	2	5	5	10
Sulfur hexafluoride [2551-62-4]	1,000	6,000	—	—
‡Sulfuric acid [7664-93-9]	—	1	—	(—)
Sulfur monochloride [10025-67-9]	C 1	C 6	—	—
Sulfur pentafluoride [5714-22-7]	C 0.01	C 0.1	—	—
Sulfur tetrafluoride [7783-60-0]	C 0.1	C 0.4	—	—
Sulfuryl fluoride [2699-79-8]	5	20	10	40
Sulprofos [35400-43-2]	—	1	—	—
Systox, *see* Demeton				
2,4,5-T [93-76-5]	—	10	—	—
Talc (containing no abestos fibers) [14807-96-6]	—	2,[(i)] Respirable dust		
Talc (containing asbestos fibers)	Use asbestos TLV-TWA.[(i)]			
§Tantalum [7440-25-7], metal & oxide dusts	—	5	—	—
TEDP, *see* Sulfotep				
Tellurium & compounds [13494-80-9], as Te	—	0.1	—	—
Tellurium hexafluoride [7783-80-4], as Te	0.02	0.2	—	—
□Temephos [3383-96-8]	—	10	—	—
□TEPP [107-49-3] — Skin	0.004	0.05	—	—
Terphenyls [26140-60-3]	C 0.5	C 5	—	—

(e) The value is for total dust containing no asbestos and < 1% free silica.

(i) These TLVs are for the respirable fraction of dust for the substance listed. The concentration of respirable dust for the application of this limit is to be determined from the fraction passing a size-selector with the characteristics defined in the "c." paragraphs of Appendix E.

(k) Based on "high volume" sampling.

(l) However, should not exceed 2 mg/m^3 respirable dust.

(n) Does not include stearates of toxic metals.

Capital letters, A, B, & D refer to Appendices; C denotes ceiling limit.

□ Identifies substances for which there are also BEIs (*see* BEI section). Substances identified in the BEI documentations for methemoglobin inducers and organophosphorus cholinesterase inhibitors are part of this notation.

■ Substance identified by other sources as a suspected or confirmed human carcinogen. See the compliation in the Appendix to the Documentation of TLVs, pp. A-5(86)–A-9(86).

• Substance for which OSHA and/or NIOSH has a Permissible Exposure Limit (PEL) or a Recommended Exposure Limit (REL) lower than the TLV.

‡ See Notice of Intended Changes.

§ STEL deleted, TWA retained.

* 1988-1989 Adoption.

Substance [CAS #]	ADOPTED VALUES TWA ppm[a)]	TWA mg/m³[b)]	STEL ppm[a)]	STEL mg/m³[b)]
1,1,1,2-Tetrachloro-2,2-difluoroethane [76-11-9]	500	4,170	—	—
1,1,2,2-Tetrochloro-1,2-difluoroethane [76-12-0]	500	4,170	—	—
•■1,1,2,2-Tetrachloroethane [79-34-5]—Skin	1	7	—	—
Tetrachloroethylene, *see* Perchloroethylene				
Tetrachloromethane, *see* Carbon tetrachloride				
Tetrachloronaphthalene [1335-88-2]	—	2	—	—
Tetraethyl lead [78-00-2], as Pb—Skin	—	0.1(m)	—	—
Tetrahydrofuran [109-99-9]	200	590	250	735
•Tetramethyl lead [75-74-1], as Pb—Skin	—	0.15(m)	—	—
Tetramethyl succinonitrile [3333-52-6]—Skin	0.5	3	—	—
□Tetranitromethane [509-14-8]	1	8	—	—
Tetrasodium pyrophosphate [7722-88-5]	—	5	—	—
Tetryl [479-45-8]	—	1.5	—	—
Thallium [7440-28-0] Soluble compounds, as Tl—Skin	—	0.1	—	—
4,4'-Thiobis(6-tert-butyl-m-cresol) [96-69-5]	—	10	—	—
Thioglycolic acid [68-11-1]—Skin	1	4	—	—
Thionyl chloride [7719-09-7]	C 1	C 5	—	—
‡Thiram [137-26-8]	—	(5)	—	—
Tin [7440-31-5] Metal	—	2	—	—
Oxide & inorganic compounds, except SnH_4, as Sn	—	2	—	—
‡ Organic compounds, as Sn—Skin	—	0.1	—	(—)
Titanium dioxide [13463-67-7]	—	10(e)	—	—
■o-Tolidine [119-93-7]—Skin	A2	A2	—	—
□Toluene [108-88-3]	100	375	150	560
Toluene-2,4-diisocyanate (TDI) [584-84-9]	0.005	0.04	0.02	0.15
■□o-Toluidine [95-53-4]—Skin	2,A2	9,A2	—	—
□m-Toluidine [108-44-1]—Skin	2	9	—	—
□p-Toluidine [106-49-0]—Skin	2,A2	9,A2	—	—
Toluol, *see* Toluene				
Toxaphene, *see* Chlorinated camphene				
Tributyl phosphate [126-73-8]	0.2	2.5	—	—
Trichloroacetic acid [76-03-9]	1	7	—	—
1,2,4-Trichlorobenzene [120-82-1]	C 5	C 40	—	—
1,1,1-Trichloroethane, *see* Methyl chloroform				
•■1,1,2-Trichloroethane [79-00-5]—Skin	10	45	—	—
•■□Trichloroethylene [79-01-6]	50	270	200	1,080
Trichlorofluoromethane [75-69-4]	C 1,000	C 5,600	—	—
Trichloromethane, *see* Chloroform				
Trichloronaphthalene [1321-65-9]—Skin	—	5	—	—
Trichloronitromethane, *see* Chloropicrin				
1,2,3-Trichloropropane [96-18-4]—Skin	10	60	—	—
1,1,2-Trichloro-1,2,2-trifluoroethane [76-13-1]	1,000	7,600	1,250	9,500
Tricyclohexyltin hydroxide, *see* Cyhexatin				
Tridymite, *see* Silica—Crystaline				
Triethylamine [121-44-8]	10	40	15	60
Trifluorobromomethane [75-63-8]	1,000	6,100	—	—
Trimellitic anhydride [552-30-7]	0.005	0.04	—	—
Trimethylamine [75-50-3]	10	24	15	36
Trimethyl benzene [25551-13-7]	25	125	—	—
Trimethyl phosphite [121-45-9]	2	10	—	—
2,4,6-Trinitrophenol, *see* Picric acid				
2,4,6-Trinitrophenylmethylnitramine, *see* Tetryl				
2,4,6-Trinitrotoluene (TNT) [118-96-7]—Skin	—	0.5	—	—
Triorthocresyl phosphate [78-30-8]—Skin	—	0.1	—	—
Triphenyl amine [603-34-9]	—	5	—	—
Triphenyl phosphate [115-86-6]	—	3	—	—
Tripoli, *see* Silica—Crystaline				
Tungsten [7440-33-7], as W Insoluble compounds	—	5	—	10
Soluble compounds	—	1	—	3
Turpentine [8006-64-2]	100	560	—	—
•Uranium (natural) [7440-61-1] Soluble & insoluble compounds, as U	—	0.2	—	0.6
n-Valeraldehyde [110-62-3]	50	175	—	—
Vanadium, as V_2O_5 [1314-62-1] Respirable dust & fume	—	0.05	—	—
Vegetable oil mists(o)	—	10(e)	—	—
•Vinyl acetate [108-05-4]	10	30	20	60
Vinyl benzene, *see* Styrene				
•■Vinyl bromide [593-60-2]	5,A2	20,A2	—	—
•■Vinyl chloride [75-01-4]	5,A1	10,A1	—	—
Vinyl cyanide, *see* Acrylonitrile				
Vinyl cyclohexene dioxide [106-87-6]—Skin	10,A2	60,A2	—	—
•■Vinylidene chloride [75-35-4]	5	20	20	80
Vinyl toluene [25013-15-4]	50	240	100	485
VM & P Naphtha [8032-32-4]	300	1,350	—	—
Warfarin [81-81-2]	—	0.1	—	—
Welding fumes (NOC+)	—	5,B2	—	—

(e) The value is for total dust containing no asbestos and < 1% free silica.
(m) For control of general room air, biologic monitoring is essential for personnel control.
(o) Except caster, cashew nut, or similar irritant oils.
Capital letters, A, B, & D refer to Appendices; C denotes ceiling limit.
□ Identifies substances for which there are also BEIs (*see* BEI section). Substances identified in the BEI documentations for methemoglobin inducers and organophosphorus cholinesterase inhibitors are part of this notation.
■ Substance identified by other sources as a suspected or confirmed human carcinogen. See the compliation in the Appendix to the Documentation of TLVs, pp. A-5(86)–A-9(86).
• Substance for which OSHA and/or NIOSH has a Permissible Exposure Limit (PEL) or a Recommended Exposure Limit (REL) lower than the TLV.
‡ See Notice of Intended Changes.
† NOC = not otherwise classified.

Substance [CAS #]	ADOPTED VALUES TWA ppm[a]	mg/m^3[b]	STEL ppm[a]	mg/m^3[b]
▪Wood dust (certain hard woods as beech & oak) .	—	1	—	—
▪ Soft wood	—	5	—	10
□Xylene [1330-20-7; 95-47-6; 108-38-3; 106-42-3] (o-, m-, p-isomers)	100	435	150	655
m-Xylene α,α'-diamine [1477-55-0]—Skin	—	C 0.1	—	—
‡ Xylidine [1300-73-8]— Skin	(2)	(10)	—	—
*Yttrium [7440-65-5] metal & compounds, as Y	—	1	—	
Zinc chloride fume [7646-85-7]	—	1	—	2
*•▪Zinc chromates [13530-65-9; 11103-86-9; 37300-23-5], as Cr	—	0.01,A1	—	—
Zinc oxide [1314-13-2] Fume	—	5	—	10
Dust	—	10[e]	—	—
Zirconium compounds [7440-67-2], as Zr	—	5	—	10

Dusts: Individual dust entries appear in the main alphabetical listing only. The separate "Dusts" section was deleted this year in favor of the single entry per substance.

Radioactivity: See Physical Agents section on Ionizing Radiation.

(e) The value is for total dust containing no asbestos and < 1% free silica.
Capital letters, A, B, & D refer to Appendices; C denotes ceiling limit.
□ Identifies substances for which there are also BEIs (*see* BEI section). Substances identified in the BEI documentations for methemoglobin inducers and organophosphorus cholinesterase inhibitors are part of this notation.
▪ Substance identified by other sources as a suspected or confirmed human carcinogen. See the compliation in the Appendix to the Documentation of TLVs, pp. A-5(86)–A-9(86).
• Substance for which OSHA and/or NIOSH has a Permissible Exposure Limit (PEL) or a Recommended Exposure Limit (REL) lower than the TLV.
* 1988-1989 Adoption.

NOTICE OF INTENDED CHANGES (for 1988-89)

These substances, with their corresponding values, comprise those for which either a limit has been proposed for the first time, or for which a change in the "Adopted" listing has been proposed. In both cases, the proposed limits should be considered trial limits that will remain in the listing for a period of at least two years. If, after two years no evidence comes to light that questions the appropriateness of the values herein, the values will be reconsidered for the "Adopted" list. Documentation is available for each of these substances.

Substance [CAS #]	TWA ppm[a]	mg/m^3[b]	STEL ppm[a]	mg/m^3[b]
Acrylic acid [79-10-7]— Skin	2	6	—	—
Cadmium [7440-43-9] and compounds, as Cd	—	0.01,A2	—	—
Caprolactam [105-60-2] Vapor & aerosol	0.25	1	—	—
Chlorine [7782-50-5]	0.5	1.5	1	3
Chloroacetone [78-95-5]— Skin	C 1	C 4	—	—
†Dibutyl phenyl phosphate [2528-36-1] — Skin	0.3	3.5	—	—
†Ethyl acrylate [140-88-5] .	5,A2	20,A2	15,A2	61,A2
Formic acid [64-18-6]	5	9	10	18
Graphite (all forms)	—	10[e]	—	—
Hexachloroethane [67-72-1]	1	10	—	—
Nitrous oxide [10024-97-2]	50	91	—	—
Ozone [10028-15-6]	C 0.1	C 0.2	—	—
†Picric acid [88-89-1]	—	0.1	—	—
Sulfuric acid [7664-93-9] .	—	1	—	3
Thiram [137-26-8]	—	1	—	—
†Tin, organic compounds, as Sn—Skin	—	0.1	—	0.2
†Xylidine (mixed isomers) [1300-73-8] — Skin	0.5,A2	2.5,A2	—	—

DELETE* THE SHORT-TERM EXPOSURE LIMITS (TLVs-STELs) FOR THE FOLLOWING SUBSTANCES:

sec-Butyl alcohol
Chlordane
Chlorinated diphenyl oxide
Chlorobromomethane
Chlorodifluoromethane
Chlorodiphenyl (42% chlorine)
Chlorodiphenyl (54% chlorine)
Chloropicrin
o-Chlorotoluene
Chlorpyrifos
Clopidol
Crufomate
Cyclonite
Isobutyl acetate
Methyl acetylene
Phosphorus oxychloride
Picloram
Propionic acid

*As per the individual documentation appearing in the *Documentation of the Threshold Limit Values and Biological Exposure Indices*, 5th ed. (1986).
(a) Parts of vapor or gas per million parts of contaminated air by volume at 25°C and 760 torr.
(b) Milligrams of substance per cubic meter of air. When entry is in this column only, the value is exact; when listed with a ppm entry, it is approximate.
(e) The value is for total dust containing no asbestos and <1% free silica.
† 1988-1989 Revision or Addition.
Capital letters A, B, & D refer to Appendices; C denotes ceiling limit.

ADOPTED APPENDICES

APPENDIX A
Carcinogens

The Chemical Substances Threshold Limit Values Committee classifies certain substances found in the occupational environment as either confirmed or suspected human carcinogens. The present listing of substances which have been identified as carcinogens takes two forms: those for which a TLV has been assigned and those for which environmental and exposure conditions have not been sufficiently defined to assign a TLV. Where a TLV has been assigned, it does not necessarily imply the existence of a biological threshold; however, if exposures are controlled to this level, we would not expect to see a measurable increase in cancer incidence or mortality.

The TLV Committee considers information from the following kinds of studies to be indicators of a substance's potential to be a carcinogen in humans: epidemiology studies, toxicology studies, and, to a lesser extent, case histories. Scientific debate over the existence of biological thresholds for carcinogens is unlikely to be resolved in the near future. Because of the long latent period for many carcinogens, and for ethical reasons, it is often impossible to base timely risk-management decisions on results from human studies.

In order to recognize the qualitative differences in research results, two categories of carcinogens are designated in this booklet:

A1 — Confirmed Human Carcinogens

and

A2 — Suspected Human Carcinogens

Exposures to carcinogens must be kept to a minimum. Workers exposed to A1 carcinogens without a TLV should be properly equipped to virtually eliminate all exposure to the carcinogen. For A1 carcinogens with a TLV and for A2 carcinogens, worker exposure by all routes should be carefully controlled to levels consistent with the experimental and human experience data. Please see the *Documentation of the Threshold Limit Values* for a more complete description and derivation of these designations.

APPENDIX B
Substances of Variable Composition

B1. *Polytetrafluoroethylene* decomposition products.* Thermal decomposition of the fluorocarbon chain in air leads to the formation of oxidized products containing carbon, fluorine and oxygen. Because these products decompose in part by hydrolysis in alkaline solution, they can be quantitatively determined in air as fluoride to provide an index of exposure. No TLV is recommended pending determination of the toxicity of the products, but air concentration should be minimal.

B2. *Welding Fumes — Total Particulate (NOC)*†
TLV-TWA, 5 mg/m³

Welding fumes cannot be classified simply. The composition and quantity of both are dependent on the alloy being welded and the process and electrodes used. Reliable analysis of fumes cannot be made without considering the nature of the welding process and system being examined; reactive metals and alloys such as aluminum and titanium are arc-welded in a protective, inert atmosphere such as argon. These arcs create relatively little fume, but an intense radiation which can produce ozone. Similar processes are used to arc-weld steels, also creating a relatively low level of fumes. Ferrous alloys also are arc-welded in oxidizing environments which generate considerable fume, and can produce carbon monoxide instead of ozone. Such fumes generally are composed of discreet particles of amorphous slags containing iron, manganese, silicon and other metallic constituents depending on the alloy system involved. Chromium and nickel compounds are found in fumes when stainless steels are arc-welded. Some coated and flux-cored electrodes are formulated with fluorides and the fumes associated with them can contain significantly more fluorides than oxides. Because of the above factors, arc-welding fumes frequently must be tested for individual constituents which are likely to be present to determine whether specific TLVs are exceeded. Conclusions based on total fume concentration are generally adequate if no toxic elements are present in welding rod, metal, or metal coating and conditions are not conducive to the formation of toxic gases.

Most welding, even with primitive ventilation, does not produce exposures inside the welding helmet above 5 mg/m³. That which does, should be controlled.

* Trade Names: Algoflon, Fluon, Teflon, Tetran.
† Not otherwise classified (NOC).

APPENDIX C
Threshold Limit Values for Mixtures

When two or more hazardous substances, which act upon the same organ system, are present, their combined effect, rather than that of either individually, should be given primary consideration. In the absence of information to the contrary, the effects of the different hazards should be considered as additive. That is, if the sum of the following fractions,

$$\frac{C_1}{T_1} + \frac{C_2}{T_2} + \ldots \frac{C_n}{T_n}$$

exceeds unity, then the threshold limit of the mixture should be considered as being exceeded. C_1 indicates the observed atmospheric concentration, and T_1 the corresponding threshold limit (*see* Example A.1 and B.1).

Exceptions to the above rule may be made when there is a good reason to believe that the chief effects of the different harmful substances are not in fact additive, but *independent* as when purely local effects on different organs of the body are produced by the various components of the mixture. In such cases the threshold limit ordinarily is exceeded only when at least one member of the series (C_1/T_1 + or + C_2/T_2, etc.) itself has a value exceeding unity (*see* Example B.1).

Synergistic action or potentiation may occur with some combinations of atmospheric contaminants. Such cases at present must be determined individually. Potentiating or synergistic agents are not necessarily harmful by themselves. Potentiating effects of exposure to such agents by routes other than that of inhalation is also possible, e.g., imbibed alcohol and inhaled narcotic (trichloroethylene). Potentiation is characteristically exhibited at high concentrations, less probably at low.

When a given operation or process characteristically emits a number of harmful dusts, fumes, vapors or gases, it will frequently be only feasible to attempt to evaluate the hazard by measurement of a single substance. In such cases, the threshold limit used for this substance should be reduced by a suitable factor, the magnitude of which will depend on the number, toxicity and relative quantity of the other contaminants ordinarily present.

Examples of processes which are typically associated with two or more harmful atmospheric contaminants are welding, automobile repair, blasting, painting, lacquering, certain foundry operations, diesel exhausts, etc.

Examples of TLVs for Mixtures

A. *Additive effects.* The following formulae apply only when the components in a mixture have similar toxicologic effects; they should not be used for mixtures with widely differing reactivities, e.g., hydrogen cyanide and sulfur dioxide. In such case the formula for Independent Effects (B) should be used.

1. General case, where air is analyzed for each component, the TLV of mixture =

$$\frac{C_1}{T_1} + \frac{C_2}{T_2} + \frac{C_3}{T_3} + \ldots = 1$$

Note: It is essential that the atmosphere be analyzed both qualitatively and quantitatively for each component present, in order to evaluate compliance or non-compliance with this calculated TLV.

Example A.1: Air contains 400 ppm of acetone (TLV, 750 ppm), 150 ppm of sec-butyl acetate (TLV, 200 ppm) and 100 ppm of methyl ethel ketone (TLV, 200 ppm).

Atmospheric concentration of mixture = 400 + 150 + 100 = 650 ppm of mixture.

$$\frac{400}{750} + \frac{150}{200} + \frac{100}{200} = 0.53 + 0.75 + 0.5 = 1.78$$

Threshold Limit is exceeded.

2. Special case when the source of contaminant is a liquid mixture and the atmospheric composition is *assumed* to be similar to that of the original material, e.g., on a time-weighted average exposure basis, all of the liquid (solvent) mixture eventually evaporates. When the percent composition (by weight) of the liquid mixture is known, the TLVs of the constituents must be listed in mg/m³. TLV of mixture =

$$\frac{1}{\frac{f_a}{TLV_a} + \frac{f_b}{TLV_b} + \frac{f_c}{TLV_c} + \ldots \frac{f_n}{TLV_n}}$$

Note: In order to evaluate compliance with this TLV, field sampling instruments should be calibrated, in the laboratory, for response to this specific quantitative and qualitative air-vapor mixture, and also to fractional concentrations of this mixture, e.g., 1/2 the TLV; 1/10 the TLV; 2 × the TLV; 10 × the TLV; etc.)

Example A.2: Liquid contains (by weight):

50% heptane: TLV = 400 ppm or 1600 mg/m³
1 mg/m³ ≡ 0.25 ppm

30% methyl chloroform: TLV = 350 ppm or 1900 mg/m³
1 mg/m³ ≡ 0.18 ppm

20% perchloroethylene: TLV = 50 ppm or 335 mg/m³
1 mg/m³ ≡ 0.15 ppm

$$\text{TLV of Mixture} = \frac{1}{\frac{0.5}{1600} + \frac{0.3}{1900} + \frac{0.2}{335}}$$

$$= \frac{1}{0.00031 + 0.00016 + 0.0006}$$

$$= \frac{1}{0.00107} = 935 \text{ mg/m}^3$$

of this mixture

50% or (935)(0.5) = 468 mg/m³ is heptane
30% or (935)(0.3) = 281 mg/m³ is methyl chloroform
20% or (935)(0.2) = 187 mg/m³ is perchloroethylene

These values can be converted to ppm as follows:

heptane: 468 mg/m³ × 0.25 = 117 ppm
methyl chloroform: 281 mg/m³ × 0.18 = 51 ppm
perchloroethylene: 187 mg/m³ × 0.15 = 29 ppm

TLV of mixture = 117 + 51 + 29 = 197 ppm, or 935 mg/m³

B. *Independent effects.* TLV for mixture =

$$\frac{C_1}{T_1} = 1; \quad \frac{C_2}{T_2} = 1 \quad \frac{C_3}{T_3} = 1; \text{ etc.}$$

Example B.1: Air contains 0.15 mg/m³ of lead (TLV, 0.15) and 0.7 mg/m³ of sulfuric acid (TLV, 1).

$$\frac{0.15}{0.15} = 1; \qquad \frac{0.7}{1} = 0.7$$

Threshold limit is not exceeded.

C. *TLV for mixtures of mineral dusts.* For mixtures of biologically active mineral dusts the general formula for mixtures given in A.2 may be used.

APPENDIX D
Some Simple Asphyxiants*

Acetylene	Hydrogen
Argon	Methane
Ethane	Neon
Ethylene	Propane
Helium	Propylene

* As defined in the Introduction.

APPENDIX E
Particle Size-Selective Sampling Criteria for Airborne Particulate Matter

For chemical substances present in inhaled air as suspensions of solid particles or droplets, the potential hazard depends on particle size as well as mass concentration because of: 1) effects of particle size on deposition site within the respiratory tract, and 2) the tendency for many occupational diseases to be associated with material deposited in particular regions of the respiratory tract.

ACGIH has recommended particle size-selective TLVs for crystalline silica for many years in recognition of the well established association between silicosis and respirable mass concentrations. It now has embarked on a re-examination of other chemical substances encountered in particulate form in occupational environments with the objective of defining: 1) the size-fraction most closely associated for each substance with the health effect of concern, and 2) the mass concentration within that size fraction which should represent the TLV.

The Particle Size-Selective TLVs (PSS-TLVs) will be expressed in three forms, e.g.,

a. *Inspirable Particulate Mass TLVs (IPM-TLVs)* for those materials which are hazardous when deposited anywhere in the respiratory tract.
b. *Thoracic Particulate Mass TLVs (TPM-TLVs)* for those materials which are hazardous when deposited anywhere within the lung airways and the gas-exchange region.
c. *Respirable Particulate Mass TLVs (RPM-TLVs)* for those materials which are hazardous when deposited in the gas-exchange region.

The three particulate mass fractions described above are defined in quantitative terms as follows:

a. Inspirable Particulate Mass consists of those particles that are captured according to the following collection efficiency regardless of sampler orientation with respect to wind direction:

$$E = 50(1 + \exp[-0.06\, d_a]) \pm 10;$$
$$\text{for } 0 < d_a \leq 100\ \mu m$$

Collection characteristics for $d_a > 100\ \mu m$ are presently unknown. E is collection efficiency in percent and d_a is aerodynamic diameter in μm.

b. Thoracic Particulate Mass consists of those particles that penetrate a separator whose size collection efficiency is described by a cumulative lognormal function with a median aerodynamic diameter of 10 μm ±1.0 μm and with a geometric standard deviation of 1.5 (±0.1).
c. Respirable Particulate Mass consists of those particles that penetrate a separator whose size collection efficiency is described by a cumulative lognormal function with a median aerodynamic diameter of 3.5 μm ±0.3 μm and with a geometric standard deviation of 1.5 (±0.1). This incorporates and clarifies the previous ACGIH Respirable Dust Sampling Criteria.

These definitions provide a range of acceptable performance for each type of size-selective sampler. Further information is available on the background and performance criteria for these particle size-selective sampling recommendations.[1]

References

1. **ACGIH:** *Particle Size-Selective Sampling in the Workplace,* 80 pp. Cincinnati, Ohio (1984).

APPENDIX B

PHYSICAL CONSTANTS OF SELECTED MATERIALS

Substance	Formula	Molecular Weight	Specific Gravity	Flash Point F: Closed Cup	Flash Point F: Open Cup	Explosive Limits (Volume Percent): Lower	Explosive Limits (Volume Percent): Upper
Acetaldehyde	CH_3CHO	44.05	0.821	−17	—	3.97	57.0
Acetic Acid	CH_3COOH	60.05	1.049	104	110	5.40	—
Acetic Anhydride	$(CH_3CO)_2O$	102.09	1.082	121	130	2.67	10.13
Acetone	CH_3COCH_3	58.08	0.792	0	15	2.55	12.80
Acrolein	$CH_2:CHCHO$	56.06	0.841	Gas		Unstable	
Acrylonitrile	$CH_2:CHCN$	53.06	0.806	—	32	3.05	17.0
Ammonia	NH_3	17.03	0.597	Gas		15.50	27.0
Amyl Acetate	$CH_3CO_2C_5H_{11}$	130.18	0.879	76	80	1.10	—
iso-Amyl Alcohol	$(CH_3)_2CHCH_2CH_2OH$	88.15	0.812	109	115	1.20	—
Aniline	$C_6H_5NH_2$	93.12	1.022	168	—	—	—
Arsine	AsH_3	77.93	2.695 (A)	Gas		—	—
Benzene	C_6H_6	78.11	0.879	12	—	1.40	7.10
Bromine	Br_2	159.83	3.119	—	—	—	—
Butane	$CH_3(CH_2)_2CH_3$	58.12	2.085	Gas		1.86	8.41
1,3-Butadiene	$(CH_2:CH)_2$	54.09	0.621	Gas		2.00	11.50
n-Butanol	$C_2H_5CH_2CH_2OH$	74.12	0.810	84	110	1.45	11.25
2-Butanone (Methyl ethyl ketone)	$CH_3COC_2H_5$	72.10	0.805	30	—	1.81	9.50
n-Butyl Acetate	$CH_3CO_2C_4H_9$	116.16	0.882	72	90	1.39	7.55
Butyl "Cellosolve"	$C_4H_9OCH_2CH_2OH$	118.17	0.903	141	165	—	—
Carbon Dioxide	CO_2	44.01	1.53	—	—	—	—
Carbon Disulphide	CS_2	76.13	1.263	−22	—	1.25	50.0
Carbon Monoxide	CO	28.10	0.968	Gas		12.5	74.2
Carbon Tetrachloride	CCl_4	153.84	1.595	Nonflammable			
Cellosolve	$C_2H_5O(CH_2)_2OH$	90.12	0.931	104	120	2.6	15.7
Cellosolve Acetate	$CH_3CO_2C_4H_9O$	132.16	0.975	124	135	1.71	—
Chlorine	Cl_2	70.91	3.214	Gas		—	—
2-Chlorobutadiene	$CH_2:CClCHCH_2$	88.54	0.958	—	—	—	—
Chloroform	$CHCl_3$	119.39	1.478	Nonflammable			
1-Chloro-1-nitropropane	NO_3ClC_3H6	139.54	1.209	144	—	—	—
Cyclohexane	C_6H_{12}	84.16	0.779	1	—	1.26	7.75
Cyclohexanol	$CH_2(CH_2)_4CHOH$	100.16	0.962	154	—	—	—
Cyclohexanone	$CH_2(CH_2)_4CO$	98.14	0.948	147	—	—	—
Cyclohexene	$CH_2(CH_2)_3CH:CH$	82.14	0.810	—	—	—	—
Cyclopropane	$CH_2CH_2CH_2$	42.08	0.720	Gas		2.40	10.40
o-Dichlorobenzene	$Cl_2C_6H_4$	147.01	1.305	151	165	—	—
Dichlorodifluoromethane	CCl_2F_2	120.92	1.486	Nonflammable			
1,1-Dichlorethane	CH_2CHCl_2	98.97	1.175	—	—	—	—
1,2-Dichloroethane (Ethylene dichloride)	$ClCH_2CH_2Cl$	98.97	1.257	56	65	6.2	15.9
1,2-Dichloroethylene	$ClCHCHCl$	96.95	1.291	43	—	9.7	12.8
Dichloroethylether	$ClCH_2CHClC_2H_5$	143.02	1.222	131	180	—	—
Dichloromethane	H_2CCl_2	84.94	1.336	—	—	—	—
Dichloromonofluoromethane	$HCCl_2F$	102.93	1.426	—	—	—	—
1,1-Dichloro-1-nitroethane	$H_3C_2Cl_2NO_3$	143.97	1.692	—	168	—	—
1,2-Dichloropropane	$CH_3CHClCH_2Cl$	112.99	1.159	59	65	3.4	14.5
Dichlorotetrafluoroethane	$CClF_2CClF_2$	170.93	1.433	Nonflammable			
Dimethylaniline	$(CH_3)_2NC_6H_5$	121.18	0.956	145	170	—	—
Dimethylsulfate	$(CH_3)_2SO_4$	126.13	1.332	182	240	—	—
Dioxane	$O(CH_2)_4O$	88.10	1.034	—	35	—	—
Ethyl Acetate	$CH_3CO_2C_2H_5$	88.10	0.901	24	30	2.18	11.4
Ethyl Alcohol	C_2H_5OH	46.07	0.789	55	—	3.28	18.95
Ethyl Benzene	$C_6H_5C_2H_5$	106.16	0.867	59	75	—	—
Ethyl Bromide	C_2H_5Br	109.98	1.430	—	—	6.75	11.25
Ethyl Chloride	C_2H_5Cl	64.52	0.921	−58	−45	3.6	14.80
Ethylene Chlorohydrin	$ClCH_2CH_2OH$	80.52	1.213	—	140	—	—

PHYSICAL CONSTANTS OF SELECTED MATERIALS (con't)

Substance	Formula	Molecular Weight	Specific Gravity	Flash Point F		Explosive Limits (Volume Percent)	
				Closed Cup	Open Cup	Lower	Upper
Ethylenediamine	$NH_2CH_2CH_2NH_2$	60.10	0.899	—	—	—	—
Ethylene Oxide	CH_2CH_2O	44.05	0.887	—	—	3.0	80.0
Ethyl Ether	$(C_2H_5)_2O$	74.12	0.713	—	—	—	—
Ethyl Formate	$HCO_2C_2H_5$	74.08	0.917	—	—	2.75	16.40
Ethyl Silicate	$(C_2H_5)_4SiO_4$	208.30	0.933	—	125	—	—
Formaldehyde	HCHO	30.03	0.815	Gas		7.0	73.0
Gasoline	CHnH(2n + 2)	86.0	0.660	−50	—	1.3	6.0
Heptane	$CH_3(CH_2)_5CH_3$	100.20	0.684	25	—	1.1	6.7
Hexane	$CH_3(CH_2)_4CH_3$	86.17	0.660	−7	—	1.18	7.4
Hydrogen Chloride	HCl	36.47	1.268 (A)	—	—	—	—
Hydrogen Cyanide	HCN	27.03	0.688	Gas		5.6	40.0
Hydrogen Fluoride	HF	20.01	0.987	Gas		—	—
Hydrogen Selenide	H_2Se	80.98	2.12	Gas		—	—
Hydrogen Sulfide	H_2S	34.08	1.189 (A)	Gas		4.3	45.5
Iodine	I_2	253.82	4.93	—	—	—	—
Isophorone	$(CH_3)_3C(CH_2)_2CCHCO$	138.20	0.923	—	205	—	—
Mesityl Oxide	$(CH_3)_2$:$CHCOCH_3$	98.14	0.857	87	—	—	—
Methanol	CH_3OH	32.04	0.792	54	60	6.72	36.5
Methyl Acetate	$CH_3CO_2CH_3$	74.08	0.928	15	20	3.15	15.60
Methyl Bromide	CH_3Br	94.95	1.732	—	—	13.5	14.5
Methyl Butanone (Isopropyl butane)	$CH_3COCH(CH_3)_2$	86.13	0.803	—	—	—	—
Methyl Cellosolve	$HOCH_2CH_2OCH_3$	76.06	0.965	107	115	—	—
Methyl Cellosolve Acetate	$CH_3OCH_2CH_2OOCCH_3$	118.13	1.007	132	140	—	—
Methyl Choloride	CH_3Cl	50.49	1.785	Gas		8.25	18.70
Methyl Cyclohexane	$CH_3(CHC_5H_{10})$	98.18	0.769	25	—	1.15	—
Methyl Cyclohexanol	$CH_3(CHC_4H_8CHOH)$	114.18	0.934	154	—	—	—
Methyl Cyclohexanone	$CH_5C_5H_9CO$	122.17	0.925	118	—	—	—
Methyl Formate	HCO_2CH_3	60.05	0.974	−2	—	4.5	20.0
Methyl Isobutyl Ketone	$CH_3COC_4H_9$	100.16	0.801	73	—	—	—
Monochlorobenzene	C_6H_5Cl	112.56	1.107	90	—	—	—
Monofluorotrichloromethane	Cl_3CF	137.38	1.494	Nonflammable			
Mononitrotoluene	$CH_3C_6H_4NO_2$	137.13	1.163	223	—	—	—
Naphtha (coal tar)	$C_6H_4(CH_3)_2$	106.16	0.85	100 − 110	—	—	—
Nickel Carbonyl	$Ni(CO)_4$	170.73	1.31	—	—	—	—
Nitrobenzene	$C_6H_5NO_2$	123.11	1.205	190	—	1.8 (200 F)	—
Nitroethane	$CH_3CH_2NO_2$	75.07	1.052	82	106	—	—
Nitrogen Oxides	NO	30.0	1.0367(A)	—	—	—	—
	N_2O	44.02	1.53	—	—	—	—
	N_2O_3	76.02	1.447	—	—	—	—
	NO_2	46.01	1.448	—	—	—	—
	N_2O_5	108.02	1.642	—	—	—	—
Nitroglycerine	$C_3H_5(ONO_2)_3$	227.09	1.601	—	—	—	—
Nitromethane	CH_3NO_2	61.04	1.130	95	112	—	—
2—Nitropropane	$CH_3CHNO_2CH_3$	89.09	1.003	—	103	—	—
Octane	$CH_3(CH_2)_6CH_3$	114.22	0.703	56	—	0.95	3.2
Ozone	O_3	48.0	1.658 (A)	—	—	—	—
Pentane	$CH_3(CH_2)_3CH_3$	72.15	0.626	−40	—	1.4	7.8
Pentanone (Methylpropanone)	$CH_3COCH_2C_2H_5$	86.13	0.816	45	60	1.55	8.15
Phosgene	O:C:Cl_2	98.92	1.392	—	—	—	—
Phosphine	PH_3	34.0	1.146 (A)	—	205	—	—
Phosphorus Trichloride	PCl_3	137.35	1.574	—	—	—	—
iso—Propanol	$(CH_3)_2CHOH$	60.09	0.785	53	60	2.02	11.80
Propane	$CH_3CH_2CH_3$	44.09	1.554	Gas		2.12	9.35
Propyl Acetate	$CH_3CO_2CH_2C_2H_5$	102.13	0.886	43	60	1.77	8.0
iso—Propyl Ether	$(CH_3)_4(CH)_2O$	102.17	0.725	−18	−15	—	—
Stibine	SbH_3	124.78	4.344 (A)	—	—	—	—
Styrene Monomer	C_6H_5HC:CH_2	104.14	0.903	90	—	1.1	6.1

PHYSICAL CONSTANTS OF SELECTED MATERIALS (con't)

Substance	Formula	Molecular Weight	Specific Gravity	Flash Point F Closed Cup	Flash Point F Open Cup	Explosive Limits (Volume Percent) Lower	Explosive Limits (Volume Percent) Upper
Sulfur Chloride, Mono	S_2Cl_2	135.03	1.678	245	None	—	—
Di	SCl_2	102.97	1.621	—	—	—	—
Tetra	SCl_4	173.89		—	—	—	—
Sulfur Dioxide	SO_2	64.07	2.264 (A)	Gas		—	—
1,1,2,2, Tetrachloroethane	$Cl_2CHCHCl_2$	167.86	1.588	—	—	—	—
Tetrachloroethylene	$Cl_2C{:}CCl_2$	165.85	1.624	Nonflammable			
Toluene	$C_6H_5CH_3$	92.13	0.866	40	45	1.27	6.75
Toluidine	$CH_3C_6H_4NH_2$	107.15	0.999	188	205	—	—
Trichloroethylene	$ClCHCCl_2$	131.40	1.466	Nonflammable			
Turpentine (Turpene)	$C_{10}H_{16}$	136.23		95	—	0.8	—
Vinyl Chloride (Chloroethane)	C_2H_5Cl	62.50	0.908	Gas		4.0	21.70
Xylene	$C_6H_4(CH_3)_2$	106.16	0.881	63	75	1.0	6.0

SOLVENT DRYING TIME

SOLVENT	Dry Time Relation	Boiling Range Deg. C	Boiling Range Deg. F.	Weight per Gal. Lbs.
Ethyl Ether, C.P	1.0	34-35	93-95	5.98
Petrolene	1.8	61-96	142-205	5.83
Carbon Tetrachloride	1.9	76	169	13.30
Acetone	2.0	55-58	133-136	6.35
Methyl Acetate	2.2	56-62	133-144	7.79
Ethyl Acetate 85-88%	2.5	74-77	165-171	7.37
Trichlorethylene	2.5	87	189	12.20
Benzol (Industrial)	2.6	79-81	174-178	7.38
Methyl Ethyl Ketone	2.7	77-82	171-180	6.95
Isopropyl Acetate 8%	2.7	84-93	183-199	7.26
Ethylene Dichloride	3.0	84	183	10.49
Solvsol 19/27	3.7	86-123	187-254	6.58
Ethylene Chloride	4.0	81-87	178-189	10.49
Propylene Dichloride	4.1	93-97	199-207	9.64
Troluoil	4.1	90-122	194-252	6.17
Methanol	5.0	64-65	147-149	6.63
Toluol (Industrial)	5.0	109-111	229-232	7.19
Methyl Propyl Ketone	5.2	101-107	214-225	6.77
V. M .& P	5.8	95-141	203-286	6.23
Perchlorethylene	6.0	121	250	13.55
Nor. Propyl Acetate	6.1	97-101	207-214	7.50
Sec. Butyl Acetate	6.5	106-135	223-275	7.13
Solox (Anhydrous)	6.5	71-78	160-172	6.80
Isobutyl Acetate 90%	7.0	106-117	223-243	7.28
Apocthinner	7.0	115-143	239-289	6.31
Ethyl Alcohol, Den. No. 1	7.7	78	172	6.64
Solox	8.0	76-78	169-172	6.73
Isopropyl Alcohol 99%	8.6	79-82	174-180	6.75
Nor. Propyl Alcohol	9.1	96-98	205-208	6.73

SOLVENT DRYING TIME (con't)

SOLVENT	Dry Time Relation	Boiling Range		Weight per Gal. Lbs.
		Deg. C	Deg. F.	
Solvsol 24/34	9.4	101-168	214-334	6.80
Nor. Butyl Acetate	9.6	110-132	230-270	7.29
Diethyl Carbonate	9.6	100-130	212-266	8.14
Methyl Butyl Ketone	9.7	114-137	237-279	6.84
Xylol (Industrial)	9.7	127-144	261-291	7.17
Monochlor Benzol	10.0	130-132	266-270	9.20
Tertiary Butyl Alcohol	11.9	82-83	180-181	6.55
Sec. Butyl Alcohol	14.0	99-100	210-212	6.85
Sec. Amyl Acetate	16.9	121-144	250-291	7.21
Amyl Acetate	17.4	105-142	221-288	7.24
Isobutyl Alcohol	17.7	107-111	225-232	6.70
Methyl Cellosolve	18.0	121-126	250-259	8.07
Butyl Propionate	18.0	124-171	255-340	7.31
Pentacetate	20.0	121-155	250-311	7.19
Turpentine	20.0	155-173	311-343	12.24
Butanol	21.0	116-119	241-246	6.79
Sec. Amyl Alcohol	25.0	105-125	221-257	6.79
2-50—W Hi-Flash Naphtha	27.5	148-187	298-369	7.18
Amyl Alcohol (Fusel Oil)	32.1	126-130	259-266	6.76
Di Isopropyl Ketone	33.9	164-169	327-336	6.75
Ethyl Cellosolve	36.2	133-137	271-279	7.77
Odorless Mineral Spirits	38.6	150-201	302-394	6.52
Ethyl Lactate	40.0	119-176	246-349	8.59
Sec. Hexyl Alcohol	41.7	157	315	6.97
Solvsol 30/40	43.2	142-199	288-390	7.06
Pentasol	45.0	112-140	234-284	6.76
Hi-Solvency Mineral Spirits	46.7	152-200	306-392	6.79
No. 380 Mineral Spirits	47.0	151-196	304-385	6.57
No. 10 Mineral Spirits	55.0	154-196	309-385	6.49
Distilled Water	60.0	100	212	8.32
Apco No. 125	60.5	162-200	324-392	6.52
Cellosolve Acetate	65.0	145-166	293-331	8.15
Sec. Butyl Lactate	73.0	172	342	8.14
Sec. Hexyl Acetate	76.5	129-158	264-316	7.19
Butyl Cellosolve	88.5	163-172	325-342	7.58
Dipentene	89.2	149-215	300-419	7.10
No. 140 Thinner	91.0	185-210	365-410	6.62
Octyl Acetate	152.5	195-203	383-397	7.20
Isobutyl Lactate	156.5	168-200	334-392	8.15
Hexalin	177.5	159-162	318-324	7.89
Solvsol 40/50	270.5	191-248	376-478	7.42
Methyl Hexalin	276.5	170-190	338-374	7.66
Butyl Lactate	339.0	185-195	365-383	8.14
Excellene	384.0	162-260	324-500	6.55
Special Heavy Naphtha	403.0	202-242	396-468	6.73
Dispersol	425.0	193-242	379-468	6.59
No. 50 Kerosene	626.7	178-256	352-493	6.76
Triethylene Glycol	Over 5200.0	276-310	529-590	9.30
Dibutyl Phthalate	Over 5200.0	195-200	383-392	8.73

Dry Time Relation:
Below 5 — Fast
5-15 — Medium
15-75 — Slow
75 over — Nil

CONVERSION FACTORS

Metric Equivalents

Length

cm = 0.3937 in
meter = 3.2808 ft
meter = 1.0936 yd
km = 0.6214 mile

in = 2.5400 cm
ft = 0.3048 m
yd = 0.9144 m
mile = 1.6093 km

Volume

cm^3 = 0.0610 in^3
m^3 = 35.3145 ft^3
m^3 = 1.3079 yd^3

in^3 = 16.3872 cm^3
ft^3 = 0.0283 m^3
yd^3 = 0.7647 m^3

Capacity

liter = 2.2046 lb of pure water at 4 C
liter = 61.0250 in^3
liter = 0.0353 ft^3

liter = 0.2642 gal (US)
in^3 = 0.0164 liter
ft^3 = 28.3162 liter
gal (US) = 3.7853 liter

Weight

gram = 15.4324 grains
gram = 0.0353 oz
kg = 2.2046 lb

grain = 0.0648 g
oz = 28.3495 g
lb = 0.4536 kg

Concentration

grain/ft^3 = 2288.1 mg/m^3
lb/1000 ft^3 = 16,017 mg/m^3
ton/sq mile = 0.35026 g/m^2

lb/acre = 0.11208 g/m^2
lb/1000 ft^2 = 4.8807 g/m^2
grain/ft^2 = 0.69725 g/m^2

Heat and Energy Units

1 boiler hp = 33,475 Btu/hr
1 ton (refrig) = 12,000 Btu/hr
Btu = 0.2520 kg-cal

1 kw-hr =

1000 whr
1.3410 hp-hr
2,655,217 ft-lb
3413 Btu

1 hp-hr =

0.7457 kw-hr
1,980,000 ft-lb
2545 Btu

1 lb water evaporated from and at 212 F

0.2844 kw-hr
0.3814 hp-hr
970.2 Btu

Temperature Equivalents

Fahrenheit, F = Rankin – 460 = 9/5 C + 32
Centigrade, C = Kelvin – 273 = 5/9 (F-32)

Pressure Equivalents

1 atmosphere =

14.696 lb/in^2 = 2116.3 lb/ft^2
33.96 ft of water = 407.52 in water
29.92 in of mercury = 760 mm mercury
234.54 oz/in^2 = 10,340 mm water

1 in water =

0.0361 lb/in^2 = 5.196 lb/ft^2
0.0735 in mercury = 1.876 mm mercury
0.002456 atmosphere = 0.5774 oz/in^2
25.4 mm water = 0.08333 ft water

1 in mercury =

0.491 lb/in^2 = 70.70 lb/ft^2
25.4 mm mercury = 7.86 oz/in^2
0.03342 atmosphere = 345.6 mm water
13.61 in water = 1.134 ft water

1 mm mercury =

0.01934 lb/in^2 = 2.789 lb/ft^2
0.3094 oz/in^2 = 0.001316 atmosphere
0.5357 in water = 0.04464 ft water
13.61 mm water = 0.03937 in mercury

1lb per in^2 =

144 lb/ft^2 = 16 oz/in^2
51.71 mm mercury = 2.036 in mercury
0.06804 atmosphere = 703.7 mm water
27.70 in water = 2.309 ft water

1 oz per in^2 =

0.0625 lb/in^2 = 9.00 lb/ft^2
1.733 in water = 0.1441 ft water
0.1272 in mercury = 3.23 mm mercury
0.00425 atmosphere = 44.02 mm water

- 1 inch of water resistance lowers density of air by 0.25 of 1%.
- 1 inch of water represents 66 feet difference in elevation at sea level to 4000 feet altitude.
- 1 inch of water represents 74 feet difference in elevation at 4000 to 6000 feet altitude.
- 1 inch of mercury represents 900 feet difference in elevation at sea level to 4000 feet altitude.
- 1 inch of mercury represents 1000 feet difference in elevation at 4000 to 6000 feet altitude.
- 1000 feet difference in elevation at sea level represents 1.11 inchs of mercury = 15.2 inches of water.
- 1000 feet difference in elevation at 4000 feet represents 1 inch of mercury = 13.6 inches of water.

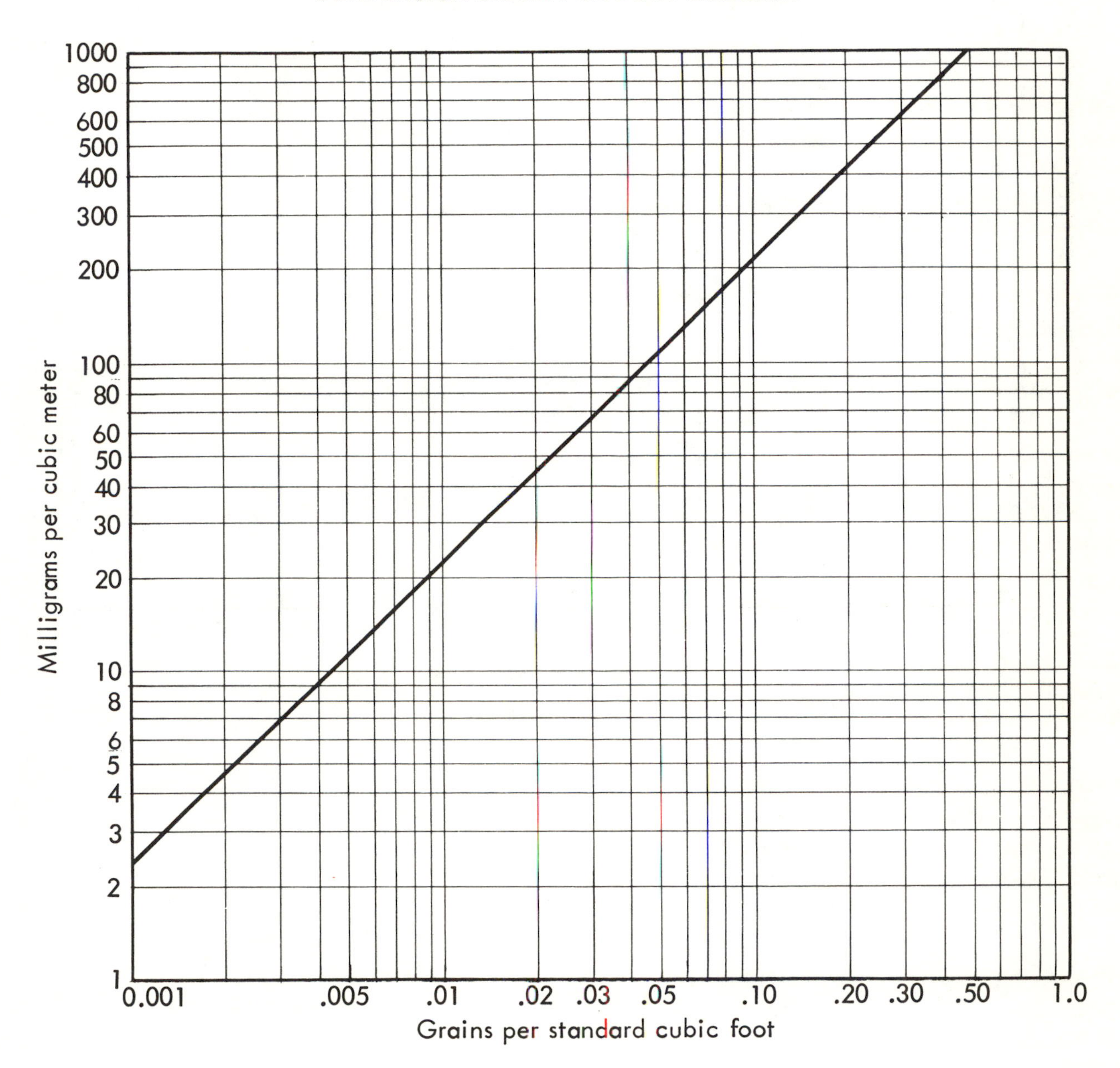

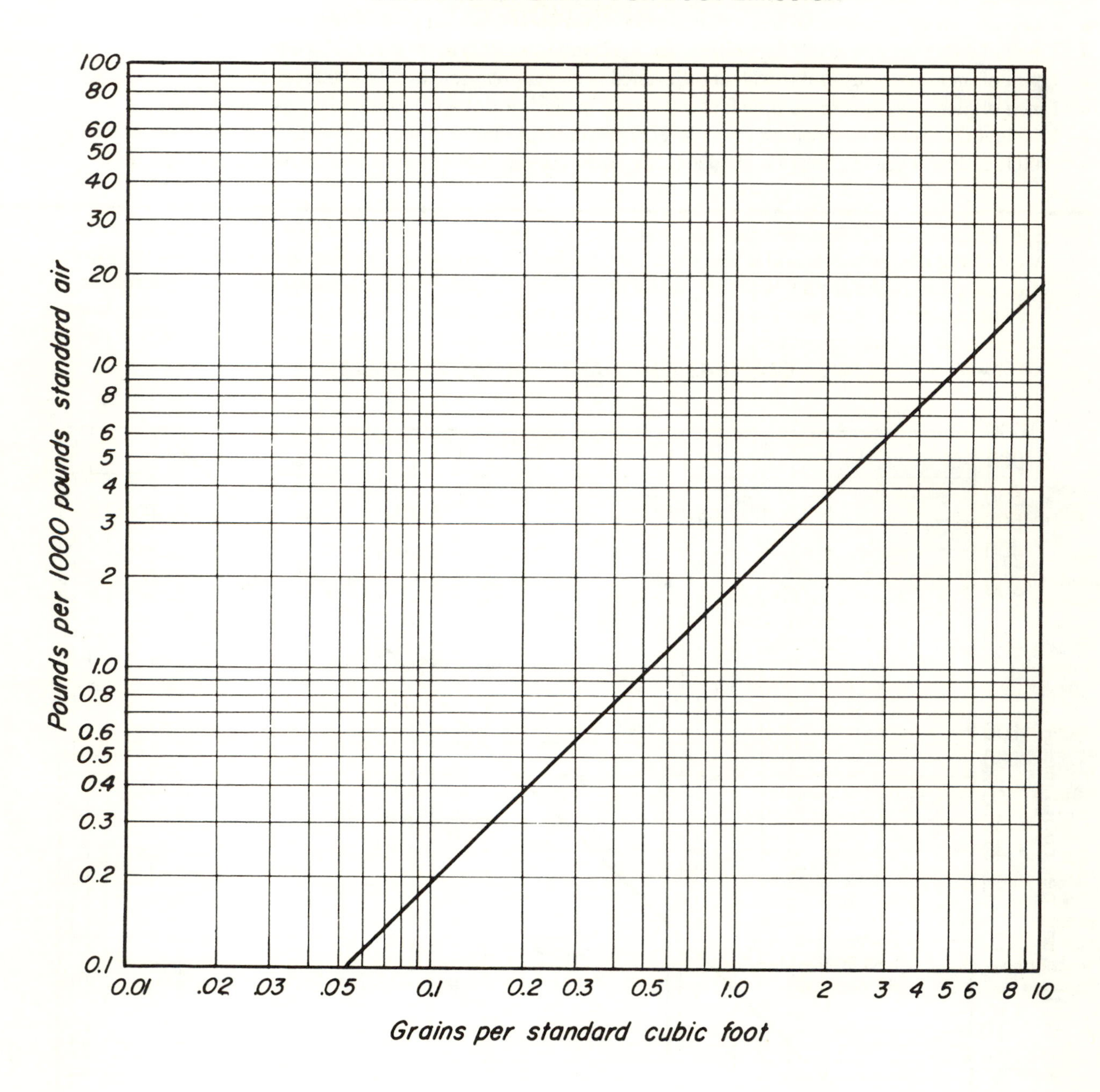
CONVERSION CHART FOR DUCT EMISSION
Pounds per 1000 pounds standard air
100
80
60
50
40
30
20
10
8
6
5
4
3
2
1.0
0.8
0.6
0.5
0.4
0.3
0.2
0.1
0.01
.02
.03
.05
0.1
0.2
0.3
0.5
1.0
2
3
4
5
6
8
10
Grains per standard cubic foot

APPENDIX C

Metric Supplement

INTRODUCTION

This supplement has been prepared for those designers who are accustomed to working in metric units. Most of the design charts and tables from Chapter 5 of the Industrial Ventilation Manual, plus the Pitot traverse point tables from Chapter 9, are included. These chapters in the Manual should be referred to for instructions in their use. No attempt was made to conform to any standard system of duct sizes, since several such systems are in use. An air density of 1.2 kg/m^3, which approximates the density of 21 C dry air at sea level, was used throughout.

GLOSSARY

ENGLISH	GERMAN	FRENCH	SPANISH
fan	der Ventilator	ventilateur	ventilador
duct	die Leitung	conduit	conducto
volume	der Rauminhalt, der Raumgehalt	volume	volumen
velocity	die Geschwindigkeit, (die Schnelligkeit)	vitesse	velocidad
air	die Luft, der Luftzug	air	aire
pressure	die Druckkraft	pression	presión
exhaust	auspumpen (verb)	épuiser (verb)	escape (de pas o vapor)
	der Auspuff (noun)	échappement (noun)	
diameter	der Durchmesser	diamètre	diàmetro
acceleration	die Beschleunigung	accélération	aceleración
friction	die Reibung, (die Friktion)	friction	fricción
density	die Dichtigkeit	densité	densidad
area	die Flächeneinhalt	aire	área
inlet	die Einlaßöffnung	entrée	estuario
slot	der Schlitzeinwurf	fente	abertura
flange	das Seitenstück; (die Flantsche)	collet	pestaña
filter	der Filter	filtre	filtrar
plenum	vollkommen ausgefüllter Raum	plenum	pleno
canopy	das Verdeck	baldaquin	dosel
ventilation	die Lüftung, der Luftwechsel	ventilation	ventilación
suction	die Zugkraft	succion	succión

TABLE MS-1. Velocity Conversion

fpm	m/s	fpm	m/s	fpm	m/s	fpm	m/s	fpm	m/s
100	0.508	2000	10.16	4000	20.32	6000	30.48	8000	40.64
150	0.762	2100	10.67	4100	20.83	6100	30.99	8100	41.15
200	1.016	2200	11.18	4200	21.33	6200	31.49	8200	41.65
300	1.524	2300	11.68	4300	21.84	6300	32.00	8300	42.16
400	2.032	2400	12.19	4400	22.35	6400	32.51	8400	42.67
500	2.540	2500	12.70	4500	22.86	6500	33.02	8500	43.18
600	3.048	2600	13.21	4600	23.37	6600	33.53	8600	43.69
700	3.556	2700	13.72	4700	23.87	6700	34.03	8700	44.19
800	4.064	2800	14.22	4800	24.38	6800	34.54	8800	44.70
900	4.572	2900	14.73	4900	24.89	6900	35.05	8900	45.21
1000	5.080	3000	15.24	5000	25.40	7000	35.56	9000	45.72
1100	5.588	3100	15.75	5100	25.91	7100	36.07	9100	46.23
1200	6.096	3200	16.26	5200	26.41	7200	36.57	9200	46.73
1300	6.604	3300	16.76	5300	26.92	7300	37.08	9300	47.24
1400	7.112	3400	17.27	5400	27.43	7400	37.59	9400	47.75
1500	7.620	3500	17.78	5500	27.94	7500	38.10	9500	48.26
1600	8.128	3600	18.29	5600	28.45	7600	38.61	9600	48.77
1700	8.636	3700	18.80	5700	28.95	7700	39.11	9700	49.27
1800	9.144	3800	19.30	5800	29.46	7800	39.62	9800	49.78
1900	9.652	3900	19.81	5900	29.97	7900	40.13	9900	50.29

TABLE MS-2. Volume Conversion

cfm	m^3/s	cfm	m^3/s	cfm	m^3/s	cfm	m^3/s	cfm	m^3/s
100	0.0471	300	0.1415	500	0.2359	1500	0.7079	3500	1.651
110	0.0519	310	0.1463	550	0.2595	1600	0.7551	3600	1.699
120	0.0566	320	0.1510	600	0.2831	1700	0.8023	3700	1.746
130	0.0613	330	0.1557	650	0.3067	1800	0.8495	3800	1.793
140	0.0660	340	0.1604	700	0.3303	1900	0.8967	3900	1.840
150	0.0707	350	0.1651	750	0.3539	2000	0.9438	4000	1.887
160	0.0755	360	0.1699	800	0.3775	2100	0.9910	4100	1.934
170	0.0802	370	0.1746	850	0.4011	2200	1.038	4200	1.982
180	0.0849	380	0.1793	900	0.4247	2300	1.085	4300	2.029
190	0.0896	390	0.1840	950	0.4483	2400	1.132	4400	2.076
200	0.0943	400	0.1887	1000	0.4719	2500	1.179	4500	2.123
210	0.0991	410	0.1934	1050	0.4955	2600	1.227	4600	2.170
220	0.1038	420	0.1982	1100	0.5191	2700	1.274	4700	2.218
230	0.1085	430	0.2029	1150	0.5427	2800	1.321	4800	2.265
240	0.1132	440	0.2076	1200	0.5663	2900	1.368	4900	2.312
250	0.1179	450	0.2123	1250	0.5899	3000	1.415	5000	2.359
260	0.1227	460	0.2170	1300	0.6135	3100	1.463	5100	2.406
270	0.1274	470	0.2218	1350	0.6371	3200	1.510	5200	2.454
280	0.1321	480	0.2265	1400	0.6607	3300	1.557	5300	2.501
290	0.1368	490	0.2312	1450	0.6843	3400	1.604	5400	2.548

TABLE MS-3. Velocity Pressure

$V = 4.043\sqrt{VP}$ DENSITY OF AIR = 1.2 Kg / m^3

VP = VELOCITY PRESSURE IN mm OF WATER V = VELOCITY IN m / sec.

VP	V	VP	V	VP	V	VP	V	VP	V	VP	V
0.1	1.28	5.1	9.13	11.0	13.41	61.0	31.58	111.0	42.59	161.0	51.30
0.2	1.81	5.2	9.22	12.0	14.00	62.0	31.83	112.0	42.79	162.0	51.46
0.3	2.21	5.3	9.31	13.0	14.58	63.0	32.09	113.0	42.98	163.0	51.62
0.4	2.56	5.4	9.39	14.0	15.13	64.0	32.34	114.0	43.17	164.0	51.77
0.5	2.86	5.5	9.48	15.0	15.66	65.0	32.59	115.0	43.35	165.0	51.93
0.6	3.13	5.6	9.57	16.0	16.17	66.0	32.84	116.0	43.54	166.0	52.09
0.7	3.38	5.7	9.65	17.0	16.67	67.0	33.09	117.0	43.73	167.0	52.24
0.8	3.62	5.8	9.74	18.0	17.15	68.0	33.34	118.0	43.92	168.0	52.40
0.9	3.84	5.9	9.82	19.0	17.62	69.0	33.58	119.0	44.10	169.0	52.56
1.0	4.04	6.0	9.90	20.0	18.08	70.0	33.82	120.0	44.29	170.0	52.71
1.1	4.24	6.1	9.99	21.0	18.53	71.0	34.07	121.0	44.47	171.0	52.87
1.2	4.43	6.2	10.07	22.0	18.96	72.0	34.30	122.0	44.65	172.0	53.02
1.3	4.61	6.3	10.15	23.0	19.39	73.0	34.54	123.0	44.84	173.0	53.18
1.4	4.78	6.4	10.23	24.0	19.81	74.0	34.78	124.0	45.02	174.0	53.33
1.5	4.95	6.5	10.31	25.0	20.21	75.0	35.01	125.0	45.20	175.0	53.48
1.6	5.11	6.6	10.39	26.0	20.61	76.0	35.24	126.0	45.38	176.0	53.63
1.7	5.27	6.7	10.46	27.0	21.01	77.0	35.48	127.0	45.56	177.0	53.79
1.8	5.42	6.8	10.54	28.0	21.39	78.0	35.71	128.0	45.74	178.0	53.94
1.9	5.57	6.9	10.62	29.0	21.77	79.0	35.93	129.0	45.92	179.0	54.09
2.0	5.72	7.0	10.70	30.0	22.14	80.0	36.16	130.0	46.10	180.0	54.24
2.1	5.86	7.1	10.77	31.0	22.51	81.0	36.39	131.0	46.27	181.0	54.39
2.2	6.00	7.2	10.85	32.0	22.87	82.0	36.61	132.0	46.45	182.0	54.54
2.3	6.13	7.3	10.92	33.0	23.22	83.0	36.83	133.0	46.62	183.0	54.69
2.4	6.26	7.4	11.00	34.0	23.57	84.0	37.05	134.0	46.80	184.0	54.84
2.5	6.39	7.5	11.07	35.0	23.92	85.0	37.27	135.0	46.97	185.0	54.99
2.6	6.52	7.6	11.15	36.0	24.26	86.0	37.49	136.0	47.15	186.0	55.14
2.7	6.64	7.7	11.22	37.0	24.59	87.0	37.71	137.0	47.32	187.0	55.28
2.8	6.76	7.8	11.29	38.0	24.92	88.0	37.93	138.0	47.49	188.0	55.43
2.9	6.88	7.9	11.36	39.0	25.25	89.0	38.14	139.0	47.66	189.0	55.58
3.0	7.00	8.0	11.43	40.0	25.57	90.0	38.35	140.0	47.84	190.0	55.73
3.1	7.12	8.1	11.51	41.0	25.89	91.0	38.57	141.0	48.01	200.0	57.17
3.2	7.23	8.2	11.58	42.0	26.20	92.0	38.78	142.0	48.18	210.0	58.59
3.3	7.34	8.3	11.65	43.0	26.51	93.0	38.99	143.0	48.35	220.0	59.96
3.4	7.45	8.4	11.72	44.0	26.82	94.0	39.20	144.0	48.51	230.0	61.31
3.5	7.56	8.5	11.79	45.0	27.12	95.0	39.40	145.0	48.68	240.0	62.63
3.6	7.67	8.6	11.86	46.0	27.42	96.0	39.61	146.0	48.85	250.0	63.92
3.7	7.78	8.7	11.92	47.0	27.72	97.0	39.82	147.0	49.02	260.0	65.19
3.8	7.88	8.8	11.99	48.0	28.01	98.0	40.02	148.0	49.18	270.0	66.43
3.9	7.98	8.9	12.06	49.0	28.30	99.0	40.23	149.0	49.35	280.0	67.65
4.0	8.09	9.0	12.13	50.0	28.59	100.0	40.43	150.0	49.51	290.0	68.85
4.1	8.19	9.1	12.20	51.0	28.87	101.0	40.63	151.0	49.68	300.0	70.02
4.2	8.29	9.2	12.26	52.0	29.15	102.0	40.83	152.0	49.84	310.0	71.18
4.3	8.38	9.3	12.33	53.0	29.43	103.0	41.03	153.0	50.01	320.0	72.32
4.4	8.48	9.4	12.40	54.0	29.71	104.0	41.23	154.0	50.17	330.0	73.44
4.5	8.58	9.5	12.46	55.0	29.98	105.0	41.43	155.0	50.33	340.0	74.55
4.6	8.67	9.6	12.53	56.0	30.25	106.0	41.62	156.0	50.49	350.0	75.63
4.7	8.76	9.7	12.59	57.0	30.52	107.0	41.82	157.0	50.66	360.0	76.71
4.8	8.86	9.8	12.66	58.0	30.79	108.0	42.01	158.0	50.82	370.0	77.77
4.9	8.95	9.9	12.72	59.0	31.05	109.0	42.21	159.0	50.98	380.0	78.81
5.0	9.04	10.0	12.78	60.0	31.32	110.0	42.40	160.0	51.14	390.0	79.84

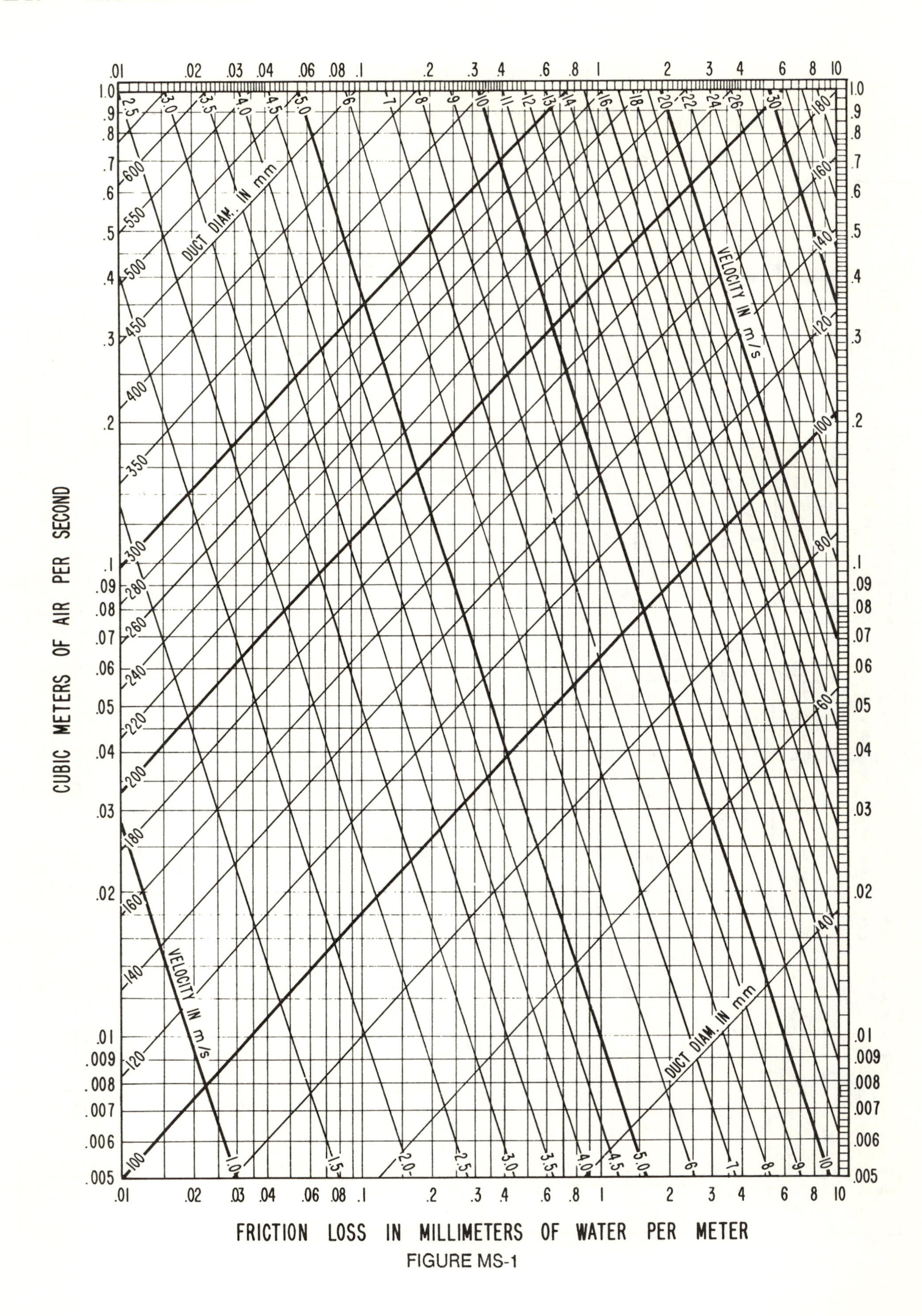
CUBIC METERS OF AIR PER SECOND
FRICTION LOSS IN MILLIMETERS OF WATER PER METER
DUCT DIAM. IN mm
VELOCITY IN m/s

FIGURE MS-1

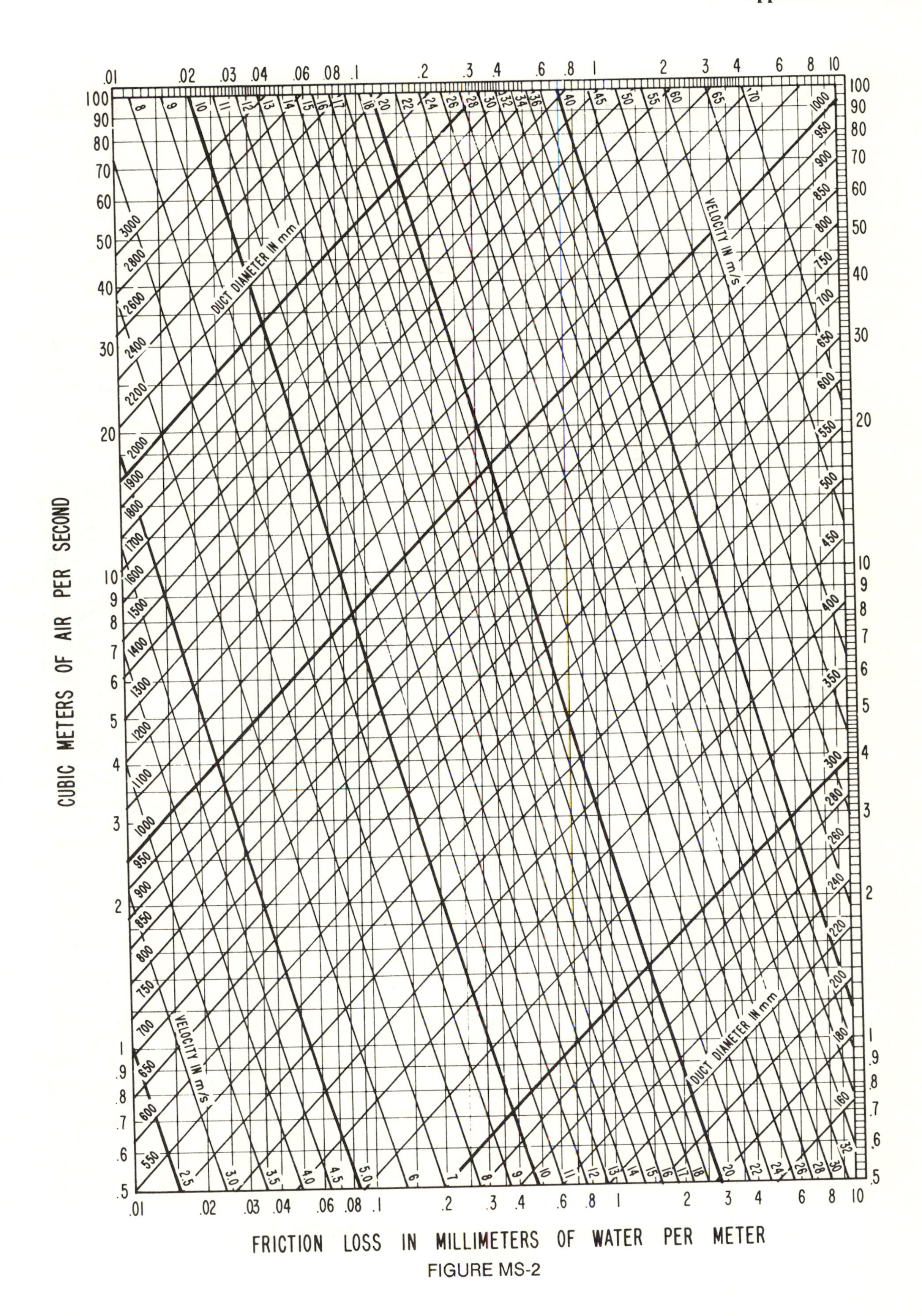
CUBIC METERS OF AIR PER SECOND
FRICTION LOSS IN MILLIMETERS OF WATER PER METER
DUCT DIAMETER IN mm
VELOCITY IN m/s

FIGURE MS-2

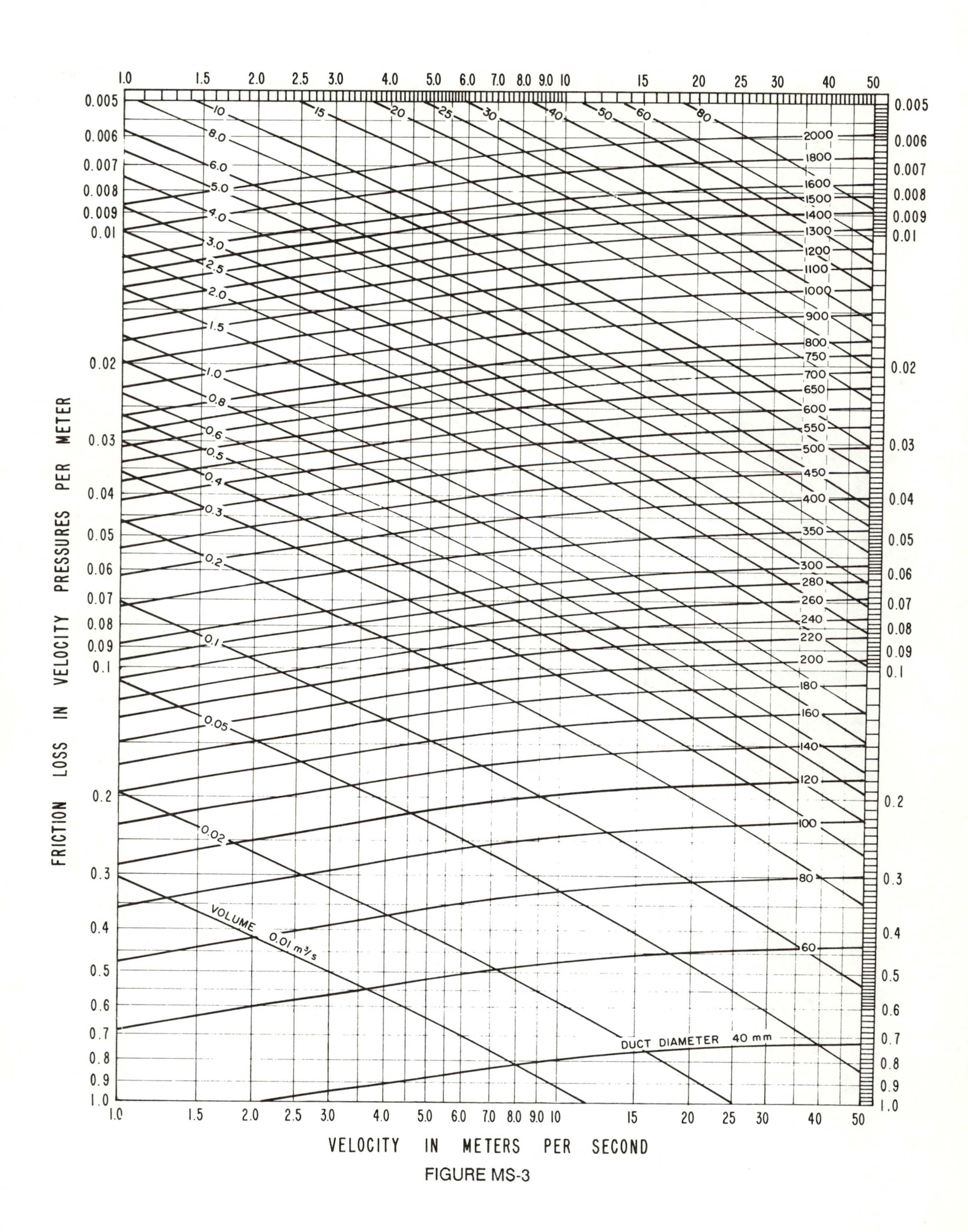

FRICTION LOSS IN VELOCITY PRESSURES PER METER
VELOCITY IN METERS PER SECOND
1.0 1.5 2.0 2.5 3.0 4.0 5.0 6.0 7.0 8.0 9.0 10 15 20 25 30 40 50
0.005 0.006 0.007 0.008 0.009 0.01 0.02 0.03 0.04 0.05 0.06 0.07 0.08 0.09 0.1 0.2 0.3 0.4 0.5 0.6 0.7 0.8 0.9 1.0
10 8.0 6.0 5.0 4.0 3.0 2.5 2.0 1.5 1.0 0.8 0.6 0.5 0.4 0.3 0.2 0.1 0.05 0.02
15 20 25 30 40 50 60 80
VOLUME 0.01 m³/s
2000 1800 1600 1500 1400 1300 1200 1100 1000 900 800 750 700 650 600 550 500 450 400 350 300 280 260 240 220 200 180 160 140 120 100 80 60
DUCT DIAMETER 40 mm

FIGURE MS-3

TABLE MS-4. Area and Circumference of Circles

DIAM. cm	AREA m²	CIRC. cm	DIAM. cm	AREA m²	CIRC. cm	DIAM. cm	AREA m²	CIRC. cm
1	0.000079	3.142	42	0.1385	131.9	122	1.169	383.3
2	0.000314	6.283	44	0.1521	138.2	124	1.208	389.6
3	0.000707	9.425	46	0.1662	144.5	126	1.247	395.8
4	0.001257	12.57	48	0.1810	150.8	128	1.287	402.1
5	0.001963	15.71	50	0.1963	157.1	130	1.327	408.4
6	0.002827	18.85	52	0.2124	163.4	132	1.368	414.7
7	0.003848	21.99	54	0.2290	169.6	134	1.410	421.0
8	0.005027	25.13	56	0.2463	175.9	136	1.453	427.3
9	0.006362	28.27	58	0.2642	182.2	138	1.496	433.5
10	0.007854	31.42	60	0.2827	188.5	140	1.539	439.8
11	0.009503	34.56	62	0.3019	194.8	142	1.584	446.1
12	0.01131	37.70	64	0.3217	201.1	144	1.629	452.4
13	0.01327	40.84	66	0.3421	207.3	146	1.674	458.7
14	0.01539	43.98	68	0.3632	213.6	148	1.720	465.0
15	0.01767	47.12	70	0.3848	219.9	150	1.767	471.2
16	0.02011	50.27	72	0.4071	226.2	152	1.815	477.5
17	0.02270	53.41	74	0.4301	232.5	154	1.863	483.8
18	0.02545	56.55	76	0.4536	238.8	156	1.911	490.1
19	0.02835	59.69	78	0.4778	245.0	158	1.961	496.4
20	0.03142	62.83	80	0.5027	251.3	160	2.011	502.7
21	0.03464	65.97	82	0.5281	257.6	162	2.061	508.9
22	0.03801	69.11	84	0.5542	263.9	164	2.112	515.2
23	0.04155	72.26	86	0.5809	270.2	166	2.164	521.5
24	0.04524	75.40	88	0.6082	276.5	168	2.217	527.8
25	0.04909	78.54	90	0.6362	282.7	170	2.270	534.1
26	0.05309	81.68	92	0.6648	289.0	172	2.324	540.4
27	0.05726	84.82	94	0.6940	295.3	174	2.378	546.6
28	0.06158	87.96	96	0.7238	301.6	176	2.433	552.9
29	0.06605	91.11	98	0.7543	307.9	178	2.488	559.2
30	0.07069	94.25	100	0.7854	314.2	180	2.545	565.5
31	0.07548	97.39	102	0.8171	320.4	182	2.602	571.8
32	0.08042	100.5	104	0.8495	326.7	184	2.659	578.1
33	0.08553	103.7	106	0.8825	333.0	186	2.717	584.3
34	0.09079	106.8	108	0.9161	339.3	188	2.776	590.6
35	0.09621	110.0	110	0.9503	345.6	190	2.835	596.9
36	0.1018	113.1	112	0.9852	351.9	192	2.895	603.2
37	0.1075	116.2	114	1.021	358.1	194	2.956	609.5
38	0.1134	119.4	116	1.057	364.4	196	3.017	615.8
39	0.1195	122.5	118	1.094	370.7	198	3.079	622.0
40	0.1257	125.7	120	1.131	377.0	200	3.142	628.3

EQUIVALENT RESISTANCE IN METERS OF STRAIGHT PIPE

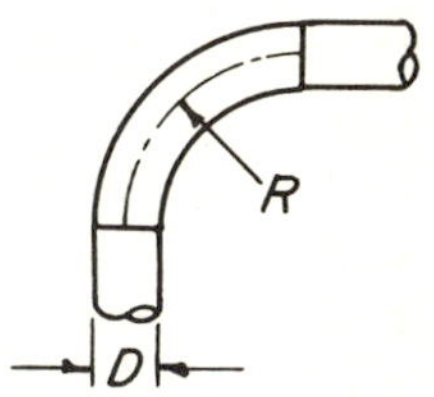

R
D

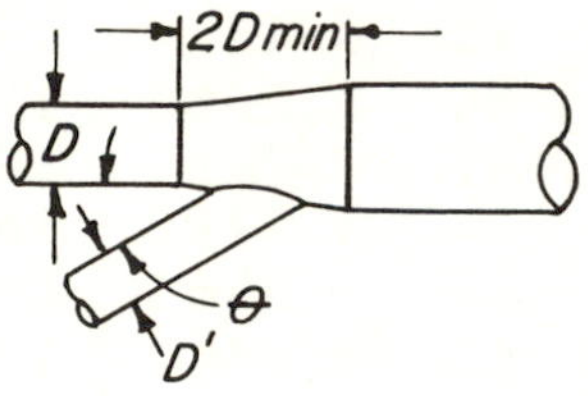

2D min
D
θ
D'

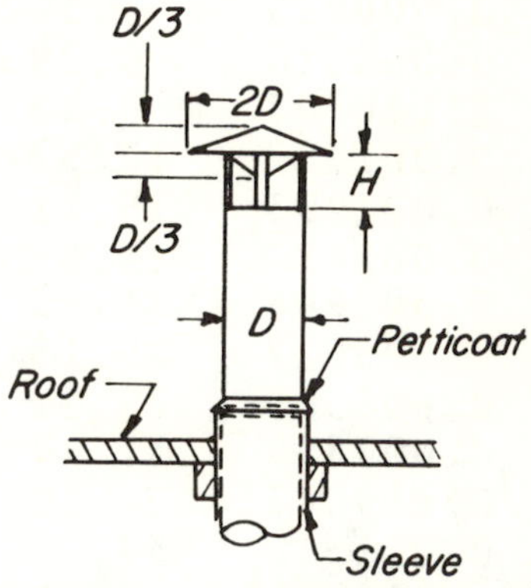

D/3
2D
H
D/3
D
Petticoat
Roof
Sleeve
Not recommended

Pipe in mm	90° Elbow * Centerline Radius			Angle of Entry		H, No of Diameters		
	1.5D	2.0D	2.5D	30°	45°	1.0H	0.75H	0.5H
75	1.4	0.9	0.7	0.5	0.9	0.3	0.5	2.0
100	2.0	1.3	1.1	0.8	1.3	0.5	0.8	3.4
125	2.6	1.7	1.4	1.1	1.7	0.6	1.1	4.4
150	3.2	2.2	1.8	1.4	2.2	0.8	1.4	5.5
175	3.9	2.6	2.2	1.7	2.6	0.9	1.7	6.6
200	4.6	3.1	2.5	2.0	3.1	1.1	2.0	7.8
250	6.0	4.0	3.3	2.6	4.0	1.4	2.6	10
300	7.4	5.0	4.1	3.2	5.0	1.8	3.2	13
350	8.9	6.0	5.0	3.8	6.0	2.1	3.8	15
400	10	7.0	5.8	4.5	7.0	2.5	4.5	18
450	12	8.1	6.7	5.2	8.1	2.8	5.2	21
500	14	9.2	7.6	5.9	9.2	3.2	5.9	23
600	17	11	9.5	7.3	11	4.0	7.3	29
700	21	14	11	8.8	14	4.8	8.8	35
800	24	16	13	10	16	5.7	10	41
900	28	19	15					
1000	32	21	18					
1200	39	26	22					
1400	47	32	26					
1600	55	37	31					
1800	64	43	36					
2000	72	49	40					

*For 60° elbows — x.67
For 45° elbows — x.5

AMERICAN CONFERENCE OF GOVERNMENTAL INDUSTRIAL HYGIENISTS

DUCT DESIGN DATA

DATE 1-70

FIGURE MS-4

TABLE MS-5. Circular Equivalents of Rectangular Ducts for Equal Fraction and Capacity

RECT. DUCT SIDE	35	40	45	50	55	60	65	70	75	80	85	90	95	100	105	110	115	120	125	130	135	140	145	150	155	160	165
100	63	67	72	76	80	84	88	91	94	98	101	104	107	109													
105	64	69	74	78	82	86	90	93	97	100	103	106	109	112	115												
110	65	70	75	80	84	88	92	95	99	102	105	109	112	115	117	120											
115	67	72	77	81	86	90	94	97	101	104	108	111	114	117	120	123	126										
120	68	73	78	83	87	91	95	99	103	107	110	113	117	120	123	126	128	131									
125	69	74	79	84	89	93	97	101	105	109	112	116	119	122	125	128	131	134	137								
130	70	76	81	86	90	95	99	103	107	111	114	118	121	124	128	131	134	137	139	142							
135	71	77	82	87	92	96	101	105	109	113	116	120	123	127	130	133	136	139	142	145	148						
140	72	78	84	89	93	98	102	107	111	115	118	122	125	129	132	135	139	142	145	147	150	153					
145	73	79	85	90	95	100	104	108	112	116	120	124	128	131	134	138	141	144	147	150	153	156	159				
150	75	80	86	91	96	101	106	110	114	118	122	126	130	133	137	140	143	146	150	153	156	158	161	164			
155	76	82	87	93	98	103	107	112	116	120	124	128	132	135	139	142	146	149	152	155	158	161	164	167	169		
160	77	83	88	94	99	104	109	113	118	122	126	130	134	137	141	144	148	151	154	157	161	164	166	169	172	175	
165	78	84	90	95	100	105	110	115	119	124	128	132	136	139	143	147	150	153	157	160	163	166	169	172	175	178	180
170	79	85	91	96	102	107	112	116	121	125	129	134	137	141	145	149	152	156	159	162	165	168	171	174	177	180	183
175	79	86	92	98	103	108	113	118	123	127	131	135	139	143	147	151	154	158	161	164	168	171	174	177	180	183	186
180	80	87	93	99	104	110	115	119	124	129	133	137	141	145	149	153	156	160	163	167	170	173	176	179	182	185	188
185	81	88	94	100	106	111	116	121	126	130	135	139	143	147	151	155	158	162	165	169	172	176	179	182	185	188	191
190	82	89	95	101	107	112	117	122	127	132	136	141	145	149	153	157	160	164	168	171	174	178	181	184	187	190	193
195	83	90	96	102	108	113	119	124	129	133	138	142	146	151	155	158	162	166	170	173	177	180	183	187	190	193	196
200	84	91	97	103	109	115	120	125	130	135	139	144	148	152	156	160	164	168	172	175	179	182	186	189	192	195	198
205	85	92	98	104	110	116	121	127	131	136	141	145	150	154	158	162	166	170	174	177	181	184	188	191	194	198	201
210	86	93	99	106	111	117	123	128	133	138	142	147	151	156	160	164	168	172	176	179	183	186	190	193	197	200	203
215	87	94	100	107	113	118	124	129	134	139	144	149	153	157	162	166	170	174	178	181	185	189	192	196	199	202	205
220	87	95	101	108	114	120	125	130	136	141	145	150	155	159	163	168	172	176	180	183	187	191	194	198	201	204	208
225	88	95	102	109	115	121	126	132	137	142	147	152	156	161	165	169	173	177	181	185	189	193	196	200	203	207	210
230	89	96	103	110	116	122	128	133	138	143	148	153	158	162	167	171	175	179	183	187	191	195	198	202	205	209	212
235	90	97	104	111	117	123	129	134	140	145	150	155	159	164	168	173	177	181	185	189	193	197	200	204	208	211	214
240	91	98	105	112	118	124	130	135	141	146	151	156	161	165	170	174	179	183	187	191	195	199	202	206	210	213	217
245	91	99	106	113	119	125	131	137	142	147	153	158	162	167	172	176	180	185	189	193	197	201	204	208	212	215	219
250	92	100	107	114	120	126	132	138	143	149	154	159	164	169	173	178	182	186	190	195	198	202	206	210	214	217	221
255	93	100	108	115	121	127	133	139	145	150	155	160	165	170	175	179	184	188	192	196	200	204	208	212	216	219	223

TABLE MS-6. Density Correction Factor

Kg/m^3 = Density Factor x 1.2

Weight of Dry Air at 21C and Sea Level = 1.2 Kg/m^3

Altitude, Meters Above Sea Level

	− 250	0	250	500	750	1 000	1 250	1 500	1 750	2000	2500	3000
Temp. E Hg	782	760	738	71 7	697	677	657	639	620	603	569	536
C. E H_2O	1 0649	1 0345	1 0048	3761	3482	9211	8947	8691	8443	8201	7739	7303
0	1.11	1.08	1.05	1.02	0.99	0.96	0.93	0.91	0.88	0.86	0.81	0.76
21	1.03	1.00	0.97	0.95	0.92	0.89	0.87	0.84	0.82	0.79	0.75	0.71
50	0.94	0.91	0.89	0.86	0.84	0.81	0.79	0.77	0.75	0.72	0.68	0.64
75	0.87	0.85	0.82	0.80	0.78	0.75	0.73	0.71	0.69	0.67	0.63	0.60
100	0.81	0.79	0.77	0.75	0.72	0.70	0.68	0.66	0.65	0.63	0.59	0.56
125	0.76	0.74	0.72	0.70	0.68	0.66	0.64	0.62	0.60	0.59	0.55	0.52
150	0.72	0.70	0.68	0.66	0.64	0.62	0.60	0.59	0.57	0.55	0.52	0.49
175	0.68	0.66	0.64	0.62	0.60	0.59	0.57	0.55	0.54	0.52	0.49	0.46
200	0.64	0.62	0.61	0.59	0.57	0.56	0.54	0.52	0.51	0.49	0.47	0.44
225	0.61	0.59	0.58	0.56	0.54	0.53	0.51	0.50	0.48	0.47	0.44	0.42
250	0.58	0.56	0.55	0.53	0.52	0.50	0.49	0.47	0.46	0.45	0.42	0.40
275	0.55	0.54	0.52	0.51	0.49	0.48	0.47	0.45	0.44	0.43	0.40	0.38
300	0.53	0.51	0.50	0.49	0.47	0.46	0.45	0.43	0.42	0.41	0.38	0.36
325	0.51	0.49	0.48	0.47	0.45	0.44	0.43	0.41	0.40	0.39	0.37	0.35
350	0.49	0.47	0.46	0.45	0.43	0.42	0.41	0.40	0.39	0.38	0.35	0.33
375	0.47	0.46	0.44	0.43	0.42	0.41	0.39	0.38	0.37	0.36	0.34	0.32
400	0.45	0.44	0.43	0.41	0.40	0.39	0.38	0.37	0.36	0.35	0.33	0.31
425	0.43	0.42	0.41	0.40	0.39	0.38	0.37	0.35	0.34	0.33	0.32	0.30
450	0.42	0.41	0.40	0.38	0.37	0.36	0.35	0.34	0.33	0.32	0.31	0.29
475	0.41	0.39	0.38	0.37	0.36	0.35	0.34	0.33	0.32	0.31	0.29	0.28
500	0.39	0.38	0.37	0.36	0.35	0.34	0.33	0.32	0.31	0.30	0.28	0.27
525	0.38	0.37	0.36	0.35	0.34	0.33	0.32	0.31	0.30	0.29	0.27	0.26

Friction loss varies directly with the density (first approximation).
See fan laws or cited references for more detail.

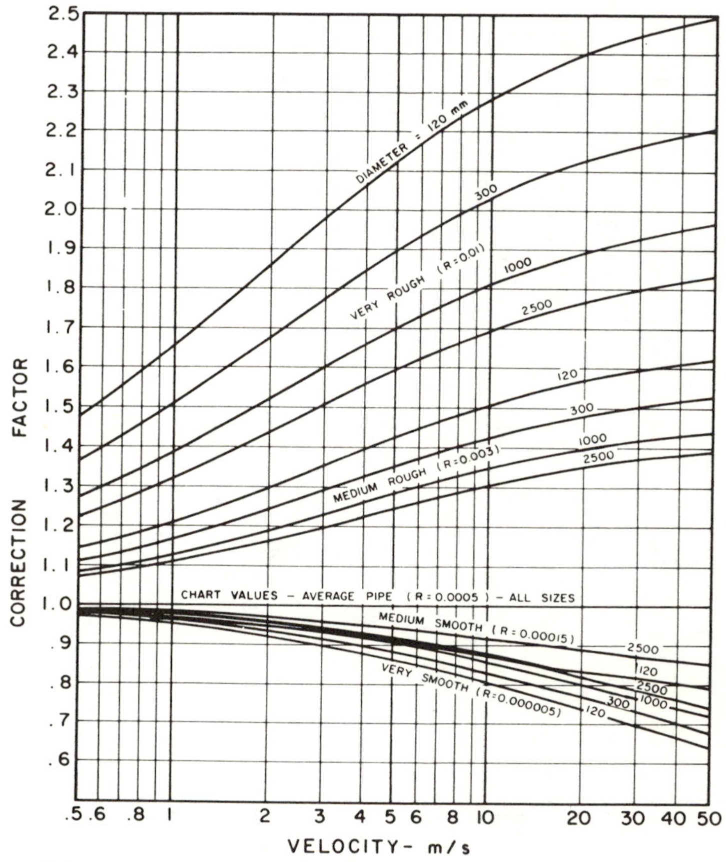

FIGURE MS-5. ROUGHNESS CORRECTION FACTOR

TABLE MS-7. Pitot Tube Traverse Distances (from wall — 10 point traverse — nearest mm)

DIAMETER mm	R_1 0.026D	R_2 0.082D	R_3 0.146D	R_4 0.226D	R_5 0.342D	R_6 0.658D	R_7 0.774D	R_8 0.854D	R_9 0.918D	R_{10} 0.974D
50	1	4	7	11	17	33	39	43	46	49
60	2	5	9	14	21	39	46	51	55	58
70	2	6	10	16	24	46	54	60	64	68
80	2	7	12	18	27	53	62	68	73	78
90	2	7	13	20	31	59	70	77	83	88
100	3	8	15	23	34	66	77	85	92	97
110	3	9	16	25	38	72	85	94	101	107
120	3	10	18	27	41	79	93	102	110	117
130	3	11	19	29	44	86	101	111	119	127
140	4	11	21	32	48	92	108	119	129	136
150	4	12	22	34	51	99	116	128	138	146
160	4	13	23	36	55	105	124	137	147	156
170	4	14	25	38	58	112	132	145	156	166
180	5	15	26	41	62	118	139	154	165	175
190	5	16	28	43	65	125	147	162	174	185
200	5	16	29	45	68	132	155	171	184	195
225	6	18	33	51	77	148	174	192	207	219
250	6	20	37	57	85	165	193	213	230	244
275	7	22	40	62	94	181	213	235	253	268
300	8	25	44	68	103	197	232	256	275	292
325	8	27	48	73	111	214	252	277	298	317
350	9	29	51	79	120	230	271	299	321	341
375	10	31	55	85	128	247	290	320	344	365
400	10	33	59	90	137	263	310	341	367	390
425	11	35	62	96	145	280	329	363	390	414
450	12	37	66	102	154	296	348	384	413	438
475	12	39	70	107	162	313	368	405	436	463
500	13	41	73	113	171	329	387	427	459	487
550	14	45	81	124	188	362	426	469	505	536
600	15	49	88	136	205	395	464	512	551	585
650	17	53	95	147	222	428	503	555	597	633
700	18	57	103	158	239	461	542	597	643	682
750	19	61	110	170	256	494	580	640	689	731
800	21	65	117	181	274	526	619	683	735	779
850	22	69	124	192	291	559	658	726	781	828
900	23	74	132	204	308	592	696	768	826	877
950	24	78	139	215	325	625	735	811	872	926
1000	26	82	146	226	342	658	774	854	918	974
1050	27	86	154	237	359	691	813	896	964	1023
1100	28	90	161	249	376	724	851	939	1010	1072
1150	30	94	168	260	393	757	890	982	1056	1120
1200	31	98	176	271	410	790	929	1024	1102	1169
1250	32	102	183	283	427	823	967	1067	1148	1218
1300	33	106	190	294	444	856	1006	1110	1194	1267
1350	35	110	198	305	462	888	1045	1152	1240	1315
1400	36	114	205	317	479	921	1083	1195	1286	1364
1450	37	118	212	328	496	954	1122	1238	1332	1413
1500	38	123	220	339	513	987	1161	1280	1377	1462

INDEX

A

Abbreviations, xi
Abbreviations, table of, xi
Abrasive Blasting
- cabinet, 10-5
- chamber, 10-33
- elevator, 10-33
- room, 10-4

Abrasive material, 4-3
Abrasive saw, 10-41
Abrasive wheel manufacture, 10-134
Absolute humidity, definition, ix
Acclimatization, 2-9
Acid dipping
- airborne contaminants, 10-72 to 10-73, 12-3
- ventilation, 10-63 to 10-77

Actual duct velocity, 5-2
Adding hoods, 8-5
Adsorber, 4-26
Aerographing — ceramic, 10-134
Aerosol, XI, 4-26
Air
- bleed-in, 5-29
- changes, 7-1, 7-4, 7-5
- cleaning devices, Chapter 4, 5-28
- contaminants, 1-2, 10-72 to 10-73, 123 to 12-17
- density, XI, 1-3, 5-11, 5-24, 5-26, 5-28
- dilution, principles, 1-2, 2-2
- distribution, 2-2, 2-4, 3-8, 7-5
  - slot, 3-8
- ejectors, 6-5
- filters, 4-2
- filtration
  - equipment, 4-28
  - straining, 4-28
- flow characteristics
  - blowing, 1-9
  - exhausting, 1-9
- flow contours, 3-7
- flow distribution, 5-3
- flow instruments, Chapter 9
- flow measurement, Chapter 9
  - discharge stacks, 9-9
- flow principles
  - assumptions 1-5
- horsepower, definition, ix
- motion, sources, 3-2
- proper introduction point, 7-5
- recirculation, 7-2
- replacement
  - principles, 1-2, 2-2, 7-2
  - quantity required, 7-2
- standard, definition, ix
- supply system, 7-2
- supply temperature, 7-5

Air-to-air heat exchangers, 7-15
Air Moving and Conditioning Association, 6-10
Air moving device, 6-2
Air stream characteristics, 6-6
Air to cloth ratio, 4-12
Air velocity instruments, 9-2, 9-12
Alignment, 6-21
Altitude
- correction for, 5-23

Aluminum furnaces, 10-134
Ambient concentration level, 4-2
Anemometer
- corrections for density, 9-16
- heated thermocouple, 9-22
- rotating vane, 9-12
- swinging vane, 9-13
- thermal, 9-16
- velometer, 9-13, 9-19, 9-27

Aneroid gauge, 9-17
- electronic, 9-17

Arrestance, 4-32
Asbestos
- bag opening, 10-131
- fiber belt conveying, 10-132
- spool winding, 10-134

Aspect ratio, 3-8
- definition, ix

Auto parking garage, 10-134
Auto spray paint booth, 10-78, 10-82
Axial fans, 6-2

B

Backdrafting, 7-2
Backstand idler, 10-42
Backward incline fans, 6-10
Baffles, 3-6
Bag filling, 10-34, 10-35, 10-134
Bag tube packer, 10-35
Bagging (paper & cloth bags), 10-134
Baking ovens, 10-79
Balanced design method, 5-6
Banbury mixer, 10-104
Band saw, 10-60, 10-90
Barbeque, 10-119
Barrel filling, 10-36, 10-134

Belt, conveyors, 10-37, 10-39
Belt driven fans, 6-10
Belt grinders, 10-42, 10-43
Belt sanders, 10-85, 10-94
Bench hoods, 10-54, 10-55, 10-69, 10-121
Bernoulli's Theorem, 1-6
Bin ventilation, 10-37
Black iron
    duct material, 8-2
Blade shape
    airfoil, 6-2
    solid, 6-2
Blast gate, definition, ix
Blast gate method, 5-6
    fan power required, 5-7
Blast gates, 5-3, 8-4
Bleed-in, 5-29
Blow (throw), definition, ix
Blowing characteristics, 1-9
Booth-type hoods, 3-2, 10-55, 10-78, 10-80, 10-81, 10-83
Brake horsepower, definition, ix
Branch entry, 5-2, 5-10, 5-32, 5-36, 5-50
Bright dip, 10-64 to 10-77
Broiler, 10-118
Brushes, 10-98
Bucket elevators, 10-38
Buffing
    automatic, 10-44, 10-45
    buffing dust, 4-3
    wheels, 10-46, 10-47
Bypass dampers, 7-10

C

Calculation sheet, 5-3, 5-11
Calendar rolls, rubber, 10-105
Calibration wind tunnel, 9-19
Canopy hoods
    high, 3-19
    low, 3-10
Capture velocity, 3-6
Capture velocity, definition, ix
Carding, asbestos, 10-134
Catalytic oxidizer, 4-26
Centrifugal collectors, 4-3
    cyclone, 4-23
    dry dynamic precipitator, 4-23
    efficiency, 4-18
    gravity separation, 4-18
    high efficiency, 4-23
    inertial separator, 4-23
    wet centrifugal, 4-18
Centrifugal fans, 6-2
Ceramic manufacture, 10-134
Charcoal broiler and barbeque, 10-119
Check out procedure
    new systems, 9-22
Chipping table, 10-53, 10-54
Chisel, pneumatic, hood, 10-99
Cleaning tanks, 10-61 to 10-77
Cleanout doors, 8-4
Clean room air flow, 10-123, 10-124
Clearance, 8-4
Coating pans, 10-134
Coefficient of entry, definition, ix, 1-7
Coefficient, hood entry, 1-7, 3-5, 3-6, 5-33
Cold header, 10-127
Collection efficiency
    pressure drop, 4-18
Collectors
    dry dynamic, 4-23
    centrifugal, high efficiency, 4-23
    comparison chart, 4-30
    cyclone, 4-23
    efficiency, 4-2
    electrostatic precipitator, 4-3, 4-7, 4-8
    fabric, 4-9 to 4-17
    unit collectors, 4-28
    wet types, 4-18
Combustibility in collectors, 4-3
Combustion flues, 7-4
Combustion products, 7-4
Combustion calculations, 7-11 to 7-15
Compound hood, 3-15
Computer design, 5-3
Concentration, 4-12
Condensate, 7-8
Condensation, 7-11
    considerations, 8-4
Cone wheels, 10-97
Configuration
    fabric, 4-12
Construction classes, 6-10
Construction
    classifications, 8-2
    hoods, 8-4
    materials, 8-2
    obstructions, 8-2
    round duct, 8-2
    specifications
        minimum, 8-2
Contaminant 7-17
    characteristics, 4-2
    concentration, 4-12
    nature of, 4-2
    shape, 4-3
    visibility, 4-3
Contaminants, air, 1-2, 2-2, 10-72 to 10-77, 7-1, 12-3
Contours
    flow, 3-7
Contraction, 5-10
Contraction losses, 5-35
Control

odors, 7-5
radioactive matrials, 3-18
Convection, ix, 3-19
Conversion factors, 13-29
Conveyor Belts, 10-37, 10-39, 10-41
Cooling tower, 4-9
Cooling tunnels (foundry), 10-134
Cooling velocity, 2-11
Core
grinder, 10-6
knockout, 10-134
molding, 10-18, 10-19
sanding, 10-134
Corrected velocity pressure, 9-3
Correction factor
different duct materials, 5-11
non-standard density, 9-26
Corrosion damage, 7-11
Corrosive applications, 6-7
Cottrell, 4-3
Crucible furnaces, 10-7, 10-20, 10-109
Crushers, 10-134
Cup wheels, 10-98
Cut-off wheel, 10-41
Cyclone collector, 4-23, 5-24

D

Dampers, 6-10
Degreasing tank, 10-61
Degree days, 7-12
Density
correction factors, 5-44, 9-29, 9-30
definition, x, 1-3
effect on flow rate, 5-24
psychrometric chart, 5-24
Density factor, 5-23, 5-26
Design procedure, 5-2
altitude correction, 5-23
temperature
moisture corrections, 5-23
Dew point temperature, 5-23, 5-26
Die casting hood, 10-108, to 10-111
Diesel Engine Exhaust, 10-114
Diffusion, 4-32
Dilution, 7-5
air distribution, 2-4
definition, 2-2
principles, 2-2
ventilation for health, 2-2
concentration buildup, 2-5
mixtures, 2-6
purging rate, 2-6
steady state concentration, 2-5
ventilation for fire and explosion, 2-7
Dip tanks, 10-63 to 10-77
Direct driven fans, 6-10
Direct combustor, 4-26
Direct-fired heater, 7-8, 7-11
bypass design, 7-11
CO concentrations, 7-11
Disc sander, 10-87
Discharge conditions, 6-21
Dishwasher hood, 10-118
Distribution
in plenum, 3-8
slot, 3-8
Doors
pressure difference across, 7-4
DOP, 4-32
Downdraft hoods, 3-20, 10-15, 10-16, 10-53, 10-54
Drafts, 7-2, 7-4
Drilling, rock, 10-115, 10-134
Drive arrangements, 6-10
mounting, 6-6
Drum sander, 10-88, 10-89
Dry box, 10-21
Dry bulb temperature, 5-23
Dry centrifugal collector, 4-3
Drying
oven, 10-79
time, solvent, 12-20 to 12-21
Duct contraction 5-10
Duct design
calculations for, 5-2
choice of method of, 5-6
data, Chapter 5
primary steps for, 5-2
principles, Chapter 5
procedure, 5-2
Duct, enlargement, 5-10
Duct, fitting losses
coefficient, 1-9
equivalent length, 1-9
Duct material
black iron, 8-2
Duct segment
definition, 5-2
Duct size
available, 5-2
Duct support, 8-2, 8-4
Dust
airborne, 3-2, 12-3 to 12-17
definition, ix
particle size range, 4-27
Dust cake, 4-9
porosity, 4-12
Dust collector, 6-6
application of, 4-1, 4-30
centrifugal efficiency, 4-23
choice
efficiency, 4-23
combined fan and, 4-18, 4-23, 6-6

(see dynamic precipitator)
comparison chart, 4-30
concentration, 4-2
cost, 4-28
cost, chart, 4-30
disposal of dust, 4-3
dry centrifugal, 4-18
efficiency, 4-2
energy considerations, 4-3
fabric, 4-2
fabrics, 4-9
multiple collectors, 4-23
pilot installation, 4-12
primary, 4-2
selection, 4-2
settling chamber, 4-18
wet type (see wet collector)
types, 4-3
unit, 4-28
Dust collector choice
effect of emission rate on, 4-2
Dust collector selection
contaminant characteristics, 4-3
gas stream characteristics, 4-3
Dust collectors, 4-2
explosion or fire considerations, 4-17
Dust disposal, 4-3
Dynamic precipitator, 4-23

E

Economic velocity
optimum, 5-29
Effective specific gravity, 3-2
Effective temperature, x, 2-11
corrected for radiation, 2-11
Efficiency, 4-2
air filters, 4-32
Egg-crate straighteners, 6-21
Ejector, 6-2
Elbow material thickness, 8-4
Elbows, 5-31, 5-36, 5-49
Electric field, 4-3
Electric furnaces, 10-8, 10-9, 10-134
Electrode, 4-3
Electroplating, 10-64 to 10-77
Electropolishing, 10-64 to 10-77
Electrostatic, 4-32
Electrostatic precipitator, 4-2, 4-3
Elevation, correction for, 5-23, 5-44
Elevators, 10-38
Employee exposures levels
evaluating, 7-17
Enclosing, 3-17
hoods, 3-2
Energy recovery, 7-15
Enlargements, 5-10
Enlargement losses, 5-10
Enthalpy, 5-24
Entrainment ratio, 3-18
Entry-hood
coefficient, 3-8
coefficient, definition, ix
losses, ix, 1-5, 3-8
Entry loss
simple hoods, 3-8
Environmental control, 7-5
Equivalent fan static pressure, 5-28
Equivalent foot method, 5-2
Equivalent length
fittings, 5-3
Equivalent pressure, 6-16
Equivalent sizes, round to rectangular, 5-42
Etching, 10-72
Ethylene glycol, 7-10
Evaluating exposure levels
equations for, 7-17
Evaporative cooling, 7-6
Evasé, 5-28
Exhaust
need for, 7-14
Exhaust stack, 5-28
Exhaust system
design
plenum system, 5-7
evaluation, 9-22
general, 1-2, Chapter 2
dilution, 1-2, 2-2
heat control, 1-2, 2-2
local, 1-2, Chapters 3, 4, 5, 6
low volume-high velocity, 10-95 to 10-103
plenum type, 5-7
Exhausting, characteristics, 1-9
Expansion, 8-4
joints, 8-4
Expansions, 5-10
Explosion venting, 4-34
Explosive atmosphere, 8-5
Explosive limit, lower, definition, x
Explosive limits, x, 12-18 to 12-20
Exterior hoods, 3-2

F

Fabric
configuration, 4-12
efficiency, 4-9
efficiency of new, 4-9
non-woven, 4-9
penetration, 4-9
permeability, 4-9
type, 4-12
Fabric collectors, 4-3
air to cloth ratio, 4-12

area required, 4-12
by-pass, 4-9
duty cycle, 4-12
filter drag, 4-12
filtration velocity, 4-12, 4-14
housing, 4-12
intermittent duty, 4-12
leak testing, 4-9
pulse-jet or fan-pulse, 4-14
reconditioning, 4-12, 4-14
reverse-air type, 4-14
reverse-jet, 4-14
shaker, 4-14
time between reconditioning, 4-9
Fabric collectors, 4-2
Fabric configurations, 4-9
Fabric drag, 4-12
Face velocity, 3-15, 3-18
Fan
alignment, 6-21
bearing check, 6-21
belt drive, 6-21
belt driven, 6-10
blade shape, 6-2
connections, 8-4
construction classes, 6-10
corrosive applications, 6-7
dampers, 6-10
desired operating point, 6-14
direct driven, 6-10
discharge conditions, 6-21
drive arrangements, 6-10
effect of temperature, 6-7
efficiency, 6-2, 6-10
energy
humidification in wet collectors, 4-17
equivalent pressure, 6-16
explosive materials, 6-7
flow rate determination for, 6-6
gas density
effect of, 6-14, 6-16
homologous, 6-16
inlet dampers, 6-10
inlet spin, 6-21
laws, 6-14
noise, 6-10
outlet dampers, 6-10
performance curve, 6-13
point of operation, 6-13
rating
non-standard conditions, 6-10
rating tables, 6-10, 6-16
required power, 5-28
rotation rate
effect of, 6-14
selection 6-6
speed
effect on flow rate, 6-14
effect on power, 6-14
effect on pressure, 6-14
Static Pressure, 5-7, 5-28, 6-6, 6-10, 6-13, 6-16
equivalent, 5-28
system curves, 6-16
system effect, 6-20
Total Pressure, 5-7, 6-6, 6-10, 6-13, 6-16
types, 6-2
axial, 6-2
propeller, 6-2
backward inclined, 6-10
centrifugal, 6-2
forward curved, 6-2
homologous, 6-13, 6-16
in-line centrifugal, 6-2
special, 6-2
tubeaxial, 6-2
vaneaxial, 6-2
window, 7-15
vibration 6-21
Fan-Collector
dry dynamic precipitator, 4-23
wet, 4-18
Fan-Collector combination, 6-6
Fan energy
humidification in wet collectors, 4-17
Fan and system curves, 6-13
Fans, Chapter 6
accessories, 6-10
corrosive applications, 6-7
density correction, 6-7
effect of temperature on, 6-7
explosive material, 6-7
gas density
effect of, 6-14, 6-16
homologous, 6-16
impeller design, 6-2
propeller, 6-2
rotation rate
effect of, 6-14
vibration, 6-21
Fan installation, 6-6
Fan static pressure, 5-7, 6-13
Feed mills, 10-128
Field meters, 9-2
Filtration velocity, 4-12
redeposition, 4-17
Filtration velocity choice
fabric collector, 4-14
Fire dampers, 8-4
Fire, dilution ventilation for, 2-7
Fitting
contraction, 5-10
Fitting losses

coefficient, 1-9
equivalent length, 1-9
Flammable atmospheres, 8-5
Flanges, 3-6
Flash points, 12-18 to 12-20
Flexible ducts, 8-5
Flour mills, 10-133
Flow rate
changes with density, 5-24
collector selection, 4-2
hood, 3-6
measurement of, 9-3
orifice, 9-19
pitot traverse method, 9-5
reduction of, 7-14
replacement, 7-14
total, 3-20
Flues, 7-4
Fluidized beds, 10-121
Forge, 10-134
Forward curved fan, 6-2
Foundry
bins, 10-37
chipping, 10-54
conveyor belts, 10-37, 10-39
cooling tunnels, 10-134
core grinder, 10-6
core knockout, 10-134
core sanding, 10-134
elevators, 10-38
grinding, 10-48 to 10-55
mullers, 10-11, 10-12
pouring, 10-13
screens, 10-40
shakeout, 10-14 to 10-16
shell core molding, 10-18, 10-19
Freeze up, 7-8
heating unit, 7-10
Friction
losses in non-fabricated ducts, 8-5
losses in non-round ducts, 5-23
Friction chart, 5-33, 5-37
Friction loss, 1-7, 5-2, 5-37
FSP, 5-7
FTP, 5-7
Fuel
conservation of consumption, 7-4
Fumes, xi, 3-2, 13-7
Furnaces
aluminum, 10-134
crucible, 10-7
electric, 10-8, 10-9, 10-134
melting, 10-7, 10-10
non-tilting, 10-7, 10-20, 10-109
rocking, 10-8
tilting, 10-10

G

Galvanized metal, 8-2
Garage ventilation, 10-111, 10-112, 10-134
Gas collectors, 4-23
Gas contaminants
control of, 4-23
Gaseous collector
types, 4-26
Gases, 3-2
concentration, 12-3 to 12-17
definition, ix
properties of, 12-18 to 12-20
General ventilation, 1-2, 2-2
Glove box, 10-21
Governmental codes, 7-11, 8-5, 9-2
Grain elevators, 10-128
Grains per cubic foot, 4-2
Granite cutting, 10-109
Gravity, specific, ix, 1-4, 12-18 to 12-20
Grilles, 7-5
Grinders
belt, 10-42, 10-43
core, 10-6
disc
horizontal single spindle, 10-49
horizontal double spindle, 10-48
vertical spindle, 10-50
jack, 10-51, 10-52
portable, 10-53, 10-54
small radial, hood, 10-100
surface, 10-59
hood, 10-98
swing, 10-55
Grinding
hoods, 10-48 to 10-55, 10-59, 10-94, 10-97, 10-98, 10-100, 10-134
table, 10-53, 10-54

H

Health, dilution ventilation for, 2-2
Heat
disorders, 2-10
modes, 2-9
adaptive mechanism, 2-9
exchanger factors, 7-15
latent, 2-14
load, 7-5
pipes, 7-15
radiant, 2-15
sensible, 2-14
tolerance, 2-10
transfer, 7-7
wheel, 7-15
stress, 2-10
indices, 2-11

WBGT, 2-11
WGT, 2-12
measurement, 2-10
air velocity, 2-11
radiant heat, 2-11
metabolic, 2-11

Heating coils
bypass dampers, 7-8
Heating costs
reduction of, 7-14
Heating degree days, 7-12
Heating equipment
replacement air, 7-8
HEPA, 4-33
High efficiency filters, 7-17
High toxicity, 3-17, 4-33
High toxicity hoods, 10-20 to 10-33
High velocity drafts, 7-2
Homologous fan, 6-16
Homologous fans, 6-13
Homologous series, 6-14
Hood design, Chapter 3
canopies, 10-107
cross drafts, 3-2, 3-17
diameter, hot process, 3-20
enclosure, 3-2, 3-6, 3-17
entry loss, 3-8, 5-33
compound hoods, 3-15
flow rate determination, 3-7
glove box, 3-17
grinding wheel, 10-48 to 10-55, 10-95, 10-96
local, 3-2
location, 3-2, 3-6
openings, 3-7
push-pull, 10-66 to 10-67
radioactive materials hoods, 3-17, 10-21 to 10-31
slots, 3-4, 3-8
static suction, 1-6
verification, 9-22
Hood entry losses, 1-5, 1-7
factors, 9-11
Hood static pressure, 1-6
interpretation of, 9-11
Hood types
bench, 10-53, 10-54, 10-57, 10-58, 10-122
booth, 10-55, 10-56, 10-78, 10-80, 10-81
canopy, 10-79, 10-107
downdraft, 10-15, 10-16, 10-53
slots, 10-63, 10-64
specific operations, 10-1 to 10-3
Hoods
rectantular, 3-8
round, 3-8
Hopper ventilation, 10-37
Horsepower
air, definition, ix
brake, definition, ix
Horsepower — fans, 6-7
Hot air column, 3-19
Hot contaminated air, 3-2
Hot processes, 3-19
Hot water heaters, 7-8
Humid volume, 5-24, 5-26
Humidification, 5-26
Humidifying efficiency, 5-25
definition, 5-27
Humidity, absolute, definition, ix
openings, 3-7
push-pull, 10-66 to 10-67
radioactive materials hoods, 3-17, 10-21 to 10-31

I

Impaction, 4-18
Impeller design, 6-2
Impingement, 4-32
Impregnating — wire, 10-135
Inclined manometer, 9-3, 9-17
Indices, heat stress, 2-11
Indirect-fired heaters, 7-8, 7-10
Inertial effects, 3-2
Infiltration, 7-2
Inlet
box, 6-21
dampers, 6-10
spin, 6-21
Integral face and bypass, 7-10
Interception, 4-32
Interpolation
fan tables, 6-10
Interpretation of measurements
hood static pressure, 9-11
Ionization, 4-3, 4-9

J

Jointer, 10-84

K

K-value, 2-2, 2-4
Kettles, varnish, 10-135
Kitchen hoods, 10-116 to 10-118

L

Laboratory hoods, 3-17, 10-23 to 10-28
Large particulates, 3-2
Latent heat, 2-14, 7-5
Lathe hood, 10-29
Laws, fan, 6-14
Lead, glaze spraying, 10-134
Leak testing, 8-2
Lift truck ventilation, 10-113
Linty material, 4-3
Loss

hood entry, 3-8
factors, 5-2
Losses
duct, 1-7
fitting, 1-9
hood, 1-5, 1-7
low velocity media, 6-13
Louvers, 7-5
Low volume-high velocity exhaust systems, 10-97 to 10-103
Lower explosive limit, definition, x
Lower explosive limits, 12-18 to 12-20

M

Magnehelic gauuge, 9-17
Manometer, x
inclined, 9-3
U-tube, 9-2
Matchers, 10-94
Material
different duct, 5-11
Material handled by fan, 6-7
Material selection
corrosive applications, 8-2
Maximum plenum velocity, 3-8
Measurement
corrections for density, 9-26
difficulties encountered, 9-25
hood static pressure, 9-11
initial for new system, 9-22
instrument selection, 9-25
number of traverse points, 9-9
periodic testing, 9-22
survey form, 9-22
velocity
in rectangular ducts, 9-9
Measurement instruments
calibration of, 9-19
Mechanical
efficiency, 6-10
refrigeration, 7-6
Melting
furnace, 10-7 to 10-10, 10-20
pot, 10-110
Metabolism, 2-9
Metal bandsaw, 10-60
Metal cleaning, 10-61 to 10-77
Metal polishing belt, 10-43
Metal shears, 10-30
Metal spraying, 10-56
Metallizing booth, 10-56
Micron, definition, x
Mill, tumbling, 10-17
Milling machine, 10-31
Minimum duct velocity, 3-17, 5-2
Miscellaneous specific operations, 10-133 to 135
Mist, 4-26
Mists, XII, 13-7 to 13-16
Mixer, 10-11, 10-12, 10-104
Mixtures, dilution ventilation, 2-6, 2-8
Moisture content, 5-24
Molders, 10-94
Muller, 10-11, 10-12
Multi-rating table, 6-10

N

National Board of Fire Underwriters, 6-7
National Fire Protection Association, 6-7
Natural draft stacks, 7-4
Nature of contaminant, 4-2
Negative pressure, 7-2
Negative pressures, 7-4
Non-circular duct
friction losses, 5-23
Non-overloading fan, 6-2
Non-standard duct material, 5-11
Non-woven fabric, 4-9
Nozzles, 9-19
push, 3-18
plenum pressure, 10-68
push jet, 3-18
Number of gores in elbows, 8-4

O

Opacity, 4-3
Open surface tanks, 10-61 to 10-77
Operating temperature, 6-6
Optimum economic velocity, 5-29
Orifices, 3-15, 5-29, 9-11, 9-19
Outboard motor test tank, 10-128
Outlet dampers, 6-10
Ovens, 10-79

P

Packaging Machine, 10-35, 10-134
Painting, spray, 10-78, 10-80 to 10-83
Paper machine, 10-134
Partial enclosure, 3-2
Particle size range, 4-27
Particulates, 7-17
Penny, 4-3
Percent saturation curves, 5-23
Perchloric acid hood data, 10-26
Performance curve
fan, 6-13
limits, 6-10
Permeability, 4-9
Physical constants, 12-18 to 12-20
Physiological principles, 2-9
Pickling, 10-63 to 10-77
Pistol range, 10-120
Pitot tube, 9-3

accuracy, 9-3
calibration of, 9-11
limitation, 9-11
modified, 9-11
traverse with, 9-5, 9-9, 9-16, 9-25. 9-27
Planers, 10-94
Plastic pipe, 8-5
Plate exchangers, 7-15
Plating, 10-63 to 10-77
Plenum, x, 3-8
Plenum pressure
push nozzle, 10-68
Plenum system, 5-7
manual cleaning design, 5-7
self-cleaning design, 5-7
Plenum velocity, 3-8
Plugging, 3-17
Pneumatic hand tools, 10-99, 10-115
Point of operation, 6-13
Polishing
belt, 10-42, 10-43
wheels, 10-44 to 10-46
Porosity of dust cake, 4-12
Portable grinding, 10-53, 10-54, 10-97 to 10-101
Pouring, 10-13
Power
required fan, 5-28
Power exhausters, 6-2
Power roof ventilators, 6-2
Precipitator
dry dynamic, 4-23
electrostatic, 4-3
high voltage, 4-3
low voltage, 4-9
pressure drop in, 4-9
wet dynamic, 4-18
Preheat coil, 7-10
Pressure
drop
collection efficiency, 4-18
hood static, 1-5
loss calculations, Chapter 5
measurements, 9-2
regain, 5-11
static, x, 1-3, 3-8, 5-2, 5-6, 5-10, 5-23, 5-27, 6-13, 9-2
total, x, 1-3
vapor, x
velocity, x, 1-3
Primary collector, 4-2
Pressure gauge calibration, 9-17
Primary filter, 7-17
Propeller fans, 6-2
Protective suits for heat, 2-15
Psychrometric chart, 2-7, 5-23, 5-26
density determination from, 5-24
density factor, 5-23
enthalpy, 5-24
high temperature, 5-46, 5-47, 5-48
humid volume, 5-24
moisture content, 5-24
relative humidity, 5-23
saturation curves, 5-23
Psychrometric principles, 5-23
Pull hood, 3-18, 3-19
Pulse-jet collector, 4-14
filtration velocities, 4-17
Push jet, 3-1, 3-18
velocity, 3-18
Push nozzle, 3-18
plenum pressure, 10-68
Push-pull, 3-1, 3-18
hood design, 10-66 to 10-67

Q

Quartz fusing, 10-135

R

Radial, 6-2
Radial saw, 10-87
Radiant heat, 7-5
engineering control, 2-15
shielding, 2-15
Radiation, x, 2-15
Radioactive, 3-17
Radioactivity, 4-33
Rating tables, 6-10, 6-16
Recirculation, 4-2, 7-1, 7-15
factors to consider, 7-16
governmental codes, 7-11
system
bypass to outside, 7-19
malfunction, 7-16
monitoring, 7-16
potential problems in, 7-19
Rectangular equivalents (ducts), 5-42
Redeposition, 4-17
Reducing air flow rate, 7-14
Regulations
governmental, 4-2
Refrigerated suits, 2-15
Reheat coil, 7-10
bypass dampers, 7-10
Relative humidity, ix, 5-23
Replacement air, 1-2, 2-2, 7-2
heating costs, 7-12
heating equipment, 7-8
reasons for, 7-2
untempered air supply, 7-15
Replacement air unit, 7-4
Respiratory
protection, 10-78, 10-80, 10-81

Resultant velocity pressure, 5-10
Reversed fan rotation, 6-21
Rifle range, 10-120
Rip saw, 10-92, 10-94
Rock drilling, 10-94, 10-115, 10-134
Rolls, rubber calendar, 10-105
Roof exhausters, 7-15
Rotary blasting table, 10-135
Rotating vane anemometer, 9-12
Routers, 10-94
Rubber
    calendar rolls, 10-105
    mill, 10-106
Run-around coils, 7-15

S

S-type pitot tube, 9-11
Safety guards, 6-10
Salt bath, 10-64 to 10-77
Sampling box, 10-33
Sanders
    belt, 10-85, 10-94
    disc, 10-86, 10-94
    drum, 10-88, 10-89
    swing, 10-94
Sash stickers, 10-94
Saws
    abrasive, 10-42
    band, 10-60, 10-90
    miscellaneous, 10-94
    radial, 10-93
    rip, 10-92, 10-94
    swing, 10-91
    table, 10-92, 10-94
Screens
    cylindrical, 10-40
    flat deck, 10-40, 10-134
Scrubber, 4-17
Sealed joints, 8-2
Seamless tubing, 8-5
Secondary filter, 7-17
Selection, fan, 6-6
Sensible heat, 2-14, 7-5, 7-6, 7-15
Service garages, 10-111, 10-112, 10-114
Settling chamber, 4-18
Shaft seal enclosure, 10-32
Shakeouut, 10-14 to 10-16
Shears, metal, 10-30
Shell core molding, 10-18, 10-19
Shot blasting room, 10-4
Side draft hood, 10-14, 10-63 to 10-65
Silver soldering, 10-135
Sizers, 10-93
Slots, 3-8, 10-36, 10-61 to 10-69
    hoods, 3-15 to 3-17
    lateral, 10-57, 10-61 to 10-64
    resistance, 3-10, 5-33
Slot velocity, x, 3-8
Sludge, 4-3
SMACNA, 8-2
Smoke, definition, x
Smoke tubes, 9-17
Soldering, 10-57, 10-130
Solvent
    degreasing, 10-61, 10-62
    drying time, 12-20 to 12-21
Solvent mixtures, dilution ventilation for, 2-6
    effective, 1-3, 1-4
Sparking, 6-10
Specific gravity, ix, 12-18 to 12-20
Specific operations, 10-1 to 10-3
Spiral wound duct, 8-2
Splitter sheets, 6-21
Spraying
    ceramic, 10-134
    metal, 10-56
    paint, 10-78 to 10-83
Spread, 7-6
Squirrel cage, 6-2
Stack
    discharge, 8-4
    exhaust, 5-28
Static pressure, x, 1-3, 5-2, 5-6, 5-10, 5-23, 5-27, 9-2
    comparison, 5-6
    correction for, 5-27
    definition, 1-3
    losses
        non-standard density, 5-26
    taps, 9-3
Static tube, 9-3
Staubscheid pitot tube, 9-11
Steam coil, 7-8
    throttling range, 7-8
Steam condensate, 7-8
Steam heaters, 7-8
Steam kettles, 10-135
Steam traps, 7-8
Stone crusher, 10-134
Stone cutting, 10-115
Streamlines, 3-7
Stripping, 10-61 to 10-77
Supply systems, 7-2
Surface grinder 10-48
Surfacing machine, 10-115
Swing grinder, 10-55
Swinging vane anemometer, 9-13
    limitations, 9-16
    measuring pressure with, 9-13
System curves, 6-16
System design
    adding hoods, 8-5
    method choice, 5-6

methods, 5-2
preliminary steps, 5-2
procedure, 5-2
testing, 8-5
System effect, 6-20
factor, 6-21
System requirement curves, 6-13

T

Table exhaust, 10-53, 10-54, 10-57, 10-69
Tailpipe ventilation, 10-111 to 10-113
Take-off, 3-8
Tank exhaust, 10-61 to 10-77
Tap
static pressure, 9-3
Tapered transitions, 8-4
Temperature
air supply, 7-5
corrections for, 5-23
corrections to density, 9-3
dew point, 5-23
dry bulb, 2-10, 5-23
map, 7-13
wet bulb, x, 2-10, 5-23
Tenoner, 10-90
Terminal air supply velocity, 7-6
Testing, 9-2
Thermal
anemometer, 9-16
maintenance, 9-16
measuring pressure with, 9-16
equilibrium, 2-8
oxidizer, 4-26
radiation control, 2-15
Threshold limit values, x, 12-3 to 12-17
Throttling range, 7-8
Throw, ix, 1-9, 7-6, See Blow
Tolerance to heat, 2-10
Total heat, 5-24
Total pressure, 9-2
Torch cutting, 10-122
Total pressure, x, 5-8
definition, 1-4
Toxic contaminant
preferred design method, 5-6
Tracer gas, 9-17
Trailer interior spray painting, 10-82
Transport velocity, definition, x
Traps, 7-8
Truck spray painting booth, 10-83
Tubeaxial fans, 6-2
Tumbling mill, 10-17
Turbulence loss, 1-7
Turn-down ratio, 7-11
Turning vanes, 6-21

U

U-tube manometer, 9-2, 9-16
liquids used in, 9-17
Unit collectors, 4-2, 4-28

V

Vane anemometer, 9-12, 9-13
Vapor, definition, x
TLVs, 12-3 to 12-17
Vapor pressure, definition, x
Vapors, 3-2
Varnish kettles, 10-135
V-belt drives, 6-10
V-belt power losses, 6-10
Velocity
capture, ix, 3-17
contours, 3-7
cooling, 2-14
correction for changes in, 5-10
definitions, ix, x, 3-3
face, 3-15, 3-18
instruments, 9-2, 9-12
jet, 3-18
minimum duct, 5-2
over person, 7-5
plenum, 3-8
pressure, 3-8, 5-42
definition, 1-4
pressure chart, 5-39, 5-40
slot, x, 3-3
transport, x, 3-17
variation with distance, 3-7
Velocity pressure
actual, 5-26
conversion to velocity, 9-11
resultant, 5-10
Velocity pressure method, 5-2
Velometer, 9-2, 9-12
Vena contracta, 3-8
Ventilation
dilution, 2-2
air distribution, 2-5
fire and explosion, 2-7
K factor, 2-2, 2-4
principles applied to, 2-2
general, 2-2
general principles, 1-2
heat control, 2-2, 2-8
hot process, 3-19
system, supply, 7-2
systems testing, 9-2
Venting of combustion products, 7-11
Vibration, 6-21, 8-4
Volumetric flow rate
definition, 1-3

Vortex, 6-21

W

Water hammer, 7-8
Water pollution, 4-17
Welding, 8-2, 10-57, 10-58
    joints, 8-4
    seams, 8-4
Wet bulb temperature, x, 2-10, 5-23, 5-26
Wet collector
    high energy 5-25
    humidification, 5-16, 7-19
    humidifying efficiency, 5-25, 5-27
Wet collectors
    air to water contact, 4-17
    centrifugal type, 4-18
    fogging type, 4-18
    humidification, 4-17
    orifice type, 4-18
    packed towers, 4-18
    pressure drop, 4-18
    spray tower/chamber, 4-17
    temperature and moisture, 4-17
    venturi, 4-18
Wind tunnel
    use of, 9-22
Window exhaust fans, 7-15
Winter temperatures, in US, 7-13
Wiper, belt, 10-41
Wire brushes, 10-98
Wire impregnating, 10-135
Woodshaper, 10-94
Woodworking machines, 10-84 to 10-94